Handbook of Elastic Properties of Solids, Liquids, and Gases

Handbook of Elastic Properties of Solids, Liquids, and Gases

Volume I   Dynamic Methods for Measuring the Elastic Properties of Solids
Arthur G. Every, Wolfgang Sachse

Volume II   Elastic Properties of Solids: Theory, Elements and Compounds, Novel Materials, Technological Materials, Alloys, and Building Materials
Moises Levy

Volume III   Elastic Properties of Solids: Biological and Organic Materials, Earth and Marine Sciences
Moises Levy

Volume IV   Elastic Properties of Fluids: Liquids and Gases
Moises Levy, Richard Raspet, Dipen N. Sinha

Handbook of Elastic Properties of Solids, Liquids, and Gases

Volume IV Elastic Properties of Fluids: Liquids and Gases

Editors-in-Chief

Moises Levy
Physics Department
University of Wisconsin-Milwaukee
Milwaukee, Wisconsin

Henry E. Bass
National Center for Physical Acoustics
University of Mississippi
University, Mississippi

Richard R. Stern
Applied Research Laboratory
Penn State University
State College, Pennsylvania

Volume Editors

Moises Levy
Physics Department
University of Wisconsin
Milwaukee, Minnesota

Richard Raspet
National Center for Physical Acoustics
University of Mississippi
University, Mississippi

Dipen N. Sinha
Los Alamos National Laboratory
Los Alamos, New Mexico

Supervising Editor

Veerle Keppens
National Center for Physical Acoustics
University of Mississippi
University, Mississippi

ACADEMIC PRESS

A Harcourt Science and Technology Company

SAN DIEGO SAN FRANCISCO NEW YORK BOSTON
LONDON SYDNEY TOKYO

This book is printed on acid-free paper. ∞

Copyright © 2001 by Academic Press

All rights reserved.
No part of this publication may be reproduced or transmitted in any form or by any means, electronic or mechanical, including photocopy, recording, or any information storage and retrieval system, without permission in writing from the publisher.

Requests for permission to make copies of any part of the work should be mailed to:
Permissions Department, Harcourt, Inc., 6277 Sea Harbor Drive, Orlando, Florida, 32887-6777

ACADEMIC PRESS
A Harcourt Science and Technology Company
525 B Street, Suite 1900, San Diego, CA 92101-4495, USA
http://www.academicpress.com

Academic Press
Harcourt Place, 32 Jamestown Road, London NW1 7BY, UK
http://www.academicpress.com

Library of Congress Catalog Card Number: 00-106894
International Standard Book Number: 0-12-445764-9

Printed in the United States of America
00 01 02 03 04 IP 9 8 7 6 5 4 3 2 1

Contents

Foreword	ix
Preface	xi
List of Contributors	xv

PART 1 ELASTIC PROPERTIES OF LIQUIDS 1

Chapter 1   Noninvasive Determination of Sound Speed and Attenuation in
Chemical Warfare Agents 3
- Abstract 3
- 1.1. Introduction 3
- 1.2. Theory 5
- 1.3. Experimental 13
- 1.4. Results and Discussion 14
- 1.5. Conclusions 16
- Appendix 18
- Acknowledgment 20
- References 20

Chapter 2   The Speed of Sound in Pure and Neptunian Water 23
- Abstract 23
- 2.1. Introduction 24
- 2.2. General Overview 25
- 2.3. Historical Survey 27
- 2.4. The Laboratory Sound Velocimeters 35
- 2.5. The Neptunian Domain 42
- 2.6. The Available Measurements of Sound Speed in Water ... 44
- 2.7. The Equations for the Speed of Sound 49
- 2.8. Discussion 56
- 2.9. Conclusion 75
- Acknowledgment 76
- Appendix: Recommended Equations for the Speed of Sound in Pure Water and Neptunian Waters ... 76
- References 78

Chapter 3   Absorption of Sound in Fresh and Sea Water 83
- Abstract 83
- 3.1. Introduction 84
- 3.2. Physical and Chemical Causes of Absorption in Liquids ... 85
- 3.3. Sound Absorption above 20 kHz 92
- 3.4. Sound Absorption at Low Frequencies 96
- 3.5. Discussion and Conclusion 108
- References 112

CONTENTS

Chapter 4   Velocity and Absorption of Sound Waves in Superfluid ^{3}He 117
- 4.1. Introduction to Normal Liquid State of ^{3}He at Very Low Temperatures 117
- 4.2. Introduction to Wave Propagations 122
- 4.3. First Sound 125
- 4.4. Second Sound in Superfluid ^{3}He A and B Phases 126
- 4.5. Fourth Sound 129
- 4.6. Spin-Entropy Wave in Superfluid ^{3}He A_1 137
- Acknowledgment 144
- References 144

Chapter 5   Elastic Properties of Superfluid Helium Four 147
- Abstract 147
- 5.1. Introduction 147
- 5.2. Acoustic Properties of Superfluid Helium 148
- 5.3. Experimental Measurement of the Sounds in Superfluid Helium .. 150
- 5.4. Results 155
- References 156

Chapter 6   Elastic Properties of Nematic and Smectic Liquids 159
- Abstract 159
- 6.1. Introduction 159
- 6.2. Elastic Energy of the Nematic Phase 161
- 6.3. Elastic Energy of the Smectic A Phase 161
- 6.4. Hydrodynamics and Sound Propagation 163
- 6.5. Behavior of the Elastic Constants Near the N–SmA Transition .. 165
- 6.6. Behavior of the Elastic Constants Near the SmA–SmC Transition and Comparison with the Specific Heat Behavior 169
- 6.7. Experiments 171
- Acknowledgments 179
- References 179

Chapter 7   Fundamental Acoustic Properties of Bubbly Liquids 183
- Abstract 183
- 7.1. Introduction 183
- 7.2. Linear Waves in Dilute Bubbly Liquids 184
- 7.3. Bubble Dynamics 191
- 7.4. Speed of Sound in Bubbly Liquids 197
- 7.5. Some Applications 201
- 7.6. Conclusions 203
- Acknowledgment 204
- References 204

Chapter 8   Acoustic Properties in Petroliferous Liquids 207
- Abstract 207
- 8.1. Introduction 207
- 8.2. Major Components of Crude Oil 208
- 8.3. Products of Crude Oil 211
- 8.4. Classification of Crude Oils 211
- 8.5. Indirect Methods of Oil Classification 212
- References 218

Chapter 9   Acoustic Properties in Rocks Saturated with Petroliferous Liquids ... 219
- Abstract 219
- 9.1. Introduction 219
- 9.2. Light Oil Saturants 220
- 9.3. Natural Heavy Oil Saturants 224
- References 230

CONTENTS

PART 2	ELASTIC PROPERTIES OF GASES	233
Chapter 10	Introduction to Elastic Constants of Gases	235
10.1.	Introduction	235
10.2.	Ideal Gas	236
10.3.	Real Gas Corrections	238
10.4.	Contribution of Transport Processes on Speed of Sound	239
10.5.	Changes in Elastic Properties Due to Relaxation Processes	241
10.6.	Measurements at Moderate Pressures	245
10.7.	Typical Results	247
10.8.	Diffusion	256
10.9.	Gases at Low Pressure	257
10.10.	Systems Not in Equilibrium	263
10.10.	Summary	263
	References	264
Chapter 11	Sound Speed as a Thermodynamic Property of Fluids	267
	Abstract	267
11.1.	Introduction	267
11.2.	Wave Equation in a Fluid	268
11.3.	Equations of State for Fluids	274
11.4.	Speed of Sound in Fluids	280
11.5.	Conclusions	289
11.6.	Tables of the Speed of Sound in Important Fluids	294
	Additional Reading	328
	References	328
Chapter 12	Acoustic Measurements in Gases: Applications to Thermodynamic Properties, Transport Properties, and the Temperature Scale	329
	Abstract	329
12.1.	Introduction and Scope	330
12.2.	Acoustic Measurements and Thermodynamic Properties of Dilute Gases	331
12.3.	Measuring the Speed of Sound	335
12.4.	Resonance Measurements of Transport Properties	340
12.5.	Acoustic Thermometry	345
12.6.	Acoustic Determination of the Universal Gas Constant R	349
12.7.	Concluding Remarks	351
12.8.	Preface to Tables	351
	Acknowledgments	369
	References	369
Chapter 13	Use of Cylindrical Acoustic Resonance to Measure the Speed of Sound in Gases	371
	Abstract	371
13.1.	Introduction	372
13.2.	Apparatus	373
13.3.	Gases	375
13.4.	Method of Operation	376
13.5.	Experimental Results and Discussion	377
13.6.	Conclusion	384
	Acknowledgment	385
	References	385
Chapter 14	Speed of Sound in Planetary Atmospheres	389
14.1.	Introduction	389
14.2.	Acoustic Measurements on Spacecraft	389
14.3.	Instrumental Considerations	391

14.4.	Speed of Sound — A Diagnostic of Composition and Transport	392
14.5.	Acoustic-Gravity Waves in High Planetary Atmospheres	394
	References	397

Chapter 15 Sound Speed in Normal Stars ... 399
    Abstract ... 399
15.1. Introduction ... 399
15.2. Determining the Sound-Speed Profiles in Stars ... 400
15.3. Sound Speed in the Sun ... 407
15.4. Implications of Helioseismology Results ... 419
    Acknowledgment ... 420
    References ... 420

Chapter 16 The Properties of Condensed Matter in White Dwarfs and Neutron Stars ... 423
    Abstract ... 423
16.1. Introduction ... 423
16.2. The Cold, High-Density Equation of State ... 425
16.3. White Dwarf and Neutron Star Structure ... 433
16.4. Observations of Condensed Matter in White Dwarfs ... 437
16.5. Observations of Condensed Matter in Neutron Stars ... 439
16.6. Thermal Properties of Matter in White Dwarfs and Neutron Stars ... 441
16.7. Concluding Remarks ... 444
    References ... 445

Index ... 449

Foreword

Dynamical measurements of the elastic properties of solids, fluids, and gases have provided a rich base of data upon which to build physical models of matter. Indeed, there are few physical processes or characterizations that do not rely heavily on the elastic properties of their constituents. More often than not, much of the required data are presented in archived journals or technical reports that are difficult to obtain and treat only a few materials. When presented with the need for the elastic properties of a particular material under specific physical conditions, the scientist, researcher, or design engineer often has to search the contents of hundreds of individual papers, some in obscure journals or books that are out of print. This process is labor intensive, time consuming, and usually frustrating. Past efforts have captured the elastic properties of materials into printed tables. Those tables are a valuable source of information about many materials of interest, but with the passage of time they cannot remain up-to-date without reissue. The *Handbook of Elastic Properties of Solids, Liquids, and Gases* contains an ambitious compendium of both the real and complex components of the elastic properties of solids, liquids and gases, as well as the modern and archival theory and techniques relating to the methods of obtaining the actual values.

The *Handbook* is divided into four volumes. Volume I provides the reader with a description of how the elastic properties of solids are measured and how measurements are analyzed. Since the quality of individual measurements depends upon the techniques used to collect the data, tables of data would be incomplete without providing the user with a basis for a critical evaluation of the accuracy of the table entries. The *Handbook* contains a comprehensive description of the experimental procedures used to collect the data presented in the *Handbook*. A collection of data, though important, does not alone provide insight into the nature of the material involved. Although much of our understanding of the mechanical properties of materials comes from measurements of elastic properties, the interpretation of these data often provides comprehensive and useful insight. When used for design purposes, knowledge of the microscopic basis for the elastic properties can guide the selection of material classes for particular applications. In addition, when presented with an entirely new material, the theory that has evolved from similar measurements can be used to help predict the properties of the new material. Volumes II and III are devoted to the theory and measurements of the elastic properties of solids, including biological, earth, and seismic materials.

Volume IV includes the elastic properties of liquids and gases. Although they are likely to be homogeneous and isotropic, their elastic properties are as difficult to predict as those for solids, particularly over the wide range in density and temperature of interest, from a few particles per cubic meter to neutron star densities and from microkelvin temperatures to 10^8 K. Even at normal temperatures and densities, subtle processes such as chemical reactions or energy transfer can complicate the determination of elastic properties and require careful measurements.

FOREWORD

There are materials that were omitted from this complication because their elastic properties were not available in the open literature, because they were overlooked, or because potential contributing authors were not available at press time. The editors hope that whenever a scientist or engineer discovers material that should have been included he or she would be willing to provide the data, appropriately referenced to the National Center for Physical Acoustics. The editors also welcome recommendations of other classes of materials that should have been included.

This project began in large part through the efforts of Dr. Logan E. Hargrove, Office of Naval Research. It was his goal to make elastic property data available to the science and engineering community and found willing collaborators in the editors. Without Dr. Hargrove, this project would never have been started and once begun would never have been completed. The editors are sincerely indebted to him for his insight and his encouragement.

HENRY E. BASS VEERLE KEPPENS MOISES LEVY RICHARD STERN

Preface

PREFACE FOR PART 1

This Volume is one of the four volumes in the *Handbook* of *Elastic Properties of Solids, Liquids and Gases*. The focus of this Part is on the elastic properties of liquids. It includes chapters on liquids ranging water to quantum liquids such as liquid helium three and four. The purpose of this Volume is to provide theory, in-depth discussions of specific experimental techniques, and data. The speed with which sound wave or sound energy travels through a liquid depends on the interaction between neighboring molecules of the substance. Therefore, the study of sound propagation through liquids provides an understanding of the elastic property B, the bulk modulus of liquids. In fact, sound speed is the only direct way to determine adiabatic compressibility of a liquid. The various chapters in this volume thus discuss sound propagation (sound speed and sound attenuation) in liquids of various kinds.

The Volume starts with a discussion in Chapter 1 of a novel noninvasive technique to determine sound speed in chemical warfare agents and their significant precursor chemicals as there are little sound propagation data available anywhere in the literature on this interesting group of chemicals. The data presented include sound speed and sound attenuation data on the most important chemical warfare agents and sound speed data of 40 precursor chemicals. In the next chapter, Chapter 2, an excellent survey of sound speed measurements in both fresh and seawater is presented. This survey contains extensive original analyses of the various laboratory measurements and equations that are complemented by further measurements. A companion chapter, Chapter 3, provides a thorough presentation of sound absorption and the various relaxation mechanisms in fresh water and seawater. We then turn our attention in Chapters 4 and 5 to quantum fluids, whose superfluid properties can only be observed at low temperatures — below 2.5 mK for He3 and below 2.17 K for He4. Although both are superfluids below their corresponding transition temperatures, the first is a Fermi liquid, while the second is a Bose-Einstein liquid. Three different phases have been discovered in superfluid He3. In addition to the usual sound wave mode called first sound, four other different modes of sound have been discovered and measured in superfluid He4. In fact, simultaneous measurements of some of these modes have been used to determine the thermodynamic properties of He4 from 1.2 K to the Lamda-transition 2.17 K. Although second sound, a thermal wave mode, was first discovered in superfluid He4, it and first sound are used extensively to obtain the elastic constants of liquid crystals, as discussed in the following chapter, Chapter 6. It includes experimental measurements and theoretical description of the elasticity of the Nematic, Smectic-A, and Smectic-C liquid crystal phases and the phase transitions between them. The next chapter, Chapter 7, gives a comprehensive theoretical analysis of propagation of linear pressure waves in a bubbly liquid. At a fundamental physics level, the peculiar physics of bubbly liquids arises from the sharp separation between the system's mass, mostly contained in the liquid phase, and the system's compressibility, mostly due to the gas. This differentiates the behavior of bubbly liquids from other, apparently similar, gas-liquid systems such as fogs or other suspensions of droplets in gases. In the final two chapters of this part, Chapters 8 and 9, the measurement of sound speed, density, and viscosity in petroleum fluids that contain various types of hydrocarbons is presented. This work is particularly important from the perspective of commercial oil exploration where the primary goal of a seismic survey is to detect and map hydrocarbons.

DIPEN N. SINHA
MOISES LEVY

PREFACE FOR PART 2

The 2nd Part of the Volume discusses the elastic properties of gases. This part covers the widest range of densities, from micrograms per cubic centimeter ($10^{-6}/cm^3$) for low density gases, to Petagrams per cubic centimeter ($10^{15}/cm^3$) for neutron stars. The corresponding velocities go from about 10 km/sec to close to the velocity of light, 3×10^5 km/sec. The lowest velocity mentioned above is not the lowest recorded velocity in matter. In the first part of this volume, it was pointed out that near the superfluid transition of He4, the velocity of second sound, third sound, and fourth sound approach zero. In addition, according to thermodynamic considerations of phase transitions the velocity of first sound also approaches zero near first-order phase transitions.

Both theoretical and experimental results on the propagation, absorption and dispersion of sound are introduced in Chapter 10. Results for low density gases are presented. The interesting result is that even for the lowest density gases, the sound velocity is within an order of magnitude of the ideal gas result. A comprehensive theoretical analysis of the sound speed in gases is presented in Chapters 11 and 12, the goal being to describe the equation-of-state of the gases sufficiently well in order to predict sound velocity to better than .01%. Chapter 11 concentrates on predicting the "thermodynamic sound speed of a fluid system at any temperature, density (or pressure), and composition." Extensive data for argon, nitrogen, air and water are presented in multiple figures. Chapter 12 concentrates on the measurement techniques required to achieve these accuracies. The authors use a double Helmholtz or Greenspan viscometer to measure the speed of sound. They also describe two elegant measurement vessels, one for making extremely accurate sound velocity measurements, and the other for measuring the Prandtl number, which is related to the ratio of the viscocity to the thermal conductivity. The first vessel is a sphere, which lends itself to exact theoretical analysis of its resonant modes, and therefore eliminates many theoretical uncertainties in the evaluation of the sound speed. The other vessel is a double resonator separated by a honeycomb-shaped array of hexagonal ducts. This is an ingenious arrangement that permits the measurement of both the viscosity of the gas and its thermal conductivity by measuring the frequencies and half-widths of the resonator's even and odd harmonics. They present data for 21 gases, including candidates for replacing refrigerants that may be harmful to the atmosphere, and also for He-Xe mixtures, which may be important for thermoacoustic prime movers and for some very reactive hexafluorides. Chapter 13 presents an inexpensive and simple cylindrical acoustic resonator that may be used to obtain sound speeds with uncertainties of less than plus or minus 0.1%.

PREFACE

In addition, sound speed data for ten different gases and extensive tables for He-Ar mixtures, important for thermoacoustic engines, are given. Chapter 14 discusses sound in the atmospheres of Earth, Mars, Venus, Saturn, and Saturn's largest moon, Titan, Uranus, and Neptune. Measurements of the speed of sound in planetary atmospheres are used to determine composition and transport processes in these atmospheres. For instance, scintillation effects prove the existence of gravity waves in high-density planetary atmospheres. Chapters 15 and 16 cover sound speed in the Sun, normal stars, white dwarfs and neutron stars. As the authors of the chapter on normal stars say "Helioseismology, or the study of solar oscillations, has provided an almost direct and very precise measurement of sound speed in the solar interior." Therefore, with the help of solar models, the sound speed profiles in normal stars may be determined. They are given as functions of radius and age for normal star masses ranging from 0.8 solar masses to 5 solar masses. The final chapter of this *Handbook* discusses the properties of matter in white dwarfs and neutron stars. These are stellar objects whose masses are comparable to those of our Sun, but whose densities may be as much as seven orders of magnitude larger for white dwarfs and fifteen orders of magnitude larger for neutron stars. It is interesting to note that our knowledge of the equation-of-state above nuclear density is not sufficient to deduce that the sound speed above nuclear density is smaller than light speed. Instead this condition is imposed as an additional constraint on candidate equations of state for nuclear star matter.

RICHARD RASPET
MOISES LEVY

List of Contributors

Numbers in parenthesis indicate the pages on which the authors' contributions begins.

SHMUEL BALBERG, PH.D. (423)
Physics Department, University of Illinois at Urbana-Champaign, Urbana, Illinois, USA

HENRY E. BASS, PH.D. (235)
National Center for Physical Acoustics (and) Department of Physics and Astronomy, The University of Mississippi, University, Mississippi, USA

SARBANI BASU, PH.D. (399)
Astronomy Department, Yale University, New Haven, Connecticut, USA

JACK DVORKIN, PH.D. (207, 219)
Geophysics Department, Stanford University, Stanford, California, USA

DANIEL G. FRIEND, PH.D. (267)
Physical and Chemical Properties Division, Chemical Science and Technology Laboratory, National Institute of Standards and Technology, Boulder, Colorado, USA

F.W. GIACOBBE (371)
Chicago Research Center/American Air Liquide, Inc., Countryside, Illinois, USA

K.A. GILLIS, PH.D. (329)
Process Measurements Division, National Institute of Standards and Technology, Gaithersburg, Maryland, USA

BILL HUBBARD, PH.D. (389)
Lunar and Planetary Laboratory, University of Arizona, Tucson, Arizona, USA

J.J. HURLY, PH.D. (329)
Process Measurements Division, National Institute of Standards and Technology, Gaithersburg, Maryland, USA

GREGORY KADUCHAK, PH.D. (3)
Electronic Materials & Devices Group, Los Alamos National Laboratory, Los Alamos, New Mexico, USA

LIST OF CONTRIBUTORS

H. KOJIMA, PH.D. (117)
Rutgers University, Department of Physics and Astronomy, Piscataway, New Jersey, USA

CLAUDE C. LEROY (23, 83)
Private Consultant in Underwater Acoustics, SANARY-sur-Mer, France

R.D. LORENZ, PH.D. (389)
Lunar and Planetary Laboratory, University of Arizona, Tucson, Arizona, USA

PHILLIPE MARTINOTY, PH.D. (159)
Laboratoire de dynamique des fluides complexes, Universite Louis Pasteur, Strasbourg Cedex, France

GARY MAVKO, PH.D. (207, 219)
Geophysics Department, Stanford University, Stanford, California, USA

J.D. MAYNARD. PH.D. (147)
Department of Physics, Pennsylvania State University, University Park, Pennsylvania, USA

J.B. MEHL, PH.D. (329)
Process Measurements Division, National Institute of Standards and Technology, Gaithersburg, Maryland, USA

ROBERT H. MELLEN, PH.D. (83)
Kildare Corporation, New London, Connecticutt, USA

MICHAEL R. MOLDOVER, PH.D. (329)
Process Measurements Division, National Institute of Standards and Technology, Gaithersburg, Maryland, USA

AMOS NUR, PH.D. (207, 219)
Geophysics Department, Stanford University, Stanford, California, USA

MANIKA PRASAD, PH.D. (207, 219)
Geophysics Department, Stanford University, Stanford, California, USA

DR. ANDREA PROSPERETTI (183)
John Hopkins University, Baltimore, Maryland, USA

STEWART L. SHAPIRO, PH.D. (423)
NCSA Senior Research Scientist, University of Illinois at Urbana-Champaign, Department of Physics Urbana, Illinois, USA

DIPEN N. SINHA, PH.D. (3)
Electronic Materials & Devices Group, Los Alamos National Laboratory, Los Alamos, New Mexico, USA

GILES WATON (83)
Laboratoire de dynamique des fluides complexes, Universite de Strasbourg, Strasbourg Cedex, France

J. WILHELM, PH.D. (329)
Process Measurements Division, National Institute of Standards and Technology, Gaithersburg, Maryland, USA

PART 1

ELASTIC PROPERTIES OF LIQUIDS

CHAPTER 1

NONINVASIVE DETERMINATION OF SOUND SPEED AND ATTENUATION IN CHEMICAL WARFARE AGENTS

Dipen N. Sinha and Gregory Kaduchak
Electronic Materials and Devices Group, Los Alamos National Laboratory, Los Alamos, New Mexico, USA

Contents

Abstract 3
   1.1. Introduction 3
   1.2. Theory 5
   1.3. Experimental 13
   1.4. Results and Discussion 14
   1.5. Conclusions 16
Appendix 18
Acknowledgments 20
References 20

ABSTRACT

Swept-Frequency Acoustic Interferometry (SFAI) is a nonintrusive liquid characterization technique developed specifically for detecting and identifying chemical warfare (CW) compounds inside sealed munitions. The SFAI technique can rapidly (<20 s) and accurately determine sound speed and sound attenuation of a liquid inside a container over a wide frequency range. From the frequency-dependent sound attenuation measurement, liquid density is determined. The theory of the technique is described. SFAI is a frequency domain technique, and we describe how accurate time domain information, equivalent to high-quality pulse-echo measurements, can also be extracted from the data. Sound speed data for 40 CW precursor chemicals are also presented.

1.1. INTRODUCTION

There are many situations in which it is necessary to determine information about the liquid contents inside sealed or otherwise inaccessible containers. In such situations, the only choice is to make measurements from outside the container without any direct physical contact between the sensor and the liquid content. The need for such noninvasive measurement is particularly crucial for highly toxic chemicals, such as the various chemical warfare (CW) agents and many of their precursor chemicals. This is necessary to address the requirements of the Chemical Weapons Convention, the treaty that calls for eventual destruction of all chemical weapons [1], where it is essential to have techniques that can be used to monitor compliance and destruction of existing stockpiles in a verifiable way. If the container has a metal wall, as is the case most often for chemical munitions, then one cannot use any of the commonly used optical, electromagnetic (e.g., microwave and radio frequency), or mass spectrometry techniques for liquid characterization or identification. Information regarding thermal characteristics of the liquid inside may be obtained through calorimetric type measurements but is insufficient to identify a chemical. A good possibility is to use nuclear techniques, with which it is possible to determine the presence of various elements in the liquid and guess at the chemical composition based on such elemental information. This approach has been quite successful in identifying CW agents in munitions. Unfortunately, such techniques require a radioactive source, a liquid nitrogen cooled detector, and a long particle counting time. A simple solution is to use ultrasound. Ultrasound can easily penetrate the walls of a container and interrogate the liquid content.

Typically, for the noninvasive characterization of any such liquid filler using ultrasound, one needs to determine various physical properties of the liquid, such as sound speed, sound attenuation, and density of the liquid. This information is often sufficient to identify and characterize various subclasses of chemicals. Popular ultrasonic techniques for the measurement of sound speed and sound attenuation in fluids primarily include resonance reverberation, pulse-echo, and ultrasonic interferometry (UI). Of these techniques, UI, developed several decades ago, stands out for its ability to provide frequency-dependent measurements and for its ability to work with small liquid samples [2–5]. Eggers and Kaatze [6] give a comprehensive review on this subject. Unfortunately, this technique requires the transducers to be in direct contact with the test fluid and precision-machined resonator cells, thereby limiting the applicability of this technique primarily to laboratory fluid characterization. Recently, Pope *et al.* [7] applied this technique for noninvasive monitoring of density and concentration of liquids inside pipes. Although the quality of their data was very poor and extremely noisy compared against typical data obtained from traditional ultrasonic measurements, it showed that it is possible to adapt the UI technique for noninvasive liquid characterization.

Sinha *et al.* [8–13] showed for the first time how to properly adapt the UI technique for noninvasive liquid characterization in general containers and called their approach Swept-Frequency Acoustic Interferometry (SFAI). With this adaptation, it is now possible to measure not only sound speed but also sound attenuation over a wide frequency range, density, and viscosity of liquids inside sealed containers. One interesting feature of the SFAI technique is that it takes advantage of the container wall to extract density and viscosity information. A combination of all these physical properties allows a unique identification of various CW agents inside munitions and other containers noninvasively.

The SFAI technique is primarily a frequency domain technique. However, in this chapter we also discuss how to derive high-quality time domain measurement, equivalent to the traditional pulse-echo measurement, from the frequency data and point out the various advantages of this approach. Although the main focus of this chapter is to present data on the sound speed of CW agents and their precursor chemicals, we also present information on how sound attenuation and density are measured noninvasively.

An important objective of this chapter is to present details about how the UI technique is adapted for noninvasive acoustic characterization of liquids. The chapter is organized as follows: In Section 1.2, we discuss the theory behind the SFAI technique and provide details on the frequency domain analysis. In this section, we also show how accurate time domain information can be derived from the frequency domain data. In Section 1.3, we describe the experimental details. The results and discussion are presented in Section 1.4. Finally, Section 1.5 presents our conclusions.

1.2. THEORY

An SFAI measurement can be described in terms of planar ultrasonic wave transmission and reflection through a multilayered system consisting of the test fluid sandwiched between symmetric layers of transducer crystal, wear plate, coupling gel, and cell (container) wall. To better understand this multilayered system, it is instructive to first start with the basic single-layer model, as shown in Figure 1.1. It shows a volume of liquid between two identical parallel piezoelectric transducers where one transducer is used as the excitation source and the one on the opposite side as the receiver. If a sine-wave electrical signal is applied to the source transducer and its frequency is stepped through in time over a frequency range with appropriate residence time at each frequency step, the signal detected by the receiver transducer is a series of regularly spaced resonance peaks, as shown in the bottom frame of Figure 1.1. The typical regularly spaced interference peaks are observed whenever an integral number of half-wavelengths span the liquid between the two transducers. The sound speed c is then derived from the frequency spacing between any two consecutive resonance peaks Δf as

$$c = 2L\Delta f, \qquad (1.1)$$

where L is the liquid path length. Eq. 1.1 provides accurate measurement in practice, in particular if the measurements are made in a frequency range far from the transducer resonance frequency. For measurements closer to the transducer resonance frequency, corrections due to the frequency-dependent transducer transfer function can be included [14, 15]. The width of the peaks (FWHM) is related to the sound attenuation in the liquid and will be discussed later.

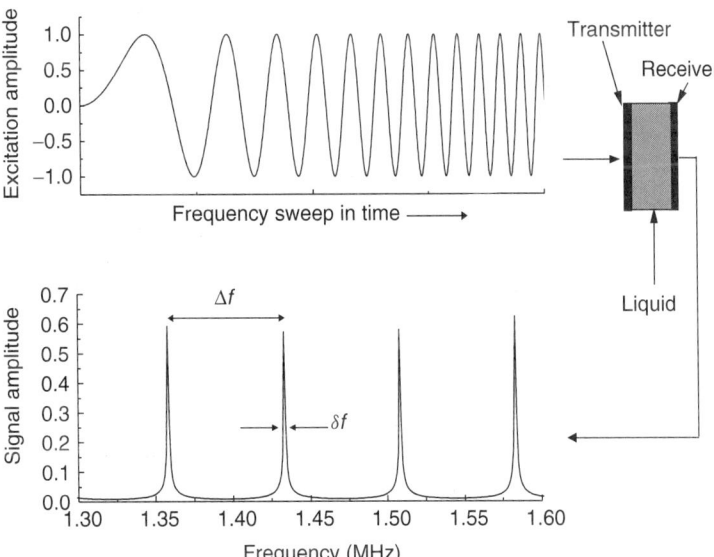

Fig. 1.1. Ultrasonic interferometry with transducers directly in contact with liquid. The top frame shows the excitation signal and the bottom frame shows the interference spectrum due to setting up of standing waves in the liquid at regular intervals.

The spectrum shown in Figure 1.1 can be easily derived from a simple sound transmission model for a single-layer bounded by infinite media and can be found in many textbooks [16] and articles [17, 18]. The solution is general purpose and can also be used for solids, such as metal plates, at near normal incidence.

In noninvasive measurements, one has to deal with container walls; therefore, the system now becomes a problem of sound transmission through a multilayered medium. For a good description of such sound transmission, refer to the excellent book by Brekhovskikh [19]. Hill and El-Dardiry [20] give simpler expressions for three layers and five layers. When all layers are solid, the situation is somewhat complicated because now both compressional and shear wave propagation through the composite medium needs to be considered. In trying to probe a liquid inside a container, the system becomes much simpler because of the presence of the liquid as the main layer. There is no need to worry about shear wave coupling except at the transducer–wall interface. In practice, it turns out that this effect is quite small and can be neglected unless very accurate liquid attenuation measurement is desired. It does not affect the compressional sound speed measurement in the fluid.

Figure 1.2 shows a typical spectrum obtained from a liquid inside a container with the transducers attached to the outside of the container. The spectrum is now much more complicated than in Figure 1.1. It now shows a composite spectrum where two sets of spectra are coupled together: one from the liquid (as in Fig. 1.1) and the other from each wall, the two walls being identical in thickness and material. The sound transmission through the wall now has a periodic structure reaching a maximum whenever standing waves set up within the thickness of the wall. This modifies the spectrum obtained from the liquid and forms the composite spectrum. Typically, the path length in the liquid is much larger than the thickness of a wall. Because the sound speed in metals is much higher than in liquids, the frequency spacing of consecutive wall peaks is quite large (see Eq. 1.1). Only a single wall resonance is shown in Figure 1.2. A closer examination of this figure reveals two important features. First, as the wall peak is approached from either side, the liquid peaks get progressively broader, indicating greater loss. This is because of the higher sound transmission through the wall near the wall peaks.

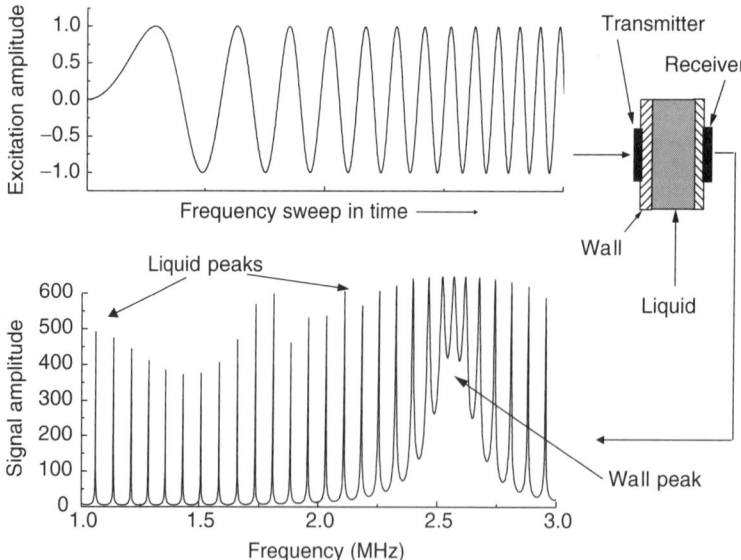

Fig. 1.2. Frequency-sweep measurements in a liquid through a container wall. The irregular variation in transmission amplitude is due to poor resolution in the data.

The second feature is the bunching of the liquid peaks near the wall peak: the liquid peak spacing gets progressively shorter as the maximum of the wall peak is approached due to the coupling between the wall and the liquid, and it is very similar to two coupled oscillators. The impedance mismatch between the wall material and the liquid plays a big role in this coupling. The composite spectrum shown in Figure 1.2 is very similar to that observed in traditional interferometry measurements [21] around the transducer crystal resonance, where the transducer is in direct contact with the liquid. The effect of the transducer alone in the traditional interferometry case is similar to the effect of the wall in the present situation, as shown in Figure 1.2. Although not obvious from the figure, the effect of this wall–liquid coupling extends through the entire frequency range but gets weaker as one moves away from the wall peak. This coupling manifests itself as a small periodic oscillation about its average value in the various observable quantities, such as liquid peak amplitude, liquid width, and liquid peak spacing as a function of frequency. The characteristic of this oscillatory behavior as a function of frequency is the same for each observable quantity. This effect is also observed in traditional interferometry. For example, Gavish *et al.* [22] allude to a fixed oscillatory pattern in the observed sound speed data. A Mathcad program that models the system shown in Figure 1.2 is provided in the Appendix. This program is derived from a straightforward plane wave 1-D propagation model through multiple layers, and it faithfully reproduces the observed behavior in the spectrum described in Figure 1.2.

In practice, the wall–liquid coupling does not affect the sound speed measurement, but it does require care in properly extracting the sound attenuation value in the liquid and will be discussed later. We would like to emphasize here that our goal is to derive sound speed and attenuation values with sufficient accuracy and resolution to be useful for practical applications; this is not meant for high-accuracy acoustic characterization. The SFAI technique shows how to derive proper sound attenuation information by subtracting the effect of the transducer–wall coupling.

It should be pointed out that for use on munitions and other containers, the SFAI technique employs a novel transducer implementation that allows all measurements to be made from only one side of the container using a dual-element transducer system instead of the two transducers on opposite sides, as shown in Figures 1.1 and 1.2.

1.2.1. Frequency Domain Analysis

To derive the physical properties of the liquid from the observed frequency spectrum, the full theory that includes all layers can be used, as presented in the Mathcad program in the Appendix. However, for rapid extraction of such properties, it is more convenient to select an appropriate measurement frequency range to avoid resonance contributions from the walls (e.g., around 4, 6, and 8 MHz in Fig. 1.3). Essentially, in the first order, this reduces the problem to a basic one-layer model, making the analysis significantly straightforward and without introducing significant errors in the measurement of sound speed and sound attenuation. This is similar to avoiding the transducer crystal resonance frequency region in traditional interferometry. The intensity transmission coefficient T for the simplified case of a single fluid layer of path length L, attenuation coefficient $\alpha_L (\alpha_L L \ll 1)$, and sound speed c_L between two identical wall boundaries can be expressed as

$$T = \frac{1}{\left(1 + \frac{1}{2}\sigma \alpha_L L\right)^2 + \frac{\sigma^2 - 4}{4}\sin^2\left(\frac{\omega}{c_L}L\right)} \qquad (1.2)$$

Here, $\sigma = z_w/z_L + z_L/z_w$, $\omega (\omega = 2\pi f)$ is the angular frequency, z_w and z_L are acoustic impedance (ρc) of wall and liquid, respectively, and ρ is the density. For most liquids inside a metal container, $\sigma \approx z_w/z_L$. T_T in Eq. 1.2 is a periodic function of $\omega L/c_L$ and

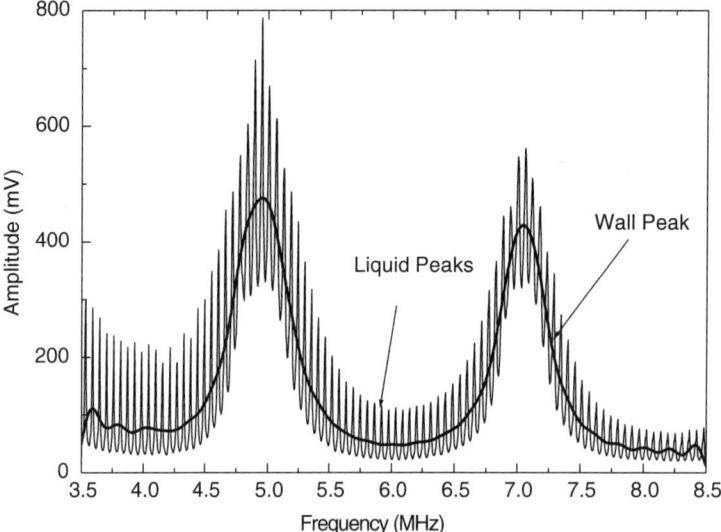

Fig. 1.3. Composite spectra of container wall and liquid. The solid line indicates the sound transmission characteristics of the wall alone. The gradual decrease in the amplitude of the liquid peaks with frequency is due to sound absorption in the liquid. Measurements made in the frequency regions near 4, 6, and 8 MHz (away from the influence of the wall) are similar to measurements without the walls.

reaches a maximum (peak) whenever the condition $2\pi f_n L/c_L = n\pi$ is satisfied, where f_n is the n-th peak frequency. From this condition, the sound speed c_L ($c_L = 2L\Delta f$) can be determined if the frequency difference between any two consecutive resonance peaks $\Delta f = f_{n+1} - f_n$ is measured. However, we see from both experiments and model predictions that Δf oscillates (>1% in most cases in practice) around the mean value over the measurement frequency range due to effects of container wall resonance modulation, a characteristic behavior of the noninvasive measurement as mentioned earlier. Therefore, c_L is obtained using the averaged frequency difference $<\Delta f>$ (e.g., between 5.7 and 6.4 MHz in Fig. 1.3)

$$c_L = 2L<\Delta f> \qquad (1.3)$$

Next, we discuss how sound attenuation and liquid density are determined from the frequency spectrum. The ratio of transmission coefficient minima, $T_{\min}$, and maxima, $T_{\max}$, can be expressed in terms of σ and α_L as [18]

$$\frac{T_{\min}}{T_{\max}} = \frac{2}{\sigma} + L\alpha(f^2) \qquad (1.4)$$

Equation 1.4 shows that both α_L and σ can be determined from a linear fit of the data of the transmission ratio factor as a function of f^2. The intercept at zero frequency is related to the acoustic impedance ratio σ. If the impedance of the wall material is known, the liquid density can be determined because the liquid sound speed is determined independently, as discussed above (see Eq. 1.3) and σ is the only unknown.

Another approach to determining the sound attenuation coefficient is to use the half-power bandwidth of observed resonance peaks. From Eq. 1.2, one can derive an inverse solution for the half-power bandwidth, δf, in terms of acoustic properties in the fluid as [13, 18]

$$\delta f = \frac{2c_L}{\pi \sigma L} + \frac{c_L \alpha_L(f^2)}{\pi} \qquad (1.5)$$

Similar to Eq. 1.4, the second term is the contribution from liquid sound absorption and is identical to the solution obtained in a previous resonator theory [23, 24] for

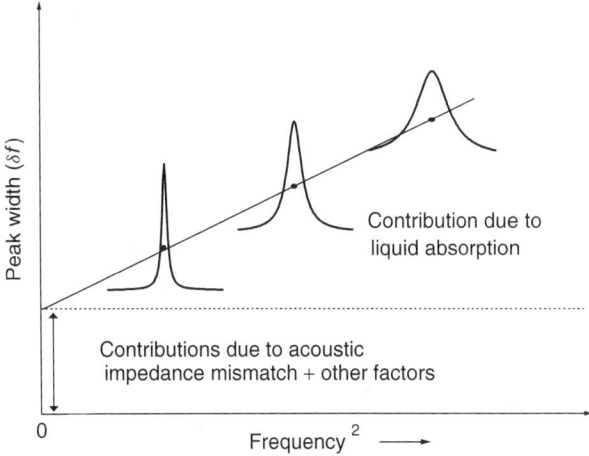

Fig. 1.4. Liquid peak width as a function of square of frequency. The slope of the line provides information on sound attenuation, whereas the intercept extrapolated to zero frequency provides a measure of acoustic impedance mismatch between the container wall and the liquid.

transducers in direct contact with the test liquid. The first term, the width extrapolated to zero frequency, δf_0, is independent of frequency and depends on σ, c_L, and L. This term results from the reflection loss at the wall–liquid interface due to the acoustic impedance mismatch and can be used to determine liquid density if the acoustic impedance of the wall is known. Figure 1.4 shows a plot of Eq. 1.5. The intercept is simply the first term in Eq. 1.5 and the slope, the second term. A similar plot is obtained from Eq. 1.4 as well. Either of these approaches can be used to determine sound attenuation and liquid density. From the first term of Eq. 1.5, since c_L is determined independently, the liquid density can be determined if the acoustic impedance of the wall material is known. Typically, one needs to know only the density of the wall material because the sound speed can be determined from the wall peak spacing (see Fig. 1.3).

In practical measurements, however, other loss mechanisms — such as diffraction loss and losses due to the transducer backing and tangential mode generation in the container — although small in magnitude, also contribute to the measured peak width, and their effects need to be included. The diffraction loss effects can be minimized by avoiding the low-frequency region, typically 1 MHz and below for most measurements. If the liquid path length inside the container is of comparable dimensions to that of typical commercially available transducers, all the measurements at higher frequencies ensure that a near-field condition is met. Being in this Fresnel region minimizes the beam divergence effects and, therefore, the diffraction loss [24, 25]. However, for accurate determination of sound attenuation and density at any frequency range and particularly for larger containers (path length) where near-field conditions are not ensured, the diffraction loss needs to be subtracted from the measured δf. We have used the standard expression [26] $\delta f_{\text{diff}} = \frac{0.147}{\pi \beta} \left(\frac{2c_L}{D}\right)^3 f^{-2}$, where D is the effective diameter of the transducer and $\beta = z_L/z_W$. Other losses can be accounted for by using a reference liquid of known acoustic properties [6, 23, 26]. Therefore, the attenuation coefficient of the test liquid α_L can be written as

$$\alpha_L = \frac{\pi}{c_L}(\delta f - \delta f_0) - \frac{\pi}{c_L}(\delta f_r - \delta f_{r0}) + \frac{c_r}{c_L}\alpha_r \qquad (1.6)$$

Here, δf is the measured half-power bandwidth of the test liquid at any selected frequency, f, and δf_0 is the extrapolated width at $f = 0$. Subscript "r" refers to the reference liquid. From Eq. 1.6, the attenuation coefficient factor $\alpha_0 = (\alpha_L/f^2)10^{17}$ np

s^2/cm can be determined with good accuracy. The density of the test liquid can be determined from

$$\rho_L = \delta f_{0c} \frac{\pi c_w \rho_w L}{2 c_L^2} \qquad (1.7)$$

Here, δf_{0c} is the extrapolated peak-width value (at $f^2 = 0$) that is corrected for miscellaneous small losses by using a reference liquid. This density determination thus takes advantage of the container wall. On the other hand, the measurement of peak width or the $T_{\min}/T_{\max}$ ratio over a wide range of frequencies allows determination of the sound attenuation in the liquid.

1.2.2. Time Domain Analysis: Traditional Pulse-Echo and Pulse-Echo Transform

Historically, the most popular approach to determining sound velocity has been the pulse-echo time-of-flight technique [27]. In the technique, a narrow electrical pulse is used to excite a transducer attached to a container or in direct contact with a liquid. The pulse generates a pressure pulse that propagates through the liquid, reflects from the opposite wall of the container, and gets detected by either the same transducer or by a second transducer. By measuring the time it took for the pulse to make the round trip travel over the known distance inside the container, it is possible to determine the sound velocity. By transforming the time domain signature to the frequency domain, it is also possible to gather other physical properties of the liquid (e.g., attenuation and density) [28, 29]. This section will describe how to do high-accuracy, high–signal-to-noise measurements in the time domain by taking advantage of the properties of the *analytic* signal and frequency domain data.

The concept of the analytic signal has been around for several decades [30]. It has received much attention by researchers in the radar and sonar community. Its use in NDE applications has been somewhat limited [31, 32]. The analytic signal is described by a real-valued band pass signal that can be represented by complex signals with one-sided spectral densities. In the time domain, the analytic signal is simply written as a real signal, $f(t)$ (which is typically a time-varying quantity measured in an experiment), and its quadrature component, $g(t)$ (phase offset by 90° from the signal $f(t)$) [33].

$$z(t) = f(t) + i g(t) \qquad (1.8)$$

This signal has an interesting property in the frequency domain, as shown by the Fourier transform of Eq. 1.8.

$$Z(\omega) = \begin{cases} 2F(\omega) & \omega > 0 \\ 0 & \omega < 0 \end{cases} \qquad (1.9)$$

The spectrum of the analytic signal is one-sided and is composed wholly of the positive frequency portion of the real part of the analytic signal. It is not within the scope of this chapter to give a detailed account of the analytic signal or the Hilbert transform. For this, the reader is referred to the references. The technique of using only the positive frequency portion of a real signal, as shown in Eq. 1.9, is the Hilbert transform.

The analytic signal cannot be appreciated until some examples are given. Consider a measured frequency signature in an experiment such as the one shown in Figure 1.5(a). The spectrum is complex and has been quadrature sampled, as discussed in Section 1.3 of this chapter. The spectrum represents a sinusoidal pulse of center frequency 2 MHz modulated with a Gaussian envelope. The Fourier transform of the spectrum is displayed in Figure 1.5(b). The resultant signal is oscillatory, and a time delay measurement can be made by measuring the location of zero crossings, amplitude extrema, or correlating the pulse with a replica. Now conduct a Hilbert transform on the spectral data in Figure 1.5(a). Erasing the negative frequency components in the spectrum

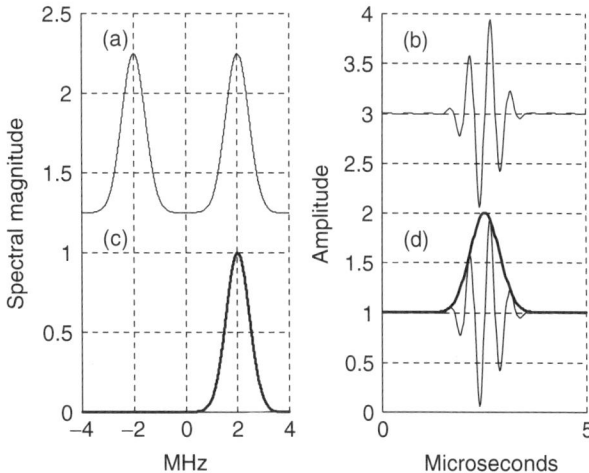

Fig. 1.5. (a) Spectrum of a 2-MHz pulse with a Gaussian envelope. (b) Fourier tranform of the spectrum in (a) displaying its time domain representation. (c) Same spectrum as in (a) except the negative frequency components have been set to zero. (d) Fourier transform of the spectrum in (c) representing the results of the Hilbert transform. The bold, solid line is the magnitude of the analytic signal. The real part of the analytic signal is the replica of the signal in (b). The plots in this figure have been offset for clarity.

yields the one-sided spectrum in Figure 1.5(c). Performing a Fourier transform on this spectrum yields the analytic signal. The real part of the analytic signal is shown by the oscillatory, solid line in Figure 1.5(d) [and is a replica of the signal seen in Figure 1.5(b)]. But the magnitude of the analytic signal is shown as the bold, solid line in Figure 1.5(d). It is the envelope of the real signal. It can be shown that this envelope is proportional to the rate-of-arrival of the total energy of the pulse at the measuring device. Thus, time-delay measurements may be made easily and effectively by locating the maxima of the envelope.

In practice, we derive the analytic signal from swept frequency data. Exactly the same type of parameters may be found from analysis of the analytic signal as from analyzing the data in the frequency domain from an SFAI frequency sweep measurement. Consider the experiment diagrammed in Figure 1.7(b). Instead of applying a frequency sweep, pulse is passed through the wall of a container, propagated through a liquid conduit, and then received after passing through another wall. The frequency spectrum for this experiment is shown in Figure 1.6. After performing the Hilbert transform on this spectral data, the resultant time signature is displayed in Figure 1.7(a). It consists of clusters of closely spaced impulses. The first cluster corresponds to the energy, which has traveled once through the liquid conduit. The closely spaced impulses within the cluster represent multiple reflections within the container wall, as shown in Figure 1.7(b). An accurate time-of-flight through the fluid is derived from the time difference between the peaks of the first impulse in each cluster (labeled A and B in the first two clusters). This subtracts out the time the pulse requires to propagate through the container wall, which is not always known for a measurement on unknown systems. A closeup of the first cluster of echoes labeled A is shown in Figure 1.7(b). Impedance mismatch information about the liquid relative to the container is found in the rate of decay of the peaks in Figure 1.7(b). Using this impedance mismatch information and the decay of the amplitude of the first peak in each cluster shown in Figure 1.7(a) (i.e., A and B), one determines the sound attenuation.

One advantage of using this particular type of time domain approach, which is derived from the frequency sweep data, is the ability to achieve much higher signal-to-noise ratios for much lower values of the excitation voltage. In the experiments,

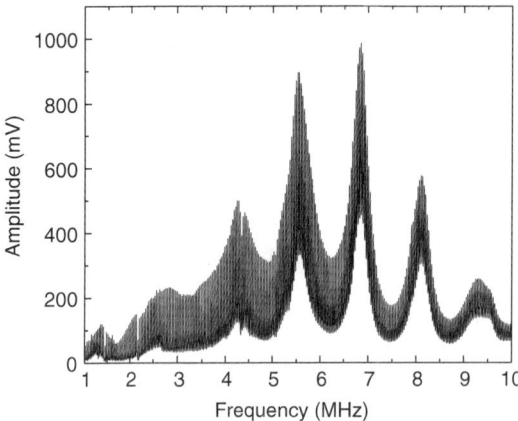

Fig. 1.6. Wide range frequency response of a stainless steel container with ethylene glycol inside.

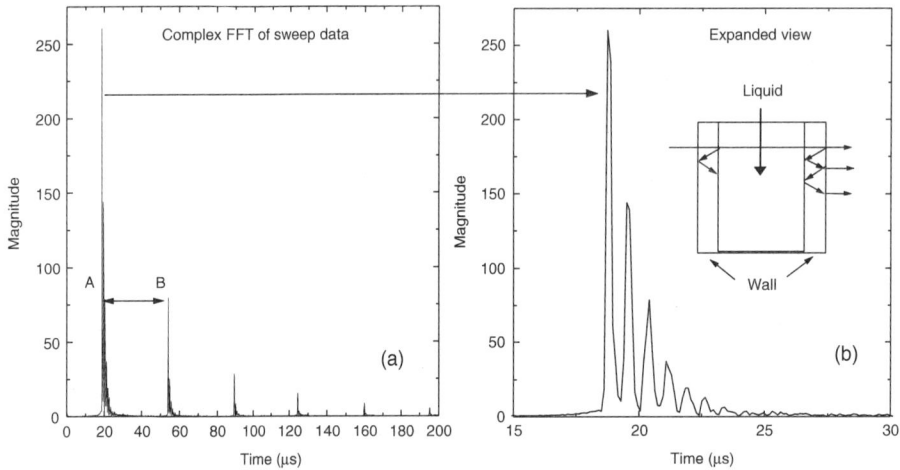

Fig. 1.7. Complex FFT of frequency-sweep data to obtain time domain information. (a) The multiple echoes through the liquid as would be obtained in a typical pulse-echo measurement. (b) An expanded view of the first set of peaks (labeled A) in (a) that clearly shows the multiple reflections in the container wall.

measurement at each frequency is made through a narrowband (~100 Hz) tracking filter, restricting the received noise to a narrow, sliding frequency window throughout the entire frequency sweep range. In contrast, the noise window for pulse-echo types of measurements would be the equivalent of the entire bandwidth of the transmitted pulse (typically on the order of several megahertz). Thus, for a 10-MHz bandwidth measurement, the SFAI technique can provide a signal-to-noise improvement of approximately five orders of magnitude when comparing the filter widths.

$$\frac{\text{filter width swept-frequency}}{\text{filter width pulse} - \text{echo}} = \frac{100 \text{ Hz}}{10,000,00 \text{ Hz}} = 10^5 \quad (1.10)$$

It should be noted that not only is there an increase in the signal-to-noise ratio by recording frequency domain data with a narrow filter, but the system being measured is allowed to reach a resonant state at each frequency. Thus, the Q of the system will increase the signal levels for a given input drive voltage.

Based on the analysis presented in this section, a single sweep measurement can be used to derive multiple physical properties of a liquid, including sound speed, sound absorption, frequency dependence of sound absorption, and liquid density. A combination of several of these properties can be used to uniquely identify all the CW compounds and their most significant precursors.

1.3. EXPERIMENTAL

Any conventional electronic system, such as Network Analyzers or Impedance Analyzers, can be used for SFAI measurements. It is also possible to design a simple system based on a computer-controlled function generator (preferably the direct digital synthesis kind) to provide a frequency sweep capability and signal processing electronics that can convert the amplitude of the received sine-wave response from the measurement system. The signal processing circuit can be either an envelope detector or an RMS-to-DC converter (available as single integrated circuit modules from various electronics vendors). However, to obtain a high signal-to-noise ratio measurement, it is advisable to use a tracking band pass (1 kHz or less bandwidth) filter. Typically, this can be built from a simple heterodyne circuit, such as an AM radio. The amplitude signal can then be digitized and recorded in the computer memory for on-screen display and data analysis.

The SFAI system that we use comprises an electronics unit that consists of a 486DX4-100 MHZ embedded PC-104 bus computer and a customized electronics Digital Synthesizer and Analyzer system, model DSA520, made by Nick Electronics, Houston, TX. This instrument is custom designed for SFAI measurements. The DSA520 system contains all electronics for sine-wave sweep signal generation, signal detection, and processing. Another version of this consists of a PC plug-in card (DSA220) that can be used with any desktop PC or any computer with a full-length ISA slot. The sweep frequency range available is 1 kHz to 10 MHz. The sweep rate can be varied from 1 to 800 frequency steps per second. With a resolution of 0.1 Hz over the entire frequency range, the system provides a frequency response directly in real time. Typical excitation voltage levels used for the measurements presented in this chapter varied between 0.1 and 0.5 V peak to peak. The maximum excitation signal available in this system is 10 V.

The transducer system consists off-the-shelf, commercially available broadband piezoelectric transducers. It is preferable that the transducers do not have delay lines attached on the front surface such that the front face can be applied directly on top of the container wall. A thin wear plate on the top surface for protection from mechanical damage is acceptable. Transducers made from 1-3 composite material seem to provide excellent data, but any broadband transducer works fine. The wider the transducer bandwidth, the better the quality of the measurement. This makes the effect of the transducer resonance characteristics less of a problem for attenuation measurement over a wide frequency range. Wide bandwidth means that the transducer frequency response is rather flat compared with a highly peaked response for narrow band transducers. It is particularly important for the time domain analysis.

The signal processing involves a heterodyne mixing technique followed by signal rectification, envelope detection, and 14-bit digitization. The digitized amplitude spectrum is then analyzed in the PC to determine sound speed, sound attenuation, and density of the sample liquid by the methods described in Sections 1.2.1 and 1.2.2. The combined measurement and analysis time is typically <30 s.

For the time domain analysis, we use a homodyne signal processing technique where the real and imaginary components of the received signal are determined. This can be used to obtain both signal amplitude and phase. The bandwidth used for this is ~100 Hz. A complex FFT is performed to derive the time domain information, as discussed in the previous section. Both the DSA220 and DSA520 systems have provisions for both heterodyne and homodyne signal processing.

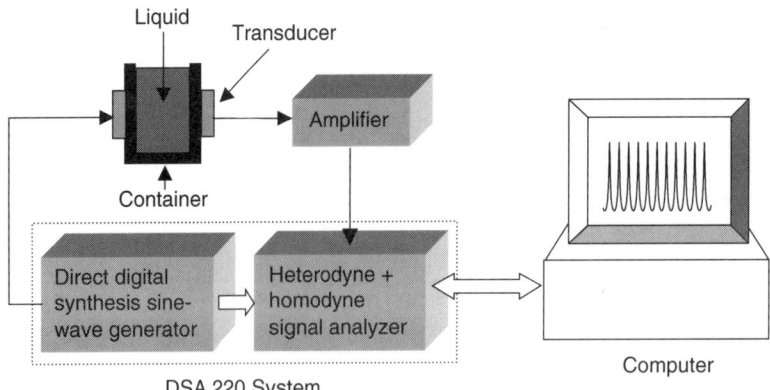

Fig. 1.8. Experimental setup. The DSA220 system is a PC plug-in card that resides inside the desktop computer. It is controlled by the computer through the ISA bus.

The experimental setup is shown schematically in Figure 1.8. The measurements on CW agents and precursors presented in this chapter were made in cylindrical glass cells used for optical measurements (Catalog Number 32-G-10, Starna Cells, Inc., Atascadero, CA). These cells were made of optical glass, with a path length of 10 mm, wall thickness of 1.25 mm, an exterior diameter of 22 mm, and a volume of 2.8 ml. The transducers used were Panametrics Contact Transducers Videoscan (part number V111-RM), 10-MHz center frequency, and element size of 13 mm. A thin film of ultrasonic gel was used for transducer coupling to the Starna cells. In addition, the transducers had a slight spring loading applied to maintain constant pressure for each measurement. A Peltier cooling–heating system was used to vary the temperature of the Starna cell and its contents and to maintain the temperature to within 1 °C. For safety reasons, all experiments were performed under a fume hood in a special laboratory. The transducer and container dimensions are such that all measurements are made in the near-field condition, which minimizes the effect due to diffraction, and the analysis is straightforward. The sound field inside the Starna cell between the two opposite transducers can be considered as plane waves confined inside a cylindrical place [24, 25] of the near-field Fresnel region. Because of the use of identical receiver and transmitter, the transmitted signal as detected by the receiver transducer is effectively integrated over the beam diameter.

1.4. RESULTS AND DISCUSSION

We show results of SFAI measurements in optically polished rectangular glass cells of liquid path length 1.0 cm and wall thickness 0.125 cm for various chemicals, including CW compounds. Sound speed can be accurately determined from Eq. 1.3. Figure 1.9 shows the sound speed measurements for glycerine, ethylene glycol, water, toluene, benzene, and isopropanol in a glass cell over a frequency range of 1–12 MHz. For these liquids, the maximum standard deviation relative to the mean values of sound speeds is 0.59%. The scatter (actually an oscillation) in the sound speed data is not due to measurement error but primarily to the wall resonance modulation effect, as discussed in Section 1.2.1., see Eq. 1.3. This oscillation is also seen in predictions from our theoretical model and is discussed in Section 1.2.

Figure 1.10 compares the sound attenuation measurement for ethylene glycol in two very different containers, a small glass cell (as in Fig. 1.8) and a cylindrical stainless steel shell with an inner diameter of 5.27 cm and a wall thickness of 0.225 cm. The resonance peak width is plotted as a function of the square of the frequency. The data

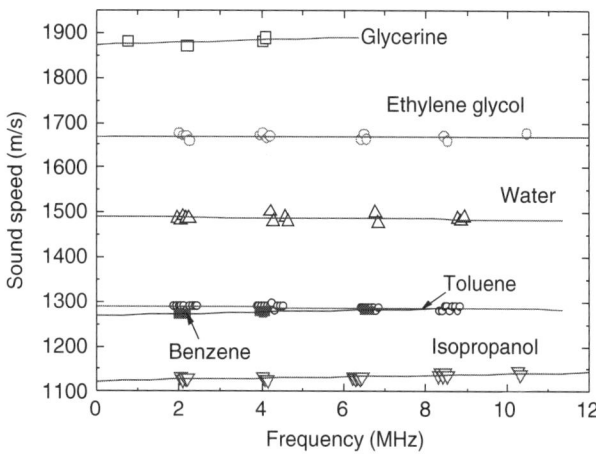

Fig. 1.9. Sound speed determination for six chemicals using Eq. 1.3.

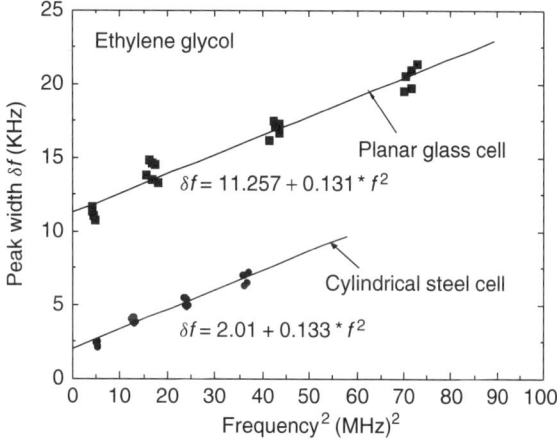

Fig. 1.10. Swept-frequency acoustic interferometry measurements on a planar glass cell and a cylindrical steel cell filled with ethylene glycol.

are consistent with Eq. 1.5. Because the acoustic impedances are very different for glass and steel, the intercepts are different. For the cylindrical container, the effect of curvature on the peak width is now included in the intercept and can be separated. It is evident in the experimental data that qualitatively the same behavior in the frequency-dependent peak width is observed for both planar and cylindrical containers. The slopes of the two lines are the same, indicating the same sound attenuation values. This demonstrates that reliable sound attenuation measurements can be made using the SFAI technique regardless of the container geometry and container wall material properties. Figure 1.11 presents the results of the noninvasive SFAI characterizations of the most common CW compounds. The data presented are for measurements made in Starna glass cells using CW agents that are more than 90% pure and at an ambient temperature of 23 °C. The density for each liquid was determined using Eq. 1.4. This forms the basis for the SFAI technique in noninvasive CW agent identification. Note that in a 3-D representation of the measured data, each liquid is clearly separated from the others, and this separation is many times larger than the accuracy of the measurement. For measurements using the Starna cells, the accuracy of sound speed measurement is approximately 0.5% and that of sound attenuation is more than 3%. Density determination is somewhat less accurate and is typically 5%. All measurements

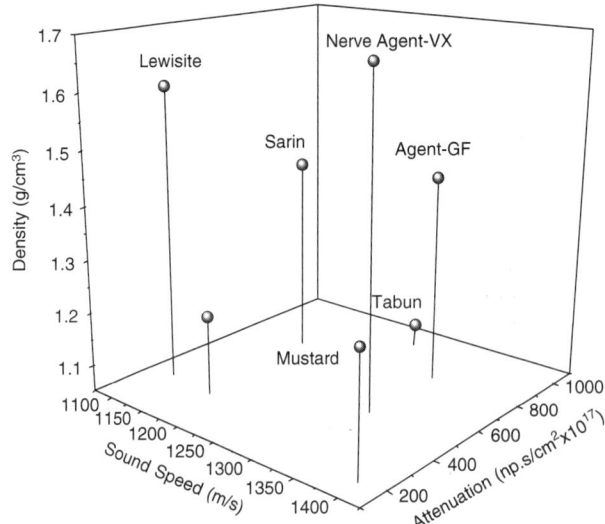

Fig. 1.11. Sound speed, sound attenuation, and density determination for seven chemical warfare agents using the swept-frequency acoustic interferometry technique using glass Starna cells of path length 1 cm and at 26.5 °C.

Table 1.1. Sound speed values of chemical warfare agents at several temperatures.

Agent	Sound speed (m/s)				Density at 25 °C (g/cm^3)
	13 °C	17.5 °C	26.5 °C	35 °C	
Tabun (GA)	1290.3	1272.8	1253.21	1220.5	1.073
Sarin (GB)	1196.77	1181.9	1154.1	1129.4	1.088
Soman (GD)	1211.79	1194.3	1172.49	1149.5	1.022
GF	1354.17	1345.7	1322.29		1.434
Sulfur mustard (HD)	1458.18	1421.8	1398.18	1367.8	1.268
Lewisite (L)	1141.82	1124.8	1107.15	1080.3	1.890
VX	1356.60	1340.10	1317.38	1283.62	1.008

were made over a temperature range between 4 and 35 °C using a specially built Peltier heater–cooler based temperature-controlled system.

The sound speed values of the CW agents and 40 precursor chemicals are presented in Tables 1.1 and 1.2.

1.5. CONCLUSIONS

In this chapter, we showed how acoustic characterization of liquids can be made noninvasively. In particular, sound speed, sound attenuation, and density can be determined over a wide frequency range. The equipment used limits the frequency range. The approach is based on an adaptation of the UI technique. In this particular adaptation, one can take advantage of the container wall itself to determine liquid density. Moreover, it is shown that if the frequency region of analysis is chosen away from the wall resonance (thickness mode), then the analysis reduces to traditional interferometry where the transducers are in direct contact with the liquid. The methodology presented is applicable for containers that are both planar and cylindrical. Although this is a frequency domain technique, the complex frequency spectrum can be Hilbert transformed to derive high signal-to-noise ratio time domain measurements. The technical approach presented here is meant

Table 1.2. Sound speed of precursor chemicals at three different temperatures.

Chemical name	Sound speed (m/s)			Density (g/cm^3)
	20 °C	25 °C	30 °C	
1, 4-Butanediol		1601.71	1576.63	1.017
1, 5 Pentanediol	1593.96	1592.15	1573.82	0.978
Sulfur monochloride 98%	1202.47	1187.60	1163.69	1.690
N, N-Diethylethanolamine 99%	1346.42	1431.56	1317.50	0.892
4-picoline	1439.52		1279.20	0.955
Diethyle phosphite 98%	1257.14	1261.98	1247.96	0.920
Phosphorus trichloride	986.78	957.46	933.10	1.574
Phosphorus oxychloride 99%	995.51	986.70	977.14	1.645
N-ethyl diethanolamine 98%	1504.13	1522.41	1490.79	1.013
Triethyl phosphite 98%	1117.02	1126.04	1093.79	0.963
Pinacolone 98%	1186.76	1180.64	1164.39	0.723
Ethylsonicotinate 98%	1413.71	1407.60	1394.29	0.993
1,4-Dichloro butane 97%	1289.43	1346.62	1269.92	1.141
3,3-Dimethyl-2-butanol 98%	1210.55	1215.84	1188.07	0.812
Triethanol amine 98%	1601.82	1622.15	1593.41	1.124
1,3-Dichloropropane	1243.37	1244.95	1193.68	1.187
N-Methyl diethanolamine	1567.91	1576.54	1559.50	1.038
Dimethyl methyl-phosphonate 97%	1378.52	1376.97	1355.02	1.151
Triisopropyl phosphite 95%	1080.70	1091.23	1059.50	0.968
Diethyl methyl phosphonate 97%	1267.73	1268.96	1244.09	1.041
2-(Diisopropylamino)-ethanol 99%	1328.84	1325.29	1309.19	0.826
1,5-Dichloropentane 99%	1293.18	1312.17	1294.63	1.101
2-Chloroethyl ether 99%	1328.11	1328.31	1308.47	1.220
Diethyl chlorothiophosphate 97%	1158.77	1182.44	1134.78	1.183
Ethyl bromoacetate 98%		1119.91	1102.26	1.503
Trimethyl phosphite	1128.10	1154.63	1106.96	1.052
2-Mercapo ethanol 98%	1485.48	1483.74	1469.46	1.114
Diethyl dithiophosphate	1278.52	1276.89	1262.66	1.247
2,2-Thiodiethanol 99%	1680.16	1683.37	1668.15	1.075
2-Chloroethanol	1356.92	1358.18	1340.48	1.202
Thionyl chloride	1040.71	1038.59	1022.87	1.803
Sulfer dichloride	1104.76	1112.68	1087.47	1.620
Dimethyl phosphite 98%	1336.94	1333.71	1318.52	1.200
Dimethyl sulfate 99%	1240.13	1239.43	1221.93	1.332
Dichloro methane 99.6%	1079.49	1096.40		1.326
Carbon disulfide 99.9%	1150.11	1145.45		1.255
Diisopropylamine	1101.63	1093.98	1073.23	0.715
Dithanolamine 99%	1722.46	1699.62	1707.18	1.097
Arsenic (III) chloride 99.99%	1021.23	1016.68	1007.52	2.150
2-Methyl-2-butane 99%	1080.72	1066.29		0.662
Pinacolyl alcohol		1230.34		0.812
O,O'-diisopropyl methylphosphonate		1184.88		1.036
Diisopropyl [hydrogen] phosphite		1186.30		0.997

primarily to derive sound speed, sound attenuation, and density values with sufficient accuracy for practical use. This is not designed for the highest quality measurements to define fluid characterization standards. Using the methodology presented in this chapter, we determined the sound speed and attenuation for all the common CW agents. This is the first such systematic measurement of these types of chemicals. We also present sound speed data on 40 precursor chemicals to these CW agents.

APPENDIX

This program analyzes in one-dimension acoustic plane wave interactions normally incident on a system consisting of five distinct regions with the following boundaries: glass–quartz, water–glass, glass–water, and quartz–glass. The program is customized for the glass Starna cells used for the measurements presented in the chapter. It can be easily adapted for other systems by simply replacing the wall thickness, density, attenuation, and sound speed values for glass with other material. The transmitted amplitude is calculated as a pressure, defined as the inverse of the absolute value of the average of components z_3 and z_4.

$d_1 := 0.01$ the thickness of the liquid region (m)
$c_0 := 5750$ the speed of sound in crystal (approximated) (m/s)
$d_2 := 0.00125$ the thickness of the Pyrex glass (m)
$c_1 := 1480$ the speed of sound in water (m/s)
$\rho_0 := 0.1$ the density of crystal (measured) (g/cm^3)
$c_3 := 5640$ speed of sound in Pyrex glass (m/s)
$\rho_1 := 1.0$ the density of water (g/cm^3)
$i := \sqrt{-1}$ complex number defined
$\rho_3 := 2.6$ the density of Pyrex glass (g/cm^3)
$j := 0.999$ the range variable (Mathcad starts at 0)
points $:= 1000$ number of data points
startf $:= 1.0$ the initial frequency in MHz
stopf $:= 5.0$ the final frequency in MHz

$$\text{stepf} := \frac{\text{stopf} - \text{startf}}{\text{points}}$$

the incremental frequency step size as determined by the number of points

$f_j := \text{startf} + \text{stepf} \cdot j$ the frequency range in MHz
$\omega_j := 2 \cdot \pi \cdot f_j \cdot 10^6$ the conversion of linear frequency to angular frequency
$\alpha_1 := 25.3 \cdot 10^{-15}$ the absorption coefficient for water
$\alpha_3 := 2.5 \cdot 10^{-17}$ the absorption coefficient for Pyrex glass

Glass–Quartz Interface

$$k0_j := \frac{\omega_j}{c_0}$$

propagation constant in quartz

$$k3_j := \frac{\omega_j}{c_3}$$

propagation constant in glass

$e1_j := \exp[-\alpha_3 \cdot (f_j)^2 \cdot 10^{12} \cdot d_2]$ the exponential dependence of the attenuation coefficient for Pyrex glass

$$z_{al} := \frac{\rho_0}{\rho_3}$$

ratio of quartz to Pyrex glass densities: equivalent to the sum of the psi field amplitude coefficients $(B + C)$

$$zr1_j := \frac{k0_j}{k3_j + i \cdot \alpha_3 \cdot (f_j)^2 \cdot 10^{12}}$$

the difference of the amplitude psi field coefficients $(B - C)$

$$zb_j := \frac{(z_{al} + zr1_j)}{2}$$

solution of the psi field amplitude coefficients B and C

$$zc_j := \frac{(z_{al} - zr1_j)}{2}$$

$\text{ang1}_j := k3_j \cdot d_2$ the phase shift in Pyrex glass through distance $d2$, and modified by the sweep in frequency

$\text{euler1}_j := \cos(\text{ang1}_j) + i \cdot \sin(\text{ang1}_j)$ the sum and difference of the complex Euler angles in Pyrex glass

$\text{euler2}_j := \cos(\text{ang1}_j) - i \cdot \sin(\text{ang1}_j)$

$zb2_j := \dfrac{zb_j \cdot \text{euler2}_j}{e1_j}$

evaluation of the psi field at an infinitesimal amount within boundary 2

$zc2_j := zc_j \cdot \text{euler1}_j \cdot el_j$

Water–Glass Interface

$k1_j := \dfrac{\omega_j}{c_1}$ propagation constant in water

$z_{\text{am}} := \dfrac{\rho_3}{\rho_1}$ equivalent to the sum of the amplitude coefficients $(E + F)$ divided by the sum of the amplitude coefficients $(B + C)$

$zr2_j := \dfrac{k3_j + i \cdot \alpha_3 \cdot (f_j)^2 \cdot 10^{12}}{k1_j + i \cdot \alpha_1 \cdot (f_j)^2 \cdot 10^{12}}$ equivalent to the difference of the amplitude coefficients $(E - F)$ divided by the difference of the coefficients $(B - C)$

$zd_j := z_{\text{am}} \cdot (zb2_j + zc2_j)$
$ze_j := zr2_j \cdot (zb2_j - zc2_j)$ imposing boundary conditions at boundary 2

$zr3_j := \dfrac{zd_j + ze_j}{2}$

$zr4_j := \dfrac{zd_j - ze_j}{2}$ solution of the psi field amplitude coefficients E and F

$\text{ang2}_j := k1_j \cdot d_1$ the phase shift in water through distance $d1$, and modified by the sweep in frequency

$\text{euler3}_j := \cos(\text{ang2}_j) - i \cdot \sin(\text{ang2}_j)$
$\text{euler4}_j := \cos(\text{ang2}_j) + i \cdot \sin(\text{ang2}_j)$ the sum and difference of the complex Euler angles in water

$e2_j := \exp[-\alpha_1 \cdot (f_j)^2 \cdot 10^{12} \cdot d_1]$ exponential dependence of the attenuation coefficient for water

$zb3_j := \dfrac{zr3_j \cdot \text{euler3}_j}{e2_j}$ evaluation of the psi field at an infinitesimal amount within boundary 3

$zc3_j := zr4_j \cdot \text{euler4}_j \cdot e2_j$

Glass–Water Interface

$z_{\text{an}} := \dfrac{\rho_1}{\rho_3}$ equivalent to the division of the amplitude coefficients $(G + H)$ by the amplitude coefficients $(E + F)$

$zr5_j := \dfrac{k1_j + i \cdot \alpha_1 \cdot (f_j)^2 \cdot 10^{12}}{k3_j + i \cdot \alpha_3 \cdot (f_j)^2 \cdot 10^{12}}$ equivalent to the difference of the amplitude coefficients $(G - H)$ divided by the difference of the coefficients $(E - F)$

$zf_j := z_{\text{an}} \cdot (zb3_j + zc3_j)$

$$zg_j := zr5_j \cdot (zb3_j - zc3_j)$$ imposing boundary conditions at boundary 3

$$zr6_j := \frac{zf_j + zg_j}{2}$$

$$zr7_j := \frac{zf_j - zg_j}{2}$$ solution of the psi field amplitude coefficients G and H

$$zb4_j := \frac{zr6_j \cdot \text{euler2}_j}{e1_j}$$ evaluation of the psi field at an infinitesimal amount within boundary 4

$$zc4_j := zr7_j \cdot \text{euler1}_j \cdot e1_j$$

Now Calculate the Input PSI Field

$$z3_j := \frac{\rho_3}{\rho_0} \cdot (zb4_j + zc4_j)$$ equivalent to the sum of the psi field amplitude coefficients $(P + A)$

$$zs_j := \frac{k3_j + i \cdot \alpha_3 \cdot (f_j)^2 \cdot 10^{12}}{k0_j}$$ equivalent to the difference of the psi field amplitude coefficients $(P - A)$

$$z4_j := zs_j \cdot (zb4_j - zc4_j)$$ imposing boundary condition at boundary 4

$$\text{trans}_j := \frac{z3_j + z4_j}{2}$$ solution of psi field coefficient P

$$\text{pressure}_j \frac{1}{|\text{trans}_j|}$$ the transmitted pressure amplitudes as a function of swept frequency

$$\text{reflected}_j := \frac{(z3_j - z4_j)}{2}$$

solution of psi field coefficient A

$$\text{backpressure}_j \frac{1}{|\text{reflected}_j|}$$

the reflected amplitudes as a function of swept frequency

ACKNOWLEDGMENTS

This work was supported by the Defense Threat Reduction Agency. We are grateful to Kendall Springer and David Lizon for support with the measurements on CW agents and precursor chemicals. We are indebted to Wei Han for developing the methodology and extensive analysis of the data. We thank Mike Keleher for many valuable discussions on CW agents, Fred Mueller for deriving the 1-D multilayer wave propagation model, and Roger Hasse for writing the Mathcad program.

References

1. Barnaby, F. (1994). The Destruction of chemical warfare agents. *Interdisc. Sci. Rev.* **19**: 190.
2. Hubbard, J.C. (1931). The acoustic resonator interferometer: I. the acoustic system and its equivalent electric network. *Phys. Rev.* **38**: 1011.
3. Fox, F. (1937). Ultrasonic interferometry for liquid media. *Phys. Rev.* **52**: 973.
4. Pethrick, R.A. (1972). The swept frequency acoustic resonant interferometer: Measurement of acoustic dispersion parameters in the low megahertz frequency range. *J. Phys. E* **5**: 571.
5. Eggers, F., and Funck, Th. (1976). Ultrasonic relaxation spectroscopy in liquids. *Naturwissenschaften* **63**: 280.
6. Eggers, F. and Kaatze, U. (1996). Broad-band ultrasonic measurement techniques for liquids. *Meas. Sci. Technol.* **7**: 1–19.
7. Pope, N.G., Veirs, D.K., Claytor, T.N., and Histand, M.B. (1992). Fluid density and concentration measurement using noninvasive in situ ultrasonic resonance interferometry, in *Proceedings IEEE 1992 Ultrasonic Symposium*, vol. 2, McAvoy, B.R., ed., p. 855.

8. Sinha, D.N., Anthony, B.W., and Lizon, D.C. (1995). Swept Frequency Acoustic Interferometry Technique for Chemical Weapons Verification and Monitoring, Third International Conference On Site Analysis, Houston, January 22–25, Los Alamos Unclassified Report: 95–610.
9. Sinha, D.N., Springer, K., Lizon, D., and Hasse, R. (1996). Applications of Swept-Frequency Acoustic Interferometry Technique in Chemical Diagnostics, Fourth International Conference On Site Analysis, Orlando, January 22–25, Los Alamos Unclassified Report: 96–449.
10. Sinha, D.N., Springer, K.N., Han, W., Lizon, D.C., and Houlton, R.B. (1997). Swept-Frequency Acoustic Interferometry for Noninvasive Chemical Diagnostics, Fifth International On Site Analysis Conference, Seattle, February 3–5, Los Alamos Unclassified Report: 96–4561.
11. Sinha, D.N., Springer, K., Han, W., Lizon, D., and Kogan, S. (1977). Applications of Swept Frequency Acoustic Interferometer for Nonintrusive Detection and Identification of Chemical Warfare Compounds, presented at the 1977 Fall American Chemical Society Meeting, Las Vegas, Nevada, September 7–11. Los Alamos Unclassified Report: 97–3113.
12. Sinha, D.N. (1997). Applications of Ultrasonic Interferometry, Proceedings of the Symposium Sponsored by the Office of Naval Research, Resonance Meeting, Asilomar Conference, Pacific Grove, CA, May 11–15. Available from National Center for Physical Acoustics, University of Mississippi. ncpa@olemiss.edu.
13. Han, W., Sinha, D.N., Springer, K., and Lizon, D. (1998). Noninvasive measurement of acoustic properties of fluids using and ultrasonic interferometry technique, in *Nondestructive Characterization of Material VIII*, Green, R.E., ed., p. 393, New York: Plenum Press.
14. Leon, H.I. (1955). Fixed path, variable frequency acoustic interferometer. *J. Acoust. Soc. Am.* **27**: 1107.
15. Behrends, R., Eggers, F., Kaatze, U., and Telgmann, T. (1996). Ultrasonic spectrometry of liquids below 1 MHz: Biconcave resonator cell with adjustable radius of curvature. *Ultrasonics* **34**: 59.
16. Kinsler, L.E., Frey, A.R., Coppens, A.B., and Sanders, J.V. (1982). *Fundamentals of Acoustics*, Third Edition, New York: John Wiley & Son.
17. Katahara, K.W., Rai, C.S., Manghnani, M.B., and Balogh, J. (1981). An interferometric technique for measuring velocity and attenuation in molten rocks. *J. Geophys. Res.* **86**: 779.
18. Guidarelli, G., Marini, A., and Palmieri, L. (1993). Ultrasonic method for determining attenuation coefficients in plate-shaped materials, *J. Acoust. Soc. Am.* **94**: 1476.
19. Brekhovskikh, L.M. (1980). *Waves in Layered Media*, New York: Academic Press.
20. Hill, R. and El-Dardiry, S.M.A. (1980). A theory for optimization in the use of acoustic emission transducers. *J. Acoust. Soc. Am.* **67**: 673.
21. Eggers, F. (1994). Analysis of phase slope or group delay time in ultrasonic resonators and its applications for liquid absorption and velocity measurements. *Acustica* **80**: 397.
22. Gavish, B., Gratton, E., Hardy, C.J., and St. Dennis, A. (1983). Differential sound velocity apparatus for the investigation of protein solutions. *Rev. Sci. Instrum.* **54**: 1756.
23. Eggers, F. and Funck, Th. (1973). Ultrasonic measurements with milliliter liquid samples in the 0.5–100 MHz range. *Rev. Sci. Instrum.* **44**: 969–977.
24. Povey, M.J.W. (1977). *Ultrasonic Techniques for Fluids Characterization*, Orlando, FL: Academic Press. Water.
25. Champion, J.V., Langton, C.M., Meeten, G.H., and Sherman, N.E. (1990). Near-field ultrasonic measurement apparatus for fluids. *Meas. Sci. Technol.* **1**: 786.
26. Labhardt, A. and Schwarz, G. (1976). A high resolution and low volume ultrasonic resonator method for fast chemical relaxation measurements. *Ber. Bunsenges. Phys. Chem.* **80**: 83–92.
27. McClements, D.J. and Fairley, P. (1991). Ultrasonic pulse echo reflectometer. *Ultrasonics* **29**: 58–62.
28. Papadakis, E.P., Fowler, K.A., and Lynnworth, L.C. (1973). Ultrasonic attenuation by spectrum analysis of pulses in buffer rods: Method and diffraction correction. *J. Acoust. Soc. Am.* **53**: 1336–1343.
29. Adamowski, J.C., Buochi, F., Simon, C., Silva, E.C.N., and Sigelmann, R.A. (1995). Ultrasonic measurement of density of liquids, *J. Acoust. Soc. Am.* **97**: 354–361.
30. Gabor, D. (1946). Theory of communication, *J. Inst. Electric. Engrs.* **93**: 429–457.
31. Gammel, P.M. (1981). Improved ultrasonic detection using the analytic signal magnitude. *Ultrasonics* **19**: 73–76.
32. Duncan, M.G. (1990). Real-time analytic signal processor for ultrasonic nondestructive testing. *IEEE Trans. Inst. Meas.* **39**: 1024–1029.
33. Stremler, F.G. (1982). *Introduction to Communication Systems*, pp. 244–252, Reading, MA: Addison-Wesley.

CHAPTER 2

THE SPEED OF SOUND IN PURE AND NEPTUNIAN WATER

C. C. Leroy
Private Consultant in Underwater Acoustics, SANARY-sur-Mer, France

Contents

Abstract		23
2.1.	Introduction	24
2.2.	General Overview	25
2.3.	Historical Survey	27
2.4.	Laboratory Sound Velocimeters	35
2.5.	The Neptunian Domain	42
2.6.	Available Measurements of Sound Speed in Water	44
2.7.	Equations for the Speed of Sound	49
2.8.	Discussion	56
2.9.	Conclusion	75
Acknowledgment		76
Appendix: Recommended Equations for the Speed of Sound in Pure Water and Neptunian Waters		76
References		78

ABSTRACT

After publication of the last equations for the speed of sound in pure water at atmospheric pressure and seawater at pressures from 0 to 100 MPa between 1972 and 1977 (Del Grosso and Mader [71], Del Grosso [6], and Chen and Millero [4]), studies on the subject passed through an empty period. Large-scale experiments carried out in the Pacific in 1987 gave rise, starting in 1992, to renewed interest in sound speed in seawater at great depths and to a number of controversies, very well summarized by Dushaw *et al.* [10]. The author, having continuously followed the measurements and equations for sound speed in water since 1960, developed equations and the concept of Neptunian waters and, in 1988, made an unpublished review on the whole subject; he felt this to be an appropriate moment for a "review and tutorial" on sound speed in pure water and seawater. This review accordingly gives a detailed historical survey of current knowledge on sound speed and describes the various methods employed to measure it, including the main laboratory velocimeters. It reviews all measurements and equations made in the past years and provides a detailed analysis of the problems,

Handbook of Elastic Properties of Solids, Liquids, and Gases, edited by Levy, Bass, and Stern
Volume IV: Elastic Properties of Fluids: Liquids and Gases
Copyright © 2001 by Academic Press
ISBN 0-12-445764-9 / $35.00 All rights of reproduction in any form reserved.

possible errors, and inaccuracies, ending with the situation at the present time. The reasons for the observed discrepancies seem to have been found, and a number of suggestions are made for future measurements and equations to model the physical reality.

2.1. INTRODUCTION

More than 40 different equations for the speed of sound in both freshwater and seawater have been proposed in the past 50 years. This wealth, to which this author contributed (Leroy [1–3]), reflects the evolution of understanding of the mechanisms affecting sound speed in water, characterized by continuous improvement in accuracy and knowledge of the various domains of the oceans and seas explored. In the post-1960 era, a crescendo of equations was proposed, apparently ceasing abruptly in 1981, but then followed in 1983 by UNESCO's adoption of the equation of Chen and Millero [4, 5]. The ingenuous observer might have thought that nothing remained to say and that no further investigation was warranted, but that is not at all the case. The state of the art is such that the validity of the Chen-Millero equation is heavily debated, and the NRL II equation developed from Del Grosso's results [6] seems rehabilitated [7, 8]. This is largely the result of the need for highly accurate sound speed values at depths in the oceans, required by acoustic tomography, and large modern experiments at sea where results were in better agreement with the NRL II equation than with UNESCO's. There remain, however, other delicate points, such as the precise values of sound speed in waters of "intermediate" salinities ($10-30 \frac{o}{oo}$). Moreover, some anomalous conditions, such as found in hot brine pools in the Red Sea, have not yet been explored.

It was stated in the previous paragraph that equations for sound speed in water apparently ceased appearing after 1981. Until now, with the exception of some European organizations, few workers have heard of a third equation proposed by this author (Leroy [9]) following a critical survey of the subject sponsored by the French Naval Laboratory at Le Brusc. This was 20 years after Leroy's original investigations. Unfortunately, the work was influenced by then-current ideas concerning corrections to be applied to Del Grosso's measurements at high pressure. A subsequent and more intensive study led to the abandonment of such corrections and to the decision to abandon publication of that equation, which incorporated some of the corrections.

A recent article (Dushaw *et al.* [10]) has an appendix giving a brief review of laboratory measurements of the speed of sound in seawater, but because the appendix was short, many points could not be covered. In contrast, the present author's survey (Leroy [9]) contained extensive original analyses of the various laboratory measurements and equations, complemented by further investigations. It seemed, therefore, that the time was ripe for a publication in English covering the whole subject. The purpose would be to recall the physics and significance of sound speed in water and to give a detailed historical survey of current knowledge. It would describe how to measure it and how the various equations of serious interest were formulated. It would also contain unpublished investigations about sound speed at intermediate salinities and about measurements under pressure. The former French Naval Laboratory of Le Brusc (now CTSN/DES/LSM), recognizing the importance of the subject, kindly agreed to support the work and its publication.

In embarking on this work this author has to state a personal regret. He had emphasized in 1968, when defining "Neptunian" waters, that the priority at the time was to investigate the speed of sound in waters where military operations were feasible. Unfortunately, in giving a list of areas of recommended coverage for measurements, he did not mention that the careful measurement of the speed of sound in pure water under pressure, even restricted to those temperature/pressure combinations encountered in the Neptunian waters, would have been very useful for later measurements by comparisons, as done by Chen and Millero. And so, Del Grosso, who originally intended to perform such measurements, abandoned the idea—a great pity since his equipment no longer exists. See, for example, Millero and Li [11].

In the following, Section 2.2 is a reminder of the physics of sound speed in freshwater and seawater and raises questions about what is meant by seawater. Section 2.3 is a historical survey of our knowledge of sound speed in water and explains all the various difficulties encountered. Section 2.4 describes the principles used in accurate sound velocimeters and reviews the main laboratory equipment. This is an essential part of this chapter because most problems in sound speed determination arise from the instrumentation and whether it was used carefully. Section 2.5 briefly recalls the limits of the Neptunian domain to understand the measurements made and/or select those important for practical equations. Sections 2.6 and 2.7 describe all the measurements and all the equations, respectively. Section 2.8 is a critical examination of the measurements and equations. They contain in particular the most important points studied in the author's 1988 review, with additions concerning the effect of pressure on equipment. Two selected equations for the speed of sound in pure water at atmospheric pressure and in Neptunian waters are recalled in the Appendix.

2.2. GENERAL OVERVIEW

Historically, sound speed was first called sound velocity. Although both terms are used, sound speed is more correct because it is a scalar quantity. In freshwater and seawater, the speed of sound obeys the theoretical law set forth in the 1700s by Sir Isaac Newton for perfect fluids, expressed by the well-known equation

$$C = \sqrt{\frac{\kappa}{\rho}} \qquad (2.1)$$

where C stands for celerity (a rarely used term, although C remains the most commonly used symbol for sound speed). κ is the adiabatic bulk modulus of the fluid, and ρ is its density. In practice, this expression is more conveniently replaced by

$$C = \sqrt{\frac{\gamma}{\chi \rho}} \qquad (2.2)$$

where γ is the ratio c_p/c_v of the specific heats at constant pressure and constant volume and χ is the isothermal compressibility. In freshwater at atmospheric pressure between 0 and 100 °C, all these quantities vary in a complex way. The overall result of these variations is an increase of sound speed with temperature up to around 74 °C, followed by a decrease [Fig. 2.1(a)]. This is unique in natural fluids because in all others the sound speed varies monotonically with temperature between solid and gaseous states. The reason for this abnormal behaviour lies in molecular arrangements and implies highly complicated physical considerations. Despite this peculiarity, freshwater, also called *pure water*, is the fluid for which it is most important to know physical properties: it is the most common on Earth and the easiest to obtain in large quantities without any possible contestation. This is why the properties of pure water continue to be studied.

Under pressure, the various properties of water vary. The general behavior for sound speed at any given temperature consists of a monotonic increase with pressure [Fig. 2.1(b)]. At the beginning of research in this area (see Sect. 2.3), sound speed could not be measured accurately, whereas such properties as compressibility and specific heat could be obtained with more or less success ..., so sound speed was calculated through equation Eq. 2.2. We have now reached a point where the situation

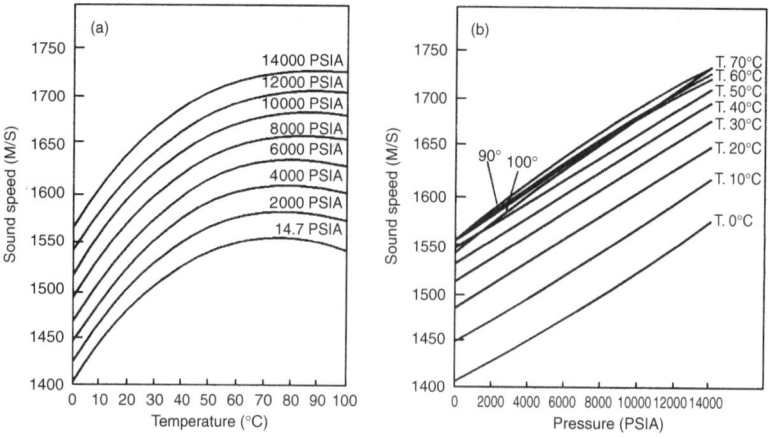

Fig. 2.1. Variation of the speed of sound in pure water as a function of temperature at different pressures (a) and as a function of pressure at different temperatures (b) (from Wilson [12]).

is reversed! Researchers concerned with the equation of state of pure water seek very accurate values of sound speed under precisely known conditions of temperature T and pressure P.

Seawater is another natural fluid of interest, but what exactly is meant by seawater raises some questions that will be dealt with in the next chapters. It was found in the early days of oceanography that water samples taken here and there in "not too exotic" seas or oceans were pure water containing various dissolved materials in relative proportions that remained practically constant wherever the sample had been taken and whatever the temperature or depth. The whole mass of dissolved matter per unit volume of the original sample was called "salinity" and was expressed in parts per thousand (conventionally, $\frac{o}{oo}$). It could vary from one sample to another, but the percentage of each "ingredient" in the dissolved matter did not vary, at least for the most important ones. Common seawater was found to have a salinity ranging from about 33 to 37 $\frac{o}{oo}$, with an average value around 35 $\frac{o}{oo}$. The measurement of salinity soon became the subject of intensive research, still in progress: Was it really imperative to measure the whole weight of dissolved materials? As the proportions of these remained invariant, would it not be simpler to measure only the most important of them, viz. the Cl− ions, by some chemical extraction from the rest that would give the so-called chlorinity, hence, salinity, from a simple rule? And, by the way, as electric conductivity was something easy to measure and depended on salinity, could it not be used to replace all other measurements, even to redefine salinity? There was soon a general consensus to name "normal" or "standard" seawater pure water containing precisely defined amounts of each chemical component, in such a way that the total salinity was exactly 35 $\frac{o}{oo}$ (we must note that this seawater is supposed to be exempt from microorganisms or undissolved gases). This "standard" seawater could be obtained artificially: for example, the so-called Copenhagen seawater was prepared early in Denmark as an available standard for use by oceanographers. One knew, however, that the various properties of real seawater could not all be simulated exactly by these compositions, but it was a good compromise.[1] By extension, it was decided that naturally existing seawater having a salinity of 33 $\frac{o}{oo}$, for example, could be simulated well enough by proper dilution with freshwater or filtered seawater samples having near-standard salinity. Conversely, natural seawater with a salinity of 37 $\frac{o}{oo}$ could be simulated by slowly evaporating near-standard filtered seawater samples. So, in the end, the word

___
[1] For a better understanding of these details, refer to one of the major books on the subject, such as *The Oceans* [13].

"seawater" was progressively taken implicitly to mean near-standard natural filtered seawater slightly diluted or evaporated, remaining in the salinity range 33–37 $\frac{o}{oo}$, and it could be studied as such as a fluid, different from pure water. The speed of sound in this seawater at atmospheric pressure was found, first by computation and later by measurements, to increase both with temperature and with salinity. With temperature, calculated and measured sound speeds were limited to realistic values and did not extend beyond 40 °C, so the variation is monotonic. For very low temperatures, an increase of 1 °C gives rise to an increase of about 4.5 m/s in sound speed, and at high temperatures the same variation increases sound speed by only about 2 m/s. The sensitivity of sound speed to salinity is of the order of 1.25 m/s per $\frac{o}{oo}$. As for pure water with T and P, there is no possibility to separate the variables T and S in one equation for sound speed in seawater at atmospheric pressure, and the same will occur for the variation with pressure. This has been studied up to about 100 MPa. The increase of sound speed with pressure is of the order of 1.6 m/s per MPa or, to give a more illustrative figure, an increase of 1 m/s in sound speed is obtained (T and S remaining constant) by an increase in depth in the sea of about 60 m. If echo sounding, sonar applications, acoustic underwater localizations and communications, acoustic tomography, etc., had not emphasized our need for an accurate knowledge of sound speed in all practically encountered waters of the world, things could have remained at that stage. Seawater would have been a special fluid derived from standard seawater within a few parts per thousand in salinity. But things were not that simple because several "special" seas have salinities far outside the 33–37 $\frac{o}{oo}$ range. Water from the Black Sea, for instance, has, at depth, a salinity of about 22 $\frac{o}{oo}$, and because it has remained stagnant for thousands of years, its exact composition can differ slightly from that of standard seawater. So the question is simple to ask, and difficult to answer: Has a sample of standard seawater, diluted as much as to reach a measured salinity of only 22 $\frac{o}{oo}$, the same properties as a sample from the Black Sea that would have also a measured salinity of 22 $\frac{o}{oo}$? It is well known also that large rivers bring to the sea large amounts of minerals such as carbonates, uncommon in the faraway oceans. Has a water sample taken in the estuary of the Amazon, with a measured salinity of 28 $\frac{o}{oo}$, the same properties as standard seawater diluted to 28 $\frac{o}{oo}$? At the other end, has Red Sea water (with $S = 42 \frac{o}{oo}$) the same properties as standard seawater evaporated to reach $S = 42 \frac{o}{oo}$? This question cannot yet be properly answered for speed of sound, and the desired accuracy is most probably the keyword in all this affair. For pure water, this accuracy is of the order of ±0.01 m/s or better for scientific reasons and because pure water is the fluid of reference. For seawater, the experts were much less demanding until recently. An accuracy of ±0.1 m/s was the utmost aim, but it was admitted that errors of ±0.3 m/s and even ±0.5 m/s could be tolerated in practical calculations. Acoustic tomography among others obliges us to reconsider these figures, and we surely need now at least ±0.1 m/s. The great difficulty to reach such a figure in measurements, especially at high pressures, and problems concerning existing natural waters have governed the evolution of sound speed studies, which we will now review.

2.3. HISTORICAL SURVEY

2.3.1. The Early Ages

The propagation of sound in water received little sustained attention until the beginning of the nineteenth century. During the preceeding 100 years, a number of scientists and philosophers had observed that sound from various sources, such as bells, pistol shots, and voices, could be heard with varying clarity and at impressive distances under water. This early pioneering work has been summarized by Hackmann [14], who quotes the experiments and observations of Hauksbee (1709), Wolff (1738), Nollet

(1740), s' Gravesande (1742), Arderon (1742), and Franklin (1762). It was discovered that, to a human ear, the sound received in water seemed to be deadened but could be heard at much greater distances than in air. Whether it had traveled faster or slower than in air was, however, unknown. At the time, Newton's law for sound speed in fluids (Eq. 2.2) could have given the clue if at least the compressibility of water had been known, but it could not yet be measured. Some inconclusive attempts to solve this intriguing problem were made by various researchers, including G.C. Lichtenberg (1772), A. Monro *secundus* (1785), and E. Perolle (1799), but in the end it was the French physicist and mineralogist François S. Beudant who made the earliest measurement in the world of sound speed in water, giving the crude value of 1500 m/s, a figure much greater than in air. His experiment was carried out in the Mediterranean near Marseilles at the dawn of the nineteenth century, operating with an underwater bell and two listeners with clocks swimming at known distances.[2]

The prize competition on the compressibility of liquids, announced in 1826 by the *Académie Royale des Sciences* of Paris, decided the Swiss physicist Jean-Daniel Colladon, who was aware of Beudant's experiment, check his results on the compressibility of water by measuring the speed of sound in Lake Geneva with the best possible accuracy, making comparison with the application of Newton's law. With the help of the mathematician Charles Sturm [18], he performed in 1826 the famous "Colladon and Sturm experiment," currently described by a number of authors [15–17, 19]: Two small boats were floated on the lake; from one a submerged bell was suspended that, when rung, served as a transmitter; from the second an ear-trumpet–like device was suspended to serve as the hydrophone receiver. A gunpowder explosion indicated when the bell was rung and thus, with the aid of chronometers, the observers were able to obtain the first value of sound speed in almost pure water—with surprising accuracy given the apparent crudity of the method. But this was only the speed of sound in lake water at 8.1 °C (1438 m/s after modern corrections of their measurement, instead of $\approx$1435 m/s).

The fact that sound could be heard in water at much greater distances than in air was at that time thought to be a possible supplement to the equipment of lighthouses, replacing foghorns or sirens to signal danger to passing ships. It seems that the idea was first put forth by Elisha Gray (who was working with Thomas A. Edison on improvements in the telephone) and Arthur J. Mondy (quoted by Lasky [17]). Their work led in the United States to the foundation in 1901 of the "Submarine Signal Company." It was later thought that Morse-coded underwater sound signals could be employed to communicate with ships, an application made possible by the invention in 1913 of the famous Fessenden oscillator as a sound transmitter.

It is usually considered that the Titanic disaster (1912) triggered research into detection methods in the sea, but the idea had arisen earlier (see e.g., Bonneycastle [20], quoted by Lasky). In any case, this disaster, plus the necessity to detect U-boats during the First World War, gave the impetus to investigate the utilization of sound echoes for submarine detection and depth sounding. This was particularly emphasized in the United Kingdom and the United States. There are good reasons, however, to believe [21] that Marti, a French Navy hydrographer, was the first to realize in 1919 an automatic device for echo sounding. His early system used an automatically recharged gun pointed toward the water. The bullet entry into the water produced a sound pulse that was reflected from the bottom and received by a waterproof microphone arrangement connected to an amplifier and an oscillogram recorder. Marti's system obviously had limited capabilities. Nevertheless, he managed to draw a map of most of the

---

[2]For references on all the preceding, see Hackmann [14]. Reviews with references on early underwater acoustics may also be found in books by Wood (*A Textbook of Sound* [15]) and Lindsay (*Acoustics—Historical and Philosophical Development*) [16]. Also of particular interest on the subject are various publications by Lasky, including *Review of Undersea Acoustics to 1950* [17].

continental shelf off the French Mediterranean coast, which was then kept secret for nearly 30 years because it showed underwater canyons, one of which leads directly to the entrance to the military harbour of Toulon! Other systems immediately followed Marti's, mainly based on the use of piezoelectric transducers, an application developed by Langevin of the properties of quartz, discovered in 1880. Piezoelectric transducers had the advantage of being able to produce narrow sound beams, thus eliminating most of the side echoes when sounding near a slope.

The problem of converting the vertical travel times measured by the new echo sounders into depths immediately arose, and in 1923 the U.S. Coast and Geodetic Survey (US C&G S) ship "Guide" undertook a cruise [22] from New London to San Diego, passing through the Panama Canal. During that cruise, many stations were devoted to comparing the results of echo soundings and wire soundings, but due to the uncertainties produced by drift, the accuracy of the vertical mean sound speed was inadequate.

Other pioneering measurements of sound speed in seawater are also worth recalling. One of these was performed in shallow water by E.B. Stephenson, operating with electrically triggered suspended explosive sources and hydrophones moored some 15 km away, radio signals being employed for the travel time measurements. This experiment took place in 1922 in the Block Island Sound [23].[3]

2.3.2. The Age of Calculated Values

Before World War II, laboratory sound velocimeters could not be contemplated because the status of electronics was not sufficiently advanced to permit the generation and processing of signals at frequencies of several megacycles. This is why, from about 1920 to 1940, scientists decided to compute the speed of sound in water from its physical properties: density, adiabatic compressibility, specific heats c_v at constant volume and c_p at constant pressure, using Newton's equation (Eq. 2.2). A first series of calculations was carried out in the 1920s but the necessary data were still incomplete or not very accurate. At the same time, semiempirical formulas for sound speed were proposed. One can cite the equation of Wood and Brown [24] and the so-called Maurer-Schumacher equation [25, 26]. Maurer had established in 1924 tables for the speed of sound in seawater of salinity 33 $\tfrac{o}{oo}$, and Heck and Service [27] at the US C&G S published in 1927 the first systematic tables based on Newton's equation. These tables were superseded the same year those prepared for the British Admiralty Hydrographic Department by Matthews [28]. In fact, these tables gave sound speed values for different T, S, and P combinations but investigated also the problem of converting vertical travel times measured in echo soundings to depths, which means evaluating average sound speeds down to various depths by taking into account the temperature and salinity structures of the oceans.

Apparently, little more was done between 1925 and the publication of the famous work by Kuwahara in 1939. However, it is of interest to note that during that period the acoustic properties of the deep sound channel, which exists in the open oceans, had been discovered. An experiment conducted in 1933 by the US C&G S with two ships [29] demonstrated that the predicted properties were exact but that the precision of the positions of the ships, hence the distance between source and receiver, was not sufficient to check accurately the average sound speed in the deep channel.

In 1939, the Japanese naval engineer Kuwahara [30] published his computation of sound speed from equation Eq. 2.2 with quite complex mathematical developments and much more accurate values of the physical constants. His results covered the following ranges: temperature, -2 to $+30\,°C$; salinity, 30 to 40 $\tfrac{o}{oo}$; and pressure, 0 to 100 MPa. Kuwahara's tables were far more precise and more complete than any

---

[3]See also Wood and Brown [24].

previously published, and his numerical computation, made by hand, has been described as "immense." At the same time, Matthews' tables were revised [31] and found to be in reasonable agreement with Kuwahara.

Based on Kuwahara, several equations were formulated to replace the tedious work of interpolation from numerical tables. Several of these equations remained for a long time unpublished in the open literature. Of particular interest are those from Stephenson and Woodsmall [32], limited in depth; the U.S. Hydrographic Office [33], for use down to 3000 m; and Mackenzie [34]. This last equation, which gives the best fit to Kuwahara's results down to great depths, arrived unfortunately a little too late: the age of sound velocimeters had already begun.

2.3.3. The Age of Sound Velocimeters and Modern Equations

2.3.3.1. The Need for Sound Velocimeters

During World War II and the subsequent years, the first sonars were investigated by various navies, and the knowledge of the in situ sound speed profile at sea in approximately the first 100 m became of great importance. Cheap sound velocimeters were invented, and the evolution of technology made their development possible. The most common velocimeters were based on the so-called "sing-around" technique. In the first equipment, a very short high-frequency sound pulse (basically, 1 MHz or more) is sent into the water by a transducer, directed toward a sequence of reflectors, used to increase the path length for limited dimensions of the instrument, and finally reaches a receiver. When the pulse is received, it triggers transmission of a new pulse, and so on. In an established situation, the repetition rate of this process, which can be measured simply by a counter, varies linearly with the speed of sound in the closed domain of the acoustic paths involved. A difficulty inherent in such cheap sound velocimeters is that they are not precise enough to make absolute measurements. The proper way to obtain sound speed C from the repetition rate F is simply calibration in pure water. By measuring F at two or three temperatures, whatever they are, provided they be measured with sufficient accuracy, the relationship between F and C can be obtained through the knowledge of the sound speed in pure water as a function of temperature.

2.3.3.2. Disagreement with Kuwahara's Results

Use of interferometry to measure sound speed in liquids had been recommended around 1930 [35, 36], and Weissler [37] proved in 1948 that such equipment could be capable of high accuracy. In 1950, Del Grosso, at the U.S. Naval Research Laboratory (NRL), made direct measurements of the speed of sound on real seawater samples by using an existing 1-MHz acoustic interferometer. His results showed for the first time that the sound speed in seawater was higher by some 3–4 m/s than predicted by Kuwahara's calculations [38, 39]. A new, more precise, 3-MHz interferometric velocimeter was then constructed at NRL, and the first results confirmed the discovery [40]. More systematic measurements were then performed with seawater samples, natural, evaporated up to a salinity of $\approx 40 \frac{o}{oo}$, or diluted down to $\approx 20 \frac{o}{oo}$, and an equation, probably the first to fit direct measurements [38], was formulated for seawater at atmospheric pressure. The 3–4 m/s difference from Kuwahara was also confirmed for seawater under pressure by Carnevale and Litovitz [41], and for pure water. With pure water, a few measurements had been made at NRL [42], and also by Owen *et al.* from Yale University in 1953 (Del Grosso [42]) and by Gucker and Haag [43], both teams using interferometers.

At that point, the U.S. National Bureau of Standards (NBS) decided to build very accurate equipment to obtain reference values of sound speed in pure water vs temperature at atmospheric pressure. This equipment was based on a principle different from interferometry, viz. the measurement of travel time of a precisely measured sound path (see Sect. 2.4). It was developed and used in 1956–1957 by Greenspann and Tschiegg [44]. The results obtained, further published in JASA [45], also indicated that Kuwahara's values were too low by some 3–4 m/s. This discrepancy was confirmed by Martin [46] and Wilson (see later) and was suspected to be due to errors in the compressibility of water [42, 47]. It was definitely identified by Del Grosso [48] as the result of too low readings given by the Amagat's pressure gauge used in the measurements.

2.3.3.3. The Discrepancy between the NBS and NOL Results and the Others

The disagreement with Kuwahara's values had then been explained, but, unexpectedly, the values obtained by Greenspann and Tschiegg (G-T) in pure water differed slightly from all the other measurements performed with interferometers, being about 0.3–0.4 m/s higher, while interferometric measurements were mutually consistent. Because the accuracy claimed in all cases was better than 0.1 m/s, this gave rise to serious worries, which became even greater when the first U.S. Naval Ordnance Laboratory (NOL) results became known. The NOL had decided to build a velocimeter based on the same principle as the NBS's, for the purpose of measuring the speed of sound in pure water and seawater at any pressure up to 14,000 psia ($\approx$96 MPa, which corresponds to the greatest usual depths). The equipment was ready in 1958 [49] and had confirmed the disagreements with Kuwahara, but the first values measured with the NOL equipment in pure water by Wilson [12] in 1959 were even higher than those of G-T.

This embarrassing affair had two different consequences, the first of which was that work at the NRL was stopped on the construction of an interferometer to measure the speed of sound under pressure. Del Grosso decided to examine thoroughly the possible sources of this discrepancy between measurements with interferometers and measurements with fixed-path velocimeters. He had already worked on problems associated with interferometers [42] and decided to reexamine and complete his analyses. Did the sound field inside the acoustical cell present the expected pattern of interference fringes, and, by the way, was the method chosen to observe those fringes correct, and was the salinity properly measured? Perhaps diffraction effects, or perturbations by waves traveling in the walls of the acoustic cell, altered the sound field in the cell more than had been expected; maybe the observation of impedance circles was not the best way to measure the fringe spacings, etc. All this research delayed considerably the final measurements made by Del Grosso. A series of internal NRL reports, such as those by Del Grosso [50–53], summarizes his conclusions. Diffraction corrections were minimized by an appropriate ratio of the crystal diameter to the acoustic cell diameter. In all, the problems with interferometers could not be held responsible for the $\approx$0.35 m/s difference sought. This affair, however, was beneficial in the end because the two "final" NRL velocimeters built afterward were indeed superb instruments.

The second consequence of the discrepancy between interferometric measurements and those with the NBS and NOL velocimeters was the construction of equipment differing from both. Basically, two kinds of equipment were built and used, with two exceptions, for measurements in pure water at atmospheric pressure at just a few values of temperature. The first category consists of velocimeters with a fused quartz buffer. The initial equipment of this kind had been studied by McSkimin [54] at the Bell Telephone Laboratory. McSkimin made new measurements in 1965 [55] and found the results to be consistent and to be about 0.4 m/s less than NBS. Subsequent velocimeters of that type, but different in detail (see Sect. 2.4), were built by Barlow and Yazgan [56, 57], Carnvale *et al.* [58], and Williamson [59]. The former was designed and used for measuring the speed of sound in pure water as a function of temperature and pressure, the others were only made for atmospheric pressure. The second type of equipment used the variable path–variable distance principle (see Sect. 2.4) with precisely positioned transmitters and receivers in a tank filled with water. Such measurements were performed by Brooks [60], Neubauer and Dragonette (from the NRL) [61], and later by Kroebel and Mahrt [62] (1976, University of Kiel). To these various measurements one must add those made in the U.S.S.R. by Ilgunas *et al.* with an interferometer, and published in 1964 [63]. All these measurements confirmed the lower values as against the too high ones of the NBS and the NOL.

In all, in 1966, although some of the previously cited measurements had not yet been made, there was a consensus, which still holds, that the NBS measurements had to be corrected by a fixed value of -0.35 m/s due to a slightly inadequate absolute calibration and/or insufficient diffraction corrections. It was also admitted that the NOL velocimeter needed corrections depending on temperature. An average correction of -0.6 m/s could be used below 30 °C. At this point, we recall that in 1960–1962 the NOL velocimeter had already been used by Wilson, in addition to the previously cited measurements in pure water, for a systematic exploration of the speed of sound in "sea" water over a wide range of temperature, pressure, and salinity (see Sect. 2.3.3.4). This whole set of values also had to be corrected, on average, by -0.45 m/s.

2.3.3.4. *The First Equations Based on Results from Velocimeters*

After the first equation developed by Del Grosso [38] for his early measurements in seawater at atmospheric pressure, Greenspann and Tschiegg [44, 45] derived a very precise equation fitting their pure water data to ± 0.06 m/s in the whole range 0–100 °C, which was first published in 1957. As it was admitted afterward that their sound speed values had to be corrected by -0.35 m/s, only the first constant term of G-T's equation needs to be corrected. Barlow and Yazgan [57] also derived an equation to fit their data. It is a "matrix form" polynomial development of the fifth degree in T, each coefficient being a polynomial of the fourth degree in P. A similar structure was used by Wilson [12] for his equation for pure water under pressure published the same year, except that he went one degree less in power for both T and P.

The values of sound speed in seawater measured at the NOL by Wilson were obtained in two subsequent series in 1959 and 1960. The first set of data [64, 65] sampled a "cubic" volume in T, S, and P, with S ranging from 33 to 37 $\frac{o}{oo}$, and a first equation, the "June" equation fitting these data, was published in [65]. A second set of measurements concerned the intermediate salinities around 10, 20, and 30 $\frac{o}{oo}$. These results were combined with the previous ones and with a few selected data from pure water to establish a second equation [66, 67], the "October" one.

2.3.3.5. *The Limitation to Neptunian Waters*

In 1967, this author, then at SACLANTCEN, became convinced that sound speed equations for practical use in underwater acoustics would gain both in simplicity and in accuracy by limitation of the data points to realistic combinations of temperature, salinity, and pressure. He then undertook systematic research of the existing combinations of T, S, and P in order to develop a final equation to cover all possible cases. He found that extraneous combinations could be found in some existing waters, such as the Dead Sea, but decided to eliminate them for a number of reasons: (1) Such waters were not normally relevant to naval operations. (2) The composition of salts in such waters could be different from that usually found, so they might need special measurements not yet made. (3) Even if measurements were later made, the risk existed of unnecessarily complicating an equation just for the sake of giving correct values in unfrequented waters. It would be better in the end to establish, if necessary, an equation appropriate to each exceptional case. The difficulty was that the "useful" waters had no specific name, so in the end this author christened "Neptunian" the waters that the ancient god of the seas could have visited, starting from his home in the Mediterranean, and traveling in water without ever emerging. This eliminated inland seas, lakes, and modern canals. Moreover, Leroy [68] wanted his equation to use depth instead of pressure. His research on this subject had proved that a universal simple formula to obtain pressure P from depth Z and latitude ϕ could be adopted for most oceans and that closed basins needed modified versions [68]. After several approaches by trial and error, he succeeded in establishing a first simplified formula [1] that fitted Wilson's second equation within better than ± 0.1 m/s. A second equation was then developed to fit directly the useful measured values of Wilson for seawater and those given at seven temperatures below 30 °C by G-T's equation for pure water. This second "merged" equation was again simple and fitted Wilson's *data* better than Wilson's *equation* itself. The two equations were published by Leroy [2, 3], and a list of desirable sound speed measurements covering the Neptunian waters was proposed.

Interesting work on Wilson's data was also done in 1970 by Frye and Pugh [69]. Following the recommendations of Leroy, they eliminated the non-Neptunian combinations but proved also by utilizing the F test that the second set of Wilson's data could not be mixed with the first. On that basis, they proposed a revised and simplified equation to replace Wilson's equations (without the -0.45 m/s correction).

2.3.3.6. *The Systematic NRL Measurements*

The systematic work of Del Grosso at the NRL started as soon as the new interferometric velocimeters were ready. The results were published progressively from 1970 to 1974 and will be examined in detail in the following sections. At this point, it is sufficient to summarize the work achieved. After a few measurements at atmospheric pressure with pure water and seawater [70], Del Grosso and Mader [71] presented in 1972 the new NRL velocimeters and the first series of systematic measurements. These followed the scheme proposed by this author for the limitations but could not yet cover the whole desirable range in T, S, and P. Six different possible polynomial equations were tried, with different degrees of accuracy, the one finally recommended being later called the NRL I equation. At the same time, extremely precise measurements of sound speed in pure water at atmospheric pressure between 3 and 95 °C were published [72, 73], and a fifth degree equation was proposed [72], with an accuracy believed to be 0.015 m/s. A publication by Del Grosso [74], gave tables for the speed of sound in seawater based on the NRL I equation, with specific emphasis on Neptunian waters. Finally Del Grosso's last published work [6] gave an equation that fitted both his seawater data from the Neptunian domain and his very accurate pure water data at atmospheric pressure. This equation is the famous NRL II equation, now under debate.

2.3.3.7. *After Del Grosso*

After the measurements of Del Grosso, only two sets of important laboratory data were obtained, one by Kroebel and Mahrt [62] at atmospheric pressure and mostly for pure water, proposing slight changes in the coefficients of DG-Mader's equation, and the second set, in the 1975–1977 period, by Millero and Kubinski (hereafter M-K) [75] and by Chen and Millero (hereafter C-M) [4]. These last measurements were made by comparing results obtained in pure water and in seawater with Nusonic sing-around velocimeters, at atmospheric pressure and under pressure. They used for sound speed in pure water under pressure recalculated values from Wilson [76]. The T-S-P domain explored by C-M was "cubic," like Wilson's, and their equation purported to fit this whole domain instead of just the Neptunian one. This equation was adopted in 1983 as the UNESCO equation and will be discussed in Section 2.8.

Finally, in the period extending from the last equation of Del Grosso to 1981, a number of scientists commented on the differences between DG's data, Wilson's data, and others, as well as the equations. Critical analyses were made, most of them demonstrating that many of Wilson's data had to be corrected. Anderson, from the U.S. Naval Undersea Research and Development Center (NURDC) [77, 78], and Lovett, from the Naval Ocean System Center (NOSC) [79], were among the main investigators in this domain. In the same period many researchers developed corrected, merged and/or simplified equations, in general following this author's real ocean limitations. Such are the equations established by Kell [80], Medwin [81], Ross [82], Coppens [83], and Mackenzie [84].

While the various laboratory measurements reviewed above were being made, a number of direct scientific measurements of the speed of sound at sea were performed with commercial velocimeters associated with in situ temperature measurements, and even salinity measurements, direct or from samples. Mackenzie (the author of the 1960 formula to fit Kuwahara's data) is certainly the most persevering scientist to have followed this track. For 20 years, he published results of measurements made at great depth with manned submersibles and made comparisons with the results of modern equations [85–89], before publishing his 1981 equation [84].

2.3.4. The Age of Doubts

The values of sound speed in seawater under high pressure measured by Wilson and Del Grosso were not in perfect agreement, and those made by Mackenzie in deep water were not accurate enough to decide where the truth was. A number of researchers had investigated this problem and some, like Lovett, tried to establish corrections to Del Grosso's results on the basis of a better validity of some selected values from Wilson. The results of Chen and Millero provided yet other values differing from both Wilson's and Del Grosso's, but this intriguing situation was not considered sufficiently critical until the need for very accurate values in acoustic tomography arose. Because a famous experiment in the Pacific Ocean had shown that direct use of the UNESCO equation did not give sufficiently accurate results whereas the NRL's second equation gave better ones, a new impetus has been given to restart the investigations. We shall briefly describe those measurements here as they are a world "premiere."

In 1987, a large instrumental set was moored for 4 months in the Northern Pacific. It consisted (Fig. 2.2) of three moored transceivers and one source and used nine existing hydrophone arrays. This provided several acoustic sensors, some of which were placed near the deep sound channel axis (at a depth of around 900 m). The transceivers were moored at distances between each other ranging from ≈ 750 to ≈ 2000 km. Each consisted of an acoustic 250-Hz broadband transmitter and a four-element receiving array. The anchoring locations had been determined with care from the operating ship, equipped with an acoustic measuring system for underwater localizations, its position being obtained with GPS. The subsequent motions of the transceivers due to currents were tracked from another acoustic measuring system. In all, the positions of the transceivers were estimated to be known in absolute coordinates to within ± 70 m. The nine hydrophone arrays were moored at distances of 1000 to 4000 km from the transceivers along two separate 2500-km lines (see Fig. 2.2). Their positions were known within ± 120 m. This equipment was used for different measurements, mostly concerned with acoustic tomography, and could also be considered as a gigantic "sound velocimeter" placed in a variable environment as the distances between sources and receivers were known with a relative accuracy of the order of 3×10^{-5}. Two different

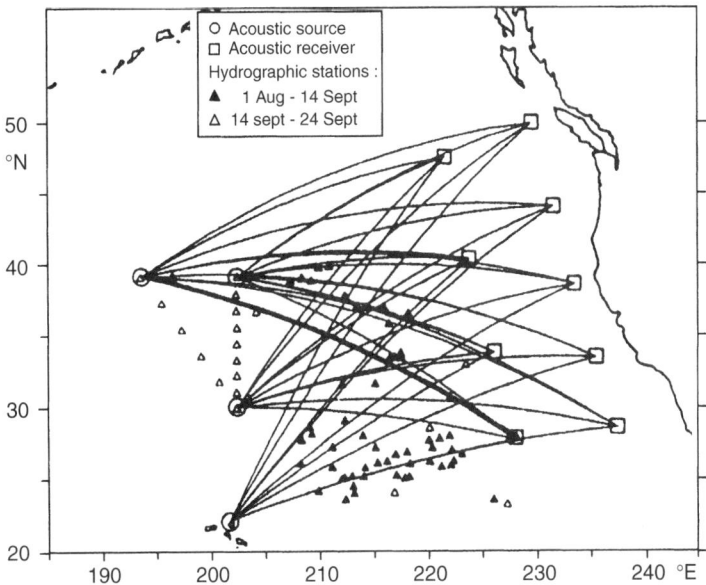

Fig. 2.2. Plan view of the 1987 acoustic experiment in the Pacific (after Spiesberger and Metzger [7]).

sets of measurements, which could be used for sound speed control, were performed. The first [7, 90] concerned the comparison of the observed and calculated travel times of the various individual arrivals from processed short phase–modulated pulses along a 3000-km track from one transmitter to one hydrophone. These arrivals correspond to refracted acoustic paths, the early ones reaching the greatest depths and undergoing a reduced number of refractions. The calculated values were obtained by the application of ray theory in a variable medium. Several hydrographic measurements provided temperature and salinity profiles along the sound track, from which the desired sound speed profiles could be obtained by the use of an appropriate equation. It is precisely at this point that it was discovered that the use of DG's NRL II equation gave much better agreement between calculated and measured travel times for the observable first eight pairs of refracted arrivals than the use of UNESCO's equation. The second set of experiments [10] dealt with sound signals received on the acoustic sensors from three of the transceivers. The well-known final "crescendo" at the end of the received energy from one pulse corresponds to the accumulation of acoustic paths around the theoretical "axial" ray, that follows the lowest sound speed in the sound channel. Comparisons between the observed and calculated travel times for this crescendo demonstrated again that good agreement was obtained by the use of DG's equation rather than UNESCO's. It must be pointed out here that these last measurements concern only a very reduced domain in T, S, and P, as only the properties of seawater at the sound channel depth are involved.

2.4. LABORATORY SOUND VELOCIMETERS

2.4.1. The Basic Methods for Absolute Measurements

The accurate measurements of sound speed in water have been made using laboratory velocimeters built on purpose, or with the aid of special transducer assemblies immersed in tanks. The accuracy achieved depends the method used and the care taken in mechanically calibrating the instrumentation. The equipment has to make precise measurements of the water temperature and, for those designed for studies under pressure, the hydrostatic pressure. It would be too cumbersome to include this equipment in the following description. The question of pressure will be discussed together with the results. One has to add that for seawater, salinity had to be measured as well, but this was done independently of the velocimeter itself, either before or after measurements, or both. This question will also be discussed in Section 2.8.

Laboratory sound velocimeters measure the sound speed of samples contained in small acoustic cells of 3–20 cm maximum dimensions. The measurements in tanks, sometimes called "free-field" measurements, use longer paths, but these do not exceed 2 m. Sound velocimeters can be divided into two categories. The first utilizes fixed acoustic paths and some method for measuring the associated travel time, the second makes use of interferometry in standing wave situations. The free-field instrumentation measures differences of times of arrival of sound pulses vs differences of path lengths. To simplify this presentation, we use terminology derived from that adopted by McSkimin [54, 55]. This leads to the following list of types and principles: type A1, fixed path with time-delay measurements; type A2, fixed path with HF "phase" measurements; type B, interferometric measurements; and type C, differential paths and delays measurements.

We describe first the type A1 and B sound velocimeters because it is with such equipment that the main corpus of sound speed data for Neptunian waters has been obtained. Type A2 velocimeters were only built for measurements of sound speed in pure water. Type C could only be used for a modest range of temperature in tap water at atmospheric pressure.

2.4.2. The Type A1 Velocimeters (Fixed-Path, Time-Delay Measurements)

An ultrasonic "pulse-type" velocimeter was developed by Carnevale and Litovitz [41] but their equipment, certainly of the A1 type, has not received as much detailed attention as those of the NBS and the NOL. These two laboratory velocimeters were developed at the NBS by Greenspann and Tschiegg (G-T) and at the NOL by Dudley Taylor, Wilson being in charge of the measurements. The basic acoustic cells used differed only in detail, the main difference residing in the fact than the NOL velocimeter could be put under internal pressure and placed in a container at the same pressure, the whole being immersed in a bath at a controlled temperature, whereas the NBS instrument had not been designed for measurements under pressure. In addition, the NBS velocimeter was not used for published measurements with seawater but only with pure water. A first NBS velocimeter was built, with a 10-cm-long cylindrical cell, and was soon replaced by a more precise one, with a 20-cm-long cell. At the NOL, one single equipment seems to have been built, although modified equipment to measure both sound speed and density had been envisaged at the start [49]. Whatever the details, the published measurements of G-T and Wilson are both from one single velocimeter. Both used an acoustic cell made of a thick cylindrical chromium or nickel steel tube, at the ends of which were precisely positioned quartz crystals, one used as a transmitter and the other as a receiver [Fig. 2.3(a)]. The mechanical distances between the two crystal mountings were measured with the greatest accuracy at a precisely known temperature. The longitudinal extension of the tubes with temperature was known from accurate measurements on similar tubes. In all, for both velocimeters, the distance between the two internal faces of the opposite crystals could be known with great accuracy at any temperature. Unfortunately, it was later proved, after the discrepancy with measurements by interferometers had been observed (see Sect. 2.3.3.3), that the absolute calibration was not sufficiently accurate.

In both the NBS and the NOL velocimeters, the transmitting crystal was periodically excited by a short electric pulse. This resulted, after an unknown extremely short time delay, in an ultrasonic sound pulse of oscillatory shape that was received by the other crystal and reflected back to the transmitting crystal, then from the latter back again to the receiving crystal. To eliminate the inaccuracy due to the unknown time delay between the electric triggering and the resulting sound pulse generated in water, the measurement consisted of adjusting the frequency of repetition of the electric pulses so that the direct acoustic arrival from one pulse would coincide with the twice reflected arrival from the preceding pulse [Fig. 2.3(b)], or the general coincidence of all arrivals. This was done by visual observation on an oscilloscope of the electric signals obtained from the receiving crystal. This is where the difference lies between what are usually called sing-around velocimeters and the scientific equipment, a fact that seems to have been lost by some modern authors. If L is the mechanical distance between the opposite faces of the crystals and T is the time elapsed between one electric pulse and the next (accurately measured by a counter on the adjustable oscillator used to generate the electric pulses), then the speed of sound C is given by $C = 2 \times L/T$. The difficulty is to decide when the desired pulses coincide on the oscilloscope. What is observable is the shape resulting from the summation of the somewhat different signals (direct and reflected) when "coincidence" is obtained. Both teams at the NBS and the NOL were conscious of this, and several ways were tried with several observers of deciding when a good result was achieved, averages then being made. It is certainly for this reason,

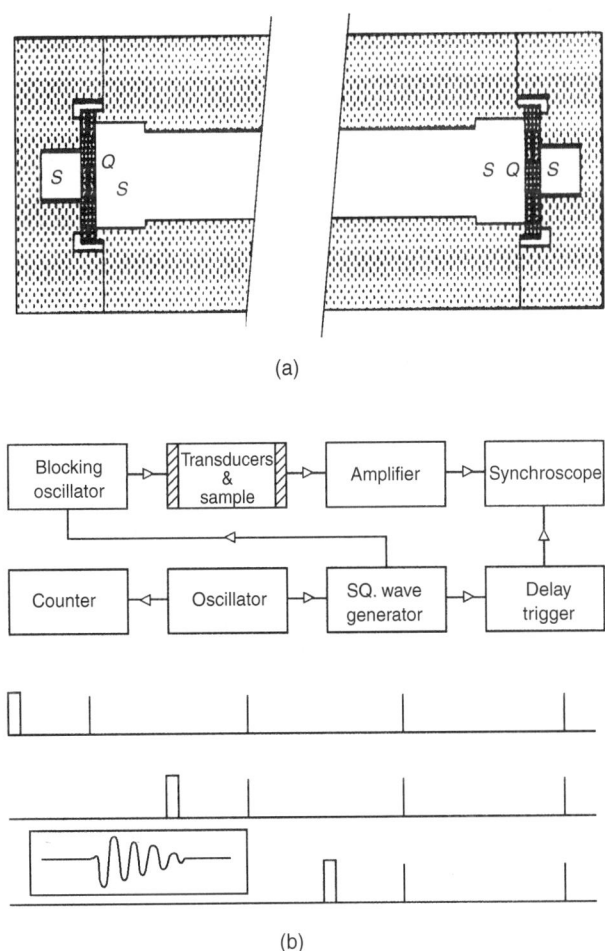

Fig. 2.3. The NBS type A1 velocimeter. (a) Schematic of the axial cut of the instrument. Q = quartz, S = spaces for the springs applied to each quartz. (b) Schematic of method (after Greenspann and Tschiegg [44]). Diagram of three successive pulses, with the thick line representing the input pulse.

plus the difficulties in mechanical calibrations, that after the NBS and NOL type A1 velocimeters, not one such equipment has later been built.

The measurements performed with these velocimeters immediately followed their construction. The G-T data were obtained in a short period of time, and just one set of measurements has been published. The main Wilson data were published in three subsequent articles within only 15 months. An additional small set of points was published later but essentially all the work of Wilson was accomplished between 1958 and 1961, and by the same equipment. Things were completely different with the interferometric velocimeter of Del Grosso.

2.4.3. The Type B Velocimeters (Interferometric Measurements)

Several important laboratories have developed interferometers for measuring the speed of sound in water, the most famous being certainly that of the NRL with Del Grosso, first associated with other scientists and then on his own. In the Soviet Union, interferometric measurements have been performed by Ilgunas et al. [63], but these are rather scarce, so we will concentrate on the work of DG and later comment on recent developments in Japan. In contrast with the type A1 velocimeters of the NBS and the NOL, directly built and used, the equipment of DG has been continuously revised, modified, and improved so that there was not a single interferometer built at the NRL but several, with variations. Finally, the main systematic measurements were performed between 1970 and 1974. From the descriptions given in 1966 [48] and 1972 [71] by DG, one can conclude that at least the final atmospheric equipment was the same, with the exception of the method used to measure displacements. The principle of the measurements is as follows. The measuring cell where water is contained is equipped with one 5-MHz transmitting quartz and one movable plane reflector facing it and parallel to it within 10 s of arc. A mechanical stirrer and a thermometer are located in another part of the cell, aside from the acoustic path between the transducer and the reflector. Still another thermometer is positioned behind the reflector. The quartz transmits permanently, so that a standing wave situation is created inside the cell. The standing wave pattern is altered when the separation between the transducer and the reflector is changed and a displacement of one half wavelength reproduces the same pattern. Del Grosso found that the most accurate way to determine that an identical pattern was achieved was the examination of the impedance of the transducer, which is controlled by the sound field at the quartz–fluid interface. A display of the impedance circles obtained during each displacement of one half wavelength enables one to count the number of wavelengths covered by displacements. One can also examine the precise position of the impedance point on any circle. The problem that needs to be solved is then the accurate measurement of the mechanical displacements. After having employed gauge rods, optical methods using a laser were found to be more accurate. The difficulty, however, lies in the realization of a precise mechanical arrangement, as the displacement measured by the laser has to be exactly the same as the displacement of the reflector. In addition to the mechanical design, Del Grosso made detailed calculations for diffraction effects, control of reflections, examinations of waves transmitted in the metal, etc. Finally, as the speed of sound had to be measured under pressure, two different velocimeters were built, one for atmospheric pressure and one for high pressures. Figure 2.4 gives a schematic view of the second velocimeter of the NRL for measurements under pressure. This has no internal stirrer but still two thermometers, and the rod supporting the reflector passes though the whole cell under pressure. This detail is dicussed in Section 2.8.

Measurements of the speed of sound in water by interferometry has recently received a new interest, apparently totally disconnected from acoustics and related to obtaining equations of state for the fluids. We shall only briefly describe the interferometric

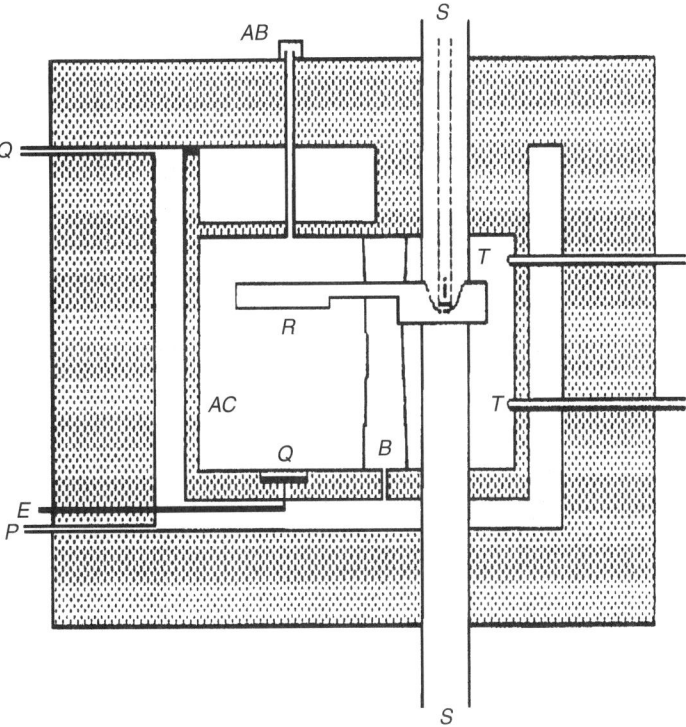

Fig. 2.4. The NRL interferometric velocimeter for under pressure measurements (schematic cut). Q is the transmitting quartz electrically connected by E; S is the main shaft, sliding up and down without rotation through the pressure jacket, thus moving the reflector (R) in the waterproof acoustic cell (AC); the displacement is measured by an interferometric laser acting through the central hole of the upper part of the shaft and pointing at the internal reflector (i); oil is maintained under pressure by P and Q, and pressure transmitted to the acoustic cell by the rubber blader (B); AB = air bleed, T = thermometers.

velocimeter of Fujii and Masui [91] in Japan, based on the use of phase-sensitive detectors, a technique studied by various researchers (see [92–94], for example). A transducer is directly applied to an ultrasonic buffer, the other end of which is one face of the acoustic cell. A movable reflector is the other face, and both are mechanically connected to a laser interferometer for measuring the displacements. The acoustical interference fringes that are controlled are those between the first reflection at the fixed interface buffer-cell and the second one at the movable reflector. Sound CW pulses are used instead of a permanent field (Fig. 2.5). These pulses are mixed with in-phase and in-quadrature signals from the electric HF prime generator, and the levels obtained in the middle of the resulting pulses are measured and used to obtain phase variations as a function of displacements. The authors claim an accuracy of 10^{-5} for their first measurements under atmospheric pressure.

2.4.4. The Type A2 Velocimeters (Fixed Path With HF "Phase" Measurements)

Although several different methods for time measurements through "phase" have been employed in these sound velocimeters, they all have one common characteristic: the acoustic cell containing the water sample is limited by two cylindrical fused quartz rods separated by a removable tube-shaped spacer made of the same material. The cell is then just like a "slice" of cylinder, about 1 to 2 in. in diameter and only 1/4–1/2 in. in length (Fig. 2.6). Several spacers of different thickness can be employed for controls. In all cases, both ends of the rods and the faces of the spacers are machined with

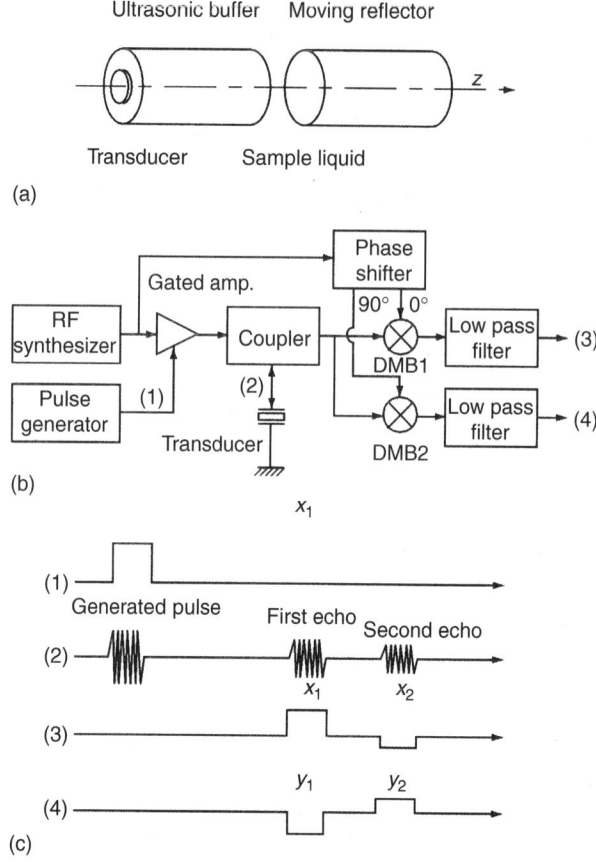

Fig. 2.5. The Japanese NRLM interferometer, using the coherent phase detection technique. Principle of measurement after Fujii and Masui [91].

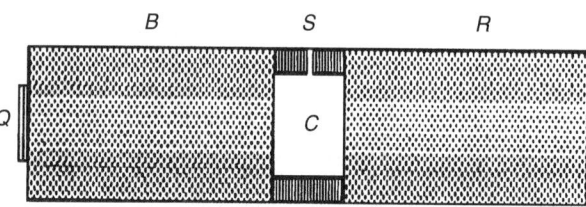

Fig. 2.6. The type A2 velocimeters. Schematic of general mechanical structure. Q = transmitting quartz, B = buffer, S = spacer, R = rod (all three in fused quartz), C = acoustic cell.

extreme accuracy to provide parallel acoustic faces, and the thickness of the spacers is measured to better than ± 1 µm at a precisely known temperature. These spacers are pressed between the two rods and a transmitting crystal is applied on the end of the first rod opposite the cell, so sound is transmitted to water through that rod acting as a buffer.

The first velocimeter of this kind was developed at the Bell Telephone Laboratory by McSkimin [54]. In this equipment only one quartz transducer was used to generate high-frequency short CW sound pulses between 60 and 100 MHz. Each sound pulse crosses the fused quartz buffer. At the interface with water in the cell the pulse is partly reflected back to the quartz, where it generates an electric signal. That part of the original acoustic pulse transmitted to water crosses the sample, is reflected by the

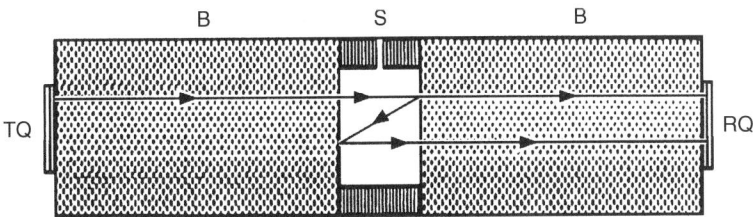

Fig. 2.7. The type A2 velocimeter of Carnvale *et al.* [58]. Schematic cut and sound paths used. TQ and RQ = transmitting and receiving quartz, B = fused quartz buffer, S = fused quartz spacer.

face of the second rod, and reaches the buffer, where again part of the energy arrives at the quartz and the rest is reflected into water, and so on. In all, the quartz receives a series of back and forth reflections from the two faces of the cell. It is now a question of adjusting both the transmitted pulses' time spacing and their frequency to observe interference patterns between overlapping pulses, from which precise travel times can be calculated for various configurations. In the velocimeter of Barlow and Yazgan [56] for measurements under pressure, a quartz ring spacer was placed inside a cylindrical container put under pressure. An accurate knowledge of the bulk modulus of the fused quartz employed for the spacer was used to correct for the changes in dimension of the acoustic cell under pressure. Unlike McSkimin's equipment, the travel time in the cell was obtained from the measurement of the total phase shift experienced by a frequency-modulated high-frequency pulse.

The type A2 velocimeter built at NAVOCEANO (Carnvale *et al.* [58]) was a variant of McSkimin's equipment in that the receiver was not the transmitting quartz but another one located at the end of the terminating rod (Fig. 2.7). What is measured is then the travel time difference between the directly received pulse and the first "echo" from one reflection on each plane of the cell. The measurement is carried out by sending pulses in pairs, adjusting their spacing and the frequency to obtain nulls in the interference between reflections from the first pulse and direct signals from the second (an artifice somewhat similar to that used in the NBS and NOL type A1 equipments). The equipment of Williamson [59] at the NASA Electronic Research Center was very similar to that of Carnvale *et al.*

2.4.5. The Type C Velocimeters (Differential Paths and Delays Measurements)

These equipments are not, strictly speaking, velocimeters but rather precise mountings of instruments installed in an open tank filled with water. The principle exploited is as follows. A transmitter and a receiving hydrophone are immersed in homogeneous water with constant temperature. The time t elapsed between a well-defined point related to the electric signal applied to the transmitter and a well-defined point of the electric signal generated by its reception on the hydrophone, at a distance corresponding to an acoustic path length l, is related to the sound speed C by $l = C(t - t_x)$, where t_x is an unknown, not measurable time delay, and l is difficult to determine with accuracy. If similar measurements are made for two path lengths l_1 and l_2 one gets

$$C = l_1/(t_1 - t_x) = l_2/(t_2 - t_x) = (l_1 - l_2)/(t_1 - t_2) \qquad (2.3)$$

so that t_x is eliminated. By using simultaneously two separate hydrophones and path lengths l_1 and l_2, the need for a reference time at transmission would be eliminated also, but the measurement of $l_1 - l_2$ would be very difficult. Instead, if only one hydrophone is used and moved by an accurately measured distance, this distance is, in free-field conditions, equal to $l_1 - l_2$. The measurement of $t_1 - t_2$, however, has to be done with a reference at transmission because sound pulses separated in time are used. Various mechanical mountings with different systems to measure displacements can be employed, and the measurements of t_1 and t_2 can be based on different techniques. The limitations of these systems, however, are as follows: (1) Measurements cannot be made under pressure. (2) The dimensions of the tank make it difficult to obtain a perfectly homogeneous medium, with temperature remaining constant within $\pm 0.001\,°C$. (3) Great care must be taken to make sure that the displacement of the immersed transducers is exactly the same as that of the reference mechanical mounting. The first of these "velocimeters" was set up by Brooks [60], who used an elongated aluminium tank ($10' \times 8'' \times 8''$) filled with distilled water at ambient room temperature only ($\approx 22\text{--}25\,°C$), and precisely located transducers with $l_2 = 2l_1$ (a fact that increases accuracy). The achievement of distances was realized by two inside mounted micrometers with indicator gages that read to 10^{-4} in. In the instrumentation used at the NRL by Neubauer and Dragonette [61], water was contained in a $10 \times 5 \times 5$ ft cypress tank. It was heated up by addition of hot water and stirring. Finally, the equipment built at the University of Kiel by Kroebel and Mahrt [62] is certainly the most precise of these type C velocimeters. The aim was to achieve an accuracy in sound speed of ± 0.01 m/s, but temperature problems reduced it to ± 0.04 m/s. Distilled water was placed in a 150-litre enameled steel tank, and the acoustic measurements were performed between panels of sound-absorbing brushes. The assembly of transducers was particularly robust, being mounted on a rigid cast iron thick platform (1.4×0.4 m). Laser interferometers optically aligned with the transducer faces were used to measure displacements. The average path lengths were of the order of 70 cm, and displacements and field corrections were established. The measurement of time differences was made by the use of a feedback system in which the received signal was used to control the signal fed to the acoustic transmitter. A modified version of the equipment using a Plexiglas tank was used for just a few measurements with seawater.

2.5. THE NEPTUNIAN DOMAIN

The reasons for specifying the Neptunian waters domain have been given in the preceding sections. The investigation carried out in 1968 by this author to determine the limits of this domain arrived at the volume in TSZ coordinates illustrated in Figure 2.8 from Leroy [1]. It is a three-dimensional representation of the volume when looking upwards toward the sea surface. At zero depth, the area indicated by the letter C corresponds to the Persian Gulf; that with letter G to intertropical zones with heavy precipitation and/or river inflow, and area F to cold, low-saline waters mostly found along the coasts of the Arctic or high-latitude seas. At depth, stalactite-looking parts of the domain are found in closed basins. They are lettered according to the list in Table 2.1. This table gives the depths of the sills making each basin communicate with the open ocean or with the next basin toward the open ocean, e.g., from the Black Sea toward the Sea of Marmara, from the Sea of Marmara toward the Sea of Thrace, and so forth. It also provides the maximum depth of each basin and the constant temperature and salinity (T_c and S_c, respectively) reached from about $1\frac{1}{2}$ times the sill depth to the greatest depth.

To facilitate measurements inside the Neptunian domain, this author gave in Table III of Refs. 2 and 3, a list of temperature and salinity ranges as a function of pressure (in kg/cm^2). Closed basins were on part b of the table but the small Sulu Sea had been omitted, as pointed out by Lovett [79], because of its importance in the problem of transforming depth to pressure. Taking into account the Sulu Sea adds the range $10\text{--}11\,°C$, $34\text{--}35\,\tfrac{o}{oo}$, 0–60 MPa.

It was pointed out by Leroy [2, 3] that at the *TSP* combinations encountered in closed basins it would be better to operate sound speed measurements from in situ

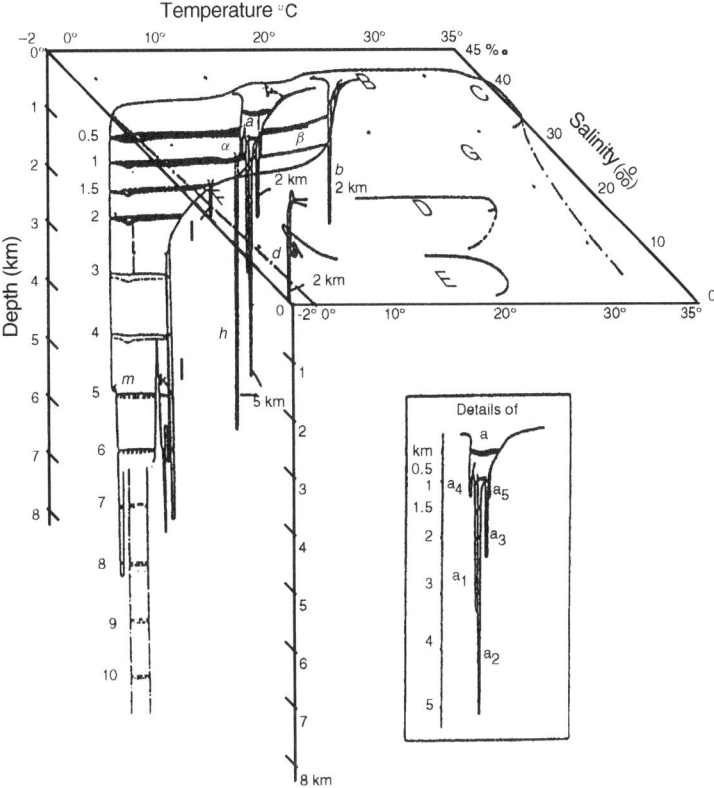

Fig. 2.8. Three-dimensional representation of the temperature-salinity-depth volume enclosed by Neptunian waters (after Leroy [1]). "Stalactites" are lettered according to Table 2.1.

Table 2.1. World's main closed basins with sill depths much lower than their maximum depth (after Leroy [2, 3]).

Name of Basin	Ref. on Fig. 2.8	Sill depth (m)	Maximum depth (m)	T_i (°C)	S_i ($\frac{o}{oo}$)
Western Mediterranean	a1	300	3600	13.0	38.4
Eastern Mediterranean	a2		5100	13.6	38.7
Sea of Crete	a3		2500	14.4	38.9
Sea of Thrace	a4		1500	12.7	38.8
Sea of Marmara	a5	55	1300	14.3	38.5
Black Sea	d	35	2200	8.8	22.0
Red Sea	b	100	2800	21.5	40.5
Baltic Sea	e	20	400	4–6	12–14
Sulu Sea	h	400	5600	10.3	34.5
Halmahera Basin	i	700	2000	7.8	34.6
"American Med." (Caribbean, etc.)	j	1600	7200	3.8	35.0
Miscelleneous East Indian basins (Celebes, etc.)	k	1400–3000	3700–7400	3.2–4.2	34.6
Sea of Japan	l	160	≈4000	0.2	34.1
Artic Basin (Polar Basin and Norvegian Sea)	m	2 × 500	≈4800	≈ −1.0	≈35.0

water samples rather than with diluted or evaporated standard water samples. Also the sound speed data from non-Neptunian waters ought to be withdrawn when establishing an accurate equation for use in underwater acoustics. We must emphasize strongly some of those "details" and comment on what has been said already. If one takes a sample of seawater at 200 m in the central Atlantic or Pacific and uses it at the *TSP* combination existing where it has been taken, this is seawater and it is Neptunian water. If this sample is brought to 1 °C under 80 MPa, it is still Neptunian water. If it is put under that same 80 MPa pressure but heated to 30 °C, it is still seawater, placed in a very uncomfortable situation indeed, but it will not be Neptunian water any more. Now, if one dilutes or evaporates this original sample so that its salinity is decreased or increased within the range 33–37 $\frac{o}{oo}$, then places it under T and P conditions existing in Neptunian waters at the achieved salinity, it can be considered Neptunian water because the relative composition of dissolved matter remains invariant in Neptunian waters within that range. On the contrary, if the sample is diluted to a measured salinity of 22 $\frac{o}{oo}$ and if it is put under 15 MPa at 8 °C, it will not exactly be Neptunian water. It will be diluted sealike salted water with TSP combinations encountered in the Black Sea, which belongs to the Neptunian domain. The real Neptunian water under these TSP conditions has a slightly different composition, because it comes from a closed basin (H_2S in particular is in excess compared with what is found in normal ocean water). But a question can be raised: Is there any risk of finding Neptunian waters from different origins having the same salinity? The practical answer is in the negative. A detailed examination of the Neptunian domain shows that the "abnormal" combinations of temperature and salinity at depth are distributed in salinity according to discrete separated values or ranges: 10–16 $\frac{o}{oo}$, Baltic Sea; $\approx$22 $\frac{o}{oo}$, Black Sea; 38–39 $\frac{o}{oo}$, Mediterranean; $\approx$40.5 $\frac{o}{oo}$, Red Sea. As for the other basins, their salinity stays in the classical 33–37 $\frac{o}{oo}$ range, it is temperature that may take unusual values, but that does not alter the salt content. The only reason "practical" was used previously is due to some near-coastal, near-surface waters affected by river inflows of different mineral content. As an example, the calcium content in the Mississippi River is twice that of the Colorado River, whereas the reverse occurs for SO_4. To take into account all those local shades of surface waters would lead us to abandon any hope of ever reaching a proper terminology. It is much more important in the end to solve the problem of knowing whether the real Black or Red Sea waters have a sound speed as given by their equivalent "quasi" Neptunian waters of the same salinity.

Some criticism must, however, be leveled at the strict definition of the Neptunian domain. Such as it is, it excludes the totally closed seas (Aral, Caspian, and Dead seas) and the lakes, where the precise speed of sound remains of minor operational importance. But the definition also eliminates the modern canals, while the Suez Canal is of great strategic interest and presents unusual salinities due to the crossing of the Bitter Lakes. On the other hand, the strict Neptunian domain does include the deep waters of the Red Sea, where hot brine pools can be found, and those of near volcanic underwater activity, as in the Pacific. It was also just mentioned that the salt composition of coastal ocean waters can be altered by the inflow of large rivers. In all, it would be too complicated and much too academic to try a definition of a "revised Neptunian domain," but additions have to made to the list given by Leroy [3] to specify the desirable measurements. These need not be all performed with equal accuracy: what the speed of sound is at some deep spots of the Red Sea is at present totally unknown, and a ± 2 m/s accuracy would probably be sufficient, whereas better than ± 0.1 m/s is wanted in other cases. A detailed investigation of the unusual waters could be carried out but is considered beyond the scope of this review.

2.6. AVAILABLE MEASUREMENTS OF SOUND SPEED IN WATER

2.6.1. Preliminary Comments

This section details the available accurate measurements of sound speed in water, fresh or sealike salted, whatever the combinations of T and P for freshwater and of T, S, and

P for seawater, regardless of whether they belong to the Neptunian domain. However, as the final aim is the analysis of data corresponding to that domain, its limits will be recalled when necessary.

When reviewing the available measurements of sound speed, one observes enormous disparities in both the number and quality of data. Some publications are limited to a few points around a tiny subdomain in TSP, among which one can find poor results or, on the contrary, very accurate controls. Other publications cover a very large domain in TSP combinations, more or less finely sampled by more or less precise measurements. It is not easy to represent by one number or one symbol the quantity and quality of a particular set of measurements. To illustrate this point let us look at different available values of sound speed in near-standard seawater ($S \approx 35 \frac{o}{oo}$) around 5 °C and under a pressure of about 40 MPa. In an NOL report, Wilson [64] gives one value: $S = 35.02 \frac{o}{oo}$, $T = 5.083$, $P = 6000$ psia, $C = 1540.27$ m/s. But Wilson notes that 10 readings were made to build up an average value at each of his TSP combinations. If we look at Del Grosso and Mader's data [71], around this chosen TSP combination we find (Table V, page 970) a sequence of 6 "data points" at $S = 34.994 \frac{o}{oo}$ and $P = 6000$ psig, each corresponding to a precise given value of temperature, varying from 5.065 to 5.011 °C, the values of C being given with three decimals. Del Grosso preferred to present all results that way instead of giving one single average value at, say, 5.08 °C. Should we say that DG-M's published results are six times more "important" than Wilson's? Certainly not! But on the other hand, one can really observe the dispersion of his measurements (which is very small, see Sect. 2.8). For about the same TSP combination again we find now from Chen and Millero [4] two different points at 399.33 dbar and $T = 5$ °C exactly. One is with a measured salinity corresponding to $S = 34.998 \frac{o}{oo}$, the other with $S = 35.002 \frac{o}{oo}$. It becomes obvious this time that we are confronted with two independent measurements with two different water samples, but C-M do not say how many observations were made to arrive at their given values.

With sound speed measurements at atmospheric pressure on pure water, different situations will again be found: Greenspann and Tschiegg [44] made temperature vary between 0.14 and 99.06 °C, and performed in that range measurements at 83 different temperatures, not listed but displayed in a graph. At each value, the reading of the repetition rate in their type A1 velocimeter that leads to sound speed was made "three times or more," apparently by different observers, to obtain averages. Del Grosso and Mader published in 1972 [72] what they called "112 new data points" of sound speed in pure water. If one looks carefully at these data (Table I in [72]), one will find that they are grouped into packets around successively 0.002, 0.012, 0.053, 0.198 °C, etc. (23 values in all), and each time 2 to 12 values are very close in temperature. We cannot simply say that there are 112 data points from DG-M and 83 from G-T. Now, if we look at Neubauer and Dragonette's (N-D) data [61], we notice that they did not attempt to make measurements around specific values of T but instead let water follow the slow increase of ambient temperature from 16.6 to 23.09 °C. This gave 45 experimental points in a small T domain, each resulting from observations by three different experimenters.

The previous paragraph clearly illustrates the great disparity of the measurements and the danger of giving a false impression of their quantity by merely quoting one number. To avoid confusion, we adopted the following rules. For each set of published measurements we shall indicate the range of temperature and a number of "measured values" in that range: It will be 83 values for G-T above but 23 only for DG-M in pure water. In addition, to show the occasional greater concentrations of measurements in a subrange of the total temperature coverage, this will sometimes be divided in an appropriate manner. If measurements inside one T range do not correspond to desired discrete values, such as for N-D, we shall mention "continuous values." To indicate afterward the amount of data, information will be provided in tables under "No. of data points." There the total will be 112 for DG-M. For G-T it will be said ">3 readings per value," and for N-D "45 with 3 observers." This way of presenting the work of each scientist is the most "honest" we could find.

In the following presentation, we deliberately separated the measurements at atmospheric pressure and those under pressure. All measurements presented by scientists as "under pressure" contain at each T-S combination one value at atmospheric pressure. For the sake of a clear distinction, we placed in the "atmospheric group" all data coming from studies under pressure that were obtained at 1 atm.

2.6.2. Sound Speed in Pure Water

The measurements of sound speed in pure water at atmospheric pressure were made between 1950 and 1976, and recently (1993) in Japan [91]. They are summarized in Table 2.2 in chronological order. This table gives the year of publication of the results; the reference number in this chapter; the investigator(s) and laboratory; the type of velocimeter employed following, the lettering adopted in Section 2.4; the temperature range(s) of the measurements; and the number of "values" and "data points" according to what was said in the previous paragraphs. The last column indicates whether an equation for pure water at atmospheric pressure only was developed from the obtained values. This will avoid another presentation when reviewing sound speed equations in the next section. The discussion of results will be made in Section 2.8. At this point we must recall that for applications to Neptunian waters temperature was limited to 35 °C.

It is unfortunate that there exist basically only two sets of measurements of sound speed in pure water under pressure [see Table 2.3(a)]. The first set is from Wilson [12] with a type A1 velocimeter and covers the range ≈ 0 to ≈ 91 °C, 2,000 to 14,000 psia (lb/in^2 in absolute value), that is ≈ 14–96 MPa. The other set is the second one from Barlow and Yazgan [57] (their first one having been considered not accurate enough) and covers the range ≈ 16 to 94 °C, 10 to 80 MPa (pressure values in the communication were transformed into psia). In both cases, the measurements were made at a few discrete values of temperature, with pressure varying by constant steps.

2.6.3. Sound Speed in Seawater

Apart from a few sporadic measurements around very limited values, like those of Kroebel and Mahrt [62] for example, the speed of sound in seawater at atmospheric pressure specifically has only been studied on a large scale at the NRL and at Miami University (the measurements from the NOL are part of Wilson's investigation as a function of the three parameters T, S, and P). These specific measurements were performed by (1) Del Grosso, first in 1950 with the interferometer existing at that time at the NRL, then in 1970 [70] and together with Mader [71] in 1972 using the NRL atmospheric pressure velocimeter and (2) Millero and Kubinski [75] in 1975, at Miami University, with a Nusonics commercial sing-around velocimeter, by comparison with results in pure water and with the use of the equation of Kell to obtain the speed of sound in pure water. Measurements at atmospheric pressure were also obtained during each of the three systematic investigations in the TSP domain performed at the NOL by Wilson, the NRL by Del Grosso and Mader, and Miami University by Chen and Millero. This time, the NRL results were obtained with the velocimeter designed to operate under pressure. Table 2.4 summarizes all the systematic measurements of sound speed in seawater at atmospheric pressure, in a manner similar to that of Table 2.2. To visualize the coverage of measurements, the TS combinations explored in these investigations are given in the three two-dimensional plots of Figure 2.9. It is clear that both Del Grosso and Millero covered the range 30–40 $\frac{o}{oo}$, 0–40 °C with a quasi uniform sampling, and Wilson concentrated his measurements on the more common

Table 2.2. Measurements of the speed of sound in pure water at atmospheric pressure.

Year	Reference no.	Investigator(s)	Laboratory	Type of equipment	Temperature range (°C)	No. of values	No. of data points	Equation
1957	44	Greenspann-Tschiegg	NBS	A1	0–100	83	> 3 readings per value	2.1
1959	12	Wilson	NOL	A1	1–91	11	10 readings per value	
1960	60	Brooks	NOTS	C	22–25	7	Several	
1964	61	Neubauer-Dragonette	NRL	C	18–23	[a]	45 with 3 observers	
1965	55	McSkimin	BTL	A2	23–30	11	$11\times \approx 13$	
					31–72	9	id	
					72–79	17	id	
1967	57	Barlow-Yazgan	Glasgow University	A2	17–93	8	Several?	
1968	58	Carnvale et al.	NOO	A2	0–40	9	Several[b]	
1969	59	Williamson	NASA ERC	A2	23–75	19	Several	
1970	70	Del Grosso	NRL	B	0–74	11	36	
1972	72	Del Grosso-Mader	NRL	B	0–10	18	94	2.3
					40–95	5	18	2.3
1976	62	Kroebel-Mahrt	Kiel University	C	3–34	20	Numerous[c,d]	
1992	91	Fujii-Masui	NRLM[e]	B	19–45	6	21	

[a] "Continuous" measurements.

[b] Different ring spacers were used.

[c] An example of 20 measurements with 16 readings is presented.

[d] The equation was a modification of the third one from Del Grosso and Mader [72] (DG-M c).

[e] Naval Research Laboratory of Metrology, Japan.

Table 2.3. Measurements of the speed of sound in water under pressure.

Year	Reference No.	Investigator(s)	Domain explored in T, S, P			No. of values	No. of data points	Equation
			T (°C)	S (o/oo)	P (MPa)			
(a) Pure Water								
1959	12	Wilson	1–91	0	12–96	77	10 readings per value	2.1
1967	57	Barlow-Yazgan	7–93	0	10–80	64	Several?	2.1
(b) Seawater								
1960	64	Wilson	3–30	33–36.5	12–96	511	10 readings per value	2.1
1960	66	Wilson	0–30	10–30	12–96	126	id	2.1
1972	71	Del Grosso-Mader	Neptunian		3–103	55	314	2.6
1977	4	Chen-Millero (a)	0–40	5–39	10–100	296	327	1.1

low-temperature values because they are found more often at great depth. One must add that if we except D-G's very first equation, the data from M-K are the only ones that have been fitted to an equation by their experimenters.

For seawater under pressure, there are very few isolated laboratory measurements of sound speed performed at limited combinations of T, S, and P. The measurements

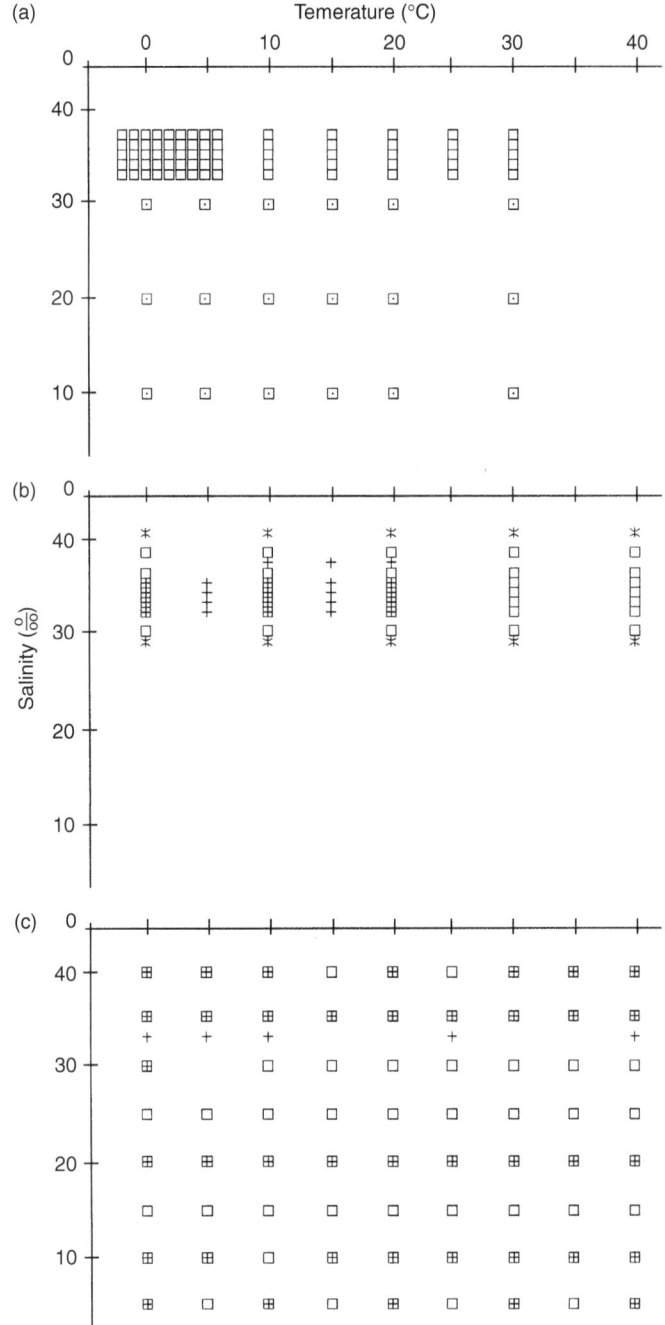

Fig. 2.9. Measurements of sound speed in seawater at atmospheric pressure. (a) By Wilson [64–66]. Dotted squares stand for measurements at intermediate salinities [66]. (b) By Del Grosso [70] and Del Grosso and Mader [71]. Stars stand for measurements at nominal salinities of $S = 30$ and $41 \frac{o}{oo}$ [70], squares for measurements with the atmospheric cell [71], and crosses for measurements with the pressure cell. (c) By Millero and Kubinsky [75] (squares) and Chen and Millero [4] (crosses).

by Carnevale and Litovitz [41] and Martin [46], for example, can be forgotten at that stage: they had supported at their time the discrepancy between Kuwahara's calculations and measurements by velocimeters, and their accuracy was limited to about 1 m/s. There remain only three large sets of data, that overwhelm the rest, viz. in chronological order: Wilson [64–67, 95], Del Grosso and Mader [71], and Chen and

Table 2.4. Measurements of the speed of sound in seawater at atmospheric pressure.

Year	Reference no.	Investigator(s)	Type of equipment	Temperature range (°C)	Salinity range ($\frac{o}{oo}$)	No. of values	No. of data points	Equation
1960	64	Wilson	A1	−2–30	33–36.5	70	10 readings per value	
1960	66	Wilson	A1	0–30	10–30	18	id	
1970	70	Del Grosso	B	0–41	30–41	15	50	
1972	71	Del Grosso-Mader	B[a]	0–40	31–39	35	179	2.2
			B[b]	0–15	33–38	20	100	
1975	75	Millero-Kubinski	[c]	0–40	5–40	81	88	2.1
1977	4	Chen-Millero	[c]	0–40	5–40	35	39	

[a]Velocimeter for measurements at atmospheric pressure.

[b]Velocimeter for measurements under pressure.

[c]Nusonics sing-around velocimeter. Comparison with pure water.

Millero [4]. The measurements of DG-M were performed with the last NRL type B velocimeter, those of Wilson with the NOL type A1 velocimeter, and those of C-M by comparison with measurements in pure water. Table 2.3(b) gives the details and extent of the measured values. The three-dimensional coverage of these measurements are illustrated by the "vertical" cuts at the constant salinities of the measurements, provided in Figures 2.10–2.12. The restriction to Neptunian waters of Del Grosso and Mader's measurements is clearly illustrated.

2.7. EQUATIONS FOR THE SPEED OF SOUND

2.7.1. General

We exclude from this exposition the equations that are not based on laboratory measurements; the early empirical simple equations and those derived from computed sound speeds must now be abandoned. Del Grosso's first equation [38] will not be discussed either. His further measurements were much more accurate than those used in the first equation, which were not included in the data collection fitted by the subsequent equations at the NRL.

The equations for sound speed considered hereafter can be clearly divided into three main categories:

(1) The "direct" equations. These are least squares best fits to a number of measurements performed by one scientist and his team, and to these measurements only. Such are the equations of Greenspann and Tschiegg, Wilson, Del Grosso and Mader, Millero and Kubinski, etc.

(2) The "revised" equations. The final aim here is increased accuracy with respect to the direct equations inside a given domain of TSP, mostly the Neptunian one. They were developed by researchers who had not themselves made measurements but used available data from others, eventually helped by direct equations, the whole with or without corrections and/or limitations. In certain instances, critical analyses and hypotheses led to corrected values of C, T, or S that were used to establish new sets of more reliable data that are fitted with equations.

(3) The "simplified" equations. These are again equations not developed by measurers and their aim is simplicity for practical use, always in Neptunian

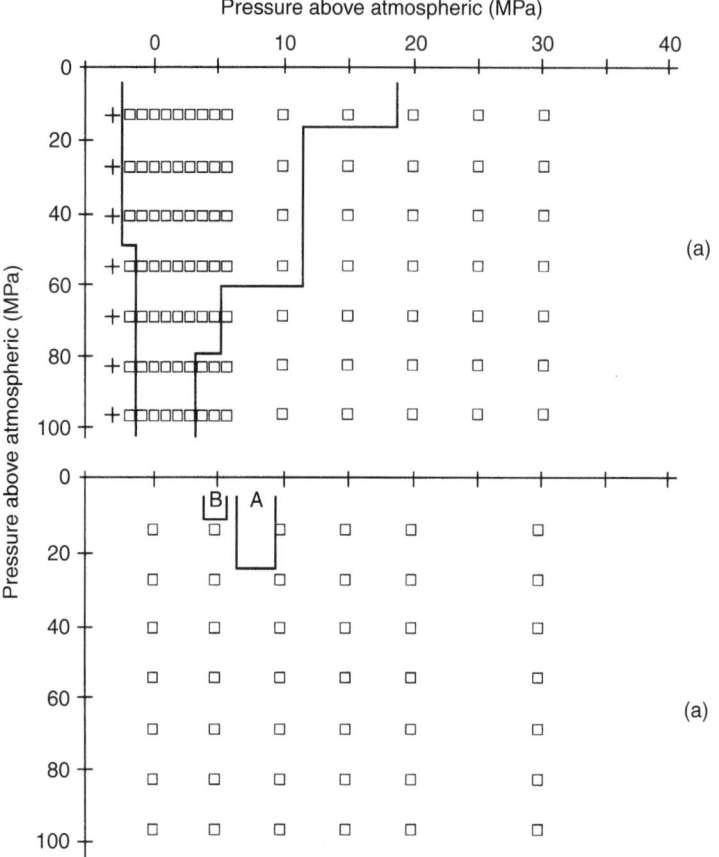

Fig. 2.10. The measurements of sound speed in seawater under pressure by Wilson. (a) Nominal salinities $S = 34$, 35, and 36 $\frac{o}{oo}$ (squares) and $S = 33$ and 36.5 $\frac{o}{oo}$ (crosses) [64]. (b) Nominal salinities $S = 10$, 20, and 30 $\frac{o}{oo}$. The Neptunian limits are for (A) the Black Sea and (B) the Baltic Sea [66].

waters: the number of terms is reduced and so is the number of significant figures used for all coefficients. This is made at the cost of reduced accuracy and a reduced domain of applicability, for example in maximum pressure, corresponding to depths of 4500 or 6000 m. It is obvious that such equations were useful before the common availability of microcomputers. This purely practical interest is now considerably reduced.

2.7.2. The Formulation of Equations

All direct equations for the speed of sound are computer-made best fits to given sets of data based on least squares methods. The first equations for pure water under pressure by Wilson [12] and Barlow and Yazgan [57] illustrate the early procedure. In both cases, the data consist of a completely filled two-dimentional grid of values. The sound speed C is then expressed by

$$C = \sum_{i=0}^{p} a_i T^i \qquad (2.4)$$

with coefficients a_i given by

$$a_i = \sum_{k=0}^{r} c_{ik} P^k \qquad (2.5)$$

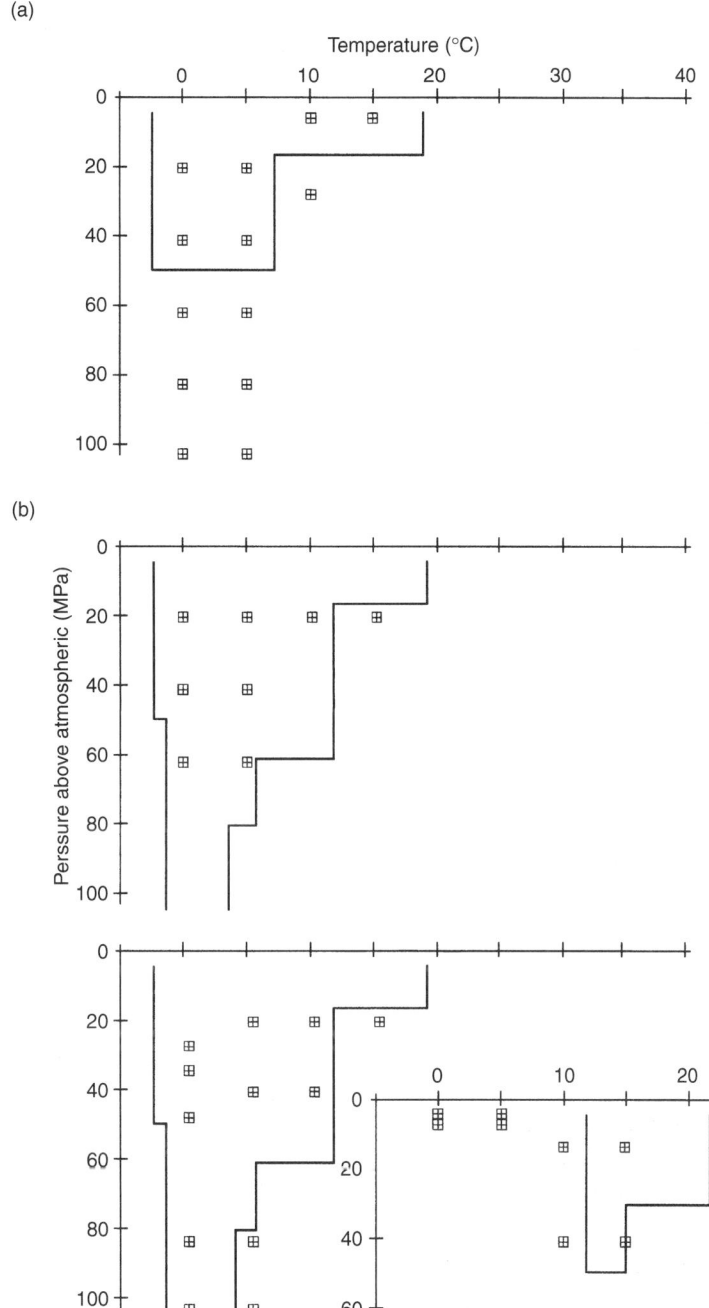

Fig. 2.11. The measurements of sound speed in seawater under pressure by Del Grosso and Mader [71]. The crosses inside the squares are to emphasize the high number of published data points and their quality. The nominal salinity values are: (a) $S = 33$ and $36 \frac{o}{oo}$, (b) $S = 34 \frac{o}{oo}$, (c) $S = 35 \frac{o}{oo}$, (d) $S = 38 \frac{o}{oo}$.

The integers p and r are set a priori, as well as the number s of significant figures adopted for all c_{ik} coefficients. The calculation searches for coefficients c_{ik} that minimize the rms value σ of the differences between the measured data and the corresponding values calculated from the equation (standard deviation). Different values for p, r and s can be tried to examine if σ is substantially reduced when increasing complexity and, conversely, if prohibitive deterioration results from

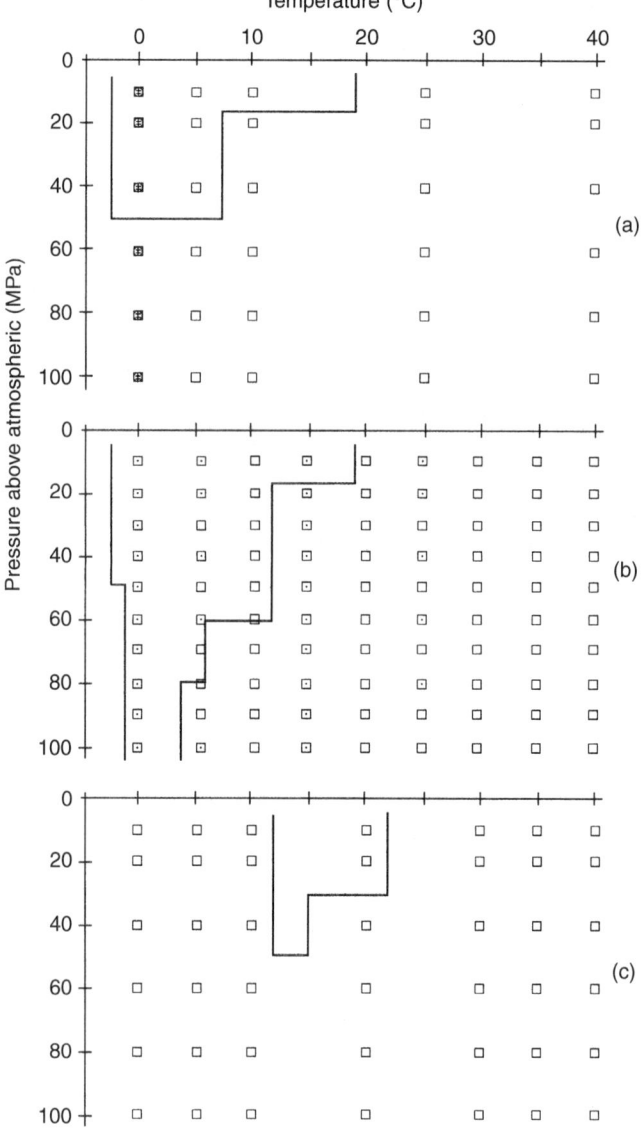

Fig. 2.12. The measurements of sound speed in seawater under pressure by Chen and Millero [4] at common and high salinities of nominal values. (a) $S = 33 \frac{o}{oo}$ (squares). The additional measurements at $S = 30 \frac{o}{oo}$ are indicated by crosses. (b) $S = 35 \frac{o}{oo}$. (c) $S = 39 \frac{o}{oo}$.

simplifications. Wilson ended up with $p = 4$, $q = 3$, and $s = 7$, that is fourth power in T, third power in P, and 7 significant figures for all the $(4+1) \times (3+1) = 20$ coefficients in a matrix form. G-T used $p = 5$ and $q = 4$, resulting in 30 coefficients, all with 7 significant figures, as for Wilson. These equations were presented as such, without any attempt at simplification, in contrast with what was later done for seawater. The reason for this is simple: the speed of sound in pure water under pressure is of interest only to a small community of physicists and need not be calculated every day. Even in the 1960s when these equations were formulated, all scientists interested could use the possibilities of a computer. This was not the case at the time for the numerous users of sound speed in the sea, with sonar or other underwater applications. In addition, a third variable, salinity, increases the complexity of equations. There we find, instead of the previous formulation:

$$a_i = \sum_{i=0}^{q} b_{ij} S^j \qquad (2.6)$$

with

$$b_{ij} = \sum_{k=0}^{r} c_{ijk} P_k \qquad (2.7)$$

that is, in all, $(p+1) \times (q+1) \times (r+1)$ coefficients. With a third power in all variables T, S, and P, this leads already to 64 coefficients. Another alternative is to regroup all constant terms and write

$$C = C_0 + \sum_{i=1}^{p} \sum_{j=1}^{q} \sum_{k=1}^{r} a_{ijk} T^i S^j P^k \qquad (2.8)$$

All modern personal computers can of course easily master such calculations at breakneck speed, but they did not exist when the equations were formulated. Even now, it does not seem advisable to systematically maintain such complexity because it is really tedious to type a large amount of long numbers when writing the program, and the risks of errors are greatly increased. For example, the equation of B-Y was published with a transcription error: the c_{44} term is given with a wrong sign, a fact pointed out by Dushaw et al. [10].

With the exception of the final C-M equation (UNESCO 83, see the next paragraph), all direct equations for the speed of sound in water under pressure were first calculated in a matrix form from the given data $C(T, S, P)$, and T, S, P, but the equation obtained was further processed by computer to get a more simple polynomial. Briefly, the procedure is as follows: In a first step, the most significant term is selected from the whole lot. Each other individual term is then added to the main term, and the one leading to the closest fit to the data is selected and incorporated in the second "provisory" equation, from which a third one will be developed. This is subsequently repeated in a stepwise regression process, where statistics are used to reject terms that do not improve the resulting equation by more than given criteria. A more detailed explanation with references is given by Del Grosso and Mader [71], IV Equation development. However, still further simplifications can be made by reducing the number of significant figures of certain coefficients, which so far have not been changed from the first common decided values. This can be done for each preserved c_{ijk} coefficient by considering the maximum value reached by the quantity $T^i S^j P^k$ multiplied by this coefficient. This last simplification was not made in the various equations presented by DG-M, and it will be discussed in the next section, but other scientists have done it in their formulations.

The UNESCO 83 equation from Chen and Millero has been developed differently because it is not derived from a number of data points given as $C(T, S, P)$, and this is a unique case. In a first step, an equation was given by M-K [75] for the speed of sound in seawater at atmospheric pressure. The data were not absolute values of $C(T, S)$ but differences between the sound speed in seawater of a certain salinity at one temperature and that in pure water at exactly the same temperature (within the accuracy of temperature stabilization). The equation was then of the form $C(T, S) = C_0(T) + f(T, S)$, the subscript zero standing for zero salinity. The function $f(T, S)$ was a computer-made polynomial best fit of the special form $f(T, S) = AS + BS^{1.5} + CS^2$, with A, B, and C functions of T. This is a unique case as M-K, then later C-M, were the only ones to use a 1.5 power term in salinity. In a second step, Wilson's data for the speed of sound in pure water under pressure [12] were transformed into differences, at each temperature, between speeds under pressure and speed at atmospheric pressure. The set of values thus obtained was found to be more centered than the original data. By means of a computer-made best fit, an equation of the form $C_0(T, P) = C_0(T) + g(T, P)$ was formulated [76]. The polynomial $g(T, P)$ was a postprocessed one. Finally, measurements were performed by C-M for the difference in sound speed with "seawater" of different salinities and temperatures and pure water at the same temperatures under similar different pressure conditions. They revised the first polynomial $f'(t)$ from M-K to a simpler one, $f'(t)$, and arrived at a final equation, the Chen and Millero equation for seawater under pressure and the UNESCO 83 equation, given in the form:

$$C(T, S, P) = C_0(T) + f'(T, S) + g(T, P) + h(T, S, P) \qquad (2.9)$$

The equation used for $C_0(T)$ was first, in the M-K equation for atmospheric pressure, the one developed by Kell [80] from various measurements. In the final complete equation of Chen and Millero, the third equation of DG-M for pure water [72] (here DG-M c) was used.

Formulation of the "revised" equations generally follows the same procedures as the usual "direct" equations, the corrected data replacing the original data. One particular treatment, however, was used by Anderson [77]. His point was not only to correct certain data from Wilson, but also to obtain an equation leading to a smooth natural variation of sound speed with pressure at great depth, for more accurate calculations of convergence zones. This leads to the best possible value for the gradient $\partial C/\partial P$, which is not the case from a blind use of least squares applied to all data points. This was done by considering constant values of temperature and salinity, as usually found at great depths. As for the "simplified" equations, concerning the Neptunian domain only, one cannot pretend that their formulation obeys strict mathematical principles. Intuition and trial and error are commonly used to guide the investigator toward a limited number of terms with a limited number of significant figures; computers are used to examine the results in detail and to furnish statistics.

2.7.3. Presentation of the Sound Speed Equations

All sound speed equations are summarized in Tables 2.5 and 2.6. They are arranged as direct and revised and simplified equations. In each case the presentation is subdivided, when necessary, into pure water, first at atmospheric pressure then under pressure, and seawater, under the same conditions. In each subcase, equations are given in chronological order. For each equation, the tables give successively the following information:

- The "type" of water, pure water or seawater, the latter meaning "sealike salted water," as discussed previously
- The maximum pressure P_{max} for the data, or claimed by the author(s). For uniformity, it has been transformed into MPa values above atmospheric, rounded to the nearest integer.
- The author(s)' name(s) in abbreviated form, given in Table 2.7.
- The number of data points used in the mathematical treatment.
- The range of validity of the equation in temperature and salinity. When all these figures are present, the equation is claimed valid inside the whole "cubic" domain described in T and S and up to the maximum pressure. When equations

Table 2.5. List of the direct equations.

P_{max} (MPa)	Equation	No. dp	Range of validity		Complexity		Standard Deviation (m/s)
			T (°C)	S ($^o/_{oo}$)	No. terms	No. sig. f.	
			Pure Water				
0	G-T	83	0–100	0	6	5–6	0.0263
0	DG-M a	112	0–95	0	6	9	0.0026
0	DG-M b	36	0–74	0	6	9	0.0025
0	DG-M c	148	0–95	0	6	9	0.0029
0	K-M	20	3–34	0	6	9	0.018
96	W-o	88	0–92	0	20	7	≈0.2
80	B-Y	72	16–94	0	30	7	≈0.1
			Seawater				
0	DG-M 1	143	0–30	31–39	7	11	0.013
0	DG-M 6	159	0–30	30–39	7	12	0.015
0	M-K	89	0–40	0–40	18[a]	5–7	0.04
0	C-M[b]	id	id	id	14[a]	3–4	0.03
96	W 1	581	−3–30	33–37	21	2–6	0.22
0	W 2	813	−3–30	0–37	23	5–6	0.30
103	DG-M 2	627	Neptunian		24	12	0.044
0	DG-M 3[c]	627	Neptunian		19	12	0.050
0	DG-M 4	627	Neptunian		12	12	0.101
0	DG-M 5	110[d]	Neptunian		17	12	0.059
0	DG-M 7	73[d]	Neptunian		12	12	0.077
0	DG-M 8	414	Neptunian		17	12	0.047
0	NRL II	662[e]	Neptunian		19	12	0.050
100	UNESCO 83	375[f] +89[g]	0–40	0–40	42	3–6	≈0.2

[a]Including the six terms for pure water.

[b]Differing from M-K only by the use of DG-M c equation instead of Kell.

[c]Equation also called NRL I.

[d]Averages from the original data points.

[e]Previous 627 dp plus 7 calculated points for pure water with a weight of 5.

[f]Data from measurements under high pressure.

[g]Data from K-M at atmospheric pressure.

were formulated to be valid only in Neptunian waters, the word "Neptunian" is indicated in the place of values, and it must be understood that the maximum pressure is the maximum Neptunian pressure. In other words, the mention of 103 MPa for all DG-M equations for Neptunian waters does not mean that the speed of sound can be given up to 103 MPa at 13 °C and 38.5 $^o/_{oo}$.

- The complexity of the equation, expressed by the number of terms used in the polynomial and by the number of significant figures found in the various coefficients (with the exception of the constant term, which necessarily needs at least five or six figures). When this number varies with the coefficient, the minimum and maximum values are given.

- The standard deviation of the equation as given by the author(s). This is different from their accuracy, which depends on the measurements and equipment.

Table 2.6. List of the revised and simplified equations.

P_{max} MPa	Equation	No. dp and/or Origin	Complexity No. terms	Complexity No. sig. f.	Standard deviation (m/s)
		Pure Water (Atm. P)			
0	Kell	a	6	5–6	0.03
		Seawater (Neptunian)			
≈100[b]	Ler 1	c	13	1–2	≈0.05[d]
≈100[b]	Ler 2	299[e]	14	1–2	≈0.2
96	And	813[f]	?	?	?
96	F-P	209	12	2–4	0.10
10[b]	Medw	g	7	2–3	≈0.3
103	Lov 1	h	19	7	0.049
103	Lov 2	662[i]	25	6–7	0.035
103	Lov 3	662	13	6–7	0.048
≈30[j]	Ross	k	13	1–4	?
≈40[b]	Copp	l	13	1–4	0.07[d]
80[b]	Mack	m	9	4	0.07[d]
92	Ler 88	h	19	3–5	n

[a]Data from Carnvale *et al*, Barlow and Yazgan, and McSkimin.

[b]Depth is used instead of pressure.

[c]"Neptunian" values given by equation Wilson 2.

[d]With respect to the equation it was derived from.

[e]292 dp from Wilson and 7 values calculated from G-T.

[f]From Wilson 1 and 2, out of which 208 dp at modified salinities.

[g]A simplification of NRL II equation.

[h]A modification to NRL II equation.

[i]The 662 points used in NRL II, corrected as a function of pressure.

[j]Equations using depth are presented after that using pressure.

[k]A merged equation derived from the results of Wilson, DG-M, M-K, and C-M.

[l]A simplification of Lovett's equation Lov 3.

[m]A simplification of NRL I equation with pressure to depth conversions.

[n]Not evaluated. Similar to NRL II.

2.8. DISCUSSION

2.8.1. Preliminary Remarks

The final objective in this study of sound speed in pure water or seawater is a set of equations that provide with specified accuracy the speed as a function of the pertinent variable(s) T, S, and P. In this context, the word accuracy is relative to the exact value, theoretically not obtainable, but only approached "at best" through measurements at a number of combinations of the variables. The first condition to arrive at a "good" equation is of course that it should be based on accurate and reliable mesurements. This involves the use of carefully constructed and calibrated accurate velocimeters, and a precise knowledge of the relevant parameters T, S, and P. It also requires care, patience, and motivation by the experimenters. Moreover, the equation resulting from the measurements will gain in accuracy if (1) the number of data used for its formulation is large enough, (2) these data cover the whole domain of applicability

Table 2.7. Key to the authors of sound speed equations.

Equation	Investigator(s)	Reference
And	Anderson	77
B-Y	Barlow and Yazgan	57
Copp	Coppens	83
C-M	Chen and Millero	4
DG-M a-c	Del Grosso and Mader	72[a]
DG-M 1-8	id	71[b]
F-P	Frye and Pugh	69
G-T	Greenspann and Tschiegg	45
Kell	Kell	80
K-M	Kroebel and Mahrt	62
Ler 1,2	Leroy	3[c]
Ler 88	Leroy	9
Lov 1-3	Lovett	79
Mack	Mackenzie	84
Medw	Medwin	81
M-K	Millero and Kubinski	75
NRL II	(published by Del Grosso)	6
Ross	Ross	82
UNESCO 83	UNESCO 83, from Chen and Millero	4
W o	Wilson	12[d]
W 1	Wilson	65[e]
W 2	Wilson	67[e]

[a] Three equations for pure water.

[b] Eight equations for seawater.

[c] Two equations (1968).

[d] Equation for pure water.

[e] Equation for seawater.

of the equation with an appropriate distribution, and (3) the equation is not forced to apply also to data too remote from this domain.

In the analysis, we concentrate on a final application to Neptunian waters on the one hand and to pure water (as being a reference) on the other. Pure water will, however, be given priority in the TP domain existing in Neptunian waters. This limits the analysis of pure water at atmospheric pressure to 35 °C. For pure water under pressure, the problems encountered with the measurements will oblige us also to take into account the high temperature measurements to see more clearly the differences between Wilson and Barlow and Yazgan. Considerations about data, their quality and scatter, their number, and their distribution will be of importance, as well as the problems related to the accuracy of the equipment and the associated measurements of the parameters T, S, and P. The number of points of measurement has already been discussed, and we made a distinction between nominal values, averaged values, and data points ("dps"). The number of values and dps are listed in Tables 2.2 to 2.4 for each set of measurements, but they will have to be reconsidered in connection with optimization to Neptunian waters.

2.8.2. Speed of Sound in Pure Water at Atmospheric Pressure

The measurements of the speed of sound at atmospheric pressure are listed in Table 2.2. It can be seen from Tables 2.5 and 2.6 that six equations are in question, five of them being "direct" equations proposed by the measurers: one by Greenspann and Tschiegg [44, 45] (G-T), three by Del Grosso and Mader [72] (DG-M), and one by Kroebel and Mahrt [62] (K-M). The data from K-M are found to be very well adapted for the formulation of their equation, but unfortunately there exists some doubt about the homogeneity of temperature inside the tank used for their measurements. For this reason, the results of DG-M will be preferred as a basis of discussion. The accuracy of their temperature measurements is better, the data are numerous, with a good distribution in the desired domain. The offset of their 1972 data (Eq. b) toward low temperatures derives from the fact that the authors wanted to investigate a possible anomaly [73] around 4 °C, where the density of water presents a minimum, and reinforced their measurements around this temperature. Another well-adapted set of measurements is that of G-T, but it is subject to a correction of −0.35 m/s, as seen in Section 2.3.3.3.

The standard deviation of residuals, usually given by the author(s) of an equation fitting data, is an acceptable estimate of the quality of both the equation and the data, as small standard deviations can only be obtained with small scatters of the data. The possible remaining sources of error are then systematic biases, as due to insufficient calibration or to a lack of appropriate corrections for different effects. With pure water at atmospheric pressure, the only source of varying error would be due to improper corrections for dilatation of the equipment. This cannot be the case in the measurements of DG-M because, at each temperature, displacements were exactly measured by a laser interferometer. In a 1988 analysis [9] this author noted that the values given by DG-M for the "scatter" of their pure water data (Ref. 72, Table V, page 1445) were much higher than the true scatter. What happened is that DG-M considered each cluster of data points around a nominal value of temperature as an individual random population of data. This is not the case, because temperature slightly varies in a quasi-monotonic way inside each cluster of dps, as can be observed from Table I in Ref. 72. This variation is not the result of errors in measurement, but corresponds to a real, gradual, slight warming-up of water inside the velocimeter cell during each set of measurements. As the final equation very well represents the general trend of variation of sound speed with temperature, one can use it to correct each data point of each cluster for temperature variations. In the minute temperature interval covered within each data cluster, it is sufficient to consider corrections obtained from the local value of the first derivative $\partial C/\partial T$. If T_o is the average temperature of a cluster of dps, and $C(T_i)$ the measured value of sound speed at T_i, its value corrected for T_o is simply given by

$$Ci(T_o) = C(T_i) + (\partial C/\partial T)_{T_o} \times (T_o - T_i) \qquad (2.10)$$

Repeating this calculation for all T_i and $C(T_i)$ results in a new set of data, all at T_o, to which classical statistics can be applied. As an example, let us consider in [65], Table I the cluster of 5 dps around the nominal value $T = 6\,°C$. They are in fact grouped around an average value $T_o = 5.9914\,°C$. By employing DG-M's Equation "c" (third column of Table III) and taking its derivative, one arrives at the following figures:

Measured T (°C)	Measured sound speed (m/s)	Sound speed corrected for T_o (m/s)
5.9892	1430.543	1430.5465
5.9902	1430.548	1430.5471
5.9902	1430.551	1430.5501
5.9922	1430.559	1430.5494
5.9952	1430.572	1430.5492

The use of four decimals for the corrected speeds is intended to demonstrate the quality of the results and to preserve the accuracy of the standard deviation. The mean corrected sound speed at temperature $T_o = 5.9914\,°C$ is 1430.5550 m/s, the same as that of the uncorrected values, for obvious reasons of symmetry plus linearity. The standard deviation of the corrected values, on the contrary, is now 0.0014 m/s only, instead of the 0.004 m/s given by DG-M in [72], Table V as the "scatter," a reduction by almost a factor 3! This reduced value of 0.0014 m/s is the real scatter of the measurements, and the figure 0.004 m/s, which corresponds to a crude calculation on the dps without correction, is artificially spoilt by the tiny increase of temperature. Having discovered this effect, this author decided to treat all original pure water data given by Del Grosso and Mader in the same way. Because the results were extremely encouraging, he later decided to examine DG-M's data for seawater, found the same phenomenon, and performed the computation for the whole set of DG-M's results.

The results of the recalculated scatters are given in Table 2.8, together with those given by DG-M. The differences between the mean measured values of sound speed and those given by equation named DG-M c at the mean temperature of each cluster of data points are also listed. For simplicity, the temperatures given in the table are the nominal ones, but calculations have been performed as in the example developed above. One can observe that, with only four exceptions in 24 values, all real scatters are two to three times smaller than given by DG-M. The maximum departure from the data of equation DG-M c remains within ±0.5 cm/s. This is quite remarkable, and DG-M's measurements were indeed much better than they had thought when writing ([72], Discussion of results) that they were "at most probably accurate to 0.015 m/s." With the restriction that will be made below in reference to Fujii and Masui, they are probably accurate to 0.005 m/s, and so is their equation in the 0–35 °C range.

As a consequence of this superb result, we shall take the DG-M c equation as a reference to examine the other results. This is done in Figure 2.13, redrawn from Leroy [9]. This figure concerns only measurements falling within roughly ±0.1 m/s from Del Grosso and Mader. It can be seen that DG-M's values lie in the middle of all others. The data points from McSkimin [55] present the greatest scatter, whereas those from Carnvale et al. [58] have the smoothest evolution. Even smoother behavior would be found with the measurements of Greespann and Tschiegg [44, 45] once translated by

Table 2.8. Sound speed in pure water at atmospheric pressure (after Del Grosso and Mader [71])[a].

T nom (°C)	Scatter (cm/s)		Cm-Cc (cm/s)	T nom (°C)	Scatter (cm/s)		Cm-Cc (cm/s)
	DG-M	Corr.			DG-M	Corr.	
0.01	0.3	0.12	0.13	4.2	0.2	0.12	0.22
0.05	0.0	0.12	0.03	4.5	0.2	0.13	0.43
0.06[b]	0.6	0.31	−0.14	5.0[b]	0.6	0.23	0.07
0.2	0.0	0.09	0.18	5.5	0.2	0.09	0.18
0.5	0.4	0.18	0.07	6.0	0.4	0.14	0.28
1.0	0.6	0.21	−0.09	8.0	0.4	0.17	0.29
2.0	0.4	0.19	−0.22	10.0	0.4	0.16	0.13
2.5	0.6	0.20	−0.32	10.0[b]	0.6	0.25	−0.51
3.0	0.2	0.09	−0.03	20.0[b]	0.2	0.20	−0.25
3.5	0.6	0.23	−0.34	25.0[b]	0.0	0.04	−0.23
3.8	0.4	0.44	−0.12	30.0[b]	0.3	0.17	−0.18
4.0	0.2	0.11	−0.14	35.0[b]	0.4	0.16	−0.11

[a]Scatter of data and differences Cm-Cc between measured and calculated values using equation DG-M c.

[b]Data are from the first set of measurements.

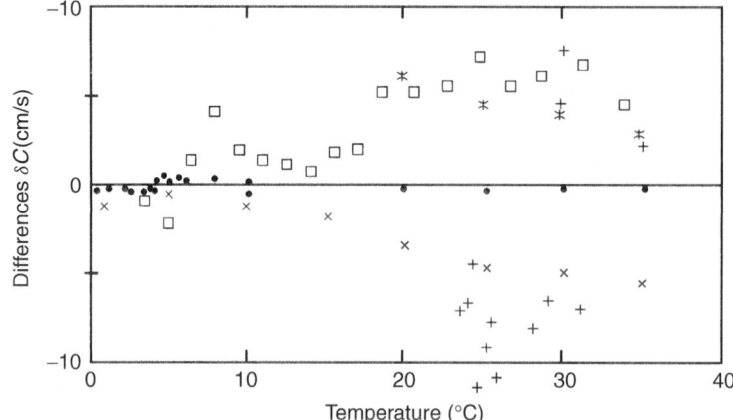

Fig. 2.13. Differences δC between measured sound speed in pure water at atmospheric pressure and the equation DG-M c from. Del Grosso and Mader [72]. Measurements are from Del Grosso [72] (black circles), Kroebel and Mahrt [62] (squares), McSkimin [55] (plus signs), Carnvale et al. [58] (x), Fujii and Masui [91] (asterisks).

-0.35 m/s. The differences between G-T -0.35 and DG-M do not exceed ± 0.01 m/s between 0 and 40 °C. The only embarrassment arises from the recent measurements of Fujii and Masui [91], found to exceed those of DG-M by some 0.06 to 0.02 m/s between 20 and 40 °C. Because the relative accuracy claimed by these investigators is better than 10^{-5} (± 0.015 m/s), this certainly requires further investigation. So far, in this context of application to Neptunian waters, we shall keep to the results of Del Grosso.

In conclusion, we must notice that the equation derived by Kell [80] from the measurements of Barlow and Yazgan [57], Carnvale et al. [58], and McSkimin [55], so different in nature and scatter, gives results that are surprisingly close to those obtained from the DG-M c equation (-0.005 to -0.08 m/s between 0 and 35 °C). This is largely a consequence of the excellent results of Carnvale et al.

2.8.3. Sound Speed in Seawater at Atmospheric Pressure

We shall divide the measurements of sound speed in seawater at atmospheric pressure into two categories. The first will concern "common" salinities, and the second the so-called "intermediate" salinities, classically ranging from 5 to 30 $\frac{o}{oo}$. Common salinities are concentrated in the 33–37 $\frac{o}{oo}$ range, but we have extended it to 30–40 $\frac{o}{oo}$ to rejoin intermediate salinities, include those encountered in the Mediterranean and to be close to those of the Red Sea. This can be justified by the fact that salinity in excess of the mean value 35 $\frac{o}{oo}$ is not too pronounced (5 ppt). The laboratory measurements of sound speed in sealike salted water at atmospheric pressure are listed in Table 2.4, and their distribution is illustrated in Figure 2.9.

2.8.3.1. Common Salinities

If we simplify common salinities to a rectangle bounded by salinities 30 and 40 $\frac{o}{oo}$ and by temperatures -3 and $+35$ °C, it can be seen from Figure 2.9 that the best coverage with measurements is achieved by Del Grosso and Mader. Wilson's data have been found to be in excess by about 0.42 m/s in the temperature range 0–30 °C, a value adopted by Lovett [79]. Lovett, however, rejected by t-tests Wilson's data at $S = 33.08$ $\frac{o}{oo}$, as did Anderson [77, 78]. It will be seen in VII D that these measurements also display an anomalous behavior with pressure, which justifies that elimination. We shall use the results of Del Grosso and Mader as a reference for comparison. Their data [71, Tables I, II, and V] show as for pure water a distribution into clusters at average values of temperature close to nominal values, temperature varying slightly in each cluster. A processing of the data points at each salinity and each average temperature, as explained for pure water, has been performed by this author [9]. The results are given in Table 2.9 for the two sets of values: (a) those obtained with the atmospheric velocimeter and (b) those from the velocimeter for measurements under pressure. Equation 6 of DG-M [71, Table 2.10] was used for establishing the derivatives because it is based on the total set of data. It can be seen that the values of scatter of each cluster are very small for the atmospheric cell, being always smaller than 0.005 m/s, again an indication of the amazing quality of the data. The overall standard deviation obtained from this set is 0.0028 m/s, a figure three times smaller than that claimed by DG-M. The scatter of data points from the pressure cell is larger and can reach 0.030 m/s with an overall standard deviation of 0.014 m/s. This could be an indication that the measurements from the pressure cell are less accurate, but the authors confess that the platinum thermometer could display "erratic behavior." In fact, the temperature range of fluctuations is larger than in the atmospheric cell set of data. Note also that, as distinct from measurements at atmospheric pressure (both with pure water and with seawater), temperature decreased on average in each cluster instead of increasing. That could be the result of the tail of the physical decrease in temperature for recovery after compression, an indication that perhaps other measurements, where temperature has not been measured in the cell but in the surrounding water of the pressure tank, have been affected. The deviations between the mean corrected values and the sound speed calculated from equation DG-M 6 remain within ±0.03 m/s for the atmospheric cell and +0.02 − 0.01 m/s for the pressure cell. This lower departure from the equation simply arises from the limitation in temperature variation (0 to 15 °C only, for restriction to Neptunian waters at depth). These remarkable results are illustrated in Figure 2.14, redrawn from Leroy [9]. Salinity has been chosen as the variable, for reasons that will become clear in the next paragraph, and the values given

Table 2.9. Speed of sound in seawater at atmospheric pressure (after Del Grosso and Mader [71])[a].

Nominal salinity $\left(\frac{o}{oo}\right)$	Temperature (°C)				
	0	10	20	30	40
(a) Measurements with the Atmospheric Cell					
30	0.3	0.4	0.3	0.2	0.5
31	0.5	0.3	0.3	0.2	0.4
33	0.3	0.2	0.1	0.3	0.3
34	0.2	0.1	0.1	0.3	0.2
35	0.3	0.2	0.2	0.2	0.3
36	0.3	0.2	0.3	0.2	0.5
37	0.3	0.2	0.1	0.2	0.3
39	0.2	0.4	0.2	0.3	0.1
41	0.4	0.2	0.3	0.2	0.2
(b) Measurements with the Pressure Cell					
33	1.6	3.0	1.4	1.6	
34	0.7	0.9	1.2	0.7	
35	1.9	1.4	1.1	1.3	
36	1.5	1.0	0.9	1.1	
38			1.0	1.1	

[a]Scatter of their data in cm/s when corrected for temperature changes.

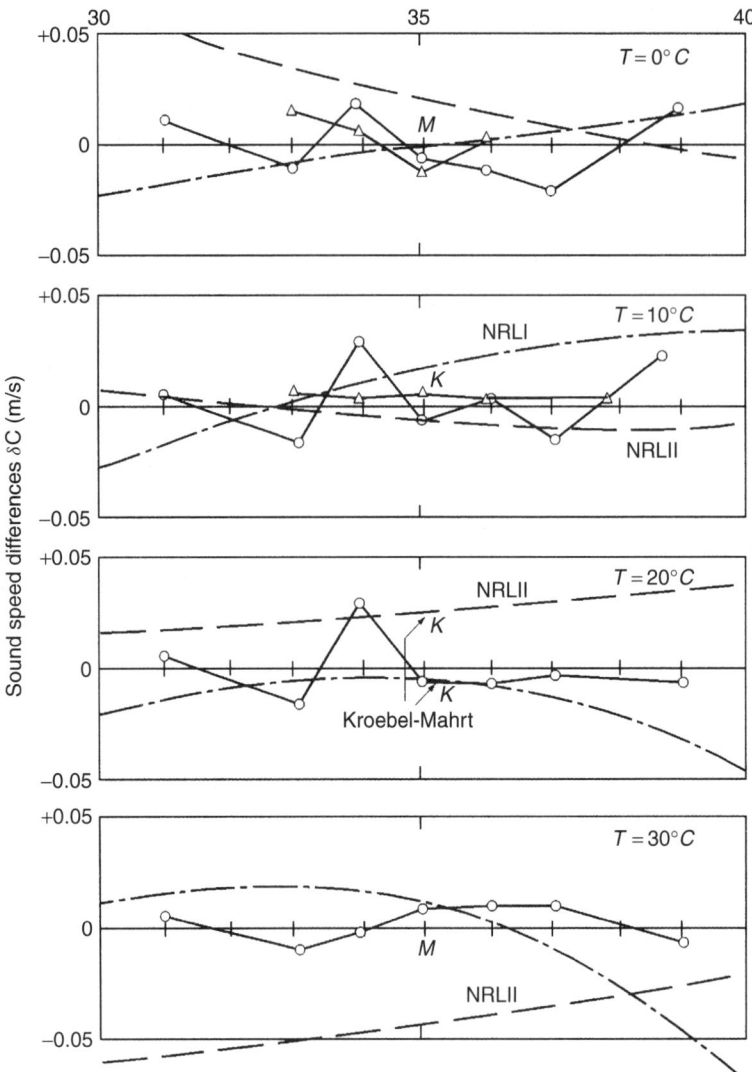

Fig. 2.14. Differences δC between measured values of sound speed in seawater at atmospheric pressure and values calculated from equation DG-M 6: Del Grosso and Mader's measurements with the atmospheric cell, (circles), idem with the pressure cell (triangles) (all from [71]).

for temperatures are, for simplicity, the nominal values. It is seen that, although the scatter of data is larger with those from the pressure cell, the average measured values are as good as those from the atmospheric cell. Most other laboratory measurements performed in the 32–40 $\frac{o}{oo}$ salinity range fell outside this figure, even Wilson's after correction. This will be illustrated in the next set of figures. The difference between equations NRL I and NRL II, used with $P = 0$, and the DG-M 6 equation is given on Figure 2.14. It is seen that they are less precise than equation DG-M 6, a fact illustrating again that an equation for a limited domain is more precise than fitting a larger domain.

2.8.3.2. *The Intermediate Salinities*

The intermediate salinities have been explored only by Wilson [66, 67] and Millero (with Kubinski [75] and with Chen [4]), all with diluted natural or standard seawater.

Del Grosso would have wanted to measure sound speed at low salinities with real Black Sea and Baltic water samples (as well as high salinities with Mediterranean and Red Sea samples) but was unable to get them. All measurements by Wilson at the intermediate salinities have been rejected by Frye and Pugh [69], on the basis of t-test results. Those at $S = 9.94 \frac{o}{oo}$ were rejected by both Anderson [77] and Lovett [79] for similar reasons, and those at $20.06 \frac{o}{oo}$ were accepted by Anderson but rejected by Lovett.

We shall base our investigation of sound speed at intermediate salinities on an original development made by Leroy [9]. The reasoning is as follows. Equation DG-M 6 was found the best to represent sound speed in seawater at atmospheric pressure between $S = 30$ and $40 \frac{o}{oo}$, but it fails outside this domain. The NRL II equation was developed to fit both seawater data at atmospheric pressure and under pressure, and pure water at atmospheric pressure, but is less accurate in the range 30–40 $\frac{o}{oo}$ than the DG-M 6 equation. In the case of pure water, NRL II again does not give exactly the very precise values of equation DG-M c; differences represent the departure of NRL II when $S = 0$. By drawing for various values of T, the differences (NRL II − DG-M 6) as a function of S between $S = 30$ and $S = 40 \frac{o}{oo}$, one obtains part of the complete "departure curves" (NRL II - best estimated values) that extend from $S = 0$ to $S = 40 \frac{o}{oo}$. These partial curves have been drawn on Figure 2.15 in thick lines. The points that the total curves should reach at $S = 0$ have also been indicated. It becomes clear that a smooth interpolation between each partial curve and the point at $S = 0$ can be made by hand and that it represents the complete departure curve. These interpolations are given in thin lines on Figure 2.15. It is obvious that they are accurate to within ±0.01 m/s, with the exception of the dashed line curve at 0 °C, which may have an uncertainty of about ±0.02 m/s. From these results Leroy derived a set of corrections to apply to the NRL II equation to obtain accurate values of sound speed in seawater at atmospheric pressure and a list of reference values of sound speed at 0 to 35 °C and 0 to 40 $\frac{o}{oo}$, reproduced in Table 2.10. These values and corrections were used for comparison with the measurements of Wilson and of Millero and Kubinsky. For the latter, no additional corrections needed to be made since their values of temperatures are multiples of 5 °C. For Wilson, on top of the −0.42 m/s correction from Lovett, an adjustment to the nominal values was made by a method similar to that used in correcting DG-M's data points, using the local values of the derivative of the NRL II

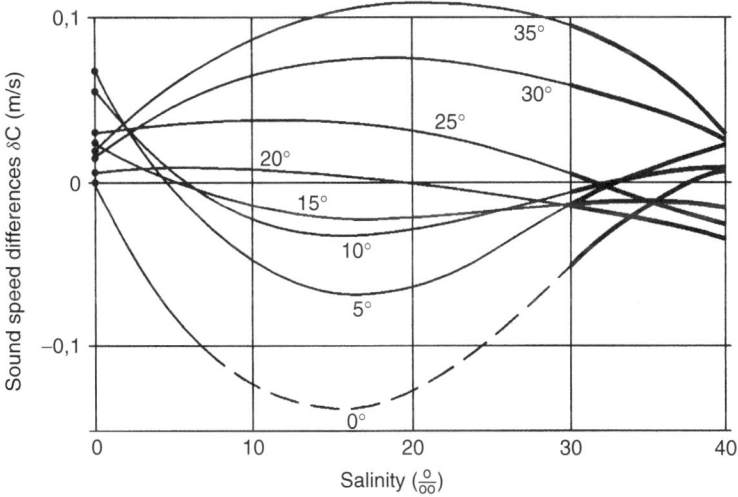

Fig. 2.15. Illustration of the method by interpolation developed by Leroy [9] for evaluating seawater sound speed at atmospheric pressure at the intermediate salinities.

Table 2.10. Reference values of sound speed in seawater at atmospheric pressure as a function of temperature and salinity (after Leroy [9]).

Salinity ($\frac{o}{oo}$)	Temperature (°)							
	0	5	10	15	20	25	30	35
0	1402.39	1426.16	1447.27	1465.93	1482.34	1496.68	1509.12	1519.81
5	1408.96	1432.43	1453.28	1471.71	1487.92	1502.08	1514.34	1524.86
10	1415.57	1438.74	1459.32	1477.51	1493.50	1507.46	1519.54	1529.90
15	1422.22	1445.08	1465.39	1483.31	1499.07	1512.84	1524.73	1534.94
20	1428.90	1451.45	1471.47	1489.13	1504.66	1518.20	1529.92	1539.98
25	1435.62	1457.84	1477.57	1494.97	1510.25	1523.57	1535.12	1545.02
30	1442.34	1464.24	1483.67	1500.80	1515.83	1528.94	1540.31	1550.05
35	1449.05	1470.66	1489.78	1506.65	1521.44	1534.33	1545.52	1555.07
40	1455.79	1477.06	1495.91	1512.50	1527.03	1539.71	1550.70	1560.08

equation. The results are given in Figure 2.16, redrawn from Leroy [9]. The two lines between 30 and 40 $\frac{o}{oo}$ at ±0.03 m/s schematize the spread of DG-M's values. Wilson's points rejected by Lovett are identified by different symbols.

Examination of this figure is quite instructive. It clearly shows the large and irregular scatter of Wilson's data in the 30–37 $\frac{o}{oo}$ range, which could not be observed on Figure 2.14 due to its scale. Wilson's values at nominal salinities 10 and 20 $\frac{o}{oo}$ are too low by some −0.3 m/s. The data from M-K present, except at 0 and 15 °C, an unexpected scatter in the range 30–40 $\frac{o}{oo}$, where the reference sound speed has not yet been obtained by interpolation. Departures are as much as −0.08 to +0.2 m/s. At intermediate salinities, M-K's data are always, with one exception only, higher than Leroy's reference. It does not seem that this is due to errors in Leroy's method, but to inaccuracies in M-K's values because (1) there is often an abnormal jump in sound speed right at $S = 30$ or 25 $\frac{o}{oo}$ for all temperatures and (2) the values of differences δC between 25 and 5 $\frac{o}{oo}$ are not compatible with the precise values at zero salinity, the jump from 5 to 0 $\frac{o}{oo}$ is too high and too systematic.

To conclude this section concerning sound speed in seawater at atmospheric pressure, we assert that the data from Del Grosso and Mader are certainly the most accurate and reliable. This conclusion is based on the unusually low values of their scatter but can be also supported by other considerations: (1) the B type velocimeter of DG-M (interferometer) is insensitive to temperature variations; (2) temperature measurements are precise to 0.001 °C and made inside the acoustic cell; and (3) salinity was measured from the sample used in the acoustic cell after recovery following measurements. As pointed out by Del Grosso, this is the best method because residual humidity in the cell can alter the salinity of a sample for which it has been measured before. (No detail about when these measurements were made was given in the quoted publications from Wilson or C-M.) No precise values for the intermediate salinities have been obtained from real Neptunian waters. Owing to the lower accuracy of M-K results, and to their higher scatter, we shall consider Leroy's interpolated values [9] as being the most accurate ones.

2.8.4. The Speed of Sound Under Pressure

The results of the analysis of sound speed in pure water or seawater at atmospheric pressure are very satisfactory. It seems that accuracies of 0.01 m/s and 0.1 m/s respectively, have indeed been reached, with the restriction that similar sound speeds would be obtained with real Neptunian water samples of low and high salinities found in closed basins, which needs to be confirmed.

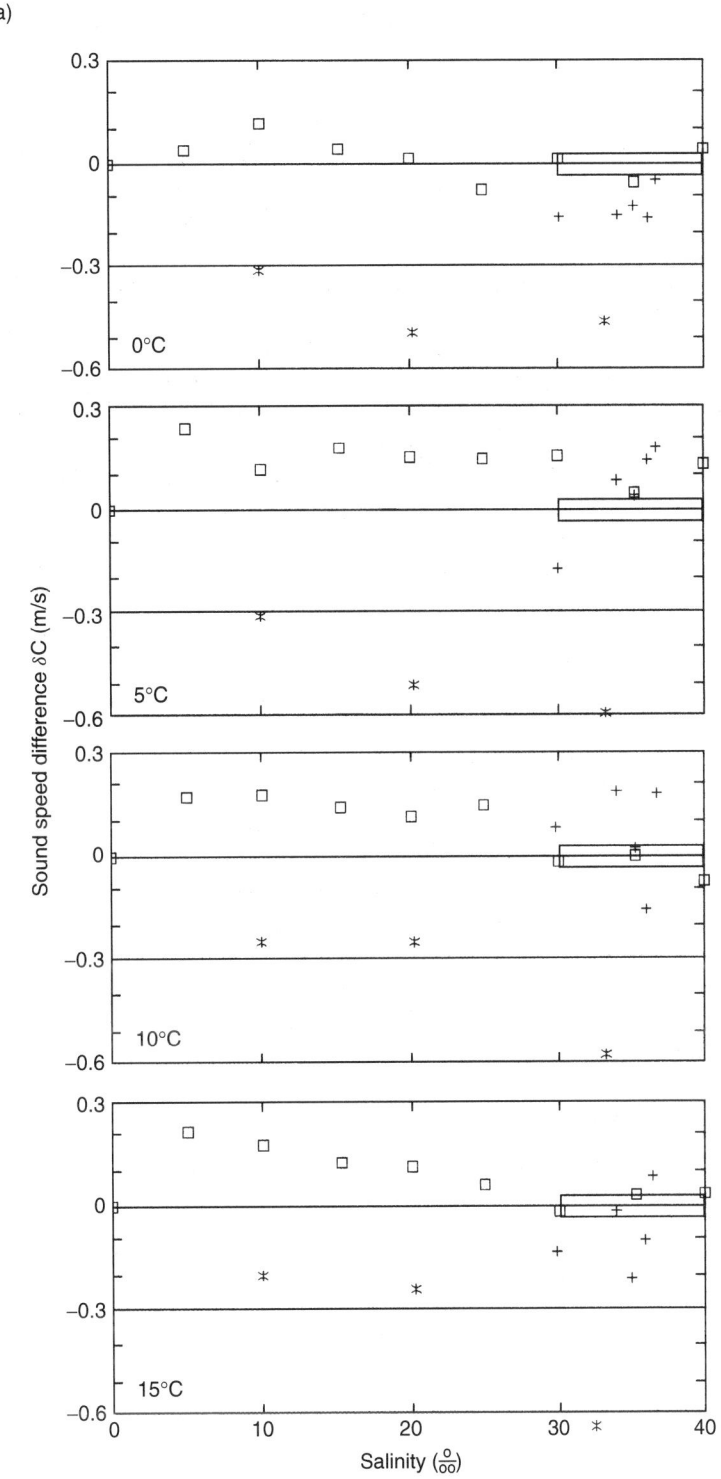

Fig. 2.16. Speed of sound in seawater at atmospheric pressure for the intermediate salinities. (a) $T = 0$ to $15\,°C$. (b) $T = 20$ to $35\,°C$. Differences in sound speed δC between measured values and the values estimated by Leroy [9]. Wilson's values [64, 66] accepted by Lovett [79] (plus signs) and rejected by Lovett (asterisks). Millero and Kubinsky's values [75] (squares).

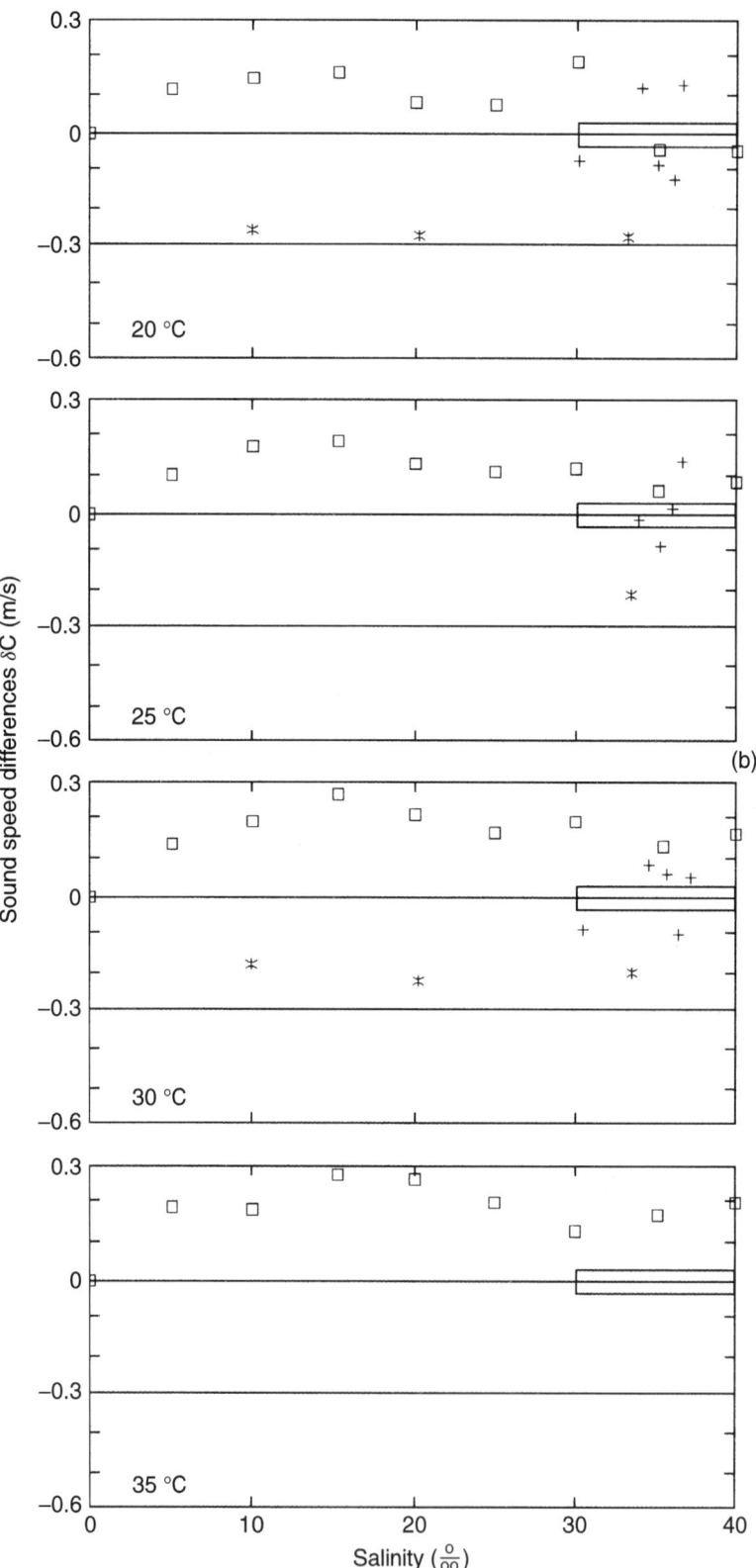

Fig. 2.16. (Continued)

In contrast with this satisfactory situation, that for high pressure values is very chaotic. Of the two sets of extended measurements in pure water under pressure by Wilson and B-Y, the first was entirely rejected by Anderson. However, it has been readjusted by Chen and Millero [76] as follows: Basically, Wilson's data were at all pressures shifted by constant values depending only on temperature. This permitted the formulation by C-M of an equation for pure water under pressure fitting corrected data with a standard deviation of only 0.08 m/s, instead of 0.2 m/s for Wilson's equation on uncorrected data. It does not mean, obviously, that the equation of C-M represents physical reality, but it seems the best for making comparisons with the results of B-Y. This is done on Figure 2.17 for all temperatures, and shows quite erratic behavior. No obvious conclusion could be obtained by looking at each separate curve in temperature, and this is why all results have been displayed on one single figure. Comments will be given later.

Since Chen and Millero's results for absolute values of sound speed in seawater under pressure depend on the exactness of Wilson's measurements, we shall concentrate on the two sets of absolute measurements, from Wilson [64–67] and Del Grosso and Mader [71]. The average values of sound speed at atmospheric pressure from the NRL velocimeter designed for under pressure measurements have been found to be in excellent agreement with those from the atmospheric equipment and with the equations from DG-M, whereas Wilson's data showed much larger scatter and irregularities. It is therefore natural to use the NRL II equation as a first reference for examining Wilson's results. The NRL I equation would be slightly more accurate for seawater at common salinities but would be less adapted to intermediate salinities as it did not fit any pure water data. If one takes Wilson's data, uniformly corrected by -0.42 m/s, and compares them with the values given by applying the NRL II equation, one finds differences of sound speeds δC as illustrated in Figure 2.18. Figure 2.18 a presents the results at $S = 35.02 \frac{o}{oo}$ and shows a progressive departure when T increases above $15\,°C$. This is perfectly normal since DG-M did not make any measurements under pressure for such temperatures. The remaining curves (below $15\,°C$) show a slight progressive departure of Wilson's values from DG-M's with increasing pressure. In the other example given in Figure 2.18b, corresponding to $S = 30.03$, the behavior of the curves at $15\,°C$ and higher is quite abnormal, and it is not surprising that data at

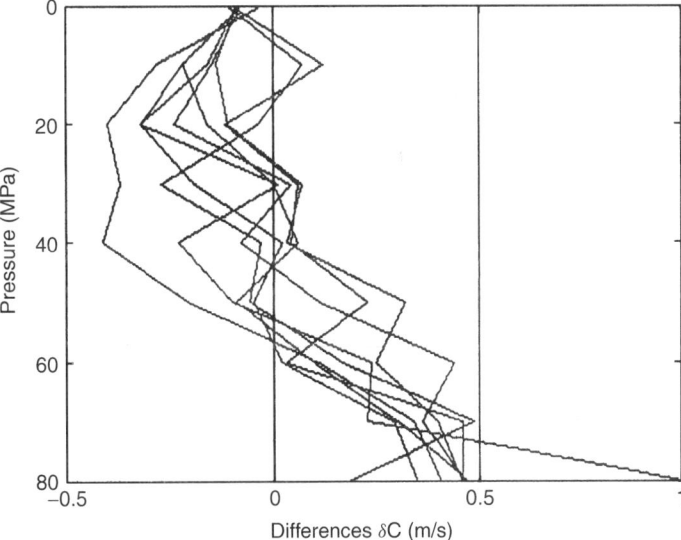

Fig. 2.17. Sound speed differences δC between Barlow and Yazgan [57] and the equation of Chen and Millero [76] for reevaluated values of Wilson [12] as a function of pressure (all temperatures).

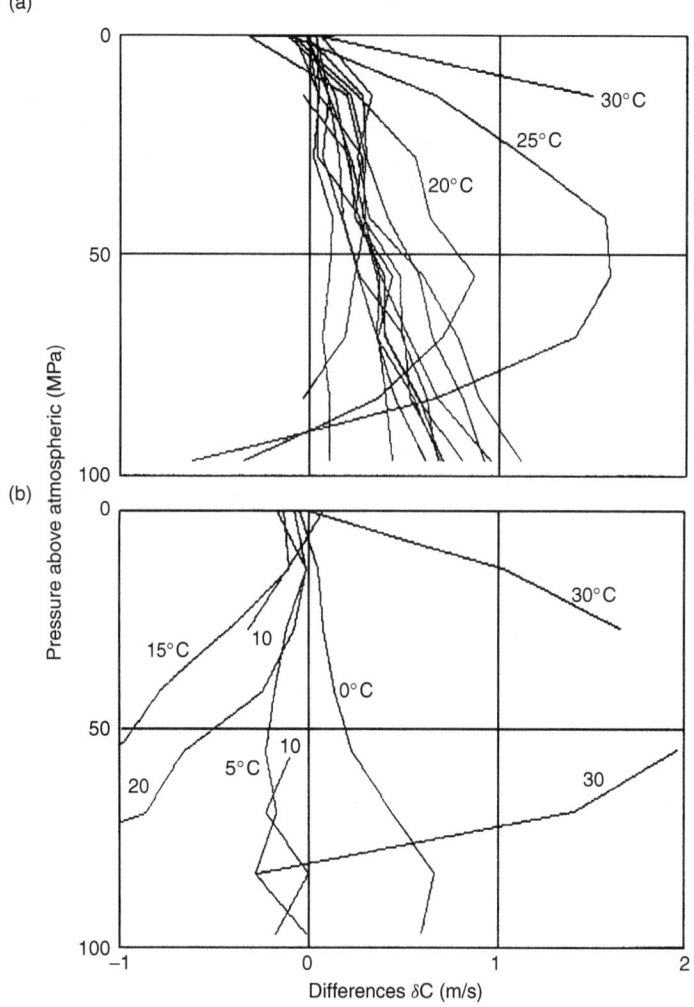

Fig. 2.18. Differences δC between the values of Wilson [64, 66] corrected by -0.42 m/s for sound speed in seawater under pressure and the NRL II equation [6]: (a) At $S = 35.02 \left(\frac{o}{oo}\right)$. (b) At $S = 30.03 \left(\frac{o}{oo}\right)$.

this salinity have been rejected by Anderson. Wilson's values at nominal S values of 20 and 10 $\frac{o}{oo}$ were found again to give totally erratic variations with pressure. This confirms the rejection of these intermediate salinity data. If we look only at the common salinity sets of values from Wilson, we can observe that the differences δC follows two distinct patterns, as illustrated in Figure 2.19, where temperatures higher than 12 °C have been ignored as not adapted to a comparison with the NRL II equation. Values of δC at all salinities except 33.08 follow the same trend of variation with pressure, as seen in Figure 2.19(a). Values at $S = 33.08$ (Fig. 2.19b) display a completely different behaviour. This is in conformity with the rejection of these data by Anderson and Lovett, but no explanation can be given for it. If we forced all differences to be null at atmospheric pressure, we should obtain curves staying within some ± 0.2 m/s of a global average curve depending only on pressure. The values given by this curve could be envisaged as possible corrections to apply to the NRL II equation. They are sampled in Table 2.11, together with values deriving from Lovett's proposed corrections [79].

Now the outstanding questions are as follows: (1) Why is it thought that Wilson's sound speed values under pressure, greater than DG-M's, are more reliable? (2) Why are B-Y's values in pure water at high pressure on average still higher than Wilson's?

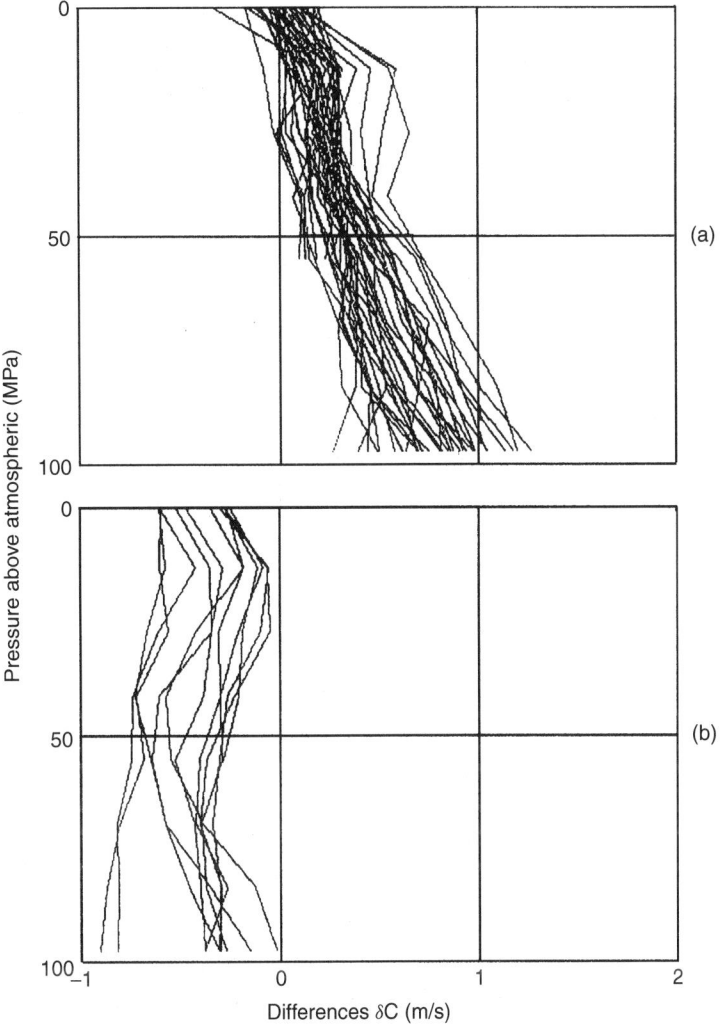

Fig. 2.19. A regrouping of all sound speed differences δC between Wilson, corrected by -0.42 m/s, and the NRL II equation. (a) Nominal salinities of 34, 35, 36, and 36.5 $\frac{o}{oo}$. (b) Nominal salinity 33 $\frac{o}{oo}$.

Table 2.11. Possible corrections δC to NRL II equation for agreement with the variation of sound speed with pressure measured by Wilson.

P (MPa)	δC (m/s)		
	a	b	c
14.0	0.39	0.17	+0.21
28.0	0.42	0.32	+0.10
42.1	0.56	0.49	+0.07
56.1	0.74	0.67	+0.07
70.2	0.93	0.93	0
84.3	1.09	1.24	−0.15
98.3	1.35	1.67	−0.32

[a]Present calculation.

[b]Values from Lovett [79].

[c]Présent — Lovett.

It is clear that errors in salinity, when they exist, cannot be taken as responsible for the trend. Errors in temperatures (on the high side by excess) could be possible in all equipment except DG-M's, where temperature was measured inside the acoustic cell. A higher temperature in the cell could result from insufficient recovering of temperature increases due to pressurizations, although Wilson waited 1 h for apparent stabilization, observed from the sound speed measured (but was it sufficient?). A much more important effect, however, is left out or not treated with great attention. It concerns the deformation of the cell with pressure. Barlow and Yazgan specifically mention [57, page 648] that they had corrected the acoustic path length for compression of the ring spacer of the velocimeter. They quote a 1.24 m/s sound speed excess at 800 bars computed from a bulk modulus of fused quartz of 5.35×10^6 lb/in^2. For a cube of side a and volume $V = a^3$ withstanding a uniform pressure P, the use of the bulk modulus k gives a volume deformation $dV/V = -k/P$. As $dV/V \approx 3\, da/a$, this finally leads to $da/a \approx -k/3P$. If one makes the assumption that for a ring of radial thickness e, height h, and mean circumference L placed under pressure, all values de/e, dh/h, and dL/L are also the third of dV/V, then the fractional reduction of height of the ring spacer, the same as the path length l reduction, gives $dl/l = -k/3P$. The application at 800 bars with the previous figure for k gives $dl/l = -7.25 \times 10^{-4}$. With a sound speed of 1700 m/s, this leads to a variation of -1.24 m/s, the value quoted by B-Y, which suggests strongly that their calculation was made on that basis. But this oversimplification can be objected to. More elaborate corrections than those of B-Y are found in the review by Mackenzie [88] of velocimeters for in situ measurements at great depths, where some account is taken of deformation in sing-around velocimeters of the TR4 and TR5 types.

It is astonishing that Wilson mentioned no correction for deformation with pressure. A number of possible sources of errors are provided by Wilson [12], such as shear and bulk viscosities of water, but not changes in path length due to deformations! With the NOL velocimeter cell, deformations can certainly not be evaluated like the reduction in size of a simple cube. Here we have a cylinder with length 10 times greater than the internal diameter. The stresses undergone are mostly due to radial compression of the cylinder wall, and longitudinal forces result only from external pressure applied to the cross section of the cylinder.

The final NRL results of sound speed in seawater under pressure deserve the greatest attention. Some "doubts" may have been expressed about their exactness, but no valuable proof of error has ever been given. Del Grosso and Mader have stated [71, page 964] that the design of the NRL velocimeter "eliminates pressure-caused distortions of pertinent acoustic geometry since the interferometer cell itself experiences identical internal and external pressures." Unfortunately the end of this statement may be the source of some confusion. Lovett [79], who quoted it, made the comment that it ignores the fact that the internal reflector shaft (see present Fig. 2.4) undergoes compression, and therefore elongation, but no conclusion could be drawn from this remark. What should have been added is that this extension starts from the mid-height of the shaft, a null point, and extends in both directions as the shaft is free to slide at both ends of the cell (O rings, not drawn on the schematic figure, are placed after to prevent leakage). When the reflector is in its lower position, it undergoes a displacement with pressure smaller than when it is in its higher position. The displacement of the reflector is then progressively greater under pressure than could be evaluated, for instance, by the number of rotations of the gear in the mechanism creating displacements. But what is measured with the final version of the DG-M interferometer is, at each pressure, the real displacement during the observation of the interference fringes. This is because the laser interferometer measuring displacement points from a fixed location toward a mirror located inside the rod at the level of the reflector (point named i on Fig. 2.4). Problems of shaft deformation are therefore of no concern, and Lovett admits it, saying that DG-M's "apparent pressure discrepancy remains unclear" [79, page 1715]. Still, Lovett made corrections to the NRL II equation on the basis of Wilson's results, whose eventual corrections for deformations are not given (and Leroy fell into the same trap in establishing his unpublished 1988 equation). It is now our opinion that DG-M's values need no correction at all, from the very principle of their measurements. The rather good agreement of results from the extensive experiments in the Pacific mentioned earlier [8, 10] when using the NRL II equation instead of the UNESCO 83 equation is another point in favor of this view. It is also generally confirmed by the various point measurements *in situ* by Mackenzie and his comments [85, 89]. To conclude, the measurements made by Del Grosso and Mader are most certainly accurate within 0.1 m/s, the NRL II equation being the best, although it could gain by some reformulation (see later). Why Wilson's results are different may be due to deformations of the velocimeter, and the discrepancies between Barlow and Yazgan and Wilson due to applied and nonapplied corrections according to the case. These deformation corrections need to be considered very seriously and calculated more scientifically than so far, probably requiring the use of finite element algorithms. This might hint at the answer to the present questions.

We cannot leave this subject without looking at the data of Chen and Millero. Basically, considering what has been said, and to remain independent of Wilson's results for pure water, the only thing that can be done is to examine, for each measured salinity, the differences of C-M's results for different temperatures as a function of pressure and to compare them with what is found when applying NRL II at the *TSP* combinations of C-M. The results are provided in Table 2.12. It is seen that, on average,

Table 2.12. Evolution of measured sound speed differences (in m/s) vs pressure for temperature changes of 5 or 10 °C at 3 salinities, according to the data of Chen and Millero [4] and the NRL II equation [6], and differences δC (m/s) between the two results[a].

Temperature change (°C) Authors S $\left(\frac{o}{oo}\right)$	P(MPa)[c]	$0^{[b]}$ to 5			5 to 10		
		C-M	DG-M	δC	C-M	DG-M	δC
33.323	0	21.63	21.66	−0.03	19.41	19.23	+0.18
	10	22.08	21.78	+0.30	19.29	19.34	−0.05
	20	21.91	21.77	+0.14	19.31	19.33	−0.02
	40	21.74	21.41	+0.33	19.06	18.87	+0.19
	60	20.50	20.70	−0.20	18.50	18.57	−0.07
	80	19.61	19.76	−0.15	18.14	17.98	+0.16
35.002	0	21.48	21.57	−0.09	19.28	19.14	+0.14
	10	21.44	21.47	−0.03	19.29	19.23	+0.06
	20	21.54	21.65	−0.11	19.22	19.22	0
	40	21.04	21.28	−0.24	19.10	18.94	+0.16
	60	20.58	20.55	+0.03	18.36	18.43	−0.07
	80	19.66	19.62	+0.04	18.05	17.82	+0.23
	100	18.74	18.62	+0.12			
39.442	0	31.65	31.24	+0.41	Temperature change 10 to 20 °C		
	10	31.54	31.34	+0.20			
	20	31.89	31.30	+0.59			

[a]The examples given have been limited to the Neptunian domain.

[b]Average of two sound speed measurements at 0 °C for C-M.

[c]Rounded to multiples of 10.

the experimental data from C-M are not in contradiction with DG-M at $S \approx 33$ and $\approx 35\ \frac{o}{oo}$. Their scatter, however, can reach ±0.3 m/s, which is much greater than with DG-M's measurements. At the highest salinity explored of 39.44 $\frac{o}{oo}$, the values no longer agree, but this is slightly outside the domain explored by DG and Mader. The overall result is finally reassuring. The main criticism of Chen and Millero's article [4] is that it does not discuss the problem of temperature stabilization inside the pressure cell, while temperature was measured in the bath outside. The accuracy of their equation obviously totally depends on Wilson's results in pure water under pressure, a fact with which Millero and Li [11] are in complete agreement.

2.8.5. Equations for Sound Speed in Water

More than 40 equations for the speed of sound in pure and seawater have already been proposed, but most scientific organizations now confine themselves to a few selected among the "great equations developed from acccurate measurements with precise velocimeters." For pure water at atmospheric pressure there exists no doubt. The equation, here called DG-M c, due to Del Grosso and Mader, is the best and is recalled in the Appendix. For pure water under pressure what to do is less clear: perhaps it is sufficient just to cite the Chen and Millero equation for the reevaluated values of Wilson, but without real confidence because of the lack of given corrections for deformations. Seawater at atmospheric pressure only is of little operational concern, but it is important for scientific investigations and as a reference. For common salinities, equation DG-M 6 is the best, but it fails for exceptional values, and one must look for alternatives. For the total Neptunian domain, a detailed scrutiny of the validity of the equations is needed.

The first thing when examining an equation supposed to be applicable for seawater under pressure is to look at what it gives at atmospheric pressure. If these results are insufficient there is no point in going further. On this simple basis, this author (Leroy [9]) examined the results of all equations with respect to the reference values provided in Table 2.10. Equations were abandoned if their maximum departure at $P = 0$ from these values exceeded 0.4 m/s. On this basis, there remain only five equations: DG-M (NRL II), C-M (the same as M-K at atmospheric pressure), Ross [82], Lovett [79], and Coppens [83]. For Lovett, only the third equation was considered, as it claimed to be the best for calculations under high pressures. Examination of these five equations at atmospheric pressure was made through figures such as illustrated in Figures 2.20 and 2.21. These give the lines of equal departures between equations and reference values, at 0, 0.1, and sometimes 0.05 m/s intervals, in the TS plane at $P = 0$. NRL II is found very good and only higher by more than 0.1 m/s in the unexplored region around 15 $\frac{o}{oo}$, as could be expected from Figure 2.15. The equation from C-M is less accurate in the intermediate salinity range despite the unusual addition of a term in S to the 1.5 power. This results from the departure of their data from the interpolated reference illustrated in Figure 2.16. Lovett equation 3, a merged and improved equation from NRL II, is less good at intermediate salinities. Finally, the simplified equations of Ross and Coppens (the latter a simplification of Ross) are quite good at atmospheric pressure.

The difficulty in examining the validity of these remaining five equations when used at pressure is directly connected with the confidence that one has in the measurements from which they are derived. However, one can already observe that the sound speed values given by both Ross and Coppen's equations are not sufficiently accurate for pressures greater than 40 MPa. This is normal, and in conformity with the aim of the authors, who were looking for simple, limited applications to the common great depths of the North Atlantic and the Mediterranean. However, an unfortunate application to the trough of Puerto Rico would lead to unacceptable errors. In all, to avoid errors due

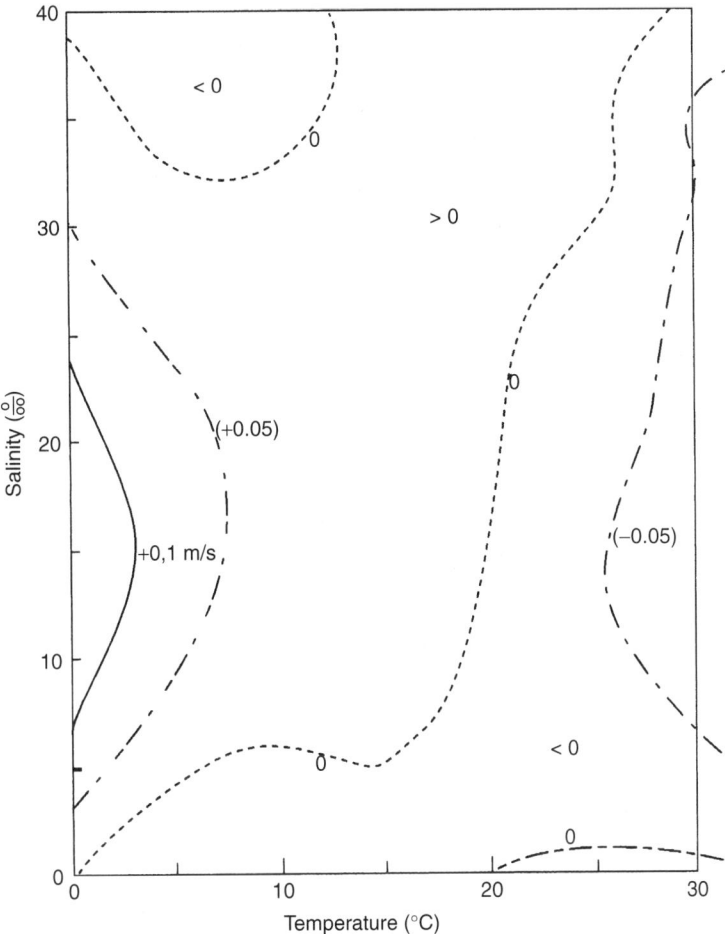

Fig. 2.20. Accuracy of the NRL II equation [6] at atmospheric pressure.

to the application of an equation outside its domain of validity, it seems now that, with the availability of personal computers, and even small programmable hand calculators, the age of simplified equations is over. As for the third equation of Lovett, it takes into account a correction to the pressure dependence of sound speed that has now been found uncertain and not justified in the end. We cannot blindly accept the equation for the moment, just as is the case for that of Chen and Millero.

Something now has to be said in favor of the formulation of C-M's equation. It has been seen that the best equation for pure water at atmospheric pressure is DG-M c, accurate to 0.01 m/s. It has also been seen that the best equation for seawater of common salinities was DG-M 6, within some 0.03 m/s, all this because in each case the equation was based on just the appropriate necessary data. Chen and Millero did the right thing in formulating their equation in a "progressive" manner but unfortunately did not pay attention to the Neptunian limitations and fitted data points far away from realistic values. However, the progressive method is quite good, and its interest has not been emphasized sufficiently. Chen and Millero employed it systematically. In fact, their equation $C(T, S, P)$ is of the form $C(T, S, P) = C(T, 0, 0) + C(T, S, 0) + C(T, 0, P) + C(T, S, P)$, where each additional term enables enlargement of the domain of validity (see Sect. 2.7). This means that, if the equation is employed without thinking about its structure, one will get (when data are excellent) the best equation in each limited case: pure water at atmospheric pressure, pure water under

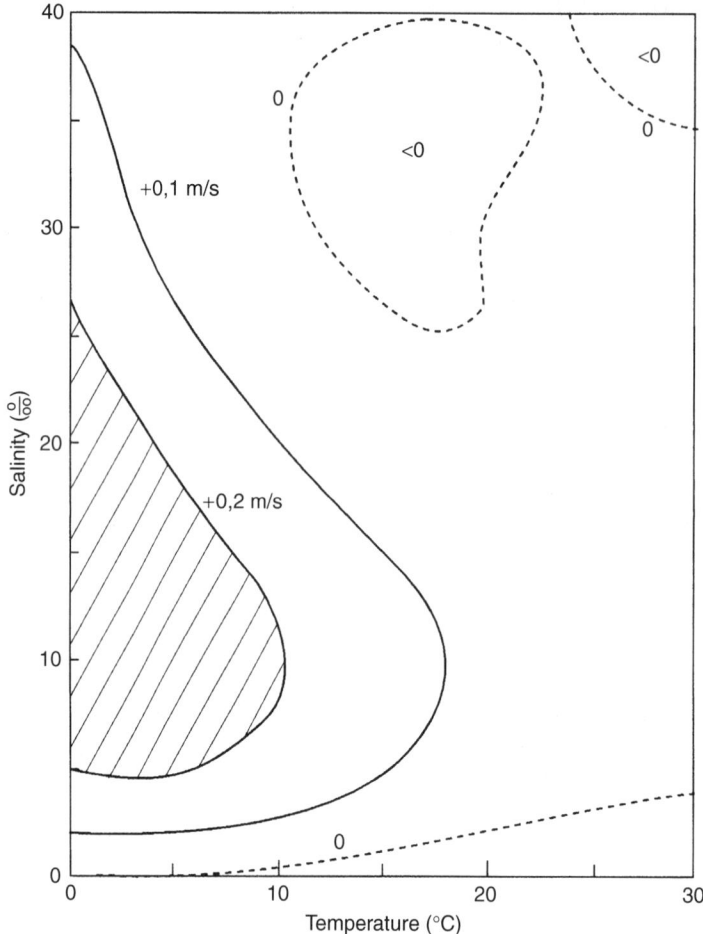

Fig. 2.21. Accuracy of the UNESCO 83 equation from Chen and Millero [4] at atmospheric pressure.

pressure, seawater at atmospheric pressure, etc. It is not at all the same when a best fit is globally made to the three-dimensional large set of data, as was the case with Wilson or Del Grosso and Mader, or others. It is obvious that the formulation made on this basis is much more complicated than a unique best fit. Equation UNESCO 83 has 42 terms and looks discouraging, but the immense advantage of an equation written that way would be a unique equation, the best in all applications, instead of a best one for pure water at $P = 0$, another best one for pure water in all conditions, and so on.

A last comment now on the complexity of an equation. The number of significant figures employed in the various coefficients can be considerably improved by remaining "reasonable" and limiting the domain of application to the Neptunian domain for seawater. A straightforward and illustrative example can be given by considering the coefficient of the T^3SP^2 term of equation UNESCO 83 [4, page 1132], given as 7.988×10^{-12} (when P is in bars). The maximum that can be reached in Neptunian waters by the product T^3SP^2 is found at the bottom of the Sulu Sea, with $T \approx 10\,°C$, $S \approx 35\,\frac{o}{oo}$, and $P \approx 60$ MPa, giving 1.25×10^8. The same term reaches as much as 2.56×10^{10} (a value 200 times greater!) when one uses $T = 40\,°C$, $S = 40\,\frac{o}{oo}$, and $P = 100$ MPa, the unrealistic values found at the extreme corner of the cubic domain explored and fitted by C-M. When remaining in the Neptunian domain, a replacement of the coefficient of T^3SP^2 from 7.988×10^{-12} to simply 8×10^{-12} results in a departure of only 0.00015 m/s. It is hoped that this kind of "pruning" will be made in the future.

2.9. CONCLUSION

Eighty years have passed since interest in sound speed in seawater became an issue. After some rudimentary sea trials, the scientific community realized that it was better to obtain sound speed from the application of Newton's law for fluids and laboratory measurements of the necessary physical properties. A number of tables were established from which equations for sound speed could be formulated. This research culminated with the work of Kuwahara in 1939 and the 1960 equation of Mackenzie. In the 1950s, advances in electronic technology initiated the age of sound velocimeters. The first measurements demonstrated that Kuwahara's tables were wrong by some 3 m/s, and this was soon explained. Contemporaneous measurements of sound speed in pure water with fixed-path velocimeters at the NBS and the NOL on the one hand, and interferometric techniques at the NRL on the other, showed discrepancies of some 0.35 m/s. This triggered the invention and development of new and more sophisticated velocimeters, but also delayed the fundamental work undertaken at the NRL by Del Grosso and his collaborators. The two final NRL velocimeters were, indeed, superb instruments, but the measurements had been eagerly awaited. The needs of underwater applications, mostly military, forced Del Grosso to concentrate first on measurements at realistic combinations of temperature, salinity, and pressure that this author had defined in 1968 as the Neptunian domain. Del Grosso, however, was not able to use, as he would have liked, real water samples from low- and high-salinity waters taken from the world's main closed basins, nor could he make measurements in pure water at high pressures as he had planned from 1956. After publication of the NRL II equation in 1974, his work on the subject ceased. A number of studies were made concerning the different behavior with pressure of sound speed in seawater given by Del Grosso and by Wilson at the NOL. The generally admitted facts were that Del Grosso's measurements at atmospheric pressure were the best, but that Wilson's trend of sound speed variation with pressure was more reliable. Measurements made by Millero's team were published in the 1975–1977 period. These were made in seawater by comparison with pure water, and Wilson's reevaluated values of sound speed in pure water under pressure were used as a reference to establish an equation by Chen and Millero. This was adopted as the UNESCO 83 equation. It gave at high pressures a progressive departure from the NRL II equation. Since that work, the situation has remained "stagnant." For nearly 15 years the scientific community seems to have decided that nothing more was to be done concerning sound speed in water. Then, very long-range accurate sound propagation measurements in the Pacific, involving deep acoustic paths, demonstrated in 1993 that the use of the NRL II equation gave better results than that of UNESCO 83. This resuscitated interest in Wilson's measurements under pressure.

From the discussion given in this review we can arrive at a number of conclusions. First, the speed of sound in pure water at atmospheric pressure is certainly known within an accuracy of 0.01 m/s, and an equation based on Del Grosso's and Mader's work preserves this accuracy. Second, the sound speed in sealike salted water at atmospheric pressure, in the complete T-S domain of Neptunian waters, is known within 0.1 m/s, but measurements with real low- and high-salinity samples from Neptunian waters are lacking. The situation of sound speed in both pure water and seawater under pressure is rather chaotic. We pointed out that this could simply be due to a lack of serious calculation of the deformation of acoustic cells under pressure. We believe also that Del Grosso and Mader's measurements by interferometry were not affected by deformations, hence that they are most probably the closest to reality. As a result, what is finally lacking at the moment is an exact knowledge of sound speed in pure water under pressure. We suggest, like Millero and Li, that this should be measured, preference being given to an interferometer technique, like that used by the NRL. However, considering the cost of such development, we suggest first that some scientific calculations should be made of the deformation of Wilson's and Barlow and Yazgan's acoustic cells under pressure, using finite element techniques if necessary. This would illuminate this obscurity and, perhaps, could avoid the reconstruction of expensive equipment. Nevertheless, it remains that measurements under pressure with Black Sea and Red Sea water samples are most advisable, if not also with real Mediterranean water.

From the numerous equations developed since the beginning of studies on sound speed in water, only a few are acceptable. For pure water, the equation called DG-M c here is without doubt the best and is sufficient for the future, provided that the discrepancy with the recent measurements of Fujii and Masui can be explained. The NRL II equation applied with $P = 0$ is at the moment the best to fit the expected values of sound speed in seawater at atmospheric pressure, but it could be slightly improved to remain within ±0.1 m/s. At high pressure in Neptunian waters, NRL II should be considered as the best for the time being, but can be simplified. These two equations are given in the Appendix.

For further development of speed equations it must be emphasized that for practical use in precise underwater applications only the data corresponding to realistic TSP combinations of the Neptunian domain should be taken into account. The coefficients of such equations should be limited to a realistic number of significant figures adapted to the extreme possible value of the term they multiply to be found in the Neptunian domain. Finally, a progressive formulation like that of Chen and Millero may be of great interest. This could lead to a unique universal equation that would be the best in all subsituations, even pure water only.

ACKNOWLEDGMENTS

The author wishes first to thank Chief Naval Engineer D. Long, former Director of the LSM Branch of the CTSN (DCN, French Navy, Toulon) for supporting this work, and to his Scientific Research Assistant J.Y. Morin, who made many helpful suggestions. Thanks are due also to the reviewers of an earlier version proposed to the *Journal of the Acoustical Society of America* for many comments and suggestions. A great part of the analysis developed in the discussion is derived from the work done in 1988, supported by the DCN, under the scientific responsibility of J.A. Roy and R. Ancey. The author is particularly indebted to his friend and former colleague, B.W. Conolly, for help in the text and suggestions. Finally, special thought must be given to V. Del Grosso, certainly the greatest scientist in this field.

APPENDIX: RECOMMENDED EQUATIONS FOR THE SPEED OF SOUND IN PURE WATER AND NEPTUNIAN WATERS

At this stage, and from the results of this review, we can only recommend two equations for calculating the speed of sound in water — one for pure water at atmospheric pressure and one for Neptunian waters — with the comment that the latter needs confirmation by accurate measurements with water samples from the four main closed basins of the world, viz., the Black Sea and the Baltic Sea on the low salinity side, and also, for safety, the Mediterranean Sea and the Red Sea for high salinities. Unfortunately, no equation can be recommended for the speed of sound in pure water under pressure.

2.A.1. Equation for Pure Water at Atmospheric Presure

This equation is the one designated by DG-M c in this review, and it has been given by Del Grosso and Mader [72, Table 2.3, page 1443]. It is, with C in m/s and T in °C:

$$C = 0.140238754 \times 10^4 + 0.503711129 \times 10^1 \, T - 0.580852166 \times 10^{-1} \, T^2 \\ + 0.334198834 \times 10^{-3} \, T^3 - 0.147800417 \times 10^{-5} \, T^4 + 0.314643091 \times 10^{-8} \, T^5$$

This equation can be considered accurate within 0.01 m/s. In that respect, the coefficients of the equation can be "pruned" without altering this accuracy (here in the whole range 0–100 °C) as follows:

$$C = 1402.3875 + 5.03711T - 5.80852 \times 10^{-2}T^2 + 3.34199 \times 10^{-4}T^3 \\ - 1.478 \times 10^{-6}T^4 + 3.14643 \times 10^{-9}T^5$$

2.A.2. Equation for Neptunian waters

This equation is the NRL II equation, developed by Anderson under Del Grosso [6] from the data of Del Grosso and Mader. In its original form C is in m/s, T is in °C, S is in p.p.t., and P is in kg/cm². It reads [6, page 1084]:

$$C_{STP} = C_{000} + C_T + C_S + C_P + C_{STP}$$

where

$$C_{000} = 1402.392$$

$$C_T = 0.501109398873 \times 10^1 T - 0.550946843172 \times 10^{-1}T^2 \\ + 0.221535969240 \times 10^{-3}T^3$$

$$C_S = 0.132952290781 \times 10^1 S + 0.128955756844 \times 10^{-3}S^2$$

$$C_P = 0.156059257041 \times 10^0 P + 0.244998688441 \times 10^{-4}P^2 \\ - 0.883392332513 \times 10^{-8}P^3$$

$$C_{STP} = -0.127562783426 \times 10^{-1}TS + 0.635191613389 \times 10^{-2}TP \\ + 0.265484716608 \times 10^{-7}T^2P^2 - 0.159349479045 \times 10^{-5}TP^2 \\ + 0.522116437235 \times 10^{-9}TP^3 - 0.438031096213 \times 10^{-6}T^3P \\ - 0.161674495909 \times 10^{-8}S^2P^2 + 0.968403156410 \times 10^{-4}T^2S \\ + 0.485639620015 \times 10^{-5}TS^2P - 0.340597039004 \times 10^{-3} \, TSP$$

The overall accuracy of this equation is believed to be 0.1 m/s, and a modernized version of this equation should use Pascals for pressure, i.e., Pa or rather MPa. With respect to the accuracy, the coefficients of the NRL II equation using MPa can be "pruned" as above in the equation for pure water. We propose the following development, with P in MPa on the basis of 1 kg/cm² = 9.80655×10^{-2} MPa:

$$C = 1402.392 + 5.0111T - 5.55095 \times 10^{-2}T_2 + 2.21536 \times 10^{-4}T^3$$
$$+ 1.3295S + 1.289 \times 10^{-4}S^2$$
$$+ 1.59136P + 2.5475 \times 10^{-3}P^2 - 9.367 \times 10^{-6}P^3$$
$$- 1.2756 \times 10^{-2}TS + 9.684 \times 10^{-5}T^2S$$
$$+ 6.477 \times 10^{-2}TP - 1.657 \times 10^{-4}TP^2 + 5.536 \times 10^{-7}TP^3$$
$$+ 2.76 \times 10^{-6}T^2P^2 - 4.467 \times 10^{-6}T^3P$$
$$- 1.681 \times 10^{-7}S^2P^2$$
$$+ 3.473 \times 10^{-3}TSP + 4.952 \times 10^{-5}TS^2P$$

with T in °C, S in $\frac{o}{oo}$, and P in MPa.

When the speed of sound has to be calculated from values of temperature T and salinity S given vs depth instead of pressure, a conversion is needed. A detailed and very accurate procedure, used by oceanographers, is given for the reverse problem (the conversion of depth Z into pressure P) by Fofonoff and Millard [5, chapter IV]. In this context, the algorithm requires the introduction of complete detailed profiles of T and S and the use of iteration to obtain P from Z instead of Z from P. An alternative and much simpler procedure is to employ an equation using depth and latitude, completed by an additional correction term depending only on the area. Such a development has been recently published [96]. One equation plus one main additional term can be used in 80% of the world's conditions. A set of 10 alternative terms is provided for use in the other cases, mostly closed basins. This new algorithm preserves the accuracy on sound speed within ±0.02 m/s.

References

1. Leroy, C.C. (1968). A simple and accurate formula for the calculation of the velocity of sound in sea water. SACLANT ASW Research Centre Technical Report 111.
2. Leroy, C.C. (1968). Development of simple equations for accurate and more realistic calculation of the speed of sound in sea water. SACLANT ASW Research Centre Technical Report 128.
3. Leroy, C.C. (1969). Development of simple equations for accurate and more realistic calculation of the speed of sound in sea water. *J. Acoust. Soc. Am.* **46**: 216–226.
4. Chen, C.T. and Millero, F.J. (1977). Sound speed in seawater at high pressures. *J. Acoust. Soc. Am.* **62**: 1129–1135.
5. Fofonoff, N.P., and Millard, R.C. (1983). Algorithms for computation of fundamental properties of seawater. *Unesco Tech. Pap. in Mar. Sci.*, No. 44.
6. Del Grosso, V.A. (1974). New equation for the speed of sound in natural waters (with comparisons to other equations). *J. Acoust. Soc. Am.* **56**: 1084–1091.
7. Spiesberger, J.L., and Metzger, K. (1991). New estimates of sound speed in water. *J. Acoust. Soc. Am.* **89**: 1697–1700.
8. Spiesberger, J.L. (1993). Is Del Grosso's sound speed algorithm correct?. *J. Acoust. Soc. Am.* **93**: 2235–2237.
9. Leroy, C.C. (1988). La vitesse du son dans les eaux neptuniennes: Synthése et mise á jour. (Speed of sound in Neptunian waters: Review and updating). Consultant Report for CERDSM No. ACS.
10. Dushaw, B.D., Forcester, P.F., Cornuelle, B.D., and Howe, B.M. (1993). On equations for the speed of sound in seawater. *J. Acoust. Soc. Am.* **93**: 255–275.
11. Millero, F.J., and Li, X. (1994). Comments on On equations for the speed of sound in seawater. *J. Acoust. Soc. Am.* **95**: 2757–2759.
12. Wilson, W.D. (1959). Speed of sound in distilled water as a function of temperature and pressure. *J. Acoust. Soc. Am.* **31**: 1067–1072.
13. Sverdrup, H.U., Johnson, M.W., and Fleming, R.H. (1970). Chemistry of sea water, in *The Oceans—Their Physics, Chemistry, and General Biology*, pp. 165–227, Englewood Cliffs NJ: Prentice-Hall.
14. Hackmann, W. (1984). *Sonar, Anti-submarine Warfare and the Royal Navy—1914-54*. London: Seek and Strike, Her Majesty's Stationery Office.
15. Wood, A.B. (1941). *A Textbook of Sound*, New York: MacMillan.

16. Lindsay, R.B. (1973). *Acoustics—Historical and Philosophical Development*, Stroudsburg PA: Douden, Hutchinson & Ross.
17. Lasky, M. (1977). Review of undersea acoustics to 1950. *J. Acoust. Soc. Am.* **61**: 283–297.
18. Colladon, J.D., and Sturm, C. (1827). Vitesse du son des liquides. *Ann. Chim. Phys.* **36**: 236–257.
19. Urick, R.J. (1975 and 1985). *Principles of Underwater Sound*, 2nd and 3rd ed., New York: McGraw-Hill.
20. Bonneycastle, C. (1838). Notes of experiments made August 22-25, 1838 with a view of determining the depth of sea by echo. *Proc. Am. Philos. Soc.* **1**: 39–41.
21. Bourcart, J. (1961). *Le Fond des Océans (The Bottom of the Oceans)*, pp. 18–19, Paris: Presses Universitaires de France, Que sais-je No. 621.
22. U.S. Coast and Geodetic Survey, Special Publication 108 (1924).
23. Stephenson, E.B. (1923). Velocity of sound in sea water. *Phys. Rev.* **21**: 181–186.
24. Wood, A.B., and Brown, U. (1923). Determination of the velocity of sound of explosion waves. *Proc. Roy. Soc. Lond. A* **103**: 284–303.
25. Maurer, J. (1924). Uber Echolotungen der Nord-Amerikanischen Marine. (On U.S. Navy Echo Rangers), pp. 75–87. Annalen der Hydrographie und maritimen Meteorologie.
26. Schumacher, A. (1924). Hydrographische Bemerkungen und Hilfsmittel zur akustischen Tiefenmessung (Hydrographic observations and aids for acoustic depth measurement), pp. 87–95.
27. Heck, N.H., and Service, J.H. (1927). Velocity of sound in sea water. U.S. Coast and Geodetic Survey. Publication No. 108.
28. Matthews, D.J. (1927). Tables of the velocity of sound in pure water and sea water for use in echo-sounding and sound-ranging. British Admiralty, Hydrographic Department, No. 282.
29. Smith, P.A. (1934). Recent acoustic work of the U.S. Coast and Geodetic Survey. Field Engineer Bulletin No. 8, 60–73.
30. Kuwahara, S. (1939). Formulas for computation of sound speed in sea water. *Hydrographic Rev.* **16**: 123–140.
31. Matthews, D.J. (1939). Tables of the velocity of sound in pure water and sea water for use in echo-sounding and sound-ranging. British Admiralty, Hydrographic Department, No. 282 2nd ed.
32. Stephenson, E.B., and Woodsmall F.I. (1941). Velocity of sound in sea water. NRL Report No. S-1722.
33. Equation presented as Original Hydrographic Office machine computation, by L. A. Sower.
34. Mackenzie K.V. (1960). Formulas for the computation of sound speed in sea water. *J. Acoust. Soc. Am.* **32**: 100–104.
35. Hubbard, J.C., and Loomis, A.L. (1928). *Phil. Mag.* **5**: 1177.
36. Klein, E., and Hershberger W.D., (1931). *Phys. Rev.* **37**: 760.
37. Weissler, A. (1948). *J. Am. Chem. Soc.* **70**: 1634.
38. Del Grosso, V.A. (1952). *Naval Research Laboratory Technical Report No. 4002*.
39. Sower, L.A. (1961). Sound velocity formulas. U.S. Naval Hydrographic Office Unpublished manuscript.
40. Weissler, A., and Del Grosso, V.A. (1951). The velocity of sound in sea water. *J. Acoust. Soc. Am.* **23**: 219–223.
41. Carnevale, E.H., and Litovitz, T.A. (1955). Pressure dependence of ultrasonic velocity in sea water. *J. Acoust. Soc. Am.* **27 (LE)**, 794–795.
42. Del Grosso, V.A. (1958). Speed of sound in pure and sea water. Conf. on Phys. and Chem. Properties of Sea Water, O.N.R. Committee on Oceanography.
43. Gucker, F.T., Jr., and Haag, R.M. (1953). *J. Acoust. Soc. Am.* **25**: 470.
44. Greenspann, M., and Tschiegg, C.E. (1957). Sound speed in water by a direct method. *J. Res. Natl. Bur. Stand.* **59**: 249–254.
45. Greenspann, M., and Tschiegg C.E. (1959). Tables of the speed of sound in water. *J. Acoust. Soc. Am.* **31**: 75–76.
46. Martin, A.V. (1957). Sur la propagation du son dans l'eau de mer (On sound propagation in seawater). *Ann. Géophysique* **13**: 307–309.
47. Beyer, R.T. (1954). *J. Mar. Res.* **13**, 113.
48. Del Grosso, V.A. (1966). Problems in the absolute measurement of sound speed. U.S.N. *J. Underwater Acous.* **16**: 597–612.
49. Wilson, W.D. (1958). Velocity in sea water: Pressure effects. Conf. NAS–NRC–600.
50. Del Grosso, V.A. (1964). Naval Research Laboratory Technical Report No. 6026.
51. Del Grosso, V.A. (1965). Naval Research Laboratory Technical Report No. 6133.
52. Del Grosso, V.A. (1966). Naval Research Laboratory Technical Report No. 6409.
53. Del Grosso, V.A. (1968). Naval Research Laboratory Technical Report No. 6852.
54. McSkimin, H.J. (1957). *IRE Trans. Ultrasonics Eng.* **5**: 25–47.
55. McSkimin, H.J. (1965). Velocity of sound in distilled water for the temperature range 20°-75°C. *J. Acoust. Soc. Am.* **37**: 325–328.
56. Barlow, A.J., and Yazgan, E. (1966). Phase change method for the measurement of ultrasonic wave velocity and a determination of the speed of sound in water. *Br. J. Phys.* **17**, 807–819.
57. Barlow, A.J., and Yazgan, E. (1967). Pressure dependence of the velocity of sound in water as a function of temperature. *Br. J. Appl. Phys.* **18**, 645–651.

58. Carnvale, A., Bowen, P., Basileo, M., and Sprenke, J. (1968). Absolute sound-velocity measurements in distilled water. *J. Acoust. Soc. Am.* **44**, 1098–1102.
59. Williamson, R.C. (1969). Echo phase-comparison technique and measurement of sound velocity in water. *J. Acoust. Soc. Am.* **45**, 1251–1257.
60. Brooks, R. (1960). Determination of the velocity of sound in distilled water. *J. Acoust. Soc. Am.* **32**: 1422–1425.
61. Neubauer, W.G., and Dragonette, L.R. (1964). Experimental determination of the freefield sound speed in water. *J. Acoust. Soc. Am.* **36**: 1685–1690.
62. Kroebel, W., and Mahrt, K.H. (1976). Recent results of absolute sound velocity measurements in pure water and sea water at atmospheric pressure. *Acustica* **35**, 154–164.
63. Ilgunas, V., Kubilyurere, O., and Yapertas, A. (1964). A high pressure interferometer for measuring ultrasonic velocity in liquids in the 1–12 MHz frequency range. *Soviet Physics—Acoust.* **10**: 44–48.
64. Wilson, W.D. (1960). Naval Ordnance Laboratory Report 6746.
65. Wilson, W.D. (1960). Speed of sound in sea water as a function of temperature, pressure and salinity. *J. Acoust. Soc. Am.* **32**: 641–644.
66. Wilson, W.D. (1960). Naval Ordnance Laboratory Report 6906.
67. Wilson, W.D. (1960). Equation for the speed of sound in sea water. *J. Acoust. Soc. Am.* **32 (LE)**: 1357.
68. Leroy, C.C. (1968). Formulae for the calculation of underwater pressure in acoustics. *J. Acoust. Soc. Am.* **44 (LE)**, 651–653.
69. Frye, H.W., and Pugh, J.D. (1971). A new equation for the speed of sound in seawater. *J. Acoust. Soc. Am.* **50 (LE)**, 384–386.
70. Del Grosso, V.A. (1970). Sound speed in pure water and sea water. *J. Acoust. Soc. Am.* **47 (LE)**, 947–949.
71. Del Grosso, V.A., and Mader, C.W. (1972). Speed of sound in seawater samples. *J. Acoust. Soc. Am.* **52**: 961–974.
72. Del Grosso, V.A., and Mader, C.W. (1972). Speed of sound in pure water. *J. Acoust. Soc. Am.* **52**: 1442–1446.
73. Del Grosso, V.A., and Mader, C.W. (1973). Another search for anomalies in the temperature dependence of the speed of sound in pure water. *J. Acoust. Soc. Am.* **53**: 561–563.
74. Del Grosso, V.A. (1973). Tables of the speed of sound in open ocean water (with Mediterranean Sea and Red Sea applicability). *J. Acoust. Soc. Am.* **53**: 1384–1401.
75. Millero, F.J., and Kubinski, T. (1975). Speed of sound in seawater as a function of temperature and salinity at 1 atm. *J. Acoust. Soc. Am.* **57**: 312–319.
76. Chen, C.T., and Millero, F.J. (1976). Reevaluation of Wilson's sound-speed measurements for pure water. *J. Acoust. Soc. Am.* **60**: 1270–1273.
77. Anderson, E.R. (1971). Sound speed in sea water as a function of realistic temperature, salinity, pressure domains. NURDC T.P. 243.
78. Anderson, E.R. (1973). Comparison of measured and computed sound speeds, in *Proc. Second STD Conference* Danielson, W.R., ed., San Diego: Plessey. pp. 19–28.
79. Lovett, J.R. (1978). Merged seawater sound-speed equations. *J. Acoust. Soc. Am.* **63**: 1713–1718.
80. Kell, G.S. *J. Chem. Eng. Data* **15**: 119.
81. Medwin, H. (1975). Speed of sound in water: A simple equation for realistic parameters. *J. Acoust. Soc. Am.* **58 (LE)**, 1318–1319.
82. Ross, D. (1978). Revised simplified formulae for calculating the speed of sound in sea water. SACLANT ASW Research Centre Memorandum No. SM-107.
83. Coppens, A.B. (1981). Simple equations for the speed of sound in Neptunian waters. *J. Acoust. Soc. Am.* **69 (LE)**: 862–863.
84. Mackenzie, K.V. (1981). Nine-term equation for sound speed in the oceans. *J. Acoust. Soc. Am.* **70**: 807–812.
85. Mackenzie, K.V. (1961). Sound speed measurements utilizing the bathyscaphe TRIESTE. *J. Acoust. Soc. Am.* **33**: 1113–1119.
86. Mackenzie, K.V. (1962). Further remarks on sound-speed measurements aboard the TRIESTE. *J. Acoust. Soc. Am.* **34 (LE)**: 1148–1149.
87. Mackenzie, K.V. (1967). Sound speed measurements aboard french bathyscaphe ARCHIMEDE in the Japan trench. *J. Acoust. Soc. Am.* **42**: 1210(A).
88. Mackenzie, K.V. (1971). A decade of experience with velocimeters. *J. Acoust. Soc. Am.* **50**: 1321–1333.
89. Mackenzie, K.V. (1981). Discussion of sea water sound-speed determinations. *J. Acoust. Soc. Am.* **70**: 801–806.
90. Spiesberger, J.L., and Metzger, K. (1991). A new algorithm for sound speed in water. *J. Acoust. Soc. Am.* **89**: 2677–2688.
91. Fujii, F. and Masui, R. (1993). Accurate measurements of the sound velocity in pure water by combining a coherent phase-detection technique and a variable path-length interferometer. *J. Acoust. Soc. Am.* **93**: 276–282.

92. Heydemann, P.L. (1971). A fringe-counting pulsed ultrasonic interferometer. *Rev. Sci. Instrum.* **42**: 983–986.
93. Wallace, P.W., and Garland, C.W. (1986). Ultrasonic velocity and attenuation measurements with a computer controlled phase-sensitive detection tecnhique. *Rev. Sci. Instrum.* **57**: 3085–3088.
94. Iwaza, I. Koizumi, H., and Suzuki, T. (1988). Automated ultrasonic measuring systems using phase-sensitive detection. *Rev. Sci. Instrum.* **89**: 356–361.
95. Wilson, W.D. (1962). Extrapolation of the equation for the speed of sound in sea water. *J. Acoust. Soc. Am.* **34 (LE)**, 866.
96. Leroy, C.C., and Parthiot, F. (1998). Depth-pressure relationships in the oceans and seas. *J. Acoust. Soc. Am.* **103**: 1346–1352.

CHAPTER 3

ABSORPTION OF SOUND IN FRESH AND SEA WATER

C. C. Leroy
Private consultant in Underwater Acoustics, SANARY-sur-Mer, France

R. H. Mellen
Kildare Corporation, New London, Connecticutt, USA

G. Waton
Laboratoire de dynamique des fluides complexes, Universite de Strasbourg, Cedex, France

Contents

Abstract		83
3.1.	Introduction	84
3.2.	Physical and Chemical Causes of Absorption in Liquids	85
3.3.	Sound Absorption above 20 kHz	92
3.4.	Sound Absorption at Low Frequencies	96
3.5.	Discussion and Conclusion	108
References		112

ABSTRACT

The absorption of sound in seawater is a major limitation to all techniques using underwater sound waves for detection, localization, or soundings. The knowledge of its value is therefore of importance to equipment designers and users when evaluating maximum operational range in a given environment.

In the 300 Hz to several MHz frequency range, the absorption of sound is due to a number of different mechanisms, three of which are predominant. These mechanisms are all of the relaxation type: the propagating sound wave disturbs the existing equilibrium of molecules or ions that pass from one type of arrangement to another and vice versa, with different time lags, thus giving rise to a tiny partial transfer of local acoustic energy into kinetic energy and heat.

In freshwater, absorption is due to ultrahigh frequency oscillations between two ways of configuring the water molecules. In seawater, a first relaxation process, predominant in the 10 to 300 kHz band, is due to ionic dissociations of the magnesium sulfate dissolved in water. Both mechanisms were discovered and studied between 1930 and 1950. A second relaxation process, predominant in the 300 Hz to 5 kHz range, was finally discovered in 1972 and could explain the ambiguous results of numerous measurements at sea and in lakes that extended from 1957 to about 1972. It is a very complex physicochemical mechanism mostly involving the boric acid dissolved in seawater. The effect of hydrostatic pressure on this second relaxation process is not yet known. We must also note that several measurements at sea would indicate another absorption excess below 300 Hz, but this is not confirmed, and diffraction or even artifacts could be at the origin of what is observed.

After an introduction of the subject in Section 3.1, and a presentation of the absorption mechanisms in general with the methods used in laboratory measurements in Section 3.2, the text is divided into two main parts—one concerning the high-frequency relaxations and absorption above 20 kHz and the other at low frequencies. The reason for this clear distinction is the fact that the experimental conditions and possibilities of measurements are entirely different.

Section 3.3 deals with the absorption at high frequencies in freshwater and in seawater as well as in solutions of magnesium sulfate in general. It gives a short historical survey and describes the measurements and the mechanisms. The equations for calculating absorption above 20 kHz are presented.

Section 3.4 treats the low frequencies (below 20 kHz). A detailed historical survey is given first that explains all the difficulties encountered and the fact that more than 15 years of research were necessary to find the boric acid relaxation effect. The techniques of measurements at sea are then presented, followed by a description of the complex mechanism of the boric acid dissociation and its relationship with absorption. The results and the equations for calculating absorption at low frequencies are presented.

Finally, Section 3.5 gives a discussion of all results and equations and makes recommendations for complementary research.

3.1. INTRODUCTION

A sound wave generated by a source S located in the sea or in freshwater is received at a point M of the medium through a number of possible acoustic paths. To each path corresponds a loss of acoustic energy between that generated by the source in the proper direction, referred to as a unit distance (usually 1 m), and that received at point M. This "transmission loss" is the result of several mechanisms, viz: (1) a geometrical spreading of the sound (usually the predominant effect); (2) losses at the boundaries due to reflections, if any; (3) losses due to scattering by inhomogeneities of the medium; and (4) losses by absorption of sound in water. This last loss is due to continuous local transformations of a small part of the acoustic energy into another form of energy (heat in the end) during the transit of the sound wave.

For continuous sound waves, or time-limited pulses of sufficient duration, acousticians employ "intensity" as a unit. It is defined by the time average of the energy flow across a unit area normal to the motion per unit of time and is found to be given at a point P_i by

$$I_i = p_i^2/\rho_i c_i \tag{3.1}$$

where p_i, ρ_i, and c_i are the local values of acoustic pressure (RMS in general), density of the medium, and speed of sound at P_i, respectively, the product $\rho_i c_i$ being the acoustic impedance.

In an ideal infinite and homogeneous medium of water (fresh or salted), an omnidirectional sound source would radiate in a spherical manner, and if there were no sound absorption, the acoustic intensity would vary with the distance r from the source as $I(r) = I_o/r^2$, I_o being the intensity at unit distance. This results simply from the conservation of the energy distributed on concentric spheres. Because the processes governing the absorption of sound are local but independent of place for a homogeneous medium, the fraction of acoustic energy lost per unit volume is a constant. The variation of intensity with distance thus follows the rule

$$I(r) = I_o/r^2 e^{-\alpha r} \tag{3.2}$$

where α is the absorption coefficient, commonly expressed in Nepers per unit distance. Physicists, however, often use Nepers per wavelength of the sound, which, as will be seen in Section 3.2, has the advantage of revealing, when making graphs, the resonant frequencies involved in the physics of the phenomena.

In underwater acoustics, losses are usually expressed in decibels. If I_o and I_r are the transmitted acoustic intensity (at unit distance from S) and the received acoustic intensity at M, respectively, then the transmission loss TL in dB between S and M is given by

$$TL = 10 \log(I_o/I_r) \tag{3.3}$$

In practice, however, acousticians in underwater sound simplify this formulation by neglecting the variations of acoustic impedance in the medium, which are extremely small compared with the variation of the acoustic pressure as the wave travels. Eq. 3.3 is thus reduced to

$$TL = 20 \log(p_o/p_r) \tag{3.4}$$

When calculating the transmission loss, the acousticians make another simplification, which is only acceptable as a first approximation, and which is discussed in Section 3.5: They assume that in the limited environment concerned with one application, the absorption is a mechanism independent of the location of the sound wave, hence the current use of "decibels per km (or kyd)" as a unit for the coefficient of absorption. In the case of a "cylindrical" spreading of the sound, often assumed to be valid in measurements of low-frequency absorption at sea, the transmission loss vs horizontal range r is then written in the form

$$TL = K + 10 \log r + ar10^{-3} \tag{3.5}$$

where r is in meters but a is in dB/km, and where K is a constant depending on the area. For further details on these matters, consult Urick [1].

The absorption of sound in water was first studied from a purely physical point of view, starting with pure water, then extending to seawater. The fact that the use of sound is the only way to transmit messages at long ranges in the sea, detect deep submerged targets, etc., considerably reinforced the need for accurate values. The experimental results also were always quite puzzling and progressively revealed the existence of various unpredicted phenomena that needed to be explained and studied in detail. In all, during half a century (roughly between 1935 and 1985), numerous scientists throughout the world spent years and years of lives on the subject, and the number and cost of experiments performed has been considerable.

The fundamental relaxation mechanisms that govern the absorption of sound in seawater is first explained in Section 3.2, followed by a brief description of the main experimental methods of measurements in laboratory. Because the possible types of experiments and ways of investigation both in laboratory and at sea are very different above and below some 20 kHz, and the main mechanisms at the origin of sound absorption are also different, the results will be presented in two clearly divided sections: Section 3.3 deals with absorption above 20 kHz and Section 3.4 with that at lower frequencies. A general discussion, mainly concerned with what still needs to be known and with improvements in practical calculations, is provided in Section 3.5.

3.2. PHYSICAL AND CHEMICAL CAUSES OF SOUND ABSORPTION IN LIQUIDS

3.2.1. General

The absorption of acoustic waves in liquids is due to viscosity and to the fact that the thermodynamic parameters of the liquid do not vary in phase when they are perturbed by sound. The absorption coefficient α varies with the characteristics and composition of each liquid, its environmental conditions (temperature and pressure), and the frequency of the sound wave. The expression for the absorption coefficient is usually divided into two parts, namely, the Stokes-Kirchhoff coefficient and the relaxation process contribution.

3.2.2. The Stokes-Kirchhoff Coefficient

The Stokes-Kirchhoff coefficient for "classical" absorption takes into account the loss by viscous friction due to the velocity gradient of the acoustic wave on the one hand and the energy dissipation by thermal conductivity resulting from the temperature differences between the warm and the cold parts of the wave on the other. It is given by

$$\alpha_{ST} = \frac{2\pi^2}{\rho c^3}\left(\frac{4}{3}\eta + \frac{\gamma-1}{\gamma C_V}\kappa\right) f^2 \qquad (3.6)$$

where f is the frequency of the wave; c, the speed of sound; ρ, the density; η, the shear viscosity; κ, the thermal conductivity; C_v, the heat capacity at constant volume; and γ, the ratio of specific heats.

3.2.3. Relaxation Processes

The heat capacity and density values depend on the structure of the liquid, which itself depends on temperature and/or pressure. The acoustic wave produces pressure and temperature variations that lead to structural changes. These are not instantaneous but require a time τ. To illustrate the effect of the structural change on ρ we reported on Figure 3.1 the variation of ρ for a very rapid change of pressure: At the beginning, ρ evolves with pressure but keeps varying after pressure has become constant; this is the relaxation process. For an acoustic wave, the propagation properties are different according to whether its period is shorter or longer than the relaxation time τ.

In the case where there exists only one relaxation process, the compressibility is given by

$$\beta(\omega) = \beta_\infty + \frac{\beta_0 - \beta_\infty}{1 + i\omega\tau} \qquad (3.7)$$

where β_∞ and β_0 are the compressibility for frequency $\omega \gg \tau^{-1}$ and for zero frequency, respectively. The compressibility of such a liquid is complex, and when an acoustic

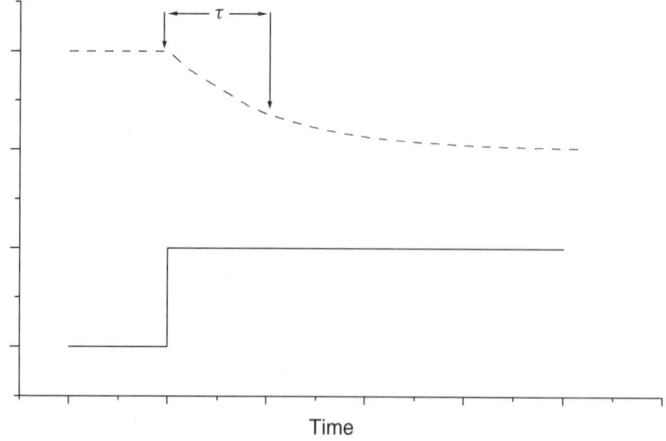

Fig. 3.1. Variation with time of the density (dashed line) after a pressure jump (solid line).

wave is propagating in it, the variation of pressure is out of phase with the density variation. The amplitude of the wave decreases while propagating, and its velocity depends on the frequency. The relationship between the wave vector k and the compressibility is

$$k^2 = \omega^2 \beta \rho_0 \qquad (3.8)$$

where ρ is the density at rest, and where k has a complex value given by

$$k = \frac{\omega}{c} - i\alpha \qquad (3.9)$$

c being the speed of sound and α the absorption coefficient at ω.

With these equations, the contribution of the relaxation process to the value of α and to the speed of sound value may be calculated to be

$$\alpha = \frac{\beta_0 - \beta_\infty}{2c\beta_0} \frac{\omega^2 \tau}{1 + \omega^2 \tau^2} \qquad (3.10)$$

$$\left(\frac{c_0}{c(\omega)}\right)^2 = 1 - \frac{\beta_0 - \beta_\infty}{\beta_0} \frac{\omega^2 \tau^2}{1 + \omega^2 \tau^2} \qquad (3.11)$$

where c_0 and c_∞ are the speed of sound at $\omega\tau \ll 1$ and at $\omega\tau \gg 1$, respectively.

3.2.4. Chemical Relaxation of Salts in Water

Seawater is an electrolytic solution in which about 50 different kinds of ions are dissolved. Certain ions join together and make up different molecules: for example, the Mg^{+2} ion associates with the SO_4^{-2} ion, giving the molecule $MgSO_4$, or with ion CO_3^{-2}, giving $MgCO_3$. All these species are present in seawater, their concentrations depending on temperature and on pressure and being able to change when an acoustic wave is passing through the medium. The association of ions and the dissociation of molecules are not instantaneous because collisions between species are needed, which are not systematically efficient in that respect. In all, the more there are collisions between species, the shorter is the relaxation time. The density of water is then changed because the volume occupied by two associated ions is not the sum of the ion volumes. Indeed, more molecules of water are associated with ions than a neutral molecule.

The chemical equilibrium is characterized by the kinetic equation

$$A^+ + B^- \underset{k_{21}}{\overset{k_{12}}{\rightleftharpoons}} AB \qquad (3.12)$$

where k_{12} is the probability per unit time that, in a volume equal to unity with only one molecule A and one molecule B, A and B join together, and where k_{21} is the probability per unit time that one molecule AB breaks up. The number of molecules built up during the time Δt is $k_{12} C_A C_B \Delta t$, and the number of molecules AB broken up is $k_{21} C_{AB} \Delta t$, where C_A, C_B, and C_{AB} are the numbers per unit volume of the species A, B, and AB, respectively.

The variation of C_{AB} is given by

$$\frac{\partial C_{AB}}{\partial t} = k_{12} C_A C_B - k_{21} C_{AB} \qquad (3.13)$$

To make the calculation easier, the variables C_A, C_B, and C_{AB} are split into static and dynamic parts as

$$C_A = \overline{C_A} + \Delta C_A(t);\, C_B = \overline{C_B} + \Delta C_B(t);\, C_{AB} = \overline{C_{AB}} + \Delta C_{AB}(t) \qquad (3.14)$$

At equilibrium, the number of molecules AB is constant and $\frac{\partial C_{AB}}{\partial t} = 0$.

This relationship gives a ratio between the different species:

$$k_{12}\overline{C_A C_B} = k_{21}\overline{C_{AB}} \quad \text{and} \quad \frac{\overline{C_A C_B}}{\overline{C_{AB}}} = \frac{k_{21}}{k_{12}} = K \tag{3.15}$$

where K is a constant that is independent of the concentrations of A, B, and AB but depends on temperature and on pressure.

When the variations of the concentrations are small, Eq. 3.13 provides the linear differential equation:

$$\frac{\partial \Delta C_{AB}}{\partial t} = k_{12}(\overline{C}_B \Delta C_A + \overline{C}_A \Delta C_B) - k_{21}\Delta C_{AB} \tag{3.16}$$

One ion A and one ion B disappear when one molecule AB appears, so the variations of the different species are linked:

$$\Delta C_A = \Delta C_B = -\Delta C_{AB} \tag{3.17}$$

The equation for the relaxation process is now:

$$\frac{\partial \Delta C_{AB}}{\partial t} = -(k_{12}(\overline{C_B} + \overline{C_A}) + k_{21})\Delta C_{AB} \tag{3.18}$$

The variation of C_{AB} corresponds to an exponential law with a time constant τ given by

$$\tau = k_{12}(\overline{C_B} + \overline{C_A}) + k_{21} \tag{3.19}$$

3.2.5. Relationship between Chemical Relaxation and Absorption of Sound

The concentrations of the different species depend on temperature and pressure, as well as the value of K. Planck's law gives the variation of K with pressure as

$$\frac{d\log(K)}{dP} = -\frac{\Delta V^0}{RT} \tag{3.20}$$

where $-\Delta V^0 = V_{AB}^0 - V_A^0 - V_b^0$, with V_{AB}^0, V_A^0 and V_B^0 being the volumes of each species.

Van't Hoff's law gives the variation of K with temperature as

$$\frac{d\log(K)}{dT} = -\frac{\Delta H}{RT} \tag{3.21}$$

where ΔH is the enthalpy of the reaction.

The compressibility is given by

$$\beta = -\frac{1}{V}\left(\frac{dV}{dP}\right) = \beta_\infty - \frac{1}{V}\left(\frac{dV}{dP}\right)_{relax} \tag{3.22}$$

The second term represents the contribution of the chemical relaxation process to the compressibility β_r. For the relaxation process previously described it is equal to

$$\beta_r = -\frac{1}{V}\left(\frac{\partial V}{\partial C_{AB}}\right)\left(\frac{\partial C_{AB}}{\partial K}\right)\left(\frac{\partial K}{\partial P}\right) \tag{3.23}$$

The first factor is equal to the change of volume induced from the fusion of ions A and B: ΔV_c. The second factor is obtained from Eq. 3.15:

$$\left(\frac{\partial C_{AB}}{\partial K}\right) = -\frac{1}{K}\left(\frac{1}{C_A} + \frac{1}{C_B} + \frac{1}{C_{AB}}\right)^{-1} \tag{3.24}$$

The third factor is given by Planck's law.

The amount of contribution of the chemical relaxation process to the compressibility thus depends on the change of volume ΔV induced by the chemical reaction on the concentration of the different species, and on the equilibrium constant K.

The absorption coefficient is equal, for the previous process to:

$$\alpha_i = \frac{\rho_0 c_0}{2RT}(\Delta V_0)^2 \left(\left(\frac{1}{C_A} + \frac{1}{C_B} + \frac{1}{C_{AB}}\right)^{-1}\right) \frac{\omega^2 \tau}{1+\omega^2 \tau^2} \qquad (3.25)$$

In the case where several relaxation processes occur, the time variation of each species is written as in Eq. 3.13. With the conservation equations, and under the assumption of weak perturbations, a set of linear differential equations is obtained, the number of relaxation times being equal to the number of relaxation processes.

The overall absorption coefficient can finally be written as

$$\alpha_r = \sum_i A_i \frac{\omega^2 \tau_i}{1+\omega^2 \tau_i^2} \qquad (3.26)$$

where A_i is the amplitude associated to the relaxation time τ_i.

3.2.6. General Formulation and Comments

For practical applications Eq. 3.26 is generally found written as

$$\alpha = \sum_i A_i \frac{f_{ri} f^2}{f_{ri}^2 + f^2} \qquad (3.27)$$

where the f_{ri} are the relaxation frequencies in hertz or, more often, in kilohertz, and f is the frequency of the sound wave in the same unit. For one process i its contribution to absorption is bounded by the limit $\alpha_{iM} = A_i f_{ri}$, asymptotically reached at infinite frequency. At $f = f_{ri}$ the contribution of the process is $\alpha(f_{ri}) = \frac{1}{2} A_i f_{ri}$, half the limiting value [Fig. 3.2(a)].

It is of interest to consider the absorption per wavelength $\alpha \lambda$. For process number i this gives

$$\alpha \lambda = \sum_i \frac{A_i}{c} \frac{f_{ri} f^2}{f_{ri}^2 + f^2} \qquad (3.28)$$

where c is the speed of sound in the medium. The variation of $\alpha \lambda$ with frequency presents a maximum at the relaxation frequency f_{ri}, starting and ending at zero for $f = 0$ and $f = \infty$ [Fig. 3.2(b)]. The coefficient of amplitude A_i of the absorption is then related to the maximum reached by $\alpha \lambda$ by

$$A_i = 2 \frac{\alpha_i \lambda_M}{c} \qquad (3.29)$$

Thus, examining the variation of $\alpha_i \lambda$ with frequency is one method to detect easily the existence of relaxation frequencies and to measure them, as well as to obtain the values of the corresponding amplitudes A_i.

3.2.7. Techniques of Laboratory Measurements

In the laboratory, the study of sound absorption in liquids can be made by different methods, all requiring the use of a container of limited dimensions. Descriptions of the various methods employed from the early measurements are found in Markham *et al.* [2] and Kurtze and Tamm [3]. Another more recent technique is the temperature jump (T-jump). We shall only describe briefly three of these methods, which have been

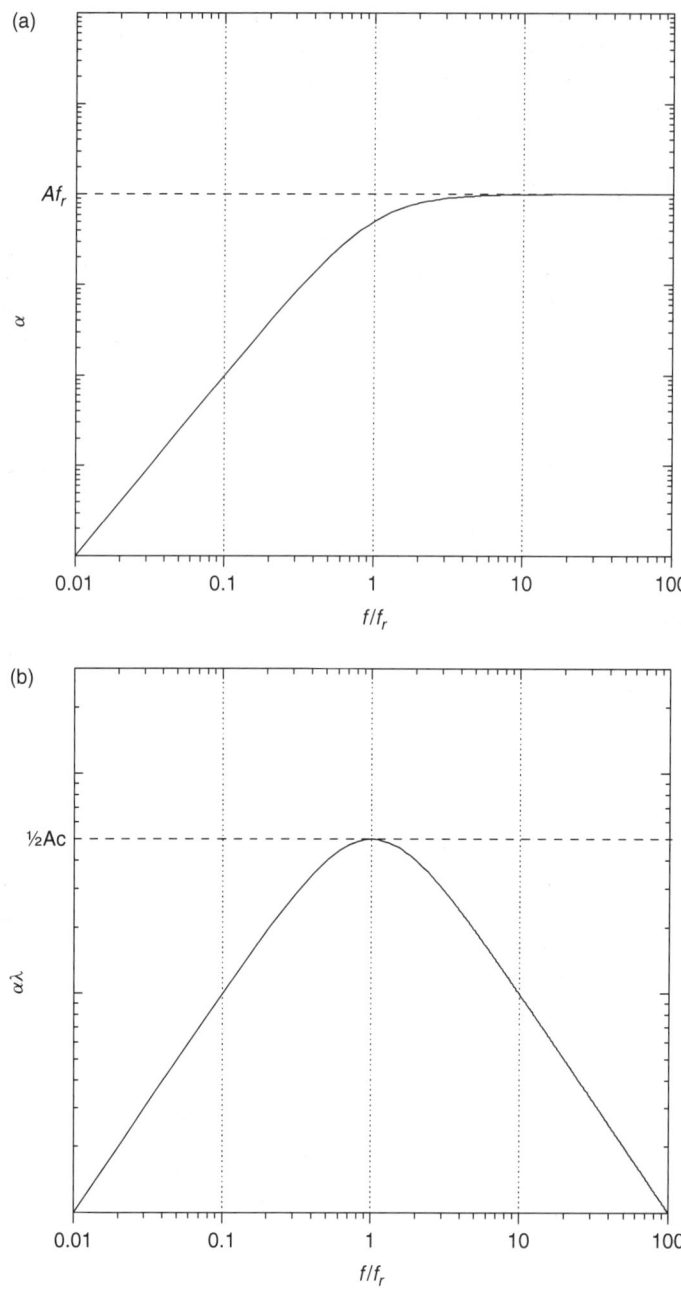

Fig. 3.2. Variation with frequency of the absorption coefficient (a) and of its product $\alpha\lambda$ by the wavelength (b) in the case of a relaxation process.

used to study the absorption of sound in freshwater and seawater. The main point to bear in mind is that measurements become more and more difficult toward the lower frequencies for two reasons: the absorption gets smaller and the wavelength λ of the sound approaches or exceeds the dimensions of the containers.

3.2.7.1. The "Direct" Method

This method is only applicable at very high frequencies (above 1 MHz) where transducers of dimensions greater than some 50 λ can be used, enabling to generate quasi-plane waves. A receiver moved at various distances from the sound source receives an energy that, ideally, is only reduced by absorption. Pulse techniques and corrections need to be used to avoid the influence of reflections and to account for the radiation pattern of the transducer.

3.2.7.2. The Resonator Method

In this method, the most commonly employed, the liquid under study is placed in a container that is excited by a tone-burst at one of its possible modes of vibration. The intensity of the sound produced in the liquid by this single excitation progressively decays with time due to absorption and to losses at the successive reflections on the walls of the container. Several variations of this method have been developed, the most frequently used for seawater, and electrolyte solutions being the spherical resonator (10–200 liters) suspended in vacuum by thin leads. The pulse is generated by a transducer fixed on the spherical shell, which is then switched in a receiving mode (Fig. 3.3 from [4]). This arrangement reduces the reflection losses that must be taken into account when making absolute measurements of absorption coefficients. In differential procedures, the rate of decay of the sound is compared with that obtained by the use of a reference liquid of known acoustic properties. The frequency range of the resonator technique extends from 30 kHz or even less to some 500 kHz, but accurate measurements at the lower modes of vibration are difficult to obtain. In all cases, the experimental search for resonances is a tedious process.

3.2.7.3. The T-jump Method

At low frequencies (below 20 kHz), the absorption of sound in seawater or in solutions of electrolytes with equivalent concentrations can no more be measured in the laboratory. However, it is still possible to detect relaxations due to dissolved components by the use of sudden perturbations applied to the liquid in a container. In the T-jump technique, a sudden increase of temperature is created by the electric discharge of a condensator. After the temperature jump, the variation of the concentration of the

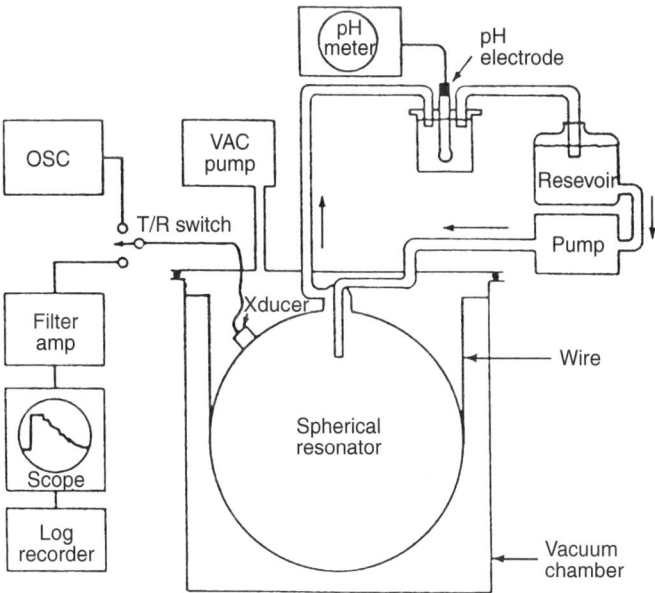

Fig. 3.3. Schematic of a resonator (after Mellen [4]).

chemical species is generally observed optically by measurements of light absorption or of fluorescent intensity, and the time lag corresponding to a relaxation process if it exists can be measured.

3.3. SOUND ABSORPTION ABOVE 20 kHZ

3.3.1. Historical Survey

As for all liquids, the absorption of sound in water was first studied from theoretical considerations. The oscillatory superimposed pressure due to the sound transiting through an elementary volume ΔV was considered as generating deformations and molecular frictions resulting in a local loss of acoustic energy in ΔV, determined by a coefficient of absorption. From data about heat conductivity and viscosity in pure water, the pioneering work of Stokes (1850), followed by that of Kirchhoff (1868), led to values for what has been called the "classical" absorption (Sect. 3.2.2.). It was found that absorption by thermal conductivity was small compared with that due to shear viscosity and that both varied as the square of frequency, but direct experimental values could not be obtained at the time by acoustical means.

It is only after the discovery of piezoelectricity and the development of high-frequency transducers based on this effect that the first precise measurements could be performed. Using various methods, mechanical, optical, but above all electrical (in particular the resonator method, see Sect. 3.2.7.) values of absorption in pure water were obtained between 1935 and 1950 that were three times greater than predicted by the Stokes-Kirchhoff model, still following a variation with frequency as f^2 inside the range of measurements, i.e., from 100 kHz to 300 MHz (Fig. 3.4). Hall proposed in 1948 [5] the following explanation: The absorption of sound is due to structural rearrangements of the water molecules, disturbed by sound and oscillating between two possible ways of piling in a relaxation manner. Although no physical experiment has ever proved entirely this theory (the relaxation frequency is too high), it is generally admitted as it gives values that correspond to measurements.

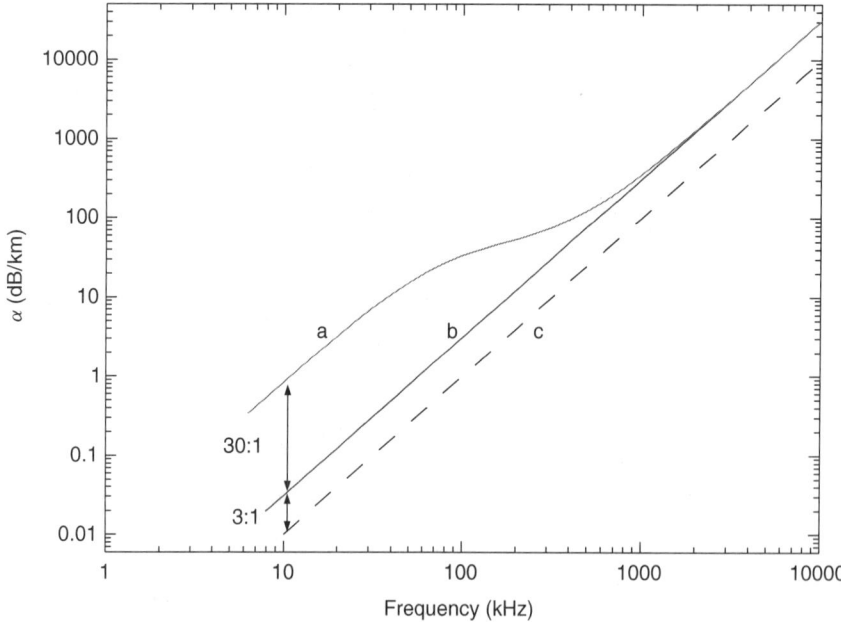

Fig. 3.4. Differences in absorption coefficient for (a) seawater, (b) pure water, and (c) a calculation from the classical Stokes-Kirchhoff theory (after Urick [1]).

In parallel with measurements of sound absorption in pure water, early measurements of absorption were performed in seawater because it is the most common liquid medium on the planet. The experimental results soon revealed that the absorption coefficient was progressively departing from that in pure water when frequency decreased below 500 kHz, reaching as much as 30 times the value given by extrapolating pure water absorption down to about 20 kHz (Fig. 3.4). Similar effects were also found in various mixtures of electrolytes, which all exhibited a behavior with frequency following a rule in $\alpha \propto A f_r f^2/(f_r^2 + f^2)$, typical of a relaxation process with resonance frequency at f_r. Sodium chloride, the main salt of seawater, was first suspected to be at the origin of this "excess" absorption (Lieberman [6]), but laboratory measurements on various solutions of electrolytes soon proved that this was not the case. Leonard *et al.* demonstrated in 1949 [7] that indeed $MgSO_4$, the magnesium sulfate, was responsible for the effect, but with some "chemical help" from NaCl. Systematic studies were then undertaken to search for the influence of salt concentration, temperature, and pressure. These lasted until about 1980, but in the meantime another absorption excess had been discovered, which would lead to a great agitation among researchers: below 20 kHz there was another departure of sound absorption from that given by pure water molecular rearrangements and by the ionic dissociation of $MgSO_4$. It is the subject of Section 3.4.

3.3.2. Sound Absorption in Freshwater

The absorption of sound in pure water has been intensively measured in laboratories by various acoustical methods during 1935–1950. As the first measurements had shown, that absorption was about three times greater than expected from the classical theory the main preoccupation of scientists at the time was to investigate whether this ratio remained constant with frequency. Not less than 23 independent researchers or teams from various countries are listed by Kurtze and Tamm [3][(1)] as having performed such measurements, all at ambient temperature around 20 °C. The overall frequency range explored extended from 100 kHz to 250 MHz. Although some measurements were not accurate enough, especially below 1 MHz, the general trend of the data, once approximately corrected for the slight temperature offsets from 20 °C (i.e., as if it followed the classical absorption), clearly demonstrated that in the whole frequency range the ratio of the absorption coefficient to the square of frequency remained constant. While at 20 °C the classical theory gave $\alpha/f^2 = 8.16 \times 10^{-15}$ Np/m/Hz^2, the experimental value was found to be $\alpha/f^2 = 25 \times 10^{-15}$.

Various hypotheses were advanced to explain the departure from the classical theory, which had to be rejected. It is finally the model developed by Hall in 1948 [5] that was accepted. In this model, the impact of the sound beam first compresses groups of water molecules, which are assumed to possess some degree of order. Then a structural compression sets in, the molecules breaking the structural bands to move into a closed packet arrangement. Absorption appears from the time lag. On this basis, and assuming an instantaneous compressibility, Hall calculated a theoretical absorption coefficient that varied as $\alpha = a f_o f^2/(f_o^2 + f^2)$, f_o being the relaxation frequency related to the time lag. As the ratio α/f^2 is found constant up to $f_M = 250$ MHz, this leads to $\alpha/f^2 = k \cong a/f_o$ with $f_o \gg f_M$. Thus, only the ratio a/f_o needs to be known, and in fact the frequency f_o is so high that it has never been possible to measure it. Hall calculated a/f_o as a function of temperature, and his values were found in good agreement with other laboratory measurements exploring the variation of absorption with temperature. These were performed by various researchers and in particular by Pinkerton [8], whose data are given in Table 3.1.

[(1)]This reference, plus the review by Markham *et al.* [2], provides a sufficient set of references for a reader who might want greater details and data on the subject.

Table 3.1. Absorption of sound in pure water as a function of temperature according to Pinkerton's measurements.

Temperature (°C)	Coefficient A of $\alpha = Af^2$ in Np/m with f in Hz ($A \times 10^{15}$)
0	56.9
5	44.1
10	36.1
15	29.6
20	25.3
30	19.1
40	14.6
50	12.0
60	10.2

The effect of pressure on sound absorption in pure water was later explored by Litovitz and Carnevale [9] at 0 and 30 °C. Absorption was found to decrease monotically with pressure in an almost similar way at both temperatures, being reduced by some 30% at 100 MPa.

In all, the absorption coefficient α_o in pure water can be written as a function of temperature t, pressure p above atmospheric, and frequency f up to 200 MHz in the following manner:

$$\alpha_o(t, p, f) = f(t)g(p)f^2 \tag{3.30}$$

where $f(t)$ and $g(p)$ are independent functions valid in the whole domain

$$0 < t < 60\,°C$$

$$0 < p < 100 \text{ MPa}$$

At atmospheric pressure (i.e., with $g(p) = 1$), and for α expressed in Np/m, frequency in Hz, and t in °C, $f(t)$ was written by Fisher and Simmons [10] as

$$f(t) = (55.9 - 2.37t + 4.77 \times 10^{-2}\,t^2 - 3.48 \times 10^{-4}t^3) \times 10^{-15} \tag{3.31}$$

The same authors also proposed for $g(p)$:

$$g(p) = 1 - 3.84 \times 10^{-4}p + 7.57 \times 10^{-8}p^2 \tag{3.32}$$

where p is in atmospheres.

To be closer to Pinkerton's absorption data, François and Garrison [11] later proposed for $f(t)$ a formulation that depended on the temperature range, but they used the decibel per kilometer as the unit of absorption, their aim being to provide equations for undersea applications.

Their formulations are:

(a) In the temperature range $0° < t \leq 20\,°C$

$$f(t) = 4.937 \times 10^{-4} - 2.59 \times 10^{-5}t + 9.11 \times 10^{-7}t^2 - 1.50 \times 10^{-8}t^3 \tag{3.33}$$

(b) In the temperature range $20° \leq t \leq 60\,°C$

$$f(t) = 3.964 \times 10^{-4} - 1.146 \times 10^{-5}t + 1.45 \times 10^{-7}t^2 - 6.5 \times 10^{-10}t^3 \tag{3.34}$$

All these last points are discussed in Section 3.5.

3.3.3. Sound Absorption above 20 kHz in Seawater

3.3.3.1. Mechanisms and Measurements

In the frequency range 10–300 kHz, the absorption of sound in seawater is mainly governed by the relaxation dissociation of $MgSO_4$ as $Mg^{++} + SO_4^{--} \leftrightarrow MgSO_4$.

This dissociation occurs in a multiple step manner and is slowed down by the presence of other electrolytes. Its study and the associated absorption measurements have been carried out in the laboratory using the resonator technique with synthetic seawater and with enriched solutions of electrolytes. The main authors of those investigations were Kurtze and Tamm; Tamm *et al.*; Wilson and Leonard; Fisher; and Glotov [3, 12–15].

The measurements in synthetic seawater yielded an absorption that is equal to that of pure $MgSO_4$ solutions of 0.014 mole/liter, a concentration smaller than that of the Mg^{++} and SO_4^{--} ions in seawater. In fact, the dissociation process of $MgSO_4$ involves other electrolytes, NaCl in particular: it was observed that solutions of magnesium sulfate enriched by the addition of NaCl only could give the same results as synthetic seawater.

The relaxation frequency f_r of the process was found to vary considerably with temperature, evolving from 35 kHz to 250 kHz in the Neptunian domain (−2 to 35 °C). This variation follows an exponential law in θ, the absolute temperature in °K, according to $f_r = ke^{-q/\theta}$, which can be reformulated as $f_r = m \, 10^{(n-q'/\theta)}$, n being an integer. A very slight dependence of f_r on salinity was also found. The coefficient of absorption itself is directly proportional to salinity and presents a slight increase with temperature (see Sect. 3.3.3.2 and Table 3.2 for the details).

It has been suspected that the dissociation of $MgSO_4$ in real seawater could be altered near the freezing point, and a few direct acoustic measurements of the absorption coefficient were made under the Arctic ice [16, 17]. No change with respect to what was given by the equations could be found above 100 kHz, but an increase of absorption by 50% was observed below 50 kHz. It is not clear, however, whether this is really due to modifications in the dissociation process or more simply to an additional loss of intensity by scatterers.

The effect of the hydrostatic pressure P on seawater absorption has been the subject of several controversies, but finally the results from laboratory and at-sea measurements proved that absorption was reduced under high pressures. The first measurements were carried out in the laboratory by Fisher on more concentrated $MgSO_4$ solutions than in the sea. In their 1962 simplified equation for absorption, Schulkin and Marsh [18] interpreted Fisher's results as a linear variation of absorption with pressure following $\alpha(P) = a(\text{atmospheric}) \times (1 - 6.54 \times 10^{-4} P)$ (atm).

Table 3.2. Equations for sound absorption due to the $MgSO_4$ relaxation.

	Schulkin-Marsh	Fisher-Simmons	François-Garrison
f_r	kHz	Hz	kHz
	$21.9 \times 10^{6-1520/\theta}$	$1.55 \times 10^7 \theta e^{-3052/\theta}$	$8.17 \times 10^{8-1990/\theta}$
			$1 + 0.0018(S - 35)$
A	Np/m	Np/m	dB/km
	2.34×10^{-6}	$5.62 \times 10^{-8} + 7.52 \times 10^{-10} T$	$\dfrac{21.44}{C}(1 + 0.025)^{(a)}$
g(P)	kg/cm²	atm	Depth in m
	$1 - 6.54 \times 10^{-4} P$	$1 - 10.3 \times 10^{-4} P + 3.7 \times 10^{-7} P^2$	$1 - 1.37 \times 10^{-4} Z + 6.2 \times 10^{-9} Z^2$

[a] C being the local speed of sound in m/s, which, according to the authors, may for this calculation be approximated simply by $C = 1412 + 3.21 T + 1.19 S + 0.0167 Z$.

This variation was suspected to be too low for real seawater conditions from the measurements made at sea by Bezdek [19], who found the coefficient of P to be between -12×10^{-4} and -14×10^{-4}. This coefficient was later corrected by François and Garrison to -11.8×10^{-4}. New laboratory measurements performed by Hsu and Fisher on MgSO$_4$ solutions with or without NaCl additions and on FLEMING seawater [20] led to the intermediate value -8×10^{-4}.

3.3.3.2. Equations

Disregarding some excessively simplistic early formulations, three different equations have been proposed to obtain the excess of absorption in seawater due to the magnesium sulfate relaxation. Basically they are all of the form

$$\alpha = ASf_r f^2 / (f_r^2 + f^2) g(P)$$

where f is the frequency in Hz or in kHz; f_r, the relaxation frequency in the same unit and given by a specific formula; S, the salinity in p.p.t.; and $g(P)$, a correction for pressure above atmospheric. The two first equations give the absorption coefficient in Nepers per meter and the third one in decibels per kilometer. According to the equation, A is a constant or a function of temperature alone or of temperature and sound speed. As for the relaxation frequency, it always varies with temperature following the physical law in $f_r = ke^{-q/\theta}$, where θ is the absolute temperature in °K. This formulation is also written as $f_r = m \, 10^{n-q'/\theta}$ with n an integer.

All three equations are presented in Table 3.2, which gives the formulas for f_r, A and $g(P)$, with their original coefficients, and specifies the units. The results are discussed in Section 3.5. The first equation is that of Schulkin and Marsh [18], proposed in 1962 on the basis of laboratory measurements and a few field data. It is oversimplified since A is considered as a constant, but it was the only equation for 15 years. The second equation is that of Fisher and Simmons [10] in 1977 based on values from Simmons at atmospheric and laboratory measurements of the effect of pressure by Fisher. The third equation, from François and Garrison [11] was published 5 years later. In this equation, the relaxation frequency is as determined by Glotov [15], the value of A is adjusted from measurements performed by Applied Physics Laboratory at the University of Washington, and the pressure correction is the one given by Bezdek [19] from at-sea measurements, thus using depth Z instead of pressure P.

3.4. SOUND ABSORPTION AT LOW FREQUENCIES

3.4.1. Historical Survey

The two relaxational mechanisms governing the absorption of sound in seawater above 20 kHz had been discovered soon after measurements had given values greater than expected from existing theories. Things went quite differently for absorption below 20 kHz, which turned out to be a real burden to researchers all over the world for more than 25 years.

The absorption of sound by seawater toward the low frequencies had become of interest in World War II and postwar sonar applications. The vast knowledge acquired during the war in the field of underwater acoustics had been compiled by Eckart in the famous classified "brown books" series, later revised and released as the "green books" [21]. From the analysis of many sonar trials and performances, it had been concluded that the absorption coefficient in seawater was about 3.5 dB/km at 20 kHz, in agreement with the effect of MgSO$_4$, but decreased at lower frequencies as $f^{1.3}$ instead of f^2. In fact, it was not sure whether this was due to absorption only or to a mixture of absorption and scattering, and the intermediate terminology "attenuation" was often used. Still, in the 1950s and even later this was of considerable importance: It was clear that high-frequency sonars were too limited in range detection due to excessive sound absorption, and that one would gain in performance by using lower frequencies at the cost of the development of much larger and expensive transducers, but how far to go? As a simple illustration, just to detect a target at a mere 15 km with a sonar to be designed for working at 5 kHz, the expected propagation loss due to "attenuation" would be about 6.6 dB according to the physics of $MgSO_4$, but about 17.3 dB if the law in $f^{1.3}$ given by Eckart was the right one. Such a difference of 10.7 dB implied, for given identical performances, a requirement to produce acoustic power nearly 12 times greater than calculated from pure theory. The confusion was indeed serious, as we find for instance Schulkin and Marsh writing in 1962 about absorption [18]: "The present authors have carried out an analysis of thousands [sic] of measurements at sea in the frequency range 2–25 kcps .... Their results agree with those of Del Grosso and Kurtze and Tamm" (i.e., with absorption due to magnesium sulfate alone!).

A few experiments, however, led to results that did not agree with this statement, but the greatest impulse certainly arose from very unusual measurements published in 1957 by Sheehy and Halley [22]. They dealt with the analysis of the noise produced by an underwater atomic explosion recorded at various remote stations across the Pacific. It sounded like endless thunder rolls due to multiple reflections from islands or entire archipelagos that could be identified from sound travel times. Due to the enormous distances involved (several thousands of kilometers), the usable frequency spectrum was reduced to 20–200 Hz, but Sheehy and Halley noticed that it varied according to the total distance covered by sound. Assuming that this variation was only due to the effect of absorption, the authors could determine in the 20–200 Hz band a set of differences of absorption coefficients $\Delta\alpha_{ij} = \alpha(f_i) - \alpha(f_j)$, where $\alpha(f_k)$ stands for the coefficient at f_k, but there was no way to obtain absolute values from the measurements. This set of differential coefficients was therefore adjusted to a law of variation with frequency in $\alpha = kf^n$ and the best-fit calculation concluded that the exponent n verified $1.4 < n < 1.5$. Adopting $n = 1.5$ led, in dB per kiloyard, to the formulation $\alpha = 0.033 f^{1.5}$, which happened to give at 20 kHz the correct value for average seawater and was not that different from Eckart's empirical law.

Schulkin [23] appears to be the first to suspect from the Sheehy and Halley measurements that the excess of absorption above that from $MgSO_4$ was due to a relaxation mechanism. His points were as follows: (1) a variation in $f^{1.5}$ is not physical and (2) the set of differential coefficients obtained by Sheehy and Halley may as well be approximated by a law in f^2 without any valuable loss of confidence; the admitted value of absorption at 20 kHz can then be reached if the frequency dependence follows a law in $f^2/(f_r^2 + f^2)$ (see Fig. 3.5). In his short paper on the subject, Schulkin suggested for f_r the value of 630 Hz and eddy viscosity as a possible cause for the mechanism.

The evidence that there existed indeed a new unknown relaxation mechanism was demonstrated for the first time at the SACLANT Center of La Spezia by Stangerup and Leroy in 1963–1964. At that time a team of scientists was exploring the peculiarities of sound propagation in the Mediterranean Sea, whose waters stay isothermal from 300 m down to the deepest waters (>5000 m). "Classical" sound propagation measurements with explosive sources (see Sect. 3.4.2) in the very asymmetrical sound channel of that sea had led to absorption values (Lallement and Waterman [24]) that were higher than those predicted by the Schulkin-Marsh equation. Stangerup noticed that the general trend of these data could be interpreted according to Schulkin's idea by a law in $f^2/(f_r^2 + f^2)$ and suggested a relaxation frequency of 1 kHz (quoted in [27]). Leroy, who feared sound leakage effects in the asymmetrical channel to alter the results of the absorption evaluation but agreed with Stangerup's interpretation, realized the interest to analyze the spectral variation with path length of the direct sound arrivals from the

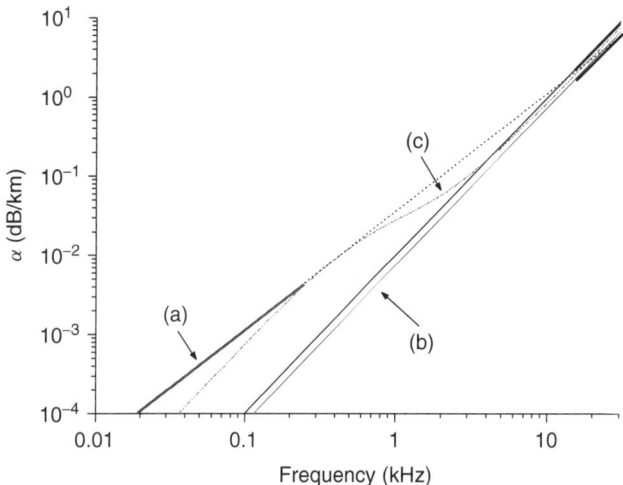

Fig. 3.5. Illustration of the uncertainty about the variation of absorption with frequency in the 1960s. (a) Results from Sheely and Halley. Thick line is from their measurements; dotted line, extrapolation. (b) Absorption due to $MgSO_4$. Thick lines are limits of measurements; thin lines, extrapolation. (c) Schulkin's hypothesis (mixed line).

shock wave signals used in the available Mediterranean experiments: these acoustic paths undergo only one refraction at depth from the constant speed of sound gradient due to hydrostatic pressure, so the spreading losses are independent of frequency, and absorption is the only factor altering the frequency spectrum. The analysis was performed from recordings of six different acoustic runs and made at 17 preselected frequencies that sampled the range 0.3–8 kHz. The results clearly demonstrated that the shape of the curve giving the absorption vs frequency was that of a relaxation mechanism that could be adjusted to $\alpha = 0.006 f^2 + 0.155 f^2/(f^2 + f_r^2)$, with $f_r = 1.7$ kHz and α in dB/km (Fig. 3.6). Leroy's results were presented in 1964 at the 67th meeting of the Acoustical Society of America and later published [25–27]. His data points were added, with those from Lallement and Waterman, by Thorp to data coming from measurements in the Atlantic and Pacific oceans, but no relaxation process was suggested by this author, and instead the formulation proposed followed a variation in $f^{1.5}$ [28]. It was only 2 years later that Thorp proposed his better-known formula in which the low-frequency absorption mechanism is expressed by $0.1 f^2/(1 + f^2)$ dB/kyd [29].

The exact cause of the mechanism could not be known from Leroy's or Thorp's results, and, even more embarrassing, serious doubts had been set forth by Marsh already in 1963 [30]: He had noticed that the few measurements of absorption performed with pure-tone low-level signals gave values lower than the others, all performed with high-level signals from explosions. As nonlinear effects occur near high-level sources there exists an internal transfer of acoustic energy from the low to the high frequencies in the shock wave sound pulse. If this transfer went on at longer ranges, the observed results could be due to artifacts, not to absorption. This, plus the fact that no physicochemical explanation had yet been given, resulted in many measurements, studies, and suggestions in the 1964–1972 period. They can be regrouped according to their purpose as follows: (1) Could nonlinear effects be at the origin of artifacts? (2) Could the "classical" measurements (see Sect. 3.4.2.) be affected by the oversimplification made in evaluating the spreading losses? (3) Was the absorption excess specific to seawater or could it be also found in freshwater? (4) Did the absorption excess depend on temperature and how? (5) What could be the possible reasons for a new relaxation effect?

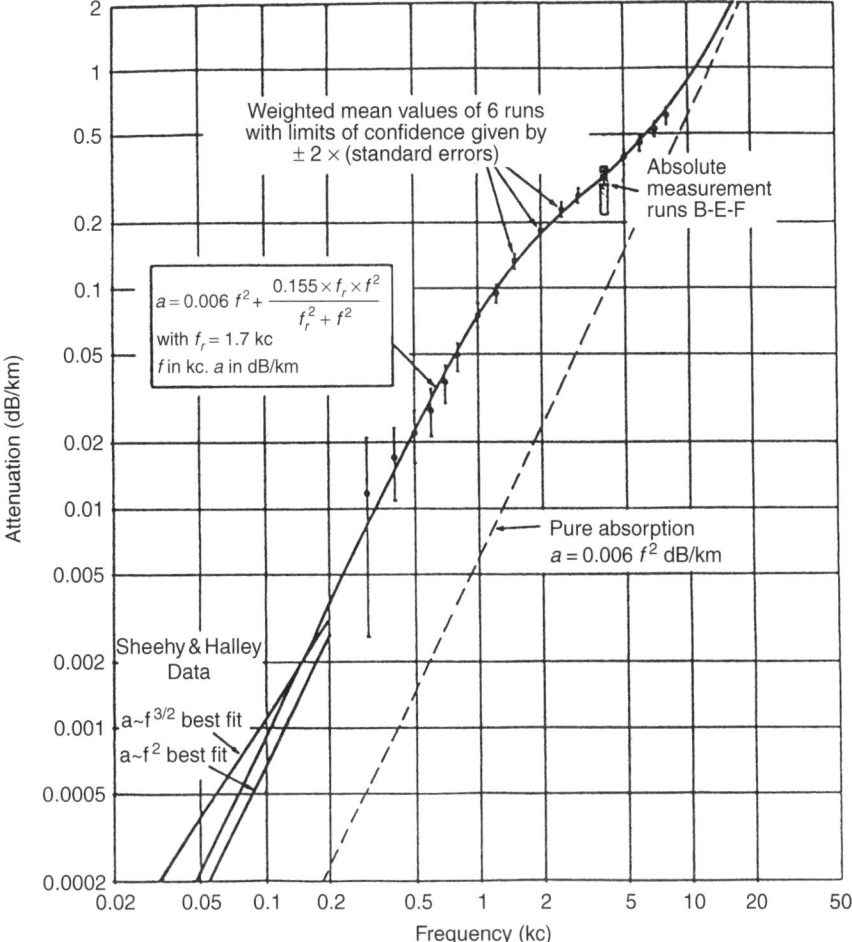

Fig. 3.6. First evidence of a low-frequency relaxation: Leroy's results from 1965 [26].

Question (1) resulted in several mathematicophysical developments [31] that in the end could not prove artifacts, a few more measurements (of limited accuracy) with pure-tone signals [32], and the very complex experiment set up by Leroy and Gerrebout in 1967. This experiment was completed by additional measurements in 1968, and the whole was reported only in 1971 by Skretting and Leroy [33]. It is described in Section 3.4.2. The results proved that there was no nonlinear effect affecting the previous results from Leroy, and they confirmed with accuracy the equation already proposed for the low-frequency relaxation in the Mediterranean.

Question (2) was raised by a number of authors (Leroy; Di Napoli and Powers; Hanna and Rost [27, 34, 35] and is commented on in Section 3.4.2. In all there were only minor and occasional errors due to the simplification.

Question (3) required sound propagation measurement in lakes. The one most adapted was, and remains, Lake Baikal, but the researchers had to be satisfied with Lake Tanganyika and Lake Superior (Browning et al. [36, 37]). The results were not very satisfactory: In the first case the absorption could be interpreted by $\alpha = 0.006 f^2 + 0.015$ (dB/km), and in the second the "attenuation" was found independent of frequency from 700 to 9000 Hz!

The influence of temperature (point [4]) was investigated by measurements in warm waters (Red Sea, Jones et al. [38]; Gulf of Aden, (Browning et al. [39]) and in Polar waters (Browning [40]). Complementary measurements of absorption were also performed in the Pacific (Defense Research Establishment Pacific, Canada, and Naval Undersea System Center [41]). In all, it was found that there was a dependence on temperature and on location, but no clear conclusion could be drawn. These measurements, however, were later of great use to establish the variation of the relaxation frequency with temperature and pH and to establish a general equation (François and Garrison [11].

Question (5), the most essential, was at the origin of several suggestions: Chemical effect of a component of beryllium (Thorp [29]), of aluminum sulfate (Knoche [42]) or others (Garland *et al.* [43]); animal effect: fish swim bladders (Weston [44]), plankton (Duykers [45]), etc., until the true origin of the low-frequency excess absorption was finally discovered in 1972 by Yeager *et al.* at Case Western Reserve University. By applying the T-jump technique (see Sect. 3.2.7) to samples of freshwater and seawater containing the same amount of HCO_3 ions, it was observed that only seawater presented a relaxation frequency in the range 1–2 kHz. As this relaxation was found to depend on the pH of the seawater samples, a systematic study was undertaken on concentrated solutions of all the acids contained in the sea, and one of the least common, the boric acid $B(OH)_3$, was finally ferreted out to be responsible for the absorption. The discovery was announced at the 84th meeting of the Acoustical Society of America and reported in various publications [46, 47].

Only a few measurements have been performed at sea after 1972, and studies were concentrated in laboratory experiments to detect the sensitivity of the mechanism to the environment. A number of sea trials, however, seemed to indicate that there was still another absorption excess below 300 Hz, but no real proof could be given due to lack of accuracy and reliability. In fact, the possible effect of diffraction is generally admitted (see Sect. 3.4.4). As for the laboratory measurements — mainly performed in the United States by Simmons, Simmons and Fisher, and Mellen *et al.* [48–57]; in USSR by Glotov [58]; in France by Mallo, Waton, and Candau [59–61]; and in China by Qiu *et al.* [62–64] — they demonstrated that the phenomena were indeed much more complicated than expected. Of particular interest in that field is the research by Mellen of the intermediate effect of carbonates in the dissociation of $B(OH)_3$, and his discovery (later confirmed by the works of Qiu *et al.*) that still other components, in particular $MgCO_3^-$, were contributing to some degree to the total absorption below 20 kHz. The dependence of absorption on pH and temperature was studied both from laboratory and sea trials, but the dependence of pressure has not yet been satisfactorily determined. Still it remains of interest in the evaluation of sonar detection capabilities of targets in convergence zones. This is discussed in Section 3.5.

3.4.2. Techniques of Measurements at Sea

The most usual method for measuring the absorption of sound at low frequencies in the oceans is very elementary and requires simple equipment. It was first developed for geophysical purposes and is already found described in a 1948 paper by Ewing and Worzel [65]. One or several hydrophones are suspended from a drifting experimental ship, in general in the deep sound channel. A second ship, or an aircraft, drops at increasing ranges explosive charges set to explode in the sound channel. Because the absorption is small at low frequencies, very large ranges must be explored, up to several hundreds of kilometers, and even thousands for the very low frequencies below some 300 Hz. The range is measured at each explosion from the time elapsed between it and its reception on board the experimental ship. The time of the explosion is known on the ship from a radio pulse automatically transmitted when a "device" associated with the launcher detects the shock wave. This device can use listening equipment of the ship, or a suspended hydrophone, or also, when aircraft are used, sonobuoys launched shortly before the explosive charges. Corrections are made to account for the travel time between the explosion and its reception by the device that triggers the radio transmission. The acoustic signals received on board the experimental ship are recorded together with the time signals and the analyses are usually performed later in the laboratory.

The explosive sources provide broadband acoustic signals, and what is received at long ranges is mostly due to the energy trapped in the sound channel. If necessary, a proper time filtering enables to reject the bottom reflected paths that, suffering reflection losses that depend on frequency, would alter the results. The useful portions of the received signals are then analyzed in frequency bands (usually thirds of octaves). The acoustic levels in each band correspond to a noncoherent energetic addition from all the acoustic paths involved in the guided propagation of sound, which avoids the unpredictable artifacts of interferences that are encountered with pure-tone signals. Thus, the variation of energy with range tends to obey that of a cylindrical spreading, to which is added a loss by absorption with a coefficient that depends on the location and the average temperature and hydrostatic pressure at the depth of the channel axis. A variant of this method is found, for example, in an experiment from Urick [66]. The receiving ship is replaced by a buoy equipped with the analysis instrumentation, the measured levels being transformed into coded signals transmitted by radio to the launching aircraft where they are recorded.

Various situations may be encountered, depending on the possible departure from the cylindrical spreading assumption. The simplest case is found when this assumption is acceptable. The energy in the filter centered around frequency f is then given in decibels as a function of range r by

$$E_r(f) = E_o(f) - H(r) - \alpha(f)r \tag{3.35}$$

$$= E_o(f) - H(r_1) - 10\log(r/r_1) - \alpha(f)r \tag{3.36}$$

where E_r and E_o are the received and transmitted energies; $H(r)$, the spreading loss up to the range r; r_1, the range beyond which a cylindrical spreading is valid; and $\alpha(f)$, the absorption coefficient at f. It is not necessary to determine r_1 and $H(r_1)$ by theoretical calculations from the environmental conditions as Eq. 3.36 can be rewritten

$$E_r(f) + 10\log(r) = K(f) - a(f)r \tag{3.37}$$

where $K(f)$ is unknown but does not depend on r. Consequently, a linear regression on the set of data $E_r(f)$ corrected by $+10\log(r)$ gives the desired coefficient $\alpha(f)$.

In some cases, however, it is not possible to assess that a real cylindrical loss is reached, but both the medium and the geometry of the measurements permit an evaluation of $H(r)$ through calculations using ray theory or, rather, mode theory. In still other cases, in particular with very long ranges and a varying medium (noticeable changes in the speed of sound profile), calculations to obtain $H(r)$ are illusory, and the absolute values a_f cannot be reached. If, however, one makes the assumption that $H(r)$ is independent of f, differential absorption coefficients $\Delta\alpha_{ij} = \alpha(f_i) - \alpha(f_j)$ can be calculated from the set of values $E(f_i)$ and $E(f_j)$. To obtain the absolute values of all $\alpha(f)$ necessitates a known value of the absorption coefficient at one of the measured frequencies (usually the highest) or an additional assumption or approximation. It can be that of a negligible value (in fact zero) at the lowest frequency or a given law of variation of absorption with frequency, in the whole range of measurements or in a limited range (a low in f^2 at the lower end for instance).

According to the experiments and their authors, quite different situations can indeed be found, and the accuracy of the results, even the degree of confidence in them, may vary considerably. It has also to be noted that sound absorption measurements in a sound channel suppose that no energy is dissipated by diffraction or lost by leakage effects. This is particularly important at very low frequencies (below some 300 Hz) and is commented on in Section 3.5.

The measurements carried out in sound channels with transducers emitting pure sound tones are much less numerous and instructive than those using explosives because only one frequency is measured at a time and because interference effects result in uncontrollable fluctuations that reduce the degree of confidence in the results.

A completely different approach for measuring low-frequency sound absorption at sea with explosives is to follow the spectral variation of the impulsive signal with range along identified single acoustic paths. The method is limited to rays refracted at depth, and as absorption varies with temperature it would give poor results in most open oceans because rays go deeper with increasing ranges and travel in layers of different temperature. The method is, however, appropriate in closed basins if one uses rays that have traveled in the deep isothermal water, but the maximum exploitable range is limited by the water depth to a few tens of kilometers only, hence the absorption at very low frequencies is too small to be measured with enough accuracy. The first measurements of this type have been performed by Leroy [25–27] from experiments that had not at all been organized for this purpose: the only usable source-receiver configuration corresponded to a common depth of 100 m, not enough for refracted rays to reach less than 10 km by traveling mostly in isothermal water. The received signals that were exploitable were in fact limited to ranges varying from 15 to 35 km on one side, and to recordings around 200 m from the source for calibration purposes on the other. Under such circumstances, the set of data used for the regression analyses giving the differential absorption coefficients (the range measurements were too crude to obtain very accurate values for the spreading losses) presented a large gap between 200 m and 15 km, and the spectrum at 200 m was of great weight in the results. This is why, when Marsh raised the doubt that nonlinear effects could be at the origin of artifacts [30], Leroy realized that if nonlinear effects had really been present at 200 m and had disappeared after 15 km, then his measurements that had revealed a relaxation law could be questionable, and he decided with Gerrebout to embark on a large and quite unique experiment, fortuitously made possible by the presence in the Mediterranean of a large oceanographic buoy anchored by 2500 m of depth. The ideas were as follows: If nonlinear effects occurred at short ranges and later disappeared, the change could be observed by continuously following the frequency spectrum variation with range from 200 m up to 40 km without any gap of signals traveling in perfectly isothermal water. Furthermore, the evolution would be different for the impulse signal due to the shock wave of the explosion and for the bubble pulse of much lower peak value and different spectrum. Finally, to be independent of any assumption such as an adjustment to a given law of variation of the absorption with frequency, results would gain in reliability from accurate calculations of the spreading losses. All this was possible with explosive charges detonated at a depth of 500 m and with a knowledge of all distances within 1 or 2%. This last requirement, hard to obtain at short ranges, was made possible with a calibrated and stable base of two suspended hydrophone arrays. In the experiment, the receiving ship was attached to the anchored buoy by a strong 2000-m-long nylon rope, and pulled on the buoy with its active rudder. After stabilization, the buoy and the ship remained steady for hours, and the hydrophone arrays suspended from each vessel kept vertical. The buoy-suspended array was connected to the receiving ship, where all signals were recorded. Explosive 500-m-deep charges launched from a second ship opening the range in the alignment buoy-receiving vessel provided signals both for the calibration of the base and for the investigation of the 2–40 km range (radio signals were transmitted at the reception on the ship of the shock wave). A small boat maneuvering along the cable between the buoy and the receiving ship also launched charges to cover the 200–2000 m range, accurately known from the arrival times and the base calibration. The experiment was entirely satisfactory, except that the noise generated by the active rudder masked the high-frequency spectrum of distant signals. This is why a complementary trial with just

THE ABSORPTION OF SOUND

two ships (the receiving one simply drifting) was performed to obtain the whole set of spectra. No nonlinear effect was observed from the analysis, and the values found from the cruder experiments of 1963 were confirmed [33]. Some comments, however, need to be made in the discussion in Section 3.5.

We must note before closing this section that measurements at sea along single paths are also of interest when searching for particular effects at the higher frequencies where the MgSO$_4$ relaxation is predominant. The experimental procedure does not need to be sophisticated, and even measurements with transducers can be performed. This has been the case in particular for verifications of the absorption coefficient in polar waters (isothermal), and for the study of the effect of hydrostatic pressure on absorption (Bezdek [19]).

3.4.3. Mechanisms of Sound Absorption in Seawater at Low Frequencies

The T-jump experiments performed by Yeager *et al.* [47] showed that the relaxation-like sound absorption observed at low frequencies from at-sea measurements is due to the chemical equilibrium or borate species. Simmons [48] restarted this work and proposed a two-step model to explain his results:

$$\mathrm{B(OH)_3 + OH^- \underset{k_1'}{\overset{k_1}{\rightleftharpoons}} B(OH)_3OH^- \underset{k_2'}{\overset{k_2}{\rightleftharpoons}} B(OH)_4^-}$$

This two-step reaction can be interpreted by a change of location of the OH$^-$ ion during the reaction. In fact, B(OH)$_3$ is a planar molecule and B(OH)$_4^-$ is a tetragonal one. The first step is then the capture of OH$^-$ by the borate, and may be a diffusion-controlled reaction, while the second step controls the structural change. The relaxation times are worked out using the kinetic equations

$$\frac{d\Delta \mathrm{B(OH)_3}}{dt} = -k_1(\mathrm{B(OH)_3}\Delta \mathrm{OH^-} + \mathrm{OH^-}\Delta \mathrm{B(OH)_3}) + k_1'\Delta \mathrm{B(OH)_3OH^-} \quad (3.38)$$

$$\frac{d\Delta \mathrm{B(OH)_4^-}}{dt} = -k_2'\Delta \mathrm{B(OH)_4^-} + k_2\Delta \mathrm{B(OH)_3OH^-} \quad (3.39)$$

and the conservation equations

$$\Delta \mathrm{B(OH)_3} + \Delta \mathrm{B(OH)_4^-} + \Delta \mathrm{B(OH)_3OH^-} = 0 \quad (3.40)$$

$$\Delta \mathrm{B(OH)_4^-} + \Delta \mathrm{B(OH)_3OH^-} + \Delta \mathrm{OH^-} = 0 \quad (3.41)$$

This system has two relaxation times:

$$\frac{1}{\tau_{1,2}} = \frac{1}{2}\left[\Sigma k \pm \sqrt{(\Sigma k - 4\Pi k)}\right] \quad (3.42)$$

where

$$\Sigma k = k_1(\mathrm{B(OH)_3 + OH^-}) + k_1' \quad (3.43)$$

and

$$\Pi k = k_1(\mathrm{B(OH)_3 + OH^-})(k_2 + k_2') + k_1'k_2' \quad (3.44)$$

The first relaxation time is diffusion controlled and is faster than the change from a planar to a tetragonal symmetry. Under the assumption that the second step is the slower ($k_1, k_1' \gg k_2, k_2'$), the relaxation times satisfy

$$\frac{1}{\tau_1} = k_1 + k_1(\mathrm{B(OH)_3 + OH^-}) + k_1' \quad (3.45)$$

$$\frac{1}{\tau_2} = k'_2 + \frac{k_2 K_1 (B(OH)_3 + OH^-)}{1 + K_1 (B(OH)_3 + OH^-)} \qquad (3.46)$$

where $\quad K_1 = \dfrac{k_1}{k'_1} = \dfrac{B(OH)_3 OH^-}{B(OH)_3 \times OH^-} \qquad (3.47)$

At rest, the concentration of $B(OH)_3 OH^-$ is low, and so is the value of the equilibrium constant K_1. The second time constant, which corresponds to the slow process, varies as $B(OH)_3$ at low borate concentrations and is constant at high borate concentrations.

Simmons has obtained values of the relaxation time that are not the same for NaCl solutions and for seawater. In seawater, NaCl is the main constituent but many other ions can make up ion-pairing with borate. The dissociation of borate is sensitive to the pH, and it is coupling with the carbonate that acts as a buffer. We reported in Table 3.3 the main links between borate and the other species. The kinetic equations set is very complicated and will not be presented here; the interested reader can refer to the works of Eigen or Mellen et al. [67, 68, 4, 51–56].

To explain the shift of the relaxation frequency, two possibilities can be considered, viz: (1) a buffer effect by the ion-pairing and the pH buffer, and (2) a catalysis effect.

(1) The buffer effect

A very simple model is presented. Taking the more simple equation, which fits with the experimental measurements at low borate concentration, we get

$$\frac{d\Delta B(OH)_3}{dt} = -k_1 (B(OH)_3 \Delta OH^- + OH^- \Delta B(OH)_3) + k'_1 \Delta B(OH)_4^- \qquad (3.48)$$

Because the relaxation time of borate is very long for an electrolyte, the other reaction may be assumed to be much faster. This complicated process may be observed at two different time scales. First, one may consider a time scale much longer than the relaxation time of the ion-pairing but shorter than the relaxation time of $B(OH)_3$. At this time scale, the value of $\Delta B(OH)_3$ is not zero but the fast processes of ion-pairing are at equilibrium for this instantaneous value of $\Delta B(OH)_3$. As when the value of $\Delta B(OH)_3$ decreases, a part of the $B(OH)_4^-$ ions are absorbed by ion-pairing, and the relation

$$\Delta B(OH)_3 = -\Delta B(OH)_4^-$$

is not valid. The concentration conservation equation may be written as

$$\Delta B(OH)_3 = -\Delta B(OH)_4^- (1 + a) \qquad (3.49)$$

where a is the ratio of $B(OH)_4^-$ ions that are absorbed by ion-pairing.

Table 3.3. Schema of reaction for boron.

$$B(OH)_3 + OH^- \rightleftharpoons B(OH)_4^- \Rightarrow \text{ion pairing with } \begin{cases} Na^+ \\ Mg^{2+} \\ Ca^{2+} \end{cases}$$

$$\text{pH process} \begin{cases} \updownarrow \\ OH^- + H^+ \rightleftharpoons H_2O \\ \updownarrow \\ CO_3^{2-} + H^+ \rightleftharpoons HCO_3^- \\ \updownarrow \\ \text{ion-pairing with} \\ \overbrace{Na^+ Mg^{2+} Ca^{2+}} \end{cases}$$

With the same reasoning for the acid-base equilibrium we obtain

$$\Delta B(OH)_3 = \Delta(OH)^-(1+b) \tag{3.50}$$

where b is the ratio of $(OH)^-$ ions that are absorbed by the other acid-base reaction.

In the second time scale (longer than the first), one can observe the relaxation process of $B(OH)_3$. The kinetic equation in this time scale is now:

$$\frac{d\Delta B(OH)_3}{dt} = -\left[k_1\left(\frac{B(OH)_3}{1+b} + OH^-\right) + \frac{k_1'}{1+a}\right]\Delta B(OH)_3 \tag{3.51}$$

Thus, the resulting effect of ion-pairing or of the acid-base reaction is to decrease the relaxation frequency of boric acid. The measurements show an opposite effect: the relaxation frequencies are higher in seawater than in NaCl solutions.

(2) The catalysis effect

This effect can be explained by a catalysis reaction between boric acid and carbonic acid.

$$B(OH)_3 + CO_3^{2-} + H_2O \underset{k_{BC}'}{\overset{k_{BC}}{\rightleftharpoons}} B(OH)_4^- + HCO_3^- \tag{3.52}$$

This reaction increases the probability that $B(OH)_3$ is transformed into $B(OH)_4^-$ because a collision with another species than OH^- is also effective.

The observed increase is then the result of two competitive processes: the buffer effect, which leads to a decrease of the relaxation frequency, and the catalysis effect, which leads to its increase. For carbonic acid, the main effect is the latter. Laboratory measurements, however, have shown that the two processes modify the relaxation time of boric acid if phosphoric acid, ammoniac, or silicic acid are added to a borate solution. To summarize, the main processes that control the relaxation frequency of the boric acid in seawater are

$$\text{direct process: } B(OH)_3 + OH^- \underset{k_1'}{\overset{k_1}{\rightleftharpoons}} B(OH)_4^-$$

$$\text{pH buffer process } \begin{cases} CO_3^{2-} + H^+ \underset{k_2'}{\overset{k_2}{\rightleftharpoons}} HCO_3^- \\ OH^- + H^+ \underset{k_w'}{\overset{k_w}{\rightleftharpoons}} H_2O \end{cases} \tag{3.53}$$

$$\text{catalysis process: } B(OH)_3 + CO_3^{2-} + H_2O \underset{k_{BC}'}{\overset{k_{BC}}{\rightleftharpoons}} B(OH)_4^- + HCO_3^-$$

The relaxation time with a fast pH buffer process is

$$\frac{1}{\tau} = k_1\left(\frac{B(OH)_3}{1+a} + OH^-\right) + k_1' + k_{BC}\left(\frac{a}{1+a}B(OH)_3 + CO^{2-}\right)$$

$$+ k_{BC}'\left(\frac{a}{1+a}B(OH)_4^- + HCO_3^-\right) \tag{3.54}$$

where

$$a = \frac{K\,HCO_3^-}{1 + K\,OH^-} \quad \text{with} \quad K = \frac{k_1 k_{BC}'}{k_1' k_{BC}}$$

If the relaxation time can be measured by T-jump experiments for slow relaxations, the amplitude of the sound absorption coefficient is very difficult to determine in laboratories because direct measurements are not possible: To estimate the value of the change of volume ΔV corresponding to the reaction, the ion-pairing and pH buffer processes cannot be neglected as this was the case for the relaxation time. Indeed, when $B(OH)_3$ ions disappear, many species of ions with a bore atom appear, and the ΔV induced is the result of several processes. This is why the value of the absorption coefficient has to be obtained from experiments made in oceans and seas with different temperatures and pH values.

Measurements by the resonator method solutions with higher concentration than in natural seawater led in 1979 to the discovery by Mellen et al. [52] (and subsequent) of a relaxation process in the 5–20 kHz frequency range due to magnesium carbonate. It was followed by the discovery of two other, less important relaxations at higher frequencies, due to ions $MgB(OH)_4^-$ and $MgHCO_3^+$. The whole was later studied by Qiu et al. with solutions at normal concentrations [62–64]. The corresponding mechanisms are not described here.

3.4.4. Results and Equations

The absorption of sound at low frequency in seawater governed by the boric acid relaxation obeys the classical law in $\alpha = Ag(P)f_r f^2/(f_r^2 + f^2)$. The variations of A and f_r with the parameters describing the environment are, however, quite different from those encountered with the $MgSO_4$ relaxation. Even more, another parameter is found, viz. the pH, a quantitative measurement of acidity. On the other hand, no noticeable effect of hydrostatic pressure has been found so far, and the general consensus is to adopt $g(P) = 1$.

As explained previously, the laboratory measurements cannot provide accurate values of the absorption coefficient in real seawater at low frequencies, but only general trends. Instead, they permit investigation of the very complex phenomena involved and provide some knowledge about the variation of the relaxation frequency with the parameters. It is finally from the measurements at sea that the most accurate values for A and f_r can be obtained, but in that respect the quality and even the validity of the measurements need to be examined seriously. Many more equations for evaluating the absorption coefficient at low frequencies have been proposed than for the higher ones, but in general their validity is confined to limited environments. The reason is simply that most experimenters have proposed a simple equation to fit the results they obtained from one particular trial or set of trials in one ocean area or one particular sea. We shall not list all those equations of limited values but instead concentrate on the results of those researchers who proposed general equations for worldwide applications, taking into account the pertinent parameters of the medium.

A special mention, however, must be made of the work by Qiu et al. [62–64]: With a large cylindrical resonator, filled at appropriate water heights with various solutions of the same concentrations as the seawater, they succeeded to perform decay time measurements at several modal frequencies lower than 10 kHz, from which they determined the relaxation frequency and the absorption coefficient for several mechanisms (see Sect. 3.4.3 and the following paragraphs).

The first publication reviewing the variation of the relaxation frequency for the absorption due to $B(OH)_3$ and the amplitude of this absorption is that of Mellen and Browning [69]. From the laboratory measurements of the relaxation frequency f_r they conclude that, for the natural seawater conditions, f_r does not depend on the pH but varies with temperature in an exponential manner. They mention as a crude approximation the very simple formulation $f_r = 10^{(T-4)/100}$, with T in °C and f_r in kHz.

From the values given by at-sea experiments, they show that, as expected from theory, the maximum $(\alpha\lambda)_r$ of the absorption per wavelength varies with the pH according to an exponential law, which is illustrated by a graph, but no equation is proposed in their publication. (Mellen later proposed [70] the value $A = 0.1*10^{(pH-8)}$).

Soon after, Fisher and Simmons [10] published results that were derived from the work of Simmons's thesis, viz:

(1) f_r depends not on pH but only on temperature:

$$f_r = 1315\theta \, e^{(-1700/\theta)} \tag{3.55}$$

where θ is the temperature in °K and f_r is in Hz

(2) for the constant values of salinity and pH $S = 35$ o/oo and pH $= 8$, A varies with temperature as

$$A = 1.03 \times 10^{-8} + 2.36 \times 10^{-10} T - 5.22 \times 10^{-12} T^2 \tag{3.56}$$

One step further is accomplished by Schulkin and Marsh [71], who propose from measurements at sea:

$$f_r = 6.1(S/35)^{\frac{1}{2}} \times 10^{(3-1051/\theta)} \text{ (kHz)} \tag{3.57}$$

$$A = 2(\alpha\lambda)_r/c \tag{3.58}$$

where

$$(\alpha\lambda)_r 10^5 = 3.1 \times 10^{(0.69pH-6)} \tag{3.59}$$

This particular variation of f_r involving the square root of salinity is in accordance with the results of Glotov [58] and those of Simmons [48].

Both the previous equations were derived respectively from linear best fits on f_r and $(\alpha\lambda)_r$ logarithmic values vs $1/\theta$ and pH in linear values. The sets of f_r and $(\alpha\lambda)_r$ values were directly taken from the authors of the equations proposed for the following oceans, seas, and areas: Atlantic, Pacific, Baffin Bay, Gulf of Aden, Mediterranean, and Red Sea.

A different and deeper approach to the problem of establishing a universal equation for the low-frequency absorption coefficient is found in the work of François and Garrison [72]. In a first step they carefully analyze all publications and retain only five at-sea measurements as being confident and rich enough in frequency data to provide good values for the B(OH)$_3$ relaxation frequency, using the $\alpha\lambda$ vs T curves. These are from the Pacific (presented by Mellen and Browning [73]), the Atlantic (selected values among Thorp's [28]), the Mediterranean (Skretting and Leroy [33]), the Red Sea and the Gulf of Aden (presented by Jones *et al.* and Browning *et al.* [38, 39]). Then, instead of using the values taken from the equations proposed by the experimenters, François and Garrison refer back to the data points that have been employed, i.e., the direct results of the measurements (either given by the authors themselves or read from graphs). Finally, as the original data of absorption values represent the sum of the absorptions due to the various effects, they subtract what was due to MgSO$_4$ in the specific environment of each measurement to get only that part of absorption that is related to B(OH)$_3$. The graphs representing $\alpha\lambda$ vs frequency provide a set of five relaxation frequencies that are more accurate than those given by the experimenters. These values of f_r are plotted on a logarithmic scale vs the sea temperature mostly encountered in the sound paths of the measurements, presented on a linear scale. Following — like Schulkin and Marsh — the assumption of proportionality of f_r to the square root of salinity, François and Garrison finally arrive at f_r in kHz in the expression:

$$f_r = 2.8(S/35)^{\frac{1}{2}} \times 10^{(4-1245/\theta)} \tag{3.60}$$

The good agreement of this formulation with the other results is confirmed by a set of graphs drawn with the data from the measurements that had not been taken into account when selecting the five basic experiments.

The coefficient A of the absorption formula is obtained from plots of the five values of $(\alpha\lambda)_r$ in log scale vs pH in linear scale. This leads, for A in dB/km to

$$A = 8.86/c \times 10^{(0.78pH-5)} \qquad (3.61)$$

A set of simplified absorption formulas was later proposed by Browning and Mellen [74]. For the boron contribution they gave

$$f_r = 0.9 \times 10^{t/70} \qquad (3.62)$$

$$A = 0.1 \times 10^{(pH-8)} \qquad (3.63)$$

And they added a term to account for the magnesium carbonate relaxation, which follows the same dependence on pH as boron, contrarily to the that of magnesium sulfate. This reads

$$f_r(\text{MgCO}_3) = 4.5 \times 10^{(t-30)} \qquad (3.64)$$

$$A(\text{MgCO}_3) = 0.03 \times 10^{(pH-8)} \qquad (3.65)$$

In the same paper, separate specific simple formulas were proposed for the Atlantic, Pacific, Sub-Arctic, Mediterranean, and Red Sea.

From their measurements at 12.4 °C, Qiu et al. proposed [63] still another formulation for the overall contribution to absorption of the four relaxational effects of $B(OH)_3$, $MgCO_3$, $MgB(OH)_4^-$, and $MgHCO_3^+$. They are not reproduced here, being only valid at a single temperature.

3.5. DISCUSSION AND CONCLUSION

The absorption of sound in pure water is greater than predicted by the classical Stokes-Kirchhoff theory by a factor of about 3, and the explanation proposed by Hall in 1948 (a structural rearrangement of the configuring of molecules) has been accepted. The frequency fr_o of the corresponding relaxation process is still unknown but far outside any range of application. From a purely scientific point of view it is obvious that we have here the subject of challenging research. From a practical point of view, however, this frequency does not need to be known since all frequencies of interest satisfy $f \ll fr_o$, which simplifies the equation for the absorption coefficient according to

$$\alpha_o(f) = \alpha_o fr_o f^2 = A_o f^2 \qquad (3.66)$$

Early laboratory measurements provided values of A_o as a function of temperature and pressure that were in good agreement with theoretical expectations. The most accurate equation for the calculation of A_o is that of François and Garrison (F-G for simplification) provided in Section 3.3.2. and reproduced in Table 3.4. The proposed polynomial for a sufficient temperature range of application is, however, divided into two parts that do not ensure the continuity of the second derivative, and a revised formulation could be proposed.

The absorption of sound in seawater is the result of that in pure water and of a number of relaxation absorptions due to ionic dissociation and reassociations of various dissolved components. Some of these phenomena are extremely complex and involve the contribution of other components, but in the end they can be identified as separate and independent effects. As a result, the coefficient for the total absorption of sound in seawater, expressed in logarithmic units (Np/m or dB/km), takes an additive form as $\alpha = \alpha_o + \Sigma\alpha_i$, where α_i stands for the contribution of chemical component number i. Fisher and Simmons, who developed the first "universal" equation, adopted the reverse order: their three-term equation reads $\alpha = \alpha_1 + \alpha_2 + \alpha_3$, where α_3 stands

Table 3.4. Total equation for sound absorption in seawater.

$\alpha = A_o f^2 + \Sigma A_i f_{ri} f^2 / (f_{ri}^2 + f^2)$ with

α in dB/km, f in kHz, T in °C, θ in °K($= T + 273.1$), S in o/oo,

P in MPa, (mega-Pascal), C (sound speed) in m/s

$A_o = f(T) \times (1 - 3.79 \times 10^{-3} P + 7.38 \times 10^{-6} P^2)$ with

at $T < 20\,°C$ $f(T) = 4.937 \times 10^{-4} - 2.59 \times 10^{-5} T + 9.11 \times 10^{-7} T^2 - 1.5 \times 10^{-8} T^3$

at $T \geq 20\,°C$ $f(T) = 3.964 \times 10^{-4} - 1.146 \times 10^{-5} T + 1.45 \times 10^{-7} T^{-2} - 6.5 \times 10^{-10} T^3$

$A_1 = \dfrac{21.44}{C}(1 + 0.025T)(1 - 8 \times 10^{-2} P)$

$f_{r1} = \dfrac{8.17 \times 10^{(8-1990/\theta)}}{1 + 0.0018(S - 35)}$

$A_2 = \dfrac{8.86}{C} 10^{(0.78\,pH - 5)}$

$f_{r2} = 2.8(S/35)^{1/2} 10^{(4 - 1245/\theta)}$

$A_3 = 0.03 \times 10^{(pH - 8)}$

$f_{r3} = 4.5 \times 10^{(T - 30)}$

for the absorption due to pure water. For α_1 and α_2, they give the contribution of $B(OH)_3$ and that of $MgSO_4$, respectively. This choice was guided by the fact that the absorption by boron is predominant at usual modern sonar frequencies, that from magnesium sulfate being lower and that from pure water almost negligible. A similar formulation was adopted by most authors of equations—François and Garrison in particular—but not all of them, especially when other relaxation effects were taken into account. In what follows we have preferred to adopt a somewhat logical and historical order: The absorption due to water itself is unavoidable and will be indicated first by the term α_o. The term α_1 will designate the absorption by $MgSO_4$, α_2 that by $B(OH)_3$, and α_3 that by $MgCO_3$. Under these conditions, any further contribution would have a subsequent number, which could be of interest if an additional term is needed to account for the phenomena at very low frequencies.

The contribution of magnesium sulfate to absorption can be considered sufficiently known: The laboratory measurements are not too difficult, and the influence of temperature, salinity, and pressure has been studied quantitatively, some measurements at sea guiding or complementing the research. Of all equations proposed, we shall recommend that of F-G [11], with the exception of their pressure correction: For an easier utilization in underwater acoustics, the equation uses depth instead of pressure, but this is no more convenient for more general physical applications. Besides, their pressure correction was based on the measurements at sea by Bezdek, and it was seen in Section 3.3.3.1 that it was better to adopt different coefficients. The modified F-G equation for a_1 is given in Table 3.4.

Below 20 kHz, various relaxation processes have been observed, but only two of them bring a noticeable contribution to absorption, viz. those of $B(OH)_3$ and $MgCO_3$. The former, very complex, is by far the more important, and its effect is predominant over that of $MgSO_4$ below some 8 kHz. On the contrary, the $MgCO_3$ relaxation dissociation can only be of some practical importance in the 10 kHz region at the highest pH values (above ≈ 8.1). As explained previously, the main corpus of data from which accurate enough equations can be obtained is the set of at-sea measurements. Under these conditions, the respective contributions of boron and magnesium carbonate cannot be separated, and the measurements made at sea give the sum of these two effects. However, from the existing laboratory measurements (Mellen *et al.*, Qiu *et al.*) it can be deduced that the Mediterranean Sea is the only exploited part of the world's waters where $MgCO_3$ has a nonnegligible effect. This is probably the main reason why a high relaxation frequency is observed [72, 74], too high if one considers it as being that of $B(OH)_3$ alone. The weight of the Mediterranean measurements when making regression analyses to obtain worldwide equations for calculating the relaxation frequency and the coefficient of absorption, as did all authors, cannot be so important as to justify a revision of those equations. Instead, the availability of more accurate measurements at sea in waters of the world with different characteristics would be of great interest.

With the exception of those made by Leroy and by Skretting and Leroy, all measurements of absorption at sea were performed using the sound channel technique, mainly with explosive sources. One consequence is that the pressure dependence of absorption cannot be studied: the average pressure is that at the channel axis, which is constant in one area, and the measurements from various areas involve different values of temperature or of pH or salinity. In fact, the acoustic energy that is measured in this technique travels via numerous paths, some of which reach depths that are far from the axis depth, which would tend to alter the results. It could be possible to restrain the analyses to the near-axis rays by a proper time filtering (those rays are the last to reach the receiver), but this improvement does not seem to have been exploited. As a general rule, one could also deplore the lack of homogeneity and systematization in the experiments. Great differences are found in their extent, number of trials and data points, number of frequencies in the analysis, accuracy or even care, and above all proper documentation. It is perfectly understandable that many early measurements have been made without coordination in the fever of research, but it can be regretted that, once the origin of the low-frequency absorption had been discovered, with a dependence on temperature and on pH, no systematic campaign to obtain more accurate values in areas of well-known different hydrologic characteristics had been undertaken.

It is also puzzling that the possibility of making accurate measurements from the analysis of signals propagated along single paths in closed basins — as Leroy did — was not apparently exploited anymore. Closed basins are not that numerous, but they have the advantage of offering a variety of different waters, e.g., the Black Sea, the Red Sea, the Sea of Japan, that of Sulu, or the Arctic Ocean. Maybe the great complexity of the experiment reported by Skretting and Leroy discouraged the researchers, but in the end this complexity was only chosen at the time to make sure once and for all that the suspected nonlinear effects were not altering the results. Measurements with just two ships, or a ship plus an aircraft or even an aircraft with buoys, as in the case of sound channel experiments, are perfectly sufficient; even more, the maximum distance to cover at sea is limited as single path cannot exceed the "limiting ray" that grazes the sea bottom.

The single path technique, however, exhibits an error that seems to have been ignored so far: with increasing ranges of reception the acoustic paths used in the analysis travel deeper and deeper. At frequency f, the energy $E_f(r)$ received at range r is not given by $E_f(r) = E_o - H(r) - a_f r$ but by the integral form $E_0 - H(r) - \int_0^L a_f(z)dl$, where $a(z)$ is the absorption at depth z and L the curvilinear length of the path. In the case of a constant speed of sound gradient as encountered in all closed basins, and of a linear variation of absorption with pressure (i.e., with depth at first approximation) it is easy to demonstrate [75] that $\int_0^L a_f(z)dl \approx \overline{a_f r}$, where a_f is the mean absorption coefficient that would be found at two thirds of the maximum depth z_M reached by the acoustic ray, this depth increasing as the square root of the range. Under such effects the linear regressions leading to absorption coefficient values are slightly biased if absorption does vary with pressure. This could explain the "negative" values mentioned in [33], and corrections could be made under the assumption of a given law of variation of absorption with pressure.

The absence of data about this pressure effect constitutes perhaps the most important lack in our knowledge of the absorption of sound at low frequencies in seawater: if there is an effect with an order of magnitude similar to that observed with $MgSO_4$

solutions, a difference of as much as 10 dB would occur on the echo level from a target detected in the first convergence zone in the Atlantic (at ≈60 km), which is not at all negligible [75].

On the other hand, absorption could increase with pressure: A theoretical estimation of the pressure effect was carried out by Qiu [76], based on laboratotry data about molecular volume changes in the dissociation steps of boric and carbonic acids. His results show an increase of $(a)_r$, noticeable only beyond 20 kg/cm^2.

Specific experiments to elucidate this question are highly advisable, and two main orientations are possible: (1) laboratory measurements like those of Qiu *et al.* with a resonator under pressure and (2) simple paths measurements at specific ranges with varying source/receiver depths, the best conditions for such experiments being found in the little Sulu Sea (5600 m deep).

In the absence of measurements about the effect of pressure on the absorption due to boron, no pressure-dependent correction will be proposed. The most reliable equation to adopt is again that proposed by François and Garrison, reexposed in Table 3.4 for the a_2 term. One must point out that an alternative simplified equation giving approximately the same results in the limit of the overall accuracy was proposed in 1998 by Ainslie and McColm [77]. This simplification, however, does not seem very helpful because the required parameters are the same and because all modern, small, hand-held calculators can easily handle the complete F-G equation.

The fact that the F-G equation for the boron contribution requires (as all others) introduction of the pH of water in the area of application has resulted in a number of studies that will not be detailed here. Of special interest is the work of Lovett, who undertook surveys of the various pH values found in world waters [78–80].

It seems advisable to add the contribution of magnesium carbonate to have a complete universal equation for the absorption of sound. As for boron, the effect of pressure is still unknown. The additional term $\alpha_3(f)$ proposed is that given by Browning and Mellen [74], reproduced in Table 3.4. As explained previously, this last term will slightly modify the result only in the case of the Mediterranean Sea.

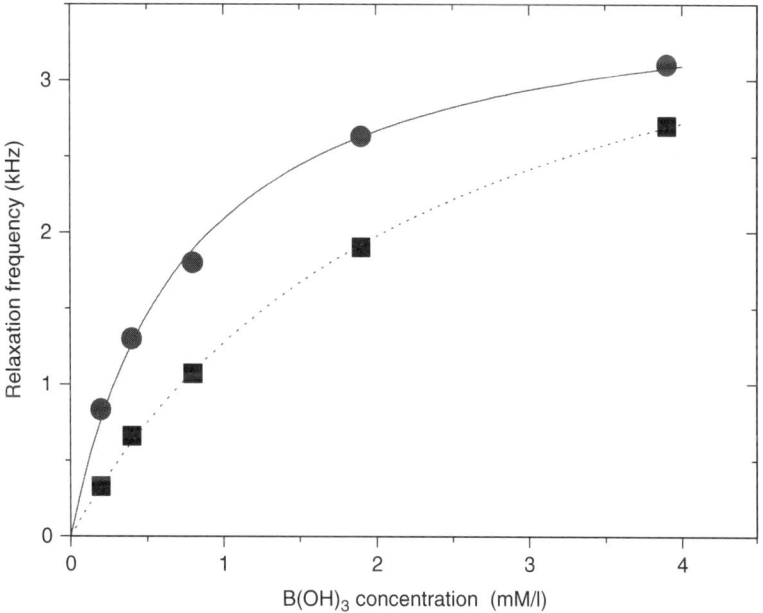

Fig. 3.7. Variation with borate concentration of the relaxation frequency. Circles indicate in synthetic seawater; squares, in 0.7 M/l NaCl (after Simmons [48]).

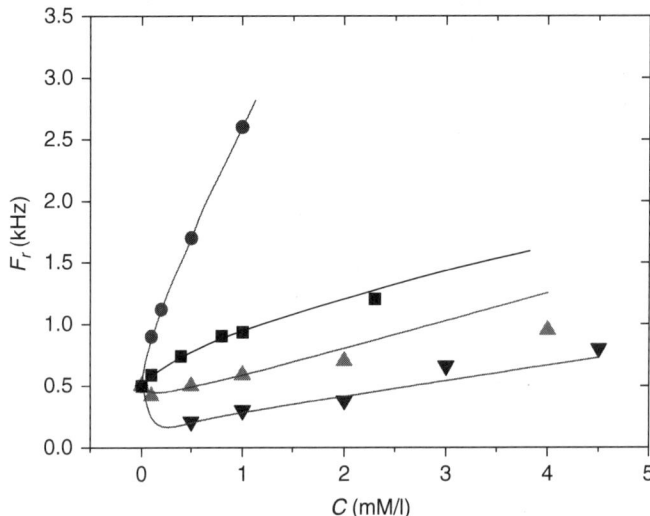

Fig. 3.8. Variation of the relaxation frequency with the concentration. Squares indicate of H_2CO_3; circles, of H_3PO_4; triangles, of $Si(OH)_4$; and inverted triangles, of NH_3 (after Mallo [59]).

Finally, as was pointed out in Section 3.4, there still exists another "excess" of sound attenuation at very low frequencies (below 500 Hz). It has been observed by very long-range at-sea measurements in sound channels, starting with experiments in the Pacific. This excess does not seem to follow a relaxation behavior with frequency. It is highly dependent on the location and is more likely to be due to scattering, if not to artifacts due to acoustic energy leakage outside the sound channels. It is not taken into account in this presentation, and the reader interested in the subject may find information and references in the article by Kibblewhite and Hampton [81].

References

1. Urick, R.J. (1975). *Principles of underwater sound*, New York: McGraw-Hill Co.
2. Markham, J.J., Beyer, R.T., and Lindsay, R.B. (1951). Absorption of sound in fluids. *Rev. Mod. Phys.* **21**: 353–411.
3. Kurtze, G., and Tamm, K. (1953). Measurements of sound absorption in water and in aqueous solutions of electrolytes. *Acustica* **3**: 33–48.
4. Mellen, R.H. (1983). Chemical sound absorption in sea water: Low-frequency relaxation mechanism. *Mar. Inc. Tech. Rep.* 335N.
5. Hall, L. (1948). The origin of ultrasonic absorption in water. *Phys. Res.* **73**: 775.
6. Lieberman, L.N. (1948). The origin of sound absorption in water and sea water. *J. Acoust. Soc. Am.* **20**: 868–873.
7. Leonard, R.W., Combs, P.C., and Skidmore, L.R. (1949). Attenuation of sound in synthetic sea water. *J. Acoust. Soc. Am.* **21**: 63.
8. Pinkerton, J.M.M. (1947). A pulse method for the measurement of ultrasonic absorption in liquids: results for water. *Nature* **160**: 128.
9. Litovitz, T.A., and Carnevale, E.H. (1955). Effect of pressure on sound propagation in water. *J. Appl. Phys.* **26**: 816–820.
10. Fisher, F.H., and Simmons, V.P. (1977). Sound absorption in sea water. *J. Acoust. Soc. Am.* **62**: 558–564.
11. François, R.E., and Garrison, G.R. (1982). Sound absorption based on ocean measurements: Part I: Pure water and magnesium sulfate contributions. *J. Acoust. Soc. Am.* **72**: 896–907.
12. Tamm, K., Kurtze, G., and Kaiser, R. (1954). Measurements of sound absorption in aqueous solutions of electrolytes. *Acustica* **4**: 380–386.
13. Wilson, O.B., and Leonard, R.W. (1954). Measurements of sound absorption in aqueous salt solutions by a resonator method. *J. Acoust. Soc. Am.* **26**: 223–226.
14. Fisher, R.H. (1958). Effect of high pressure on sound absorption and chemical equilibrium. *J. Acoust. Soc. Am.* **30**: 442–448.

15. Glotov, V.P. (1964). Calculation of the relaxation time from the degree of dissociation of magnesium sulfate in fresh and sea water as a function of temperature. *Sov. Phys. Acoust.* **10**: 33–38.
16. Garrison, G.R., François, R.E., and Pence, E.A. (1975). Sound absorption measurements at 10–60 kHz in near-freezing sea water. *J. Acoust. Soc. Am.* **58**: 608–619.
17. Garrison, G.R., Early, E.W., and Wen, T. (1976). Additional sound absorption measurements in near-freezing sea water. *J. Acoust. Soc. Am.* **59**: 1278–1283.
18. Schulkin, M., and Marsh, H.W. (1962). Sound absorption in sea water. *J. Acoust. Soc. Am.* **34**: 864–865.
19. Bezdek, F.H. (1973). Presssure dependence of the relaxation frequency associated with $MgSO_4$ in the ocean. *J. Acoust. Soc. Am.* **54**: 1062–1065.
20. Hsu, C.C., and Fisher, F.H. (1983). Effect of pressure on sound absorption in synthetic seawater and in aqueous solutions of $MgSO_4$. *J. Acoust. Soc. Am.* **74**: 564–569.
21. Eckart, C., ed. (1969). *Physics of sound in the sea*, NDRC Technical Reports.
22. Sheehy, M.J., and Halley, R. (1957). Measurement of the attenuation of low-frequency underwater sound. *J. Acout. Soc. Am.* **29**: 464–469.
23. Schulkin, M. (1963). Eddy viscosity as a possible acoustic absorption mechanism in the ocean. *J. Acoust. Soc. Am.* **35**: 253–254.
24. Lallement, B., and Waterman, H. (1963). *A bottom study with explosive sources and 7-kc pulses*, SACLANT ASW RC.
25. Leroy, C.C. (1964). *Sound attenuation between 200 and 10,000 cps measured along single paths*, 67th ASA Meeting, Paper E3.
26. Leroy, C.C. (1965). *Attenuation of low-frequency sound in fresh water*, SACLANT ASW RC Tech. Report 43.
27. Leroy, C.C. (1967). Sound propagation in the mediterranean sea, In *Underwater Acoustics*, vol. 2, Abers, V.M. ed., pp. 228–235, New York: Plenum Press.
28. Thorp, W.H. (1965). Deep-ocean sound attenuation in the sub- and low- kilocycle-per-second region. *J. Acoust. Soc. Am.* **38**: 648–654.
29. Thorp, W.H. (1967). Analytic description of the low-frequency attenuation coefficient. *J. Acoust. Soc. Am.* **42**: 270.
30. Marsh, H.W. (1963). Attenuation of explosive sounds in sea water. *J. Acoust. Soc. Am.* **35**: 1837.
31. Marsh, H.W., Mellen, R.H., and Konrad, W.L. (1965). Anomalous absorption of pressure waves from explosions in sea water. *J. Acoust. Soc. Am.* **38**: 326–338.
32. Webb, D.C., and Tucker, M.J. (1970). Transmission characteristics of the SOFAR channel. *J. Acoust. Soc. Am.* **48**: 767–769.
33. Skretting, A., and Leroy, C.C. (1971). Sound attenuation between 200 Hz and 10 kHz. *J. Acoust. Soc. Am.* **49**: 276–282.
34. Di Napoli, F.F., and Powers, M.R. (1971). *Fast Field Program (F.F.P.) and attenuation loss in Hudson Bay*, 82nd ASA Meeting, Paper NG.
35. Hanna, J.S., and Rost, P.V. (1977). The influence of range-dependent environments on low-frequency volume attenuation measurements in the sea. *J. Acoust. Soc. Am.* **61**: 369–374.
36. Mellen, R.H., Browning, D.G., and Simmons, V.P. (1979). Acoustic attenuation in Lake Tanganyika. *Nature* **277**: 374–375.
37. Browning, D.G., Jones, E.N., Mellen, R.H., and Thorp, W.H. (1968). Attenuation of low-frequency sound in fresh water. *Science* **162**: 1120–1121.
38. Jones, E.N., Browning, D.G., Thorp, W.H., and Mellen, R.H. (1971). *Sound attenuation in the red Sea*, NUSC Tech. Rep. 4101.
39. Browning, D.G., Jones, E.N., and Thorp, W.H. (1973). *Low frequency sound attenuation in the Gulf of Aden*, NUSC Rep. 4501.
40. Browning, D.G. (1971). *Profect CANUS: Sound propagation measurements in the hudson bay*, NUSC Rep. 4221.
41. Browning, D.G., and Thorp, W.H. (1974). *Attenuation of low-frequency sound in the ocean*, NUSL Research Program 1957–1972, NUSC Report 4581.
42. Knoche, W. (1972). *Durch Aluminium sulfa-complex shallabsorption in seewasser*, 71th Congress of Physical Chemistry, Hambourg.
43. Garland, F., Patel, R.C., and Atkinson, G. (1973). Simulation of sound absorption spectra of seawater systems. *J. Acoust. Soc. Am.* **54**: 996–1003.
44. Weston, D.E. (1966). Fish as a possible cause of low-frequency acoustic attenuation in deep water. *J. Acoust. Soc. Am.* **40**: 1558.
45. Duykers, R.L.B. (1971). *Low frequency sound attenuation in open waters*, 81st ASA Meeting, Paper Y2.
46. Fisher, F.H., Yeager, E., Bressel, R., and Miceli, J. (1972). *Origin of the relaxational sound absorption in sea water below 1 kHz*, 84th ASA Meeting, Paper MM8.
47. Yeager, E., Fisher, F.H., Miceli, J., and Bressel, R. (1973). Origin of the low-frequency sound absorption in sea water. *J. Acoust. Soc. Am.* **53**: 1705–1707.

48. Simmons, V.P. (1975). *Investigation of the 1-kHz Sound Absorption in Sea Water*. PhD Thesis, San Diego: University of California.
49. Simmons, V.P., and Fisher, F.H. *Sound absorption at low frequency in sea water and aqueous solutions of boric acid*, (unpublished).
50. Fisher, F.H. (1979). Sound absorption in sea water by a third chemical relaxation. *J. Acoust. Soc. Am.* **65**: 1327–1329.
51. Mellen, R.H., Browning, D.G., and Simmons, V.P. (1979). Sound absorption in sea water: A third chemical relaxation. *J. Acoust. Soc. Am.* **65**: 923–925.
52. Mellen, R.H., Browning, D.G., and Simmons, V.P. (1979). Acoustical absorption by $MgCO_3^0$ ion-pair relaxation. *Nature* **279**: 705–706.
53. Mellen, R.H., Simmons, V.P., and Browning, D.G. (1980). Low-frequency sound absorption in sea water: A borate-complex relaxation. *J. Acoust. Soc. Am.* **67**: 341–342.
54. Mellen, R.H., Browning, D.G., and Simmons, V.P. (1980). Investigation of chemical sound absorption in seawater by the resonator method: Part I. *J. Acoust. Soc. Am.* **68**: 248–267.
55. Mellen, R.H., Browning, D.G., and Simmons, V.P. (1981). Investigation of chemical sound absorption in seawater: Part II. *J. Acoust. Soc. Am.* **69**: 1660–1662.
56. Mellen, R.H., Browning, D.G., and Simmons, V.P. (1981). Investigation of chemical sound absorption in seawater: Part III. *J. Acoust. Soc. Am.* **70**: 143–148.
57. Mellen, R.H., Browning, D.G., and Simmons, V.P. (1983). Investigation of chemical sound absorption in seawater: Part IV. *J. Acoust. Soc. Am.* **74**: 987–993.
58. Glotov, V.P. (1976). Influence of minute quantities of beryllium and boron on the bulk viscosity of water and on the low frequency attenuation of sound in the ocean. *Sov. Phys. Acoust.* **22**: 67–68.
59. Mallo, P. (1984). *Etude des mécanismes de relaxation chimique dans l'eau de mer en relation avec la propagation des ondes acoustiques de trés basse fréquence dans les océans* (Study of chemical relaxation mechanisms related to low fequency sound absorption in the oceans). PhD Thesis, Strasbourg. France: University of Strasbourg.
60. Waton, G., and Candau, S. (1983). *Revue sur les mesures d'absorption acoustique d'ondes sonores de basse fréquence dans l'eau de mer—Partie II: expériences en laboratoire* (A review of low frequency sound absorption measurements in seawater—Part II: Laboratory experiments). *Rev. du Cethedec* **75**: 49–89.
61. Waton, G., and Mallo, P. (1984). *L'absorption acoustique en basse fréquence dans l'eau de mer: Etude des mécanismes de relaxation* (Low frequency acoustic absorption in sea-water: Study of relaxation mechanisms. *Rev. du Cethedec* **78**: 45–52.
62. Qiu, X.F., Jiang, J., and Wan, S. (1984). Investigation of the mechanism of sound absorption by the boric acid relaxation in sea water. *Chinese J. Acoust.* **3**: 51–63.
63. Qiu, X.F., Jiang, J., and Wan, S. (1988). Measurement of the pH dependence of low frequency sound absorpyion in sea water. *Chinese J. Acoust.* **7**: 295–301.
64. Qiu, X.F. (1991). A cylindrical resonator method for the investigation of low-frequency sound in sea water. *J. Acoust. Soc. Am.* **90**: 3263–3270.
65. Ewing, M., and Worzel, J.L. (1948). Long-range sound transmission. *Geol. Soc. Am.* Memo 27.
66. Urick, R.J. (1963). Low-frequency sound attenuation in the deep ocean. *J. Acoust. Soc. Am.* **35**: 1413–1422.
67. Eigen, M., and de Maeyer, L. (1963). Relaxation methods In *Technique of Organic Chemistry*. New York: Wiley Interscience.
68. Eigen, M. (1964). Proton transfer, acid-base catalysis and enzymatic hydrolysis, Part I: Elementary processes. *Angewandte Chem.* **3**: 1–172.
69. Mellen, R.H., and Browning, D.G. (1977). Variability of low-frequency sound absorption in the ocean: pH dependence. *J. Acoust. Soc. Am.* **61**: 704–706.
70. Mellen, R.H. (1981). Chemical sound absorption in the sea, In *Underwater Acoustics and Signal Processing*, pp. 71–80, Denmark: Reidel Publishing Co.
71. Schulkin, M., and Marsh, R.W. (1978). Low-frequency sound absorption in the ocean. *J. Acoust. Soc. Am.* **63**: 43–48.
72. François, R.E., and Garrison, G.R. (1982). Sound absorption based on ocean measurements: Part II: Boric acid contribution and equation for total absorption. *J. Acoust. Soc. Am.* **72**: 1879–1890.
73. Mellen, R.H., and Browning, D.G. (1976). Low-frequency attenuation in the Pacific Ocean. *J. Acoust. Soc. Am.* **59**: 700–702.
74. Browning, D.G., and Mellen, R.H. (1987). Attenuation of low-frequency sound in the sea: Recent result, in *Progress in Underwater Acoustics*, Merklinger, H.M., ed., pp. 403–410, New York: Plenum Publishing Co.
75. Leroy, C.C., and Martinez, C. (1981). *Effet combiné de la rempérature et de la pression sur l'absorption du son dans les couches profondes* (Combined effects of temperature and pressure on sound absorption at depth). 8th GRETSI Congress, Nice.
76. Qiu, X.F. (1991). Effect of ocean environmental factors on sound absorption by boric acid relaxation in sea water. *Acta Oceanologica Sinica* **10**: 271–280.

77. Ainslie, M.A., and McColm, J.G. (1998). A simplified formula for viscous and chemical absorption in sea water. *J. Acoust. Soc. Am.* **103**: 1671–1672.
78. Lovett, J.R. (1979). Geographic variation of low-frequency sound absorption in the Pacific Ocean. *J. Acoust. Soc. Am.* **65**: 253–254.
79. Lovett, J.R. (1980). Geographical variation of low-frequency sound absorption in the Atlantic, Indian and Pacific Oceans. *J. Acoust. Soc. Am.* **67**: 338–340.
80. Brewer, P.G., Glover, D.M., Goyet, C., and Shafer, D.K. (1995). The pH of the North Atlantic Ocean: Improvements to the global model for sound absorption in seawater. *J. Geophys. Res.* **100**: 8761–8776.
81. Kibblewhite, A.C., and Hampton, L.D. (1980). A review of deep ocean sound attenuation data at very low frequencies. *J. Acoust. Soc. Am.* **67**: 147–158.

CHAPTER 4

VELOCITY AND ABSORPTION OF SOUND WAVES IN SUPERFLUID ^{3}He

H. Kojima
Department of Physics and Astronomy, Rutgers University, Piscataway, New Jersey, USA

Contents

4.1.	Introduction to Normal Liquid State of ^{3}He at Very Low Temperatures	117
4.2.	Introduction to Wave Propagations	122
4.3.	First Sound	125
4.4.	Second Sound in Superfluid ^{3}He A and B Phases	126
4.5.	Fourth Sound	129
4.6.	Spin-Entropy Wave in Superfluid ^{3}He A_1	137
Acknowledgments		144
References		144

4.1 INTRODUCTION TO NORMAL LIQUID STATE OF ^{3}He AT VERY LOW TEMPERATURES

^{3}He is an isotope of the more common ^{4}He. The natural abundance of ^{3}He in atmosphere is only about 1 part in 10^{11}. Amounts needed for experiments can be produced only by nuclear reactions such as neutron bombardment of Lithium-6 to produce tritium and subsequent tritium decay to ^{3}He. ^{3}He became available for macroscopic experiments only in the late 1950s. The liquid ^{3}He has been found tremendously rich in its fascinating properties at low temperatures. Among the diverse physical phenomena of ^{3}He, this chapter touches on the experiments on low-frequency sound waves in liquid ^{3}He at very low temperatures.

Schematic pressure-temperature phase diagrams of ^{3}He in zero and 50 koe external magnetic field are shown in Figure 4.1. Note the logarithmic scale of temperature. The boiling point under atmospheric pressure is 3.2 K. The critical temperature and pressure are 3.3 K and 1.2 bar, respectively. The ^{3}He remains in its liquid state down to absolute zero temperature because the quantum mechanical zero point motion prevents solidification. The vapor pressure is a standard thermometer useful down to about 500 mK. The liquid ^{3}He solidifies at pressure above 33.4 bar at zero temperature, but the freezing pressure has a characteristic minimum around 300 mK. The liquid ^{3}He remains in the "normal" liquid phase at temperatures down to a phase transition

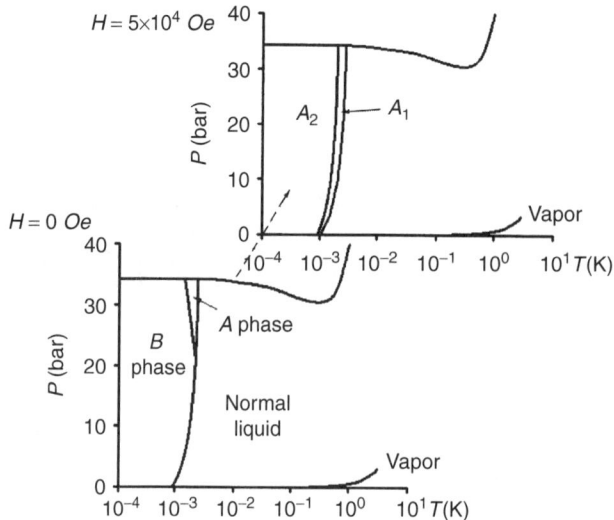

Fig. 4.1. Schematic pressure-temperature phase diagrams of ^{3}He in zero and 50 koe magnetic field.

temperature, T_c, which ranges from 1 mK at zero pressure to 2.5 mK at the solidification pressure. There are three distinct phases of superfluid ^{3}He that appear depending on temperature, pressure, and magnetic field. In zero magnetic field at "high" temperatures near the melting pressure, the superfluid phase between T_c and T_{AB} is known as A phase. At low temperatures below T_c and T_{AB} and at all pressures, B phase becomes stable. The T_{AB} line intersects the T_c line at the polycritical point pressure of 21.5 bar. When an external magnetic field is applied, the transition temperature T_c splits into two, T_{c1} and T_{c2} ($<T_{c1}$). The phase between these two new transition temperatures is known as A_1 phase. The A phase in magnetic field (A_2 phase) now extends down to zero pressure and at fields greater than about 5 koe, the B phase no longer exists. The purpose of this chapter is to describe experiments to study the nature of the superfluid ^{3}He phases using sound wave propagations as probes.

The observed properties of normal liquid ^{3}He are very different from those of liquid ^{4}He [1, 2]. That the odd number of neutrons "buried" deep in the nucleus influences the observed macroscopic properties is a testament to the importance of quantum mechanics in the low-temperature behavior of ^{3}He. Some of the thermal, magnetic, and transport properties of the normal liquid ^{3}He at low temperatures are summarized. The characteristic simple temperature dependences indicated below by Eqs. 4.1 and 4.2 are observed only at a temperature lower than about 50 mK, where quantum mechanical effects become dominant. The measured specific heat is proportional to temperature T [3]:

$$\Gamma_V = \gamma RT \tag{4.1}$$

where R is the gas constant and γ is a pressure-dependent constant. Owing to the odd nucleon number, ^{3}He nucleus carries a magnetic moment. At high temperatures, the magnetic susceptibility of liquid ^{3}He obeys the Curie law: $\chi \propto 1/T$. It becomes constant (χ_0) at low temperatures. The temperature dependences of the specific heat and the magnetic susceptibility are the same as those for metals at low temperatures where the conduction electrons dominate these properties. The shear viscosity (η) [4], thermal conductivity (K) [5], and spin diffusion (D) [5] coefficients depend on temperature as

$$\eta = \eta_o/T^2, \quad K = K_o/T, \quad D = D_o/T^2 \tag{4.2}$$

where η_o, K_o, and D_o are pressure-dependent constants. The pressure-dependent coefficients are tabulated in Table 4.1 at the vapor pressure, 18 bar, and the melting pressure.

Table 4.1. Selected properties and parameters of liquid ^{3}He.

Property		P(bar)	
	0	18	34.4
V(cm^3/mole)	36.8	28.9	25.5
V_F(m/s)	59	40	32
γ (K^{-1})	2.8	3.8	4.6
χ_o(10^{-8})	3.8	8.4	11.2
C_1(m/s)	183	345	422
m^*/m	2.8	4.5	5.9
F_0	9.3	48.3	87.1
F_1	5.4	10.6	14.4
Z_0	−2.8	−3.0	−3.0
η_0 (poisem K^2)	1.8	1.3	0.9
K_0(erg/scm)	35	17	11
D_0(10^{-7} cm^{2}K^2/s)	15	3.2	1.5
τT^2(10^{-6} smK^2)	1.2	0.9	0.7
T_c(mK)	0.929	2.177	2.486
$(T_{c1} - T_{c2})/H$(μK/10 koe)	14	48	63
$\Delta\Gamma/\Gamma$	1.46	1.80	1.97
$(\rho_s/\rho)/(1 - T/T_c)$	0.9	0.7	0.6

The observed *temperature* dependence of the properties cited in Table 4.1 may be understood by regarding the liquid ^{3}He as a noninteracting gas of identical spin 1/2 particles obeying Pauli exclusion principle and Fermi-Dirac statistics [6]. Statistical mechanics methods lead to the distribution function, giving the number of particles with energy ε as

$$f(\varepsilon) = \{e^{(\varepsilon-\mu)/k_BT} + 1\}^{-1} \tag{4.3}$$

k_B is Boltzmann's constant and μ is the chemical potential. In the limit of zero temperature, the distribution function $f(\varepsilon) = 1$ in the range $0 \leq \varepsilon \leq \mu$ and is zero otherwise. This value of μ is called the Fermi energy $E_F = (\hbar^2/2m)(3\pi^2N/V)^{2/3}$, where N/V is the particle number per unit volume. The Fermi momentum is defined as $p_F = (2mE_F)^{1/2}$. The density of states per unit volume and unit energy at the Fermi surface is given by

$$\left(\frac{dn}{d\varepsilon}\right) = \frac{m}{\pi^2\hbar^2}\left(\frac{3\pi^2N}{V}\right)^{1/3} \tag{4.4}$$

In the "degenerate" Fermi gas temperature range, $k_BT \ll E_F$, the distribution function differs little from the zero temperature limit except for the energy range of k_BT centered around E_F. Some particles below the Fermi energy are excited into those above the Fermi energy. Considerations of these "excitations" lead to the observed temperature dependence of the specific heat and the magnetic susceptibility(of the nuclear spins):

$$\Gamma_0 = \frac{\pi^2 k_B^2}{3}\left(\frac{dn}{d\varepsilon}\right)T \quad \text{and}$$
$$\chi_0 = \left(\frac{g\hbar}{2}\right)^2 \left(\frac{dn}{d\varepsilon}\right) \tag{4.5}$$

where g is the gyromagnetic ratio [7]. Elementary kinetic theory [6] applied to the Fermi gas shows the transport coefficients may be written as shear viscosity, $\eta = (1/3)\rho\tau v$; thermal conductivity, $K = (1/3)\rho\tau v^2 C_V$; and self diffusion, $D = (1/3)\tau v^2$. Here, τ is the mean time between collisions and is proportional to $(1/T)^2$. The mean velocity of the particles is v and may be taken as the Fermi velocity, $v = p_F/m$. Inserting the temperature dependence of τ and C_V gives temperature dependence of the coefficients in agreement with observations.

Though the observed temperature dependence can be explained, the Fermi gas theory is inadequate to explain the magnitude of the multiplying coefficients. Inadequacy may be illustrated by the velocity of sound. Applying Eq. 4.10 to the Fermi gas gives the velocity of sound as

$$C_{10}^2 = \frac{p_F^2}{3m^2} \tag{4.6}$$

Evaluating Eq. 4.6 for liquid ^{3}He at the vapor pressure gives 95 m/s, which differs substantially from the measured value of 183 m/s. Similar discrepancies between the Fermi gas predictions and the measurements are found in γ and χ_o. Landau [8] formulated the Fermi liquid theory to incorporate interactions between ^{3}He atoms to account for the discrepancies. The first feature of the Landau Fermi liquid theory is to recognize that, as a ^{3}He atom moves about in the liquid, a "cloud" of other surrounding atoms is dragged along. So the ^{3}He atom becomes a "quasi-particle" with an effective mass parameter m^*. The mass m in the expression for Fermi energy is replaced by m^*. The number of quasi-particles is equated to the total number of particles. The second feature is that the theory introduces an effective interaction between two quasi-particles. The interaction energy may depend on the momentum and spin of the quasi-particles. The two-particle interaction is expressed in terms of an infinite set of phenomenological "Landau parameters," F_l and Z_l. Only a few of these parameters may be determined by fitting the experiments to the theory according to

$$\begin{aligned}
\Gamma &= \frac{m^*}{m}\Gamma_o, \\
\frac{m^*}{m} &= 1 + \frac{F_1}{3}, \\
\chi &= \frac{m^*}{m}\frac{\chi_0}{1+\frac{Z_0}{4}}, \quad \text{and} \\
C_1^2 &= C_{10}^2 \frac{1+F_o}{1+\frac{F_1}{3}}
\end{aligned} \tag{4.7}$$

Measurements of the specific heat, the magnetic susceptibility, and the velocity of sound are used to determine m^*, F_0, F_1, and Z_0. The parameter F_2 may be determined from the difference in first and zero sound velocities [9].

About the time the Fermi liquid theory by Landau came out, Bardeen, Cooper and Schrieffer [10] published their celebrated BCS theory of superconductivity in which a formation of pairs of electrons (Fermi particles) leads to Bose-Einstein condensation-like condensation into a superfluid state. The BCS theory was soon applied to ^{3}He to predict its transition to superfluid state. The search was on to discover superfluid ^{3}He. As the minimum achievable temperature was lowered but no transition observed, theoretical estimates were revised. It was not until 1971 that the superfluid phases below 2.6 mK were discovered by Osheroff, Richardson and Lee [11] in a pressurization experiment on ^{3}He at the melting pressure. By means of nuclear magnetic resonance, they [12] showed soon afterwards that the signatures they observed were related to the liquid phase of ^{3}He. Tremendous amount of activity followed these pioneering experiments worldwide in both experimental and theoretical work [13, 14].

Like superconductors, liquid ^{3}He has a jump, $\Delta\Gamma$, in the specific heat at T_c in going from the normal phase to A, B, and A_1 phases. The measured specific heat jump [15] to the normal liquid specific heat ratio, $\Delta\Gamma/\Gamma$ is shown in Table 4.1. Perhaps the most dramatic evidence for superfluidity (and superconductivity) is that a circulating current once created can last forever. Existence of long-lasting mass currents in A and B phases has been experimentally demonstrated [16, 17]. In striking contrast with superfluid ^{4}He, the superfluid mass density of A phase is anisotropic. This was demonstrated in an elegant torsion pendulum experiment to measure the superfluid mass by applying a magnetic field that reoriented the anisotropy axis direction [18]. The superfluid density varies with temperature (T) just below T_c as $\rho_s/\rho = b(1 - T/T_c)$, thus ρ_s increases as T is lowered. The value of b is listed [19] in Table 4.1. Some of the most interesting effects are seen in nuclear magnetic resonance experiments. For example, the transverse magnetic resonance frequency in A phase departs from the Larmor frequency below T_c and increases as the temperature is lowered [12]. In addition to the transverse magnetic resonance, there exists a longitudinal magnetic resonance in A phase [20]. In all the ^{3}He superfluid phases, the shear viscosity of the normal fluid component decreases rapidly just below T_c, producing a kink in the temperature dependence and providing a convenient indicator of T_c.

Theoretical understanding of the experimental results in the previous paragraph can be made in terms of coherence effects between ^{3}He nuclear magnetic moments [13, 21]. The theory is complex, and only a rough picture of the basic ideas are given here. According to the BCS theory of "conventional" low-temperature superconductors, bound pairs of electrons are formed. The pair formation results from an indirect attractive force between a passing electron and a "cloud" of positive charge of the lattice ions that its partner electron traveling in the opposite direction. Being pointlike particles, the electron pair favors a greater overlap of wave functions and therefore the s-wave or $L = 0$ state. To preserve the quantum mechanical requirement of antisymmetry of the total wave function on particle exchange, the spin quantum number is taken as $S = 0$. The formation of a pair results in a reduction of energy 2Δ, where Δ is called the energy gap. In the superfluid phases of ^{3}He, pairs of ^{3}He atoms are also formed to reduce the overall energy. The force responsible for the attraction is thought to be the nuclear magnetic moment of a ^{3}He atom, which attracts a cloud of other moments, which in turn attracts its partner. Unlike the pair of electrons, there is now a hard core repulsion between ^{3}He atoms. Overlapping of wave function may be reduced by p-wave pairing state or $L = 1$ state. The antisymmetry requirement forces the spin pairing to be triplet or $S = 1$ state. That is, the ^{3}He pairs are magnetic and circulate around each other to carry angular momentum. According to the rules of quantum mechanics, the state with $S = 1$ may have a spin projection quantized as $S_z = +1$, 0, and -1. The A phase is thought to be described by the Anderson-Brinkman-Morel (ABM) state [22] containing pairs with $S_z = +1$ and -1. The B phase is thought to be described by the Balian-Werthamer (BW) state [23] containing $S_z = \pm 1, 0$. The A_1 phase is perhaps the simplest in that it contains pairs with only $S_z = -1$ (see [7]).

It should be emphasized that the paring in superfluid ^{3}He in the spin triplet ($S = 1$) and p wave ($L = 1$) state is a very different superfluid from the conventional superconductors ($S = 0, L = 0$). The conventional superconductor pair wave function may be specified by a complex number of given amplitude and a phase. The superfluid ^{3}He requires a 3×3 matrix with nine complex numbers. It is said that the pair wave function of superfluid ^{3}He possesses an internal structure. The superfluid phases have, therefore, intrinsically anisotropic properties. The energy gap Δ is, in general, anisotropic, though Δ of B phase turns out to be isotropic. The state of anisotropy may be specified by two quantization axis directions, **s** and **l** vectors, respectively, of the spin and the orbital components of the wave function. The superfluid densities of A and A_1 phases are anisotropic relative to the **l** vector. In a tensor notation, the superfluid density at temperatures near T_c or T_{c1} may be given as [13]

$$[\rho_s]_{\alpha\beta} = \rho_{s\|}(2\delta_{\alpha\beta} - l_\alpha l_\beta) \tag{4.8}$$

where $\rho_{s\parallel}$ is the superfluid density along **l** direction. Note that the superfluid density measured perpendicular to the **l** direction is two times that measured parallel to it. The parallel component of the superfluid density increases linearly with the reduced temperature near T_c as

$$\rho_{s\parallel} = \rho \frac{6}{5}\left(\frac{m}{m^*}\right)\left(\frac{\Delta\Gamma/\Gamma}{1.42}\right)\left(1 - \frac{T}{T_c}\right) \quad (4.9)$$

where ρ is the total density. The factor (m/m^*) comes from the "molecular field" correction [13] in the Fermi liquid theory and reduces the superfluid density considerably. The factor 1.42 is the ratio $\Delta\Gamma/\Gamma$ expected from the BCS theory in the "weak coupling" limit.

The physics of achieving low temperatures is a fascinating subject in itself. The developments in ^{3}He–^{4}He dilution refrigerator, Pomeranchuk cooling, and adiabatic demagnetization techniques all contributed to the low-temperature research discussed in this chapter. The readers are referred to other books and articles on these and other aspects of low-temperature production and measurements [24–27].

4.2. INTRODUCTION TO WAVE PROPAGATIONS

Let us now consider the propagation of "sound" modes in these superfluids. The considerations will be limited to the *hydrodynamic* regime in which the liquid remains in local thermodynamic equilibrium. The sound frequency ω is low enough to satisfy $\omega\tau \ll 1$, where τ is the characteristic quasi-particle collision time. The sound propagation at high frequencies, where $\omega\tau > 1$, is called zero, or collisionless, sound [28]. The high-frequency zero sound has been applied extensively to study superfluid ^{3}He, but the topic is outside the scope this chapter and the reader is referred to other articles [29–33].

The sound propagations in the isotropic superfluid ^{4}He are well described by the two-fluid model equations of superfluid. There are numerous publications on this and other aspects of superfluid ^{4}He physics [1,34,35]. However, a brief summary is useful for introducing terminologies. The two-fluid model assumes that the liquid is composed of two *interpenetrating* fluids of normal and superfluid components. The normal and superfluid components have mass densities, ρ_n and ρ_s, and velocity fields, $\mathbf{V}_n$ and $\mathbf{V}_s$, respectively. The total density is the sum, $\rho = \rho_n + \rho_s$. The superfluid component carries zero entropy and can flow with zero viscosity. The normal component carries the total entropy of the liquid and has finite viscosity. The total mass current is given by $\rho_n\mathbf{V}_n + \rho_s\mathbf{V}_s$. In the approximation of neglecting all dissipations, the two-fluid model hydrodynamics consist of the equations of conservation of mass, conservation of entropy, conservation of momentum, and the equation of motion of superfluid component [35]. These eight form a complete set of equations for the variables pressure, temperature, $\mathbf{V}_n$, and $\mathbf{V}_s$. Assuming traveling waves (wave vector k and frequency ω) of the form $\sim \exp(ikx - i\omega t)$ for the variables, the equations may be solved for propagating sound modes in the bulk. One mode is the first sound in which both the superfluid and normal fluid move in phase: $\mathbf{V}_n = \mathbf{V}_s$. The first sound is the same as the ordinary acoustic wave, which exists in ordinary liquids, and its velocity is given by:

$$C_1^2 = \frac{1}{\rho K_s} \quad (4.10)$$

where K_s is the adiabatic compressibility. The other mode is the second sound in which the normal and superfluid components move out of phase such that there is no mass current: $\rho_n\mathbf{V}_n + \rho_s\mathbf{V}_s = 0$. This mode is accompanied by temperature (and entropy) oscillation and is unique to superfluid. Its propagation velocity is given by

$$C_2^2 = \left(\frac{\rho_s}{\rho_n}\right)\frac{S^2 T}{C_p} \tag{4.11}$$

where S is the entropy per unit mass, C_p is the specific heat at constant pressure; and T is the temperature. Consider next the superfluid being placed in so narrow (straight, for simplicity) channels that the normal component is immobilized by viscous forces. The superfluid component in this situation can still flow in such a "superleak." The propagation of pressure and density oscillation in the superfluid component in a superleak is called the fourth sound and its velocity is given by:

$$C_4^2 = \left(\frac{\rho_s}{\rho}\right) C_1^2 \tag{4.12}$$

To account for the attenuation of sound modes, dissipative currents and forces must be included in the hydrodynamic equations. The dissipation-related quantities are the shear viscosity of the normal component (η), the bulk viscosity coefficients (ζ_1, ζ_2, ζ_3 and ζ_4 [$= \zeta_1$ by Onsager relations]), and the thermal conductivity (K). The shear viscosity enters as a coefficient to a combination of spatial derivatives of $\mathbf{V}_n$ and the bulk viscosity coefficients, ζ_1 and ζ_2, enter as coefficients to the spatial derivatives of $(\mathbf{V}_n - \mathbf{V}_s)$ and $\mathbf{V}_n$, respectively, in the dissipative part of the stress tensor in the conservation of the momentum equation. The bulk viscosities, ζ_3 and ζ_4, enter as coefficients to the spatial derivatives of $(\mathbf{V}_n - \mathbf{V}_s)$ and $\mathbf{V}_n$, respectively, in the equation of motion of the superfluid component. Solving for the propagating modes again gives the attenuation coefficient α in $k = \omega/C + i\alpha$, where C is the velocity in the no-dissipation limit. The attenuation coefficient for the first sound is

$$\alpha_1 = \frac{\omega^2}{2\rho C_1^3}\left(\frac{4}{3}\eta + \zeta_2\right) \tag{4.13}$$

and that for the second sound is

$$\alpha_2 = \frac{\omega^2}{2\rho C_2^3}\left[\frac{\rho_s}{\rho_n}\left\{\frac{4}{3}\eta + \zeta_2 + \rho^2\zeta_3 - 2\rho\zeta_1\right\} + \frac{K}{C_V}\right] \tag{4.14}$$

A small term related to K is neglected in Eq. 4.13. If the shear viscosity is independently measured, the bulk viscosity ζ_2 may be determined by measuring α_1. If the thermal conductivity can also be measured independently, the measurement of α_2 allows determination of only a combination of ζ_1 and ζ_3. An Onsager relation requires $\zeta_2\zeta_3 > \zeta_1^2$.

The foregoing description of sound propagations in superfluid ^{4}He is thought to be applicable to the isotropic ^{3}He B phase. It is modified for the anisotropic A phase [36, 13]. The superfluid density is written as a tensor, and the total mass current becomes

$$\rho V_\alpha = [\rho_n]_{\alpha\beta} V_{s\beta} + [\rho_s]_{\alpha\beta} V_{n\beta} \tag{4.15}$$

where α and β stand for x, y, and z spatial directions. Neglecting all dissipative effects, and assuming a spatially uniform $\mathbf{l}$ vector "texture," the linearized two-fluid model hydrodynamic equations for the anisotropic superfluid are (summation over repeated indices are assumed)

$$\frac{\partial\rho}{\partial t} + [\rho_s]_{\alpha\beta}\frac{\partial}{\partial x_\alpha}V_{s\beta} + [\rho_n]_{\alpha\beta}\frac{\partial}{\partial x_\alpha}V_{n\beta} = 0 \tag{4.16}$$

$$\frac{\partial\rho S}{\partial t} + \rho S\frac{\partial V_{n\alpha}}{\partial x_\alpha} = 0 \tag{4.17}$$

$$[\rho_s]_{\alpha\beta}\frac{\partial}{\partial t}V_{s\beta} + [\rho_n]_{\alpha\beta}\frac{\partial}{\partial t}V_{n\beta} = -\delta_{\alpha\beta}\frac{\partial}{\partial x_\beta}P \qquad (4.18)$$

$$\frac{\partial}{\partial t}V_{s\beta} = -\frac{1}{\rho}\frac{\partial P}{\partial x_\beta} + S\frac{\partial T}{\partial x_\beta} \qquad (4.19)$$

Here P is the pressure and S is the entropy per unit mass. The first three of these represent the conservation of mass, entropy, and momentum, respectively. The fourth gives the equation of motion for the superfluid component. The hydrodynamic equation becomes very complex when spatial variations in **l** are included [21].

Looking for solution of all oscillatory variables in the form $\exp(ikx - i\omega t)$ in Eqs. 4.16–4.19 gives two propagating modes as before. One is of course the first sound, which remains unchanged from the isotropic superfluid with velocity given by Eq. 4.10. The second sound propagation is anisotropic with the velocity given by [36]

$$C_2^2(\hat{q}) = [\rho_s]_{\alpha\beta}[\rho_n]_{\alpha\beta}^{-1}\frac{S^2 T}{C_p}\hat{q}_\alpha \hat{q}_\beta \qquad (4.20)$$

where $\hat{q}$ is the propagation direction. At temperatures near T_c, the second sound velocity propagating along **l** is $\sqrt{2}$ times that along perpendicular to it. The second sound propagation would be a convenient method for measuring the anisotropic effects if the **l** vector can be specified or measured independently. Turning to the fourth sound propagation, setting $V_n = 0$ and combining Eqs. 4.16 and 4.19 leads to a good approximation:

$$C_4^2(\hat{q}) = \frac{[\rho_s]_{\alpha\beta}}{\rho}C_1^2 \hat{q}_\alpha \hat{q}_\beta \qquad (4.21)$$

Again the velocity is anisotropic. If the angle between $\hat{q}$ and **l** is θ, the velocity near T_c is given by

$$C_4^2 = \frac{\rho_{s\parallel}}{\rho}(2 - \cos^2\theta)C_1^2 \qquad (4.22)$$

Inclusion of dissipation into the hydrodynamics of anisotropic superfluid is a complicated affair even if the **l** texture is assumed to be uniform. Dissipative coefficients in general must now be considered as having parallel and perpendicular (to **l**) components, $v^\parallel$ and $v^\perp$, where v stands for previously introduced coefficients, η, ζ_1, ζ_2, and K. In addition, new $\eta_L^{\parallel,\perp}$ coefficients to the spatial derivative coupling **l** and $\mathbf{V}_n$. As before, ζ_3 and ζ_4 enter as dissipative coefficients in the equation of motion of the superfluid component. So there are 12 coefficients in all. The attenuation of first sound A phase is given by Eq. 4.13 by replacing the term η by an effective viscosity given by [21]

$$\bar{\eta} = \eta_L^\parallel \cos^4\theta + (3\eta^\parallel - \eta_L^\parallel)\sin^2\theta\cos^2\theta + \tfrac{1}{4}(3\eta^\perp + \eta_L^\parallel)\sin^4\theta \qquad (4.23)$$

where θ is the angle between the propagation direction and **l**. The bulk viscosity ζ_2 is small. If the **l** texture is uniform and θ can be varied, measurements of the attenuation of first sound in A phase, in principle, can be used to determine the shear viscosity components. In practice, it is difficult as the **l** texture is typically not uniform. The attenuation of second sound in A phase is given by Eq. 4.14 by replacing η by the effective viscosity above and the thermal conductivity K by a thermal conductivity tensor. The attenuation of fourth sound in the ideal limit, $\mathbf{V}_n = 0$, is given by [37]

$$\alpha_4 = \frac{\omega^2 \rho_s \zeta_3}{2C_4^3} \qquad (4.24)$$

In the superfluid ^{3}He A_1 phase the hydrodynamic equations must be modified to account for the fact that the superfluid mass flow is at the same a totally polarized spin flow [38, 39]. The A_1 hydrodynamics is complex in its full generality. It simplifies in the long wavelength limit where the variations in **l** and spin quantization direction can be neglected. The conservation of mass, entropy, and momentum remain the same as before. To these, the conservation of spin is added:

$$\frac{\partial \sigma}{\partial t} + \frac{\partial}{\partial x_\alpha}\left[\sigma_n V_{n\alpha} + \frac{\hbar}{2m}[\rho_s]_{\alpha\beta} V_{s\beta}\right] = 0 \qquad (4.25)$$

where σ is the total spin density and σ_n the normal component spin density. The superfluid spin density is $(\hbar/2)$ times (ρ_s/m), the superfluid particle number density. The new equation of motion of the superfluid component is

$$\frac{\partial}{\partial t} V_{s\beta} = -\frac{1}{\rho}\frac{\partial P}{\partial x_\beta} + S\frac{\partial T}{\partial x_\beta} - \frac{\hbar}{2m}\frac{\partial \omega_\beta}{\partial x_\beta} \qquad (4.26)$$

where $\omega = g(gS/\chi - \mathbf{H})$ and χ is the magnetic susceptibility, **S** is the spin density, and **H** is the external magnetic field. The first sound is again given by Eq. 4.10. The second sound takes on an interesting form:

$$C_{se}^2(\hat{q}) = \left[\frac{\rho_s}{\rho_n}\right]_{\alpha\beta}\left[\left(\frac{\rho}{\chi}\right)\left(\frac{\gamma\hbar}{2m}\right)^2 + \frac{S^2 T}{C_p}\right]\hat{q}_\alpha \hat{q}_\beta \qquad (4.27)$$

Note that the second term in the bracket is precisely the thermal entropy wave, Eq. 4.11. The first term originates in the stiffness against spin compression of the polarized superfluid spins. The subscript *se* represents the spin and entropy wave nature of the second sound in A_1 phase. The spin stiffness term overwhelmingly dominates the entropy term. The fourth sound velocity A_1 phase is given by Eq. 4.21. The effects of magnetic terms are small compared with the bulk compressibility. The attenuation of the spin-entropy wave is discussed later in Section 4.6.5.

A "third" sound is another wave propagation familiar from ^{4}He physics [40]. This "sound" is analogous to the water surface wave that propagates on thin superfluid films in equilibrium with its vapor of ^{4}He. The films are thin enough that the normal component is locked to the substrate. The two-dimensional geometry of the sample gives a distinct advantage in interpretation of third sound results. The readers are referred to the recent article on the observation of third sound on films of superfluid ^{3}He [41].

4.3. FIRST SOUND

4.3.1. Velocity of First Sound

First sound is the ordinary hydrodynamic acoustic pressure-density wave propagation whose velocity is related to the density and the compressibility by Eq. 4.10. It propagates in both the normal and superfluid phases. In the degenerate Fermi liquid temperature range, the density and the compressibility are those of the quasi-particle excitations. Given the value of F_1^S parameter determined from the measurement of specific heat, the measurement of velocity of sound serves as an important method to determine the F_0^S parameter.

4.3.2. Attenuation of First Sound

The attenuation coefficient of first sound in B phase in the hydrodynamic limit is given by Eq. 4.13. The shear viscosity term dominates over the bulk viscosity term and the attenuation is found to vary with temperature as $\alpha_1 \propto T^{-2}$. Note that the sound propagation transforms to the zero sound mode at temperatures below which $\omega\tau > 1$. In the crossover temperature region where $\omega\tau \sim 1$, the attenuation passes through a maximum for a fixed frequency [30, 42].

Care must be exercised in applying Eq. 4.13, which is applicable to the attenuation in bulk free of wall effects. Assuming the usual bulk hydrodynamics boundary condition (valid in the limit of small mean free-path length compared with the enclosing wall dimensions) that the fluid velocity tangential to the wall vanishes, the wall exerts a viscous force on the fluid. For a cylindrical resonator with radius R, the attenuation coefficient owing to the wall losses is

$$\alpha_w = \frac{\omega \delta_v}{2RC_1} \tag{4.28}$$

where $\delta_v = (2\eta/\rho\omega)$ is the viscous penetration depth. The total attenuation would be the sum of α_w and α_1. At frequencies less than about 50 kHz, the wall contribution dominates and the attenuation is proportional to $1/T$. The wall effect needs further examination since the mean free path becomes extremely large in the normal and superfluid ^{3}He at low temperatures and the bulk hydrodynamics boundary condition is no longer valid. The theoretical treatment of first sound propagation in a resonator was extended to the nonhydrodynamic range by Nagai and Wölfle [43]. They modified the usual boundary condition of vanishing tangential velocity at the wall to the one of a "slip" boundary condition (see the section on fourth sound for discussion). In the limit that δ_v is small compared with both R and the wavelength, they found that the wall contribution was replaced by

$$\alpha_{ws} = \frac{ReZ(\omega)}{RC_1\rho} \tag{4.29}$$

where $ReZ(\omega)$ represents the real part of the surface impedance of the resonator wall. The impedance $Z(\omega)$ is a function of shear viscosity and slip length. Experiments [44, 45] on the propagation of sound were carried out in the frequency range 20–300 kHz. The measured attenuation was analyzed in terms of this theory to extract the shear viscosity. The extracted viscosity was consistent with those measured by torsion oscillator and vibrating wire methods. The effects of frequency can be more conveniently examined with the acoustic method than the other methods.

4.4. SECOND SOUND IN SUPERFLUID ^{3}He A AND B PHASES

4.4.1. Introduction

An important prediction of the two-fluid model of superfluid ^{4}He was a new (second) "sound" mode [46]. The observation [47] of the second sound in accordance with its predicted temperature dependence gave important support for the two-fluid model. The velocity of second sound is related to the two-fluid model parameters by Eq. 4.11 in the isotropic B phase and by Eq. 4.20 in the anisotropic A phase. If other thermodynamic quantities are known, the second sound velocity may be used to determine the (anisotropic) superfluid component density. If the superfluid component density and the temperature are accurately known, the second sound serves as a novel method for measuring the entropy. The second sound velocity in ^{4}He is typically around 20 m/s, except near its transition temperature. In contrast, the second sound velocity in ^{3}He B phase is only on the order of 5 cm/s. The small velocity poses a challenge in observing second sound in ^{3}He.

The attenuation coefficient of second sound in B phase is given by Eq. 4.14. For a typical frequency of 10^3 Hz and for the same viscous effects in the square bracket, the attenuation of second sound in ^{3}He is 10^7 times that in ^{4}He. If there is to be any hope, an experiment would have to be done at low frequencies around 1 Hz. If the attenuation of second sound can be measured, it may be combined with viscosity coefficient (from a separate experiment) to evaluate the thermal conductivity. The (diffusive) thermal conductivity in superfluid is difficult to determine by the conventional temperature gradient in the presence of heat flow since convective heat flow easily masks the diffusive effect.

4.4.2. Peshkov Transducer

The first observation [47] of second sound in ^{4}He was made by generating it with heater and detecting the subsequent temperature oscillation with a sensitive bolometer. These techniques are not likely to work in ^{3}He as the thermal boundary resistance between the bolometer and liquid ^{3}He becomes prohibitively large. To avoid heating effects, some mechanical means of producing and detecting second sound would be preferred. The transducer we chose to employ was invented by Peshkov and examined theoretically by Liu [48]. The commonly used oscillating superleak transducer (OST) [49, 50] is not suitable for ^{3}He since the driving efficiency is much reduced owing to thermal shorting by the high diffusive heat flow in the normal fluid in the pores of thin superleak membrane.

The Peshkov transducer (PT) consists of a rigid superleak and a flexible nonporous membrane behind it. In the PT operation, the motion of the membrane creates a superfluid component motion at the face of the superleak. The superleak may be designed to reduce the nuisance thermal diffusive current by increasing its path length. Our version of PT is shown in Figure 4.2. The superleak was a 15-mm-diameter and 2-mm-thick glass with a built-in array of parallel capillaries of diameter 25 μm [51]. The porosity of the array was 50%. The membrane was a 1-μm-thick nickel shim placed 0.5 mm from the superleak. A fixed electrode was placed about 50 μm away from the membrane. The fixed electrode was drilled with 1-mm-diameter holes, which allowed sufficient flow to eliminate pressure buildup. A DC-biased voltage applied between the electrode and the membrane-induced superfluid component flows through the glass capillary array to induce relative motion with the normal component.

One PT was fitted onto one end of a cylindrical resonator chamber (diameter = 13 mm, length = 19 mm). A second PT constructed in the same manner was fitted onto the other end and served as a detector. In the detector side, the membrane acts as a microphone. The changes in the capacitance between the membrane and the fixed electrode was measured with a capacitance bridge and a lock-in detector. The whole resonator assembly was immersed in liquid ^{3}He, which was cooled by an adiabatic demagnetization apparatus. Two holes (2 mm in diameter) made in the wall of the resonator served as thermal conduction paths as well as pressure release port. Second

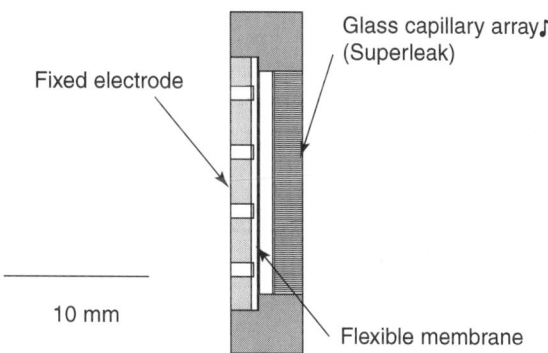

Fig. 4.2. Peshkov transducer.

sound was searched by looking for resonant response in one PT while sweeping the frequency of the voltage applied to the other PT.

4.4.3. Observation and Analysis

Observations of the purely thermal second sound were indeed made in superfluid ^{3}He B by the resonance techniques based on PT [52]. The observed velocity was small as expected and the resonance was observed in the frequency range 0.5–2.5 Hz. The signal strengths were quite weak owing to a large attenuation as expected. A large spurious fourth sound resonance also made the observation difficult. The thermal contact holes in the resonator wall turned out to have significant effects on the resonance spectrum [53]. In the cylindrical resonator, plane wave resonance occurs at $f_m = mC_2/2L$, where L is the length of the resonator. Modes up to $m = 3$ were observed.

The velocity of second sound determined from resonance frequency is shown as a function of temperature in Figure 4.3. The temperature range of measurement was limited by the cooling apparatus. The lines represent the second sound velocity computed using Eq. 4.11 and are consistent with the measurement. In the experiment at 24.4 bar, the second sound signal abruptly disappears at a temperature close to the B to A transition temperature. Apparently, the second sound was not well formed in the A phase. Recall that the velocity of propagation is anisotropic in A (see Eq. 4.20). It is likely that a complicated interplay between the superflow and **l** vector prevented a

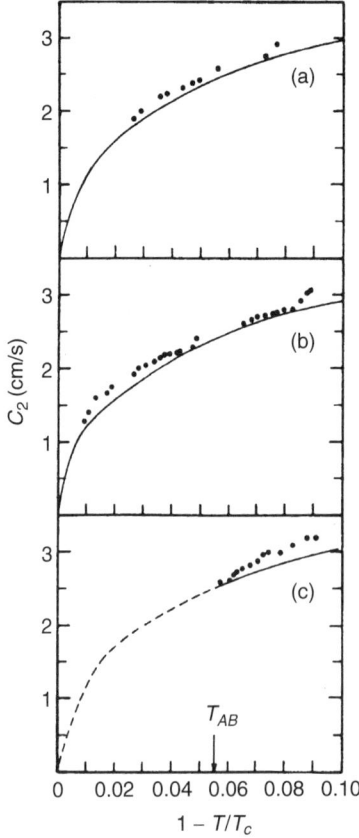

Fig. 4.3. Measured (dots) velocity of second sound in superfluid ^{3}He B as a function of reduced temperature at a pressure of (a) 18.2 bar, (b) 21.3 bar, and (c) 24.4 bar. The lines were computed from Eq. 4.21.

propagation of well-defined second sound. The role of anisotropy is explored in detail in the spin-entropy wave propagation discussion that follows.

The quality factor of the observed resonances ranged between 2 and 3 compared to about 4 expected from Eq. 4.14. The expected frequency dependence could not be verified. There is much room for improvement in the measurement of attenuation of the second sound. It is important as the attenuation of second sound can be analyzed to determine the thermal conductivity. A systematic measurement over wider temperature and pressure ranges is highly desirable. The mysterious disappearance of the signal in A phase may be related to instability in the **l** texture in the cell induced by the superflow accompanying second sound. A combination of magnetic field and wall geometry is likely needed for stabilizing the texture.

4.5. FOURTH SOUND

4.5.1. Introduction

Fourth sound can be propagated in a porous structure ("superleak") in which the superfluid component is free to move, but the normal component is immobilized by the viscous force against the pore walls. Consider an oscillatory (frequency f) motion of normal fluid along the axis of a cylindrical tube of radius R. The fluid velocity varies from zero at the wall to that at the axis with a characteristic "viscous penetrating length" given by $\delta_v = (\eta/\pi\rho_n f)^{1/2}$. The pore radius must be small enough that $R < \delta_v$ to "lock" the normal component. Since the viscosity of liquid ^{3}He is considerable at low temperatures (see Table 4.1), it is relatively easy to satisfy the superleak condition. Taking the values at 18 bar and T_c, we have $\delta_v \sim 0.8/f^{1/2}$ (cgs units), which give $\delta_v \sim 250$ µm a typical frequency of 1 kHz.

In practice, a superleak is often made by the complicated (not simple straight tubes considered in the introduction) pores in a packed powder. To account for the multiple scattering of sound by the powder, an index of refraction, n, is introduced. The measured velocity of fourth sound, C^*, is written as

$$C_4^* = \frac{C_4}{n} \qquad (4.30)$$

where C_4 would be given by Eq. 4.12 or 4.21. The irregular geometry in the pores produces local variations in the order parameter and the **l** texture, hence the superfluid density. The fourth sound would then sense the average over the distribution of the superfluid density. Thus, Eqs. 4.11 and 4.8 are modified as

$$C_4^{*2} = \frac{1}{n^2} \left\langle \frac{\rho_s}{\rho} \right\rangle C_1^2 \qquad (4.31)$$

where the angular bracket indicates an average of the superfluid density over the pore spaces. The averaging is discussed in more detail in Section 4.5.3.

The requirement that a fluid be a superfluid for fourth sound to propagate means that its observation is a convincing proof for superfluidity. The fourth sound velocity gives a simple and direct measure of the superfluid density. The confining environment of superleak needed for propagation of fourth sound makes a natural setting for studying the effects of the solid wall boundaries and restricted geometries on the superfluid density and the interaction between quasi-particles and the boundaries. These applications of fourth sound are described in the next three sections.

4.5.2. Search for Fourth Sound: Proof of Superfluidity

The fourth sound can propagate only in a superfluid. Thus, observation of fourth sound constitutes a convincing proof of superfluidity in a new fluid state. This was the primary motivation for the search for fourth sound in the then newly discovered [11] state of liquid ^{3}He cooled to ultra-low temperatures below about 3 mK. An apparatus for observing fourth sound needed to contain simultaneously a superleak for immobilizing the normal component, a refrigerant for cooling the sample liquid ^{3}He, and a thermometer. In our first fourth sound cell, cerium magnesium nitrate (CMN) was utilized to the fullest extent to meet the three requirements [54]. Ground polycrystalline CMN were sifted through a wire mesh to select grains with diameters of 37 μm or less and was packed to 80% packing fraction into a cylindrical tube to make a superleak. Demagnetization of CMN provided sufficient cooling power to reach the low temperatures needed. The magnetic susceptibility of the paramagnetic CMN acted as a convenient thermometer in good thermal contact with ^{3}He. Each end of the cylindrical cell was fitted with a condenser transducer. The transducers were made by stretching a Mylar foil (6-μm thick) over a stationary electrode. The construction of the transducers is essentially identical to that for first sound.

Fourth sound was observed by pulse and resonance methods. In the pulse method, a step voltage was applied to the drive transducer to look for response in the detector transducer. In the resonance method, a sinusoidal voltage was applied to the drive transducer to measure the response of the detector transducer as a function of frequency. Spurious signals may arise from certain mechanical resonance in the structural parts of the cell such as the Mylar foil, the electrical lead, supporting structure.

For any potential signal to qualify as fourth sound in origin, (1) it must occur only below the transition temperature; (2) the velocity derived from it must be temperature dependent and be vanishing toward the transition temperature; (3) given a fundamental mode frequency in the resonance method, higher harmonics should be present at integer multiples; and (4) when the transition temperature is varied by changing the liquid pressure, the temperature variation of the velocity should reflect the change in transition temperature. In our cell, the transition temperature could be detected by observing the change in the warmup rate owing to the jump in specific heat at T_c.

It is often the case with fourth sound experiments that even though the liquid temperature is measured to be lower than the transition temperature, fourth sound signal cannot be found in the initial several adiabatic demagnetization cooling. This is probably caused by residual heat leaks that take a long time to dissipate through the small heat paths in the superleak. So it was discouraging to find no signs of fourth sound signal in the initial cool down runs. Thus, after about three demagnetizations, it was great excitement to see signal that eventually satisfied all verification tests for fourth sound. It was observed with both of the methods mentioned. Thus, superfluidity in the newly discovered state of liquid ^{3}He was convincingly proven experimentally [54]. Fourth sound was also observed by an ingenious mechanical oscillator method [55].

4.5.3. Velocity of Fourth Sound and Superfluid Density Fraction

An important application of fourth sound is the measurement of superfluid fraction, as implied by Eq. 4.12. An example of the results of superfluid fraction determined from fourth sound experiment [56] is shown in Figure 4.4 as a function of the reduced temperature, $1 - T/T_c$. In this experiment, three fourth sound cylindrical resonators (each with length = 14 mm and diameter = 8 mm) were simultaneously mounted onto a copper nuclear adiabatic demagnetization cooling apparatus. Each resonator was packed with a distinct superleak packed with different powder, as shown in Table 4.2. The powders were alumina polishing powders of manufacturer-specified nominal grain size.

In the experiment, fourth sound resonance frequency of a particular mode (m) in each resonator is tracked as the temperature is varied. The resonant frequency is converted to the velocity of sound from the relation, $C_4^* = f_m/2mL$, where L is the length of

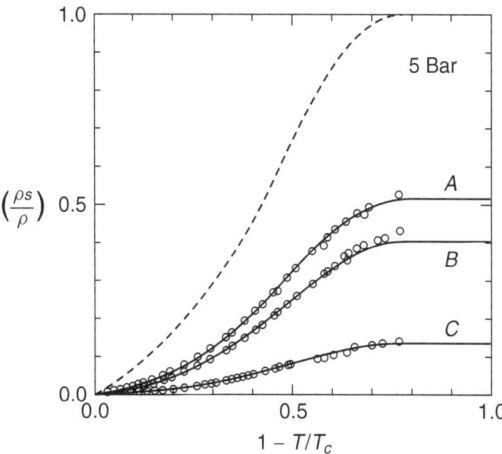

Fig. 4.4. Superfluid fraction.

Table 4.2. Pertinent parameters of resonators in Figure 4.4.

Resonator	Nominal grain size (µm)	Porosity (%)	n	R_{av}
A	3.0	68.7	1.17	0.42
B	1.0	75.1	1.09	0.30
C	0.3	79.8	1.11	0.11

the resonator). To convert the measured speed of sound to superfluid fraction, the index of refraction must be known for the porous medium. If the superleak is made of parallel straight tubes of length L connecting one transducer to the other, the index of refraction n in Eq. 4.30 is unity. If, on the other hand, the tubes are bent around such that the path length from one transducer to the other is increased by a factor. The index of refraction n represents this factor for the torturous path presented to fourth sound by the packed powder in the superleak. The magnitude of n depends on unknown details of the powder such as the particle size distribution, shape, and packing pattern. Fortunately, in our case, it can be measured from a separate experiment in the same packed powder in superfluid ^{4}He, whose fourth sound velocity in the $n = 1$ limit is well-known. The ratio of the velocity in Eq. 4.12 to the measured velocity gave the index of refraction [57] and is given in Table 4.2 for each resonator. This index of refraction was used to compute the superfluid density fraction in superfluid ^{3}He.

The dashed line in Figure 4.4 represents the bulk superfluid density as measured by Parpia *et al.* [58]. It is clear that the average superfluid density measured by fourth sound is suppressed from that in bulk. The smaller the average pore size, the greater the suppression. For a given resonator, the relative suppression increases as the temperature is increased toward the transition temperature and as the pressure is decreased. The superfluid to normal fluid transition temperatures appear to be same as that in bulk in Figure 4.4. However, precision measurements showed systematically greater suppression of transition temperature in smaller pores [59]. All of these qualitative features can be understood in terms of the effects of the boundary walls on ^{3}He superfluidity. The lines through the data points in Figure 4.4 indicate fitted superfluid fraction based on the following analysis described below.

Now let us turn to the effect of solid boundary walls on the superfluid component density. Far away from the walls, the superfluid does not feel their effects, and the superfluid density is the bulk value. Near to the wall boundary, the local superfluid density is affected depending on boundary conditions. Theoretically, the boundary condition is determined by how the quasi-particles scatter off of the boundary [60, 61]. If the scattering is a mirrorlike specular one, then the superfluid density is not depressed. If the scattering is completely diffusive, the superfluid density vanishes at the wall. The boundary may present a partially diffusive scattering resulting in a partial depression of superfluid density. It seems reasonable to assume that the powder grains used in our experiment present diffuse scattering surfaces. Fourth sound provides a nice technique of examining the boundary conditions at the walls on the superfluid density. Care must be exercised in interpretation as the pore structure is in general complicated.

The local superfluid density recovers from the wall according to $\sim e^{-x/\xi}$, where x is the distance from the wall surface and $\xi(T)$ is the temperature-dependent "healing" length. The healing length according to the BCS theory diverges near T_c according to $\xi(T) = \xi_s/\{(5/3)(1 - T/T_C)\}^{1/2}$. If the pore is filled with superfluid ^{3}He-B phase, the suppression of the superfluid density may be written as

$$\frac{\rho_{sp}}{\rho_{sb}} = 1 - k\xi(T)/R \qquad (4.32)$$

where ρ_{sp} is the average superfluid density in the pore and ρ_{sb} is the bulk superfluid density, R is the radius for a cylindrical pore or the distance between parallel walls for a slab pore, and k is a constant of about 2.4.

Our powder packing was modeled as interconnected cylindrical pores, each contributing to the total superfluid density according to Eq. 4.32. The distribution of pore sizes was measured by the Mercury intrusion method. The effective superfluid density in our resonator was calculated by summing over the pore size distribution. This method was adapted from ^{4}He studies [62]. The scale of the healing length ξ_s was left as an adjustable parameter to achieve the best fit for each resonator. The solid lines in Figure 4.4 show the fitted average superfluid fraction. The fit is seen to be excellent at all pressures in all resonators. The best-fit healing length ξ_s for each resonator is shown in Figure 4.5 at all pressures measured. Note that the values for all resonators are consistent with each other. There is fairly large pressure dependence in the low pressure range, but it tends to be constant in the high-pressure range. The line in Figure 4.5 is evaluated from the BCS expression for the healing length given by $\xi_s = 0.021 \hbar V_F/k_B T_C$. The magnitude as well as the pressure dependence of ξ_s are in fair agreement with that obtained from the fit but with significant deviation at the high pressure side. At higher pressures and at low temperatures the ratio $\xi(T)/R$ becomes smaller and the superfluid density suppression should become smaller. The observed suppression of superfluid density in each resonator is consistent with the expectation.

Let us first look at the zero temperature limit of the superfluid fraction. In bulk it approaches unity, but in confined geometries it remains less than unity in the zero temperature limit. If the pore size is made sufficiently small, the confined ^{3}He may never become superfluid. To study this critical size effect, available data at all pressures studied are combined to plot the observed superfluid fraction in the zero temperature limit, $\rho_s(T=0)$, plotted as a function of R_{av}/ξ_s in Figure 4.6 for all resonators. Data from different resonators overlap smoothly on this plot. If a linear extrapolation is made, superfluid fraction vanishes at the critical radius $R_{avC} = \xi_s$. The critical size was considered theoretically [63] and found to be $R_C = 1.5\xi_s$ for cylindrical pore with diffuse scattering wall. The average pore radius gives a good indication of the critical radius but more studies are needed in this direction. By choosing the pore radius appropriately, superfluidity would be totally suppressed at zero temperature below a critical pressure but emerge as a superfluid above the critical pressure.

The relative suppression of superfluid fraction is greatest near the transition temperature, but the detail is difficult to discern in Figure 4.4. The transition temperature region was studied in more detail in a separate experiment in which three resonators (I

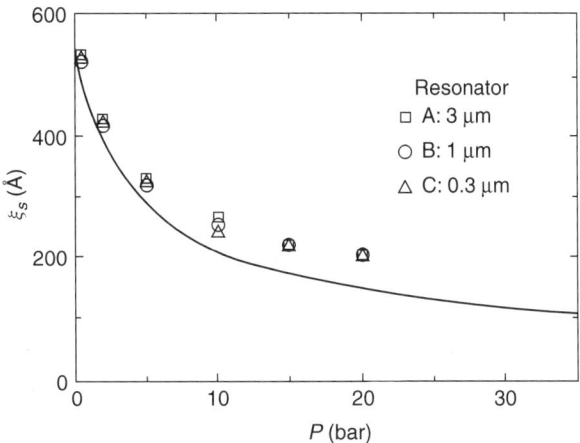

Fig. 4.5. Pressure dependence of healing length. Symbols are obtained from fits to data. The line is the BCS formula.

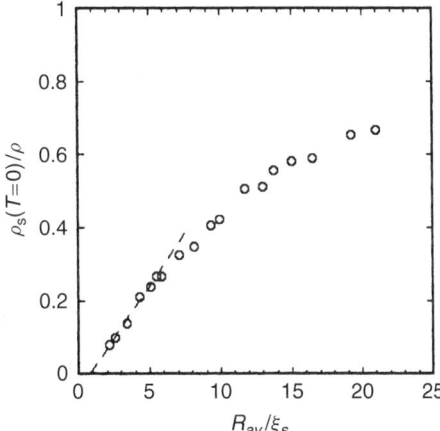

Fig. 4.6. Zero temperature limit of superfluid fraction vs R_{av}/ξ_s.

[nominal powder size = 1 μm, Rav = 1.9 μm], II [0.3 μm, 0.44 μm], and III [0.05 μm, 0.18 μm]) were again mounted simultaneously in an adiabatic demagnetization apparatus [59]. The results are shown in Figure 4.7. The transition temperature in bulk is T_C^0. Extrapolation of superfluid fraction (and fourth sound signal amplitude) to zero (indicated by arrows) shows that there is a systematically greater depression of transition temperature in the resonators with finer powder. The depression of the transition temperature is plotted as circles in Figure 4.8 as a function of the average pore radius as determined by Mercury intrusion porosimetry. Theoretical analyses on the depression of transition temperature was carried out [63, 64]. If the bounding surfaces present totally diffusive scattering to quasi-particles, the depression of T_C may be solved by setting the left side of Eq. 4.32 to zero:

$$1 - \frac{T_C}{T_C^0} = \left(\frac{3}{5}\right)\left(\frac{k\xi_s}{R}\right)^2 \qquad (4.33)$$

The line in Figure 4.8 is evaluated using Eq. 4.33, with ξ_s set to 200 Å as read off from Figure 4.5 and R set to the average pore radius. The magnitude of the depression is correlated fairly well with the average pore radius.

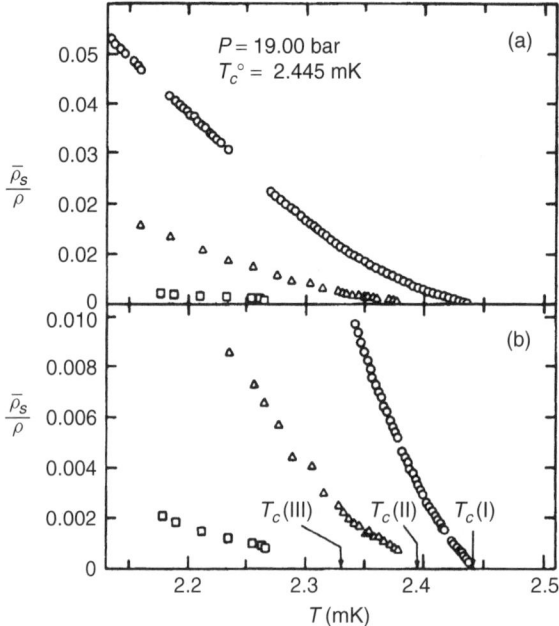

Fig. 4.7. Depression of superfluid transition temperature. (b) Expanded view of (a). The pressure is 19 bar and the bulk transition temperature is 2.442 mK.

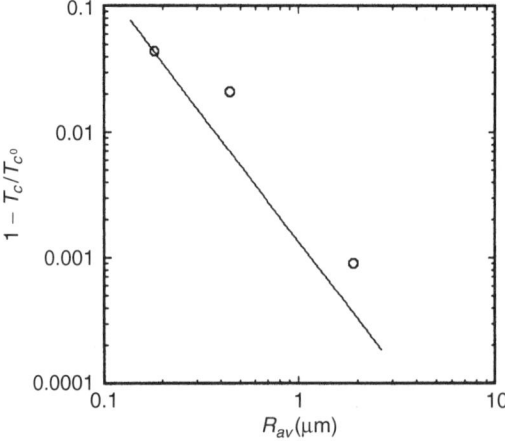

Fig. 4.8. Fractional depression of transition temperature as a function of pore size.

4.5.4. Attenuation of Fourth Sound

If the normal component is totally immobilized by viscous interaction with the walls within the superleak, the attenuation of fourth sound would be governed by the bulk viscosity ζ_3 according to Eq. 4.24. Evaluating Eq. 4.24 with the theoretical values [65] of ζ_3 gives much smaller attenuation than measured. To the extent that the normal component moves relative to the superleak owing to imperfect locking, the viscous stress leads to additional dissipation. According to the boundary condition of the ordinary hydrodynamics, the tangential velocity of the normal component vanishes at the wall surface. Consider fourth sound propagation in a superleak made of cylindrical tubes of radius R. The attenuation of fourth sound in this case may be expressed in terms of the quality factor Q_H of fourth sound resonance at angular frequency ω:

$$\frac{1}{Q_H} = \frac{\rho_n^2 R^2 \omega}{8\rho_s \eta} \qquad (4.34)$$

Since the normal fluid shear viscosity is large in liquid ^{3}He, this formula would imply quality factors of several 10^3 in pores with radius 0.5 μm. This hydrodynamic limit turns out again inadequate to explain the fourth sound attenuation in ^{3}He.

The mean free path length, $l \sim V_F \tau \sim 40$ μm/T^2(mK), is ~ 8 μm at T_c at 18 bar. The condition for validity of bulk hydrodynamics, $l <$ pore diameter, is clearly violated. In this case, the usual "sticky" boundary condition, $\mathbf{V}_{n\text{ tangential}} = 0$, is modified to one of a "slip" boundary condition where $\mathbf{V}_{n\text{ tangential}} = 0$. Jensen, Smith, and Wölfle [66] showed that the slip contribution to the attenuation is given by

$$\frac{1}{Q_S} = \frac{\rho_n^2 \kappa \zeta R \omega}{8\rho_s \eta} \qquad (4.35)$$

where κ is a numerical constant of about 6 and ζ is the slip length. The slip length is defined as the distance *behind* the wall at which the fluid velocity tends to zero when extrapolated from the fluid velocity in the bulk [67]. Unlike in the sticky boundary condition, the fluid flow with finite velocity at the wall. Within the accuracy of about 15%, the slip length is proportional to the mean free path l and may be written in terms of the specularity S as

$$\zeta = 0.7\, l \left(\frac{1+S}{1-S} \right) \qquad (4.36)$$

In the limit of completely diffusive scattering surface, S is zero. Note that as the specularity approaches unity, the slip length diverges as expected.

The attenuation of fourth sound in superfluid ^{3}He should be analyzed in terms of Eqs. 4.35 and 4.36. An example of the measured temperature dependence of the dissipation factor, $1/Q$, is shown in Figure 4.9 for a resonator packed with 3-μm powder at liquid pressure of 5 bar [68]. The dots are the Q_S^{-1} computed from Eqs. 4.35 and 4.36 by setting $S = 0$ with the pore radius adjusted to 0.67 μm. The average pore radius as determined by Mercury intrusion porosimetry is 0.42 μm. Qualitative temperature dependence of the measurement is reproduced but there is large deviation at the low temperature end. Experiments with well-defined superleak geometry are needed. Assuming that the formulation of the fourth sound attenuation based on slip boundary conditions is correct, its measurement can be used to compute the specularity S.

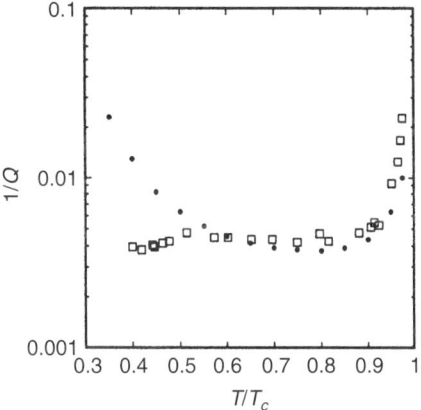

Fig. 4.9. 1/Q vs temperature. Dots are from theory and squares from experiment.

4.5.5. Boundary Condition on Superfluid ^{3}He as Altered by ^{4}He Coverage

In the discussions on the suppression of superfluid fraction in the previous section, it was more or less assumed that the surfaces surrounding ^{3}He were rough such that incident quasi-particles scatter off diffusely and that coherence of superfluid pairs was broken in the process. This led to the boundary condition that the local superfluid density vanishes at the wall. What property of the surface determines the diffusivity of scattering was left unanswered. Is the roughness of the surface important? Is it the magnetic impurities at the solid surface that change the diffusivity? In this section, experiments to probe such questions as these using fourth sound will be described.

When a small amount of ^{4}He is mixed in with ^{3}He, it preferentially coats the solid surfaces in the apparatus. This occurs because of smaller zero point motion of the heavier ^{4}He atoms. The ^{4}He coverage of surfaces gives us a handle on altering the nature of the surface scattering by the deposition of nonmagnetic inert layer. If the coverage is sufficiently increased, the ^{4}He layer changes from normal solid/fluid to superfluid, which should again change the scattering condition. If sufficiently high pressure is applied, the ^{4}He layer should solidify. The liquid/solid transition should again change the scattering process and hence the boundary condition. All of these considerations are motivations for studying the fourth sound propagation in porous media whose surfaces are "painted" with varying amounts of ^{4}He.

A systematic study of the influence of ^{4}He coverage on fourth sound was carried out by Kim et al. [68]. The main results were as follows. At any applied liquid pressure below the freezing pressure of ^{4}He, up to a critical coverage of about 1.5 to 2 monolayers of ^{4}He, the measured superfluid density did not change. The smaller the powder size, the greater was the critical coverage. The observed presence of critical coverage was consistent with the torsion oscillator experiments of Freeman et al. [69] and Tholen and Parpia [70]. When the ^{4}He coverage was increased beyond the critical thickness, the superfluid density was found to increase sharply toward the bulk value. Before the bulk superfluid density could be achieved, the thermal contact between the nuclear demagnetization apparatus and the sample liquid ^{3}He became so poor that it could not be cooled below the superfluid transition temperature. As the pressure was increased at a given ^{4}He coverage greater than the critical coverage, the relative enhancement of the superfluid density decreased as the freezing pressure (24 bar) of ^{4}He was approached. At the ^{4}He freezing pressure, the superfluid density enhancement vanished and reverted to the "bare" powder value.

The unique feature of fourth sound measurement was that it allowed determination of the specularity as the ^{4}He coverage was increased. The specularity was derived from the measurement of quality factor as follows:

$$\frac{1+S}{1-S} = \frac{(\rho_s/Q\omega\rho_n^2)_{\text{He-4}}}{(\rho_s/Q\omega\rho_n^2)_{\text{bare}}} \qquad (4.37)$$

where bare refers to the quantity with pure ^{3}He and ^{4}He refers to that with ^{4}He coverage. As it seemed reasonable, it was assumed here that the constant κ, the slip length ζ, and the pore radius R did not depend on ^{4}He coverage. The specularity determined in this manner at $T/T_C = 0.47$ and at 5 bar is shown in Figure 4.10. The results in Figure 4.10 demonstrate that the critical thickness of ^{4}He needed for enhancement of superfluid density coincides with that for specularity. However, whether the superfluid fraction reaches the bulk value when the specularity reaches unity could not be answered. In this limit, fourth sound loses its usefulness as the attenuation diverges.

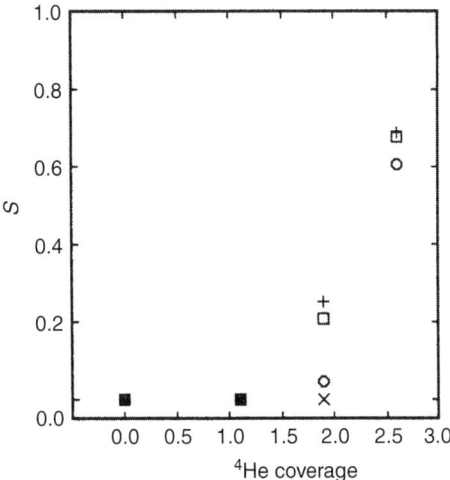

Fig. 4.10. Specularity measured by fourth sound. Symbols refer to different powders in the resonators.

4.5.6. Fourth Sound in Clean Geometry, Aerogel, Magnetic Field, and A_1

To avoid the index of refraction effects as well as the complications arising from the unknown pore space geometry, the simplest would be to construct a fourth sound superleak from "clean" straight parallel cylindrical pores or rectangular channels of well-defined dimension. If the pore radius is increased to 10 μm, the quality factor of the resonator of Figure 4.9 decreases substantially to about 18, in the mid-temperature range and at the same frequency. The difficulty is to arrange a regular pore structure with dimensions of order 10 μm opening and of order 1 cm length. There have been several attempts [71, 72], but observation of plane wave fourth sound resonance in a clean pore structure has proven elusive.

Effects of magnetic field on the fourth sound propagation were studied by Daly [73]. It was found that the superfluid density and the transition temperature were depressed by applied magnetic field. More recently, propagation modes that couple liquid ^{3}He to the tenuous structure of Aerogel were studied [74]. Two types of modes with different velocities were found and analyzed to extract the superfluid density in the Aerogel porous structure. Fourth sound propagation has not been studied in the A_1 phase. Stern and Liu [75] considered it theoretically and found that the attenuation is dominated by the surface magnetic relaxation. Fourth sound may provide a convenient tool in studying magnetic coupling at the superleak surfaces. Another application of fourth sound, which has not been realized in ^{3}He but extensively investigated in ^{4}He, is the Doppler-shift fourth sound propagation in the presence of persistent superfluid flow [76]. Recently, the Doppler-shift of fourth sound in B phase was predicted to exhibit unusual behavior [77].

4.6. SPIN-ENTROPY WAVE IN SUPERFLUID ^{3}HE A_1

4.6.1. Introduction

When an external magnetic field is applied on liquid ^{3}He, the superfluid transition T_c splits into two, creating a new phase called A_1 phase. In this section, the second sound, or more appropriately, the spin-entropy wave propagation in the A_1 phase, is described. The spin-entropy wave is a hybrid between the (dominant) longitudinal spin

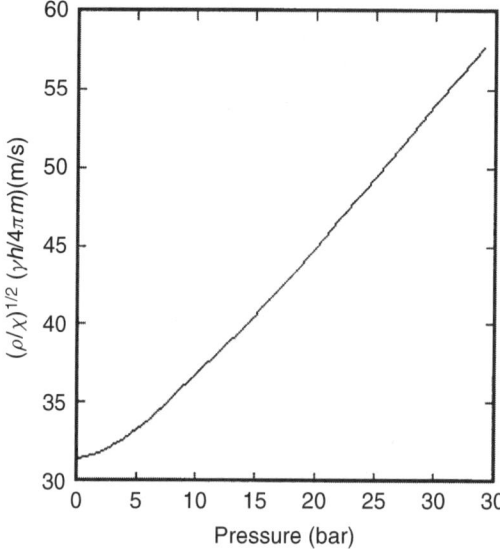

Fig. 4.11. Pressure dependence of "spin stiffness." See text for definition.

wave and entropy wave propagations. The wave propagation arises primarily from the stiffness in the spin density. In contrast to the purely thermal second sound wave in B phase, the propagation velocity is almost 3 orders of magnitudes greater. This makes the spin-entropy wave a much easier and friendlier tool for studying superfluid phenomena.

The anisotropic velocity of spin-entropy wave is given by Eq. 4.27. The pressure dependence of the temperature independent "spin stiffness" factor, $(\rho/\chi)^{1/2}(\gamma\hbar/2m)$, is shown in Figure 4.11. The stiffness factor should be multiplied by (ρ_s/ρ_n) to compute the temperature-dependent velocity.

The first observation of the spin-entropy wave was made at the melting pressure in a magnetic field of 9 kOe [78]. The observations were made with pulse techniques using OSTs. As expected, the spin-entropy wave propagation occurred only in the A_1 phase and it disappeared abruptly just below T_{c2}. The observed velocity was consistent with Eq. 4.27. The experimental observation of the spin-entropy wave consistent with theory provided a convincing support for the theory advanced by Liu [38].

4.6.2. Oscillating Superleak Transducer (OST)

An OST is an efficient mechanical generator/detector of second sound [49, 50]. Its active element is a superleak membrane whose motion induces relative motion between the normal and superfluid components. The motion is usually capacitively driven by a DC-biased voltage applied between the membrane and a fixed electrode. The most popular superleak membrane for OSTs is a Nuclepore filter sheet (polycarbonate), which can be obtained in a variety of pore sizes. One side of the sheet is coated with thin ($\sim$1000 Å) film of metal (e.g., silver) to serve as an electrode. In our experiments, the pore diameter of the Nuclepore membrane ranged from 1 to 8 μm. The pore diameter should be larger than the healing length at the temperature of interest to avoid size effects. A schematic of a typical resonator is shown in Figure 4.12. The resonator body was a cylinder (diameter = 8 mm and length = 12.5 mm). The Nuclepore membranes were clamped between the resonator body and the casing for the fixed electrodes by a set of screws (not shown) around the perimeter. The membranes were lightly stretched around the rim of the resonator body during the assembly process. The spacing between

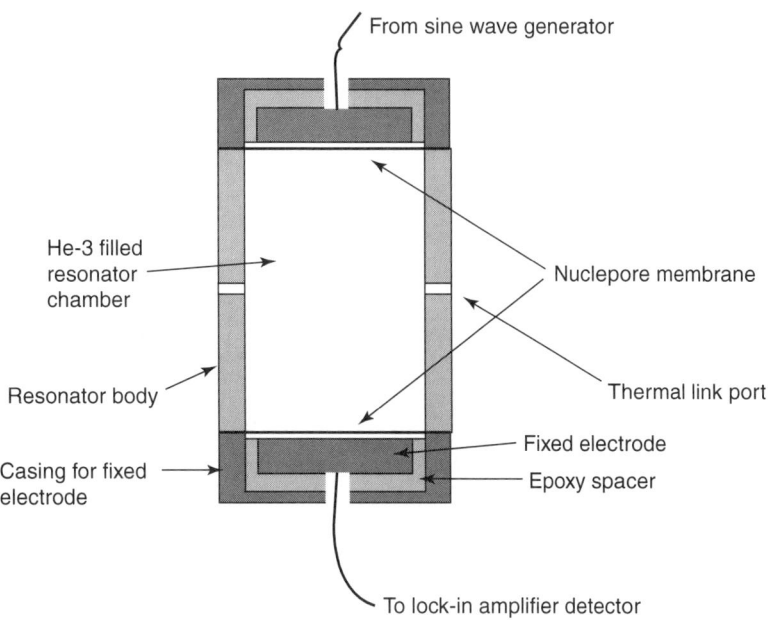

Fig. 4.12. Schematic of spin-entropy wave resonator.

the membrane and the fixed electrode was set to about 50 μm. The capacitance was typically 5 pF.

4.6.3. Superfluid Density

Since the superfluid density is a tensor quantity, the anisotropy orientation effects must be taken into account in interpretation of experiments. The spatial dependence of the **l** vector texture is determined by that which minimizes the total free energy [21]. Let us consider the case when the external magnetic field **H** is applied along the axis (**z** direction) of the cylinder in Figure 4.12. The wave-vector **q** of the spin-entropy wave propagation is also directed along **z**. In this case, the Zeeman magnetic energy of the Cooper pairs is minimized when its spin is directed antiparallel to the magnetic field. The magnetic dipolar energy of the pair is minimized if the spin direction and the **l** vector are perpendicular to each other. In this way, the spins of the pair spend at least part of the time in the smallest interaction energy alignment of "head" to "tail" orientation. The **l** vector must lie in the **x − y** plane. Within the **x − y** plane, the **l** vector texture forms in such a way to minimize the bending while satisfying the boundary condition that **l** be perpendicular to the resonator wall. The superflow that comes with the spin-entropy wave propagation would tend to make **l** parallel to **q**, but the associated kinetic energy is much smaller than dipolar energy for this to occur. Immediately adjacent to the transducer faces, the **l** vector is along **z**, but the direction heals into **x − y** plane within a short distance of order μm. Thus, the spin-entropy wave senses the perpendicular (to **l**) component of the superfluid density, $[\rho_s]_\perp$.

The measurements of spin-entropy wave in the configuration discussed in the previous paragraph has been extended to magnetic fields as high as 15 tesla and to liquid pressures down to 10 bar. The superfluid density was computed from the j-th plane wave resonance frequency f_j:

$$\left[\frac{\rho_s}{\rho_n}\right]_\perp = \left(\frac{f_j}{2jL}\right)^2 \bigg/ \left[\left(\frac{\rho}{\chi}\right)\left(\frac{\gamma\hbar}{2m}\right)^2\right] \qquad (4.38)$$

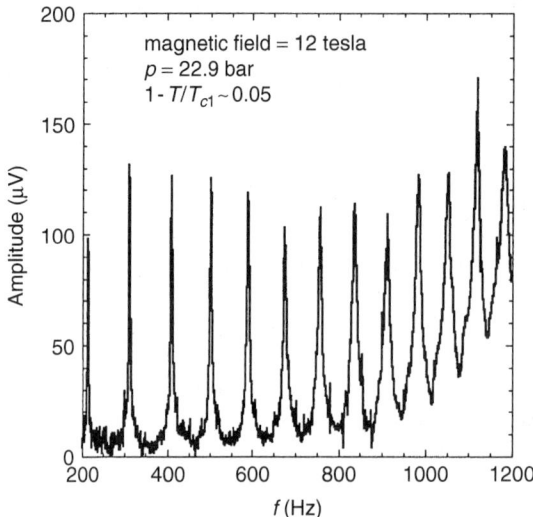

Fig. 4.13. Resonance amplitude spectrum of spin-entropy wave.

where L is the length of the resonator. As high as 15-th mode was observed. An example of resonance spectrum is shown in Figure 4.13. The gradual decrease in the difference in resonant frequency of adjacent modes was caused by a temperature rise during the sweep. The detected signal amplitude together with the amplifier gain, transducer capacitance, the bias circuit parameters, and the porosity of the Nuclepore membrane gives an estimated superfluid velocity between 5 and 20 μm/s. Though this is much smaller than the critical velocities observed in A_1 phase [79], it can produce explosive texture transitions under appropriate conditions (see last paragraph of this section).

The results for the data at 22.9 bar are summarized in Figure 4.14 by plotting the $[\rho_s/\rho]_\perp$ at T_{c2} for all fields (from 13 to 120 kOe) of measurements. To convert to the reduced temperature, $1 - T_{c2}/T_{c1}$, the following measured values [80, 81] were used: $T_{c1} - T_{c2} = 5.2$ (μK/kOe), $T_{c1} - T_c = 3.2$ (μK/kOe), and $T_c = 2.315$ mK. Plotted in this manner, the $[\rho_s/\rho]_\perp$ increases strictly linearly with the reduced temperature.

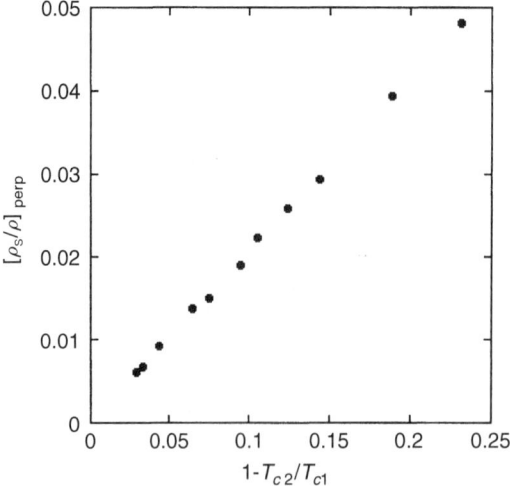

Fig. 4.14. Perpendicular component of superfluid fraction in A_1 phase at $p = 22.9$ bar.

To compare with theory, the measured superfluid fraction should be corrected for the molecular field effects to extract the "bare" superfluid density: $[\rho_s/\rho]_{\text{bare}} = [\rho_s/\rho]_\perp \{1 + F_{1S}/3\}/\{1 + [\rho_s/\rho]_\perp F_{1S}/3\}$, where F_{1S} ($= 11.8$ at 22.9 bar) is a Landau Fermi liquid parameter. We find $[\rho_s/\rho]_{\text{bare}} = a(1 - T_{c2}/T_{c1})$ with $a = 0.95$. The expected value of a is unity from BCS theory applied to ^{3}He.

4.6.4. Nonlinear Response

The perpendicular component of the superfluid density could be measured as described in the previous section by aligning the sound propagation direction parallel to the magnetic field. Consider now the configuration where the propagation (along the cylinder axis) is directed at 90° to the magnetic field. Minimizing the Zeeman and dipolar energies requires that the **l** vector lies in the plane parallel to the propagation direction. In the absence of superflow, the equilibrium **l** texture is set by minimizing the bending energy and satisfying the boundary condition as before. Thus, this configuration would lead to a measurement of some combination of both the parallel and perpendicular components of the superfluid density. The superflow associated with the spin-entropy wave *cannot* be neglected. Interesting and unexpected nonlinear effects were observed in the spin-entropy wave resonance spectra [82].

An example of response of $j = 1$ mode is shown in Figure 4.15. The drive transducer was driven sinusoidally with a constant amplitude. The in-phase (upper graph) and quadruture-phase (lower graph) response were detected by a lock-in amplifier. The response signal was very different depending on the direction of frequency sweep up

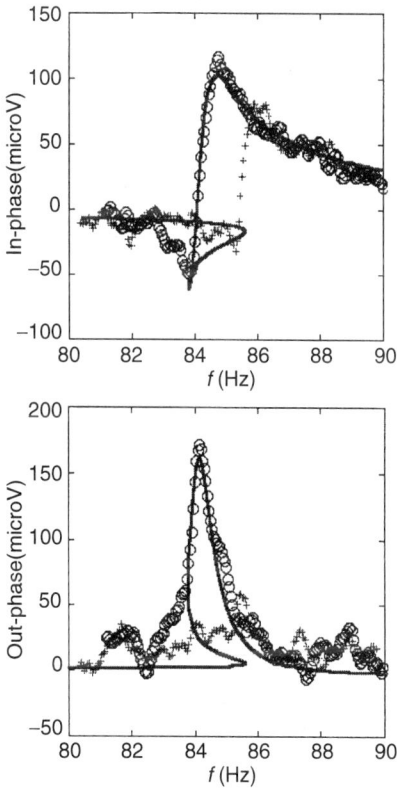

Fig. 4.15. Example of nonlinear response (crosses indicate frequency sweep up; circles, frequency sweep down). The smooth line is theory (see text). The magnetic field was directed perpendicular to the spin-entropy wave propagation direction. The temperature was near T_{c2}. $H = 40$ kOe. $P = 22.9$ bar.

(crosses) or down (circles)! The response was unexpectedly hysteretic and nonlinear. By contrast, the response in the **q**∥**H** configuration behaved linearly at all drive levels unless a critical velocity was exceeded. The behavior as shown in Figure 4.15 occurred at very low drive levels, where no critical velocity effects were expected. The frequency at which the sharp change in response occurred decreased as the drive level was increased. To add more surprise, the response tended to revert to *linear* response at *higher* drive levels.

There is a simple mechanical model [83] that can reproduce most of the observed spin-entropy wave resonance response. Consider a harmonically driven damped oscillator whose spring is "soft" and nonlinear. Let the spring constant be k_1 at low drive amplitude but it decreases smoothly to and remains constant at k_2 ($<k_1$) at high amplitude. As the drive frequency is decreased toward the resonance frequency, the amplitude increases and so the resonance frequency decreases. The resonance tends to run away from the driving frequency toward lower frequency. Dissipation prevents this runaway effect to continue and eventually jumps down to the low-frequency response. The analytical response may be solved for the response at the drive frequency and is shown as the solid lines in Figure 4.15 by adjusting the form of variation of the spring constant. The double-valued solution can explain the hysteretic spectrum dependent on the direction of frequency sweep.

It is the velocity of spin-entropy wave, C_{se}^2, that plays the role of the stiffness of the spring in the above mechanical model. Recall that C_{se} depends on the relative orientation between **q** (and hence $\mathbf{V}_s$) and **l** (constrained to lie in the plane parallel to **q**) via the dependence on the superfluid density. The wave vector **q** is fixed by the resonator. In the absence of superflow, the equilibrium **l** texture (shown by the top panel in Figure 4.16) is determined by that which minimizes free energy associated with the bending of **l** vector. The presence of superflow contributes kinetic energy to the free energy and affects the **l** texture (lower two panels of Figure 4.16). Since the superfluid density tensor in the parallel direction is smaller, the kinetic energy contribution favors the alignment of $\mathbf{V}_s$ parallel to **l**. The final **l** texture is determined by the competition between the bending and kinetic energies while satisfying the boundary condition at the walls.

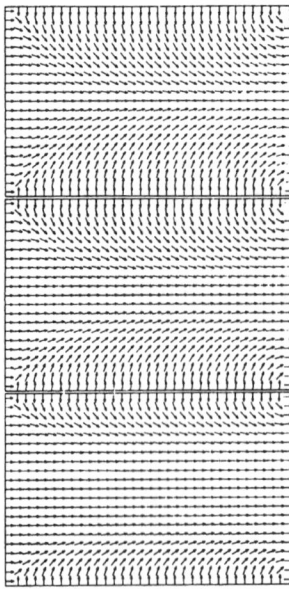

Fig. 4.16. Calculated texture by minimizing free energy. The superfluid velocity amplitude is 0, 4, and 8 μm/s in the top, middle, and bottom panels, respectively.

The calculated **l** textures in the mid-plane of the cylindrical resonator are shown in Figure 4.16 for the given values of the superfluid velocities. The curvature at the wall is neglected. Since the relaxation time for **l** texture is expected to be long compared with the period of the spin-entropy wave, the time-averaged superflow determines the texture. The spin-entropy wave propagation induces the relatively small superfluid velocity needed to alter the texture pattern. It is this superflow velocity-dependent texture and hence C_{se} that results in the observed nonlinearity. In the large flow limit, the texture saturates to $\mathbf{l} \| \mathbf{V}_s$ essentially throughout the resonator and is no longer dependent on the velocity. This is the origin of the mysterious return to linear behavior at high drive levels.

In the experiments just described, the same resonance of spin-entropy was used to create as well as to observe the nonlinear effects. To separate the driver transducers from the detector transducers and to simplify the geometry, a rectangular resonator was built with six independent transducers attached to each face [84]. The magnetic field was applied along **z** as before. The idea was to create large superflow in the **y** direction by one set of transducers and to measure their effects by observing resonance in the **x** direction. It was observed that when a threshold drive level in the **y** direction was exceeded, large, sudden, and hysteretic jumps in the resonance frequency were observed. Numerical simulations indicated that these were signatures of first-order, nonequilibrium, superfluid flow-driven textural transitions. A cubic centimeter or so of ultra-low cold liquid ^{3}He in a large magnetic field provides us a fascinating laboratory for studying flow-induced phase transitions.

4.6.5. Attenuation of Spin-Entropy Wave

The attenuation of spin-entropy wave was considered by Grabinski and Liu [85]. In the limit of low frequencies and fixed **l** texture, they found the attenuation proportional to the square of the frequency as

$$\alpha = \frac{\omega^2}{2\rho C_{se}^3}\left[D + \frac{\rho_s}{\rho_n}\left\{\frac{4}{3}\eta + \zeta_2 + \rho^2\zeta_3 - 2\rho\zeta_1\right\} - \left(\frac{2T\sigma}{\rho C_V}\right)\left(\frac{2m}{\hbar}\right)\zeta_4 + \frac{K}{C_V}\right] \quad (4.39)$$

which is similar to Eq. 4.14 except the terms involving D, the spin diffusion coefficient, and ζ_4, the dissipative coefficient to the spin current in response to temperature gradient. All terms except those involving D, η, and ζ_3 are small and can be neglected. Since the spin-entropy wave is primarily spin-density oscillation in nature, it is the spin diffusion, not entropy diffusion, that dominates the attenuation.

The attenuation of spin-entropy wave was recently measured in a cylindrical resonator at 22.9 bar in magnetic fields up to 11 tesla applied parallel to the propagation direction [86]. Measurements were made in the frequency range between 50 Hz and 2.2 kHz such that the quadratic dependence of α on frequency could be observed. The attenuation coefficient was extracted by fitting the measured resonance spectra. The viscous losses at the wall contribute to attenuation as $\alpha_v = (\rho_s/R\rho C_{se})(\pi\eta f/\rho_n)^{1/2}$, where R is the radius of the resonator. This was subtracted from the total measured attenuation to compute the bulk attenuation, $\alpha_b = \beta\omega^2$.

The measured temperature dependence of β is plotted in Figure 4.17 as a function of reduce temperature, $t = 1 - T/T_{c1}$. The symbols refer to data from the several different magnetic fields indicated. The dotted line through the data points is a guide to the eye. The major contributions to the theoretical attenuation are indicated: spin diffusion (dashed) [5], shear viscosity (dash-dotted), and bulk viscosity ζ_3 (dotted) [65]. The sum of these three is shown by the solid line. There is significant deviation between the theory and the experiment as the transition temperature is approached. The spin diffusion coefficient D was taken as a constant equal to that in the normal state at $T = T_{c1}$ since there was no theory nor measurement of D in the superfluid state. The

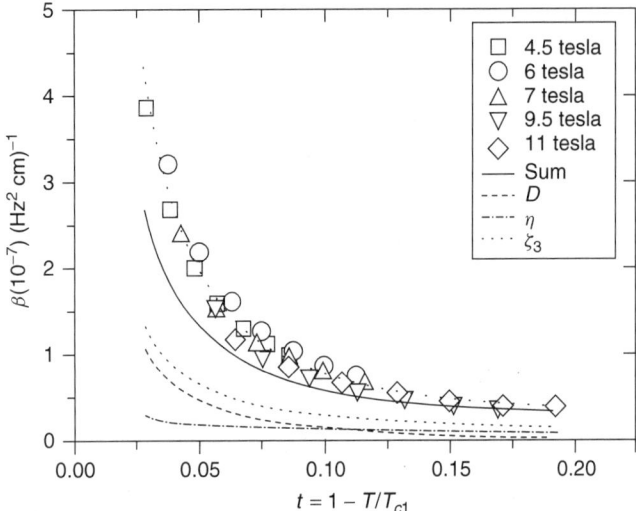

Fig. 4.17. Attenuation of spin-entropy wave.

difference between theory and experiment of the attenuation might be caused by a greater spin diffusion coefficient in the A_1 phase. The experiment should be extended to measure pressure dependence.

ACKNOWLEDGMENTS

I gratefully acknowledge the contributions of my colleagues with whom I had the pleasure of doing and discussing the experiments presented: Marina Bastea, Tim Chainer, Jim Coleman, Michael Grabinski, Keiichi Ichikawa, Q. Jiang, D. Kim, Takao Kodama, Singteh Lu, Yukio Morii, Yuichi Okuda, Doug Paulson, Seth Putterman, Rene Ruel, the late Izzy Rudnick, Taro Sato, P. deVegvar, and the late John Wheatley. The research was supported in part by National Science Foundation grant DMR9510306.

References

1. Wilks, J. (1967). *Liquid and solid helium*. Oxford: Clarendon Press.
2. Wilks, J., and Betts, D.S. (1987). *An introduction to liquid helium*. Oxford: Clarendon Press.
3. Greywall, D.S. (1983). *Phys. Rev.* **B27**: 2747.
4. Parpia, J.M., Sandiford, D.J., Berthodl, J.E., and Reppy, J.D. (1978). *Phys. Rev. Lett.* **40**: 565.
5. Wheatley, J.C. (1970). *Progress in low temperature physics*. Vol. 6, Gorter, C.J., ed., Amsterdam: North-Holland.
6. Reif, F. (1965). *Fundamentals of statistical and thermal physics*. New York: McGraw-Hill.
7. The gyromagntic ratio of ^{3}He is negative: $g = -2.0 \times 10^4$ (gauss s)$^{-1}$. So the spin vector is directed antiparallel to the magnetic moment vector.
8. Landau, L.D. (1957). *Soviet Phys. JETP.* **3**: 920.
9. Engel, B.N., and Ihas, G.G. (1985). *Phys. Rev. Lett.* **55**: 955.
10. Bardeen, J., Cooper, L.N., and Schrieffer, J.R. (1957). *Phys. Rev.* **108**: 1175.
11. Osheroff, D.D., Richardson, R.C., and Lee, D.M. (1972). *Phys. Rev. Lett.* **28**: 885. They were awarded the Nobel prize in 1996 for their discovery. It is interesting to note that the title of the original discovery paper refers to transitions in the solid phase in the apparatus, which contained both liquid and solid ^{3}He.
12. Osheroff, D.D., Gully, W.J., Richardson, R.C., and Lee, D.M. (1972). *Phys. Rev. Lett.* **29**: 920.
13. Leggett, A.J. (1975). *Rev. Mod. Phys.* **47**: 331.
14. Wheatley, J.C. (1975). *Rev. Mod. Phys.* **47**: 415.
15. Greywall, D.S. (1986). *Phys. Rev.* **B33**: 7520.

16. Pekola, J.P., Simola, J.T., Nummila, K.K., Lounasmaa, O.V., and Packard, R.E. (1984). *Phys. Rev. Lett.* **53**: 70.
17. Gammel, P.L., Ho, T.-L., and Reppy, J.D. (1985). *Phys. Rev. Lett.* **55**: 2708.
18. Berthold, J.E., Giannetta, R.W., Smith, E.N., and Reppy, J.D. (1976). *Phys. Rev. Lett.* **37**: 1138.
19. Parpia, J.M., Wildes, D.G., Saunders, J., Zeise, E.K., Reppy, J.D., and Richardson, R.C. (1985). *J. Low Temp. Phys.* **61**: 337.
20. See review by Lee, D.M., and Richardson, R.C. (1978). In *The Physics of Liquid and Solid Helium*, Part II, Bennemann, H., and Ketterson, J.B. eds., New York: Wiley-Interscience.
21. Vollhardt, D., and Wölfe, P. (1990). *The Superfluid Phases of Helium 3*. London: Taylor and Francis.
22. Anderson, P.W., and Morel, P. (1961). *Phys. Rev.* **123**: 1911. Anderson, P.W., and Brinkman, W.F. (1973). *Phys. Rev. Lett.* **30**: 1108.
23. Balian, R., and Werthamer, N.R. (1963). *Phys. Rev.* **131**: 1553.
24. White, G.K. (1979). *Experimental techniques in low temperature physics*. Oxford: Clarendon.
25. Lounasmaa, O.V. (1974). *Experimental principles and methods below 1 K*. London: Academic.
26. Betts, D.S. (1989). *An introduction to millikelvin technology*. Cambridge: Cambridge Univeristy Press.
27. Pobell, F. (1995). *Matter and methods and low temperatures*. Berlin: Springer.
28. Landau, L.D. (1957). *Soviet Phys. JETP* **5**: 101.
29. Keen, B.E., Mathews, P.W., and Wilks, J. (1963). *Phys. Lett.* **5**: 5.
30. Abel, W.R., Anderson, A.C., Black, W.C., and Wheatley, J.C. (1965). *Physics* **1**: 337.
31. Wölfle, P. (1978). *Progress in low temperature physics*, vol. VIIA, Brewer, D.F., ed., p. 191, Amsterdam: North Holland.
32. Halperin, W.P., and Varoquaux, E. (1990). *Helium Three*. Amsterdam: Elsevier Science.
33. Zhao, Z., Adenwalla, S., Sarma, B.K., and Ketterson, J.B. (1992). *Adv. Phys.* **41**: 147.
34. Khalatnikov, I.M. (1965). *An introduction to the theory of superfluidity*. Reading, Mass: Addison-Wesley.
35. Putterman, S.J. (1974). *Superfluid hydrodynamics*, Amsterdam: North-Holland.
36. Saslow, W.M. (1973). *Phys. Rev. Lett.* **31**: 870.
37. Wölfle, P. (1977). *J. Low Temp. Phys.* **26**: 659.
38. Liu, M. (1979). *Phys. Rev. Lett.* **43**: 1740.
39. See Grabinski, M. and Liu, M. (1990). *J. Low Temp. Phys.* **78**: 247 for a complete treatment.
40. Atkins, K.R. and Rudnick, I. (1970). *Progress in low temperature physics*, vol. VI, p. 37. Amsterdam: North-Holland.
41. Schechter, A.M.R., Simmonds, R.W., Packard, R.E., and Davis, J.C. (1998). *Nature* **396**: 554.
42. Rudnick, I. (1980). *J. Low Temp. Phys.* **40**: 287.
43. Nagai, K. and Wölfle, P. (1981). *J. Low Temp. Phys.* **42**: 227.
44. Eska, G., *et al.*, (1980). *Phys. Rev. Lett.* **44**: 1337, Eska, G., Neumaier, K., Schoepe, W., Uhlig, K. and Wiedemann, W. (1983). *Phys. Rev.* **B27**: 5534.
45. Kodama, T. and Kojima, H. (1981). *Phys. Lett.* **87A**: 103.
46. Tisza, L. (1940). *J. Phys. (Paris)* **1**: 164; Landau, L.D. (1941). *J. Phys. (Moscow)* **5**: 71.
47. Peshkov, V. (1946). *J. Phys. (Moscow)* **10**: 389.
48. Liu, M. (1984). *Phys. Rev.* **B29**: 2833.
49. Sherlock, R.A. and Edwards, D.O. (1970). *Rev. Sci. Instrum.* **41**: 1603.
50. Liu, M. and Stern, M.R. (1982). *Phys. Rev. Lett.* **48**: 1842.
51. Galileo Electro-Optics Corp., Mass: Sturbridge.
52. Lu, S.T. and Kojima, H. (1985). *Phys. Rev. Lett.* **55**: 1677.
53. Henjes, K. and Liu, M. (1988). *J. Low Temp. Phys.* **71**: 97.
54. Kojima, H., Paulson, P.N., and Wheatley, J.C. (1974). *Phys. Rev. Lett.* **32**: 141.
55. Yanof, A.W. and Reppy, J.D. (1974). *Phys. Rev. Lett.* **33**: 631.
56. Ichikawa, K., Yamasaki, S., Akimoto, H., Kodama, T., Shigi, T., and Kojima, H. (1987). *Phys. Rev. Lett.* **58**: 1949.
57. Shapiro, K.A. and Rudnick, I. (1965). *Phys. Rev.* **137**: 1383.
58. Parpia, J.M., Wildes, D.G., Saunders, J., Zeise, E.K., Reppy, J.D., and Richardson, R. (1985). *J. Low Temp. Phys.* **61**: 337.
59. Chainer, T., Morii, Y., and Kojima, H. (1984). *J. Low Temp. Phys.* **55**: 353.
60. Ambegaokar, V., deGennes, P.G., and Rainer, D. (1974). *Phys. Rev.* **A9**: 2676.
61. Buchholtz, L.J. (1986). *Phys. Rev.* **B33**: 1579.
62. Kriss, M. and Rudnick, I. (1970). *J. Low Temp. Phys.* **3**: 339.
63. Kjäldman, L.H., Kurkijärvi. J., and Rainer, D. (1978). *J. Low Temp. Phys.* **33**: 577.
64. Barton, G. and Moore, M.A. (1975). *J. Low Temp. Phys.* **21**: 489.
65. Einzel, D. (1984). *J. Low Temp. Phys.* **54**: 427.
66. Jensen, H.H., Smith, H., and Wölfe, P. (1983). *J. Low Temp. Phys.* **51**: 81.
67. Jensen, H.H., Smith, H., Wölfle, P., Nagai, K., and Bisgaard, T.M. (1980). *J. Low Temp. Phys.* **41**: 473.
68. Kim, D., Nakagawa, M., Ishikawa, O., Hata, T., Kodama, T., and Kojima, H. (1993). *Phys. Rev. Lett.* **71**: 1581.

69. Freeman, M.R., Germain, R.S., Thuneberg, E.V. and Richardson, R.C. (1988). *Phys. Rev. Lett.* **60**: 596.
70. Tholen, S.M. and Parpia, J. (1992). *Phys. Rev. Lett.* **68**: 2810.
71. Kojima, H., Paulson, D.N. and Wheatley, J.C. (1975). *J. Low Temp. Phys.* **21**: 283.
72. Akimoto, H. and Ishimoto, H., private communication.
73. Daly, K. (1988). *J. Low Temp. Phys.* **71**: 231.
74. Golov, A., Geller, D.A., Parpia, J.M. and Mulders, N. (1999). *Phys. Rev. Lett.* **82**: 3492.
75. Stern, M.R. and Liu, M. (1982). *Physica* **109A**: 2099.
76. Kojima, H., Veith, W., Guyon, E. and Rudnick, I. (1972). *J. Low Temp. Phys.* **8**: 187.
77. Kenis, A.M., Nepomnyashchy, Y.A., Mann, A. and Revzen, M. (1999). *Phys. Rev. Lett.* **82**: 584.
78. Corruccini, L.R. and Osheroff, D.D. (1980). *Phys. Rev. Lett.* **45**: 2029.
79. Ruel, R. and Kojima, H. (1987). *J. App. Phys.* **26**: 125.
80. Israelsson, U.E., *et al*. (1984). *Phys. Rev. Lett.* **53**, 1943. Remeijer, P. *et al.* (1998). *J. Low Temp. Phys.* **111**: 119, reported that $T_{c1} - T_{c2}$ has a quadratic dependence on magnetic field.
81. Greywall, D.S. (1986). *Phys. Rev.* **B33**: 7520.
82. Bastea, M., Kojima, H. and deVegvar, P.G.N. (1996). *Phys. Rev. Lett.* **76**: 2766.
83. Landau, L.D. and Lifshitz, E.M. (1959). *Mechanics* London: Pergomon Press.
84. deVegvar, P.G.N., Ichikawa, K., and Kojima, H. (1999). *Phys. Rev. Lett.* **83**: 1806.
85. Grabinski, M. and Liu, M. (1990). *J. Low Temp. Phys.* **78**: 247.
86. Sato, T., Coleman, J.J., deVegvar, P.G.N., Kojima, H., and Okuda, Y., to be published.

CHAPTER 5

ACOUSTIC PROPERTIES OF SUPERFLUID HELIUM FOUR

J. D. Maynard
Department of Physics, Pennsylvania State University, University Park, Pennsylvania, USA

Contents

Abstract 147
    5.1. Introduction 147
    5.2. Acoustic Properties of Superfluid Helium 148
    5.3. Experimental Measurement of the Sounds in Superfluid Helium 150
    5.4. Results 155
References 156

Abstract

Superfluid helium four has remarkable acoustic properties, as it can sustain four distinct modes of sound propagation. The velocities of the different sound modes, as functions of temperature and pressure, are uniquely related to the thermodynamic properties of the superfluid.

5.1. INTRODUCTION

Superfluid helium (He II) is interesting acoustically because it has more than one characteristic sound velocity [1, 2]. The different sound velocities are not simply due to numerically different elastic constants as for different modes in a solid, but rather are due to fundamentally different acoustic restoring forces. The existence of more than one sound mode makes it possible to use acoustic measurements to determine all of the equilibrium thermodynamic properties of He II with an order of magnitude improvement in precision and consistency over other measurement techniques. To this end, high-precision measurements of the different sound velocities have been made simultaneously in one experiment [3], and the data have been analyzed to obtain the density, thermal expansion coefficient, normal fluid fraction, entropy, specific heats, and compressibility as continuous functions of temperature and pressure [4]. The results cover the temperature range from 1.2 K to the lambda transition temperature T_λ and the pressure range from the saturated vapor pressure (SVP) to 25 bar.

Prior to the sound velocity measurements, the existing thermodynamic data for He II had a typical precision of 2–10%. In regions where different measurements overlapped, there were discrepancies as large as 10–15%. In particular, data at elevated pressures was inaccurate, and in some cases it was unavailable. However, by using high-resolution acoustic resonators, the sound velocities are measured to 0.1–0.2% and the thermodynamic properties are determined to typically 0.3%. Furthermore, since the sound velocities are measured simultaneously in one experiment, the resulting thermodynamic functions are completely self-consistent and suffer no problems due to different temperature or pressure calibrations among separate experiments.

The sound velocities determine directly the important thermodynamic derivatives that are integrated to give all of the thermodynamic quantities. The constants of integration can be found from the expansion coefficient $\beta_p = -(1/\rho)(\partial\rho/\partial T)_p$ along the SVP line and the entropy and density at a single point at SVP. Only one thermodynamic quantity must be found by numerically taking a derivative, a procedure that usually yields imprecise results. However, by using a Maxwell relation, this derivative can be calculated from the sound data in two independent ways. Hence the sound velocities contain an intrinsic cross-check on the only weak calculation.

In the experiment, the sound velocity data were taken as functions of temperature in 50 mK steps and as functions of pressure in 1 bar steps, corresponding to more than 400 points in the He II $P-T$ plane.

5.2. ACOUSTIC PROPERTIES OF SUPERFLUID HELIUM

When cooled to a temperature below 2.172 K, liquid ^{4}He becomes a superfluid, entering a macroscopic quantum mechanical state, and is designated He II. In this state it can appear to flow with absolutely no friction and may be driven not only with mechanical forces such as pressure, but also with thermal forces such as heat.

The model that accounts for the unusual properties of superfluid helium is the two-fluid model [2], in which He II is pictured as consisting of two independent, interpenetrating fluid components: a normal fluid component and a superfluid component. Each component has its own mass density ρ_n and ρ_s for normal and superfluid density, respectively) and its own velocity field ($\vec{V}_n$ and $\vec{V}_s$). The total fluid density is given by the sum of the component densities: $\rho = \rho_n + \rho_s$. The normal component is an ordinary fluid; it has finite entropy (i.e., carries heat) and viscosity. The superfluid component has absolutely no entropy and no viscosity; it carries no heat and, below a critical velocity, moves without friction. The relative amounts of the normal and superfluid components (ρ_s/ρ or $\rho_n/\rho = 1 - \rho_s/\rho$) are thermodynamic functions of temperature and pressure.

With a thermohydrodynamic system consisting of two fluids, it is possible to have more than one type of propagating sound wave, as illustrated in Figure 5.1. Figure 5.1(a) shows the spatial variation of part of a sound wave in which both fluid components are moving together. In the center of the wave there will be a buildup of fluid (or an increase in the total density) and an increase in pressure determined by the compressibility. The increase in pressure provides the restoring force for the wave motion just as in an ordinary sound wave; in superfluid helium this is called *first sound* and the velocity of propagation is given approximately by

$$C_1^2 = \left(\frac{\partial p}{\partial \rho}\right)_S = \gamma \left(\frac{\partial p}{\partial \rho}\right)_T \qquad (5.1)$$

where $\gamma = C_p/C_V$ is the specific heat ratio.

It is also possible to make a sound wave in which the two fluid components move in opposite directions, as shown in Figure 5.1(b). With this type of flow there would

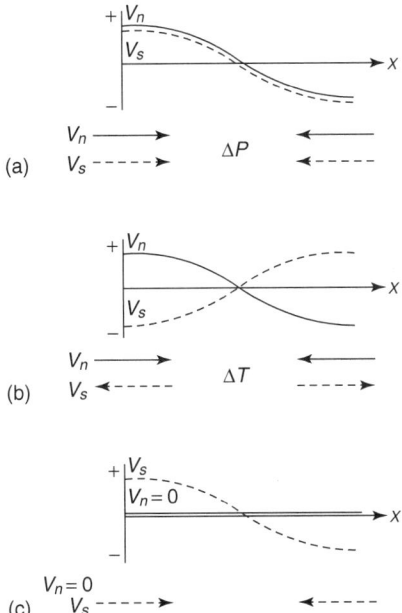

Fig. 5.1. Sound waves in superfluid helium four. (a) Fluid components moving together. (b) Fluid components moving in counterflow. (c) Normal fluid clamped, with only the superfluid component moving.

be negligible change in the density and pressure and hence there would be no pressure restoring force. However, there would be a change in the relative amounts of the normal and superfluid components, and this corresponds to a temperature change. Furthermore, since the normal fluid carries entropy and the superfluid does not, there would also be a change in the specific entropy. Hence Figure 5.1(b) represents a temperature or entropy wave; it is called *second sound* and has a velocity of propagation given approximately by

$$C_2^2 = \left[\left(\frac{\rho_n}{\rho}\right)^{-1} - 1\right] S^2 \left(\frac{\partial S}{\partial T}\right)_p^{-1} \tag{5.2}$$

where S is the entropy per unit mass. Whereas first sound has a pressure restoring force and a velocity involving $\left(\frac{\partial P}{\partial \rho}\right)$, second sound has a thermodynamic restoring force and its velocity is related to $(\partial S/\partial T)_p = C_p/T$.

Figure 5.1(c) shows another possible flow pattern for the two fluid components. In this case, there must be some environmental condition clamps the normal fluid motionless through its viscosity. The superfluid component, having no viscosity, is still free to flow and form a sound wave. Several different types of sound are possible depending on the dominant type of restoring force for the superfluid component. One normal-fluid-clamping environment, called a superleak, consists of a porous medium such as a packed powder or "Vycor" glass, with pore sizes of 10 nm to 1 μm. In such small pores, any ordinary fluid with even the smallest known viscosity would be immobilized. In the superleak the flow of the superfluid results in predominantly pressure waves, and the sound mode called *fourth sound* propagates with velocity given approximately by

$$C_4^2 = \left(1 - \frac{\rho_n}{\rho}\right) C_1^2 \tag{5.3}$$

Being a pressure wave, fourth sound is analogous to first sound, and the factor of $(1 - \rho_n/\rho)$ in the velocity expression reflects the condition that the normal fluid is clamped.

In special "pressure-release" situations, the normal fluid may be clamped as in Figure 5.1(c), but the dominant restoring force is thermal, as for second sound. This is *fifth sound* [5, 6], with a propagation speed given by

$$C_5^2 = \left(\frac{\rho_n}{\rho}\right) C_2^2 \tag{5.4}$$

Experimentally it is quite difficult to clamp the normal fluid while providing pressure release conditions. For this reason fifth sound has limited application in studying superfluid helium.

The superfluid sound mode called *third sound* is a surface wave that propagates in a thin adsorbed helium film [2, 7]. This velocity of this mode depends on the substrate material and the thickness of the film, and thus it is not a mode characteristic of the superfluid. Third sound is not considered further in this chapter.

5.3. EXPERIMENTAL MEASUREMENT OF THE SOUNDS IN SUPERFLUID HELIUM

The experimental cell used to measure the sound velocities simultaneously [3] is shown in Figure 5.2. It consists of a brass cylinder drilled with four holes to make four cylindrical acoustic resonators. Drive and pickup sound transducers seal the ends of the resonators. The sound velocities are determined from the resonant frequencies corresponding to fitting an integral number of half wavelengths in the length of the resonator. Two of the resonators were empty and were used for measuring first and second sound. The other two were packed with fine powders for measuring fourth sound. One of the packed powders had a pore size of 1 μm, which just clamped the normal fluid. The other powder had pore sizes of 0.01 μm, which produced measurable quantum mechanical size effects. The data from this resonator are not discussed here.

All of the transducers except for the second sound resonator were capacitive devices using aluminized, electreted Teflon for the oscillating membrane. The second sound transducers were similar but used a porous membrane. Such a membrane drives the normal fluid component but permits the inviscid superfluid component to flow through the pores. This results in the counterflow necessary to drive second sound.

All of the resonators had quality factors of at least 300. The resonant peaks could easily be determined to within a fraction of its full width at half maximum and hence the sound velocities could be determined to within 0.1–0.2%.

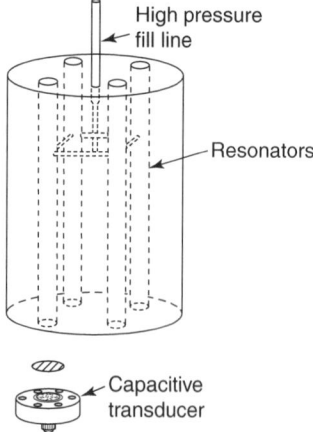

Fig. 5.2. The experimental cell, with four resonators for first sound, second sound, and fourth sound in two different superleaks.

Table 5.1. Measured velocities of first, second, and fourth sounds[a].

T	SVP			$p=1$			$p=2$		
	C_1	C_2	C_4	C_1	C_2	C_4	C_1	C_2	C_4
1.20	237.4	18.55	234.0	245.0	18.14	241.5	252.3	17.76	248.5
1.25	237.6	18.78	232.8	244.8	18.41	240.2	252.1	18.07	247.2
1.30	236.8	19.03	231.2	244.6	18.70	238.7	252.0	18.39	245.6
1.35	236.5	19.30	229.4	244.4	19.00	236.8	251.8	18.70	243.7
1.40	236.1	19.58	227.2	244.0	19.28	234.6	251.5	19.00	241.4
1.45	235.6	19.84	224.7	243.7	19.55	232.0	251.2	19.27	238.6
1.50	235.2	20.07	221.7	243.2	19.78	228.9	250.8	19.51	235.4
1.55	234.6	20.25	218.3	242.7	19.96	225.3	250.3	19.68	231.7
1.60	234.0	20.37	214.3	242.2	20.07	221.2	249.8	19.79	227.3
1.65	233.4	20.41	209.8	241.6	20.11	216.4	249.2	19.82	222.3
1.70	232.7	20.36	204.6	240.9	20.05	210.9	248.5	19.75	216.5
1.75	231.9	20.19	198.6	240.1	19.87	204.5	247.7	19.56	209.7
1.80	231.0	19.89	191.7	239.2	19.56	197.2	246.8	19.23	202.0
1.85	230.0	19.43	183.7	238.2	19.08	188.7	245.8	18.73	193.0
1.90	228.8	18.78	174.4	237.0	18.40	178.8	244.6	18.02	182.5
1.95	227.4	17.89	163.5	235.6	17.47	167.2	243.2	17.04	170.1
2.00	225.8	16.69	150.4	234.0	16.20	153.3	241.7	15.69	155.2
2.05	224.0	15.00	134.4	232.2	14.43	136.1	239.8	13.78	136.4
2.10	221.8	12.39	112.6	230.0	11.75	112.4	237.6	10.91	110.0
2.15	219.2	7.99	75.1	227.4	6.91	69.7	234.8	4.53	51.5

T	$p=3$			$p=4$			$p=5$		
	C_1	C_2	C_4	C_1	C_2	C_4	C_1	C_2	C_4
1.20	259.2	17.44	255.1	265.7	17.16	261.4	272.0	16.91	267.4
1.25	259.0	17.77	253.8	265.6	17.51	260.0	271.9	17.28	265.9
1.30	258.9	18.11	252.1	265.5	17.86	258.2	271.8	17.64	264.0
1.35	258.7	18.44	250.1	265.3	18.20	256.1	271.6	17.99	261.8
1.40	258.5	18.75	247.7	265.1	18.52	253.6	271.4	18.31	259.1
1.45	258.2	19.02	244.8	264.9	18.79	250.6	271.1	18.58	255.9
1.50	257.9	19.25	241.5	264.5	19.02	247.0	270.8	18.80	252.2
1.55	257.4	19.42	237.5	264.1	19.19	242.9	270.4	18.96	247.9
1.60	256.9	19.53	232.9	263.6	19.28	238.1	269.9	19.04	242.8
1.65	256.3	19.54	227.6	263.0	19.28	232.5	269.3	19.04	236.9
1.70	255.6	19.46	221.5	262.3	19.19	226.0	268.7	18.93	230.1
1.75	254.9	19.26	214.4	261.6	18.97	218.5	267.9	18.69	222.2
1.80	254.0	18.91	206.2	260.7	18.60	209.9	267.0	18.29	213.1
1.85	252.9	18.38	196.7	259.6	18.04	199.8	266.0	17.70	202.4
1.90	251.8	17.63	185.5	258.5	17.24	187.9	264.7	16.85	189.9
1.95	250.4	16.59	172.3	257.1	16.13	173.8	263.3	15.67	174.8
2.00	248.8	15.15	156.2	255.4	14.59	156.5	261.7	14.01	155.9
2.05	246.9	13.10	135.6	253.5	12.37	133.7	259.7	11.59	130.7
2.10	244.7	9.92	105.4	251.2	8.79	98.5	257.3	7.41	87.9

T	$p=6$			$p=7$			$p=8$		
	C_1	C_2	C_4	C_1	C_2	C_4	C_1	C_2	C_4
1.20	278.0	16.70	273.2	283.8	16.52	278.6	289.3	16.35	283.9
1.25	277.9	17.08	271.5	283.6	16.89	276.9	289.1	16.73	282.0
1.30	277.7	17.45	269.5	283.5	17.27	274.7	289.0	17.11	279.7
1.35	277.6	17.80	267.1	283.3	17.62	272.2	288.8	17.46	277.0
1.40	277.4	18.12	264.3	283.1	17.94	269.2	288.5	17.77	273.9

continues

Table 5.1. (continued)

T	C_1	C_2	C_4	C_1	C_2	C_4	C_1	C_2	C_4
1.45	277.1	18.39	261.0	282.8	18.21	265.7	288.2	18.03	270.2
1.50	276.8	18.60	257.1	282.4	18.41	261.6	287.9	18.23	265.8
1.55	276.3	18.75	252.5	282.0	18.55	256.8	287.5	18.36	260.8
1.60	275.9	18.82	247.2	281.5	18.61	251.2	287.0	18.40	254.9
1.65	275.3	18.80	241.0	281.0	18.57	244.7	286.4	18.35	248.1
1.70	274.6	18.67	233.9	280.3	18.42	237.2	285.8	18.18	240.2
1.75	273.9	18.41	225.6	279.6	18.14	228.5	285.0	17.87	231.1
1.80	273.0	17.99	215.9	278.6	17.69	218.3	284.1	17.38	220.4
1.85	271.9	17.36	204.6	277.6	17.01	206.4	282.9	16.66	207.7
1.90	270.7	16.46	191.3	276.3	16.05	192.2	281.6	15.64	192.5
1.95	269.2	15.19	175.1	274.8	14.70	174.8	280.1	14.18	173.8
2.00	267.5	13.40	154.6	273.0	12.76	152.4	278.2	12.08	149.2
2.05	265.5	10.74	126.2	270.9	9.80	120.0	275.9	8.73	111.5
2.10	262.9	5.54	70.4						

	$p = 9$			$p = 10$			$p = 11$		
T	C_1	C_2	C_4	C_1	C_2	C_4	C_1	C_2	C_4
1.20	294.7	16.20	288.9	299.8	16.06	293.7	304.8	15.92	298.3
1.25	294.5	16.57	286.9	299.6	16.43	291.6	304.6	16.29	296.1
1.30	294.3	16.95	284.5	299.4	16.81	289.0	304.4	16.67	293.4
1.35	294.1	17.31	281.6	299.2	17.16	286.0	304.1	17.02	290.2
1.40	293.8	17.62	278.3	298.9	17.46	282.5	303.8	17.32	286.5
1.45	293.5	17.87	274.4	298.5	17.71	278.3	303.4	17.56	282.1
1.50	293.1	18.06	269.8	298.1	17.89	273.5	303.0	17.72	277.1
1.55	292.7	18.17	264.5	297.7	17.99	268.0	302.6	17.81	271.2
1.60	292.2	18.20	258.4	297.2	18.00	261.5	302.1	17.80	264.5
1.65	291.6	18.13	251.2	296.6	17.91	254.1	301.5	17.69	256.6
1.70	291.0	17.94	243.0	296.0	17.69	245.4	300.8	17.45	247.5
1.75	290.2	17.60	233.3	295.1	17.32	235.2	299.9	17.05	236.8
1.80	289.2	17.07	222.0	294.2	16.76	223.2	298.9	16.44	224.1
1.85	288.1	16.31	208.6	293.0	15.94	209.0	297.7	15.55	208.9
1.90	286.7	15.21	192.4	291.5	14.76	191.6	296.1	14.28	190.2
1.95	285.0	13.64	172.1	289.7	13.07	169.6	294.2	12.45	166.1
2.00	283.0	11.34	144.8	287.6	10.53	138.9	291.8	9.61	131.2
2.05	280.5	7.44	99.5	284.8	5.71	80.8	288.4	1.79	30.1

	$p = 12$			$p = 13$			$p = 14$		
T	C_1	C_2	C_4	C_1	C_2	C_4	C_1	C_2	C_4
1.20	309.6	15.80	302.8	314.3	15.68	307.0	318.8	15.56	311.1
1.25	309.5	16.16	300.4	314.1	16.04	304.6	318.7	15.92	308.6
1.30	309.2	16.54	297.6	313.9	16.41	301.6	318.5	16.29	305.5
1.35	308.9	16.88	294.2	313.6	16.75	298.1	318.1	16.62	301.8
1.40	308.5	17.18	290.3	313.2	17.04	293.9	317.7	16.90	297.4
1.45	308.2	17.41	285.7	312.8	17.26	289.1	317.3	17.11	292.3
1.50	307.7	17.56	280.4	312.3	17.40	283.5	316.8	17.24	286.4
1.55	307.3	17.63	274.3	311.9	17.45	277.1	316.3	17.27	279.7
1.60	306.8	17.60	267.2	311.3	17.40	269.6	315.8	17.20	271.8
1.65	306.2	17.47	258.9	310.7	17.24	260.9	315.1	17.01	262.7
1.70	305.4	17.20	249.3	310.0	16.94	250.8	314.3	16.68	252.0
1.75	304.6	16.76	238.0	309.0	16.47	238.8	313.4	16.16	239.3
1.80	303.5	16.10	224.5	307.9	15.76	224.5	312.2	15.39	224.0
1.85	302.1	15.15	208.3	306.5	14.73	207.1	310.6	14.28	205.3
1.90	300.5	13.78	188.1	304.6	13.24	185.2	308.6	12.66	181.4

Table 5.1. (continued)

| 1.95 | 298.4 | 11.77 | 161.5 | 302.3 | 11.03 | 155.6 | 305.9 | 10.19 | 147.9 |
| 2.00 | 295.7 | 8.54 | 120.8 | 299.3 | 7.22 | 106.3 | 302.4 | 5.36 | 83.4 |

	$p=15$			$p=16$			$p=17$		
T	C_1	C_2	C_4	C_1	C_2	C_4	C_1	C_2	C_4
1.20	323.2	15.45	315.1	327.4	15.34	318.9	331.6	15.23	322.5
1.25	323.2	15.81	312.5	327.5	15.70	316.2	331.7	15.59	319.8
1.30	322.9	16.17	309.2	327.3	16.06	312.8	331.6	15.95	316.3
1.35	322.6	16.50	305.3	327.0	16.38	308.7	331.2	16.26	311.9
1.40	322.2	16.77	300.7	326.5	16.64	303.8	330.8	16.51	306.8
1.45	321.7	16.97	295.3	326.0	16.82	298.2	330.2	16.68	300.9
1.50	321.2	17.08	289.2	325.5	16.91	291.7	329.7	16.75	294.0
1.55	320.7	17.09	282.0	324.9	16.91	284.2	329.1	16.72	286.1
1.60	320.1	17.00	273.8	324.3	16.79	275.5	328.4	16.57	276.9
1.65	319.4	16.78	264.1	323.6	16.54	265.3	327.7	16.29	266.2
1.70	318.6	16.41	252.8	322.7	16.12	253.3	326.8	15.83	253.4
1.75	317.6	15.84	239.3	321.7	15.50	238.9	325.6	15.15	238.1
1.80	316.3	15.01	223.0	320.2	14.60	221.5	324.0	14.16	219.3
1.85	314.5	13.80	202.8	318.3	13.28	199.5	321.8	12.72	195.2
1.90	312.2	12.02	176.5	315.7	11.32	170.2	318.8	10.53	162.2
1.95	309.3	9.23	138.0	312.2	8.07	124.6	314.7	6.56	105.5

	$p=18$			$p=19$			$p=20$		
T	C_1	C_2	C_4	C_1	C_2	C_4	C_1	C_2	C_4
1.20	335.7	15.13	326.0	339.6	15.03	329.5	343.5	14.93	332.7
1.25	335.8	15.49	323.2	339.9	15.39	326.5	343.8	15.29	329.7
1.30	335.7	15.84	319.6	339.8	15.73	322.7	343.8	15.63	325.7
1.35	335.4	16.14	315.0	339.5	16.02	318.0	343.5	15.91	320.8
1.40	334.9	16.38	309.7	339.0	16.25	312.3	342.9	16.12	314.8
1.45	334.3	16.53	303.4	338.3	16.38	305.7	342.2	16.23	307.8
1.50	333.7	16.58	296.2	337.7	16.42	298.1	341.5	16.24	299.7
1.55	333.1	16.53	287.8	337.0	16.33	289.3	340.8	16.13	290.4
1.60	332.4	16.35	278.1	336.3	16.12	279.0	340.1	15.88	279.6
1.65	331.6	16.02	266.7	335.5	15.75	266.9	339.2	15.47	266.7
1.70	330.7	15.52	253.2	334.5	15.19	252.4	338.1	14.85	251.3
1.75	329.4	14.77	236.7	333.0	14.38	234.8	336.5	13.95	232.2
1.80	327.6	13.70	216.3	331.0	13.19	212.6	334.2	12.65	207.9
1.85	325.1	12.11	189.8	328.1	11.43	183.1	330.8	10.67	174.7
1.90	321.6	9.62	152.1	324.0	8.54	138.8	325.9	7.18	120.6
1.95	316.7	4.16	71.7						

	$p=21$			$p=22$			$p=23$		
T	C_1	C_2	C_4	C_1	C_2	C_4	C_1	C_2	C_4
1.20	347.3	14.84	335.9	351.0	14.75	338.9	354.6	14.66	341.8
1.25	347.7	15.20	332.7	351.4	15.10	335.6	355.1	15.01	338.4
1.30	347.7	15.52	328.6	351.5	15.42	331.3	355.1	15.32	333.8
1.35	347.3	15.79	323.4	351.1	15.67	325.8	354.7	15.56	328.0
1.40	346.7	15.98	317.0	350.4	15.85	319.1	353.9	15.72	320.9
1.45	346.0	16.08	309.7	349.6	15.93	311.3	353.1	15.77	312.6
1.50	345.2	16.07	301.1	348.8	15.89	302.3	352.2	15.70	303.1
1.55	344.5	15.93	291.3	348.0	15.71	291.9	351.4	15.50	292.1
1.60	343.7	15.64	279.8	347.2	15.38	279.7	350.6	15.12	279.2

continues

Table 5.1. (continued)

T	C_1	C_2	C_4	C_1	C_2	C_4	C_1	C_2	C_4
1.65	342.8	15.18	266.2	346.3	14.87	265.2	349.5	14.54	263.7
1.70	341.6	14.49	249.6	344.9	14.11	247.4	348.0	13.71	244.5
1.75	339.8	13.50	228.9	342.8	13.02	224.8	345.6	12.49	219.8
1.80	337.1	12.05	202.0	339.6	11.40	194.9	341.7	10.67	186.1
1.85	333.0	9.80	164.1	334.8	8.77	150.5	335.8	7.49	132.4
1.90	327.1	5.23	92.2						

	$p = 24$			$p = 25$		
T	C_1	C_2	C_4	C_1	C_2	C_4
1.20	358.1	14.58	344.6	361.5	14.50	347.2
1.25	358.6	14.92	341.0	361.9	14.84	343.3
1.30	358.6	15.21	336.1	361.9	15.11	338.2
1.35	358.1	15.44	329.9	361.3	15.32	331.6
1.40	357.2	15.58	322.4	360.4	15.45	323.7
1.45	356.3	15.61	313.7	359.4	15.46	314.5
1.50	355.4	15.52	303.7	358.5	15.33	303.9
1.55	354.6	15.27	292.1	357.6	15.04	291.7
1.60	353.8	14.85	278.4	356.8	14.57	277.2
1.65	352.6	14.21	261.8	355.5	13.85	259.4
1.70	350.9	13.28	241.0	353.4	12.82	236.9
1.75	348.0	11.93	213.8	350.0	11.31	206.5
1.80	343.3	9.84	175.3	344.2	8.86	161.5
1.85	336.0	5.73	105.4	334.7	1.50	33.8

[a] Temperature T is in Kelvin and pressure p is in bar. SVP is saturated vapor pressure.

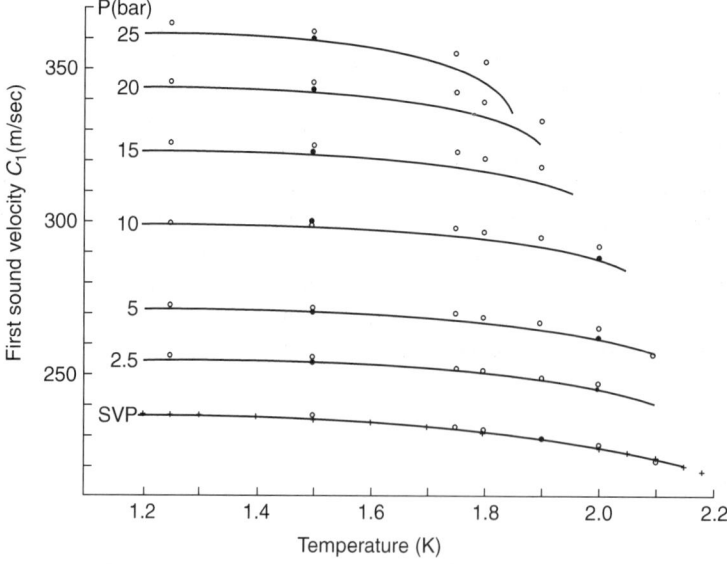

Fig. 5.3. Velocity of first sound in He II. Solid lines, this experiment; solid circles, Vignos and Fairbank [8]; open circles, Atkins and Stasior [10]; crosses at SVP, Chase [9].

The experiment was conducted in a standard liquid helium cryostat. Temperatures below that of liquid helium at atmospheric pressure were achieved by pumping on a bath of liquid helium. The temperature was regulated with a carbon resistor thermometer and heater in a feedback circuit. The temperature was found by measuring the vapor pressure of the helium bath and using the T58 scale (Brickwedde, Van Dijk,

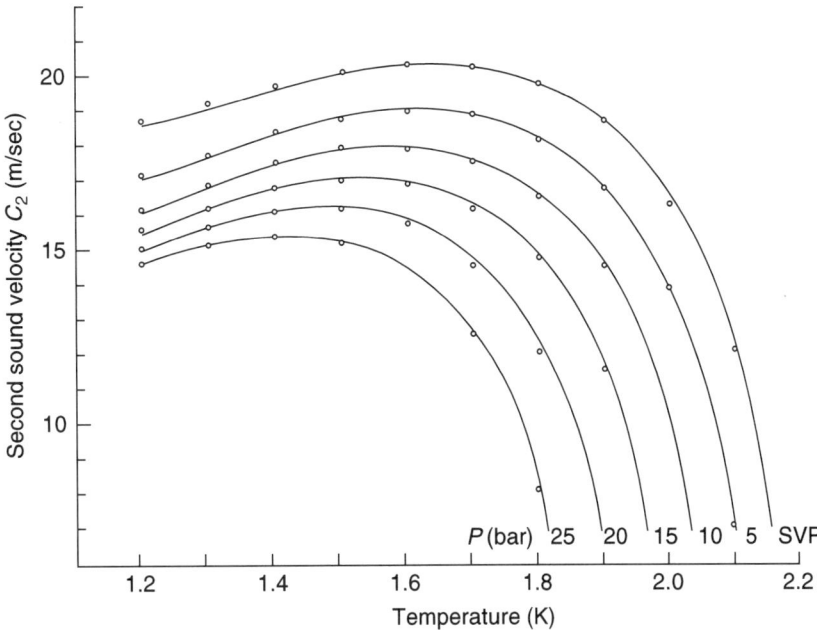

Fig. 5.4. Velocity of second sound in He II. Solid lines, this experiment; open circles, Maurer and Herlin [11]. Data points of this reference at elevated pressures have been shifted down by 2%.

Durieux, Clement, and Logan, 1960). The vapor pressure was measured with a quartz Bourdon gauge calibrated above 3 mm Hg with a mercury manometer and cathetometer and below 3 mm Hg with a calibrated capacitance gauge. The temperature could be determined to within 1 mK over the entire temperature range. The pressure in the resonators was measured with a capacitance gauge factory calibrated to 0.1% accuracy.

The resonance frequencies for the resonators were measured in parallel with three separate data acquisition channels. The first and second sound resonances had sufficiently high Qs and low backgrounds so that their resonance frequencies could be tracked automatically using wave analyzers in a phase-locked loop. The fourth sound resonances had high Qs, but the background prevented automatic tracking. The fourth sound resonances were each found in turn using a spectrum analyzer with a cathode-ray tube display. The analyzer could sweep a frequency range and locate the resonance in a matter of seconds. Once the fourth sound resonance was found, a computer was signaled, and all the resonant frequencies together with the cell temperature and pressure were simultaneously stored and tabulated.

5.4. RESULTS

The measured velocities of first, second, and fourth sounds, as a function of temperature and pressure, are presented in Table 5.1.

Comparisons of first and second sound velocities with previous measurements are found in Figures 5.3 and 5.4. The previous measurements of first sound are from Vignos and Fairbank [8], Chase (1958) [9], and Atkins and Stasior [10]. The previous measurements of second sound are from Maurer and Herlin [11].

A convenient method of viewing fourth sound is to plot $\rho_s/\rho = (C_4/C_1)^2$ vs T/T_λ, where T_λ is a function of pressure. In such a plot, all the data falls on a universal curve, as shown in Figure 5.5.

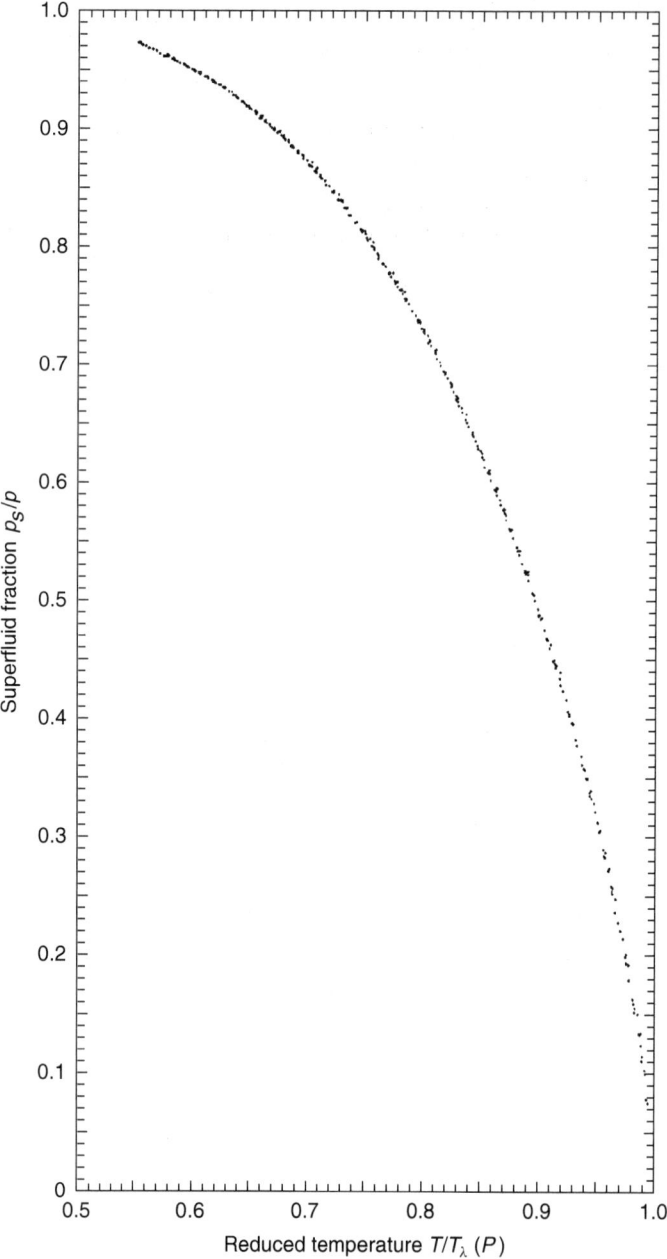

Fig. 5.5. Plot of $\rho_s/\rho \simeq (C_4/C_1)^2$ data points vs reduced temperature T/T_λ for all pressures above SVP. The ~400 data points giving the appearance of a single line are a striking manifestation of the Law of Corresponding States.

References

1. Rudnick, I. (1970). Selected problems in sound propagation in liquid helium. Proceedings of the Batsheva Conference, Technion, Haifa, Isreal (Gordon Breach), pp. 251–283.
2. Putterman, S.J. (1974). *Superfluid Hydrodynamics*. Amsterdam: North Holland (and references cited therein).
3. Heiserman, J., Hulin, J.P., Maynard, J.D., and Rudnick, I. (1976). Precision sound-velocity measurements in He II. *Phys. Rev.* **B14**: 3862–3867.
4. Maynard, J. (1976). Determination of the thermodynamics of He II from sound-velocity data. *Phys. Rev.* **B14**: 3868–3891.

5. Jelatis, G.J., Roth, J.A., and Maynard, J.D. (1979). Observation of fifth sound in a planar superfluid 4He film. *Phys. Rev. Lett.* **42**: 1285–1288.
6. Rosenbaum, R., Williams, G.A., Heckerman, D., Marcus, J., Scholler, D., Maynard, J.D., and Rudnick, I. (1979). Surface tension sound in superfluid helium films adsorbed on alumina powder. *J. Low Temp. Phys.* **37**: 663–678.
7. Rudnick, I., Kagiwada, R.S., Fraser, J.C., and Guyon, E. (1968). Third sound in adsorbed superfluid films. *Phys. Rev. Lett.* **20**: 430–433.
8. Vigonos, J.H., and Fairbank, H.A. (1966). Sound measurements in liquid and solid He3, He4, and He3-He4 mixtures. *Phys. Rev.* **147**: 185–197.
9. Chase, C.E. (1958). Propagation of ordinary sound in liquid helium near the lambda point. *Phys. Fluids* **1**: 193–200.
10. Atkins, K.R., and Stasior, R.A. (1953). First sound in liquid helium at high pressures. *Can. J. Phys.* **31**: 1156–1164.
11. Maurer, R.D., and Herlin, M.A. (1951). Pressure dependence of second sound in liquid helium: II. *Phys. Rev.* **81**: 444–447.

CHAPTER 6

ELASTIC PROPERTIES OF NEMATIC AND SMECTIC LIQUID CRYSTALS

P. Martinoty
Laboratoire de Dynamique des Fluides Complexes, Université Louis Pasteur, Strasbourg Cedex, France

Contents

Abstract		159
6.1.	Introduction	159
6.2.	Elastic Energy of the Nematic Phase	161
6.3.	Elastic Energy of the Smectic A Phase	161
6.4.	Hydrodynamics and Sound Propagation	163
6.5.	Behavior of the Elastic Constants Near the N–SmA Transition	165
6.6.	Behavior of the Elastic Constants Near the SmA–SmC Transition and Comparison with the Specific Heat Behavior	169
6.7.	Experiments	171
Acknowledgments		179
References		179

ABSTRACT

This chapter deals with the elastic constants associated with the Nematic, Smectic-A, and Smectic-C phases of liquid crystals, along with their drastic changes in behavior at the Nematic–Smectic-A and Smectic-A–Smectic-C phase transitions. It includes a theoretical description of the particular elasticity of these phases, as well as a survey of the various theories involved at these phase transitions. Because most of the elastic constants discussed here are deduced from first and second sound experiments, a brief reminder of the hydrodynamic theories of Smectic phases is also given. Results obtained with techniques allowing measurements to be taken at frequencies higher or smaller than ultrasonic frequencies are also presented to have an overall view of their behavior as a function of frequency.

6.1. INTRODUCTION

Between the solid and liquid phases, some organic compounds made up of rod-shaped molecules show a succession of intermediate phases called mesomorphic or liquid crystal phases. The most frequently encountered are the Nematic and Smectic phases.

The Nematic phase is characterized by a common direction of molecule alignment, designated as the director n, and by the absence of any positional order of the centers of gravity (Fig. 6.1); it can thus be described as an oriented fluid. This orientational order can be drastically modified by the application of an electric field. This property is the basis of the devices that have completely changed the display technology.

The Smectic phase keeps the orientational order of the Nematic phase but also possesses a positional order of the centers of gravity, the molecules being arranged in parallel layers. As many as 13 Smectic phases are listed in the literature (Smectics A, B, C, D, E, F, G, H, I, J, K, O, and Q), with numerous subcategories. In fact, many of them are not Smectic phases but three-dimensional crystals with weak cohesion energies. From a qualitative point of view, there exist only three types of Smectic phases: Smectic-A (SmA), Smectic-C (SmC), and Hexatics.

In the SmA phase, the molecules are parallel to the normal to the layers, whereas in the SmC phase, they are tilted with respect to this normal (Fig. 6.1). The Hexatic phases are differentiated from the previous two phases by the presence of a long-range bond orientational order with sixfold symmetry. There are several types of SmA phases depending on whether the interplanar spacing is equal to the molecular length (SmA$_1$), twice the molecular length (SmA$_2$), or intermediate between one and two molecular lengths (SmA$_d$). All these phases have the same macroscopic symmetry.

Typical examples of molecules giving mesomorphic phases are provided by octyloxy-cyanobiphenyl (8OCB) and terephtal-bis-p-p'-butylaniline (TBBA), whose chemical formulae and phase sequences are given in Figure 6.2. The SmA phase is of the A$_1$-type for TBBA and A$_d$-type for 8OCB.

The lamellar structure of the Smectic phases gives them original properties; for example, they display solid-type elasticity perpendicular to the layers and Nematic-type elasticity (curvature elasticity) within the plane of the layers. One is therefore dealing with systems that, depending on the direction in which they are examined, behave either like a one-dimensional solid or like a two-dimensional liquid. The Smectic phases thus belong to the so-called "low-dimensional" systems.

This chapter deals essentially with the study of the elastic properties of the N, SmA, and SmC phases. It begins with a reminder of the elastic energy of the N and SmA phases, leading to the definition of the elastic constants characterizing the phases under consideration. The sound propagation equations are then given. In particular, it is shown that the ultrasound velocity, isotropic in the N phase, becomes anisotropic in the SmA and SmC phases and that the three elastic constants, which characterize the elasticity of these phases, can be deduced from this anisotropy. The theoretical predictions on the behavior of the velocity and the elastic constants in the vicinity of the N–SmA

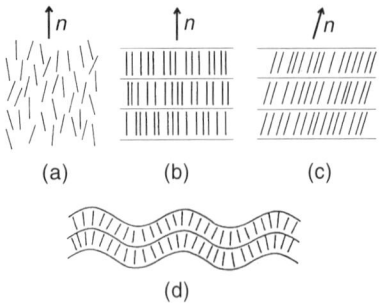

Fig. 6.1. Schematic representation of the Nematic (a), Smectic A (b), and Smectic C (c) phases and also the layer undulation mode (d), which is the main deformation of the Smectic phases. The direction of average molecular alignment is specified by the director **n**.

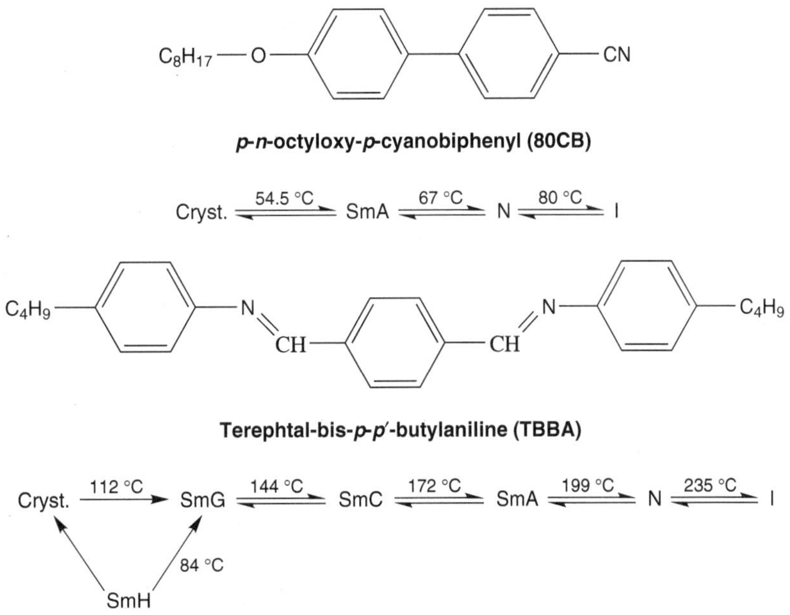

Fig. 6.2. Examples of rodlike molecules giving liquid crystal phases.

and SmA–SmC transitions are then summarized. Finally, the behavior of the elastic constants deduced from the experiments carried out by first and second sound, dynamic compression, and Rayleigh- and Brillouin-scattering is presented and discussed.

Because of limited space, some interesting studies for which elastic measurements in the hydrodynamic regime exist have been omitted; among them are those concerned with the behavior of the elastic constants near the SmA-HexB transition [1], the SmC-HexI transition [2], and the SmA-crystal B transition [3]. General background material can be obtained from several books and review articles [4–7].

6.2. ELASTIC ENERGY OF THE NEMATIC PHASE

In most practical cases, the director **n** presents spontaneous or induced distortions. Except for the vicinity of the core of the disclinations, the orientational variations of the director are only remarkable over distances L that are greater than the molecular lengths (typically $L \geq 1$ μm). Therefore, they can be described by a continuum theory. Frank [8] showed that the elastic deformation energy of Nematic phases included three contributions, due to three elementary deformation types (Fig. 6.3). The elastic energy of any deformation is written (per unit volume):

$$F = \frac{K_1}{2}(\text{div } \mathbf{n})^2 + \frac{K_2}{2}(\mathbf{n}.\text{rot } \mathbf{n})^2 + \frac{K_3}{2}(\mathbf{n} \times \nabla \times \mathbf{n})^2 \qquad (6.1)$$

Elastic constants K_1, K_2 and K_3 have the dimension of a force (energy/length). Their value can be estimated from the ratio of $k_B T$ (molecular energy scale) to a molecular length a, i.e., about 5×10^{-12} N (taking $k_B T = 5 \times 10^{-21}$ J and $a = 1$ nm, with k_B the Boltzmann constant). We shall see that the experiments confirm this order of magnitude.

6.3. ELASTIC ENERGY OF THE SMECTIC A PHASE

The elastic energy of the SmA phase is a function of the different invariants that are obtained for the symmetry operations of this phase. If the distortion energy is taken to

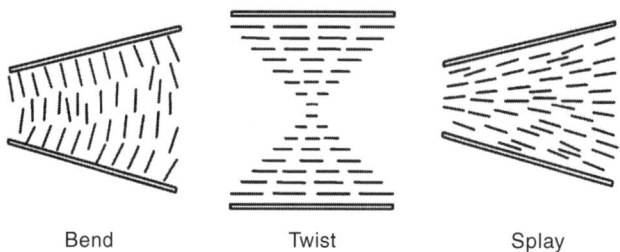

Bend   Twist   Splay

Fig. 6.3. Schematic representation of the three curvature distortions of a Nematic liquid crystal.

be only a function of the displacement u of the Smectic layers along the axis z, and the deformations are weak, then this energy is written (per unit volume):

$$F = \frac{1}{2}\bar{B}\left(\frac{\partial u}{\partial z}\right)^2 + \frac{K_1}{2}\left(\frac{\partial^2 u}{\partial x^2} + \frac{\partial^2 u}{\partial y^2}\right)^2 \tag{6.2}$$

which is the so-called "Landau-Peierls" form.

The first term is a solid-type elasticity term that represents the layer-compression energy. B is therefore an isothermal layer compression modulus, and its dimension is that of an energy per unit volume. The second term represents the layer-undulation energy; it is comparable to the K_1 term of the Frank-Oseen energy. Indeed, assuming that the molecules remain normal to the layers, then **n** is no longer an independent variable, and $\delta \mathbf{n} = (-\partial u/\partial x, -\partial u/\partial y)$. The other two deformations, of torsion and flexion, are precluded by the layer-structure of the material. This description brings out the originality of Sm-A phases: solid-type elasticity in the direction normal to the layers, and Nematic-type elasticity in the plane of the layers.

The Landau-Peierls description also enables the mean value of the layer fluctuations to be estimated; this is given by

$$\langle u^2 \rangle = \frac{k_B T}{2\pi(K_1 \bar{B})^{1/2}} \ln\left(\frac{L}{d}\right) \tag{6.3}$$

where L is the sample size and d is the interlayer spacing. This equation shows that the mean square of the layer displacement diverges with sample size. The notion of one-dimensional order is thus not totally rigorous, since fluctuations of the positional order are sufficiently great to prevent the establishment of any long-range positional order. This result, which is valid for all lamellar phases, is known as the Landau–Peierls instability.

The Landau-Peierls description does not take account of the density variation that accompanies the layer compression. This coupling was introduced by De Gennes [9], who writes the free energy as follows:

$$F = \frac{1}{2}A\left(\frac{\partial \rho}{\rho_o}\right)^2 + \frac{1}{2}B\left(\frac{\partial u}{\partial z}\right)^2 + C\left(\frac{\partial \rho}{\rho_o}\right)\left(\frac{\partial u}{\partial z}\right) + \frac{K_1}{2}\left(\frac{\partial^2 u}{\partial x^2} + \frac{\partial^2 u}{\partial y^2}\right)^2 \tag{6.4}$$

where $\partial \rho / \rho_o$ represents the variation of density compared with density ρ_o at rest. In this expression, A represents the compressibility; B, as previously, the elastic constant associated with layer compression; and C a term describing the density variation that goes along with the layer compression. We shall see in Section 6.4 that the coupling between variables $\partial \rho / \rho_o$ and $\partial u/\partial z$ is necessary to describe the dynamic behavior of the Smectic phases. In the case of a static deformation, $\partial \rho / \rho_o$ takes on the value $-C/A(\partial u/\partial z)$, and Eq. 6.4 reduces to Eq. 6.2 with $\bar{B} = B - C^2/A$.

The free energy Eq. 6.4 supposes the director to remain normal to the layers. If this requirement is suppressed, u and **n** become independent variables. In this case, F

is written

$$F = F_{Frank} + \frac{1}{2}A\left(\frac{\partial \rho}{\rho_o}\right)^2 + \frac{1}{2}B\left(\frac{\partial u}{\partial z}\right)^2 + C\left(\frac{\partial \rho}{\rho_o}\right)\left(\frac{\partial u}{\partial z}\right)$$
$$+ \frac{1}{2}B_\perp\left[\left(\frac{\partial u}{\partial x} + n_x\right)^2 + \left(\frac{\partial u}{\partial y} + n_y\right)^2\right] \quad (6.5)$$

where $\partial u/\partial x + n_x$ and $\partial u/\partial y + n_y$ represent the deviation of **n** relative to the normal to the layers and $B_\perp$ is the associated elastic constant. Equation 6.5 represents the most general form of the free distortion energy of a SmA.

Finally, the elastic constants A, B, and C can be related to the usual elastic modulus c_{ij} by writing the free energy F as a function of the strains $x_1 = \partial u/\partial x$, $x_2 = \partial u/\partial y$, and $x_3 = \partial u/\partial z$. This gives

$$F = \tfrac{1}{2}C_{11}(x_1 + x_2)^2 + \tfrac{1}{2}C_{33}x_3^2 + C_{13}x_3(x_1 + x_2) \quad (6.6)$$

with

$$C_{11} = A \quad C_{13} = A - C \quad C_{33} = A + B - 2C \quad (6.7)$$

Equation 6.6 is nothing other than the elastic energy of a uniaxial crystal that does not support shears.

6.4. HYDRODYNAMICS AND SOUND PROPAGATION

The hydrodynamic properties of SmA phases were first described by De Gennes [9]. This theory shows that the nature of the modes can change according to the orientation of wave-vector **q** in relation to the optical axis. The theory more especially predicts the existence of two propagative modes: an acoustic mode, called the first sound, comparable to that of an ordinary liquid, and, a "shearlike" mode, known as the second sound. This mode comes from the layer fluctuations u at a density that is almost constant. Martin, Parodi, and Pershan (MPP) [10] subsequently completed the De Gennes theory by including the friction and the permeation effects.

If we call θ the angle between the normal to the layers and the direction of sound propagation, the velocities of the first and second sound, respectively, are given by

$$\rho V_1^2 = A - 2C\cos^2\theta + B\cos^4\theta \quad (6.8)$$
$$\rho V_2^2 = B\cos^2\theta \sin^2\theta \quad (6.9)$$

where A, B, and C are the elastic constants appearing in the free energy F (Eq. 6.5), which is now taken at constant entropy since the sound modes are adiabatic. In deriving Eqs. 6.8–6.9 the condition $A \gg B, C$ was used. It should be noted that the second sound mode is replaced for $\theta = 90°$ by a layer-undulation mode, and for $\theta = 0°$ by the permeation mode [10]. The layer-undulation mode is sketched in Figure 6.1. This mode is associated with the curvature forces, as in nematics. There is no change in layer spacing. The permeation mode is characterized by the fact that the molecules can flow with a velocity V_z, which is different from the velocity $\partial u/\partial t$ of the Smectic layers.

Equation 6.8 shows that elastic constants A, B, and C can be deduced from the velocity measurements taken for three different orientations of the compound relative to the direction of sound propagation, for example, for $\theta = 0$, 45 and 90°. Since the first-order elastic constants of the SmC phase are identical to those of the SmA phase, the propagating modes are the same in the two cases. Only the dissipation and the behavior of the nonpropagative modes should differ.

In the Nematic phase, B and C are zero; the velocity is isotropic and is given by

$$\rho V_1^2 = A \qquad (6.10)$$

The absence of a long-range order resulting from the Landau-Peierls instability indicates that the layer fluctuations are very marked. A rigorous description of the Smectic phases therefore requires the use of the exact expression of the strain tensor e_{ij}, the trace of which is

$$\frac{\partial u}{\partial z} + \frac{1}{2}(\nabla u)^2 \qquad (6.11)$$

This description introduces into the free energy additional terms that will modify the behavior of elastic constants B, C, and K_1. The calculation, made by Grinstein and Pelcovits [11], shows that B and C tend toward 0 for long wavelengths and that K_1 tends toward infinity. While it is fundamental as a concept, the correction introduced by the layer-fluctuation effects is not very significant in practice, since it is a logarithmic correction, which has never actually been observed. On the other hand, the absence of a long-range order profoundly modifies the dynamic behavior of the Smectic phases. In particular, the α/f^2 ratio, where α and f represent the damping and frequency of the ultrasound wave, respectively, diverges as $1/f$ at low frequencies f instead of being constant as in conventional hydrodynamics. A typical example of this behavior is given by Figure 6.4. The reader interested in these spectacular effects may consult Mazenko et al. [12] for the theory and [13–16] for the experiments.

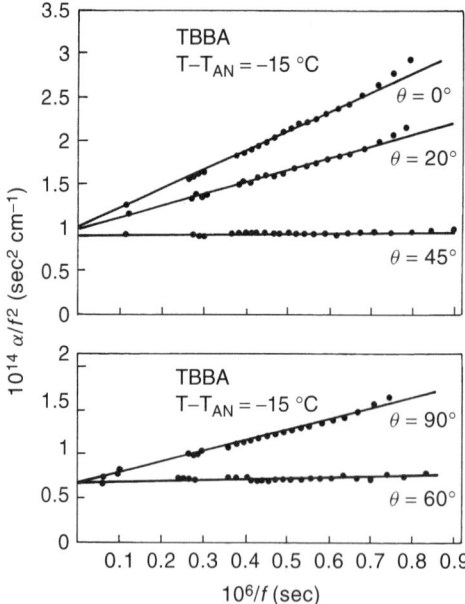

Fig. 6.4. Variation of α/f^2 as a function of $1/f$ in the Smectic A phase of TBBA for various values of angle θ between the normal to the Smectic layers and the direction of sound propagation. The behavior observed is described by the relationship $\alpha/f^2 = a + b/f$, in which a is the conventional damping term and b/f is the anharmonic contribution. The angular variation of the anharmonic contribution shows the existence of an angle θ_0 for which the anharmonic effects do not contribute to the damping. This angle is given by $\cos^2\theta_0 = C/B$ where B and C are elastic constants of the Smectic phase.

6.5 BEHAVIOR OF THE ELASTIC CONSTANTS NEAR THE N–SmA TRANSITION

We will see in the experimental section that the behavior of the elastic constants of the SmA phase is, in fact, essentially governed by the effects associated with the N–SmA and SmA–SmC phase transitions. This result leads us, therefore, to recall the theoretical predictions on the behavior of these elastic constants at both transitions.

6.5.1. Static Behavior of Sound Velocity and of Elastic Constants A, B, C, and $B_\perp$

6.5.1.1. SmA Order Parameter and Free Energy

As seen previously, the SmA phase is characterized by a periodic density modulation $\rho(z)$ along the direction perpendicular to the layers. $\rho(z)$ can be written as a Fourier expansion $\rho(z) = \rho_o + \rho_1 \cos(q_o z - \phi)$, where φ is an arbitrary phase and $q_o = 2\pi/d$ with d the interlayer spacing. Since ρ_1 is zero in the N phase and different from zero in the SmA phase, the ρ_1-term is a natural choice for the N–SmA order parameter. Since it has two components (the phase and the modulus), like the one describing the superfluid transition, the N–SmA transition could belong to the 3DXY universality class of the λ transition in liquid He. In the following, the order parameter will be written as $\psi(\mathbf{r}) = \rho_1(\mathbf{r})e^{i\varphi(\mathbf{r})}$, where the phase $\varphi(\mathbf{r})$ is simply related to the layer displacement $u(\mathbf{r})$ by $\varphi(\mathbf{r}) = -q_o \cdot u(\mathbf{r})$. By including fluctuations of ψ and of the director $\delta \mathbf{n}_\perp$, the free energy per unit volume reads:

$$F = \frac{a}{2}|\psi|^2 + \frac{b}{4}|\psi|^4 + \frac{c_\parallel}{2}|\nabla_\parallel \psi|^2 + \frac{c_\perp}{2}|(\nabla_\perp - iq_o \delta \mathbf{n}_\perp)\psi|^2 \quad (6.12)$$

where $a = a_o(T - T_{AN})$ in the mean field approximation. Because of the Nematic anisotropy, the gradient terms exhibit anisotropic coefficients ($c_\parallel \neq c_\perp$) along directions parallel and perpendicular to the director $\mathbf{n}$. As a result, there are two correlation lengths $\xi_\parallel$ and $\xi_\perp$ that are given by $\xi_\parallel^2 = c_\parallel/a$ and $\xi_\perp^2 = c_\perp/a$. As remarked by De Gennes [17], Eq. 6.12 bears a striking resemblance to the Landau-Ginzburg energy for the normal-superconductor transition, ψ playing the role of the superconducting order parameter and $\delta \mathbf{n}$ the role of the vector potential.

Since the director is introduced into the free energy, it is also necessary to take into account the Frank–Oseen elastic energy of the Nematic. By also introducing the density variation $\partial \rho/\rho_o$, the total free energy is written as

$$F = \frac{a}{2}|\psi|^2 + \frac{b}{4}|\psi|^4 + \frac{c_\parallel}{2}|\nabla_\parallel \psi|^2 + \frac{c_\perp}{2}|(\nabla_\perp - iq_o \delta \mathbf{n}_\perp)\psi|^2 + \frac{1}{2}A\left(\frac{\partial \rho}{\rho_o}\right)^2$$

$$+ \frac{c_o}{2i}\left(\frac{\partial \rho}{\rho_o}\right)(\psi^* \nabla_\parallel \psi + \psi \nabla_\parallel \psi^*) + \frac{K_1}{2}(div\, \delta \mathbf{n}_\perp)^2 \quad (6.13)$$

$$+ \frac{K_2}{2}[\mathbf{n_o} \cdot rot\, \delta \mathbf{n}_\perp]^2 + \frac{K_3}{2}[\mathbf{n}_o \times \nabla \times \delta \mathbf{n}_\perp]^2$$

If one assumes ψ_o to be spatially uniform in the SmA phase, the gradient terms of F become

$$F_g = \frac{c_\parallel}{2}q_o^2 \psi_o^2 \left(\frac{\partial u}{\partial z}\right)^2 + \frac{c_\perp}{2}q_o^2 \psi_o^2 \left[\left(\frac{\partial u}{\partial x} + n_x\right)^2 + \left(\frac{\partial u}{\partial y} + n_y\right)^2\right]$$

$$+ C_o q_o \psi_o^2 \left(\frac{\partial \rho}{\rho_o}\right)\left(\frac{\partial u}{\partial z}\right) \quad (6.14)$$

Comparison of this expression with that of the SmA phase (Eq. 6.5) gives

$$B = q_o^2 c_\parallel \psi_o^2 \quad C = q_o c_o \psi_o^2 \quad B_\perp = q_o^2 c_\perp \psi_o^2 \tag{6.15}$$

This theory predicts that the N–SmA transition could be second order. However, Halperin and Lubensky have shown that the coupling between the Smectic order parameter and the director fluctuations leads to a third-order term in the free energy expansion, thus indicating that the N–SmA transition is instrinsically first order [18]. However the discontinuities at the transition are small and the transition is weakly first order. We will see in Section 6.7.2.1 that ultrasound velocity and damping in TBBA exhibit 3DXY-type fluctuations even far from the transition, despite the fact that it is first order.

6.5.1.2. Critical Behavior of the Elastic Constants

The static behavior of velocity (or of elastic constant $A = \rho V^2$) in the vicinity of a second-order phase transition can be calculated from thermodynamic relationships. The Pippard-Buckingham-Fairbank relation [19] shows that

$$V(T) - V(T_c) \sim 1/C_p \tag{6.16}$$

This equation, which is only valid within the limit $T \to T_c$ and $\omega = 0$, applies in the N phase as well as in the SmA phase for $\theta = 90°$. It shows that a pretransitional rise of the specific heat results in a dip in the zero-frequency sound velocity.

The thermal behavior of the elastic constants B, C, and $B_\perp$ can be deduced from Eq. 6.15. In the mean field approximation, only the ψ_o^2 term is a function of temperature, and the three elastic constants vanish with an exponent of 1. Renormalization group calculations show that the fluctuations of the SmA order parameter modify the mean field behavior to give nonclassical behavior. By using the anisotropic scaling laws [20], one finds

$$B \sim (T_{AN} - T)^{2\nu_\perp - \nu_\parallel} \quad C \sim (T_{AN} - T)^{\nu_\parallel + \eta_\perp \nu_\perp} \quad B_\perp \sim (T_{AN} - T)^{\nu_\parallel} \tag{6.17}$$

where $\nu_\parallel$ and $\nu_\perp$ are the exponents associated with the correlation lengths parallel and perpendicular, respectively, to the director $\xi_\parallel \sim |T_{AN} - T|^{-\nu_\parallel}$, $\xi_\perp \sim |T_{AN} - T|^{-\nu_\perp}$. $\eta_\perp$ is an exponent associated with the Fisher equality $\gamma = (2 - \eta_\perp)\nu_\perp$, where γ is the susceptibility exponent.

The helium analogy (3DXY model) predicts that $\nu_\parallel = \nu_\perp = \nu = 0.66$ and $\eta_\perp = \eta = 0.04$. B, C, and $B_\perp$ therefore vanishes with a critical exponent of 0.66. The dislocation-loops melting theory, which is based on the topological defects of the SmA phase, indicates two possibilities [21]. One is an isotropic critical point ($\nu_\parallel = \nu_\perp$), which belongs to the 3DXY universality class but with an inverted specific heat amplitude ratio [22]. The other is an anisotropic critical point with $\nu_\parallel = 2\nu_\perp$ [23]; B then becomes constant in the vicinity of the transition, while C and $B_\perp$ continue to decrease. It should, however, be noted that the values of the critical exponents have not been calculated for this model.

6.5.2. Sound Dispersion

6.5.2.1. Critical Fluctuations of the Smectic Order Parameter

These fluctuations, which exist on each side of the transition, modify the elastic constants of the compound, and consequently the sound velocity. The theories put forward to describe the behavior of the N phase couple these fluctuations either with the director ($\psi - \mathbf{n}$ coupling) or with the density ($\delta\rho - \psi^2$ coupling). These two couplings lead to critical behaviors of the ultrasound velocity, which are anisotropic in the former case and isotropic in the latter. The $\psi - \mathbf{n}$ and $\delta\rho - \psi^2$ couplings also exist in the SmA phase. As for the N phase, they give rise to critical behaviors of velocity. The exponents associated with these behaviors are identical to those determined in the N phase as a result of scaling hypotheses, but the amplitude of the critical effects may be very different. The calculations are presented in Sections 6.5.2.1.1 and 6.5.2.1.2.

6.5.2.1.1. Coupling between the Smectic Order Parameter Fluctuations and the Director ($\psi - \mathbf{n}$ Coupling)

Swift and Mulvaney [24] have shown that this coupling leads to an anisotropic increase in velocity, which is given by

$$V^2 = V_{\text{reg}}^2 + h(\omega)\cos^4\theta \tag{6.18}$$

where θ is the angle between the direction of sound propagation and the normal to the layers and $h(\omega)$ is a temperature and frequency-dependent function. This equation shows that the $\psi - \mathbf{n}$ coupling has no incidence on the behavior of $V^2(90°)$ and thus on A.

Evaluating $h(\omega)$ in the high-frequency ($\omega\tau \gg 1$) regime shows that

$$h(\omega\tau \gg 1) \sim \omega^{(2\nu_\perp - \nu_\parallel)/z\nu_\perp} \tag{6.19}$$

where z is the exponent associated with the critical relaxation time τ, which is defined as $\tau \sim \xi_\perp^z$.

In this regime, the Smectic order parameter fluctuations are frozen, and the N phase presents properties similar to those of a SmA phase. Identifying Eq. 6.18 with Eq. 6.8, which gives the velocity associated with the SmA phase, shows that it is possible to define an effective elasticity coefficient B_{eff}, which is given by

$$B_{\text{eff}} \sim \omega^{(2\nu_\perp - \nu_\parallel)/z\nu_\perp} \tag{6.20}$$

This result indicates that the second sound mode can propagate in the N phase when $\omega\tau \gg 1$.

The influence of the $\psi - \mathbf{n}$ coupling on the behavior of B has not been calculated in the SmA phase. Transposing the result obtained in the N phase (Eq. 6.20) shows that B should, in the critical ($\omega\tau \gg 1$) regime, present a rigidity modulus given by $B(\omega) \sim \omega^{(2\nu_\perp - \nu_\parallel)/z\nu_\perp}$. The empirical formula

$$B = B_0 t^\varphi + g\omega^y \frac{(\omega\tau)^{(2-y)}}{1 + (\omega\tau)^{(2-y)}} \tag{6.21}$$

which takes account of high- and low-frequency asymptotic behaviors, and of the fact that B is an $\omega\tau$ scaling function, makes it possible to analyze the measurements of B in the nonhydrodynamic regime. In this formula, $y = \varphi/z\nu_\perp$, $\varphi = 2\nu_\perp - \nu_\parallel$, and g is a constant.

6.5.2.1.2. Coupling between the Smectic Order Parameter Fluctuations and Density Fluctuations ($\psi^2 - \delta\rho$ Coupling)

As shown by Swift and Mulvaney [25], the coupling between density and order parameter fluctuations leads to an isotropic decrease in velocity, which is written

$$V^2(\omega) = V_{\text{reg}}^2 - \delta V^2(\omega) \tag{6.22}$$

where δV^2 is given by the following integral:

$$\delta V^2(\omega) = \frac{k_B T}{4\kappa_s^2} \int \frac{dq^3}{(2\pi)^3} \left[\frac{1}{\chi(\mathbf{q})}\left(\frac{\partial\chi(\mathbf{q})}{\partial p}\right)_S\right]^2 \frac{2\tau^{-2}(\mathbf{q})}{\omega^2 + 4\tau^{-2}(\mathbf{q})} \tag{6.23}$$

which is independent of the propagation direction. A similar result was obtained by Kiry and Martinoty [26] using a dynamic specific heat model. $\chi(\mathbf{q})$ is the generalized susceptibility, and $\tau^{-1}(\mathbf{q})$ is the relaxation frequency of the order parameter fluctuations. $\tau(\mathbf{q})$ is related to $\chi(\mathbf{q})$ by $\tau(\mathbf{q}) = \gamma_3 \chi(\mathbf{q})$, where γ_3 is a viscosity. $K_s = \partial \rho / \partial p$ is the compressibility.

Evaluating this integral in the hydrodynamic ($\omega \tau \ll 1$) regime shows that

$$\delta V^2(\omega = 0) \sim (T - T_{AN})^{-\bar{\alpha}} \tag{6.24}$$

which means that the velocity anomaly reflects that of specific heat.

It should be noted that if $\bar{\alpha}$ is positive, then $\delta V^2(\omega = 0)$ diverges, and Eq. 6.22 shows that $V^2(\omega)$ becomes negative at the transition, which is not physically acceptable. It must, however, be pointed out that Eq. 6.22 is only valid within the $\delta V(\omega)/V(\omega) \ll 1$ limit, which is the hypothesis on which the theory is based. It is possible to do away with this divergence problem by considering the critical behavior of the compressibility $\kappa_3 \sim 1/\rho V^2$ instead of that of the elastic constant $A = \rho V^2$.

6.5.2.2. Relaxation of the Smectic Order Parameter Modulus

This mechanism, predicted by Landau and Khalatnikov [27] for the λ transition in helium, assumes that the order parameter modulus is shifted from its equilibrium position by the ultrasound wave, then relaxes in a finite time τ toward its equilibrium value. This effect, which shows up in the low-temperature phase where the mean value is nonzero, should appear for all the phase transitions. The influence of this relaxation on ultrasound velocity is given by

$$V^2(\omega) = V^2(0) + [V^2(\infty) - V^2(0)] \frac{\omega^2 \tau^2}{1 + \omega^2 \tau^2} \tag{6.25}$$

where $V(0)$ and $V(\infty)$ are the low-frequency ($\omega \tau \ll 1$) and high-frequency ($\omega \tau \gg 1$) limits of velocity.

The strength of the relaxation term (i.e., $V^2(\infty) - V^2(0)$) has been calculated by Liu [28] for the N–SmA transition and is given by

$$V^2(\infty) - V^2(0) = \frac{1}{\chi \rho} \left[\left(\beta_\parallel - \beta_\perp + \frac{\partial \psi_o}{\partial \nabla_\parallel u} \right) \cos^2 \theta + \beta_\perp - \rho \frac{\partial \psi_o}{\partial \rho} \right]^2 \tag{6.26}$$

χ is the susceptibility; $\chi = 1/a$ for $T > T_{NA}$ and $\chi = -1/2a$ for $T < T_{NA}$. $\beta_\parallel$ and $\beta_\perp$ are transport coefficients associated with the dynamic coupling between ψ_0 and the thermodynamic variables. Since θ represents the angle between the direction of sound propagation and the director, Eq. 6.26 shows that the relaxation of the order parameter modulus is anisotropic.

In the N phase, $\partial \psi_o / \partial \Delta_\parallel u$ and $\partial \psi_o / \partial \rho$ are zero, and Eq. 6.26 is reduced to

$$V^2(\infty) - V^2(0) = (\chi \rho)^{-1} [(\beta_\parallel - \beta_\perp) \cos^2 \theta + \beta_\perp]^2 \tag{6.27}$$

This equation shows that the relaxation effect continues to exist in the high-temperature phase, owing to the presence of the dynamic terms $\beta_\parallel$ and $\beta_\perp$. This effect, which does not exist for the λ transition in helium, is difficult to observe, since the quantity $\beta_\parallel - \beta_\perp$, which reflects the amount of Smectic order induced by the ultrasound wave in the N phase, is weak. As we shall see later, this effect is quite negligible for TBBA. To facilitate the discussion that follows, we shall assume that $\beta_\parallel = \beta_\perp = 0$.

The relaxation equations of the three elastic constants can be determined from Eq. 6.26 in the $\beta_\parallel = \beta_\perp = 0$ hypothesis. By evaluating the term $\chi^{-1}(\partial \psi_o / \partial \Delta_\parallel u)^2$, it is possible to show that B and C are not influenced by relaxation of the order parameter modulus [29], unlike A, which is given by

$$A(\omega) = A(0) + \frac{1}{\chi}\rho^2 \left(\frac{\partial \psi_o}{\partial \rho}\right)^2 \frac{\omega^2 \tau^2}{1 + \omega^2 \tau^2} \qquad (6.28)$$

Evaluating the term $\chi^{-1}(\partial \psi_o/\partial \rho)$ shows that the relaxation strength $A(\infty) - A(0)$ varies as $(T_{AN} - T)^{-\bar{\alpha}}$, where $\bar{\alpha}$ is the specific heat exponent.

6.6. BEHAVIOR OF THE ELASTIC CONSTANTS NEAR THE SmA–SmC TRANSITION AND COMPARISON WITH THE SPECIFIC HEAT BEHAVIOR

As seen previously, the SmC phase differs from the SmA phase by a tilt of the director in relation to the normal to the layers. The SmC order parameter can thus be described by the tilt angle θ_T and the azimuthal angle ϕ, or equivalently by the complex number $\psi = \theta_T e^{i\varphi}$. The SmA–SmC transition therefore is expected to be in the 3DXY universality class [30].

6.6.1. Influence of Fluctuations of the SmC Order Parameter on the Elastic Constants

The influence of the fluctuations of the SmC order parameter on the behavior of the elastic constants of the SmA phase has been calculated by Andereck and Swift [31]. This calculation shows that the critical effects are anisotropic and depend on the angle θ between the direction of sound propagation and the normal to the layers.

6.6.1.1. Energy of the SmA Phase and Jump of the Elastic Constants at the SmA–SmC Transition

According to Andereck and Swift, in the vicinity of the SmA–SmC transition, the energy of the SmA phase is written as follows:

$$F = F_\psi + F_{el} + F_v + F_{\psi\rho} + F_{\psi u} \qquad 6.29$$

F_ψ is a Landau-Ginzburg type energy, with the same form as that of Eq. 6.12, in which $\delta \mathbf{n}_\perp = 0$. F_{el} is the De Gennes elastic energy, already introduced (Eq. 6.4). F_v is a contribution of kinetic origin resulting from the displacement of the mass centers

$$F_v = \tfrac{1}{2}|\mathbf{v}|^2 \qquad (6.30)$$

$F_{\psi\rho}$ and $F_{\psi u}$ are two contributions that quadratically couple the order parameter with the density variations and the layer spacing gradient, respectively. $F_{\psi\rho}$ and $F_{\psi u}$ are given by

$$F_{\psi\rho} = \tfrac{1}{2}\gamma_\rho \delta\rho |\psi|^2 \qquad (6.31)$$

$$F_{\psi u} = \tfrac{1}{2}\gamma_u (\nabla_z u)|\psi|^2 \qquad (6.32)$$

where γ_ρ and γ_u are phenomenological constants to be determined experimentally. Both these terms are very important because they are at the source of the anisotropy of the critical effects on velocity (and damping), the degree of anisotropy depending on the γ_u/γ_ρ ratio. In these equations, the equilibrium density ρ_o was assumed to have a value of 1 unit.

The jumps of the elastic constants at the transition can be deduced from the free energy Eq. 6.29 and are given by

$$\nabla A = A(T_{AC}^-) - A(T_{AC}^+) = -\gamma_\rho^2/2b$$
$$\nabla B = B(T_{AC}^-) - B(T_{AC}^+) = -\gamma_u^2/2b \qquad (6.33)$$
$$\nabla C = C(T_{AC}^-) - C(T_{AC}^+) = -\gamma_\rho\gamma_u/2b$$

6.6.1.2. Critical Behavior of the Elastic Constants and Ginzburg Criterion

Andereck and Swift's calculation was made within the framework of a mean field type theory, based on the energy given by Eq. 6.29, which is taken at constant entropy. The calculation shows that the elastic constants A, B, and C are modified by fluctuations of the order parameter and that their abnormal parts are given by

$$\delta A = A_{\text{reg}} - A = \gamma_\rho^2 I_2(\omega)$$
$$\delta B = B_{\text{reg}} - B = \gamma_u^2 I_2(\omega) \qquad (6.34)$$
$$\delta C = C_{\text{reg}} - C = \gamma_u \gamma_\rho I_2(\omega)$$

where $I_2(\omega)$ is a temperature- and frequency-dependent function. Index "reg" indicates the value that the elastic constants would have in the absence of any fluctuations.

Evaluating the abnormal parts δA, δB, and δC of the elastic constants in the hydrodynamic regime shows that they follow the same thermal behavior:

$$\delta A \sim \delta B \sim \delta C \sim (T - T_{AC})^{-\bar{\alpha}} \qquad (6.35)$$

where $\bar{\alpha}$ is the specific heat exponent.

Modification of the elastic constants implies a modification in the velocity, this latter being linked to the elastic constants by Eq. 6.8. Replacing A, B, and C by their modified values, $V^2(\theta)$ can be rewritten as

$$V^2(\theta) = V_{\text{reg}}^2(\theta) - \Delta(V^2(\theta)) \qquad (6.36)$$

where V_{reg} is the regular part of the velocity and

$$\Delta(V^2(\theta)) = (\gamma_\rho - \gamma_u \cos^2\theta)^2 I_2(\omega) \qquad (6.37)$$

is the critical part associated with the order parameter fluctuations.

This equation shows that velocity $V(\theta)$ decreases in the vicinity of the SmA–SmC transition and that this decrease can be more or less important, according to the angle θ considered. For example, when $\gamma_u > \gamma_\rho$, the decrease associated with $V(0°)$ must be much greater than that associated with $V(90°)$.

To determine the nature of a second-order transition, the Ginzburg criterion is generally used, giving the width ΔT_c of the critical region. This width is obtained by writing that the T_{AC}-jump of the elastic constants is equal to the contribution of the fluctuations ($\delta A = \Delta A$, $\delta B = \Delta B$, $\delta C = \Delta C$) and is given by

$$\Delta T_C = \left(\frac{k_B T_C}{4\pi c_\perp \sqrt{c_\parallel}}\right)^2 \frac{b^2}{2a_o} \qquad (6.38)$$

where a_o is the coefficient of the ψ^2 term in the free energy (see Eq. 6.12).

6.6.2. Influence of Fluctuations of the SmC Order Parameter on the Specific Heat

Benguigui and Martinoty [32, 33] have remarked that the energy given by Eq. 6.29 is not suitable for analysis of the specific heat and that it is the Gibbs free energy $G = F - x_i X_i$ (the x_i's and X_i's are the strains and the stresses respectively) that is relevant. As shown in [33], G is given by

$$G = \tfrac{1}{2}a|\psi|^2 + \tfrac{1}{4}b^*|\psi|^4 \qquad (6.39)$$

where

$$b^* = b + \frac{2C\gamma_u\gamma_\rho - A\gamma_u^2 - B\gamma_\rho^2}{2(AB - C^2)} \qquad (6.40)$$

Since γ_u and γ_ρ are related to the jump of the elastic constant at T_{AC}, b^* can be written as

$$b^* = b(1 + \chi) \qquad (6.41)$$

with

$$\chi = \frac{\Delta(AB - C^2)}{(AB - C^2)} = \frac{\Delta(C_{11}C_{33} - C_{13}^2)}{(C_{11}C_{33} - C_{13}^2)} \qquad (6.42)$$

The value of χ can be determined experimentally, since the jump of the elastic constants can be deduced from the velocity measurements.

The width of the critical region, calculated from the Ginzburg criterion, is given by

$$\Delta T_C = \left(\frac{k_B T_C}{4\pi c_\perp \sqrt{c_\parallel}}\right)^2 \frac{b^{*2}}{4a_o} \qquad (6.43)$$

This equation is the key result since it shows that the behavior of the specific heat depends, via the b^* term, on the parameters that couple the SmC order parameter to the density variation and the layer spacing gradient.

6.7. EXPERIMENTS

The study of liquid crystals, and more especially their use for technological applications, requires preparation of perfectly oriented samples. Such alignments can be obtained by anchoring the director to the glass surfaces bearing the sample. The surface orientation propagates within the sample by virtue of proximity effects, and a Nematic monocrystal (typical thickness: 40 μm) can be obtained. Planar orientation (**n** parallel to the surface of the glass slides) is usually obtained by rubbing, in a given direction, a thin layer of polymer (polyimide, for example), deposited on the glass surface, whereas for the homeotropic orientation, (**n** perpendicular to the surface of the glass slides), a surfactant (silane, for example) is deposited on the slides. A strong magnetic field is generally used to orient large volumes. A monodomain SmA sample can be obtained by taking the sample into its Nematic state, then letting it cool gently within the field. This method is applied for ultrasound studies, whereas sidewall effects are generally used for second sound and light-scattering studies. The two methods are sometimes used simultaneously.

6.7.1. Curvature Elasticity in the Nematic Phases: The Freedericksz Transition

Competition between the orientation imposed by a sidewall and that given by an external field (be it magnetic or electric) can give rise to an important transition, known as the Freedericksz transition [34].

Let us, for example, consider the case of planar alignment and apply a magnetic field **H** perpendicular to the slides (Fig. 6.5). If the magnetic anisotropy χ_a is positive, the molecules will tend to align according to the field. A distortion will occur for a threshold field:

$$H_c = \frac{\pi}{d}\left(\frac{K_1}{\chi_a}\right)^{1/2} \tag{6.44}$$

beyond which the director, initially perpendicular to **H**, will orient itself parallel to **H**. H_c can be determined by optically detecting the distortion of the director, since this effect alters the refractive index of the sample. Measuring H_c thus enables K_1 to be determined, if χ_a is known. In the event that the magnetic field is applied perpendicular to the director, and parallel to the slides, constant K_2 is obtained. The homeotropic geometry in which the magnetic field **H** is applied perpendicular to the director gives K_3. Typical values of K_1, K_2, and K_3 are $K_1 = 7 \; 10^{-12}$ N, $K_2 = 4 \; 10^{-12}$ N, and $K_3 = 17 \; 10^{-12}$ N for p-azoxyanisole (PAA) at $T = 120\,°C$ [4].

The Freedericksz transition is thus a simple method for determining the Frank constants of the Nematic; it is also at the source of the Nematic displays, the orientational switch of the director in this case being the result of an electric field **E**.

6.7.2. Elastic Constants Deduced from the Velocity of the First Sound

Equation 6.8 shows that elastic constants A, B, and C can be deduced from velocity measurements taken as a function of the angle θ between the direction of sound propagation and the normal to the layers. This method assumes that the velocity displays marked anisotropy and that the measurements correspond to the hydrodynamic regime, two conditions that are not respected in common compounds, i.e., at ambient temperatures, such as 8OCB, or 8CB, which are A_d Smectics with weak B, displaying relaxation effects in the MHz domain. These effects are due to the flexibility of the end-chains of the molecules [35] and to the critical effects associated with the N–SmA transition. These drawbacks are very considerably reduced by using monolayer (SmA_1), high-temperature compounds.

From an experimental point of view, high temperatures present a serious disadvantage. Indeed, the technical problems raised by this type of study are compounded by those of the chemical deterioration of the compounds, which decompose rapidly on contact with the air at high temperatures. Using high-temperature compounds thus requires the elaboration of a special experimental procedure enabling the sample to be kept in a high state of purity throughout an experiment. This method, which is described in detail by Collin *et al.* [14], consists essentially in taking measurements under an inert atmosphere, any trace of air dissolved in the sample having been eliminated by degassing it within the cell itself.

These experimental difficulties explain why experiments providing reliable quantitative information are very scarce in the Smectic phases of liquid crystals. The data

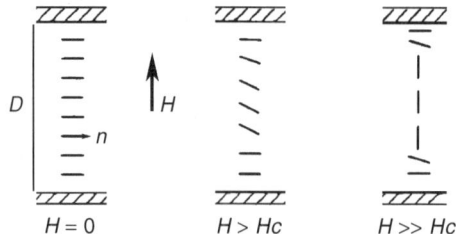

Fig. 6.5. Freedericksz transition.

presented in Figs. 6.6 to 6.11 concern TBBA, the chemical formula of which is shown in Figure 6.2. This is a high-temperature ($T_{NA} \approx 199\,°C$) compound, with wide N and SmA phases, a high layer compression modulus B, and low viscosity.

Figure 6.6 gives the behavior of the velocity measured at 1.2 MHz for $\theta = 0, 45$, and $90°$. Except for the vicinity of the N–SmA or SmA–SmC transitions, these measurements are within the hydrodynamic regime [36]. Examination of the figure shows the velocity to be a function of angle θ in the SmA phase, while it is independent of this angle in the N phase. This anisotropy, a result of the stratified structure of the SmA phase, gradually disappears on approaching the N phase. In addition, the velocity displays a marked decrease in the vicinity of the N–SmA transition. This anomaly reflects the existence of pretransitional effects that start to be visible in the N phase about $10\,°C$ before the transition. The velocity is also a function of angle θ in the SmC phase. However, velocity $V(0°)$, which is the highest velocity in the SmA phase, can be seen to become the lowest in the SmC phase. This remarkable inversion stems from the fact that the pretransitional effects are much more marked for $V(0°)$ than for $V(45°)$ and $V(90°)$. According to Eq. 6.37, this anisotropy of the critical effects indicates that $\gamma_u > \gamma_\rho$.

These measurements, analyzed with Eq. 6.8, make it possible to determine the thermal behavior of elastic constants A, B, and C, assuming $\rho = 10^3$ kg/m^3. Figure 6.7 shows that A displays a critical decrease at each phase transition. Absence of any linear variation in the N phase indicates that the range of the critical effects associated with the N–SmA and N–I transitions is large. Figures 6.8 and 6.9 show that B and C decrease strongly on approaching the N–SmA and SmA–SmC transitions. The arc-shaped curves indicate that the thermal behavior of each constant is essentially governed by the critical effects associated with both transitions.

6.7.2.1. Behavior of the Elastic Constants in the Vicinity of the N–SmA Transition

Despite the large quantity of studies that have been carried out for more than 20 years, this transition is still not understood, one of the key questions being why the exponents $\nu_\parallel$ and $\nu_\perp$, associated with the correlation lengths parallel and perpendicular to the director, respectively, are nonuniversal and different ($\nu_\parallel - \nu_\perp \approx 0.13$), a result

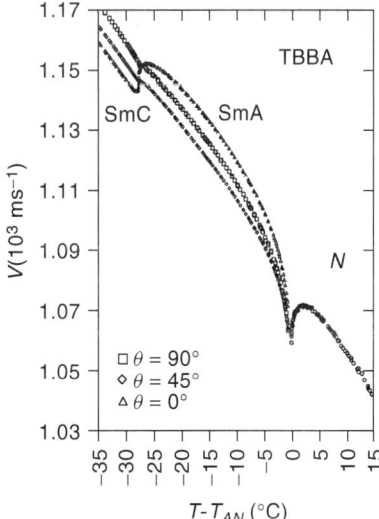

Fig. 6.6. Temperature dependence of the velocity measured at 1.2 MHz in the Nematic, Smectic A, and Smectic C phases for $\theta = 0, 45$, and $90°$.

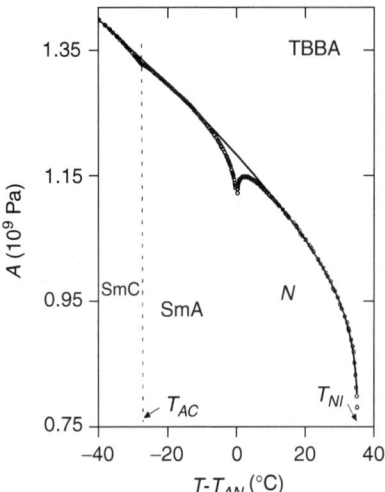

Fig. 6.7. Temperature dependence of the elastic constant A. The solid line represents the background term.

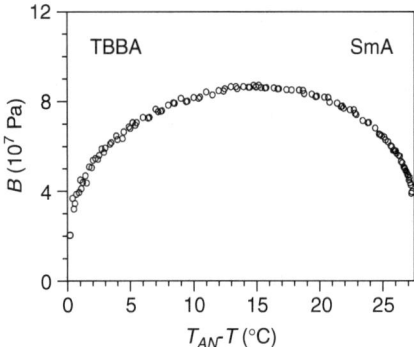

Fig. 6.8. Temperature dependence of the elastic constant B in the Smectic A phase.

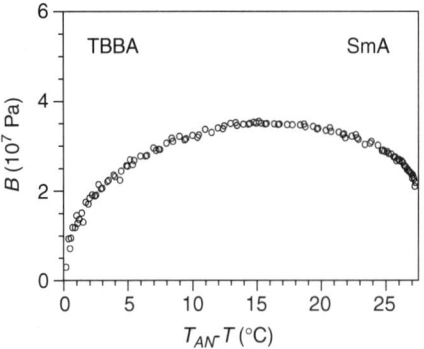

Fig. 6.9. Temperature dependence of the elastic constant C in the Smectic A phase.

that cannot be accounted for either by the 3DXY model or by the anisotropic critical point. In addition, this transition is always first order, as predicted by Halperin and Lubensky [18] and first demonstrated experimentally by Cladis *et al.* [37].

Since the velocity measurements show no anisotropy in the N phase, the critical effects come from the ψ^2-$\delta\rho$ coupling, which predicts that the elastic constant

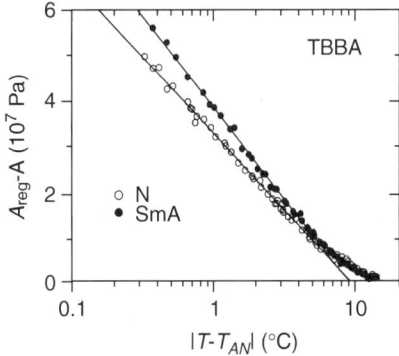

Fig. 6.10. Temperature dependence of the critical part of A near the N–SmA transition. The linear behavior indicates a logarithmic singularity ($\bar{\alpha} \approx 0$).

A behaves as $|T - T_{AN}|^{-\bar{\alpha}}$ (see Sect. 6.5.2.1.2).

Figure 6.10 shows the critical behavior of A, obtained after subtracting the regular term, plotted as a solid line in Figure 6.6. The linear variation observed in both phases up to within $\approx 6\,°C$ of the transition indicates that the critical behavior of A is near to a logarithmic singularity. This result shows that $\bar{\alpha} \approx 0$ and that the critical fluctuations are already developed in a temperature domain very far from the transition, despite it being first order. Analysis of the damping measurements leads to the same conclusion [36]. A more detailed analysis shows that A displays a 3DXY preasymptotic behavior characterized by an exponent $\bar{\alpha} = -0.007$ and a scaling law correction.

Figure 6.11 shows that the elastic constants B and C display power law behaviors ($B \sim (T_{AN} - T)^{\varphi_B}$, $C \sim (T_{AN} - T)^{\varphi_C}$) but the values of exponents φ_B and φ_C ($\varphi_B = 0.31$, $\varphi_C = 0.40$) are very far from those of the 3DXY model ($\varphi_B = \varphi_C = 0.66$).

The values of φ_B and φ_C suggest that the measurements of B and C are within a critical regime, between the 3DXY regime ($\varphi_B = \varphi_C = 0.66$) and the regime associated with the anisotropic critical point ($\varphi_B = 0$, $\varphi_C = \nu_\parallel$ with $\nu_\parallel$ unspecified). Such a regime has recently been predicted by Andereck and Patton [38] to explain why exponents $\nu_\parallel$ and $\nu_\perp$ are different and nonuniversal. This regime is dependent on the anisotropic character of the coupling between the $\delta \mathbf{n}$ fluctuations of the director and the Smectic order parameter, ψ. Since φ_B and φ_C are expressed as a function of $\nu_\parallel$ and $\nu_\perp$ (see Eq. 6.17), the fact that B and C have exponents that are very far from all the theoretical predictions suggests that these two elastic constants are under the influence of this coupling, unlike A, which involves the density coupling. This idea is currently under testing by doping TBBA with molecules that are nonmesogenic but that have a formula close to that of TBBA.

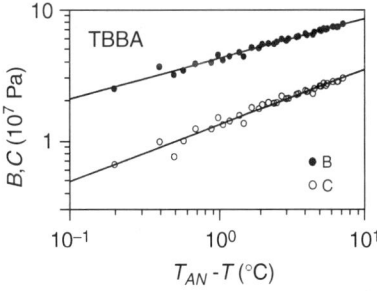

Fig. 6.11. Temperature dependence of B and C near the N–SmA transition.

6.7.2.2. Behavior of the Elastic Constants in the Vicinity of the SmA–SmC Transition, and Comparison with the Specific Heat Behavior

Although the SmA–SmC transition is expected to be in the 3DXY universality class, most of the experiments exhibit behavior that has been described by the Landau theory (no fluctuations) with an unusually large sixth order term [39]. As we shall see, measurements of the elastic constants as well as the subsequent theoretical developments have completely changed this description of the transition.

As shown in Figures 6.7 to 6.9, the elastic constants present pretransitional effects above the SmA–SmC transition, which are completely inconsistent with the Landau theory. Analysis of these effects shows that the three elastic constants follow the same thermal behavior ($\delta A \sim \delta B \sim \delta C \sim |T - T_{AC}|^{-\bar{\alpha}}$) with a specific heat exponent $\bar{\alpha}$ close to zero, indicating a 3DXY-type behavior [40]. Analysis of the damping associated with the high-frequency regime leads to the same conclusion [41].

The presence of marked pretransitional effects, which are still visible 10 °C above T_{AC}, is actually a surprising result, since the specific heat measurements taken on TBBA [42] and on most of the compounds studied display a Landau-type behavior characterized by a total absence of pretransitional effects in the SmA phase. This apparent contradiction between the behavior of the specific heat and that of the elastic constants was explained by Benguigui and Martinoty [32], showing that the Ginzburg criterion, which gives the width ΔT_c of the critical region, depends on the observable considered (see Sect. 6.6.2). This explanation rests on Eqs. 6.38 and 6.39, which show that

$$\frac{(\Delta T_C)_{C_p}}{(\Delta T_C)_{C_{ij}}} = \frac{1}{2}\left(\frac{b^*}{b}\right)^2 \tag{6.45}$$

For TBBA, $b/b^* \approx 10$ [32]. As a result, the width of the critical region for ultrasound velocity is much greater than that for specific heat ($\Delta T_{US} \approx 10$ °C and $\Delta T_{C_p} \approx 0.1$ °C). This indicates that the SmA–SmC transition of TBBA is not of the Landau mean field type, as the heat capacity behavior might suggest [42], but is really critical. The same conclusion was deduced in [43] for the SmA–SmC transition of the compound $\overline{8}S5$.

Equation 6.44 also makes it possible to show that critical-type behaviors can be observed for specific heat. Indeed, since the width of the critical region is proportional to $(b^*)^2$, a small variation of b^* can produce a large variation of ΔT_C. If b^* is very small, ΔT_C will also be very small, and the behavior of the specific heat will be of the Landau type. If b^* is large enough, ΔT_C will have a high value, and deviations from the Landau theory will appear. This explains why Gaussian or 3DXY critical behaviors have been observed for certain compounds [44–48].

6.7.3. Layer-Compression Modulus B Deduced from Second Sound and Dynamic Compression Experiments

For the previous reasons, it is not possible from ultrasound measurements to determine the hydrodynamic behavior of the elastic constants of common Smectics, i.e., ambient temperature, bilayer ones. This behavior can be determined by using frequencies below ultrasound frequencies, but the techniques used only enable layer compression modulus B to be determined. The behavior of the latter has been determined by means of experiments using undulation instability [49, 50], Rayleigh-scattering [51, 52], second sound [53–58], and dynamic compression techniques [59]. However, the experimental situation is very complicated and has only been clarified recently. We shall begin by comparing and discussing the results obtained by second sound and dynamic compression.

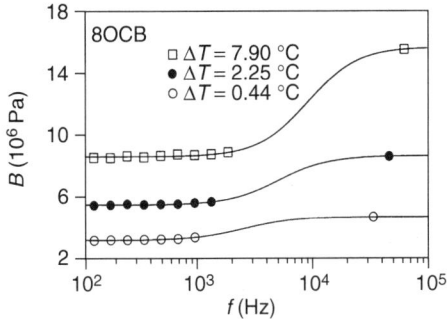

Fig. 6.12. Variation of B with frequency showing the existence of a relaxation mechanism between dynamic compression measurements ($<10^3$ Hz) and second sound ones ($>10^4$ Hz). The solid lines are the fits with a single-time relaxation mechanism (from [59]).

Figure 6.12 shows the variation of B as a function of frequency for various temperatures in the case of 8OCB. Measurements of B at frequencies above 10^4 Hz were taken using the second sound technique [56] and those below 10^3 Hz with the dynamic compression technique [59]. B can be seen to remain constant at frequencies below 10^3 Hz, down to a frequency that depends on sample quality. Below this frequency, which can be a few Hz in the most favorable cases, B decreases strongly due to the presence of defects in the sample. Above 10^3 Hz, B increases, owing to a nonhydrodynamic relaxation, and then saturates in the second sound frequency range as demonstrated by Rouillon et al. [58]. This increase in B, which is strongly temperature dependent, comes along with a decrease in the viscosity η, which goes from ≈ 100 Pa.s (10^3 Hz) to ≈ 0.1 Pa.s (second sound). The high value of the low-frequency viscosity indicates that the relaxation is of the Maxwell type with a relaxation frequency $\tau^{-1} = \Delta B / \Delta \eta$, where ΔB and $\Delta \eta$ are the relaxation strengths for B and η, respectively. Since η is only slightly temperature dependent, this relation shows that $\tau^{-1} \sim \Delta B$.

Figure 6.13 shows that B varies with temperature according to a simple power law ranging over 4 decades (full circles) with an exponent $\varphi_B = 0.36$ comparable to that observed for TBBA. Comparison with the behavior of B deduced from the second sound measurements (open circles) shows the existence of two relaxation mechanisms, one of the Maxwell type just mentioned and the other critical in nature, and associated with the $\omega\tau \ll 1 \to \omega\tau \gg 1$ change in regime occurring near the transition. The latter has been analyzed with Eq. 6.21 by imposing the φ_B value previously determined ($\varphi_B = 0.36$). Figure 6.14 shows the behavior of the $\delta B/\omega^{\varphi_B}$ ratio as a function of the reduced variable ωt^{-1} where $t = (T_{AN} - T)/T_{AN}$. The solid line is the fit to Eq. 6.21. The value of τ_o deduced from this fit ($\tau_o \approx 5\,10^{-10}$ s) leads to a value of $\omega\tau \approx 10$ for $T - T_{AN} = 0.1\,°C$

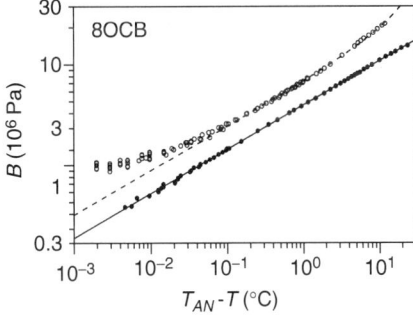

Fig. 6.13. Temperature dependence of B near the N–SmA transition. The upper curve corresponds to the second sound measurements and the lower curve to dynamic compression measurements (from [59]).

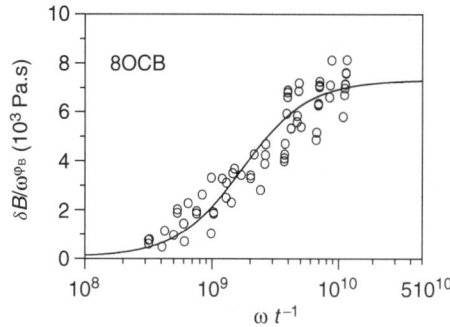

Fig. 6.14. Variation of the $\delta B/\omega^{\varphi_B}$ ratio as a function of the reduced variable ωt^{-1}. The solid line corresponds to the scaling function given by Eq. 6.21 with $\varphi_B = 0.36$ (from [59]).

and $f = 1$ MHz. This $\omega\tau$ value explains why no pretransitional effects on velocity and damping are observed at ultrasonic frequencies in this compound.

The discovery of relaxation effects in the frequency range 10^3- to 10^4-Hz rules out the previous picture held by many authors for the behavior of B. Indeed, these relaxation effects show that the second sound measurements do not belong to the hydrodynamic regime, unlike what is stated by Ricard and Prost [53], and subsequently taken up systematically [56, 57]. This statement had led to interpretation of the saturation effect observed by second sound [56] as support for the dislocation loops melting model, whereas it is a mere effect of frequency. They also explain the slight decrease in B observed by second sound when frequency diminishes, which had been interpreted [58] as the signature of anharmonic effects on the behavior of B. Finally, it should be noted that variations in B according to a power law have also been found for 8CB and the 6OCB–8OCB mixtures [60], which, at second sound frequencies, display a saturation behavior [55, 56] similar to that observed for 8OCB. This shows again that the latter is not due to the anisotropic critical point.

6.7.4. Elastic Constants Deduced from Rayleigh and Brillouin Scattering

Rayleigh-scattering experiments enable the thermal variation of the B/K_1 ratio to be determined. The results obtained for 8OCB show that B/K_1 follows a simple power-law with an exponent of ≈ 0.33 [51]. These measurements, taken in the domain of frequencies below 10^3 Hz, correspond to the hydrodynamic region in which B does not vary with frequency. They are not, therefore, affected by defects as had been suggested [53] to explain why these measurements are systematically lower than those deduced from second sound data. Comparison with the dynamic compression measurements shows that $K \sim 10^{-11}$ N, which is very close to the measurement taken in the N phase [61]. Finally, it should be pointed out that Rayleigh-scattering experiments also make it possible to determine elastic constant $B_\perp$, which holds the molecules perpendicular to the layers. The results obtained for 8OCB [51] show that $B_\perp$ follows a simple power law with an exponent of ~ 0.5, which is different from that of B. This result led to the notion of anisotropic scaling [20]. As for TBBA (see Sect. 6.7.2.1), it is possible that these exponents are crossover exponents associated with $\psi - \delta \mathbf{n}$ coupling.

Elastic constants A, B, and C can also be deduced from Brillouin scattering experiments [62, 63]. These experiments are carried out at frequencies that are typically in the gigahertz domain. They do not, therefore, correspond to the hydrodynamic regime, since nonhydrodynamic relaxations (relaxation of the director, the Smectic order parameter, the dipolar moment, or the chains) occur below this frequency. These relaxation effects increase the values of the elastic constants by more than one order of magnitude. A typical example is given by the layer compression modulus B of 8CB, which at T_{AN}-$T \sim 4\,°C$ goes from $8\ 10^6$ Pa (dynamic compression [60]) up to $5\ 10^8$ Pa (Brillouin scattering [63]).

ACKNOWLEDGMENTS

I would like to acknowledge the help of D. Collin who significantly contributed to the work made in our laboratory. I would also like to thank him and D. Rogez for partial disclosure of some of their results prior to publication.

References

1. Collin, D., Reys, V., Gallani, J.L., Benguigui, L.G., Poeti, G., and Martinoty, P. (1998). Sound propagation and damping in the vicinity of the smectic-A–hexatic-B phase transition of 4-propionyl-4′-n-heptanoyloxyazobenzene. *Phys. Rev. E.* **58**: 630–642.
2. Rogez, D., Gallani, J.L., and Martinoty, P. (1998). Strong critical fluctuations near a strongly first-order Smectic C–Hexatic F phase transition. *Phys. Rev. Lett.* **80**: 1256–1259.
3. Ali, A.H., and Benguigui, L. (1991). Elastic constants at the smectic A–smectic B phase transition of the compound 5O.8. *Liq. Cryst.* **9**: 741–750.
4. De Gennes, P.G., and Prost, J. (1995). *The Physics of Liquid Crystals*. 2nd ed., Oxford:Clarendon Press.
5. Chandrasekhar, S. (1994). *Liquid Crystals*. 2nd ed., Cambridge:Cambridge University Press.
6. Demus, D., Goodby, J.W., Gray, G.W., Spiess, H.W., and Vill, V., eds. (1998). *Handbook of Liquid Crystals*. Wiley-VCH.
7. Miyano, K., and Ketterson, J.B. (1979). Sound propagation in liquid crystals, in *Physical Acoustics* Mason, W.P., and Thurston, R.N., eds., vol. XIV, pp. 93–178. New York:Academic Press.
8. Frank, F.C. (1958). On the theory of liquid crystals. *Disc. Faraday Soc.* **25**: 19–28.
9. De Gennes, P.G. (1969). Conjectures sur l'état smectique. *J. Phys.* **30**: 65–71.
10. Martin, P.C., Parodi, O., and Pershan, P.S. (1972). Unified hydrodynamic theory for crystals, liquid crystals, and normal fluids. *Phys. Rev. A* **6**: 2401–2420.
11. Grinstein, G., and Pelcovits, R.A. (1981). Anharmonic effects in bulk smectic liquid crystals and other "one-dimensional solids." *Phys. Rev. Lett.* **47**: 856–859.
12. Mazenko, G.F., Ramaswamy, S., and Toner, J. (1983). Breakdown of conventional hydrodynamics for smectic A, hexatic B, and cholesteric liquid crystals. *Phys. Rev. A* **28**: 1618–1635.
13. Gallani, J.L., and Martinoty, P. (1985). Ultrasonic study of the breakdown of conventional hydrodynamics in the smectic-A phase of terephthal-bis-p-p'-butylaniline (TBBA). *Phys. Rev. Lett.* **54**: 333–336.
14. Collin, D., Gallani, J.L., and Martinoty, P. (1986). Nonconventional hydrodynamics of smectic-A phases of liquid crystals: An experimental study of various compounds with the use of ultrasounds. *Phys. Rev. A* **34**: 2255–2264.
15. Collin, D., Gallani, J.L., and Martinoty, P. (1987). Abnormal sound damping in the smectic-C phase of TBBA: Evidence for anharmonic effects. *Phys. Rev. Lett.* **58**: 254–257.
16. Baumann, C., Marcerou, J.P., Prost, J., and Rouillon, J.C. (1985). Frequency behavior of the second-sound damping in the smectic-A phase of octylcyanobiphenyl: Evidence for the divergence of viscosities. *Phys. Rev. Lett.* **54**: 1268–1270.
17. De Gennes, P.G. (1972). An analogy between superconductors and smectic A. *Solid State Comm.* **10**: 753–756.
18. Halperin, B.I., and Lubensky, T.C. (1974). On the analogy between smectic A liquid crystals and superconductors. *Solid State Comm.* **14**: 997–1001.
19. Barmatz, M., and Rudnick, I. (1968). Velocity and attenuation of first sound near the λ point of helium. *Phys. Rev.* **170**: 224–268.
20. Lubensky, T.C., and Chen, J.H. (1978). Anisotropic critical properties of the de Gennes model for the nematic to smectic A phase transition. *Phys. Rev. B* **17**: 366–376.
21. Lubensky, T.C. (1983). The nematic to smectic A transition: A theoretical overview. *J. Chim. Phys.* **80**: 31–43.
22. Dasgupta, C., and Halperin, B.I. (1981). Phase transition in a lattice model of superconductivity. *Phys. Rev. Lett.* **47**: 1556–1560.
23. Nelson, D., and Toner, J. (1981). Bond-orientational order, dislocation loops, and melting of solids and smectic A liquid crystals. *Phys. Rev. B* **24**: 363–387.
24. Swift, J., and Mulvaney, B.J. (1980). Anisotropic dispersion and attenuation of sound near the nematic-smectic-A phase transition. *Phys. Rev. B* **22**: 4523–4526.

25. Swift, J., and Mulvaney, B. J. (1979). Sound attenuation and dispersion near the nematic–smectic-A phase transition of a liquid crystal. *J. Phys. Lett.* **40**: 287–290.
26. Kiry, F., and Martinoty, P. (1978). Ultrasonic attenuation in CBOOA near the nematic–smectic A transition. *J. Phys.* **39**: 1019–1035.
27. Landau, L.D., and Khalatnikov, I.M. (1954). On the anomalous absorption of sound near a second-order phase transition point. *Dokl. Akad. Nauk. SSSR* **96**: 469–472.
28. Liu, M. (1979). Hydrodynamic theory near the nematic–smectic-A transition. *Phys. Rev. A* **19**: 2090–2094.
29. Martinoty, P., Sonntag, P., Benguigui, L., and Collin, D. (1994). New interpretation of second-sound measurements at the nematic-smectic A transition: Nonhydrodynamic behavior of the layer-compression elastic constant. *Phys. Rev. Lett.* **73**: 2079–2082.
30. De Gennes, P.G. (1972). Sur la transition smectique A–smectique C. *C. R. Acad. Sci. Paris* **274**: 758–760.
31. Andereck, B.S., and Swift, J. (1982). Propagation and attenuation of sound near the smectic-A–smectic-C phase transition in liquid crystals. *Phys. Rev. A* **25**: 1084–1091.
32. Benguigui, L.G., and Martinoty, P. (1989). The Smectic A–Smectic C transition: Mean-field and critical behaviors. *Phys. Rev. Lett.* **63**: 774–777.
33. Benguigui, L.G., and Martinoty, P. (1997). Characterizing the nature of the Smectic A–Smectic C and Smectic A–Smectic C* transitions. *J. Phys. II* **7**: 225–228.
34. Freedericksz, V., and Zolina, V. (1931). Uber die Doppelbrechung dünner anisotrop-flüssiger Schichten im Magnetfelde und diese Schicht orientierenden Kräfte.*Z. Kristallogr.* **79**: 255–267.
35. Candau, S., Martinoty, P., and Zana, R. (1975). Ultrasonic investigation of rotational isomerisms on mesomorphic compounds. *J. Phys. Lett.* **36**: L13–15.
36. Sonntag, P., Collin, D., and Martinoty, P. (2000). Dynamic scaling of ultrasonic damping near the nematic–smectic A transition of TBBA. *Phys. Rev. Lett.* **84**: 1950–1953.
37. Cladis, P.E., van Saarloos, W., Huse, D.A., Patel, J.S., Goodby, J.W., and Finn, P.L. (1989). Dynamical test of phase transition order. *Phys. Rev. Lett.* **62**: 1764–1767.
38. Andereck, B.S., and Patton, B.R. (1994). Anisotropic renormalization of thermodynamic quantities above the nematic–smecticA phase transition. *Phys. Rev. E* **49**: 1393–1402.
39. Huang, C.C., and Viner, J.M. (1982). Nature of the smectic A–smectic C phase transition in liquid crystals. *Phys. Rev. A* **25**: 3385–3388.
40. Collin, D., and Martinoty, P., to be published.
41. Collin, D., Gallani, J.L., and Martinoty, P. (1988) Abnormal behavior of sound velocity and damping in the vicinity of the smectic-A–smectic-C transition in terephthal-bis-p-p'-butylaniline (TBBA). *Phys. Rev. Lett.* **61**: 102–105.
42. Das, P., Ema, K., and Garland, C.W. (1989). Calorimetric study of the smectic A–smectic C transition in TBBA. *Liq. Cryst.* **4**: 205–208.
43. Collin, D., Moyses, S., Neubert, M.F., and Martinoty, P. (1994). Critical behavior of sound damping in the vicinity of the smectic A–smectic C transition in $\overline{8}$S5. *Phys. Rev. Lett.* **73**: 983–986.
44. Delaye, M. (1979). Coherence length and angular susceptibility divergences above a smectic A to smectic C phase transition observed by Rayleigh scattering. *J. Phys. Paris* **40**: 350–355.
45. Galerne, Y. (1981). Interferometric measurements at a smectic A–smectic C phase transition. *Phys. Rev. A* **24**: 2284–2286.
46. Ema, K., Watababe, J., Takagi, A., and Yao, H. (1995). Critical behavior of heat capacity at the smectic-C_α^*–smectic-A transition of the antiferroelectric liquid crystal methylheptyloxycarbonylphenyl octyloxybiphenyl carboxylate (MHPOBC). *Phys. Rev. E* **52**: 1216–1219.
47. Stoebe, T., Reed, L., Veum, M., and Huang, C.C. (1996). Nature of the smectic-A-smectic-C transition of a partially perfluorinated compound. *Phys. Rev. E* **54**: 1584–1591.
48. Ema, K., and Yao, H. (1998). Crossover from XY critical to tricritical behavior of heat capacity at the smectic-A–chiral–smectic-C liquid-crystal transition. *Phys. Rev. E* **57**: 6677–6684.
49. Clark, N.A. (1976). Pretransitional mechanical effects in a smectic-A liquid crystal. *Phys. Rev. A* **14**: 1551–1554.
50. Ribotta, R. (1974). Mesure de la longueur de pénétration dans un smectique-A au voisinage d'une transition smectique-nématique du deuxième ordre. *C.R. Acad. Sc. Paris* **279**:B, 295–296.
51. Birecki, H., Schaetzing, R., Rondelez, F., and Litster, J.D. (1976). Light-scattering study of a smectic A phase near the smectic A–nematic transition. *Phys. Rev. Lett.* **36**: 1376–1379.
52. Fromm, H.J. (1987). High resolution study of the compression modulus B in the vicinity of the nematic-smectic A transition in 6OCB/8OCB mixtures. *J. Phys.* **48**: 647–650.
53. Ricard, L., and Prost, J. (1981). Critical behaviour of second sound near the smectic A–nematic phase transition. *J. Phys.* **42**: 861–873.
54. Fisch, M.R., Pershan, P.S., and Sorensen, L.B. (1984). Absolute measurement of the critical behavior of the smectic elastic constant of bilayer and monolayer smectic A liquid crystals on approaching the transition to the nematic phase. *Phys. Rev. A* **29**: 2741–2750.

55. Fisch, M.R., Sorensen, L.B., and Pershan, P.S. (1982). Anomalous temperature dependence of the elastic constant B at the smectic-to-nematic phase transition in binary mixtures of hexyloxycyanobiphenyl-octyloxycyanobiphenyl (6OCB-8OCB). *Phys. Rev. Lett.* **48**: 943–946.
56. Benzekri, M., Claverie, T., Marcerou, J.P., and Rouillon, J.C. (1992). Non-vanishing of the layer compressional elastic constant at the smectic A-to-nematic phase transition: A consequence of Landau-Peierls instability? *Phys. Rev. Lett.* **68**: 2480–2483.
57. Beaubois, F., Claverie, T., Marcerou J.P., Rouillon, J.C., Nguyen, H.T., Garland, C.W., and Haga, H. (1997). Influence of nematic range on birefringence, heat capacity and elastic modulus near a nematic-smectic A phase transition. *Phys. Rev. E* **56**: 5566–5574.
58. Rouillon, J.C., Benzekri, M., Claverie, T., Marcerou, J.P., Nguyen, H.T., and Prost, J. (1994). Experimental evidence for the breakdown of conventional elasticity in smectics A. *Liquid Cryst.* **16**: 1065–1072.
59. Martinoty, P., Gallani, J.L., and Collin, D. (1998). Hydrodynamic and non-hydrodynamic behavior of layer-compression modulus B at the nematic–smectic A phase transition in 8OCB. *Phys. Rev. Lett.* **81**: 144–147.
60. Rogez, D., Collin, D., and Martinoty, P., to be published.
61. Karat, P., and Madhusudana, N.V. (1978). Orientational order and elastic constants of some cyanobiphenyls. *Mol. Cryst. Liq. Cryst.* **47**: 21–28.
62. Liao, Y., Clark, N.A., and Pershan, P.S. (1973). Brillouin scattering from smectic liquid crystals. *Phys. Rev. Lett.* **30**: 639–641.
63. Gleed, D.G., Sambles, J.R., and Bradberry G.W. (1988). A method for indirectly measuring the second sound velocity in smectic A liquid crystals. *Liq. Cryst.* **3**: 1689–1697.

CHAPTER 7

FUNDAMENTAL ACOUSTIC PROPERTIES OF BUBBLY LIQUIDS

Andrea Prosperetti[1]

Department of Mechanical Engineering, Johns Hopkins University, Baltimore, MD, USA

Contents

Abstract	183
7.1. Introduction	183
7.2. Linear Waves in Dilute Bubbly Liquids	184
7.3. Bubble Dynamics	191
7.4. Speed of Sound in Bubbly Liquids	197
7.5. Some Applications	201
7.6. Conclusions	203
Acknowledgment	204
References	204

ABSTRACT

This chapter begins by describing three different derivations of the dispersion relation for the propagation of linear pressure waves in a bubbly liquid. Some fundamental properties of the forced oscillations of gas bubbles are described next to develop a complete model. The theoretical predictions are shown to be mostly in good agreement with data. The only discrepancies that are found arise in a frequency region near and above the resonance frequency of the bubbles, where attenuation is very large, of the order of several dB/cm. Applications to the propagation of sound through bubbly layers, to the generation of underwater sound by the collective oscillations of bubble clouds, and to the propagation of weak shock waves in a bubbly liquid are briefly described.

7.1. INTRODUCTION

Bubbly liquids are frequently encountered both in natural phenomena and in technology. Breaking wind waves inject bubbles into the upper layers of the ocean [1–6],

---

[1] Also: Department of Applied Physics, Twente Institute of Mechanics, and Burgerscentrum, University of Twente, AE 7500 Enschede, The Netherlands.

Handbook of Elastic Properties of Solids, Liquids, and Gases, edited by Levy, Bass, and Stern
Volume IV: Elastic Properties of Fluids: Liquids and Gases
Copyright © 2001 by Academic Press
ISBN 0-12-445764-9 / $35.00
All rights of reproduction in any form reserved.

where a variety of convective processes can transport them to depths of several tens of meters [7–10]. In volcanic eruptions, the rapid ex-solution of dissolved gases causes the acceleration of a magma to velocities of hundreds of meters per second [11–13]. The injection of bubbles at the bottom of oil wells is sometimes used to help lift the liquid to the surface [14]. Some gas-liquid contactors in the chemical industry operate in the dispersed bubble regime to help maximize the contact area of the two phases [15, 16]. In aquaculture, the aeration of ponds and tanks is important to maintain an adequate supply of oxygen to fish [17], and a similar need arises in water treatment to facilitate the action of aerobic microorganisms [18–20]. As a final example, one may cite the use of bubbles as acoustic contrast agents in medicine [21, 22].

From a fluid mechanics perspective, bubbly liquids may be considered as a particular instance of multiphase systems, i.e., systems in which two (or more) phases — a gas and a liquid — simultaneously occupy the domain of interest. Although in general such systems are notoriously difficult to describe, the limitation to acoustic properties introduces significant simplifications.

It should be stated at the outset that we will mostly deal with situations in which the gas is finely dispersed in small bubbles (up to a size of, say, a few millimeters) with a relatively small concentration by volume — the so-called gas volume fraction or void fraction β. For example, in oceanic bubble plumes one encounters characteristic values $\beta \sim 10^{-6}$ at a depth of a few tens of meters, while right under a breaking wave β can be as high as, say, 10% [23, 24]. In some applications, β can be as high as 30%, although at higher concentrations small bubbles can coalesce in large gas masses that alter fundamentally the nature of the phenomena. When coalescence is inhibited by surface-active agents or by a very low surface tension, the formation of foams is possible with gas volume fractions of 95% or more (see e.g., [25]). We explicitly exclude such systems from the present considerations.

At a fundamental physical level, the peculiar physics of bubbly liquids arises from the sharp separation between of the system's mass — mostly contained in the liquid phase — and the system's compressibility — mostly due to the gas. For this reason, bubbly liquids behave very differently from other, apparently similar, gas-liquid systems such as fogs or other suspensions of droplets in gases. For example, a most striking difference between the two systems is that, while the sound speed in a fog is always bracketed by the values of the sound speed in the individual pure phases, in bubbly liquids it is easy to encounter situations in which the sound speed is significantly lower than that in the gas alone.

7.2. LINEAR WAVES IN DILUTE BUBBLY LIQUIDS

A good way to gain an appreciation of the physical nature of bubbly liquids is to consider several alternative approximate formulations for the propagation of linear waves in such systems. We shall consider here only dilute systems. The dense case is much more difficult, and its understanding is still rather incomplete. Some comments are found at the end of this section.

7.2.1. Wood's Method

The earliest calculation of the speed of sound c_m in a bubbly liquid can be found in Wood's treatise [26], the first edition of which was published in 1911. With some later additions, the argument is essentially the following.

The starting point is the standard acoustic relation

$$\frac{1}{c_m^2} = \frac{d\rho_m}{dP} \tag{7.1}$$

where ρ_m denotes the mixture density and P the mixture pressure. The density of a representative volume ΔV of mixture is given by

$$\rho_m = \frac{\Delta M_m}{\Delta V} = (1-\beta)\rho_L + \beta\rho_G \qquad (7.2)$$

where M is mass and the indices L, G refer to the pure liquid and gas components. On taking the derivative indicated in Eq. 7.1 one has

$$\frac{1}{c_m^2} = \frac{1-\beta}{c_L^2} + \frac{\beta}{c_G^2} - (\rho_L - \rho_G)\frac{d\beta}{dP} \qquad (7.3)$$

where $c_{L,G}^2 = dp_{L,G}/d\rho_{L,G}$ are the speed of sound in the individual components and one has implicitly assumed that $P \simeq p_{L,G}$. By definition, the volume fraction is the ratio of the part ΔV_G of ΔV occupied by the gas to ΔV and, if the ΔV contains ΔN bubbles, all with the same volume υ, we have $\Delta V_G = \upsilon \Delta N$ and

$$\beta = \frac{\upsilon \Delta N}{\Delta V} = \upsilon n \qquad (7.4)$$

where n is the bubble number density. From this relation we derive

$$\frac{d\beta}{dP} = \frac{d}{dP}\left(\frac{\Delta V_G}{\Delta V_L + \Delta V_G}\right) = \frac{1-\beta}{\Delta V}\frac{dV_G}{dP} - \frac{\beta}{\Delta V}\frac{dV_L}{dP} \qquad (7.5)$$

Let us define ΔV as always containing the same amount of liquid and, furthermore, assume that the amount of gas contained in each bubble is a constant so that $\rho_G d\upsilon = -\upsilon\rho_G$. Then Eq. 7.3 becomes

$$\frac{1}{c_m^2} = \frac{\beta^2}{c_G^2} + \frac{(1-\beta)^2}{c_L^2}\left(1 + \frac{\rho_G}{\rho_L}\frac{\beta}{1-\beta}\right) - \rho_L \frac{\beta(1-\beta)}{\upsilon}\frac{d\upsilon}{dP}$$
$$- \beta(1-\beta)\frac{\rho_L - \rho_G}{\Delta N}\frac{d\Delta N}{dP} \qquad (7.6)$$

Formally the right-hand side is found to approach the correct limits of $1/c_L^2$ and $1/c_G^2$ as $\beta \to 0$ and $\beta \to 1$, respectively. Actually, as will be clear from the following, this relation is only valid for small β. Retaining only the most significant terms in this quantity and disregarding ρ_G with respect to ρ_L, we then find

$$\frac{1}{c_m^2} \simeq \frac{1}{c_L^2} - \rho_L\frac{\beta(1-\beta)}{\upsilon}\frac{d\upsilon}{dP} - \beta(1-\beta)\frac{\rho_L}{\Delta N}\frac{d\Delta N}{dP} \qquad (7.7)$$

If the bubbles move with the liquid, ΔN is a constant and

$$\frac{1}{c_m^2} \simeq \frac{1}{c_L^2} - \rho_L\frac{\beta(1-\beta)}{\upsilon}\frac{d\upsilon}{dP} \qquad (7.8)$$

If, on the other hand, the bubbles are free to move with respect to the liquid, under the action of the pressure gradient that develops as the acoustic wave propagates and ignoring any drag effect, they will acquire a velocity three times that of the liquid $\mathbf{u}_L$ [27, 28]. The change in the number of bubbles contained within ΔV is then

$$d\Delta N = n\int_{\Delta V}(\mathbf{u}_{rel}dt)\cdot\mathbf{n}dS \qquad (7.9)$$

where $\mathbf{u}_{rel} = 2\mathbf{u}_L$. Since ΔV is defined to contain the same liquid mass,

$$d\Delta V = \int_{\Delta V}(\mathbf{u}_L dt)\cdot\mathbf{n}dS \qquad (7.10)$$

and, therefore, $d\Delta N = 2nd\Delta V$. Approximating ΔV by ΔV_G we thus have

$$\frac{1}{\Delta N}\frac{d\Delta N}{dP} = -\frac{2\beta}{1+2\beta}\frac{1}{\upsilon}\frac{d\upsilon}{dP} \qquad (7.11)$$

On substituting into Eq. 7.7 the result is

$$\frac{1}{c_m^2} \simeq \frac{1}{c_L^2} - \rho_L \frac{\beta(1-\beta)}{(1+2\beta)}\frac{d\upsilon}{dP} \simeq \frac{1}{c_L^2} - \rho_L\beta(1-3\beta)\frac{d\upsilon}{dP} \qquad (7.12)$$

In analogy with the case of waves in reacting media, the expressions Eqs. 7.8 and 7.12 are sometimes referred to as the *equilibrium* and *frozen* sound speeds, respectively. Since $d\upsilon/dP < 0$ it can be seen that the equilibrium sound speed is slightly smaller than the frozen sound speed, which is a consequence of the fact that bubbles are repelled by regions of high pressure so that the compressibility of the mixture decreases.

If the bubbles compress isothermally $Pd\upsilon = -\upsilon dP$ and Eq. 7.8 becomes

$$\frac{1}{c_m^2} \simeq \frac{1}{c_L^2} + \rho_L\frac{\beta(1-\beta)}{P} \qquad (7.13)$$

A comparison of this relation with some low-frequency air–water data of Gibson [29][2] is shown in Figure 7.1, where the dashed line is c_L. For many liquids, $c_L \sim 10^3$ m/s, $\rho_L \sim 10^3$ kg/m^3, and, at normal pressure, conditions $P \sim 10^5$ Pa; the two terms in the right-hand side are then of comparable magnitude for $\beta \sim 10^{-4}$, while the second one dominates for larger values of β. In these conditions one may then use the well-known approximation

$$c_m^2 \simeq \frac{P}{\rho_L\beta(1-\beta)} \qquad (7.14)$$

For an air–water system at 100 kPa, c_m reaches a minimum for $\beta = 50\%$, where it has the value 20 m/s; for $\beta = 0.1\%$, c_m is close to c_G. These numerical values are sufficient to illustrate the very profound effect that the presence of even a small amount of free

[2] Unfortunately, Gibson's data are very hard to read from his figure; we only show here the ones that we can deduce with some confidence after digitizing the original figure.

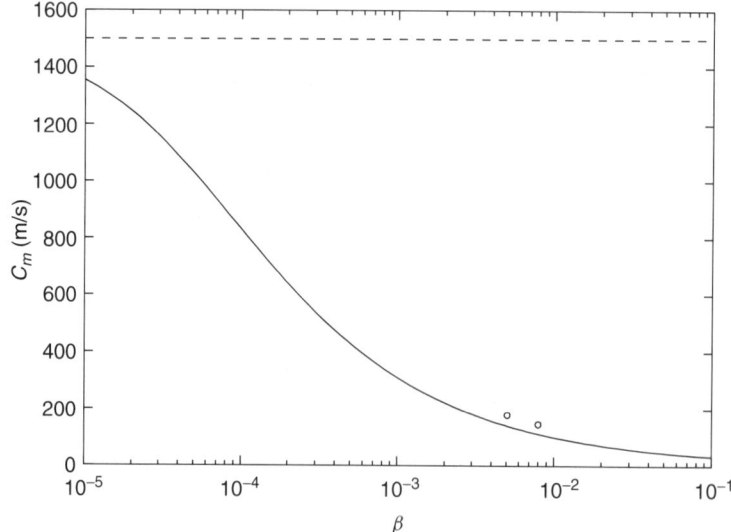

Fig. 7.1. The low-frequency limit Eq. 7.13 of the speed of sound in bubbly water as a function of the gas volume fraction β; the symbols are data of Gibson [29] and the dashed line is the speed of sound in water with no bubbles.

gas bubbles has on the acoustic properties of a liquid–gas mixture. Relating bubble volume and mixture pressure by the isothermal relation evidently ignores the fact that to compress a bubble, the liquid must acquire a certain amount of momentum. We thus see that the previous results can only describe the situation at very low frequency, as indeed is shown below in Section 7.4.

The previous argument is readily generalized to the more common case of a size distribution. Let $\mathcal{N}(a)\,da$ be the number of bubbles with equilibrium radius between a and $a + da$. Then we have

$$\beta = \int_{a_m}^{a_M} \left(\frac{4}{3}\pi a^3\right) \mathcal{N}(a)\,da \tag{7.15}$$

where $a_{m,M}$ are the minimum and maximum values of the equilibrium radius in the distribution. The same steps can be traced with this generalized expression for β with the same results.

7.2.2. An "Effective Medium" Approach

The previous derivation suffers from some obvious shortcomings. In the first place, the very expression Eq. 7.1 that lies at its heart presupposes that the bubbly mixture behaves as an effective medium that can be characterized by a well-defined equation of state expressed completely in terms of suitable mean, or effective, variables. For the development to make sense, it is necessary that the "representative volume" used in the derivation be small enough that it can be considered as homogeneous, but also large enough as to contain sufficiently many bubbles that the notion of average behavior is meaningful. Evidently, these two constraints imply the separation of scales (to be refined later):

$$\text{mean bubble distance} \ll \text{scale of } \Delta V \ll \text{wavelength of sound} \tag{7.16}$$

The questions of whether the bubbles are free to move with respect to the liquid, or indeed compress isothermally, do not admit an obvious a priori answer.

A simple justification for the use of Eq. 7.1 can be given as follows, neglecting for simplicity the compressibility of the liquid that, as noted previously, is only significant at very low volume fractions. In a low-viscosity liquid, the liquid-bubbles system can be approximately treated by means of potential flow theory. Consider N bubbles immersed in an "incident" flow characterized by a harmonic potential ϕ_∞ (this may be considered as the potential in the absence of the bubbles). If the bubbles are characterized purely in terms of their monopoles, the flow potential is modified to

$$\phi(\mathbf{x}, t) = \phi_\infty(\mathbf{x}, t) + \sum_{\alpha=1}^{N} \frac{\dot{v}_\alpha}{4\pi|\mathbf{x} - \mathbf{y}_\alpha|} \tag{7.17}$$

where the dot denotes the time derivative and $\mathbf{y}_\alpha$ is the position of the α-th bubble. If N is large, the summation can be approximately replaced by a volume integral over the bubbly region:

$$\phi(\mathbf{x}, t) = \phi_\infty(\mathbf{x}, t) + \int d^3y \int da \frac{\dot{v}}{4\pi|\mathbf{x} - \mathbf{y}|} \mathcal{N}(a; \mathbf{y}) \tag{7.18}$$

On taking the Laplacian of this expression we have

$$\nabla^2 \phi = \int da\, \dot{v}\, \mathcal{N}(a; \mathbf{x}) \tag{7.19}$$

Arguing as before, we may assume that v depends on the external pressure field so that or, using the acoustic relation $P = -\rho_m \partial\phi/\partial t$,

$$\dot{v} = -\rho_m \frac{dv}{dP} \frac{\partial^2 \phi}{\partial t^2} \tag{7.20}$$

so that Eq. 7.19 becomes

$$\nabla^2\phi + \left(\rho_m \int da \frac{dv}{dP}\mathcal{N}(a;\mathbf{x})\right)\frac{\partial^2\phi}{\partial t^2} = 0 \qquad (7.21)$$

and, again assuming isothermal behavior,

$$\nabla^2\phi - \frac{1}{c_m^2}\frac{\partial^2\phi}{\partial t^2} = 0 \qquad (7.22)$$

with c_m given by

$$\frac{1}{c_m^2} = -\rho_m \int da \frac{dv}{dP}\mathcal{N}(a;\mathbf{x}) = -\rho_L(1-\beta)n\frac{dv}{dP} \qquad (7.23)$$

where the second form is applicable to the case of a monodisperse mixture; clearly this relation reduces to Eq. 7.14 with a suitable assumption on the bubble pressure–volume relation.

An important point apparent from this argument, but not from the previous one, is that for the result to be valid it must be possible to represent the bubbles by simple monopoles. Evidently this approximation is only valid provided the interbubble distance is much greater than the radius, which introduces the necessity for a further separation of scales in the inequalities Eq. 7.16.

This derivation is a simplified version of that presented in [30], where the liquid compressibility was also included and further additional points are carefully made. The approximation of the sum in Eq. 7.17 by the integral in Eq. 7.18 presupposes a limit process with the number density $f \to \infty$ and the bubble radius $a \to 0$ so as to keep fa constant. As a consequence, in this limit, $\beta \sim a^2 \to 0$ and is accordingly very small. Since β is evidently also the probability that any point $\mathbf{x}$ is in the gas phase, it was unnecessary in the previous derivation to explicitly assume that the point $\mathbf{x}$ was in the liquid.

It may be noted that Eqs. 7.19 and 7.20 are equivalent to the (linear) system:

$$\nabla \cdot \mathbf{u} = \int da\, \dot{v}\mathcal{N}(a;\mathbf{x}), \quad \rho_m \frac{\partial \mathbf{u}}{\partial t} + \nabla P = 0 \qquad (7.24)$$

provided the mean velocity field $\mathbf{u}$ is irrotational.

7.2.3. Foldy's Multiple Scattering Theory

In a classic paper in the theory of multiple scattering published in 1945, Foldy [31] presented a derivation of the previous result which is still of considerable interest.

He starts by noting that, if the bubble radius is small compared with the acoustic wavelength (which, as is seen below, is usually the case), the field scattered by the α-th bubble has essentially a monopole nature and may be written (omitting a factor $\exp i\omega t$)

$$\phi^\alpha_{\text{scatt.}}(\mathbf{x}) = F^\alpha E(\mathbf{x}, \mathbf{y}_\alpha)\phi^\alpha_{inc}(\mathbf{y}_\alpha) \qquad (7.25)$$

where F^α is the scattering amplitude of the bubble, ϕ^α_{inc} is the field incident on it, and

$$E(\mathbf{x}, \mathbf{y}_\alpha) = \frac{\exp(-ik|\mathbf{x} - \mathbf{y}_\alpha|)}{|\mathbf{x} - \mathbf{y}_\alpha|} \qquad (7.26)$$

with $\kappa = \omega/c_L$ the wavenumber in the pure liquid. The incident field, in its turn, is given by the total field minus the contribution of the α bubble:

$$\phi^\alpha_{inc.} = \phi - \phi^\alpha_{\text{scatt.}} \qquad (7.27)$$

If, as before, ϕ_∞ denotes the potential of the external field incident on the bubbly liquid region, we may thus write the following multiple scattering relations:

$$\phi(\mathbf{x}) = \phi_\infty + \sum_{\alpha=1}^{N} F^\alpha E(\mathbf{x}, \mathbf{y}_\alpha) \phi_{inc}^\alpha(\mathbf{y}_\alpha) \tag{7.28}$$

and

$$\phi_{inc.}^\alpha(\mathbf{y}_\alpha) = \phi_\infty + \sum_{\beta \neq \alpha} F^\beta E(\mathbf{y}_\alpha, \mathbf{y}_\beta) \phi_{inc}^\beta(\mathbf{y}_\beta) \tag{7.29}$$

The first relation states that the total field is the result of the superposition of ϕ_∞ and the field scattered by all the bubbles, and is thus similar to Eq. 7.17 used earlier; the second equation states that the field incident on the bubble α is ϕ_∞ plus the field scattered by all the other bubbles. Note that, although not explicitly indicated, the fields appearing in Eqs. 7.28 and 7.29 depend on the position of all the bubbles.

While Eqs. 7.28, and 7.29 can nowadays be solved exactly up to a number of bubbles unthinkable in Foldy's time (see, e.g., Watanabe [32]). Foldy introduces a probabilistic argument by means of an ensemble of bubble configurations characterized by a probability density[3] $P(\mathbf{y}_1, \mathbf{y}_2, \cdots \mathbf{y}_N)$. On calculating the ensemble average of Eq. 7.28 we find

$$<\phi>(\mathbf{x}) = \phi_\infty + \frac{1}{N!} \sum_{\alpha=1}^{N} \int d^3 y_1 \int d^3 y_2 \int \cdots \int d^3 y_N F^\alpha E(\mathbf{x}, \mathbf{y}_\alpha) \phi_{inc}^\alpha P(\mathbf{y}_1, \mathbf{y}_2, \ldots \mathbf{y}_N) \tag{7.30}$$

where the factor $N!$ depends on the normalization of P and is due to the indistinguishibility of the bubbles (see, e.g., Batchelor [33]). Since the bubbles are all equal, each one of them makes the same contribution to the sum and we may therefore write

$$<\phi>(\mathbf{x}) = \phi_\infty + \frac{1}{(N-1)!} \int d^3 y_1 \int d^3 y_2 \int \cdots \int d^3 y_N F^1 E(\mathbf{x}, \mathbf{y}_1) \phi_{inc}^1 P(\mathbf{y}_1, \mathbf{y}_2, \ldots \mathbf{y}_N) \tag{7.31}$$

The introduction of the one-particle distribution function (or particle number density)

$$n(\mathbf{y}_1) = P(\mathbf{y}_1) = \frac{1}{(N-1)!} \int d^3 y_2 \int \cdots \int d^3 y_N P(\mathbf{y}_1, \mathbf{y}_2, \ldots \mathbf{y}_N) \tag{7.32}$$

and of the conditional probability

$$P(\mathbf{y}_2, \ldots \mathbf{y}_N | \mathbf{y}_1) = \frac{P(\mathbf{y}_1, \mathbf{y}_2, \ldots \mathbf{y}_N)}{P(\mathbf{y}_1)} \tag{7.33}$$

enables us to rewrite Eq. 7.31 as

$$\langle \phi \rangle(\mathbf{x}) = \phi_\infty + \int d^3 y_1 F^1 n(\mathbf{y}_1) E(\mathbf{x}, \mathbf{y}_1) \langle \phi_{inc}^1 \rangle_1(\mathbf{y}_1) \tag{7.34}$$

where

$$<\phi_{inc}^1>_1(\mathbf{y}_1) = \frac{1}{(N-1)!} \int d^3 y_2 \int \cdots \int d^3 y_N \phi_{inc}^1 P(\mathbf{y}_2, \ldots \mathbf{y}_N | \mathbf{y}_1) \tag{7.35}$$

is the field incident on bubble 1 averaged over all possible positions of the other bubbles subject to the condition that bubble 1 is located at $\mathbf{y}_1$. In principle, this quantity should

---

[3] Foldy allowed P to depend also on scatterer parameters such as, in the present case, the bubble radius; for simplicity, here we assume all scatterers to be equal. Also, we do not indicate the time dependence explicitly. The treatment that follows is slightly more general than Foldy's in that the N-particle probability is not factorized into the product of N one-particle probabilities.

be evaluated by taking the conditional average of Eq. 7.29 written for $\alpha = 1$, but this procedure would lead to the appearance of the two-particle average $\langle \phi_{inc}^1 \rangle_1 (\mathbf{y}_1, \mathbf{y}_2)$ and so forth. Foldy writes: "We therefore must resort to the approximation of replacing the external field acting on the [first] scatterer averaged over all configurations of the other scatterers by the average field which would exist at the position of the [first] scatterer when the scatterer is not present. This last average field would differ only by a term of the order $1/N$ from $\langle \phi \rangle (\mathbf{y}_1)$ so that, if N is large, and if the above approximation is valid, which it appears to be on physical grounds, we may substitute $<\phi>(\mathbf{y}_1)$ for $\langle \phi_{inc}^1 \rangle_1 (\mathbf{y}_1)$ under the integral sign in the above equation thus obtaining the integral equation

$$\langle \phi \rangle(\mathbf{x}) = \phi_\infty + \int d^3 y_1 E(\mathbf{x}, \mathbf{y}_1) n(\mathbf{y}_1) F^1 <\phi>(\mathbf{y}_1) \tag{7.36}$$

for the configurational average of the wave equation." To solve the equation one now applies the operator $\nabla^2 + \kappa^2$, of which E is the Green's function, to find

$$(\nabla^2 + \kappa^2)\langle \phi \rangle = -4\pi n(\mathbf{x}) F <\phi>(\mathbf{x}) \tag{7.37}$$

from which one finds the Helmholtz equation with the effective wave number

$$\kappa^2 = \kappa^2 + 4\pi n(\mathbf{x}) F \tag{7.38}$$

This result is readily extended to a distribution of bubble sizes, in which case it becomes

$$\kappa^2 = \kappa^2 + 4\pi \int da F(a) \mathcal{N}(a; \mathbf{x}) \tag{7.39}$$

From the known laws of monopole scattering and the well-known relation between the potential and the pressure valid in acoustics, it is easy to show that

$$F = \frac{1}{4\pi} \omega^2 \rho_L \frac{dv}{dP} \tag{7.40}$$

with which Eq. 7.39 becomes

$$\kappa^2 = \kappa^2 + \omega^2 \rho_L \int da \frac{dv}{dP}, \mathcal{N}(a; \mathbf{x}) \tag{7.41}$$

This result agrees with Eq. 7.21 provided the liquid compressibility is neglected, so that $\kappa = 0$ and $\rho_m \simeq \rho_L$.

7.2.4. Some Comments on the Theory

A modern version of Foldy's theory has been provided by Caflisch et al. [30], who point out the very interesting fact that Foldy's approach as recast, e.g., in the form of Eq. 7.24, leads to a model that in the dilute limit is valid also when the bubble undergoes large-amplitude, nonlinear motion. The same conclusion can be drawn by using the argument of Duraiswami and Prosperetti [34], who use a singular perturbation approach in which the disturbance field of the inclusions matches the outer field to give rise to an effective medium description. As a consequence of this validity beyond the linear domain, the theory can be used to describe the propagation of weak shock waves in a bubbly liquid, as is described in Section 7.5.

A general theory applicable to the case of a flowing bubbly liquid is described by Zhang and Prosperetti [35]. Explicit results that extend Foldy's to this situation are also presented in that work.

The only rigorous step beyond the dilute limit has been taken by Sangani [36], who was able to account not only for higher-order effects in the gas volume fraction, but also for the effect of distortions of the spherical shape. With the neglect of surface tension effects, his result is:

$$\frac{\kappa^2}{\omega^2} = (1 - \beta\lambda_\upsilon)\left[\frac{1 - \beta\lambda_p}{c_L^2} - n\rho_L \frac{d\upsilon}{dP}\lambda_p\right] \quad (7.42)$$

The term λ_υ arises from the difference between the translatory acceleration of the bubble and that of the liquid and is given by

$$\lambda_\upsilon = 3 + O(\beta\omega_r^2) \quad (7.43)$$

where $\omega_r = \omega/\omega_0$ with ω_0 the natural frequency of the bubble. With the neglect of dissipative effects, the expression of λ_p is

$$\lambda_p = \frac{1}{1 - \omega_r^2}\left[1 - \left(\frac{\omega_r^2}{1 - \omega_r^2}\right)^{3/2}\left(i\sqrt{3\beta} + \frac{3}{2}\beta\log\beta\right) + O\left(\beta\frac{\omega_r^2}{1 - \omega_r^2}\right)\right] \quad (7.44)$$

When $\omega \ll \omega_0 \lambda_p \simeq 1$ and Eq. 7.42 would essentially coincide with Eq. 7.12. However, it is remarkable that, as ω_r increases, the perturbation expansion does not proceed in integral powers of β, requires the terms $\sqrt{\beta}$ and $\beta\log\beta$.

It might seem at first sight that the present problem could be cast in the language of the theory of multiple scattering for which a very large literature exists. Actually, this is not so. For example, by means of multiple scattering arguments, Waterman and Truell [37] (as well as many others) refined the original argument of Foldy and in particular, by means of an argument based the contribution of Fresnel zones, argued that the error of Foldy's result is of the order of

$$\frac{4\pi n |F|^2}{k} = \frac{n\sigma_s}{k} \quad (7.45)$$

where σ_s is the scattering cross section of a single bubble given, as shown in the next section, by

$$\sigma_s = \frac{4\pi a^2 \omega^4}{(\omega_0^2 - \omega^2)^2 + 4b^2\omega^2} \quad (7.46)$$

with b the (effective) damping parameter of the bubble. A comparison with the leading-order corrections in Eq. 7.42 shows that the estimate Eq. 7.45 really refers to the error term in the definition Eq. 7.44 of λ_p, but misses the dominant terms of order $\sqrt{\beta}$ and $\beta\log\beta$ as well as the correction λ_υ which arise from the fluid mechanics of the problem. The bubbles are subjected to significant pressure-radiation forces, and their motion under the action of the sound wave induces a (nearly incompressible and, therefore, nonacoustic) flow that affects the behavior of the other bubbles. Furthermore unlike, for example, the case of electromagnetic scattering, when the number density increases the bubbles soon cease to be in the far field of each other and a simple monopole approximation becomes inaccurate. To these considerations one may add that, again unlike the case of electromagnetic scattering, linearity is a strongly limiting assumption in the case of a bubbly liquid.

7.3. BUBBLE DYNAMICS

While it is true that surface tension tends to maintain bubbles spherical, distortions of the spherical shape are also commonly encountered due to the action of gravity, flow, or bubble–bubble interactions. In a linear approximation, the bubble shape can be represented as a superposition of spherical harmonics with time-dependent amplitudes.

Each one of these modes is capable of inducing flow in the bubble neighborhood and to scatter sound.

The near-field effects are essentially incompressible, and it is found that the n-th mode induces a flow disturbance that decays as $(\alpha/r)^{n+1}$ with distance r from the bubble center. The monopole term, $n = 0$, is evidently the dominant contributions already a small number of radii away from the bubble. If the gas volume fraction is β, the mean interbubble distance is, approximately, $\alpha\beta^{-1/3}$ and, therefore, the flow disturbance to which a bubble is subjected due to the action of a neighboring bubble is of the order of $\beta^{(n+1)/3}$. At distances comparable with the wavelength or greater, the field decays as α/r for all modes, but the radiation efficiency compared with the monopole is proportional to $(k\alpha)^{2n}$ and is therefore very small. We may thus conclude that, in a dilute bubbly system, the monopole or volume mode is the dominant disturbance induced by the bubble in the surrounding liquid.

The response dv/dP of the bubble volume to the pressure disturbance features prominently in the results given in the previous section and it is now necessary to consider whether the simple isothermal approximation used there is justified. This problem has been treated extensively in the literature (see, e.g., Devin [38] and Prosperetti [39, 40]). Suffice it to say that if one assumes that the bubble is immersed in a pressure field given, in the absence of the bubble, by

$$P_0 + P_A \exp(i\omega t) \tag{7.47}$$

and if one writes

$$R(t) = \alpha[1 + X(t)] \tag{7.48}$$

it is found that the dimensionless radial oscillation amplitude X is

$$X = -\frac{1}{\omega_0^2 - \omega^2 + 2ib\omega} \frac{P_A}{\rho_L \alpha_2} \exp(i\omega t) \tag{7.49}$$

where

$$\omega_0^2 = \frac{p_0}{\rho_L \alpha^2}\left(\text{Re}\,\Phi - \frac{2\sigma}{p_0 \alpha}\right) \tag{7.50}$$

$$b = \frac{p_0}{2\rho_L \alpha^2 \omega}\text{Im}\,\Phi + \frac{2\mu_L}{\rho_L \alpha^2} + \frac{\omega^2 \alpha}{2c_L} \tag{7.51}$$

Here σ is the surface tension coefficient, μ_L is the liquid viscosity, and p_0 is the equilibrium pressure in the bubble given by

$$p_0 = P_0 + \frac{2\sigma}{\alpha} \tag{7.52}$$

The complex function Φ is given by

$$\Phi = \frac{3\gamma\eta^2}{\eta^2 + 3(\gamma-1)\left[(1-i)\eta\coth\left(\frac{1}{2}(1+i)\eta\right) + 2i\right]} \tag{7.53}$$

in which γ is the ratio of the gas specific heats and

$$\eta = \alpha\left(\frac{2\omega}{D_G}\right)^{1/2} \tag{7.54}$$

with D_G the thermal diffusivity proportional to the ratio of the bubble equilibrium radius to the thermal penetration length in the gas. This result is obtained assuming that the bubble contents behave as a single perfect gas, that the bubble surface is isothermal, and that the pressure inside the bubble is spatially uniform and a function of time only. The first assumption is trivially true in a linearized theory. As for the second one, it can easily be removed (see, e.g., Prosperetti [39]), but it is found that, due to the large specific heat per unit volume of the liquid with respect to the gas, the associated error is negligible. A justification for the spatial uniformity of the pressure can be found in the fact that, in typical applications, the wavelength of sound in the gas is much greater than the bubble radius.

Although the oscillation amplitude in Eq. 7.49 has the typical structure of the response of a damped harmonic oscillator with natural frequency ω_0 and damping parameter b, this appearance is deceptive as both ω_0 and b are found to depend on the frequency. At a frequency ω they give the position of the pole of the response in the complex plane and may therefore be interpreted as the "effective" or "apparent" natural frequency and damping parameter at the frequency ω. If the gas pressure-volume relationship were approximated by a polytropic relation of the type $pV^\zeta =$ constant, one would find

$$\omega_0^2 = \frac{p_0}{\rho_L a^2} \left(3\zeta - \frac{2\sigma}{p_0 a} \right) \tag{7.55}$$

from which we see that the equivalent polytropic index at the frequency ω is given by

$$\zeta = \tfrac{1}{3} \mathrm{Re}\,\Phi \tag{7.56}$$

A graph of this relation as a function of η defined in Eq. 7.54 is shown in Figure 7.2, where the dashed lines correspond to $\zeta = \gamma$. Clearly the bubble behaves isothermally at low frequency and adiabatically at high frequency, as expected.[4] The asymptotic behaviors for small and large η (i.e., for small and large frequencies) are, respectively,

$$\zeta \simeq 1 + \frac{\gamma-1}{90\gamma}\left(\frac{1}{7} - \frac{\gamma-1}{10\gamma}\right)\eta^{(4)}, \quad \zeta \simeq \gamma\left[1 - \frac{3(\gamma-1)}{\eta}\right] \tag{7.57}$$

[4] The statement about the high-frequency limit is correct only provided the sound wavelength in the gas is still much greater than the bubble radius. If this is not the case, the situation is more complex (see Prosperetti [39]), but it is doubtful that the spherical model would be accurate at such high frequencies.

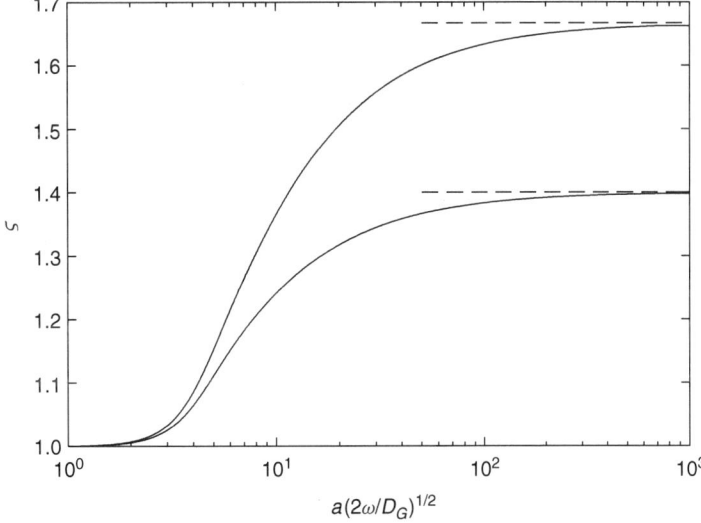

Fig. 7.2. The effective polytropic index ζ defined in Eq. 7.56 as a function of $\eta = a\left(\frac{2\omega}{D_G}\right)^{1/2}$ defined in Eq. 7.54 for $\gamma = \tfrac{5}{3}$ and $\gamma = \tfrac{7}{5}$; the dashed lines correspond to $\zeta = \gamma$. The bubble behaves isothermally at low frequency ($\zeta \simeq 1$) and adiabatically at high frequency ($\zeta \simeq \gamma$).

Figure 7.3 shows in normalized form and as a function of the equilibrium radius α, the true natural frequency of an air bubble in water at normal pressure and temperature,[5] i.e., the value ω_r of ω such that $\omega_0(\omega_r) = \omega_r$. The dashed and dotted lines are found by taking $\zeta = 1$ and $\zeta = \gamma$, respectively, and therefore correspond to isothermal and adiabatic behavior of the bubble. An easily memorized and useful rule of thumb is that, for water and air at standard temperature and pressure, $\alpha \omega_r / 2\pi \simeq 3$ mm × kHz.

The three terms in the expression Eq. 7.51 correspond, respectively, to the thermal, viscous, and acoustic damping of the bubble oscillations. The thermal contribution is proportional to Im Φ; a graph of Im Φ/η^2 vs η is shown in Figure 7.4 for $\gamma = \frac{5}{3}$ and $\gamma = \frac{7}{5}$. The asymptotic behaviors for small and large η are, respectively,

$$\mathrm{Im}\,\Phi \simeq \frac{\gamma - 1}{10\gamma}\eta^2, \quad \mathrm{Im}\,\Phi \simeq 9\gamma \frac{\gamma - 1}{\eta}\left(1 - \frac{2}{\eta}\right) \qquad (7.58)$$

The dashed lines in the Figure 7.4 show the leading order terms of these approximations for the two values of γ. Figures 7.5, 7.6, and 7.7 show the total damping Eq. 7.51 and its three components as a function of frequency for $\alpha = 1$, 0.1, and 0.01 mm, respectively, for air bubbles in water at normal pressure and temperature as a function of the driving frequency $\upsilon = \omega/2\pi$. Here the solid line is the total damping, the dashed line is the thermal contribution, the dotted line is the acoustic contribution, and the long-dashed line (out of scale in Figs. 7.5 and 7.6 and only visible for the smallest bubble in Fig. 7.7) is the viscous contribution. In these figures, the small circles denote the natural frequency of Figure 7.3.

With these results, we can now calculate $d\upsilon/dP$ provided a connection can be established between the mixture pressure and the liquid pressure "felt" by the bubble. The simplest assumption, which is essentially embodied in the quotation from Foldy's paper given previously after Eq. 7.35 and is therefore referred to as the *Foldy approximation*, is to equate the mixture pressure P with Eq. 7.47. With this approximation

---

[5] Since the quantity shown here does not include damping effects, the actual maximum amplitude for driven oscillations will occur at a slightly different value of the frequency.

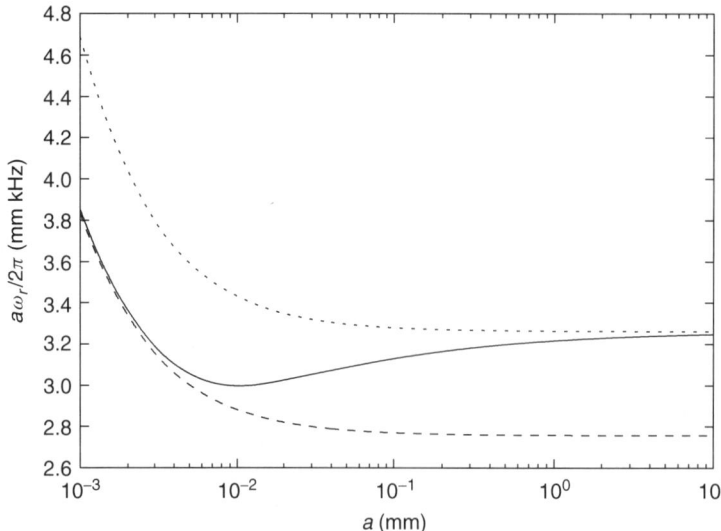

Fig. 7.3. Normalized natural frequency $\alpha \omega_r / 2\pi$ of an air bubble in water at normal pressure and temperature as a function of the equilibrium radius α. The dotted and dashed lines are hypothetical results for adiabatic and isothermal bubbles.

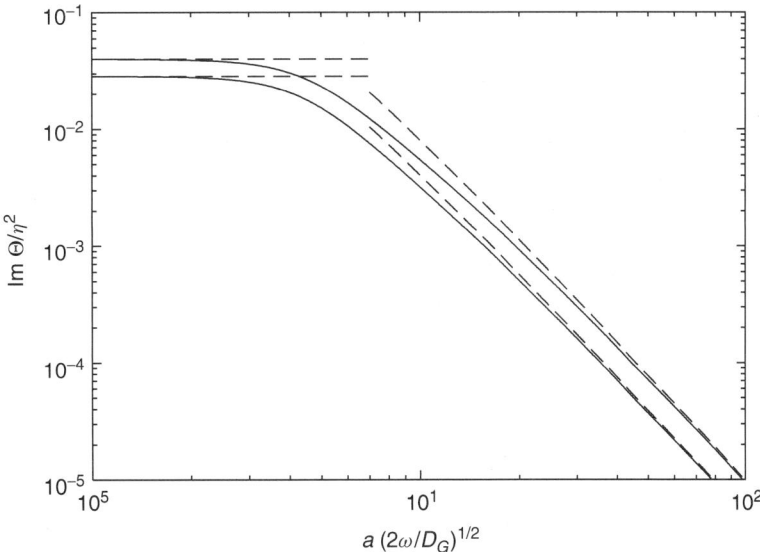

Fig. 7.4. A graph of Im Φ/η^2 vs η defined in Eq. 7.54 for $\gamma = \frac{5}{3}$ and $\gamma = \frac{7}{5}$. The dashed lines are the approximations Eq. 7.58.

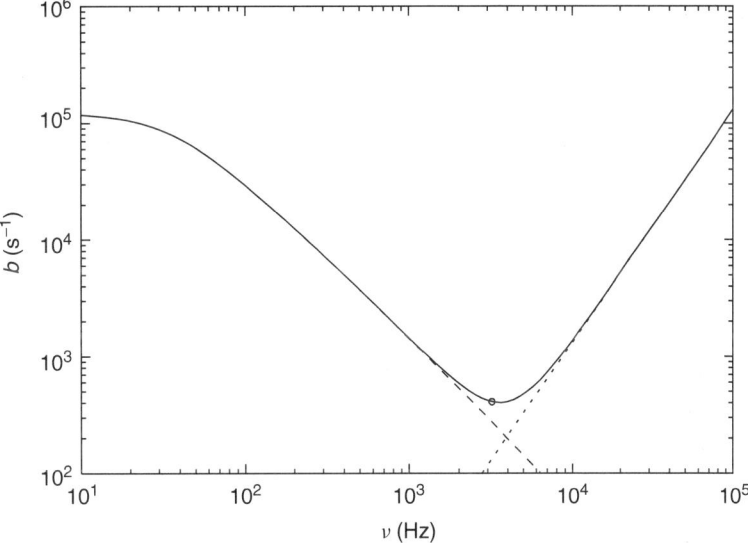

Fig. 7.5. Damping of a 1-mm-radius air bubble in water at normal pressure and temperature (natural frequency, 3.22 kHz) as function of the driving frequency v. The solid line is the total damping, the dashed line is the thermal contribution (first term in Eq. 7.51), and the dotted line is the acoustic contribution (last term in Eq. 7.51); the viscous contribution is too small to show on the scale of the graph. The circle denotes the natural frequency of Figure 7.3.

we therefore have

$$\frac{dv}{dP} = \frac{dv/dt}{dP/dt} = \frac{3\alpha^3 X}{P_A} = -\frac{3\alpha}{\rho_L} \frac{1}{\omega_0^2 - \omega^2 + 2ib\omega} \quad (7.59)$$

As mentioned previously, Foldy's ansatz, originally introduced on purely heuristic terms, has been well justified in subsequent works (see, e.g., Caflisch *et al.* [30], Duraiswami and Prosperetti [34], and Zhang and Prosperetti [35]).

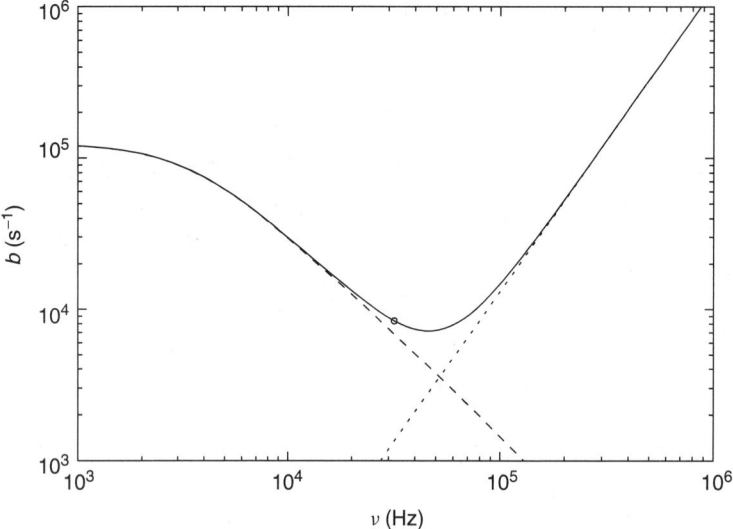

Fig. 7.6. Damping of a 0.1-mm-radius air bubble in water at normal pressure and temperature (natural frequency, 31.30 kHz) as function of the driving frequency v. The solid line is the total damping, the dashed line is the thermal contribution (first term in Eq. 7.51), and the dotted line is the acoustic contribution (last term in Eq. 7.51); the viscous contribution is too small to show on the scale of the graph. The circle denotes the natural frequency of Figure 7.3.

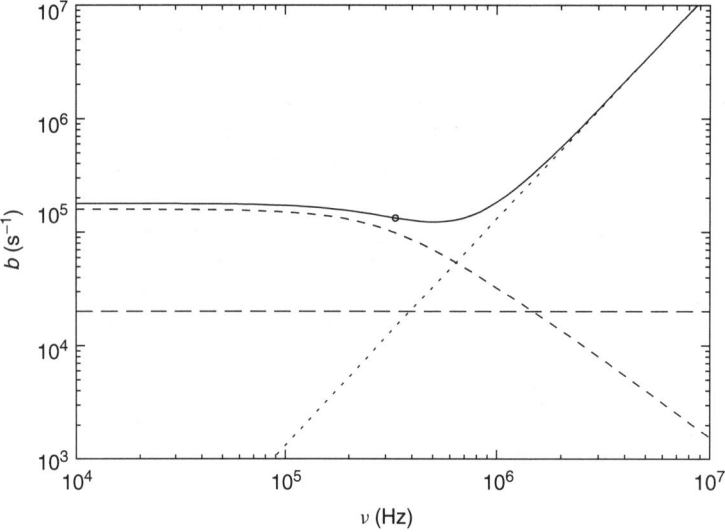

Fig. 7.7. Damping of a 0.01-mm-radius air bubble in water at normal pressure and temperature (natural frequency, 299.8 kHz) as a function of the driving frequency v. The solid line is the total damping, the dashed line is the thermal contribution (first term in Eq. 7.51), the dash-dot line is the viscous contribution (second term in Eq. 7.51), and the dotted line is the acoustic contribution (last term in Eq. 7.51). The circle denotes the natural frequency of Figure 7.3.

Due to the dependence of the bubble response on the frequency, the previous results are essentially a frequency-domain formulation. A time-domain formulation is complicated by the fact that, unless the spectrum of the wave is such that either one of the approximate results, Eqs. 7.57 or 7.58, can be used, the normalized radius perturbation X cannot be expressed in closed form in terms of P. A general formulation for this case is presented in Watanabe and Prosperetti [41].

In conclusion, we note that from the expression Eq. 7.49, it is easy to deduce the scattering amplitude F:

$$F = \frac{\omega^2 \alpha}{\omega_0^2 - \omega^2 + 2ib\omega} \qquad (7.60)$$

from which the expression Eq. 7.46 for the scattering cross section follows through the relation $\sigma_s = 4\pi |F|^2$.

7.4. SPEED OF SOUND IN BUBBLY LIQUIDS

On substituting Eq. 7.59 into Foldy's result, Eq. 7.41, we find the following expression for the dispersion relation in a bubbly liquid

$$\frac{\kappa^2}{\omega^2} = \frac{1}{c_L^2} + \int \frac{4\pi a \mathcal{N}(a; \mathbf{x})}{\omega_0^2 - \omega^2 + 2ib\omega} \qquad (7.61)$$

A plane wave perturbation has the structure $\exp i(\omega t \pm \kappa x)$, with κ real in an initial value problem (and, therefore, ω complex) or ω real in a boundary-value (or signaling) problem and, therefore, κ complex. We focus on the latter situation as it is the most common.

A graph of the (real part of the) phase velocity as given by Eq. 7.61 for a typical case is shown in Figure 7.8 redrawn from Cheyne et al. [42]. Here the gas volume fraction is 1% and the bubble radius is 1.11 mm. Agreement between theory (solid line) and data is excellent, even in the region above the bubble resonance frequency (near 2.90 kHz), where c_m grows appreciably above c_L. In general, the trend of the curve with increasing frequency is qualitatively similar to the phase speed of light in matter near a region of anomalous dispersion (see, e.g., Feynman et al. [43, pp. 31–36]). This behavior can be understood in the following terms. At very low frequencies, the bubbles respond

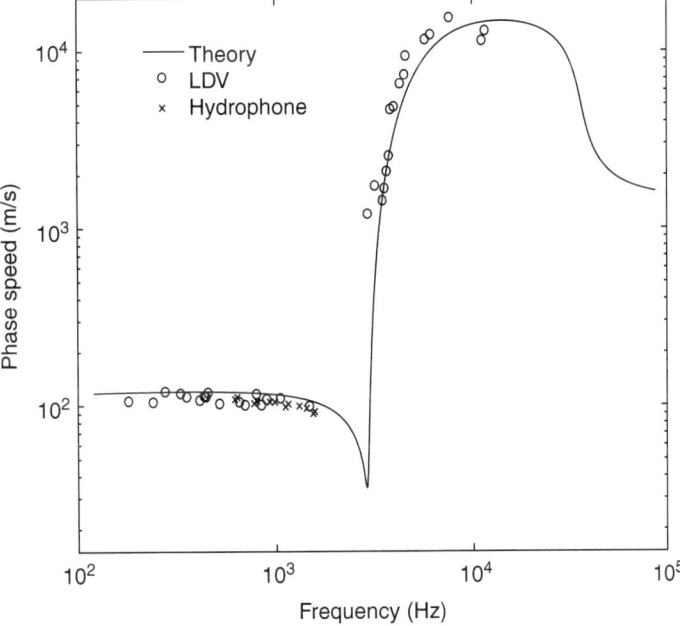

Fig. 7.8. Phase velocity in an air–water system at standard pressure and temperature as given by Eq. 7.61 compared with the data of Cheyne et al. [42]; the gas volume fraction is 1% and the radius of the bubbles, 1.11 mm (natural frequency, 2.90 kHz).

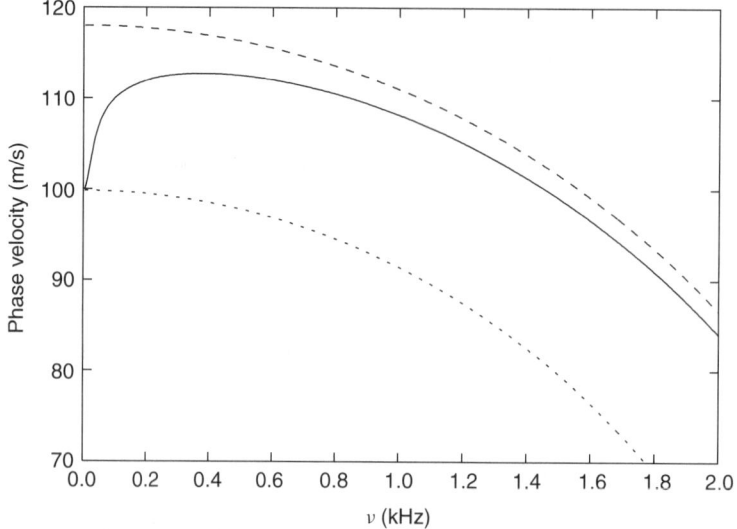

Fig. 7.9. The behavior of the phase velocity in an air–water system for the case of the previous figure at low frequencies. The solid line is the complete result; the upper dashed line and the lower dotted line are hypothetical results if the bubbles were to behave completely adiabatically or isothermally, respectively.

isothermally and Re c_m is essentially given by Eq. 7.13. As the frequency increases, at first the speed of sound tends to increase as the bubbles become less isothermal and the mixture therefore less compressible. This trend is clear from Figure 7.9, which is an enlargement of the low-frequency region of Figure 7.8; here, the upper dashed line is obtained by taking $\zeta = \gamma$ in Eq. 7.55 and would therefore describe the behavior of the bubbly liquid if the bubbles compressed adiabatically at all frequencies; the dotted line is the corresponding result for isothermal bubble behavior. As the external frequency approaches the resonance frequency, the bubbles are easily compressed, the mixture becomes very "soft," and the speed of sound falls dramatically. Just above resonance, the bubble response becomes inertia dominated and its phase changes: now to an increase of the external pressure corresponds an increase of the bubble volume, and vice versa. As a consequence, the mixture becomes much stiffer than the pure liquid and the speed of sound correspondingly dramatically increases. As the frequency continues to increase, the amplitude of the bubble response decreases and the mixture behaves more and more as a pure liquid with no bubbles.

Although the article by Cheyne et al. [42] is based on a very accurate measurement procedure, it unfortunately presents only a very limited set of data. To find additional data we have to turn to older literature, and in particular to the excellent article by Silberman [44], whose data have been compared with the theory described before by Commander and Prosperetti [45]. Figures 7.10 and 7.11 show two typical examples for which, once again, theory and experiment closely agree.

Silberman also reported data on attenuation. Figures 7.12 and 7.13 show a comparison of theory and experiment for $\beta = 0.0377\%$ and 1%. The quantity plotted is the attenuation coefficient A defined by

$$A = -20(\log_{10} e)\mathrm{Im}\frac{k^2}{\kappa} \simeq -8.6859\mathrm{Im}\frac{k^2}{\kappa} \qquad (7.62)$$

(with $k = \omega/c_L$ the wave number in the pure liquid as before) and is expressed in dB/cm. It is apparent that, while theory and experiment agree at the lower frequencies, the predicted attenuation near and above the resonance frequency appears to be greater than the measured value. Whether this discrepancy is to be imputed to the data

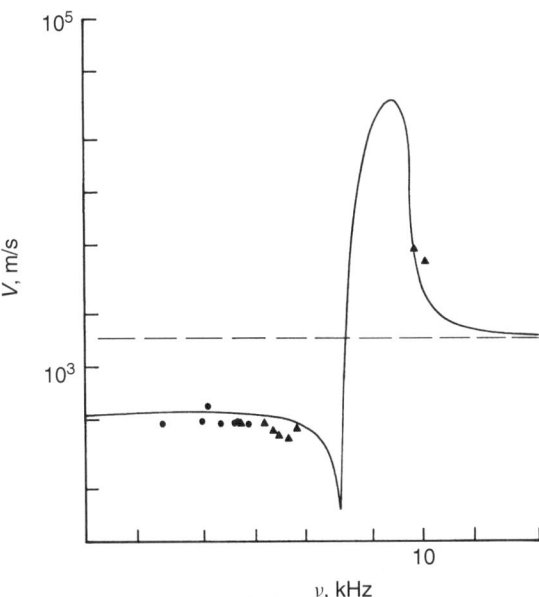

Fig. 7.10. Phase velocity in an air–water system at standard pressure and temperature as given by Eq. 7.61 compared with data by Silberman [44]; the gas volume fraction is 0.0377% and the radius of the bubbles, 1.07 mm (natural frequency, 3.01 kHz) (from [45]).

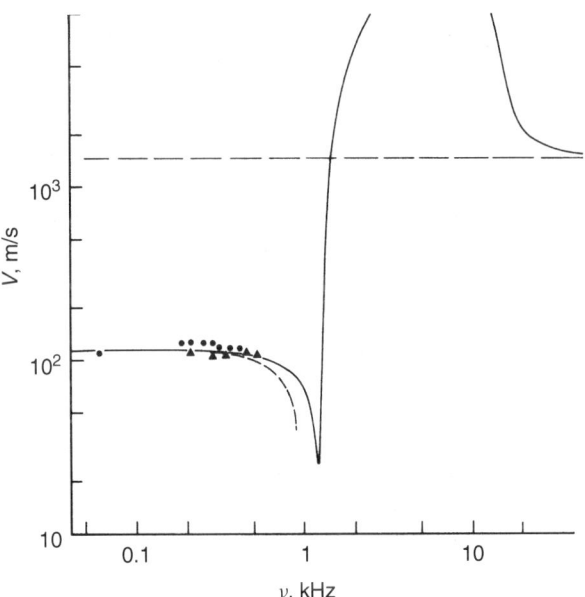

Fig. 7.11. Phase velocity in an air–water system at standard pressure and temperature as given by Eq. 7.61 compared with data from Silberman [44]; the gas volume fraction is 1%. The radius of the bubbles is 2.68 mm (natural frequency, 1.21 kHz; solid line and circles) and 3.41 mm (natural frequency, 0.949 kHz; dashed line and triangles) (from [45]).

(which are prone to inaccuracies in this region of very large attenuation), or to the theory is yet to be resolved. With reference to the estimate Eq. 7.45 given earlier, it could be that, near resonance, the bubble-scattering cross section becomes so large that Foldy's approximation is inaccurate. On the other hand, the more sophisticated theory of Sangani [36] mentioned earlier does not seem to come much closer to the

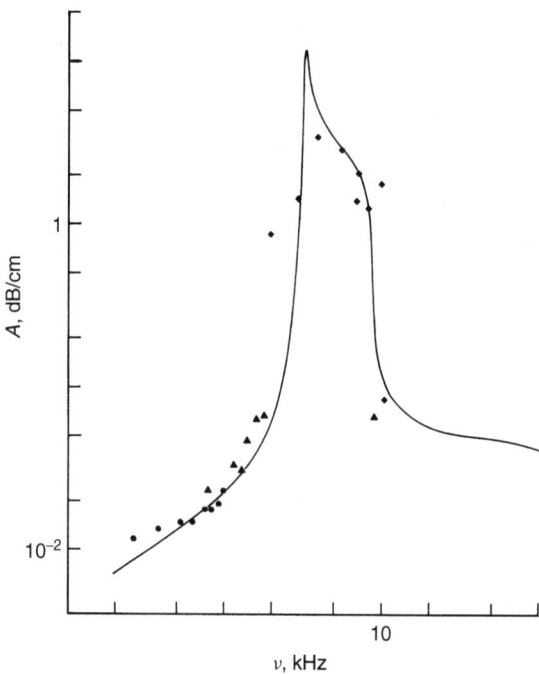

Fig. 7.12. Attenuation coefficient defined in Eq. 7.62 vs frequency for the case of Figure 7.10.

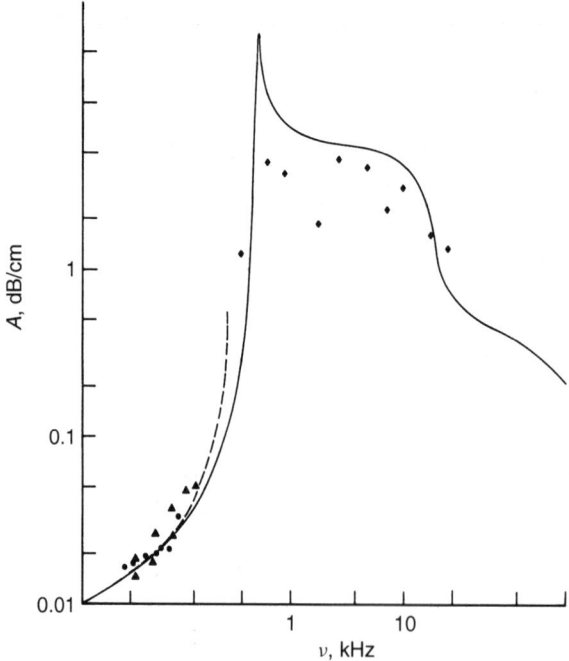

Fig. 7.13. Attenuation coefficient defined in Eq. 7.62 vs frequency for the case of Figure 7.11; the diamonds are for bubbles with a radius of 2.60 mm (natural frequency, 1.24 kHz).

data. Feuillade [46] has developed a correction to Eq. 7.61 that does show an improved agreement with data. However, in view of the heuristic nature of his argument, it is not clear whether this correction does embody the proper physical mechanisms that seem to be missing from the theory. As of now, therefore, it would seem that this question must be left unresolved.

In the cases considered thus far the bubbly mixtures contained nearly uniform-size bubbles. The literature contains a few data with a spectrum of bubble sizes, although the quality of these data is not as good as those of Silberman or Cheyne *et al.* Commander and Prosperetti [45] show some comparisons of theory and data that support the validity of the integration over the bubble size distribution in Eq. 7.61.

7.5. SOME APPLICATIONS

In many situations, bubbles occur in limited regions of space, such as under breaking waves in the ocean, in the descending portion of Langmuir circulation cells, and so forth. In such situations one would be interested in the reflection and attenuation of sound by these regions of bubbly liquid. The simplest situation that can be analyzed is the propagation of plane waves through a plane bubbly layer surrounded by the pure liquid. We take the plane of incidence as the (x, y) plane, with the source of waves at $x \to \infty$, $y \to -\infty$; the bubbly layer extends for $0 \leq x \leq L$ and is infinite in the other two directions. If the pressure perturbation in the half-space where the source is located $(x > L)$ is written as

$$\exp i[k_x(x - L) - k_y y] + A_- \exp i[-k_x(x - L) - k_y y] \tag{7.63}$$

with $R = |A_-|^2$ the reflection coefficient and, on the other side of the layer, as

$$A_+ \exp i(k_x x - k_y y) \tag{7.64}$$

with $T = |A_+|^2$ the transmission coefficient, one finds

$$A_- = \frac{i}{2} \frac{k_x/\kappa_x - \kappa_x/k_x \sin \kappa_x L}{\cos \kappa_x L + \frac{1}{2} i (k_x/\kappa_x + \kappa_x/k_x) \sin \kappa_x L} \tag{7.65}$$

$$A_+ = \frac{1}{\cos \kappa_x L + \frac{1}{2} i (k_x/\kappa_x + \kappa_x/k_x) \sin \kappa_x L} \tag{7.66}$$

where $\kappa_x = \sqrt{\kappa^2 - k_y^2}$, with κ given by Eq. 7.61. Note that, due to dissipation, $R + T < 1$.

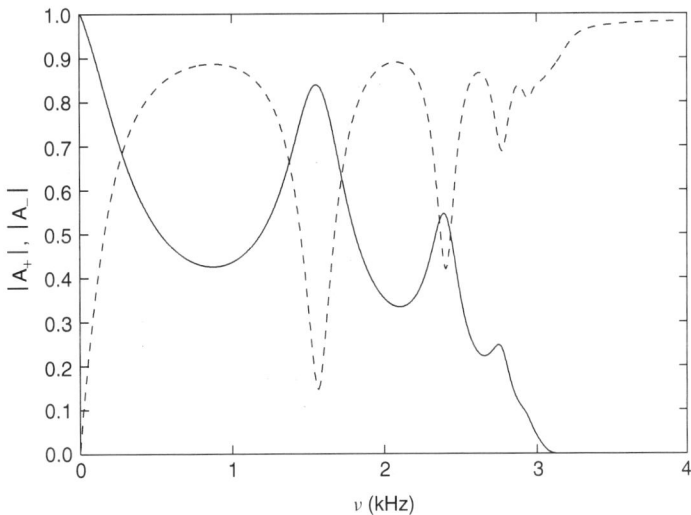

Fig. 7.14. Transmission $|A_+|$ (solid line) and reflection $|A_-|$ (dashed line) amplitudes through a 100-mm-thick bubble layer as a function of frequency; the gas volume fraction is 0.1% and the radius of the bubbles is 1 mm (natural frequency, 3.22 kHz).

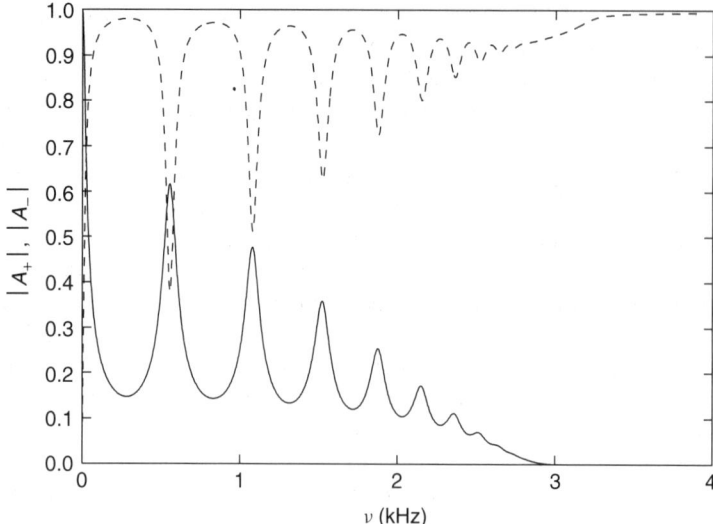

Fig. 7.15. Transmission $|A_+|$ (solid line) and reflection $|A_-|$ (dashed line) amplitudes through a 100-mm-thick bubble layer as a function of frequency; the gas volume fraction is 1% and the radius of the bubbles is 1 mm (natural frequency, 3.22 kHz).

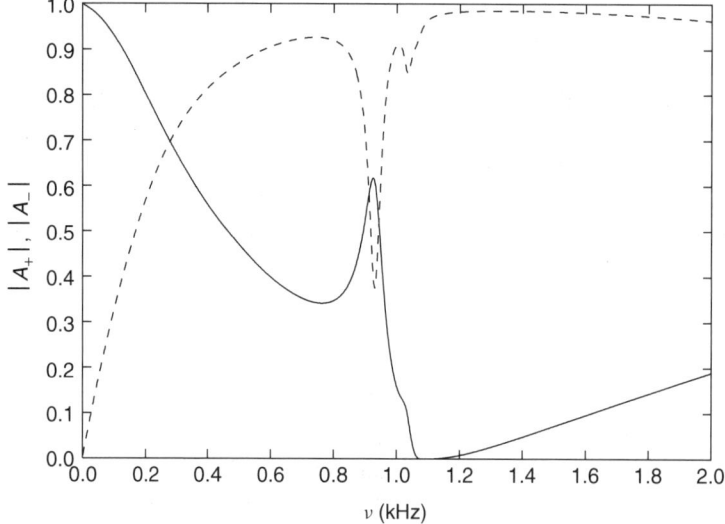

Fig. 7.16. Transmission $|A_+|$ (solid line) and reflection $|A_-|$ (dashed line) amplitudes through a 100-mm-thick bubble layer as a function of frequency; the gas volume fraction is 1% and the radius of the bubbles is 3 mm (natural frequency, 1.08 kHz).

Some illustrative results are shown in Figures 7.14 to 7.16. Below the bubble resonance frequency one encounters a series of maxima and minima that, except for attenuation, are familiar from the analogous problem in optics. It can readily be shown that the maxima of the transmission coefficient are in the neighborhood of the normal modes of the bubbly layer, which are given by

$$\cot \frac{1}{2}\kappa_x L = -i\frac{\kappa_x}{k_x}, \quad \cot \frac{1}{2}\kappa_x L = -i\frac{k_x}{\kappa_x} \qquad (7.67)$$

these two expressions giving the modes symmetric and antisymmetric about the layer's mid-plane, respectively. It is readily established that these equations correspond approximately to an even and odd number of half wavelengths in the thickness L. As a consequence, as the bubble resonance is approached, Re c_m decreases and the spacing between maxima and minima decreases. Past resonance, Re c_m increases dramatically and so does the spacing. At very high frequencies, the effect of the bubble becomes negligible and essentially the entire wave is transmitted, save for a small attenuation.

The fact that regions of bubbly liquid possess eigenmodes points to the possibility that, if suitably excited, they can act as sound sources. This fact has an important bearing on the explanation of underwater sound in the ocean. It has long been known that a substantial fraction of low-frequency oceanic ambient sound (between 50 and 100 Hz and 1 or a few kHz) exhibits a strong dependence on wind speed, with a particularly strong increase around 12 knots, which roughly corresponds to the onset of extensive wave breaking [47, 48]. This sound has been explained as being due to the free collective oscillations of the bubble clouds generated by breaking waves. This hypothesis, suggested independently by Carey and Browning [49] and Prosperetti [50], has by now been amply proven experimentally both in the field [2, 51, 52] and in the laboratory [53–55].

A simple order-of-magnitude estimate of the frequency range that may be expected is readily found by noting that since $|c_m| \ll c_L$ at low frequencies, a cloud with a size of order L will have a fundamental resonance frequency of the order of

$$\frac{c_m}{L} \sim \frac{1}{L}\sqrt{\frac{P}{\rho_L \beta}} \qquad (7.68)$$

where c_m has been approximated by Eq. 7.14. With $L \sim 1$ m, $P = 100$ kPa, $\rho_L \sim 10^3$ kg/m^3, and $\beta \sim 1\%$, we find a result of the order of 100 Hz.

As already mentioned in Section 7.2, it is possible to show that when the gas volume fraction is small, the model of Eq. 7.24 can be extended to the nonlinear regime by simply accounting for the nonlinear behavior of the bubbles. This procedure is valid because, in a dilute mixture, the effect of the nonlinearity is confined to the immediate neighborhood of the bubbles. Thus, the mixture can be described by Eq. 7.24 while the bubble response is described by the well-known Rayleigh-Plesset equation of bubble dynamics (or one of its variants):

$$R\frac{d^2R}{dt^2} + \frac{3}{2}\left(\frac{dR}{dt}\right)^2 = \frac{1}{\rho_L}\left[p_i - P(t) - \frac{2\sigma}{R} - 4\frac{\mu_L}{R}\frac{dR}{dt}\right] \qquad (7.69)$$

in which p_i is the bubble internal pressure and, again by virtue of the Foldy approximation, $P(t)$ is the mean pressure in the mixture. A remarkable phenomenon that very successfully can be explained on the basis of such a theory is the propagation of weak shock waves in a bubbly liquid [41, 56].

7.6. CONCLUSIONS

In the present chapter we described the propagation of linear pressure waves in a bubbly liquid containing a relatively small number of gas bubbles. It has been shown that theory and data are mostly in good agreement up to a gas volume fraction of a few percent, except for an excessive attenuation predicted near the resonance frequency of the bubbles. The theory can be extended to the nonlinear regime where it is capable of explaining, for example, the phenomena associated to the propagation of weak shock waves in bubbly liquids.

In all the cases considered here, except for the motion induced by the wave, the bubbly liquid was at rest. We have not therefore dealt with the propagation of sound waves in a flowing bubbly liquid. Several models to describe this situation at small gas volume fractions have been proposed, starting with the well-known one of van

Wijngaarden [57]. The interested reader may consult the works of Park *et al.* [58] and Zhang and Prosperetti [35].

ACKNOWLEDGMENT

This work has been supported by the Office of Naval Research.

References

1. Zendel, L., and Farmer, D. (1991). Organized structures in subsurface bubble clouds: Langmuir circulation in the open ocean. *J. Geophys. Res.* **96C**: 8889–8900.
2. Ding, L., and Farmer, D.M. (1994). On the dipole acoustic source level of breaking waves. *J. Acoust. Soc. Am.* **96**: 3036–3044.
3. Lamarre, E., and Melville, K.W. (1994). Void-fraction measurement and sound-speed fields in bubble plumes generated by breaking waves. *J. Acoust. Soc. Am.* **95**: 1317–1328.
4. Lamarre, E., and Melville, K.W. (1994). Sound-speed measurements near the ocean surface. *J. Acoust. Soc. Am.* **95**: 3605–3616.
5. Terrill, E., and Melville, W.K. (1997). Sound-speed measurements in the surface-wave layer. *J. Acoust. Soc. Am.* **102**: 2607–2625.
6. Deane, G.B., and Stokes, M.D. (1999). Air entrainment processes and bubble size distributions in the surf zone. *J. Phys. Oceanogr.* **29**: 1393–1403.
7. Thorpe, S., (1982). On the clouds of bubbles formed by breaking wind-waves in deep water, and their role in air-sea gas transfer. *Philos. Trans. R. Soc. London* **A304**: 155–210.
8. Farmer, D.M., and Vagle, S. (1989). Waveguide propagation of ambient sound in the ocean-surface bubble layer. *J. Acoust. Soc. Am.* **86**: 1897–1908.
9. Wallace, D.W.R., and Wirick, C.D. (1992). Large air-sea gas fluxes associated with breaking waves. *Nature* **356**: 694–696.
10. Farmer, D.M., McNeil, C.L., and Johnson, B.D. (1993). Evidence for the importance of bubbles in increasing air-sea gas flux. *Nature* **361**: 620–623.
11. Fujita, E., Ida, Y., and Oikawa, J. (1995). Eigenoscillations of a fluid sphere and source mechanism of harmonic volcanic tremor. *J. Volcanol. Geotherm. Res.* **69**: 365–378.
12. Alidibirov, M.A., and Dingwell, D.B. (1996). Magma fragmentation by rapid decompression. *Nature* **380**: 146–148.
13. Mader, H.M., Phillips, J.C., Sparks, R.J.S., and Sturtevant, B. (1996). Dynamics of explosive degassing of magma: Observations of fragmenting two-phase flows. *J. Geophys. Res.* **101**: 5547–5560.
14. Brown, K.E. (1980). *The Technology of Artificial lift Methods*, vol. 2a, Tulsa: Petroleum Publishing Co.
15. Biń, A.K. (1993). Gas entrainment by plunging liquid jets. *Chem. Eng. Sci.* **48**: 3585–3630.
16. Pan, Y., Dudukovic, M.P., and Chang, M. (2000). Numerical investigation of gas-driven flow in 2-D bubble columns. *A.I.Ch.E. J.* **46**: 434–449.
17. Boyd, C.E., and Watten, B.J. (1989). Aeration systems in aquaculture. *Rev. Aquatic Sci.* **1**: 425–472.
18. Jokela, P., and Keskitalo, P. (2000). Plywood mill water system closure by dissolved air flotation treatment. *Water Sci. Technol.* **40**: 33–41.
19. Chen, H.J., Liao, C.M., and Yang, P.Y. (1999). Development of a prefabricated treatment plant for diluted pig wastewater. *J. Environ. Sci. Heal.* **B34**: 1009–1021.
20. Gander, M., Jefferson, B., and Judd, S. (2000). Aerobic MBRs for domestic wastewater treatment: A review with cost considerations. *Sep. Purif. Technol.* **18**: 119–130.
21. Cachard, C., Bouakaz, A., and Gimenez, G. (1996). In vitro evaluation of acoustic properties of ultrasound contrast agents: Experimental set-up and signal processing. *Ultrasonics* **34**: 595–598.
22. Bouakaz, A., De Jong, N., and Cachard, C. (1998). Standard properties of ultrasound contrast agents. *Ultrasound Med. Biol.* **24**: 469–472.
23. Monahan, E.C., and Lu, M. (1990). Acoustically relevant bubble assemblages and their dependence on meteorological parameters. *IEEE J. Oceanic Eng.* **15**: 340–349.
24. Asher, W.E., Karle, L.M., Higgins, B.J., Farley, P.J., Monahan, E.C., and Leifer, I.S. (1996). The influence of bubble plumes on air-seawater gas transfer velocities. *J. Geophys. Res. Oceans* **C101**: 12027–12041.
25. Kraynik, A.M. (1988). Foam flows. *Ann. Rev. Fluid Mech.* **20**: 325–357.
26. Wood, A.B. *A Textbook of Sound*, 3nd ed, London: Bell and Sons. 1955.
27. van Wijngaarden, L. One-dimensional flow of liquids containing small gas bubbles. *Ann. Rev. Fluid Mech.* **4**: 369–396.
28. Landau, L.D., and Lifshitz, E.M. (1959). *Fluid Mechanics*. New York: Pergamon.
29. Gibson, F.W. (1970). Measurement of the effect of air bubbles on the speed of sound in water. *J. Acoust. Soc. Am.* **48**: 1195–1197.

30. Caflisch, R.E., Miksis, M.J., Papanicolaou, G.C., and Ting, L. (1985). Effective equations for wave propagation in bubbly liquid. *J. Fluid Mech.* **153**: 259–273.
31. Foldy, L.L. (1945). The multiple scattering of waves. *Phys. Rev.* **67**: 107–119.
32. Watanabe, M. (1995). *Topics in Bubbly Liquid Flows and Cavitation*. PhD thesis, Johns Hopkins University.
33. Batchelor, G.K. (1972). Sedimentation in a dilute dispersion of spheres. *J. Fluid Mech.* **52**: 245–268.
34. Duraiswami, R., and Prosperetti, A. (1995). Linear pressure waves in fogs. *J. Fluid Mech.* **299**: 187–215.
35. Zhang, D.Z., and Prosperetti, A. (1994b). Ensemble phase-averaged equations for bubbly flows. *Phys. Fluids* **6**: 2956–2970.
36. Sangani, A.S. (1991). A pairwise interaction theory for determining the linear acoustic properties of dilute bubbly liquids. *J. Fluid Mech.* **232**: 221–284.
37. Waterman, P.C., and Truell, R. (1961). Multiple scattering of waves. *J. Math. Phys.* **2**: 512–537.
38. Devin, C. (1959). Survey of thermal, radiation, and viscous damping of pulsating air bubbles in water. *J. Acoust. Soc. Am.* **31**: 1654–1667.
39. Prosperetti, A. (1977). Thermal effects and damping mechanisms in the forced radial oscillations of gas bubbles in liquids. *J. Acoust. Soc. Am.* **61**: 17–27.
40. Prosperetti, A. (1991). The thermal behaviour of oscillating gas bubbles. *J. Fluid Mech.* **222**: 587–616.
41. Watanabe, M., and Prosperetti, A. (1994). Shock waves in dilute bubbly liquids. *J. Fluid Mech.* **274**: 349–381.
42. Cheyne, S.A., Stebbings, C.T., and Roy, R.A. (1995). Phase velocity measurements in bubbly liquid using a fiber optic laser interferometer. *J. Acoust. Soc. Am.* **97**: 1621–1624.
43. Feynman, R.P., Leighton, R.B., and Sands, M. (1963). *The Feynman Lectures on Physics*, vol. I, Reading: Addison-Wesley.
44. Silberman, E. (1957). Sound velocity and attenuation in bubbly mixtures measured in standing wave tubes. *J. Acoust. Soc. Am.* **29**: 925–933.
45. Commander, K.W., and Prosperetti, A. (1989). Linear pressure waves in bubbly liquids: Comparison between theory and experiments. *J. Acoust. Soc. Am.* **85**: 732–746.
46. Feuillade, C. (1996). The attenuation and dispersion of sound in water containing multiply interacting air bubbles. *J. Acoust. Soc. Am.* **99**: 3412–3430.
47. Wenz, G.M. (1962). Acoustic ambient noise in the ocean: Spectra and sources. *J. Acoust. Soc. Am.* **34**: 1936–1956.
48. Perrone, A.J. (1969). Deep-ocean ambient noise spectra in the northeast Atlantic. *J. Acoust. Soc. Am.* **46**: 762–770.
49. Carey, W.M., and Browning, D. (1988). Low frequency ocean ambient noise: Measurement and theory, in *Sea Surface Sound*, Kerman, B.R. ed., pp. 361–376, Dordrecht: Kluwer.
50. Prosperetti, A. (1988). Bubble dynamics in oceanic ambient noise, in *Sea Surface Sound*, Kerman, B.R., ed., pp. 151–171, Dordrecht: Kluwer.
51. Hollett, R.D. (1994). Observations of underwater sound at frequencies below 1500 Hz from breaking waves at sea. *J. Acoust. Soc. Am.* **95**: 165–170.
52. Deane, G.B. (1997). Sound generation and air entrainment by breaking waves in the surf zone. *J. Acoust. Soc. Am.* **102**: 2671–2689.
53. Lu, N.Q., Prosperetti, A., and Yoon, S.W. (1990). Underwater noise emissions from bubble clouds. *IEEE J. Oceanic Eng.* **15**: 275–281.
54. Kolaini, A., Roy, R.A., and Crum, L.A. (1991). An investigation of the acoustic emissions from a bubble plume. *J. Acoust. Soc. Am.* **89**: 2452–2455.
55. Nicholas, M., Roy, R.A., Crum, L.A., Oğuz, H.N., and Prosperetti, A. (1994). Sound emission by a laboratory bubble cloud. *J. Acoust. Soc. Am.* **95**: 3171–3182.
56. Kameda, M., Shimaura, N., Higashino, F., and Matsumoto, Y. (1998). Shock waves in a uniform bubbly flow. *Phys. Fluids* **10**: 2661–2668.
57. van Wijngaarden, L. (1968). On the equations of motion for mixtures of liquid and gas bubbles. *J. Fluid Mech.* **33**: 465–474.
58. Park, J.W., Drew, D.A., and Lahey, R.T.Jr. (1998). The analysis of void wave propagation in adiabatic monodispersed bubbly two-phase flows using an ensemble-averaged two-fluid model. *Int. J. Multiphase Flow* **24**: 1205–1244.

CHAPTER 8

ACOUSTIC PROPERTIES IN PETROLIFEROUS LIQUIDS

Manika Prasad, Amos Nur, Gary Mavko, Jack Dvorkin
Geophysics Department, Stanford University, Stanford, California, USA

Contents

Abstract 207
8.1. Introduction 207
8.2. Major Components of Crude Oil 208
8.3. Products of Crude Oil 211
8.4. Classification of Crude Oils 211
8.5. Indirect Methods of Oil Classification 212
References 218

ABSTRACT

Knowledge about acoustic properties of petroliferous fluids is important to evaluate seismic response of hydrocarbon bearing rocks. We provide here an overview of the major components of crude oil and their uses, along with acoustic properties of pure fluids and multicomponent mixtures of different oils.

8.1. INTRODUCTION

In commercial exploration, the preliminary goal of a seismic survey is to detect and map hydrocarbons. To make more reliable predictions about subsurface pore fluids, a systematic study of their seismic properties is required. Theoretical and empirical relations between seismic velocity on one hand and composite rock containing fluids on the other require information about the solid components of the rocks and the fluids in its pores. A wide variety of pore fluids can occur naturally in sedimentary rocks. Petroleum, a complex mixture mainly of hydrogen and carbon, is the fluid of maximum commercial interest.

Density, viscosity, and seismic velocity are the most interesting properties derived from the seismic surveys. Density and viscosity are useful to characterize the oil and evaluate its flow properties; seismic velocity is important to help differentiate petroleum fluids from other pore fluids such as gas and brine.

In this chapter, after a brief introduction about the major groups of hydrocarbons, we give density, viscosity, and compressional wave velocity data for representative members of these groups. The velocity data are derived from Wang (1988) [1] and Wang et al. (1987, 1990) [2, 3].

Petroleum fluids can be divided in to four main hydrocarbon series: (1) n-Alkanes or paraffins; (2) cycloalkanes or napthenes; (3) aromatic series; and (4) asphalts, asphaltenes, and resins.

Depending on the major composition, petroleum is classified as paraffinic, napthenic, aromatic, or asphaltic. Alkanes, or paraffins, originate mainly from terrestrial plant organic matter. Marine organic matter is a source for napthene, or cycloalkane compounds. Bacterial degradation can change the composition of a crude oil to contain only aromatics, asphalts, and resins (Tissot and Welte, 1984 [4]).

8.2. MAJOR COMPONENTS OF CRUDE OIL

Major components of crude oils are mentioned here. The data are compiled from Hunt (1979) [5], Tissot and Welte (1984) [4], Wang (1988) [1], and Taylor et al. (1990) [6]; (1998) [7]; and Gallant and Yaws (1993) [8].

8.2.1. n-Alkanes (or Paraffins)

They are composed of open chains of carbon atoms whose remaining valence (or bonding capacity) is satisfied by hydrogen atoms (Fig. 8.1). The first and last carbon atoms are bonded to one additional hydrogen atom to satisfy the bonding capacity of all carbon atoms. The compounds here form a homologous series of C_nH_{2n+2}. (*Note: Homologous series are a set of compounds in which the members are separated by a constant amount.*) Compounds in this series are saturated hydrocarbons, since bonding capacity of all constituent atoms is satisfied. They are relatively unreactive, hence their name paraffin (Latin: parum affinum = low affinity). At room conditions, n-Alkanes C_1–C_4 are gases, C_5–C_{16} are liquids, and those above C_{16} are solids.

n-Alkanes from C_1 to C_{40} have been found in crude oils; they dominate the gasoline fraction of crude oils. Methane (CH_4) is the main component of natural gas; pentane (C_5) to pentadecane (C_{15}) are components of gasoline or petrol. Paraffin wax (n-Alkanes above C_{17}) in crude oil will make it more viscous.

Oils that contain at least 50% saturated hydrocarbons are called paraffinic. On an average, crude oils contain 15–20% n-Alkanes. The paraffin content can be as high as 35% in very paraffinic oils or can be reduced to as low as 0% due to biodegradation by aerobic bacteria. Paraffins are the principal hydrocarbon in most deeply buried reservoirs originating from terrestrial plant organic matter (Table 8.1).

8.2.2. Alkenes (or Olefins)

Open chain (and sometimes cyclic) unsaturated hydrocarbon compounds with the general formula C_nH_{2n} are called alkenes. The general structure of alkenes is shown in Fig. 8.2. They are very reactive and, except for small quantities of n-hexene, n-heptene, and n-octene, these compounds are rarely found in crude oil.

8.2.3. Napthenes (or Cycloalkanes/Cycloparaffins)

They are formed by atoms of carbon joined together in a ring (see Fig. 8.3). The remaining carbon valence is satisfied by hydrogen atoms or substituent groups giving

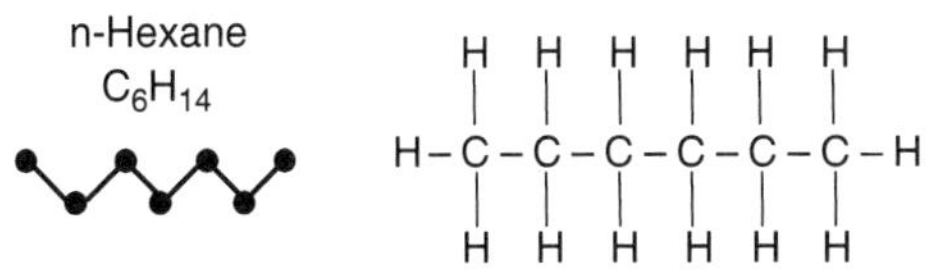

n-Hexane
C_6H_{14}

H H H H H H
| | | | | |
H–C–C–C–C–C–C–H
| | | | | |
H H H H H H

Fig. 8.1. Schematic diagram of n-Hexane showing the carbon–carbon and carbon–hydrogen bonds.

Table 8.1. Physical properties of a homologous series of n-alkanes (paraffins), 1-Alkenes (olefins), and Cycloparaffins (napthenes). The series is arranged in increasing carbon numbers for each member.

Name	Formula	Molecular weight	Melting point (°C)[a]	Boiling point (°C)[b]	Density at 20°C[c]	°API gravity	Viscosity Gas (μpoise)	Viscosity Liquid (cpoise)
\multicolumn{9}{c}{n-Alkanes (Paraffins)}								
n-Hexane	C_6H_{14}	86.06	−95	69	0.659	81.6		
N-Heptane	C_7H_{16}	100.07	−91	98	0.684	74.1		
n-Octane	C_8H_{18}	114.08	−57	126	0.703	68.7		
n-Decane	$C_{10}H_{22}$	142.10	−30	174	0.73	61.3	47.46	0.872
n-Undecane	$C_{11}H_{24}$	156.11	−26	196	0.74	58.7		
n-Dodecane	$C_{12}H_{26}$	170.12	−12	216	0.749	56.5	41.05	1.407
n-Tetradecane	$C_{14}H_{30}$	198.14	6	253	0.763	53.1	36.58	2.128
n-Pentadecane	$C_{15}H_{32}$	212.15	10	270	0.769	51.7		
n-Hexadecane	$C_{16}H_{34}$	226.16	18	287	0.773	50.5	33.77	2.83
n-Octadecane	$C_{18}H_{38}$	254.18	30	317	0.782	48.6		3.87
n-Docosane	$C_{22}H_{46}$	310.22	44	369	0.794	45.8		
n-Octacosane	$C_{28}H_{58}$	394.28	62	429	0.807	43.1		
n-Hexatriacontane	$C_{36}H_{74}$	506.36	74	493	0.817	40.9		
\multicolumn{9}{c}{1-Alkenes (Olefins)}								
1-Hexene	C_6H_{12}	84.06	−140	64	0.673	77.2		
1-Heptene	C_7H_{14}	98.07	−119	94	0.697	70.2		
1-Octene	C_8H_{16}	112.08	−101	122	0.715	65.2		
1-Decene	$C_{10}H_{20}$	140.10	−66	171	0.741	58.4	53.01	0.762
1-Undecene	$C_{11}H_{22}$	154.11	−49	192	0.75	56.1		
1-Dodecene	$C_{12}H_{24}$	168.12	−35	213	0.758	54.1	49.02	1.210
1-Tetradecene	$C_{14}H_{28}$	196.14	−13	251	0.775	51.0	45.73	1.841
1-Hexadecene	$C_{16}H_{32}$	224.16	4	274	0.783	48.8	43.19	2.711
1-Octadecene	$C_{18}H_{36}$	252.18	16	314	0.789	47.0	41.06	3.535
1-Eicosene	$C_{20}H_{40}$	280.2	28	341	0.795	45.6	39.30	4.665
\multicolumn{9}{c}{Cycloparaffins (Napthenes)}								
Cyclohexan	C_6H_{12}	84.06	−18	81	0.779	49.1	70.881	0.880
Cycloheptan	C_7H_{14}	98.07	−12	118	0.811			
Cyclooctane	C_8H_{16}	112.08	12	151	0.834			

[a] At 760 mm Hg.

[b] At 760 mm Hg; unit: g/cm^3.

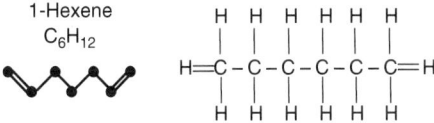

1-Hexene
C_6H_{12}

Fig. 8.2. Schematic diagram of 1-Hexene showing the carbon–carbon and carbon–hydrogen bonds.

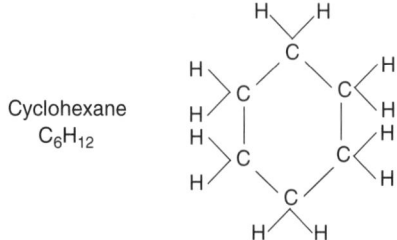

Fig. 8.3. Schematic diagram of Cyclohexane showing the cyclic carbon–carbon and carbon–hydrogen bonds.

the compounds a general formula C_nH_{2n}. Like alkanes, compounds of this series are saturated hydrocarbons. The smallest compound in this series is cyclopropane (C_3H_6). Most common napthene compounds found in crude oils are cyclopentane and cyclohexane, occasionally cycloheptane (C_5, C_6, and C_7 compounds, respectively). The methyl derivative, compounds with methylene (CH_3^-) substitutes, is more abundant in crude oil than the parent cyclic compound. Crude oils can contain up to 50% napthene compounds, usually those with higher molecular weights. Crude oils from marine organic materials are rich in cyclic saturated napthene compounds.

The relative percentage of napthenes with 1, 2, 3, 4, and 5 rings is used to assess maturity level of a crude oil. Tetra and pentacycloalkanes are most abundant in young and immature crude oils. Napthene compounds are important constituents of commercial petroleum-derived solvents.

8.2.4. Aromatic Hydrocarbons

Aromatics (Greek aroma = fragrance) are all derivatives of the benzene ring with the general formula C_nH_{2n-6R}, where R is the number of rings (see Fig. 8.4). Aromatics are usually concentrated in the heavy fraction of crude oils, for example, in gas oils, lubricating oils, and the residuum. Their contribution to crude oil is usually below 15%.

The source of the organic matter determines to a great extent the type of organic fluids that are generated. According to Tissot *et al.* (1974) [9], Tissot and Welte (1984) [4], Hunt (1979) [5], and Taylor *et al.* (1990) [6], the following classification can be made:

- Lacustrine clay deposits consisting of algae, zooplanktons, and bacteria yield mostly paraffins, with some napthenes and aromatics. They contain a large amount of volatile matter and have a very high oil yield.
- Marine marls and carbonates consisting of plankton, bacteria, and some kerogen types yield mostly napthenes and aromatics and have a high sulfur content. Oil yield is high.
- Deltaic and paralic clays consisting of terrestrial plant rests (similar to coal) have a low yield of oils that are rich in oxygenated functional and heteroatomic groups and in polycyclic aromatics.

Benzene
C_6H_{12}

Fig. 8.4. Schematic diagram of a benzene ring. The inner ring represents the carbon–carbon bonds with carbon atoms at each corner. The outer ring represents hydrogen positions.

As a result of bacterial degradation, composition of the crude oil can change to consist of only aromatics, asphalts, and resins.

8.3. PRODUCTS OF CRUDE OIL

A number of products are either distilled or otherwise extracted from crude oils. Main products derived from fractional distillation along with their uses are as follows:

1. *Gas* — Mainly methane (CH_4) with decreasing amounts of ethane (C_2H_6), propane (C_3H_8), and normal and isobutane (C_4H_{10}). Uses: As fuel and in chemical industry.
2. *Liquefied petroleum gases* — Normal and branched isomers of propane (C_3H_8) and butane (C_4H_{10}). These gases can be liquefied by pressure, making them easier to handle. Uses: As fuel.
3. *Ligroin* — Normal and branched isomers of pentane (C_5H_{12}) and hexane (C_5H_{14}) with a boiling range up to 70 °C. Uses: As fuel.
4. *Gasolene* — Hydrocarbon compounds containing carbon atoms from C_7 to C_{11} with a boiling range between 70 and 180 °C. Uses: As fuel.
5. *Kerosene* — Hydrocarbon compounds with carbon atoms between C_{11} and C_{15} with a boiling range between 180 and 270 °C. Aromatics content is higher (10–40%) in kerosene than in the average crude oil. Uses: Mainly as a jet fuel.
6. *Gas oil* (boiling range between 275 and 400 °C) — Consists of hydrocarbon compounds with carbon atoms between C_{15} and C_{25}. Uses: As jet and diesel fuels.
7. *Lubricating oil* — Hydrocarbon compounds with carbon atoms between C_{26} and C_{40} mainly from the paraffin and cycloparaffin series. Uses: As a lubricant.
8. *Residuum* — As the name implies, what is left over after fractional distillation of crude oil to produce the previous seven products is called residuum. It consists of asphalts, asphaltenes, and resins. Uses: For wood preservatives, roofing, and road asphalts.

By using methods other than fractional distillation, paraffin waxes, asphalts, and resins can also be extracted from crude oils.

8.4. CLASSIFICATION OF CRUDE OILS

A physical classification of the crude oil is necessary for assessment of the type of hydrocarbons, pure or mixed, and for its ability and ease of flow through the reservoir rocks during extraction. Such measurements are essential for an economic evaluation of the quality of an oil find.

The two main physical criteria for crude oil classification are density or specific gravity and viscosity.

8.4.1. Density or Specific Gravity

The most commonly measured parameter of crude oils is their °API gravity. °API gravity is an arbitrary measure of density that allows construction of linearly scaled hydrometers. It is defined as

$$°\text{API} = \frac{141.5}{\rho_0} - 131.5 \tag{8.1}$$

where $\rho_0 = \text{SpGr}\frac{60}{60}°\text{F}$, the specific gravity at 60 °F, is the ratio of the oil density to water density, both measured at 60 °F. Note that in this equation, the lighter the oil, the higher its °API gravity; and water has an °API gravity of about 10. °API gravity is an important factor in determining the price of a crude oil.

8.4.2. Viscosity

This property is essential to assess the flow properties of an oil find. It is used to calculate the flow through the host reservoir rock and to calculate the rate of production of oil in a reservoir. There are two methods for measuring viscosity. The kinematic viscosity, based on Poiseulle's Law, is calculated by measuring the flow through a capillary tube, and the Saybolt viscosity, which is a measure of the time required for an oil to flow through a standard orifice according to ASTM Test D-88. Viscosity is strongly dependent on temperature and is lower at higher temperatures.

Table 8.1 gives the major physical properties of some representative members of the n-Alkane, alkene, and cycloparaffin series.

8.5. INDIRECT METHODS OF OIL CLASSIFICATION

To detect and distinguish between amount and type of oil saturation, indirect methods of seismic exploration are most commonly used. These methods yield P-wave velocity, density, and acoustic impedance ($=$ velocity $\times$ density), which can then be used for reservoir evaluations. The aim of laboratory studies is to establish relations between, for example, carbon number and seismic velocity. Such studies enable identification of oil type and quality from indirect measurements of seismic impedance, velocity, and density and enhance the prediction strength of seismic methods.

P-wave velocity and density are measured in the laboratory to be used as input for the seismic evaluation schemes. The most commonly used measurement technique is pulse transmission. A basic pulse transmission measurement setup consists of a pulse generator and receiver, an oscilloscope, two acoustic transducers, and a pressure vessel. The fluid to be measured is sealed in a jacket with a transducer at each end and immersed in a pressure vessel. One transducer is excited by the pulse generator and generates acoustic waves that propagate through the fluid. The acoustic waves propagated through the fluid are received by the second transducer. They are amplified by the receiver and displayed on the oscilloscope. The oscilloscope also receives a trigger from the pulse generator to synchronize both instruments. Acoustic velocity (V_p) is calculated from length measurement of the fluid sample (L) and time required by the acoustic wave to travel through the fluid (t) using the formula $V_p = L/t$. The measurements are repeated at different pressure and temperature steps.

8.5.1. Acoustic Velocity in Single Compounds

Table 8.2 gives the results of velocity measurements in various n-Alkanes as a function of temperature, Table 8.3 gives the same data for alkenes, and Table 8.4 for napthenes. Physical properties of these compounds are given in Table 8.1. There is a good negative linear relation between temperature (T) and velocity (V_p); $V_p = a - bT$, so that velocity at any temperature can be predicted for these compounds. The linear regression coefficients, "a" and "b," are also given in Tables 8.2–8.4 for the n-Alkanes, 1-Alkenes, and Napthenes, respectively. In all cases, high values of the coefficient of determination, R^2, show that velocity predictions with temperature have a fairly high accuracy for the different oils. The regression coefficient "a" is the initial velocity at a reference temperature, and "b" is a function of the molecular weight (M) of the organic compound. It can be calculated as

$$b = 0.306 - 7.6/M.$$

Table 8.2. Acoustic properties of a homologous series of 1-alkenes (olefins) as a function of temperature[a]. The linear regression coefficients, a and b, are also given for each member, along with the coefficient of determination (R^2). The data are organized as increasing carbon numbers.

n-Alkanes (Paraffins)

C_6H_{14}		C_7H_{16}		C_8H_{18}		$C_{10}H_{22}$		$C_{11}H_{24}$		$C_{12}H_{26}$		$C_{14}H_{30}$		$C_{15}H_{32}$		$C_{16}H_{34}$		$C_{18}H_{38}$		$C_{22}H_{46}$		$C_{28}H_{58}$		$C_{36}H_{74}$	
T	V_p	T	V_p	T	V_p	T	V_p	T	V_p	T	V_p	T	V_p	T	V_p	T	V_p	T	V_p	T	V_p	T	V_p	T	V_p
−10	1224	−8	1270	−8	1314	−5	1342	−2	1357	4	1359	10	1370			23	1343								
−5	1204	0	1235	−6	1305	10	1281	0	1348	13	1327	22	1324			27	1328	28	1340						
0	1184	7	1203	−2	1289	22	1241	6	1323	22	1288	27	1304	14	1364	33	1304	29	1338						
3	1171	12	1178	5	1253	27	1224	14	1296	28	1261	30	1295	25	1323	36	1288	38	1302	44	1316				
6	1160	18	1157	12	1228	38	1178	18	1280	37	1220	36	1267	33	1290	45	1258	48	1263	48	1301	61	1301	74	1275
12	1131	28	1109	18	1200	49	1137	31	1223	45	1194	42	1248	48	1236	52	1234	58	1228	55	1278	67	1275	80	1249
19	1102	36	1078	24	1177	56	1108	44	1180	54	1157	49	1220	57	1205	62	1192	68	1194	65	1242	74	1244	85	1232
26	1070	43	1047	31	1147	64	1076	55	1135	65	1118	59	1186	71	1152	73	1152	77	1161	72	1217	83	1215	90	1211
34	1035	50	1019	41	1105	69	1059	68	1086	73	1086	71	1139	82	1111	83	1118	90	1111	82	1182	90	1188	95	1197
40	1005	59	979	52	1059	78	1023	76	1055	83	1051	83	1091	89	1085	93	1081	93	1104	93	1142	96	1169	101	1172
45	983	65	951	61	1020	89	976	87	1008	92	1011	90	1064	91	1075	104	1041	104	1060	104	1105	101	1148	110	1144
50	960	73	917	74	965	95	951	93	989	100	978	103	1016	102	1038	110	1016	114	1027	111	1077	108	1126	117	1121
55	937	85	863	86	916	106	908	101	955	110	940	110	986	112	1000	119	982	122	996	120	1047	115	1102	124	1097
61	910	92	833	102	846	118	859	112	910	119	907	120	951	121	967							120	1082	132	1076
64	894	95	820	113	797			121	875																
68	882			119	776																				

Linear Regression Coefficients: $V_p = a - bT$

a	b	a	b	a	b	a	b	a	b	a	b	a	b	a	b	a	b	a	b	a	b	a	b	a	b
1184	4.469	1234	4.352	1279	4.237	1326	3.927	1349	3.899	1373	3.932	1408	3.812	1415	3.704	1426	3.729	1441	3.650	1472	3.545	1518	3.640	1523	3.430

Coefficient of Determination, R^2

| 0.9996 | 0.9999 | 0.9999 | 0.9995 | 0.9998 | 0.9995 | 0.9998 | 0.9999 | 0.9997 | 0.9998 | 0.9999 | 0.9986 | 0.9974 |

Carbon Numbers

| 6 | 7 | 8 | 10 | 11 | 12 | 14 | 15 | 16 | 18 | 22 | 28 | 36 |

[a] At 760 mm Hg.

Table 8.3. Acoustic Properties of a Homologous series of Napthenes (Cycloparaffins) as a Function of Temperature[a]. The linear regression coefficients, a and b, are also given for each member, along with the coefficient of determination (R^2). The data are organized as increasing carbon numbers.

1-Alkenes (Olefins)

C_6H_{12}		C_7H_{14}		C_8H_{16}		$C_{10}H_{20}$		$C_{11}H_{22}$		$C_{12}H_{24}$		$C_{14}H_{28}$		$C_{16}H_{32}$		$C_{18}H_{36}$		$C_{20}H_{40}$	
T	V_p	T	V_p	T	V_p	T	V_p	T	V_p	T	V_p	T	V_p	T	V_p	T	V_p	T	V_p
−12	1241			−12	1323					−4	1380	0	1397						
−8	1222			−4	1291					0	1365	13	1349						
−4	1203	−8	1261	0	1275	−8	1355	0	1345	8	1335	23	1312	10	1385				
2	1175	0	1226	5	1249	0	1324	8	1320	15	1309	29	1288	22	1341	17	1381		
8	1147	8	1191	12	1220	12	1277	12	1293	22	1282	36	1261	30	1309	23	1361		
11	1133	11	1180	19	1188	22	1238	16	1281	28	1261	43	1236	38	1279	30	1336	32	1356
18	1102	17	1155	31	1136	28	1213	22	1257	36	1229	49	1215	46	1248	41	1293	35	1343
24	1074	25	1120	44	1084	36	1181	34	1207	47	1187	55	1191	49	1240	45	1279	48	1291
30	1048	33	1085	53	1047	47	1137	47	1153	51	1174	64	1160	57	1211	56	1238	56	1266
33	1031	42	1047	64	999	60	1087	57	1118	60	1139	70	1140	63	1191	69	1191	67	1225
39	1004	53	1000	73	960	72	1039	62	1103	64	1124	78	1111	70	1162	80	1155	78	1187
47	968	63	958	84	914	84	993	75	1052	76	1076	86	1081	75	1148	69	1174	82	1174
50	954	74	910	92	878	96	943	87	1007	86	1040	96	1046	79	1135	92	1111	96	1123
56	927	84	870	108	814	108	896	100	956	96	1003	104	1014	89	1097	100	1087	110	1073
63	896	92	833	115	784	120	849	114	896	105	969	110	993	100	1059	110	1049	117	1046
								121	868	117	924	119	959	110	1020	120	1011	120	1038
														119	987				

Linear Regression Coefficients: $V_p = a - bT$

a	b	a	b	a	b	a	b	a	b	a	b	a	b	a	b	a	b	a	b
1185	4.604	1226	4.265	1271	4.254	1324	3.959	1344	3.914	1365	3.776	1395	3.659	1418	3.618	1441	3.578	1468	3.593

Coefficient of Determination, R^2

| 0.9999 | 0.99996 | 0.9999 | 0.99998 | 0.999 | 1 | 0.9999 | 0.9997 | 0.9997 | 0.9997 |

Carbon Numbers

| 6 | 7 | 8 | 10 | 11 | 12 | 14 | 16 | 18 | 20 |

[a] At 760 mm Hg.

Table 8.4. Physical properties of a homologous series of n-Alkanes (paraffins), 1-alkenes (olefins), and cycloparaffins (napthenes). The series is arranged in increasing carbon numbers for each member.

Napthenes (cycloparaffins)					
C_6H_{12}		C_7H_{14}		C_8H_{16}	
T	V_p	T	V_p	T	V_p
7	1337	−2	1456		
12	1313	7	1416		
17	1289	14	1390	14	1427
22	1265	23	1345	22	1395
27	1237	29	1311	27	1374
32	1213	31	1304	38	1328
37	1190	39	1267	47	1288
42	1166	45	1241	58	1243
47	1141	54	1200	67	1206
53	1111	64	1157	77	1163
60	1077	74	1111	87	1121
64	1064	86	1054	95	1086
72	1023	95	1017	101	1060
79	986	106	967	110	1023
80	982	108	959	120	984
Linear Regression Coefficients: $V_p = a - bT$					
a	b	a	b	a	b
1370	4.853	1447	4.539	1487	4.211
Coefficient of Determination, R^2					
0.9997		0.9997		0.9999	
Carbon Numbers					
6		7		8	

Figures 8.5–8.7 show plots of the velocity data for the n-Alkanes, alkenes, and napthenes, respectively. The linear trend of velocity with temperature is evident. Also plotted in these figures, for example, is a linear regression fit for the lightest member of the series with the lowest carbon number. In each case, the coefficient of determination R^2 is greater than 0.999.

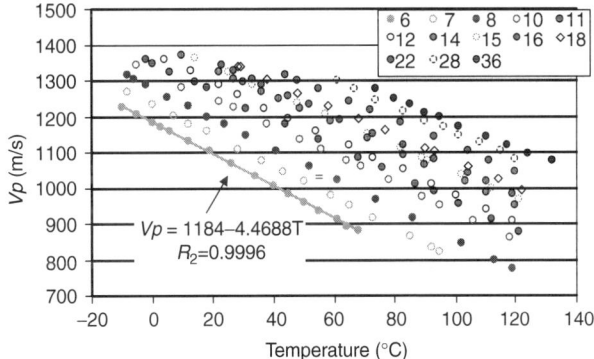

Fig. 8.5. Velocity variation in various alkanes as a function of temperature. The velocity decreases monotonously as temperature increases. The data are coded by the carbon number of the oils. The smaller carbon numbers are for lighter oils with high API gravity. At same temperature, the lighter oils (smaller carbon numbers) have lower velocity. For the lightest oil, carbon number 6, a linear regression is fitted through the data marked by a solid line. The regression parameters show goodness of the fit, with a coefficient of determination, $R^2 = 0.9996$.

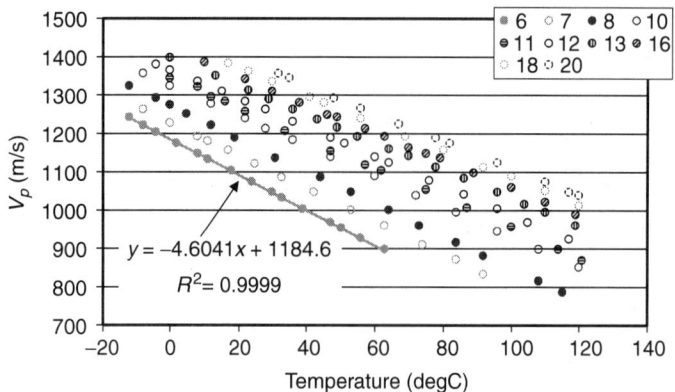

Fig. 8.6. Velocity variation in various alkenes as a function of temperature. The velocity decreases monotonously as temperature increases. The data are coded by the carbon number of the oils. The smaller carbon numbers are for lighter oils with high API gravity. At same temperature, the lighter oils (smaller carbon numbers) have lower velocity. For the lightest oil, carbon number 6, a linear regression is fitted through the data marked by a solid line. The regression parameters show goodness of the fit, with a coefficient of determination, $R^2 = 0.9999$.

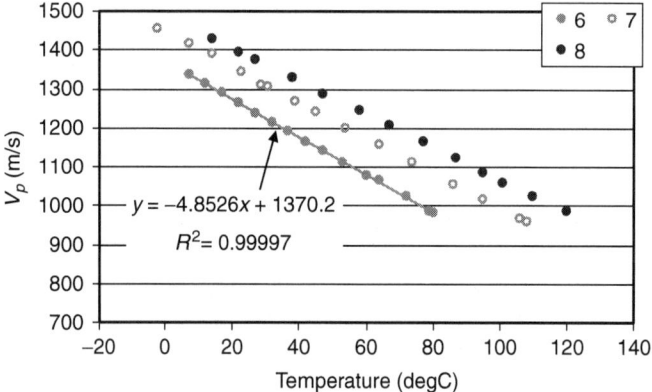

Fig. 8.7. Velocity variation in various napthenes as a function of temperature. The velocity decreases monotonously as temperature increases. The data are coded by the carbon number of the oils. The smaller carbon numbers are for lighter oils with high API gravity. At same temperature, the lighter oils (smaller carbon numbers) have lower velocity. For the lightest oil, carbon number 6, a linear regression is fitted through the data marked by a solid line. The regression parameters show goodness of the fit, with a coefficient of determination, $R^2 = 0.9997$.

8.5.2 Acoustic Velocity in Oil Mixtures

Crude oils are a mixture of numerous types of oils with very different physical properties. The physical properties of the mixture vary proportional to the volume fraction of each component making up the crude oil. Acoustic velocities in a binary mixture and in a more complex, multicomponent mixture are given in Tables 8.5 and 8.6, respectively. The acoustic velocity in mixtures can be calculated as the fractional average of the components.

$$V_P = \sum_{i=1}^{n} X_i V_i \qquad (8.3)$$

where, X_i is the volume fraction of the oil in the mixture and V_i is its acoustic velocity. Velocities calculated using this formula are also given in Tables 8.5 and 8.6. The difference between measured and calculated velocities is less than 0.5% both in the binary and in the more complex multicomponent mixtures. Figure 8.8 shows the

Table 8.5. Measured and Calculated Acoustic Properties of a Binary Mixture of 1-Decene and 1-Octadecene[a]. The difference between measured and calculated velocities is less than 0.5%.

Volume Fraction (%)		Vmeasured (m/s)	Vcalculated (m/s)	Difference (%)
$C_{10}H_{20}$	$C_{18}H_{36}$			
0	100	1369	1370	0.04
10	90	1358	1357	−0.07
20	80	1348	1345	−0.25
29.4	70.6	1336	1333	−0.23
40	60	1321	1320	−0.10
50	50	1307	1307	0.01
60	40	1298	1295	−0.25
70	30	1286	1282	−0.29
80	20	1275	1270	−0.41
90	10	1260	1257	−0.21
100	0	1247	1245	−0.17

[a] At temperature-20 °C.

Table 8.6. Measured and Calculated Acoustic Properties of a Multicomponent Mixture of n-Alkanes, 1-Alkenes, and Cycloparaffins[a]. The difference between measured and calculated velocities is less than 0.5%

Composition	V_p at 20 °C (m/s)	Volume Fraction (%)					
C_8H_{18}	1194			20			4
$C_{10}H_{22}$	1247	20		20			8
$C_{12}H_{26}$	1294	20		20			8
$C_{14}H_{30}$	1333	20		20			8
$C_{16}H_{34}$	1354	40		20			12
C_8H_{16}	1188				10		2
$C_{10}H_{20}$	1247				10	10	6
$C_{12}H_{24}$	1290		30		20		10
$C_{14}H_{28}$	1321		30		10	10	10
$C_{16}H_{32}$	1349		40		50	20	20
C_6H_{12}	1276					10	2
C_7H_{14}	1358					20	4
C_8H_{16}	1402					30	6
Vmeasured		1319	1326	1286	1312	1343	1309
Vcalculated		1316.4	1322.9	1284.4	1308.1	1346.4	1313.6
Difference (%)		−0.20	−0.23	−0.12	−0.30	0.25	0.35

[a] At temperature-20 °C.

variation in velocity in the binary mixture as a function of volume percent of the lighter oil, 1-decene ($C_{10}H_{20}$). The velocity decreases monotonously as the volume fraction of the lighter oil fraction increases. The difference between calculated (solid line) and measured (symbols) velocities is minimal.

Velocity in oils varies as functions of not only temperature but also pressure. The velocity variations as a function of temperature and pressure can be calculated using the formulae derived by Wang (1988) [1], Wang *et al.* (1990) [3], and Batzle and Wang

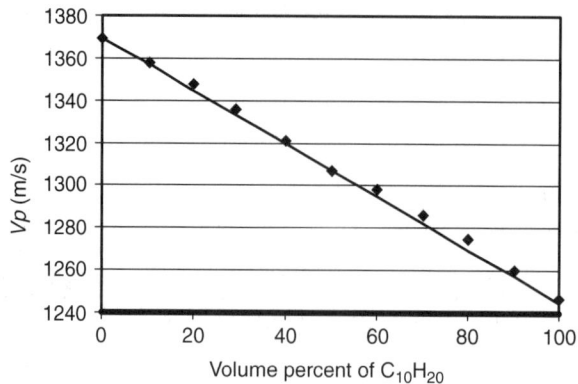

Fig. 8.8. Velocity variation in a binary mixture as a function of volume percent of the lighter oil, 1-decene ($C_{10}H_{20}$). The velocity decreases monotonously as the volume fraction of the lighter oil fraction increases. The difference between calculated (solid line) and measured (symbols) velocities is less than 0.5%.

(1992) [10]:

$$V = 2096 \left(\frac{\rho_0}{2.6 - \rho_0}\right)^{0.5} - 3.7T + 4.64P + 0.0115[4.12(1.08\rho_0^{-1} - 1)^{0.5} - 1]TP \tag{8.4}$$

or, in terms of API

$$V = 15450(77.1 + \text{API})^{-0.5} - 3.7T + 4.64P + 0.0115(0.36\text{API}^{0.5} - 1)TP \tag{8.5}$$

where P = pressure and T = temperature.

References

1. Wang, Z. (1988). Wave velocities in hydrocarbons and hydrocarbon-saturated rocks with applications to EOR monitoring. PhD dissertation. Stanford: Stanford University SRB Project v. 34.
2. Wang, Z., and Nur, A. (1987). Velocities in hydrocarbons and hydrocarbon-saturated rocks and sands. Expanded Abstracts of the 57th Annual International Society of Exploration Geophysicists Meeting and Exposition; SEG Abstracts; Vol. 57; pp. 1–4.
3. Wang, Z., Nur, A., and Batzle, M. (1990). Acoustic velocities in petroleum oils. *J of Petroleum Technology*; Vol. 42, no. 2: 192–201.
4. Tissot, B., and Welte, D.H. (1984). *Petroleum Formation and Occurrence: A New Approach to Oil and Gas Explaoration*. Berlin: Springer Verlag.
5. Hunt, J.M. (1979). *Petroleum Geologu and Geochemistry*, 615 pp., San Francisco: W.H. Freeman and Co.
6. Taylor, G.H., Teichmueller, M., Davis, A., Diessel, C.F.K., Littke, R., and Robert, P. (1990). *Organic Petrology*. Berlin: Gebrueder Borntraeger.
7. Taylor, G.H., Teichmueller, M., Davis, A., Diessel, C.F.K., Littke, R., and Robert, P. (1998). *Organic Petrology*. Berlin: Gebrueder Borntraeger.
8. Gallant, R.W., and Yaws, C.L. (1993). Physical Properties of Hydrocarbons and other Chemicals. Houston: Gulf Publishing Company.
9. Tissot, B., Durrand, B., Espitalie, J., and Combaz, A. (1974). Influence of nature and diagenesis of organic matter in formation of petroleum. *AAPG Bull.* **58**: 499–506.
10. Batzle, M., and Wang, Z. (1992). Seismic properties of pore fluids. *Geophysics* **57**: 1396–1408.

CHAPTER 9

ACOUSTIC PROPERTIES IN ROCKS SATURATED WITH PETROLIFEROUS LIQUIDS

Manika Prasad, Amos Nur, Gary Mavko, Jack Dvorkin
Geophysics Department, Stanford University, Stanford, California, USA

Contents

Abstract	219
9.1. Introduction	219
9.2. Light Oil Saturants	220
9.3. Natural Heavy Oil Saturants	224
References	230

ABSTRACT

Acoustic properties of rocks depend also on the properties of fluids that are present in their pore space. The effect of pore fluids is twofold: On one hand, it increases bulk modulus of the composite rock proportional to its fraction of total volume. On the other hand, viscosity and chemical reactions with the frame materials additionally affect seismic wave velocity. We provide here seismic wave velocity data for typical rocks that can be reservoirs for petroleum fluids and gases. The *P*- and *S*-wave velocity data are given as functions of temperature and pressure.

9.1. INTRODUCTION

Rocks are complex composites made of various minerals that are cemented together to make up the frame. In the pore space within this frame, fluids — for example, water, oil, gas, or their mixtures — can be stored. Seismic exploration is aimed at finding and identifying the type of fluids that are trapped in the pore space of reservoir rocks. Seismic wave velocity information that is derived from seismic exploration is an important tool in reservoir characterization. It can be used to predict rock type and reservoir properties. It is usually the initial and sometimes the primary source of information about reservoir properties. To make more reliable predictions about subsurface pore fluids, a systematic study of the seismic properties of typical reservoir rocks containing petroliferous fluids as saturants is required.

In addition to *in situ* state of stress and pore pressures, temperature, and frequency of measurements, seismic velocity in such rocks is dependent on the frame and pore fluid properties (King, 1966 [1]; Nur and Simmons, 1969 [2]; Kuster and Toksöz, 1974 [3]; O'Connell and Budiansky, 1974, 1977 [4, 5]; Tittmann et al., 1984 [6]; Murphy, 1985 [7]; Jones, 1986 [8]; Vo-Thanh; 1990 [9]; Clark, 1992 [10]). The effect of pore fluids is twofold: On one hand, it increases bulk modulus of the composite rock proportional to its fraction of total volume. On the other hand, viscosity and chemical reactions with the frame materials additionally affect seismic wave velocity. The change in bulk modulus of the composite rock due to fluid bulk modulus can be predicted by the Gassmann relation (Gassmann, 1951 [11]) and the Biot theory (Biot, 1956a, b [12, 13]).

In this chapter, we give seismic wave velocity data for typical rocks that can be reservoirs for petroleum fluids and gases. Data used here are derived from Wang (1988 [14]) and Wang et al. (1987, 1988, 1990 [15–17]) and Gregory (1976 [18]).

9.2. LIGHT OIL SATURANTS

Seismic wave velocity varies with oil saturant and as a function of temperature. In addition to weakening the rock, the effect of temperature is to change the viscosity of the saturant. Here, velocity variations as a function of temperature are presented for consolidated reservoir-type rocks saturated with air, water, and various pure hydrocarbons. Pure hydrocarbon members belonging to two components of crude oil, alkanes and alkenes (n-Heptane, 1-Decene, 1-Tetradecene, and 1-Octadecene), were used as saturants. Table 9.1 presents the physical properties of the hydrocarbons used as saturants and the porosity of the rocks. The two rocks presented have extreme values of porosity. Boise sandstone has high porosity and permeability, whereas Beaver sandstone has low porosity and pore connectivity.

9.2.1. Velocity in Consolidated Rocks

The P- and S-wave velocity data are given in Table 9.2 for Boise and in Table 9.3 for Beaver sandstone. The data are collected at 15 MPa effective pressure. Figure 9.1

Table 9.1. Physical properties of the light oil saturants and of the rocks.

		Properties of the hydrocarbon saturants				
Name	Formula	Purity (%)	Molecular weight	Melting point (°C)[a]	Boiling point (°C)[a]	Density at 20°C[b]
n-Heptane	C_7H_{16}	99.00	100.07	−91.00	98.00	0.68
1-Decene	$C_{10}H_{20}$	96.00	140.10	−86.00	171.00	0.74
1-Tetradecene	$C_{14}H_{28}$	95.00	198.14	−13.00	251.00	0.78
1-Octadecene	$C_{18}H_{36}$	90.00	252.18	16.00	314.00	
	Properties of the rock samples					
Name	Porosity (%)			Pore connectivity		
Boise sandstone	27			High		
Beaver sandstone	7			Low		
Ottawa sand	37			High		

[a] At 760 mm Hg.

[b] At 760 mm Hg; unit: g/cm^3.

Table 9.2. Acoustic properties (V_p, V_s) as functions of temperature in high-porosity, high-permeability Boise sandstone saturated with light oils. The data is given for dry, water, and light oil saturated conditions in the rock at 15 MPa effective pressure.

Boise sandstone (at 15 MPa effective pressure)

Air saturated			Water saturated			C_7H_{16} saturated			$C_{10}H_{20}$ saturated			$C_{14}H_{28}$ saturated			$C_{18}H_{38}$ saturated		
T(°C)	V_p (m/s)	V_s (m/s)	T(°C)	V_p (m/s)	V_s (m/s)	T(°C)	V_p (m/s)	V_s (m/s)	T(°C)	V_p (m/s)	V_s (m/s)	T(°C)	V_p (m/s)	V_s (m/s)	T(°C)	V_p (m/s)	V_s (m/s)
20.8	3404	2201	21.90	3547	2120	21.1	3490	2154	21.30	3534	2182	21.00	3583	2174	21	3582	2175
32.40	3394	2191	33.30	3532	2108	30.7	3465	2141	32.00	3502	2147	30.80	3529	2150	37.00	3547	2147
52.70	3372	2167	40.70	3517	2087	47.5	3423	2112	48.40	3466	2120	43.10	3499	2128	57.60	3488	2118
77.80	3345	2134	56.60	3483	2053	51	3413	2107	64.60	3425	2099	50.10	3484	2119	73.50	3455	2094
92.30	3314	2107	76.10	3445	2027	60.6	3390	2093	75.60	3395	2078	66.40	3451	2096	80.20	3442	2083
110.80	3288	2088	96.20	3394	2008	66.8	3376	2087	90.70	3356	2055	87.70	3409	2077	94.00	3413	2088
124.40	3263	2071	114.90	3349	1988	82.6	3349	2077	105.80	3318	2042	106.90	3368	2055	114.10	3372	2051
			126.60	3318	1977	95.2	3313	2058	119.60	3285	2028	124.20	3327	2037	124.20	3345	2039
						110.4	3291	2083	122.70	3276	2020						
						129.4	3244	2024									

Table 9.3. Acoustic properties (V_p, V_s) as functions of temperature in low-porosity, low-permeability Beaver sandstone saturated with light oils. The data is given for dry, water, and light oil saturated conditions in the rock at 15 MPa effective pressure.

Beaver sandstone (at 15 MPa effective pressure)

Air saturated			Water saturated			C_7H_{16} saturated			$C_{10}H_{20}$ saturated			$C_{14}H_{28}$ saturated			$C_{18}H_{36}$ saturated		
T(°C)	V_p (m/s)	V_s (m/s)	T(°C)	V_p (m/s)	V_s (m/s)	T(°C)	V_p (m/s)	V_s (m/s)	T(°C)	V_p (m/s)	V_s (m/s)	T(°C)	V_p (m/s)	V_s (m/s)	T(°C)	V_p (m/s)	V_s (m/s)
21.40	4814	3270	22.40	5233	3376	21.40	5183	3419	21.40	5215	3425	20.90	5238	3417	20.9	5255	3419
40.40	4789	3238	35.30	5213	3343	34.50	5121	3352	30.80	5164	3385	34.50	5186	3394	32.60	5200	3396
56.60	4764	3191	53.30	5174	3298	41.80	5091	3326	38.60	5134	3360	53.40	5125	3326	50.50	5142	3333
75.70	4740	3149	74.20	5117	3227	60.20	5021	3263	48.20	5110	3322	67.60	5085	3292	68.30	5094	3290
91.90	4724	3130	91.20	5069	3181	80.50	4963	3214	59.00	5081	3289	80.40	5036	3264	81.10	5066	3265
108.30	4708	3108	109.80	5015	3148	105.10	4879	3159	73.90	5046	3260	93.50	4998	3228	97.00	5011	3229
114.1	4700	3097	124.20	4979	3119	122.70	4794	3117	84.10	5008	3240	108.10	4951	3204	123.90	4922	3173
128.0	4685	3083							96.90	4970	3217	124.20	4896	3169			
									102.10	4951	3209						
									122.30	4869	3155						

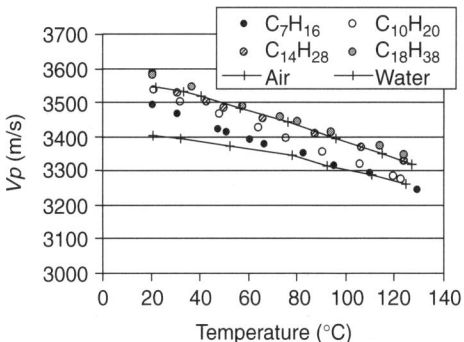

Fig. 9.1. *P*-wave velocity variations in high-porosity, high-permeability Boise sandstone saturated with light oil saturants as a function of temperature.

shows V_p variations and Figure 9.2 shows V_s variations as a function of temperature in high-porosity Boise sandstone saturated with various saturants. Figure 9.3 shows V_p variations and Figure 9.4 shows V_s variations as a function of temperature in low-porosity Beaver sandstone saturated with various saturants. *P*-wave velocity increases as the molecular weight of the hydrocarbons saturating the rock increases; *S*-wave velocity is almost unaffected. In all cases, there is a monotonous decrease in velocity with temperature.

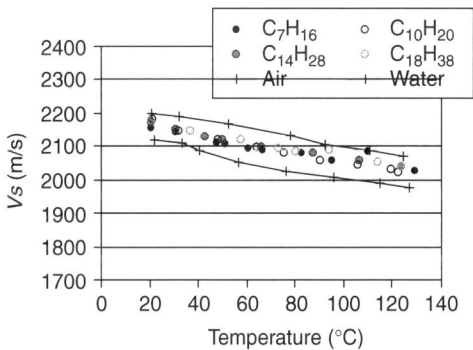

Fig. 9.2. *S*-wave velocity variations in high-porosity, high-permeability Boise sandstone saturated with light oil saturants as a function of temperature.

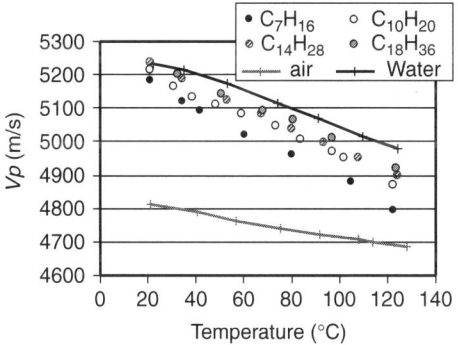

Fig. 9.3. *P*-wave velocity variations in low-porosity, low-permeability Beaver sandstone saturated with light oil saturants as a function of temperature.

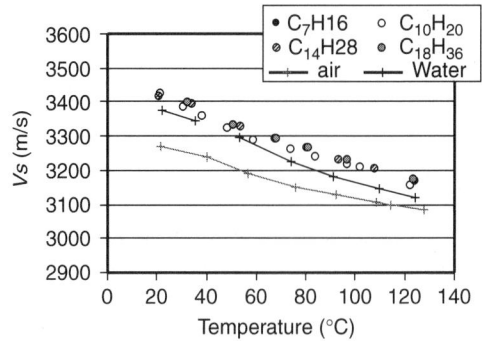

Fig. 9.4. *S*-wave velocity variations in low-porosity, low-permeability Beaver sandstone saturated with light oil saturants as a function of temperature.

Table 9.4. P-wave velocity (V_p) as function of temperature in unconsolidated Ottawa sand saturated with light oils. The data is given for dry, water, and light oil saturated conditions in the rock at 15 MPa effective pressure.

Ottawa sand (at 15 MPa effective pressure)							
Air saturated		Water saturated		C_7H_{16} saturated		$C_{18}H_{36}$ saturated	
$T(°C)$	V_p(m/s)	$T(°C)$	V_p(m/s)	$T(°C)$	V_p(m/s)	$T(°C)$	V_p(m/s)
21.50	1494	22.00	1943	20.80	1840	21.30	1933
30.40	1482	29.40	1936	36.00	1800	34.50	1891
46.60	1467	41.50	1931	50.50	1767	53.40	1841
60.60	1454	45.60	1946	69.10	1730	62.60	1808
71.40	1439	52.30	1926	83.70	1682	82.10	1750
79.00	1432	62.40	1911	98.80	1647	90.60	1731
94.70	1422	75.00	1898	112.00	1608	110.10	1684
107.20	1416	88.30	1887	126.40	1523	124.00	1650
123.30	1413	98.20	1866				
		108.20	1851				
		124.10	1825				

9.2.2. Velocity in Unconsolidated Sand

Compressional wave velocity data for unconsolidated Ottawa sand saturated with the different saturants is given in Table 9.4. Figure 9.5 shows V_p variations as a function of temperature in high-porosity, unconsolidated Ottawa sand saturated with various saturants. The data are collected at 15 MPa effective pressure. Although sand has lower velocity than the sandstones, the overall behaviors with respect to temperature and molecular weight of hydrocarbons are the same.

9.3. NATURAL HEAVY OIL SATURANTS

Seismic velocities are markedly different in rocks saturated with heavy oils than in those saturated with light oils. Since the viscosity of the oils is strongly dependent on their temperature, velocity variations in rocks and sand saturated with a heavy oil, tar, and wax as a function of temperature are given here. For comparison, data for a pure hydrocarbon of the alkene family with high molecular weight, Eicosene, is also mentioned. Table 9.5 presents the physical properties of the hydrocarbons used as

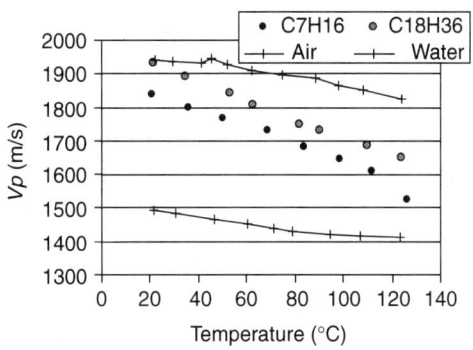

Fig. 9.5. *P*-wave velocity variations in unconsolidated Ottawa sands saturated with light oil saturants as a function of temperature.

Table 9.5. Physical properties of the heavy oil saturants, wax, alkene, crude oil, and tar, along with physical properties of the rocks they saturated.

Properties of the natural heavy oil saturants				
Name	Viscosity	Melting point (°C)[a]	Composition	Density at 20°C[b]
Parowax	Solid at room conditions	45–65	$C_{18}H_{36}$ to $C_{30}H_{62}$ normal paraffins	0.89
Eicosene	4.662 cpoise	28	Alkene $C_{20}H_{40}$ $CH_3(CH_2)_{17}CH=CH_2$	0.76
Crude oil	46 Pas		Light to very heavy hydrocarbons	0.92
Tar	760 Pas		Mainly heavy hydrocarbons	0.98
Properties of the rock samples				
Name	Porosity (%)		Pore connectivity	
Boise sandstone	27		High (902 mD)	
Massillon sandstone	22		High-medium (740 mD)	
Ottawa sand	37		High	
Gulf Coast Sand	22		n/a	

[a] At 760 mm Hg. [b] At 760 mm Hg; unit: g/cm^3.

saturants and the porosity of the rocks. The two rocks presented have high values of porosity. Both sandstones have high pore connectivity.

9.3.1. Velocity in Consolidated Rocks

The *P*- and *S*-wave velocity data are given in Table 9.6 for Boise and in Table 9.7 for Massillon sandstone and Table 9.8 for Gulf Coast Sandstone. The data are collected at 15 MPa effective pressure. Figure 9.6 shows V_p variations and Figure 9.7 shows V_s variations as a function of temperature in the high-porosity Boise sandstone saturated with crude oil. Figure 9.8 shows V_p variations and Figure 9.9 shows V_s variations as a function of temperature in Massillon sandstone saturated with various saturants. *P*-wave velocity increases as the molecular weight of the hydrocarbons saturating the rock increases; *S*-wave velocity is similarly affected. In all cases, there is an exponential decrease in velocity with temperature. The increase in shear velocity with addition of

Table 9.6. Acoustic properties (V_p, V_s) as functions of temperature in high-porosity, high-permeability Boise sandstone saturated with crude oil at 15 MPa effective pressure.

Boise sandstone		
Crude oil		
T	V_p	V_s
22	3.581	2.130
30	3.540	2.100
41	3.487	2.080
49	3.394	2.070
62	3.225	2.050
72	3.148	2.020
80	3.123	2.000
88	3.093	1.970
100	3.081	1.950
109	3.074	1.950
118	3.068	1.940
127	3.058	1.940

Table 9.7. Acoustic properties (V_p, V_s) as functions of temperature in low-porosity, low-permeability Massillon sandstone saturated with heavy oils. The data is given for different heavy oil saturants from Table 9.5. and for water saturated conditions in the rock at 15 MPa effective pressure.

Massillon light sandstone (at 15 MPa effective pressure)											
Parowax			Water			Eicosene			Crude oil		
T	V_p	V_s	T	V_p	V_s	T	V_p	V_s	T	V_p	V_s
23	4.025	2.593	20	3.421	2.190	21	3.639	2.408	22	3.423	2.230
27	4.021	2.547	25	3.402		25	3.623	2.388	29	3.397	2.220
31	3.922	2.477	35	3.403	2.157	28	3.562	2.330	35	3.380	2.210
40	3.800	2.402	42	3.422		30	3.391	2.283	44	3.351	2.190
48	3.705	2.322	48	3.409	2.140	38	3.322	2.252	48	3.334	2.190
62	3.233	2.212	58	3.400		42	3.301	2.229	59	3.264	2.170
80	3.176	2.185	68	3.393	2.110	48	3.260	2.212	71	3.215	2.150
90	3.165	2.169	72	3.370		56	3.238	2.182	84	3.182	2.120
100	3.152	2.155	81	3.343	2.094	64	3.218	2.156	97	3.156	2.110
113	3.128	2.150	100	3.349	2.082	71	3.196	2.130	103	3.149	2.100
128	3.114	2.141	109	3.325		78	3.174	2.120	121	3.138	2.090
136	3.105	2.132	118	3.312	2.049	88	3.153	2.109	130	3.130	2.090
			123	3.286		106	3.125	2.096			
						122	3.099	2.080			

the hydrocarbons is mainly due to the fact that they have a nonzero shear modulus that increases the shear modulus of the composite rock. As temperature increases, the wax starts to melt and its shear modulus decreases to zero; the shear velocity of the composite rock saturated with wax approaches that of the rock with water. Figure 9.10 shows V_p variations and Figure 9.11 shows V_s variations with pressure in high-porosity dry, water, and crude oil saturated Gulf coast sandstone.

ACOUSTIC PROPERTIES IN ROCKS SATURATED WITH PETROLIFEROUS LIQUIDS

Table 9.8. Acoustic properties (V_p, V_s) and density as functions of pressure in high-porosity Gulf Coast sandstone. The data is given for dry, water, and crude oil saturated conditions.

	Gulf coast sandstone								
	Air			Water			Crude oil		
Pressure (MPa)	Rho (g/cm³)	V_p (m/s)	V_s (m/s)	Rho (g/cm³)	V_p (m/s)	V_s (m/s)	Rho (g/cm³)	V_p (m/s)	V_s (m/s)
7	2.0569	2708	1881	2.2712	3301	2036	2.1279	3277	1935
34	2.0636	3560	2348	2.2752	3781	2246	2.1325	3392	2010
69	2.0696	3833	2527	2.2797	3927	2367	2.1381	3473	2070

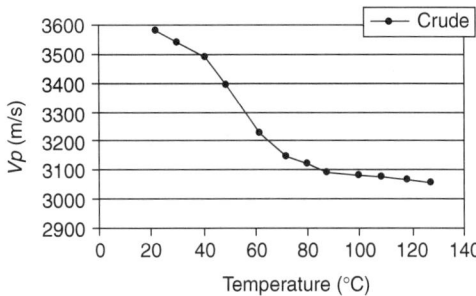

Fig. 9.6. P-wave velocity variations in high-porosity, high-permeability Boise sandstone saturated with crude oil as a function of temperature.

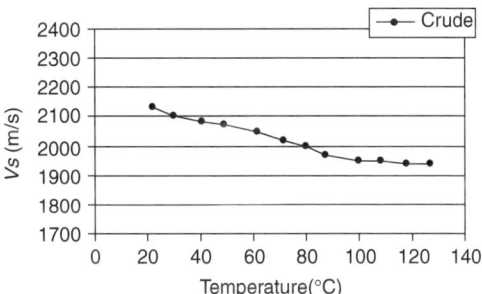

Fig. 9.7. S-wave velocity variations in high-porosity, high-permeability Boise sandstone saturated with crude oil as a function of temperature.

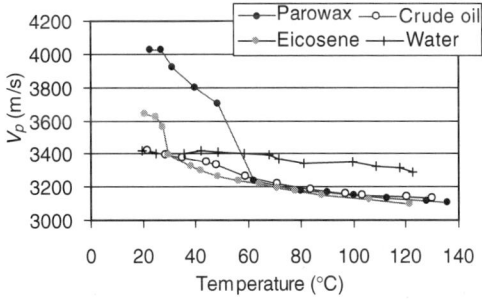

Fig. 9.8. P-wave velocity variations in high-porosity, high-permeability Massillon sandstone saturated with various heavy oil saturants as a function of temperature.

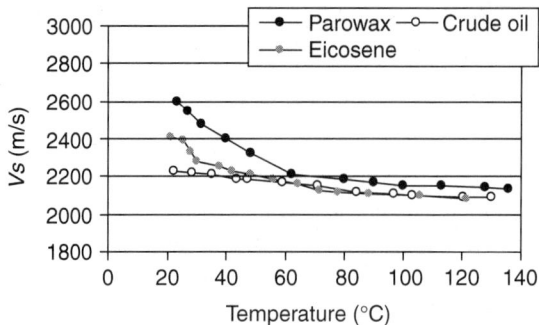

Fig. 9.9. *S*-wave velocity variations in high-porosity, high-permeability Massillon sandstone saturated with various heavy oil saturants as a function of temperature.

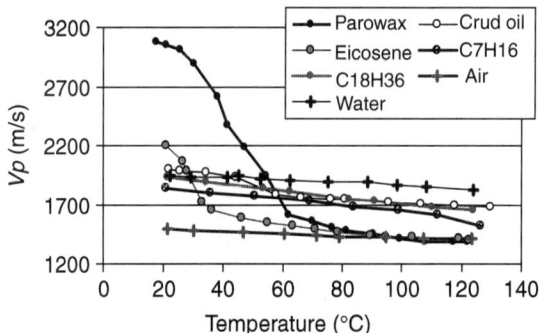

Fig. 9.10. *P*-wave velocity variations in high-porosity dry, water, and crude oil saturated Gulf Coast sandstone as a function of pressure.

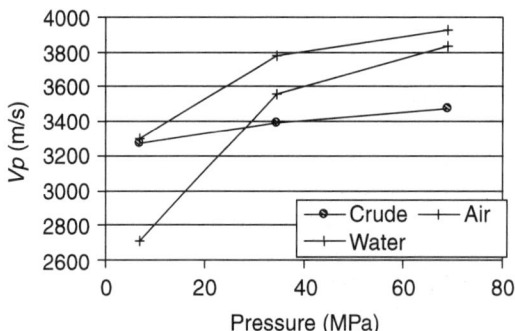

Fig. 9.11. *S*-wave velocity variations in high-porosity dry, water, and crude oil saturated Gulf Coast sandstone as a function of pressure.

9.3.2. Velocity in Unconsolidated Sand

Compressional wave velocity data for unconsolidated Ottawa sand saturated with the different natural oil saturants are given in Table 9.9. Figure 9.12 shows V_p variations as a function of temperature in high-porosity unconsolidated Ottawa sand saturated with various saturants. The data are collected at 15 MPa effective pressure. The behavior is similar to that of the consolidated rocks. There is an exponential decrease in velocity

ACOUSTIC PROPERTIES IN ROCKS SATURATED WITH PETROLIFEROUS LIQUIDS

with temperature increase. The main reason is due to the fact that the heavy oil acts as a cement, giving additional rigidity to the sand. As temperature increases, the wax melts and the sediment characteristics resemble those of water-saturated sands.

The temperature effects on *P*-wave velocity are summarized in Table 9.10. There is almost 50% velocity change with temperature in the heavy oil (Eicosene) and the wax. The velocity change for sand saturated with these hydrocarbons is correspondingly high. Velocity change for the sandstones saturated with the wax is lower.

Table 9.9. P-wave velocity (V_p) as functions of temperature in unconsolidated Ottawa sand saturated with heavy oils. The data is given for Eicosene, crude oil, and parowax saturated conditions in the rock at 15 MPa effective pressure.

Ottawa clean sand (at 15 MPa effective pressure)					
Eicosene		Crude oil		Parowax	
T	V_p (km/s)	T	V_p (km/s)	T	V_p (km/s)
21	2.203	22	1.995	17	3.081
26	2.065	26	1.989	21	3.055
28	1.991	34	1.972	25	3.019
33	1.718	44	1.929	30	2.888
36	1.658	58	1.789	38	2.609
47	1.586	66	1.765	41	2.372
55	1.549	76	1.753	47	2.188
64	1.519	81	1.744	54	1.948
71	1.489	96	1.718	62	1.608
78	1.464	107	1.713	70	1.565
90	1.438	118	1.698	77	1.511
95	1.431	130	1.686	81	1.478
104	1.422			90	1.448
119	1.409			99	1.411
123	1.401			108	1.392
				117	1.391
				122	1.383

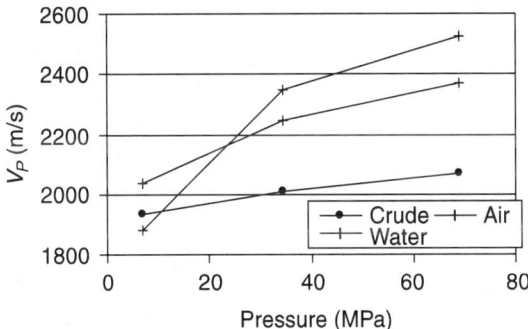

Fig. 9.12. *P*-wave velocity variations in unconsolidated dry, water saturated, and various heavy oil saturated Ottawa sand as a function of temperature.

229

Table 9.10. Comparison of velocities in sandstone and sands saturated with various heavy oils and oil–water mixtures. Velocity values and the percent change due to temperature are given. Maximum change with temperature is observed in a sand saturated with wax.

Sample	V_p (km/s) at 22 °C	V_p (km/s) at 122 °C	$\Delta V_p / V_p$ (%)
Parowax	2.010	1.181	41.2
Parowax + sand	3.044	1.387	54.4
Parowax + sandstone	4.028	3.120	22.5
1-Eicosene	1.571	1.101	29.9
1-Eicosene + sand	2.165	1.227	43.3
Durri heavy crude	1.502	1.235	17.8
Heavy crude + sand	1.994	1.691	15.2
Street ranch tar	1.676	1.312	21.7
Tar + sand	2.252	1.531	32.0
10.7% tar + 89.3% sand	2.375	1.500	36.8
10.7% tar + sand + water	2.436	1.780	26.9
Sand + gas	1.494	1.414	5.4
Sand + water	1.943	1.831	5.8

References

1. King, M.S. (1966). Wave velocities in rocks as a function of changes in overburden pressure and pore fluid saturants. *Geophysics* **31**: 50–73.
2. Nur, A., and Simmons, G. (1969). The effect of viscosity of a fluid phase on the velocity in low porosity rocks. *Earth Plan. Sci. Lett.* **7**: 99–108.
3. Kuster, G.T., and Toksöz, M.N. (1974). Velocity and attenuation of seismic waves in two-phase media, Part 1: Theoretical formulations. *Geophysics* **39**: 587–606.
4. O'Connell, R.J., and Budiansky, H. (1974). Seismic velocities in dry and saturated cracked solids. *J. Geophys. Res.* **79**: 5412–5426.
5. O'Connell, R.J., and Budiansky, H. (1977). Viscoelastic properties of fluid-saturated cracked solids. *J. Geophys. Res.* **82**: 5719–5735.
6. Tittmann, B.R., Bulau, J.R., and Abdel-Gawad, M. (1984). The role of viscous fluids in the attenuation and velocity of elastic waves in porous rocks, *In Physics and Chemistry of Porous Materials: AIP Conf. Proc.*, Johnston, D.L. and Sen, P., eds., vol. 107: pp. 131–143.
7. Murphy, W.F. (1985). Effects of microstructure and pore fluids on the acoustic properties of granular sedimentary materials. PhD dissertation, Stanford, CA: Stanford University SRB Project v. 16.
8. Jones, T.D. (1986). Pore fluids and frequency-dependent wave propagation in rocks. *Geophysics* **51**: 1939–1953.
9. Vo-Thanh, D. (1990). Effects of fluid viscosity on shear-wave attenuation in saturated sandstones. *Geophysics* **55**: 712–722.
10. Clark, V.A. (1992). The properties of oil under in-situ conditions and its effect on the seismic properties of rocks. *Geophysics* **57**: 894–901.
11. Gassmann, F. (1951). Über die Elastizität poröser Medien. *Vierteljahresschr. Naturforsch. Ges. Zürich* **96**: 1–23.
12. Biot, M.A. (1956a). Theory of propagation of elastic waves in a fluid-saturated porous solid, I: Low-frequency range. *J. Acoust. Soc. Am.* **28**: 168–178.
13. Biot, M.A. (1956b). Theory of propagation of elastic waves in a fluid-saturated porous solid, II: Higher frequency range. *J. Acoust. Soc. Am.* **28**: 179–191.
14. Wang, Z. (1988). Wave velocities in hydrocarbons and hydrocarbon-saturated rocks with applications to EOR Monitoring. PhD dissertation, Stanford, CA: Stanford Universiry SRB Project v. 34.
15. Wang, Z., et al. (1987). Velocities in hydrocarbons and hydrocarbon-saturated rocks and sands: Expanded Abstracts of the 57th Annual International Society of Exploration Geophysicists Meeting and Exposition **57**: 1–4.

16. Wang, Z., and Nur, A. (1988). Effect of temperature on wave velocities in sands and sandstones with heavy hydrocarbons. *SPE Reservoir Engineering* **3**: 158–164.
17. Wang, Z., and Nur, A. (1990). Wave velocities in hydrocarbon-saturated rocks: Experimental results. *Geophysics* **55**: 723–733.
18. Gregory, A.R. (1976). Fluid saturation effects on dynamic elastic properties of sedimentary rocks. *Geophysics* **41**: 895–921.

PART 2

ELASTIC PROPERTIES OF GASES

CHAPTER 10

INTRODUCTION TO ELASTIC CONSTANTS OF GASES

Henry E. Bass
National Center for Physical Acoustics,
The University of Mississippi, University, Mississippi, USA

Contents

10.1.	Introduction	235
10.2.	Ideal Gas	236
10.3.	Real Gas Corrections	238
10.4.	Contribution of Transport Processes on Speed of Sound	239
10.5.	Changes in Elastic Properties Due to Relaxation Processes	241
10.6.	Measurements at Moderate Pressures	245
10.7.	Typical Results	247
10.8.	Diffusion	256
10.9.	Gases at Low Pressure	257
10.10.	Systems Not in Equilibrium	263
10.11.	Summary	263
References		264

10.1. INTRODUCTION

Although gases are compressible and exert a restoring force when compressed, it is not common to address the elastic properties of gases. More often the speed of sound or attenuation in a gas are discussed. This slightly different terminology arises from the fundamental physics. Gases are a collection of individual particles that are not bound together but are usually forced to be in the same volume as other particles through the existence of walls or other constraints. A description of the elastic properties of a gas, then, is an attempt to assign a group property to a collection of individual gases. Such a description must rely on a statistical treatment [1]. The most common starting point for such a treatment is the Boltzmann Equation. There are, however, regimes where a rigorous consideration of the Boltzmann Equation is not required, conditions under which the gas behaves as a continuum. Such regimes span most of the experimental space. The major exception where a particle description becomes important is at very low pressures where the sample volume does not contain a sufficient number of particles to provide a statistical average.

Handbook of Elastic Properties of Solids, Liquids, and Gases, edited by Levy, Bass, and Stern
Volume IV: Elastic Properties of Fluids: Liquids and Gases
Copyright © 2001 by Academic Press
ISBN 0-12-445764-9 / $35.00   All rights of reproduction in any form reserved.

Many measurements of elastic properties of gas are made using acoustics. In that case, the appropriate volume over which averages are made is that defined by the scale of the acoustic wavelength. When the acoustic wavelength becomes comparable to the mean free paths between collisions, the continuum description of gases is no longer appropriate. A measure of this condition is the frequency to pressure ratio. Frequency is inversely proportional to wavelength, and mean free path is inversely proportional to gas pressure, so when that ratio is very large, the wavelength approaches the mean free path. The exact value of f/P where a particle description is required depends on the mass of the particles but, in general, is in the neighborhood of 10^{11} Hz/atm. At higher f/P ratios, solutions to Boltzmann's Equation that maintain some of the particle nature of the gas are available [2]. The characteristics of gases at this extreme condition are appropriate for astronomical applications, interstellar gas clouds, for example.

In those f/P regimes most often encountered, gases can be treated as a continuum. This continuum is isotropic so the stress tensor is simple compared to many solids. A standard form for the stress tensor is [3]

$$\sigma_{ij} = (-P + \eta' \nabla \bullet u)\delta_{ij} + 2\eta \varepsilon_{ij} \tag{10.1}$$

where P is the hydrostatic pressure, η is the coefficient of viscosity, η' is the so-called second coefficient of viscosity, u is the displacement vector, δ_{ij} is zero if $i \neq j$ and 1 if $i = j$, and ε is the strain. Note that the shear terms exist only when there is a change in strain. This is different from a solid where a static shear stress can exist. Although this stress tensor looks simple, the individual terms can become quite complex. The following sections deal with the individual terms; the final section revisits the very high f/P regime.

Much of what follows comes from an excellent book by Robert Beyer and Stephen Letcher [4]. Professor Beyer has been kind enough to allow this author to make extensive use of his work.

10.2. IDEAL GAS

An ideal gas has no internal losses, so η' and η are both zero. Since the gas is isotropic, we can further simplify the mathematics by assuming motion in only the x direction. Recognizing that we will be interested in determining the elastic constants from the speed of sound, we will write and solve Newton's second law. From the stress gradient,

$$F_i = \partial \sigma_{ij}/\partial x_j \tag{10.2}$$

This force causes an acceleration

$$\rho_o \partial^2 u_i/\partial t^2 = \partial \sigma_{ij}/\partial x_j \tag{10.3}$$

where ρ_o is the mass density. For the simple case of no shear,

$$\rho_o \partial^2 u_i/\partial t^2 = -\partial p/\partial x \tag{10.4}$$

In the same notation, the equation of continuity (or conservation of mass) can be written as [4]

$$-\partial(\rho u)/\partial x = \partial \rho/\partial t \tag{10.5}$$

The first law of thermodynamics can be written in the form

$$\Delta Q = dU - \Delta W \tag{10.6}$$

where ΔQ is the heat added (per mole) to the system in an infinitesimal process, dU is the corresponding increase in the internal energy of the system, and ΔW is the work done on the system during the same process. Assume that no heat enters or leaves the system during the process so that $\Delta Q = 0$. Replace ΔW with

$$\Delta W = -PdV = M(P/\rho_o^2)d\rho \tag{10.7}$$

where M is the gram molecular weight of the gas, $dU = C_v dT$, where C_v is the heat capacity per mole at constant volume.

In the general case, the equation of state is an expression of the form

$$P = P(\rho, T) \tag{10.8}$$

For an ideal gas, $P = \rho_o RT/M$, where R is the gas constant (per mole). In many cases, however, the more general form of Eq. 10.8 is required. The most notable cases are gases at high pressure where the gas behaves less ideally and those cases where even small deviations from ideal gas behavior are important. Both of these exceptions are treated in this chapter.

With the simplifications shown previously, the velocity of propagation of acoustic plane waves becomes

$$c_0^2 = \omega^2/k^2 = (\partial P/\partial \rho)_T + (MP_0/\rho_0^2 C_v)(\partial P/\partial T)_\rho \tag{10.9}$$

where a distinction is now made between P, the local hydrostatic pressure, and P_o, the ambient pressure in the absence of the acoustic disturbance. The same distinction is made for the densities ρ and ρ_o. By making a number of thermodynamic transformations, it is possible to show that to terms in first order, Eq. 10.9 is indeed equivalent to the general expression for the square of the sound velocity in a nondissipative medium, namely, $c_0^2 = (\partial p/\partial \rho)_s$. In the particular case of an ideal gas,

$$(\partial P/\partial \rho)_T = P_0/\rho_0 \text{ and } (\partial P/\partial T)_\rho = P_0/T = \rho_0 R/M \tag{10.10}$$

so that

$$c_0^2 = (P_0/\rho_0)(1 + R/C_v) = \gamma(P_0/\rho_0) = \gamma RT/M \tag{10.11}$$

In terms of the isothermal bulk modulus, B_T, the sound velocity is

$$c_o^2 = \gamma B_T/\rho_o \tag{10.12}$$

The isothermal bulk modulus is the quantity that would be measured in a static experiment and serves as a basis for comparing static and dynamic results.

At this point it is useful to examine some of the approximations that have been made to arrive at Eqs. 10.11 and 10.12. The first assumption made was that the acoustic wavelength is much greater than the mean free path. Relaxation of that assumption will be made in the section on rarified gases. The second assumption was that the gas is nonviscous, actually, that there are no time-dependent components of the elastic constants. Several physical processes violate this assumption. Polyatomic gases do have a measurable bulk viscosity. In addition, energy is lost due to thermal conduction and diffusion. For polyatomic molecules, relaxation processes associated with relaxation of rotational and vibrational degrees of freedom can introduce a significant source of absorption. This additional absorption may be introduced through the bulk viscosity η', but that approach is not taken here. Since absorption and velocity dispersion are related through a Kramers-Kronig relation [5], the velocity is also affected by these processes. The frequency-dependent specific heat gives rise to a frequency dependence in the thermal conduction. And, since thermal conduction and viscosity are related, including transport into the equations of motion could become complicated. To simplify the mathematics, each of the contributions to the time-dependent elastic properties of gases is treated separately in the following sections. The variation of the transport properties with frequency is a second-order effect that can be assumed to be small if the absorption due to those terms is small compared with the speed of sound.

In the derivation of Eqs. 10.11 and 10.12, it was assumed that the gas obeys the ideal gas law. This is true for monatomic gases at low pressures, but real gases exhibit deviations from ideal behavior that affect the measured speed of sound. Nonideal gas contributions to the speed of sound will be treated in the next section. The additional approximation that has been made is that the gas behaves linearly. Extreme pressures will be treated in separate chapters. But the result of dynamically changing the stress by large amounts leads to interested nonlinear elastic behavior not unlike that observed for solids. Some comments will be offered in a later section.

10.3. REAL GAS CORRECTIONS

Real gas corrections to the speed of sound can be deduced by introducing a modified equation of state. The virial equation of state in the form

$$PV = RT(1 + B(T)/V + C(T)/V^2 + \cdots) \qquad (10.13)$$

can be used as a correction to ideal gas law in most cases. The disadvantage of Eq. 10.13 is that connections to the underlying physics are obscure. The coefficients can be computed from a knowledge of the interaction potential between molecules [6]. Much of the physics can be recovered by writing the equation of state in the form

$$(P + a/V^2)(V - b) = RT \qquad (10.14)$$

In Eq. 10.14, b represents the volume occupied by individual molecules, volume that is not available to other molecules. The term "a" accounts for attractive forces between molecules.

Bhatia [7] expresses the correction to the equation of state in the form of Eq. 10.13 with $C(T) = 0$ and then writes for the specific heat at constant volume,

$$C_V = C_{Vi} - (R/V)(2B'T + B''T^2) \qquad (10.15)$$

where $B' = dB/dT$, $B'' = dB'/dT$, and C_{Vi} is the constant volume specific heat in the ideal gas limit. Neglecting powers of B' and B'' higher than first order, Bhatia finds that for this nonideal gas, the sound velocity can be written as

$$c = c_i(1 - gP) \qquad (10.16)$$

where

$$-g = B/RT + (\gamma_i - 1)B'/R + (\gamma_i - 1)^2 TB''/(2R\gamma_i) \qquad (10.17)$$

and $\gamma_i = (1 + R/C_{Vi})$.

Near standard temperature and pressure, this leads to a small correction in the speed of sound. Bhatia [7] gives values of g for CH_4 and CCl_4 as 0.000424 and 0.04422, respectively. The measured values of c at atmospheric pressure are [8] about 420 m/s (25 °C) and 145 m/s (97 °C). This suggests a nonideal gas correction of 1 part in a million and 300 parts per million, respectively. The best measurements could be expected to yield an accuracy of 10 parts in a million, so for CH_4 an ideal gas correction would be required only at very high pressures. For CCl_4 such a correction should be made even at atmospheric pressures. For gases where a large correction is required, measurements of the speed of sound can be used to experimentally determine B.

10.4. CONTRIBUTION OF TRANSPORT PROCESSES ON SPEED OF SOUND

Up to this point, the contributions of viscous drag, thermal conduction, and, in the case of gas mixtures, diffusion have been ignored. It is most simple to visualize the contributions of these effects to absorption. Since that absorption and velocity dispersion are related through a Kramers-Kronig relation, we could solve for absorption then transform to get velocity. Relating velocity dispersion to elastic moduli is then straightforward. In practice it is easier to solve for a complex propagation constant and get the velocity directly.

Using the general form of the stress tensor, Eq. 10.1, the equation of motion, Eq. 10.4, becomes the Stokes-Navier equation. For the one-dimensional case,

$$\partial/\partial t(\rho \partial u/\partial t) = -\partial P/\partial x + (\eta' + 4\eta/3)\partial^2/\partial x^2(\partial u/\partial t) \tag{10.18}$$

where η is the coefficient of shear viscosity. The quantity η' is known as the bulk viscosity and corresponds to the viscous drag that would be experienced in a pure volume dilatation, in which no shearing motions can occur.

The nature and value of η' forms one of the most interesting problems in the historical development of ultrasonic wave propagation. It was assumed by Stokes that η' was identically zero, and, to a large extent, this assumption marks the difference between classical and modern theories of ultrasonic absorption and dispersion. The bulk viscosity can be used to account for dispersion and absorption due to relaxation of internal molecular modes. This chapter calculates the effects directly. We set $\eta' = 0$.

A question often raised at this point is why the shear viscosity should enter into the description of the motion of a plane wave of large extent, since no shearing motions are immediately obvious. The answer to this question lies in the fact that one cannot restrict all the motion to one direction without doing it as a combination of deformations in all the coordinate directions. The analysis has been given by Kittel [9] for the elastic case.

Consider a unit cube of material [Fig. 10.1(a)] that is to be stretched in the x direction to a strain ε_{11}, with no net change in either the y or z directions. This will be done in three steps:

1. Uniform strain by an amount $\varepsilon_{11}/3$ [Fig. 10.1(b)].
2. Extension in the x direction to a total strain $2\varepsilon_{11}/3$, with simultaneous compression to zero in the y direction, preserving a constant volume in the process [Fig. 10.1(c)].

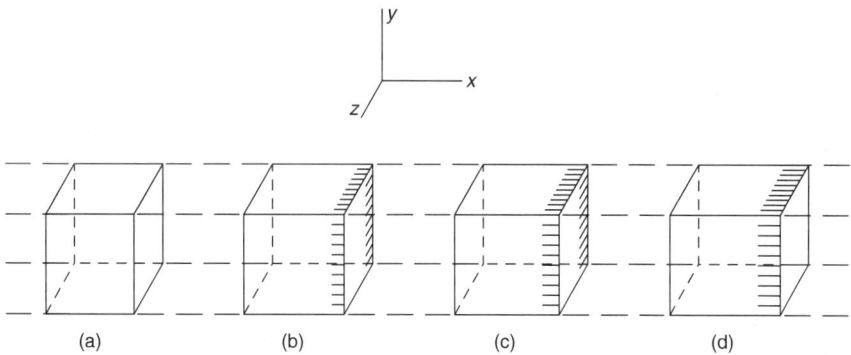

Fig. 10.1. Steps showing that a longitudinal dilation involves both compression and shear moduli. See [4] for description.

3. Extension in the x direction to a total ε_{11} and simultaneous compression in the z direction to zero, again preserving a constant volume [Fig. 10.1(d)].

In the first step, the stresses are uniform and given by

$$\sigma_{11} = B\frac{\Delta V}{V} = \frac{B}{3}(\varepsilon_{11} + \varepsilon_{22} + \varepsilon_{33}) = B\varepsilon_{11} \tag{10.19}$$

where B is the bulk modulus, since $\varepsilon_{11} + \varepsilon_{22} + \varepsilon_{33}$ in the present case. In the second step, we have a process that can be described as a shear of angle $2\varepsilon_{11}/3$, so that the stress in the x direction will be $\mu \cdot 2\varepsilon_{11}/3$, where μ is the shear modulus. The situation in the third step is identical to that in the second except that y is replaced by z. Hence, the total stress in the x direction will be equal to $(B + \frac{4}{3}\mu)$. To proceed to a discussion of the effect of the shear viscosity on the propagation of sound, we replace Eq. 10.4 by Eq. 10.18 with $\eta' = 0$, and make the linear approximation, thus obtaining

$$\partial v/\partial t = -\rho_0^{-1}\partial P/\partial x + \tfrac{4}{3}(\eta/\rho_0)\partial^2 v/\partial x^2 \tag{10.20}$$

or, for harmonic waves of frequency ω,

$$(\omega - \tfrac{4}{3}i(\eta/\rho_0)k^2)v - (k/\rho_0)P = 0 \tag{10.21}$$

where v is $\partial u/\partial t$. This leads to the result

$$k^2/\omega^2 = [c_0^2 + i\tfrac{4}{3}(\eta/\rho_0)\omega]^{-1} \tag{10.22}$$

The presence of the imaginary term on the right side of Eq. 10.22 makes k complex. Setting $k = k_r - i\alpha$, α is the amplitude absorption coefficient while k_r is the real wave number, equal to ω divided by the phase velocity c.

The substitution of $k_r - i\alpha$ in Eq. 10.22 leads to the equations

$$k_r^2 - \alpha^2 = \frac{\omega^2 c_0^2}{c_0^4 + (16/9)(\eta^2/\rho_0^2)\omega^2} \tag{10.23}$$

$$2k_r\alpha = \frac{(4/3)(\eta/\rho_0)\omega^3}{c_0^4 + (16/9)(\eta^2/\rho_0^2)\omega^2} \tag{10.24}$$

In virtually all cases, $k_r^2 \gg \alpha^2$ (and hence $(16/9)(\eta^2\omega^2/\rho_0^2) \ll c_0^4$) so that, to an excellent approximation

$$k_r = \omega/c_0$$
$$k_r\alpha = \tfrac{2}{3}(\eta\omega^3/\rho_0 c_0^4) \text{ or } \alpha = \tfrac{2}{3}(\eta\omega^2/\rho_0 c_0^3) \tag{10.25}$$

In oxygen, at $20°$ and at atmospheric pressure, α is $1.68 \times 10^{-13}\nu^2$ in cgs units so $\alpha/k_r = 0.88 \times 10^{-8}\nu$ and the approximation $k_r^2 \gg \alpha^2$ is valid to frequencies on the order of 10^8 Hz at 1 atm of pressure.

Viscosity results from the transport of momentum from one element of the gas to another. This collisional transport of momentum is always accompanied by a transport of energy. Phenomenologically, this energy transport can be described by heat flow. To include heat conduction into the wave propagation calculations, a term can be added to the energy conservation equation (Eq. 10.6), which represents the heat removed per second per mole $(\partial Q'/\partial t)$. From the Fourier's law for heat conduction,

$$\partial Q/\partial t = (M\kappa/\rho_0)\partial^2\theta/\partial x^2 \tag{10.26}$$

where κ is thermal conductivity and θ is the temperature. Repeating the derivation in the usual way [4],

$$\frac{k^2}{\omega^2} = \frac{1}{c_0^2}\frac{1-i\zeta}{1-i(\zeta/c_0^2)(\partial p/\partial\rho)_T} \tag{10.27}$$

where

$$\zeta = \kappa k^2 / \rho_0 \omega c_v \tag{10.28}$$

As written, Eq. 10.27 is not an explicit solution for k^2, since k is also contained in ζ. The quantity ζ is small, however, and it is customary to replace k in ζ by ω/c_0 so that $\zeta = k\omega/\rho_0 c_0^2 c_v$. Setting $k = k_r - i\alpha$, for the case of thermal conduction (after some algebra),

$$k_r^2 = \omega^2/c_0^2$$

$$2k_r\alpha = \frac{\omega^2}{c_0^2}\left[1 - \left(\frac{\partial p}{\partial \rho}\right)_T \frac{1}{c_0^2}\right]$$

$$\alpha = \frac{\omega^2 \kappa}{2c_v \rho_0 c_0^3}\left[1 - \frac{1}{c_0^2}\left(\frac{\partial p}{\partial \rho}\right)_T\right] \tag{10.29}$$

For an ideal gas (in the sense that $B = 0$), $(1/c_0^2)(\partial P/\partial \rho)_T = 1/\gamma$, and

$$\alpha = \frac{\omega^2 \kappa}{2c_p \rho_0 c_0^3}(\gamma - 1) \tag{10.30}$$

where $c_p = \gamma c_v$ is the heat capacity per unit mass (at constant pressure).

10.5. CHANGES IN ELASTIC PROPERTIES DUE TO RELAXATION PROCESSES

Studies of acoustic absorption and dispersion have proven to be an effective tool in developing a physical description of molecular interactions in gases. Experimentally, absorption or velocity dispersion is measured as a function of frequency/pressure. The measured quantities are computed using the theory outlined below that requires some insight into the particular physical processes that give rise to the absorption or dispersion (e.g., vibrational relaxation). Computed values are compared to experiment, and the assumed physical process or interaction parameters are varied until agreement is achieved. The theoretical model that gives rise to predicted absorption and dispersion, which agree with experiment, is assumed to correctly represent the interactions taking place at the microscopic level.

Some of the microscopic processes studied acoustically can also be studied using optical techniques. In those cases, the physical model and numerical values for the parameters describing the interactions are found to be consistent provided a proper theoretical link between microscopic model and measured quantities is used.

For the purpose of developing an understanding of relaxation processes, first consider a gas made up of diatomic molecules. The individual molecules are free to move transitionally in three directions, rotate about two perpendicular axes (actually three but the third has zero moment of inertia so has no energy), and vibrate along the bond joining the atoms. Some energy is associated with each of these allowed motions.

Translational motion can typically be considered nonquantized. Any energy is allowed. As the molecules translate, they collide, exchanging energy with their collision partners. At atmospheric pressure, assuming a hard sphere molecular model, a single molecule suffers about 10^{11} collisions/second. A single collision is typically sufficient to transfer translational energy from one molecule to another. However, a certain period of time is required to randomize energy associated with excess velocity in a particular direction. This time is often referred to as the translational relaxation time. Kohler following Maxwell [11] associates viscosity with this relaxation time, writing $\tau_{tr} = \eta/P = 1.25\tau_c$, where P is gas pressure and τ_c is the time between collisions. As pressure is lowered, the rate of collisions decreases proportionately. At 1 torr

(1/760 atm), the translational relaxation time is about 10^{-8} sec; at 1 millitorr, 10^{-6} sec. These time regimes can be effectively studied using ultrasonics.

Unlike translational motion, rotation and vibration are noticeably quantized. During a collision, a change in rotational or vibrational state can only occur if the change in energy of another state is sufficient to allow at least a one-quantum jump. For rotational energy transfer, the spacing between energy levels is given by $2(J+1)B$, where J is the rotational quantum number, $B = \hbar^2/2I$ and I is the effective moment of inertia. If one assumes J is the most probable value (from a Boltzmann distribution), the value of $2(J+1)B$ for a typical molecule (say N_2) in units of kT is about 1 K. This means that in a gas above 1 K, essentially all collisions will have sufficient translational energy to cause multiple changes in J. As a result, rotation rapidly equilibrates with translation.

An exception is hydrogen, which has much larger rotational energy level spacing due to the small moment of inertia. On average, as many as 350 collisions [12] may be necessary to transfer a quantum of rotational energy in H_2. At a pressure of 1 atm, this gives a relaxation time of about 2×10^{-8} sec. It should be noted that a given collision does or does not transfer a quantum of energy. The 350-collision average means that only 1 collision in 350 has the proper geometry and energy to cause a transfer of one quantum of rotational energy. The number of collisions necessary, on average, to transfer one quantum of energy is referred to as the collision number Z. When rotational energy is involved, the subscript "rot" is typically added (Z_{rot}). The inverse of this dimensionless quantity is the probability of transferring a quantum in a collision (P_{rot}). Since rotational energy level spacings are unequal, a $1 \to 2$ transition should be more probable than a $2 \to 3$ transition. These events are distinguished by using the symbols P_{rot}^{1-2} or P_{rot}^{2-3}.

Generally speaking, the probability for transferring a quantum of energy in a collision decreases rapidly with the size of the quantum transferred. Since vibrational levels are much more widely spaced than rotational energy levels, vibrational relaxation times are much longer than rotational. Vibrational levels in a single vibrational mode are approximately equally spaced. This means that energy can be exchanged between levels (i.e., the vibrational quantum number goes up in one molecule and down in the other) with very little energy exchanged between vibration and translation. The result is that such v-v exchanges take place very rapidly. In this case, the vibrational relaxation time is controlled by the time it takes energy to transfer between translation and the lowest lying vibrational level. Since this energy level varies greatly for different molecules, so do the probabilities of vibrational energy transfer during a collision. During $N_2 - N_2$ collisions, Z_{10} is near 1.5×10^{11}, so the relaxation time is near 1.5 sec [13] (Z_{10} is the number of collisions needed to transfer energy from the lowest vibrational level to translation). Large molecules have vibrational energy levels that are very close together. A molecule such as C_2H_6 requires only 100 collisions to transfer a quantum of vibrational energy from the first vibrational level into translation [12]. This very wide range of relaxation times presents interesting experimental challenges.

To this point, there has been no attempt to rigorously define relaxation times in terms of energy transfer probabilities. In fact, such a relation is possible in a simple form for only the few cases where gases exhibit a single relaxation time. Nevertheless, the simple case provides valuable insight into the behavior of more complex systems and deserves a detailed description.

Consider the case where the population of a vibrational state is excited to an energy E_v, which is greater than the energy $E_v(T_{tr})$, which it would have in Boltzmann equilibrium with translation. In this case, the excess vibrational energy will equilibrate with translation according to a standard relaxation equation,

$$-\frac{dE_v}{dt} = \frac{1}{\tau}[E_v - E_v(T_{tr})] \qquad (10.31)$$

The return to equilibrium occurs due to energy transfer during individual collisions.

The rate of energy transfer k_{10} is defined as the rate at which molecules go from the first excited state to the ground state due to collisions at a pressure of 1 atm. This rate is just the collision frequency, M, times the probability of energy transfer, $P^{1 \to 0}$, times the mole fraction of molecules in the first excited state, x_1. During some collisions, the reverse process will occur, that is, some molecules in the ground state will become excited at a rate k_{01}. In equilibrium, equal numbers of molecules go in both directions, so

$$k_{10}x_1 - k_{01}x_0 = 0 \tag{10.32}$$

As explained previously, energy is quickly transferred from the first excited level of the vibrational mode to higher levels of the mode by v-v exchanges. Assuming quantum mechanical laws hold for probabilities of energy exchanges between vibrational levels of a harmonic oscillator, Landau and Teller [15] showed

$$-\frac{dE_v}{dt} = k_{10}(1 - e^{-h\nu/kT})[E_v - E_v(T_{tr})] \tag{10.33}$$

where ν is the vibrational frequency of the relaxing mode. By comparison with Eq. 10.31

$$\tau = \frac{1}{k_{10}(1 - e^{-h\nu/kT})} \tag{10.34}$$

The relationship between the relaxation time and ultrasonic absorption and dispersion may be understood by noting that the relaxation process makes the specific heat of the gas time (or frequency) dependent. This time dependence can be obtained from the energy relaxation equation. Consider, as previously, that the specific heat of a simple gas can be divided into translational, rotational, and vibrational contributions. For now, assume that the translational and rotational energies both equilibrate rapidly enough to follow any acoustically induced temperature variations. In this case, the effective specific heat can be written as

$$(C_v)_{\text{eff}} = C_v^\infty + C' \frac{dT'}{dT_{tr}} \tag{10.35}$$

where C_v^∞ is the sum of rotational and translational specific heats, C' is the relaxing specific heat, and T' is the instantaneous temperature of the relaxing mode (in this case, vibration). From the energy relaxation equation (Eq. 10.33), for small periodic variations in T_{tr} and T_v about their equilibrium values,

$$(C_v)_{\text{eff}} = C_v^\infty + \frac{C'}{1 + i\omega\tau} \tag{10.36}$$

where ω is the angular frequency of the acoustic wave.

The acoustic propagation constant can be written in the form

$$\frac{k^2}{\omega^2} = \left(\frac{1}{c} - \frac{i\alpha}{\omega}\right)^2 = \frac{\rho_0 \kappa_T}{\gamma_{\text{eff}}} \tag{10.37}$$

where c is the acoustic velocity, α is the attenuation, ρ_0 is the equilibrium density, κ_T is the compressibility, and

$$\gamma_{\text{eff}} = [(C_v)_{\text{eff}} + R]/(C_v)_{\text{eff}} \tag{10.38}$$

with R the gas constant. For this simple single relaxation, assuming $\alpha/\omega \gg 1/c$ [16],

$$\alpha\lambda = \pi \left(\frac{c}{c_0}\right)^2 \varepsilon \frac{\omega\tau_s}{1 + (\omega\tau_s)^2} \tag{10.39}$$

and
$$\left(\frac{c_0}{c}\right)^2 = 1 - \frac{\varepsilon\omega^2\tau_s^2}{1+(\omega^2\tau_s^2)} \qquad (10.40)$$
where
$$\varepsilon = \left(\frac{c_\infty^2 - c_0^2}{c_\infty^2}\right) \qquad (10.41)$$

and where λ is the wavelength, c_0 is the speed of sound for $\omega\tau_s \ll 1$, and c_∞ is the speed of sound at frequencies much greater than the relaxation frequency. The adiabatic relaxation time τ_s, is related to the isothermal relaxation time τ used earlier by

$$\tau_s = (C_v + R)/(C_v^\infty + R)\tau \qquad (10.42)$$

The relaxation frequency, f_r, defined as the frequency at which the maximum absorption per wavelength occurs, is related to τ_s by

$$f_r = 1/2\pi\tau_r = c_\infty/c_0(1/2\pi\tau_s) \qquad (10.43)$$

In the case of polyatomic gases or mixtures of relaxing diatomic gases, the different relaxing modes can be coupled together by $v-v$ exchanges. Such complex or multiple relaxation processes exhibit the general behavior given by Eqs. 10.39 and 10.40, but the magnitude of the absorption and dispersion and the relaxation frequencies can take on new meaning. In the case of multiple relaxing internal energy modes, Eqs. 10.39 and 10.40 take on the form [18]

$$\left(\frac{c}{c^\infty}\right)^2 = 1 + \sum_j \frac{\delta_j k_s/k_s^\infty}{1+(\omega\tau_{s,j})^2} \qquad (10.44)$$

$$\alpha\lambda\left(\frac{c}{c^\infty}\right)^2 = -\pi\sum_j \frac{\delta_j k_s/k_s^\infty}{1+(\omega\tau_{s,j})^2}\omega\tau_{s,j} \qquad (10.45)$$

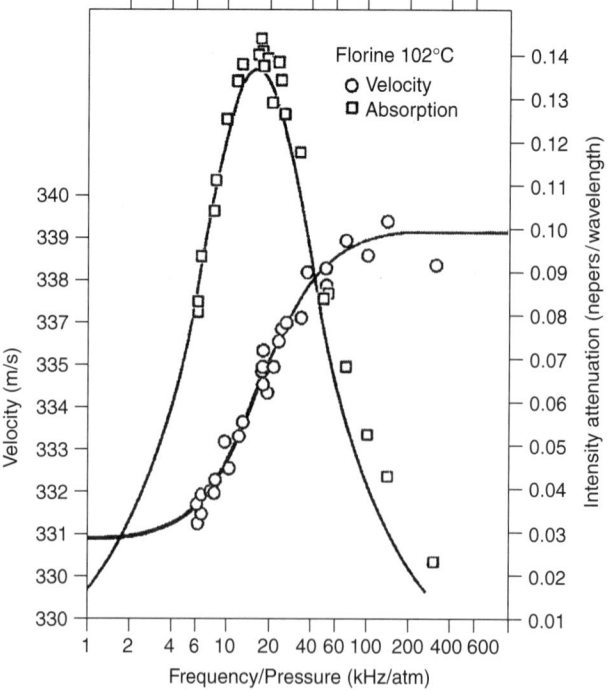

Fig. 10.2. Typical curves for absorption per wavelength and velocity dispersion to a single relaxation process. The example here is Fl_2 at $102\,°C$. The figure compares curves representing the previous theory with measured values [17].

where $\delta_j k_s / k_s^\infty$ is a relaxing adiabatic compressibility (negative) and j indicates that there might be more than one relaxation process. In these complex cases, $\tau_{s,j}$ can no longer be related to a single energy transfer reaction and $\delta_j k_s / k_s^\infty$ can no longer be related to relaxing energy of a specific mode. Instead, the various modes and reaction pathways are coupled. The sums in Eqs. 10.44 and 10.45 are over eigenvalues of the energy transfer matrix, which simultaneously accounts for all reactions. Eqs. 10.44 and 10.45 is used to calculate the sound absorption in moist air as a function of frequency and temperature in a later section. The American National Standard for sound absorption in the atmosphere is now based on these equations.

Not only can Eqs. 10.44 and 10.45 be used to calculate sound absorption and velocity, but the reverse process is also possible, i.e., the transition rates can be extracted from measured values of absorption and velocity. However, the number of possible relaxation paths multiplies rapidly with an increase in the number of relaxing modes, and the identification of specific rates becomes a tedious process and usually involves some assumptions. It has been done for only a few special cases.

A further complication arises from chemical reactions. For a reversible chemical reaction with heat of reaction ΔH, ΔH enters into the relaxation equations in a manner similar to ΔE for vibrational relaxation. A major difference is that chemical reactions allow the possibility that the number density of molecules can change. Such changes bring about additional relaxation absorption and dispersion.

10.6. MEASUREMENTS AT MODERATE PRESSURES

Before modern techniques were available for measuring transients, most ultrasonic absorption and velocity measurements were made with ultrasonic interferometers [19–22]. This instrument continues to be used with refined precision and modern methods of control and measurement [23]. Both velocity and absorption can be measured with the interferometer. A column of gas or liquid of varying length forms the load on a quartz crystal vibrating at its resonant frequency. The loading effect of the gas or liquid column increases whenever the length of the gas column is a whole multiple of a half wavelength of the sound. This loading effect is reflected in the driving circuit of the crystal. The separation between the peak values as the two path lengths changes determines the sound wavelength, and the variation of the magnitude of the peaks with length of the column allows the determination of the absorption coefficient.

Another continuous-wave method for measuring velocity and absorption in gases uses a source and receiver whose separation can be varied mounted in a tube so as to avoid standing waves [24, 25]: see Fig. 10.3. This method has been used to make measurements at audible frequencies and reduced pressures and, therefore, is capable of measurements over the f/p range where many of the interesting relaxation processes occur in gases and gas mixtures. The absorption is determined from the decrease in sound amplitude at the receiver as the source-receiver separation is increased. The wavelength of the sound is equal to the distance between points of equal phase in the received sound. This change in phase is easily observed by comparing the receiver and source signals.

With the development of the capability for generating and measuring tone bursts, it has been possible to use the pulse-echo technique [26, 27]. In this case, the sound velocity is determined from time of flight and the absorption from the variation of the tone burst amplitude with source and receiver separation.

At lower frequency-to-pressure ratios, resonant tubes are used [13]. The velocity is determined from the resonant frequency and the absorption from the width of the resonant peak.

When relaxation effects are studied, absorption measurements are designed to provide absorption as a function of frequency divided by pressure over a range of

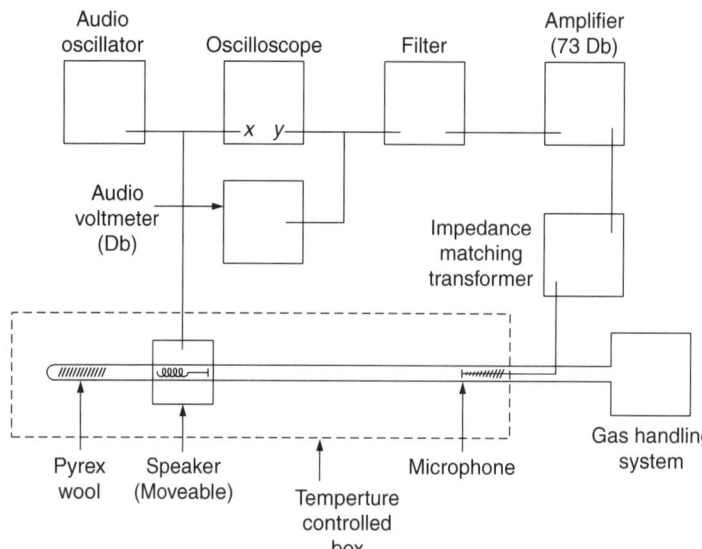

Fig. 10.3. Schematic diagram of the apparatus for measuring the attenuation of sound [24, 25].

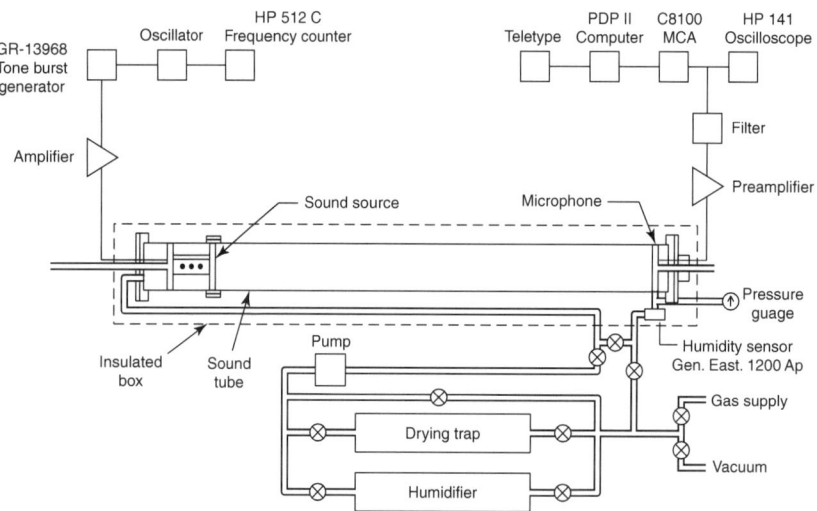

Fig. 10.4. Diagram of experimental system for measuring sound absorption, using pulse-echo.

values, where the period of the sound wave is approximately equal to the relaxation time. When studying relaxation in molecular N_2, this requires measurements of very small absorption (less than 1 dB/m) at f/p values as low as a fraction of a hertz per atmosphere. On the other hand, measurements of absorption in UF_6 involve very large attenuation (∼100 dB/m) at f/p values as high as 10 MHz/atm [28].

To separate the effects due to vibrational relaxation, the measured absorption must be corrected for (1) viscous and thermal losses in the body of the gas, (2) viscous and thermal losses to the measuring chamber walls, (3) radiation or leakage losses from the chamber, (4) spreading losses, and (5) losses due to rotational relaxation.

These different effects are generally small enough to be additive. The quantity usually reported is the absorption coefficient, α, in terms of which the amplitude of the plane wave is written as

$$A = A_0 e^{-\alpha(x_2 - x_1)} \tag{10.46}$$

Plotting log A vs x gives α as the slope in nepers/m. The measured α is due to a combination of the losses listed previously. Careful design of the experimental apparatus is necessary if accurate corrections are to be made for wall and/or spreading losses.

A common geometry for sound absorption measurements is a cylindrical tube. This method generally avoids losses c and d (due to spreading and leakage) above and involves well-known corrections for a and b. In this case, the total absorption is given by

$$\alpha_{\text{tot}} = \alpha_{cl} + \alpha_{\text{rot}} + \alpha_{tu} + \alpha_v \qquad (10.47)$$

where α_{cl} (classical absorption) is the absorption due to viscosity and thermal conductivity, α_{rot} is absorption due to relaxation processes not of interest to the study (rotational relaxation, for example). α_{tu} is the tube absorption, and α_v is the absorption due to the relaxation process of interest. The term α_{cl} can be computed from known or measured values of viscosity and thermal conductivity [29]; it is proportional to f^2. The term α_{rot} generally is estimated based on other studies. At frequencies well below the rotational relaxation frequency it also is proportional to f^2. The tube absorption, α_{tu}, depends upon the viscosity, thermal conductivity, and tube radius (r).

Only plane waves will be present in the tube if the sound frequency is below the cutoff frequency for the first nonplane mode, i.e., $f < 0.586$ c/d, where c is the sound speed and d is the tube diameter. Plane waves can be maintained for higher frequencies if the transducer generating the waves fills the tube and is carefully maintained perpendicular to the tube axis and if its surface vibrates as a piston [31]. Interferometers used for high-frequency measurements in liquids and gases will generally satisfy these conditions. As the wavelength becomes small compared with the tube diameter, wall losses become negligible.

In the previous discussion, relaxation absorption and dispersion of sound resulted when a finite time was required for the passage of the energy of translation into the vibrational energy of the gas molecules. Relaxation mechanisms can also be studied in special cases using the reverse processes. Laser light of a particular frequency is used to excite a vibrational mode. This excess vibrational energy then relaxes into translational energy generating an increase in translational temperature and pressure. If the laser light is chopped, a sound wave is generated. This phenomena is referred to as the "optoacoustic effect" and the device as a "spectrophone" [32]. The relaxation time can be determined from the phase relations between modulated laser light and the resulting sound pressure.

The primary attraction of optoacoustic measurements is the ability to excite a specific internal mode and observe how energy from the mode makes its way to translation. Experimentally, only a limited number of internal modes can be excited with available lasers, which limits the list of systems that can be studied.

10.7. TYPICAL RESULTS

For the purposes of this chapter, specific studies will be selected that illustrate the physics involved. Three cases will be treated. The first is the halogen family of diatomic molecules, specifically, F_2, Cl_2, Br_2, and I_2. This example illustrates the functional forms presented earlier for absorption due to relaxation processes, gives typical values for probability of energy transfer, $P^{1 \to 0}$, and provides some indication of how $P^{1 \to 0}$ varies with temperature, molecular weight, and vibrational frequency. The second example is a more complex molecule, SO_2, in mixtures with Ar and O_2. This example illustrates the complexity of the relaxation process when different vibrational modes exchange energy.

Shields used an acoustic traveling wave tube similar to that developed by Angora [24] to measure acoustic absorption in F_2, Cl_2, Br_2, and I_2 as a function of frequency, pressure, and temperature [17, 33]. Results for Cl_2 at five different temperatures are plotted in Figure 10.5. The absorption and velocity in F_2 at $102\,°C$ are shown in Figure 10.2. The curves drawn through the experimental points were calculated using the theory discussed previously.

Several features of these curves are interesting. First, when more than one point is plotted for the same value of f/p, it means that absorption was measured at two different pressures but at different frequencies so that the ratio was the same. The agreement between such measured values confirms that the classical and tube corrections were being made correctly and that the relaxation absorption varied as f/p. Second, note that the relaxation time decreases as temperature increases (the relaxation frequency increases). This means that although the density is lower for a given pressure, the probability of energy transfer increases more rapidly than the collision frequency decreases. Finally, note that the maximum absorption increases as the temperature increases. This is a result of increasing vibrational specific heat. The Plank-Einstein relation predicts that

$$C' = R(\theta/T)^2[e^{\theta/T}/(e^{\theta/T} - 1)^2] \tag{10.48}$$

where C' is the vibrational specific heat, θ is the vibrational temperature, and T is the gas temperature.

Results of Shields' study are summarized in Figure 10.6, which shows the probability for deexciting the first excited vibrational level in a collision.

The log of this probability is plotted vs $T^{-1/3}$. This type of plot was suggested by an early theory by Landau and Teller [15]. They predicted an approximately linear relationship between $\log P^{1\to 0}$ vs $T^{-1/3}$. From similar measurements on a great many other gases, one can expect the following trends for v-t transitions:

(1) The log of the transition probability decreases roughly linearly with $T^{-1/3}$ for a particular molecular collision pair. The reasons for this dependence are complex but can be thought of in terms of the speed molecules are traveling when they collide. The more rapidly they are moving, the more energy they carry into the collision. As the temperature increases (move to the left in

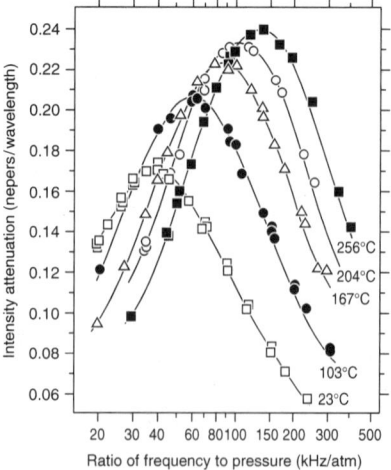

Fig. 10.5. Relaxation absorption coefficient per wavelength vs log (frequency/pressure) for chlorine. The solid curves are theoretical absorption with values of relaxation amplitude and frequency adjusted to give the best fit of the experimental points.

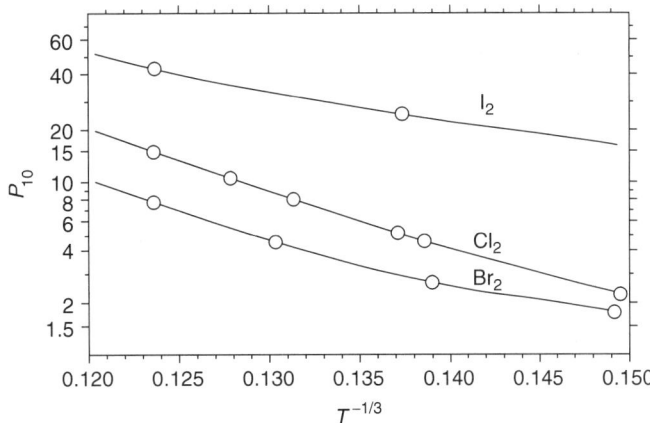

Fig. 10.6. Log of collision efficiency vs absolute temperature to the minus one-third power. The values of the ordinate should be multiplied by 10^{-5} for Cl_2 and by 10^{-4} for Br_2 and I_2.

Figure 10.6), the average speed of each molecule increases; hence, there is a greater probability of energy transfer.

(2) The transition probability is very sensitive to the amount of energy that must be transferred between vibration and translation in the transition, increasing rapidly as this energy decreases. This can be thought of the same as the temperature dependence. When two molecules collide with translational energy less than a quantum of vibration, the probability of a transition is classically zero.

(3) The transition probability is very sensitive to the time involved in the molecular collisions increasing rapidly as this time decreases. This means that the transition probability increases with increasing temperature [see trend (1)] and with decreasing molecular mass. This effect is due to the requirement that momentum as well as energy be conserved in the collision.

(4) For polyatomic molecules, bending vibrations are more easily excited than stretching vibrations (as one might expect from the geometry of the collision process). There are larger collision cross sections for excitation of a bending mode.

(5) Water vapor and a few other molecules with low moments of inertia, and, therefore, high rotational velocities, are very efficient in shortening relaxation times when added to gases, even in small amounts. For such mixtures, the relaxation time is much less sensitive to temperature change and can even increase with temperature. This effect is attributed to the coupling between vibration and rotation in such collisions [34–38].

(6) The probability of the transition is sensitive to the nature of the intermolecular potential. Figure 10.7 shows some simple potentials. These can give reasonable results for atom–diatom collisions and for molecules that have electron clouds that are reasonably spherical. For these cases, the probability of a transition depends on the depth of the potential and the steepness of the repulsive part of the potential.

As an example of the use of sound absorption measurements to determine the relaxation scheme in a polyatomic gas mixture consider the case of SO_2/O_2 mixtures [39]. SO_2 has three vibrational modes. These can be classified as a bending mode, with a fundamental vibrational frequency of 518 cm^{-1}; a symmetrical stretching mode, and

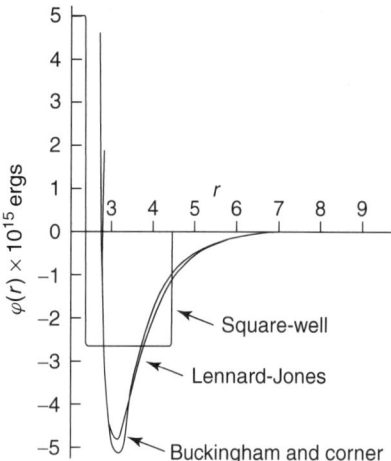

Fig. 10.7. Potential energy of interaction.

an asymmetrical stretching mode, with frequencies of 1151 and 1361 cm^{-1}, respectively. O_2 has a vibrational frequency of 1580 cm^{-1}. SO_2 was one of the first gases in which the sound absorption vs frequency curve evidenced more than one relaxation time. Figure 10.8 shows measurements in three different SO_2/O_2 mixtures at 500 °K. Twelve different transition rates were adjusted to make the theoretical curves simultaneously fit these data and the data for pure SO_2, SO_2/Ar mixtures [40, 41], and pure O_2 [42].

As an illustration of the kind of information that can be obtained from such studies, the following conclusions from the SO_2/O_2 and SO_2/Ar measurements are listed:

(1) The relaxation in SO_2 is primarily a series process. Translational energy flows into the bending mode and from there is shared with the symmetrical stretch through a two quantum for one exchange and from symmetric stretch to the asymmetrical stretch with a one for one exchange.

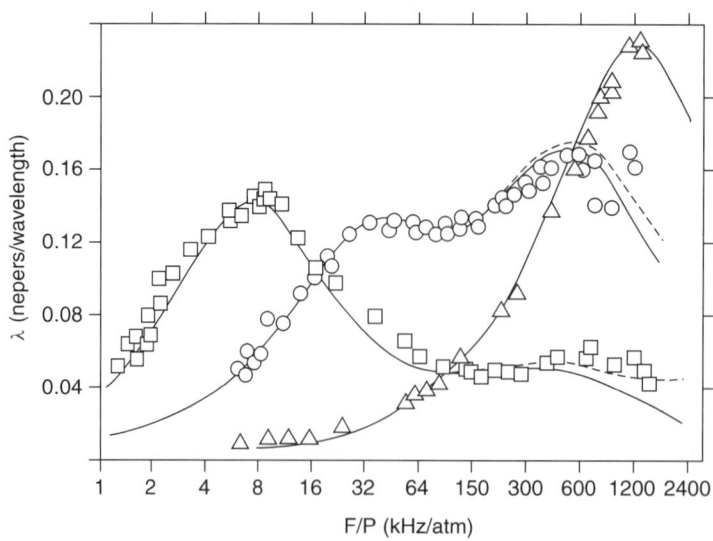

Fig. 10.8. Relaxation absorption in 75% SO_2/25% O (△), 20% SO_2/80% O_2 (○), and 5% SO_2/95% O_2 (□) at 500 °K. The solid curve is the theoretical curve for the series relaxation process.

INTRODUCTION TO ELASTIC CONSTANTS OF GASES

(2) Argon is only about one tenth and O_2 about one fourth as effective as another SO_2 molecule in deexciting the bending vibration in SO_2.

(3) The first excited level of the O_2 vibration exchanges energy in vibration-to-vibration exchange primarily with the first excited level of the asymmetric stretch mode of SO_2.

(4) The transition rate for the SO_2 vibration to vibration transfer increases with temperature faster than the vibration-to-translation rates. This is a peculiar property of SO_2 and has been attributed to its dipole moment.

An excellent example of combining knowledge of microscopic property measurements to determine elastic properties of a gas can be seen in the treatment of the absorption and velocity of sound in air. Combining absorption due to viscosity and heat conduction,

$$\alpha = [\omega^2/(2\gamma Pc)][4\mu/3 + (\gamma - 1)\kappa/(\gamma c_v)] \quad (10.49)$$

in Np m^{-1}. The Eucken expression [43]

$$\kappa = (15R\mu/4)[4c_v/(15R) + 3/5] \quad (10.50)$$

in J (kg mol)$^{-1}$ K^{-1} kg m^{-1} sec^{-1}, can be used to relate κ and μ for real air. For dry air over the temperature range 225–370 °K, the ratio c_v/R has the value $5/2 \pm 0.4\%$. For moist air, over the same temperature range, the c_v/R ratio becomes increasingly larger than the ideal gas value as the concentration of water vapor increases, but only by a small amount. For example, at a temperature of 370 °K and 100% relative humidity, $c_v/R = 2.565$ or just 2.6% more than the ideal gas value [44]. Therefore, if we assume $c_v/R = 5/2$, the computed value for κ cannot be more than 1% different from the value calculated using the correct value of c_v/R. This assumption yields $\kappa = 19R\mu/4$. For an ideal gas, $c_p = c_v + R$. Thus, with $c_v/R = 5/2$, the ratio $c_p/R = 7/2$. Substituting $\kappa = 19R\mu/4$ into Eq. 10.49 and using $(\gamma - 1)R/(\gamma c_v) = 4/35$ with $\gamma = c_p/c_v = 1.4$ gives

$$\alpha = [2\pi^2 f^2/(\gamma Pc)](1.88\mu) \quad (10.51)$$

in Np m^{-1} where f is now used to denote frequency in Hz.

For mixtures of gases, there is, in general, an additional term due to internal diffusion. It has been shown, however, that for air, this term is small and will be ignored here.

The classical absorption given by Eq. 10.51 changes with temperature due to changes in the speed of sound c, and the coefficient of viscosity μ. For an ideal gas (and air at STP is close to ideal), c varies as $T^{1/2}$. The coefficient of viscosity for dry air can be written in the form of Sutherland's equation [43]

$$\mu = \frac{\beta T^{1/2}}{1 + (S/T)} \quad (10.52)$$

in kg m^{-1} sec^{-1}, where B and S are empirical parameters. For air, the parameters $B = 1.458 \times 10^{-6}$ kg m^{-1} sec^{-1} K$^{-1/2}$ and $S = 110.4$ K are standard values [45] that provide a satisfactory fit to experimental data and permit reasonably accurate calculations of μ for the range of temperatures between 273.15 and 313.15 K (0° and 40 °C).

Using the standard values of γ, R, and M for air from the Society of Automotive Engineers [46] and defining a reference temperature T_0 to be 293.15 K gives

$$c = 343.23(T/T_0)^{1/2} \quad (10.53)$$

in m sec^{-1}, and

$$\mu = 7.318 \times 10^{-3} \frac{(T/T_0)^{3/2}}{T + 110.4} \qquad (10.54)$$

in kg m^{-1} sec^{-1}, so Eq. 10.51 becomes

$$\alpha_{cl} = 5.578 \times 10^{-9} \frac{T/T_0}{T + 110.4} f^2/(P/P_0) \qquad (10.55)$$

in Np m^{-1}, where P_0 is the reference pressure of 1.01325×10^5 N m^{-2} or 1 atm, the pressure P is in N m^{-2}, and the temperature T is in Kelvin.

It was shown earlier that for a single relaxing degree of freedom

$$\alpha = -\frac{\delta K_s/K_s^\infty}{2c} \frac{\omega^2 \tau'_{vs}}{1 + (\omega \tau'_{vs})^2} \qquad (10.56)$$

in Np m^{-1}, where K_s^∞ is the instantaneous adiabatic compressibility (in units of N m^{-2}) and equal to Pc_p^∞/c_v^∞; δK_s is the relaxing compressibility; and τ'_{vs} is the relaxation time at the partial pressure of the reactants in the mixture, at constant volume and under adiabatic conditions. For a single relaxing degree of freedom, $(\delta K_s/K_s^\infty) = -Rc'/[c_p^\infty(c_v^\infty + c')]$, where c' is the specific heat of the relaxing mode in J (kg mol)$^{-1}$ K^{-1}; c_p^∞ is the specific heat at constant pressure at frequencies $\gg 1/(2\pi\tau'_{vs})$; and c_v^∞ is the specific heat at constant volume under the same conditions. For frequencies $< 1/(2\pi\tau'_{vs})$, the specific heat at constant volume c_v equals $c_v^\infty + c'$.

Oxygen, nitrogen, and carbon dioxide all have rotational degrees of freedom and hence a rotational specific heat c' for rotational relaxation equal to the universal gas constant R. Water vapor has three rotational degrees of freedom and a rotational specific heat of $3R/2$; however, since the mole fraction of water vapor in air is at most 8% in the temperature range 273–313 °K [44], the rotational specific heat of air can be closely approximated by R.

Each rotational energy level relaxes with a different relaxation time, and the resultant absorption as a function of frequency can be quite complex. However, the principal constituents of air (oxygen, nitrogen, and water vapor) have rotational energy levels that are closer together than the average thermal energy. As a result, the rotational relaxation process for air behaves as though the rotational energy levels were continuous and can be described by a single relaxation time for isometric and adiabatic conditions $\tau'_{vs,\text{rot}}$. At all frequencies less than 10 MHz, $\tau'_{vs,\text{rot}}$ is $\ll \omega^{-1}$. For this case, and the value of $\delta K_s/K_s^\infty$ discussed previously, with $c_v = c_v^\infty + c' = c_v^\infty + R$ and $R/c_v = \gamma - 1$, Eq. 10.56 becomes

$$\alpha_{\text{rot}} = \{[\pi R(\gamma - 1)]/(cc_p^\infty)\}(f^2/f_{r,\text{rot}}) \qquad (10.57)$$

in Np m^{-1}, where

$$f_{r,\text{rot}} = 1/(2\pi\tau'_{vs,\text{rot}}) \qquad (10.58)$$

in Hertz is the rotational relaxation frequency.

Now consider again Eq. 10.56. Write

$$\alpha = f^2 \left(\frac{c}{c_o^2}\right) \left(\frac{c_\infty^2 - c_o^2}{c_\infty^2}\right) \frac{2\pi^2 \tau_{ps}}{1 + \omega^2 \tau_{ps}^2} \qquad (10.59)$$

where c_0 is the low frequency limit of sound speed. Experimentally, one finds that τ for rotational relaxation is very small so that below 10 MHz, $\omega^2 \tau_{ps}^2 \approx 0$. Then

$$\alpha = \text{const} \cdot f^2 \cdot \tau_{ps} \qquad (10.60)$$

If we write $\tau_{ps} = \text{const} \cdot Z_{\text{rot}}/P$, where Z_{rot} is the rotational collision number, we find that

$$\alpha_{\text{rot}} = [2\pi^2 f^2/(\gamma P c)]\mu[\gamma(\gamma - 1)R/(1.25 c_p^o)]Z_{\text{rot}} \tag{10.61}$$

in Np m^{-1}. Again, assuming that $\gamma = 1.4$ and $c_p^o/R = 7/2$, the term in brackets becomes 0.128, a value that is independent of temperature since the specific heat of rotation does not change with temperature. Comparing Eqs. 10.51 and 10.61, we can now write

$$\alpha_{\text{rot}}/\alpha_{cl} = 0.128 Z_{\text{rot}}/1.88 = 0.0681 Z_{\text{rot}} \tag{10.62}$$

For dry air, the rotational collision number Z_{rot} has been measured near room temperature by Greenspan [47] and at higher temperatures by Bass and Keeton [48]. A summary of these experimental results can be written in the form

$$Z_{\text{rot}} = 61.1 \exp(-16.8 T^{1/3}) \tag{10.63}$$

over the temperature range 293–690 °K.

Acknowledging the presence of water vapor but ignoring the relaxation of the small water-vapor rotational specific heat, the rotational collision number can be written as

$$Z_{\text{rot}} = \{[X(N_2 + O_2)/Z_{\text{rot}}(N_2 + O_2) + X(H_2O)/Z_{\text{rot}}(N_2 + O_2 + H_2O)]\}^{-1} \tag{10.64}$$

where $X(N_2 + O_2)$ is the mole fraction of nitrogen plus the mole fraction of oxygen; $Z_{\text{rot}}(N_2 + O_2)$ is the rotational collision number for dry air; $X(H_2O)$ is mole fraction of water; and $Z_{\text{rot}}(N_2 + O_2 + H_2O)$ is the number of H_2O collisions required for N_2 and O_2 to establish rotational equilibrium. This latter quantity can take on values from ∞ to 1 (probability 0–1). The resultant Z_{rot} will be most dependent on $X(H_2O)$ if $Z_{\text{rot}}(N_2 + O_2 + H_2O) = 1$. Since $Z_{\text{rot}}(N_2 + O_2) \cong 5$, when $X(H_2O) < 0.02$, the rotational collision number for the mixture can change by no more than 2%, so $Z_{\text{rot}} \cong (N_2 + O_2)$; hence, we are justified in ignoring the effect of water vapor on the rotational collision number.

Combining Eqs. 10.62 and 10.63 with Eq. 10.55 gives the combined absorption due to classical absorption factors and rotational relaxation α_{cr} as

$$\alpha_{cr} = 5.578 \times 10^{-9} \frac{T/T_0}{T + 110.4} \frac{[1 + 4.16 \exp(-16.8 T^{-1/3})]f^2}{P/P_0} \tag{10.65}$$

Evaluating Eq. 10.65 for various temperatures indicates that a simplified empirical equation of the form

$$\alpha_{cr} = 1.83 \times 10^{-11} \frac{(T/T_0)^{1/2} f^2}{P/P_0} \tag{10.66}$$

in Np m^{-1}, is within 2% of Eq. 10.65 for temperatures between 213 and 373 °K.

The relaxation strengths of the two vibrational relaxation processes important in air are near to those one would expect for a simple single relaxation of N_2 and O_2. With $\lambda = c/f$ and $\omega = 2\pi f$, Eq. 10.56 can be written as

$$\alpha_{vib,j} = \frac{\pi s}{c} f \frac{f^2/f_{r,j}}{1 + (f/f_{r,j})^2} \tag{10.67}$$

in Np m^{-1}, where $j = 0$ or N; $s_j = c_j'/R[c_p^\infty(c_p^\infty + c_j')]$; and $f_{r,j} = 1/(2\pi \tau_{vs,j})$. The relaxation strength s_j can be related to the particular atmospheric constituent and the temperature by using the Planck-Einstein relation [47],

$$c_j'/R = \frac{X_j(\theta_j/T)^2 e^{-\theta_j/T}}{(1 - e^{-\theta_j/T})^2} \tag{10.68}$$

where X_j is the mole fraction of the component considered; i.e., 0.20948 for oxygen and 0.78084 for nitrogen [45]; θ_j is the characteristic vibrational temperature (2239.1 °K for oxygen and 3352.0 °K for nitrogen). For the temperature range 0–40 °C, c'_j is small with respect to c_j^∞, and thus $c_j^\infty \cong c_p$ and $c_v^\infty + c'_j = c_v$. Therefore, $s_j \cong (c'_j/R)(R^2/c_p c_v)$. Using the same reasoning that led to the approximations described in developing Eq. 10.51, the quantity $(R^2/c_p c_v)$ can be set equal to $(\gamma - 1)R/c_p$, which, with $\gamma c_v = c_p$, was shown to be equal to 4/35. Thus, with Eq. 10.68, Eq. 10.67 can be written as

$$\alpha_{\text{vib},j} = \frac{4\pi X_j}{35c} \frac{(\theta_j/T)^2 e^{-\theta_j/T}}{(1 - e^{-\theta_j/T})^2} \frac{f^2/f_{r,j}}{1 + (f/f_{r,j})^2} \quad (10.69)$$

in Np m^{-1}. Using Eq. 10.56, the absorption due to vibrational relaxation can be computed if the $f_{r,j}$ frequencies for oxygen and nitrogen are known.

The frequencies of maximum absorption for oxygen and nitrogen vibrational relaxation have been computed by using the general theory described earlier [50]. To a good approximation,

$$f_{r,O} = (P/P_0)\{24 + 4.41 \times 10^4 h[(0.05 + h)/(0.391 + h)]\} \quad (10.70)$$

in hertz, where h is the mole fraction of water vapor in percent. The various constants in Eq. 10.69 can be determined from the general theory or from experimental measurements in air.

The frequency of maximum absorption for nitrogen is theoretically less difficult to determine than that for oxygen, since it is dominated by direct vibration-translation (V-T) deexcitation of the excited nitrogen molecules by water vapor (or a one-step V-V transfer to H$_2$O). Carbon dioxide provides an alternate relaxation path at very low water-vapor concentrations. The form of the relaxation frequency is

$$f_{r,N} = (P/P_0)(T/T_0)^{-1/2}[9 + 350h \exp\{-6.142[(T/T_0)^{-1/3} - 1]\}] \quad (10.71)$$

in hertz.

An alternative model for nitrogen relaxation gives

$$f_{r,N} = (P/P_0)(9 + 200h) \text{ (Hz)} \quad (10.72)$$

The contribution to atmospheric absorption due to vibration relaxation processes can now be determined from Eqs. 10.69–10.72 if the percent mole fraction of water vapor is known. By Avogadro's law, the percent mole fraction is equal to the ratio of the partial pressure of water vapor P_w to the atmospheric pressure P of the sample volume of moist air, in percent. Thus,

$$h = 100 P_w/P \quad (10.73)$$

in percent. Introducing the saturation vapor pressure of pure water over liquid water P_{sat}, Eq. 10.73 can be written as

$$h = (100 P_w/P_{\text{sat}})(P_{\text{sat}}/P) \quad (10.74)$$

or

$$h = h_r(P_{\text{sat}}/P) \quad (10.75)$$

in percent, where h_r is by definition the relative humidity for a given sample of moist air under pressure P and at a temperature T, in percent, a quantity that is usually available from experiments. For convenience, Eq. 10.75 is written in terms of the standard atmospheric pressure P_0 as

$$h = \frac{h_r(P_{\text{sat}}/P_0)}{P_0} \quad (10.76)$$

in percent. The quality P_{sat} can be determined from standard references or computed from the relation

$$\log_{10}(P_{sat}/P_0) = 10.79586[1-(T_{01}/T)]$$
$$-5.02808\log_{10}(T/T_{01})$$
$$+1.50474\times 10^{-4}(1-10^{-8.29692[(T/T_{01})-1]})$$
$$+0.42873\times 10^{-3}(10^{4.76955[1-T_{01}/T]}-1)$$
$$-2.2195983 \tag{10.77}$$

where T is the air temperature, T_{01} is the air triple-point temperature (273.16 °K). Once h is known, the relaxation frequencies can be completed. Combining all the absorption terms,

$$\alpha = f^2\left\{1.83\times 10^{-11}(P_0/P)(T/T_0)^{1/2} + (T_0/T)^{5/2}\right.$$
$$\times\left(1.278\times 10^{-2}\frac{e^{-2239.1/T}}{f_{r,o}+(f^2/f_{r,o})}\right)$$
$$\left.+\left(1.069\times 10^{-1}\frac{e^{-3352/T}}{f_{r,N}+(f^2/f_{r,N})}\right)\right\} \tag{10.78}$$

where f is the acoustic frequency in Hz, P is the atmospheric pressure, P_0 is the reference atmospheric pressure (1 atm), T is the atmospheric temperature in K, T is the atmospheric temperature in K, T_0 is the reference atmospheric temperature (293.15 °K), $f_{r,o}$ is the relaxation frequency of molecular oxygen (Eq. 10.70), and $f_{r,N}$ is the relaxation frequency of molecular nitrogen (Eq. 10.71 or 10.72).

Figures 10.9(a) and 10.9(b) show the relative contributions of the different relaxation mechanisms. Figure 10.9(c) shows comparison between theory and experiment.

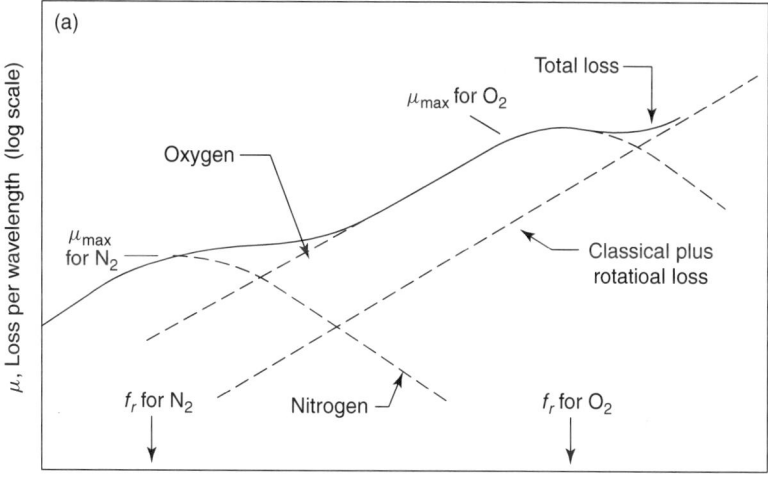

Fig. 10.9(a). Components and general behavior of total air absorption in air in terms of loss per wavelength: $\mu \sim \dfrac{2(f/f_r)}{1+(f/f_r)^2}$.

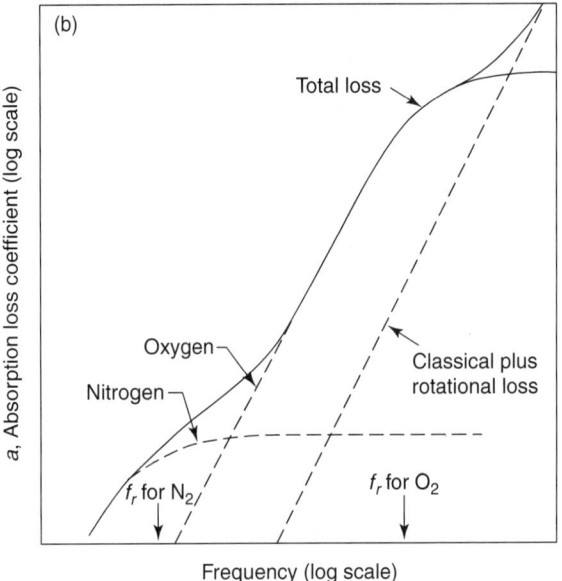

Fig. 10.9(b). Components and general behavior of total air absorption in air $\alpha \sim \dfrac{(f/f_r)^2}{1+(f/f_r)^2}$.

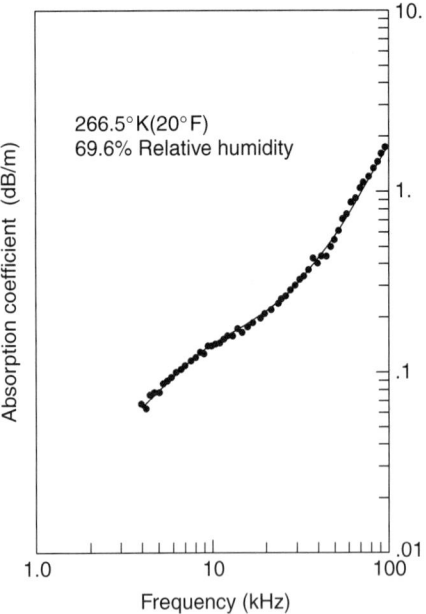

Fig. 10.9(c). Total free-field sound absorption in air at 266.5 °K and 69.9% relative humidity; points represent experimental data; solid line calculated using the computational technique described here.

10.8. DIFFUSION

There is yet another source of sound absorption that must be considered in gas mixtures. When there is local pressure or temperature gradient, less massive molecules with their higher thermal speeds move toward a condition of equilibrium more rapidly than heavier molecules. The diffusion due to the pressure gradient is accompanied by preferential diffusion of the lighter molecules due to the thermal gradient. The result is an additional absorption due to this diffusion given by [51]

$$\alpha = \frac{2\pi^2 f^2 \gamma X_1 X_2 PD_{12}}{Pc^3} \left[\frac{M_2 - M_1}{M} + \frac{(\gamma - 1)k_T}{\gamma D_{12} X_1 X_2} \right]^2 \quad (10.79)$$

in Np m^{-1}, where X_1, X_2 and M_1, M_2 are the mole fractions and molecular weights of gases 1 and 2, respectively; M is the molecular weight of the mixture of gases 1 and 2; D_{12} is the concentration diffusion coefficient and the product of pressure P and D_{12} is in units of N sec^{-1}; and k_T is the thermal diffusion coefficient in m^2 sec^{-1}.

The calculation of absorption due to the combination of pressure and thermal diffusion terms requires that PD_{12} and the ratio k_T/D_{12} be known. These quantities can be calculated from kinetic theory [43], but experimental values are sparse. For air, the term PD_{12} is largest when considering O_2/H_2O collisions; however, in this case, not only is the product $X_1 X_2$ small, but the ratio $k_T/(X_1 X_2)$ is also small. For O_2/N_2 collisions, X_1 and X_2 are both relatively large in air, but the differences in molecular weights $(M_2 - M_1)$ and kT are both small. Ignoring relaxation, Bauer [1] has shown that for air, 99.5% of the total classical absorption can be attributed to the viscosity and thermal conduction. Hence, additional absorption due to diffusion need not be considered for air. For mixtures where $M_2 - M_1$ is large, however, this situation could change dramatically.

10.9. GASES AT LOW PRESSURE

Early in this chapter, it was assumed that the f/P ratio in the gas is sufficiently large that the gas can be considered a continuum and the Navier-Stokes Equation is applicable. This description is adequate when the mean distance between collision is very small compared with a wavelength. When this condition is no longer met, treating the gas as a continuous media is no longer accurate. One must resort to a microscopic description of molecular energy and momentum transfer even for the transfer of translational energy. In this case, the governing equation is the Boltzmann Equation.

Very low frequency sound waves naturally occur in the very low density regions of space. The general trend towards favoring low frequency (long wavelength propagation) still applies giving rise to wavelengths that might be measured in light years. The interaction between particular that support wave propagation can be gravitational, or in the case of plasmas, very strong electromagnetic interactions. As a result, the wave equation is complex and the wave speeds not well understood. G. Bertin and C. C. Lin [52] have recently written an excellent book that summarizes current knowledge. For the discussion that follows, we will assume that interactions are described as hard sphere collisions. Hopefully, future versions of the handbook will include more comprehensive descriptions of this intriguing and evolving field.

Sound propagation in dilute gases has been investigated by Greenspan [2] (Helium at 1 MHz; He, Ne, Ar, Kr, X at 11 MHz), Meyer and Sessler [53] (Ar at 100 and 200 kHz), and Hassler [54] (Ar at 100 and 200 kHz). These data are plotted in Figure 10.10 on the universal abscissa $\eta f/p = \omega \tau_c / 2\pi$. One sees that the rare gases obey the Kirchhoff-Stokes theory for $\omega \tau_c < 0.2$.

In Figure 10.10, only those measurements that correspond to the conditions "inside" the gas are included. In dilute gases the experimental conditions are very unfavorable: the small density and the large absorption reduces the signal so much that measurements are taken more and more in the very neighborhood of the sound transmitter, i.e., at

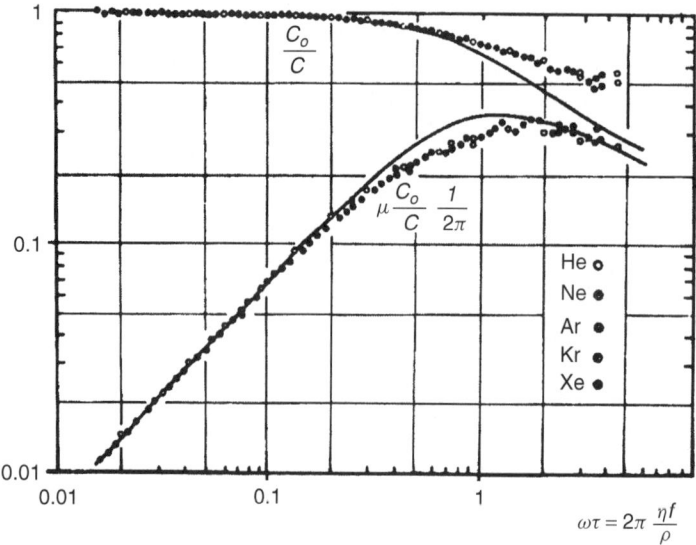

Fig. 10.10. Sound propagation in monatomic gases (Greenspan); Kirchhoff-Stokes theory. Courtesy of Academic Press, Inc., New York.

the "margin" of the gas. Here, where the molecules coming from the sound transmitter have not suffered a collision with other molecules, special phenomena show up.

The most detailed description of a system containing a single monatomic ideal gas in terms of a distribution function $f(c, r, t)$ is presented in [1]. Some of the main features of that article are included here for reference. Note that in this section, f represents the number of molecules in the differential space dr at r, which possess at t a velocity within dc of c. f can be understood as the density of particles in space/momentum hyperspace.

The thermodynamic variables and the irreversible fluxes are given by the moments of the distribution function. The zero moment yields the number density

$$n = \int f\, dc \tag{10.80}$$

The first moment yields the mean (macroscopic) velocity

$$v = \int c(f/n)\, dc \tag{10.81}$$

The kinetic energy relative to a frame moving with v defines the kinetic temperature,

$$\frac{3}{2}kT = \frac{m}{2}\int (c-v)^2 (f/n)\, dc \tag{10.82}$$

Internal friction means transport of a component of the linear momentum $m(c_\mu - v_\mu)$ with the speed $(c_\nu - v_\nu)$ ($\mu, \nu = x, y, z$). In tensor notation that establishes the pressure tensor

$$\underline{\underline{P}} = m \int (\underline{c}-\underline{v})(\underline{c}-\underline{v}) f\, dc \tag{10.83}$$

The trace of Eq. 10.83 is the threefold static pressure $3p = 3nkT$. The heat conduction stems from the transport of kinetic energy (relative to $\underline{v}$) $m(\underline{c}-\underline{v})^2/2$ in the direction $(\underline{c}-\underline{v})$. Therefore, the transport of kinetic energy is represented by q

$$q = \frac{m}{2} \int (\underline{c}-\underline{v})(\underline{c}-\underline{v})^2 f\, dc \tag{10.84}$$

where q is the trace of the third tensorial moment. Higher moments do not possess a particular thermodynamic meaning and therefore have no names. If we succeed in finding an equation for $f(c, r, t)$, the friction tensor and the heat current can be calculated by insertion of the solution into Eqs. 10.83 and 10.84. Together with the conservation equation, which also represents equations between the moments, the dispersion equation for sound propagation is obtained. For that purpose, the moments may be linearized: since f contains members proportional to the sound amplitude, we can substitute $(c - v)$ by c in Eqs. 10.82 and 10.83. In the case of Eq. 10.84, the same procedure would include convective energy transport,

$$q = \frac{m}{2} \int \underline{c}^2 c f dc - \frac{5}{2} nkT \underline{v} = \frac{m}{2} \int \left(c^2 - \frac{5}{2\beta} \right) \underline{c} f d\underline{c} \tag{10.85}$$

where $5/2 nkT$ denotes the enthalpy per unit volume.

A complete description of momentum and energy transport requires a knowledge of the distribution f. The equation for f is the Boltzmann equation. We report in the following some of its properties; more details can be gathered from the literature (e.g., [55, 56]).

The local variation of the distribution function with time is affected (neglecting external fields and walls) (1) by the flight of the molecules—when the time dt has elapsed, dr now contains molecules formerly-cdt apart; (2) by collisions between two molecules that throw these out of or into the velocity range dc at c; and (3) by the action of external fields, which we assume to be absent.

If one neglects the duration of a collision against the mean free time (or the range of the intermolecular potential against the mean free path), the changes (1) and (2) are independent of each other,

$$\frac{\partial f}{\partial t} = -\underline{c} \, \text{grad} \, f + \left. \frac{\partial f}{\partial t} \right|_{\text{collision}} \tag{10.86}$$

In elastic collisions of identical particles (velocities $\underline{c}$ and $\underline{c}_1$), the motion of the center of mass and the amount of the velocity difference is maintained, thus the velocity difference $\underline{c} - \underline{c}_1 = g$ is only rotated by angle X into a new direction (unit vector e'). The velocities of both particles $\underline{c}'$ and $\underline{c}'_1$ after the collision are, therefore,

$$\frac{\underline{c}'}{\underline{c}'_1} = \frac{1}{2}(\underline{c} + \underline{c}_1) \pm \frac{1}{2}|\underline{c} - \underline{c}_1|\underline{e}' \tag{10.87}$$

The probability of such rotation can be expressed in a differential cross section $\sigma(g, X)$, which depends on the intermolecular potential. Now we can formulate the loss of molecules out of dc,

$$d\underline{c} \left. \frac{\partial f}{\partial f} \right|_{\text{loss}} = d\underline{c} \int g \sigma f(\underline{c}) f(\underline{c}_1) d\underline{e}' d\underline{c}_1 \tag{10.88}$$

g stems from the proportionality between collision rate and velocity difference, $f(\underline{c}) f(\underline{c}_1) d\underline{c} d\underline{c}_1$ is the density of the colliding and the target molecules; the integration over the angle $(d\underline{e}')$ takes into account all products of the collision, the integration over $d\underline{c}_1$ considers all velocities of the target molecules. To calculate the gain we have to look for collisions that force the colliding molecules into $d\underline{c}$ at $\underline{c}$. If, in addition, the target molecule changes its velocity into $d\underline{c}_1$ at $\underline{c}_1$, then, for molecules without spin, we look for the "inverse" collisions, which we obtain from Eq. 10.88 by time reversal. Since the equations of motion of the molecules are reversible, those "inverse" collisions possess the same probability σ. By multiplication with the distribution function of colliding and target molecule at those $\underline{c}'$ and $\underline{c}'_1$ values compatible with Eq. 10.88, and integration overall possibilities $(d\underline{e}')$ and final velocities of the target molecules $(d\underline{c}_1)$, we obtain the production rate of molecules at c,

$$d\underline{c}\left.\frac{\partial f}{\partial t}\right|_{\text{gain}} = d\underline{c}\int g\sigma(\underline{c}')\int (\underline{c}_1')d\underline{e}'d\underline{c}_1 \qquad (10.89)$$

The loss-gain balance is the Boltzmann Equation in the absence of external fields

$$\frac{\partial f}{\partial t} + \underline{c}\text{ grad }f = \int g\sigma(f'f_1' - f'f_1')d\underline{e}'d\underline{c}_1 \qquad (10.90)$$

As usual, the argument of f is abbreviated, i.e., $f_1 = f(\underline{c}_1)$, $f' = f(\underline{c}')$. $f'f_1'$ therefore corresponds to Eq. 10.87. The nonlinear integro-differential equation for f is highly complex, and for our purpose linearization is allowed.

Equation 10.90 possesses one simple (and the only exactly known) solution, the Maxwell distribution:

$$f_M(\underline{c}) = n\left(\frac{\beta}{\pi}\right)^{3/2}\exp(-\beta(\underline{c}-\underline{v})^2), \quad \beta = m/2kT \qquad (10.91)$$

with number density n, mean velocity v, and temperature T constant in time and space. For the Maxwell distribution $(f'f_1' - f\,f_1)$ vanishes identically. However, there are systems with n, v, $T = F(r, t)$ for which Eq. 10.90 is a good zero-order approximation. That is the case if the left-hand side of Eq. 10.89 is so small that a minute deviation of the distribution function from the Maxwellian, if inserted into the then nonvanishing collision integral, is sufficient to satisfy Eq. 10.89. Since the collision integral implies the division of the distribution function by a mean free time (in the sense of a dimensional analysis), we can think of two cases.

(1) The variations of n, T, and v in time and space are small within a mean free time or a mean free path, respectively, while the total variation may be considerable. Then we suppose the solution to be a small deviation from a local Maxwellian Eq. 10.91 with n, v, $\beta = n$, v, $\beta(r, t)$ and have

$$f = f_M(\underline{r}, t)[1 + \phi(\underline{c}, \underline{r}, t)] \qquad (10.92)$$

(2) The amplitudes of the variables n, v, T are small. Then even with "steep" gradients and "fast" variations the deviation from an absolute Maxwellian (constant in time and space) will be small,

$$f = f_0[1 + \Phi(\underline{c}, \underline{r}, t)] \qquad (10.93)$$

That is the case in the propagation of small-amplitude sound. We therefore introduce Eq. 10.93 into Eq. 10.89 and obtain the linearized Boltzmann Equation by neglecting terms quadratic in Φ, because $f_0'f_{01}' = f_0 f_{01}$, the linearized Boltzmann Equation

$$f_0\left\{\frac{\partial \Phi}{\partial t} + \underline{c}\text{ grad }\Phi\right\} = \int g\sigma \int_0\int_{01}[\Phi(\underline{c}_1) + \Phi(\underline{c}') - \Phi(\underline{c}_1) - \Phi(\underline{c})]d\underline{e}'d\underline{c}_1 \qquad (10.94)$$

The collision integral on the right-hand side of Eq. 10.94 represents the rate of change of deviation from the Maxwellian distribution $f_0\Phi(\underline{c})$ at the velocity $\underline{c}$. It depends on the shape of that deviation and consists of four parts: encounters with Maxwellian-distributed target molecules diminish an excess of the colliding molecules $f_0\Phi(\underline{c})$ at $\underline{c}$ (last term of Eq. 10.94; an excess in other velocity ranges consistent with Eq. 10.87; $\Phi(\underline{c}')$; increases in the excess at c (second term). The collisions of the Maxwellian-distributed part of the colliding molecules against target molecules distributed according to $f_0(1 + \Phi)$ produce a gain (first term) and a loss (third term) at $\underline{c}$. In general, the rate of change of the relative deviation from the Maxwellian, $\Phi(\underline{c})$, will be different at different velocities. Such relative deviations, which change at every velocity with a rate proportional to $\Phi(\underline{c})$, conform to the eigenfunctions of the collision operator. They decay under retention of their shape.

Consider a region near the sound transmitter in which most of the molecules come directly from the oscillating surface without a binary collision in between. The Boltzmann Equation is

$$\frac{\partial f}{\partial t} + c_x \frac{\partial f}{\partial x} = 0 \qquad (10.95)$$

A solution is any distribution function that contains the combination $c_x - x/t$. Specific solutions are found by applying the boundary conditions at the transmitter. The simplest case is the emission of a Maxwellian from a transmitter surface located at the origin, facing to the right, oscillating according to

$$w = \hat{w} \sin \omega t' \qquad (10.96)$$

Then the molecules with direction to the right (coming from the transmitter) have a distribution in the laboratory frame [58]

$$f_+(x=0) = n_0 \left(\frac{\beta}{\pi}\right)^{3/2} \exp[-\beta(c_x - \hat{w} \sin \omega t')^2 - \beta c_y^2 - \beta c_z^2] \qquad (10.97)$$

Molecules flying to the left have no harmonic history, they shall be Maxwellian

$$f_-(x=0) = n_0 \left(\frac{\beta}{\pi}\right)^{3/2} e^{-\beta c^2} \qquad (10.98)$$

Half-range expansions of the form of Eqs. 10.97 and 10.98, but with more complicated f_+, and other assumptions regarding momentum accommodation at the oscillating surface have been introduced by Maidnik et al. [57] For small amplitudes, $\hat{w}$, Eq. 10.97 may be linearized

$$f_+(x=0) = n_0 \left(\frac{\beta}{\pi}\right)^{3/2} \exp[-\beta c_x^2 - 2\beta \hat{w} c_x \sin \omega t' - \beta c_y^2 - \beta c_z^2] \qquad (10.99)$$

or

$$f(x=0) = f_0[1 + U(c_x) 2\beta \hat{w} c_x \sin \omega t'] \qquad (10.100)$$

$U(x)$ is the unit step. The propagator $\delta(t' - t + x/c_x)$ moves the distribution function (100) from the point $x = 0$ to x,

$$f = f_0[1 + U(c_x) 2\beta \hat{w} c_x e^{j\omega(t - x/c_x)}] \qquad (10.101)$$

The received pressure is the flux of linear momentum per unit time and area, if we assume full accommodation (molecules that collide with the surface take on the momentum of that surface). It contains a part oscillating with ω, the amplitude of which is

$$\Delta P_{xx} = 2\beta \hat{w} e^{j\omega t} \int f_0 c_x^3 e^{-j\omega x/c_x} dc = 2\beta \hat{w} e^{j\omega t} n_0 \left(\frac{\beta}{\pi}\right)^{3/2} \\ \times \int_0^\infty e^{-\beta c_x^2} c_x^3 e^{-j\omega x/c_x} dc_x \qquad (10.102)$$

Higher harmonics, expected by Kahn and Mintzer [58], are not present. The integral (Eq. 10.102) cannot be solved in closed form. First we discuss the situation close to the transmitter.

(1) *Short distance to the transmitter.* For small values of x, the factor $\exp(-j\omega x/c_x)$ can be expanded for almost all molecules except the very slow ones. The integration can then be performed. If we force the result into an exponential representation again, we obtain (with the mean velocity $c_m = (4/\pi\beta)^{1/2}$),

$$\frac{\Delta P_{xx}}{p} \approx \frac{\sqrt{2}}{c_m}\hat{w}e^{j\omega t}$$

$$\times \exp\left[-j\frac{\omega x}{c_m} + \left(\frac{1}{2} - \frac{2}{\pi}\right)\frac{\omega^2 x^2}{c_m^2} + j\left(\frac{1}{3} - \frac{2}{3\pi}\right)\frac{\omega^3 x^3}{c_m^3} + \ldots\right]$$

(10.103)

The phase is no longer a linear function of the distance. From the phase term

$$\exp\left(-j\int_0^x k(x')dx'\right)$$

$$= \exp\left[-j\frac{\omega x}{c_m} + \left(\frac{1}{2} - \frac{2}{\pi}\right)\frac{\omega^3 x^2}{c_m^2} + j\left(\frac{1}{3} - \frac{2}{3\pi}\right)\frac{\omega^2 x^3}{c_m^3} + \ldots\right]$$

(10.104)

we gain differentiation as

$$k(x) = \frac{\omega}{c_m}\left[1 - j\left(\frac{4}{\pi} - 1\right)\frac{\omega x}{c_m} - \left(1 - \frac{2}{\pi}\right)\frac{\omega^2 x^2}{c_m^2} + \ldots\right] \quad (10.105)$$

This result is readily interpreted. The phase velocity near the transducer is the phase of the molecules. The phase velocity increases with increasing distance, and we notice an absorption proportional to the distance. That behavior results from the fact that in the region $\omega x/c_x \ll 1$, almost all molecules contribute to the propagation, while for greater values of x the phase factor $\exp(-j\omega x/c_x)$ begins to oscillate for the slow molecules. Fewer molecules support propagation further from the transducer, and these are faster in the mean.

(2) *General solution.* Because

$$\Delta P_{xx} = \Delta\hat{P}_{xx}e^{j\omega t}\exp\left(-j\int_0^x k(x')dx'\right) \quad (10.106)$$

we obtain $k(x)$ from Eq. 10.104 as

$$k(x) = j\frac{\partial}{\partial x}\ln P_{xx} = \int_0^\infty c_x^2 e^{-\beta c_x^2} e^{-j\omega x/c_x}dc_x / \int_0^\infty c_x^3 e^{-\beta c_x^2} e^{-j\omega x/c_x}dc_x$$

(10.107)

A relation similar to Eq. 10.107 but with modified accommodation and without the factor c_x^2 resulting from the weighting of the momentum transport, has been given by Meyer and Sessler [53] and solved by expansion techniques.

There are two standard methods of solution of the linearized Boltzmann Equation and its application to the problem of sound propagation in rarefied gases. These methods differ considerably in approach but they yield the same class of solutions, the so-called normal solutions, and the same expressions for the pressure tensor $P_{\mu v}$ and the heat current q_v. Bauer [1] discusses them in the Chapman-Enskog version and applies them to the Boltzmann Equation linearized about an absolute Maxwellian, Eq. 10.95. Then the calculation gains in simplicity, and the method becomes clear (the original method

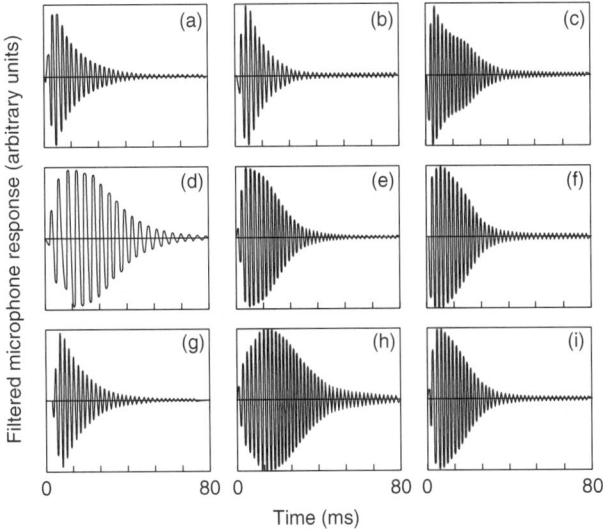

Fig. 10.11. Filtered microphone response. In moving from the central graph (e), only one of the four variables changes. For (c), (e), and (f), the sound frequency increases; (d) is the fundamental of the 60-cm tube; (e) is the fundamental for the 30-cm tube. Corresponding frequencies are approximately 340, 680, and 1360 Hz. For (b), (e), and (h) the pressure increases from 20 to 40 to 80 torr. For (a), (e), and (i) the H_2 concentration increases from 0 to 10% to 20%. For (g), (e), and (c) the energy per mole deposited by the discharge in the gas increases from 3700 to 7000 to 9000 J/mol. If all of this energy were to go into vibrations, the corresponding initial vibrational temperatures would be 1600, 2100, and 2400°K.

was based on the nonlinear Boltzmann Equation itself or a linearization about a local Maxwellian). In principle, the method is an iteration: a lower-order approximation $\phi^{(i-1)}$ is inserted into the differential part; in doing that the integro-differential equation is converted into an ordinary integral equation for a higher-order solution $\phi^{(i)}$. The dispersion relations obtained by that method and also the pertinent literature have been listed by Greenspan [2].

10.10. SYSTEMS NOT IN EQUILIBRIUM

To this point the treatment has included only those systems that are in equilibrium on a scale large compared with a mean free path. Modern research is much more involved with propagation in systems far removed from equilibrium. Thermoacoustics occurs when sound propagates through a region where there is a temperature gradient. Thermoacoustics occurs when sound propagates through a region in which the gas is in close thermal contact with a solid surface with an imposed temperature gradient. Under the correct phasing conditions, heat from the solid matrix is converted to mechanical work and a net gain in acoustic energy or intensity occurs. The stack amplifies the incident acoustic wave. A system in equilibrium could not do this without violating the third law of thermodynamics but when a temperature gradient is present, energy flow occurs, and acoustic gain can result. There are other cases where negative absorption has been observed. Shields [59] has observed gain for a wave propagating through a gas vibrationally excited by an electric discharge. He can predict and observe negative gain.

10.11. SUMMARY

A discussion of the elastic properties of gases quickly involves processes that store energy in the gas. These processes affect the dynamic elastic properties but have no noticeable effect on the static properties. Such behavior is also observed in solids but not to the extent of liquids. Not mentioned in any of the above discussion are nonlinear effects. Gases do exhibit nonlinearity as do solids. In most cases the nonlinear behavior is straightforward and involves no new properties of the gas. For a good discussion of the nonlinear behavior of gases, see [60].

References

1. Bauer, H.J. (1972). Influences of transport mechanisms on sound propagation in gases. *Adv. Mol. Relaxation Proc.* **2**: 319.
2. Greenspan, M. (1965). In *Physical Acoustics IIA*, pp. 1–45, New York: Academic Press.
3. Bhatia, A.B. (1967). *Ultrasonic Absorption: An Introduction to the Theory of Sound Absorption and Dispersion in Gases, Liquids, and Solids*, pp. 50–51, Oxford: Clarendon Press.
4. Beyer, R.T. and Letcher, S.V. (1969). *Physical Ultrasonics*, pp. 91–96, New York: Academic Press.
5. O'Donnell, M., Jaynes, E.T., and Miller, J.G. (1981). Kramers-Kronig relationship between ultrasonic attenuation and phase velocity. *J. Acoust. Soc. Am.* **69**.
6. Hirschfelder, J.O., Curtiss, C.F., and Bird, R.B. (1954). *Molecular Theory of Gases and Liquids*, pp. 147–148, New York: John Wiley & Sons.
7. Bhatia, A.B. (1967). Ultrasonic absorption: an introduction to the theory of sound absorption and dispersion in gases, liquids, and solids, pp. 14–15, Oxford: Clarendon Press.
8. Landolt-Bornstein. (1967). *Numerical Data and Functional Relationships in Science and Technology: Molecular Acoustics, vol 5*, pp. 8–9, New York: Springer-Verlag.
9. Kittel, C. (1948). *Rept. Progr. Phys.* **11**: 205.
10. Beyer, R.T. and Letcher, S.V. (1969). *Physical Ultrasonics*, pp. 359–360, New York: Academic Press.
11. Herzfeld, K.F. and Litovitz, T.A. (1959). *Absorption and Dispersion of Ultrasonic Waves*. New York: Academic Press.
12. Winter, T.G. and Hill, G.L. (1967). High-temperature ultrasonic measurements of rotational relaxation in hydrogen, deuterium, nitrogen, and oxygen. *J. Acoust. Soc. Am.* **42**: 848.
13. Zuckerwar, A.J. and Griffin, W.A. (1980). Resonant tube for measurement of sound absorption in gases at low frequency/pressure ratios. *J. Acoust. Soc. Am.* **68**: 218.
14. Hill, G.L. and Winter, T.G. (1968). The effect of temperature on the rotational and vibrational relaxation times of some hydrocarbons. *J. Chem. Phys.* **49**: 440.
15. Landau, L. and Teller, E. (1936). Zur theorie der schall dispersion. *Phy. Z. Soviet-Union* **10**: 34.
16. Beyer, R.T. and Letcher, S.V. (1969). *Physical Ultrasonics*, chapter 5, New York: Academic Press.
17. Shields, F.D. (1962). Thermal relaxation in fluorine. *J. Acoust. Soc. Am.* **34**: 271.
18. Bass, H.E., Sutherland, L.C., Piercy, J., and Evans, L. (1984). Absorption of sound by the Atmosphere. *Phys. Acoust.* **VXII**, 161.
19. Hubbard, J.C. (1931). The acoustic resonator interferometer: the acoustic system and its equivalent electric network. *Phys. Rev.* **38**: 1011.
20. Counter, J.V. (1958). Ultrasonic dispersion in oxygen. *J. Acoust. Soc. Am.* **30**: 297.
21. Steward, E.S. and Stewart, J.L. (1952). Rotational dispersion in the velocity, attenuation, and reflection of ultrasonic waves in hydrogen and deuterium. *J. Acoust. Soc. Am.* **24**: 194.
22. Lagemann, R.T. (1952). The use of glass parts in ultrasonic interferometers. *J. Acoust. Soc. Am.* **24**: 86.
23. Jacobs, B., Carain, R., Olson, J.R., and Aunne, R.C. (1990). Vibrational relaxation in $s_i f_\gamma$ using a computer controlled ultrasonic interferometer. *J. Acoust. Soc. Am.* **88**: 2812–2815.
24. Angora, F.A. (1953). Attenuation of sound in a tube. *J. Acoust. Soc. Am.* **25**: 336.
25. Shields, F.D. (1959). Measurement of thermal relaxation in cO_2 extended to 300 °C. *J. Acoust. Soc. Am.* **31**: 248.
26. Bass, H.E., Winter, T.G., and Evans, L.B. (1971). Vibrational and rotational relaxation in sulfer dioxide. *J. Chem., Phys.* **54**: 644.
27. Shields, F.D., Bass, H.E., and Bolen, L.N. (1977). Review of acoustical patents. *J. Acoust. Soc. Am.* **62**: 236–253.
28. Cravens, D., Shields, F.D., Bass, H.E., and Breshears, W.D. (1979). Vibrational relaxation of UF_6: ultrasonic measurements in mixtures with argon and N_2. *J. Chem. Phys.* **71**: 2797–2802.
29. Shields, F.D. and Lagemann, R.T. (1957). Tube corrections in the study of sound absorption. *J. Acoust. Soc. Am.* **29**: 470.
30. Shields, F.D. and Faughn, J. (1969). Sound velocity and absorption in low-pressure gases confined to tubes of circular cross section. *J. Acoust. Soc. Am.* **46**: 158. Shields, F.D. (1975). An acoustical method for determining the thermal and momentum accommodation coefficients of gases on solids. *J. Chem. Phys.* **62**: 1248.
31. Shields, F.D. and Faughn, J. (1969). Sound velocity and absorption in low-pressure gases confined to tubes of circular cross section. *J. Acoust. Soc. Am.* **46**: 158. Shields, F.D. (1975). An acoustical method

for determining the thermal and momentum accommodation coefficients of gases on solids. *J. Chem. Phys.* **62**: 1248.

32. Bass, H.E. and Yan, H.X. (1983). Pulsed spectrophone measurements of vibrational energy transfer in CO_2. *J. Acoust. Soc. Am.* **74**: 1817.
33. Shields, F.D. (1960). Sound absorption in the halogen gases. *J. Acoust. Soc. Am.* **32**: 180.
34. Lewis, J.W.L. and Lee, K.P. (1965). Vibrational relaxation in carbon dioxide/water vapor mixtures. *J. Acoust. Soc. Am.* **38**: 813.
35. Shields, F.D. and Burks, J.A. (1968). Vibrational relaxation in CO_2/D_2O mixtures. *J. Acoust. Soc. Am.* **43**: 510.
36. Shields, F.D. (1969). Sound absorption and velocity in H_2S and CO_2/H_2S mixtures. *J. Acoust. Soc. Am.* **45**: 481.
37. Shields, F.D. and Carney, G.P. (1970). Sound absorption in D_2S and CO_2/D_2S mixtures. *J. Acoust. Soc. Am.* **47**: 1269.
38. Bass, H.E. and Shields, F.D. (1974). Vibrational relaxation and sound absorption in O_2/H_2O mixtures. *J. Acoust. Soc. Am.* **56**: 856.
39. Anderson, B., Shields, F.D., and Bass, H.E. (1974). Vibrational relaxation and sound absorption in O_2/H_2O mixtures. *J. Acoust. Soc. Am.* **56**: 856.
40. Shields, F.D. (1967). Vibrational relaxation in SO_2 and SO_2/Ar mixtures. *J. Chem. Phys.* **46**: 1063.
41. Shields, F.D. and Anderson, B. (1971). More on vibrational relaxation in SO_2/Ar mixtures. *J. Chem. Phys.* **55**: 2636.
42. Shields, F.D. and Lee, K.P. (1963). Sound absorption and velocity measurements in oxygen. *J. Acoust. Soc. Am.* **35**(2): 251.
43. Hirschfelder, J.O., Curtiss, C.F., and Bird, R.B. (1958). Molecular theory of gases and liquids. New York: Wiley.
44. National Bureau of Standards, Talks of Thermodynamic and Transport Properties of Air, Argon, Carbon Dioxide, Carbon Monoxide, Hydrogen, Nitrogen, Oxygen and Steam, (Pergamon, New York, 1960).
45. U.S. Standard Atmosphere, 1962, U.S. Gov. Print. Off. Washington, DC.
46. Society of Automotive Engineers, Committee A-21, Standard Values of Atmospheric Absorption as a Function of Temperature and Humidity for Us in Evaluating Aircraft Flyover Noise, SAE Aerosp. Recomm. Pract. ARP 86 (6), (August, 1964).
47. Greenspan, M.J. (1959). Rotational relaxation in nitrogen, oxygen and air. *J. Acoust. Soc. Am.* **31**: 155.
48. Bass, H.E. and Keeton, R.G. (1975). ultrasonic absorption in air at elevated temperatures. *J. Acoust. Soc. Am.* **58**: 110.
49. Holman, J.P. (1969). Thermodynamics. New York: McGraw-Hill, **169**.
50. Evans, L.B., Bass, H.E., and Sutherland, L.C. (1972). Atmospheric absorption of sound: theoretical predictions. *J. Acoust. Soc. Am.* **51**: 1565.
51. Cottrell, T.L. and McCoubrey, J.C. (1961). Molecular energy transfer in gases. Butterworth: London.
52. Bertin, G. and Lin, C.C. In *Spiral Structure in Galaxies, A Density Wave Theory* (MIT Press, 1996, Cambridge, Massachusetts).
53. Meyer, E. and Sessler, G. (1957). *Z. Phys.* **149**: 15.
54. Hassler, H. (1968). *Acoustica* **20**: 271.
55. Walsman, L., Handbach der Physik XII, pp. 345 (Springer, Heidelberg, 1958).
56. Grad, H., Handbach der Physik XII, pp. 205–295 (Springer, Heidelberg, 1958).
57. Maidnik, G. and Heckl, M. (1965). *Phys. Fluids* **8**: 266.
58. Kahn, D. and Mintzer, D. (1965). *Phys. Fluids* **8**: 1090.
59. Douglas Shields, F. (1987). Propagation of sound in vibrationally excited n_2/mixtures. *J. Acoust. Soc. Am.* **81**: 87–92.
60. Hamilton, M.F. and Blackstock, D.T. (1998). Nonlinear acoustic. New York: Academic Press.

CHAPTER 11

SOUND SPEED AS A THERMODYNAMIC PROPERTY OF FLUIDS

Daniel G. Friend
Physical and Chemical Properties Division, Chemical Science and Technology Laboratory, National Institute of Standards and Technology, Boulder, Colorado, USA

Contents

Abstract		267
11.1	Introduction	267
11.2	Wave Equation in a Fluid	268
11.3	Equations of State for Fluids	274
11.4	Speed of Sound in Fluids	280
11.5	Conclusions	289
11.6	Tables of the Speed of Sound in Important Fluids	294
Additional Reading		328
References		328

ABSTRACT

In this Chapter, we review the principles of sound propagation in fluid systems. From a study of the hydrodynamic equations, sound propagation is shown to be a wave phenomenon. The speed of sound then can be derived at any state point from a knowledge of the thermodynamic surface of the fluid of interest. Several model equations of state are reviewed, and it is shown how the speed of sound can be obtained for a variety of systems. We then focus on several fluids of particular interest, and show the behavior of the sound speed over a wide range of the temperature and pressure variables. Tabulated values of the speed of sound are given for argon, nitrogen, water, and air based on the current standard reference thermodynamic surfaces.

11.1. INTRODUCTION

In this chapter, we discuss the propagation of sound in fluids and provide information about the thermodynamic speed of sound over substantial ranges of the state variables for a variety of fluids. In the context of this chapter, we consider sound to arise from a small periodic and isentropic (constant entropy) perturbation of the local equilibrium in a fluid, which, as we shall see, gives rise to a standard wave equation. The systems under consideration include both pure fluids and mixtures in the liquid, vapor, and supercritical states. Thus the range in temperature is from the melting line to very high temperatures (a dissociation limit) and the range in pressure is from very low values (below which the continuum approximation would not be valid) to the solidification locus (at least in principle).

The major theme of the discussion stems from the relation

$$w = \left\{ \frac{\rho}{M_U M_r} \left[2 \frac{\partial A}{\partial \rho} \bigg|_T + \rho \frac{\partial^2 A}{\partial \rho^2} \bigg|_T - \rho \frac{\left(\frac{\partial^2 A}{\partial \rho \partial T}\right)^2}{\frac{\partial^2 A}{\partial T^2}\big|_\rho} \right] \right\}^{1/2} \quad (11.1)$$

which expresses the sound speed, w, in terms of derivatives of the molar Helmholtz free energy A and the state point defined by the temperature T and the molar density ρ. In Eq. 11.1, M_U is the molar mass constant (0.001 kg/mol) and M_r is the relative molar mass of the fluid. Although Eq. 11.1 may not be the simplest expression for the sound speed, it serves to emphasize that this thermodynamic quantity is easily calculable if the Helmholtz free energy of a fluid system is known. In fact, this expression is also appropriate for fluid mixtures, in which case it is understood that the partial derivatives are to be taken at constant composition.

In the next section, we outline how Eq. 11.1 can be obtained. Although this thermodynamic relationship is straightforward and exact, a few caveats must be given regarding its use. First there are issues as to the suitability of the "thermodynamic" sound model in any application: for instance, this requires the isentropic approximation, and thus there are formal restrictions on the amplitude and frequency of the propagated disturbance. In fact these issues are seldom significant when considering sound in most fluid media under most conditions. The second caveat, however, is more essential: there are only a few fluid systems for which the Helmholtz free energy surface is sufficiently well known to allow sound speeds to be accurately calculated. In fact, these surfaces are generally empirical correlations of experimental data, and only when accurate sound speeds are included in the primary data used to determine the thermodynamic surface, are the resultant calculated sound speeds obtained with low uncertainty. Experimental determination of sound speeds, using state-of-the-art techniques as discussed in other chapters, usually produces more accurate values for this quantity. However, more theoretical or predictive models for the Helmholtz free energy surface may be used to obtain useful approximations for the sound speed for systems that have not been thoroughly characterized through experimental measurements.

Once we have completed a derivation of Eq. 11.1 in the next section, we explore some of the limits of this approach, and discuss the application to various model systems and to real fluids. Table 11.1 included at the end of this chapter, provides reliable speed of sound values for a variety of important fluid systems.

11.2. WAVE EQUATION IN A FLUID

From a standard understanding of thermodynamic conventions, it may be difficult to see how sound propagation can be considered a thermodynamic property. Despite a straightforward parsing of the name "thermodynamics," typical thermodynamic studies deal with unperturbed systems. Other responses to perturbing stresses, such as flow associated with shear stresses, heat flux related to thermal gradients, and diffusive flows caused by composition variations, are not considered thermodynamic in origin, and the separate study of steady-state kinetic theory may be required to obtain viscosity, thermal conductivity, and diffusion coefficients. The measurement of thermodynamic properties, as with most other physical quantities, may require a perturbing probe (i.e., a temperature change to determine heat capacities), but, in general, thermodynamic processes are equilibrium phenomena. As we will see, it is the linear, reversible, and small amplitude nature of the wave process that serves to connect sound propagation with fluid thermodynamics.

11.2.1. Hydrodynamic Equations

When fluids are considered in the continuum approximation, there are three conservation laws that can be used to determine the response to perturbing influences in an otherwise equilibrated system. These hydrodynamic equations enforce the conservation of mass (or the number of molecules of each species present), the conservation of momentum, and the conservation of energy.

The continuity equation,

$$\frac{\partial \rho}{\partial t} + \nabla \bullet (\rho \mathbf{u}) = 0 \tag{11.2}$$

ensures that the molar density (at a given fixed infinitesimal volume in space) can change only with a flow of molecules into or out of the volume; here, the partial derivative is with respect to time t, and $\mathbf{u}$ represents the macroscopic velocity of fluid flow at the given point. We use the Euler convention in which the frame of reference is fixed in space, so that properties such as ρ are functions of position ($\mathbf{r}$) and time and represent microscopic averages within a volume element centered on $\mathbf{r}$ in this fixed reference frame. The Navier-Stokes equation

$$\left(\frac{\partial}{\partial t} + \mathbf{u} \bullet \nabla\right) \mathbf{u} = \frac{\mathbf{F}}{m} - \frac{1}{M_U M_r \rho} \nabla(P - \eta_B \nabla \bullet \mathbf{u}) + \frac{\eta}{M_U M_r \rho} \nabla^2 \mathbf{u} \tag{11.3}$$

is equivalent to Newton's second law and balances changes in momentum with forces acting on the fluid element. The operator $\left(\frac{\partial}{\partial t} + \mathbf{u} \bullet \nabla\right)$ is the material derivative or streaming derivative, and the left side of the equation describes changes in momentum due directly to fluid flow. The forces that act to change the momentum include the external force $\mathbf{F}$, here divided by the mass m contained in the infinitesimal volume element under consideration, the force on the element due to gradients of pressure P, and forces associated with the viscous nature of any fluid: here, η_B is the bulk viscosity and η is the usual shear viscosity; the pressure tensor has been decomposed into its parts, and a linear relationship between the stress and shear has been assumed [1]. Finally, the conservation of energy can be written in the form

$$\left(\frac{\partial}{\partial t} + \mathbf{u} \bullet \nabla\right) T = -\frac{R}{C_v}(\nabla \bullet \mathbf{u})T + \frac{\lambda}{\rho C_v} \nabla^2 T \tag{11.4}$$

where T is the absolute temperature, R is the gas constant, C_v is the molar isochoric heat capacity, and λ is the coefficient of thermal conductivity in a linear, Fourier's law, relationship between thermal gradients and heat flux.

Equations 11.2 to 11.4 are often simplified in the basic development of a wave equation formalism. In particular, we will first consider an ideal fluid with no viscous effects or thermal conduction. In the absence of external forces, the Navier-Stokes relationship given by Eq. 11.3 then reduces to the Euler equation,

$$\left(\frac{\partial}{\partial t} + \mathbf{u} \bullet \nabla\right) \mathbf{u} = -\frac{1}{M_U M_r \rho} \nabla P \tag{11.5}$$

Small disturbances of an equilibrium state can cause sound propagation, and in the derivation of the wave equation, second order effects in the derivatives can be ignored;

because the equilibrium state corresponds to zero macroscopic velocity, any quadratic terms in **u** will also be dropped. These approximations are almost always quite satisfactory in typical acoustic situations, and we will provide some quantitative discussion below.

To lowest order, Eq. 11.5 can then be linearized and written as

$$\nabla P + M_U M_r \rho_o \frac{\partial \mathbf{u}}{\partial t} = 0 \tag{11.6}$$

where we have now introduced the notation x_o to denote the time and spatially invariant equilibrium value of a quantity. (In this notation, a physical quantity can be represented by $x = x_o + x_A$, where x_A is then the acoustic variable; i.e., P_A is the acoustic pressure or the deviation of the pressure caused by a perturbation from its equilibrium value.) To the same order, Eq. 11.2 is written as

$$\frac{\partial \rho}{\partial t} + \rho_o \nabla \cdot \mathbf{u} = 0 \tag{11.7}$$

Taking the time derivative of Eq. 11.6 and the divergence of Eq. 11.7 and subtracting the results gives

$$M_U M_r \frac{\partial^2 \rho}{\partial t^2} - \nabla^2 P = 0 \tag{11.8}$$

Equation 11.8 forms the basis of the wave equation when we recall that the pressure and density can be related through the thermodynamic equation of state or Helmholtz energy surface. For a pure fluid, the pressure can be considered as a function of the density and a second intensive thermodynamic variable (and for mixtures, the composition must also be given); without specifying the second variable, we write $P = P(\rho, \alpha)$, so that derivatives of the pressure can be written in terms of derivatives of the density and the parameter α. Again we drop some higher order terms and can write Eq. 11.8 as

$$M_U M_r \frac{\partial^2 \rho}{\partial t^2} - \frac{\partial P}{\partial \rho}\bigg|_\alpha \nabla^2 \rho - \nabla \cdot \left(\frac{\partial P}{\partial \alpha}\bigg|_\rho \nabla \alpha \right) = 0 \tag{11.9}$$

If we can identify an appropriate variable α which remains constant through the perturbing acoustic signal (i.e., $\nabla \alpha \equiv 0$), then Eq. 11.9 reduces to a standard propagating wave equation with wave speed $[(1/M_U M_r)\partial P/\partial \rho|_o]^{1/2}$. The earliest theoretical acoustic derivations of the wave propagation, developed by Newton, essentially assumed that the temperature is constant as the disturbance travels through the system. Although this may seem like a reasonable approximation, it is incorrect and leads to an incorrect value of the speed of sound.

To lowest order, the situation can be further analyzed by examining the energy equation, Eq. 11.4, recalling that we are considering only the ideal fluid with $\lambda = 0$. We can then write

$$\frac{\partial T}{\partial t} - \frac{RT}{C_V} \nabla \cdot \mathbf{u} = 0 \tag{11.10}$$

and use Eq. 11.7 to obtain

$$\frac{\partial}{\partial t}\left[\ln T - \frac{R}{C_V} \ln \rho \right] = 0 \tag{11.11}$$

where we have been consistent in keeping terms of lowest order and have assumed that the isochoric heat capacity is constant to this order. The time invariance of the quantity in the brackets means that $T\rho_V^{-R/C}$ is constant during the perturbation. Using the ideal gas equation of state, this leads to the constancy of $P\rho_{VV}^{-(C+R)/C}$, which is a common expression for an adiabatic process in an ideal gas. (Recall that the isobaric head capacity for an ideal gas is given by $C_P = C_V + R$, and the exponent of ρ is then γ, the ratio of heat capacities which has a value of 5/3 for an ideal monatomic gas [2].)

Thus, at least in this limit, the perturbing force leads to an adiabatic process which implies that the entropy is constant during passage of an acoustic disturbance. This turns out to be generally a good approximation, and we choose entropy, S, for the variable α in Eq. 11.9. With $\nabla S = 0$ then, we can write the wave equation

$$M_U M_r \frac{\partial^2 \rho}{\partial t^2} - \left. \frac{\partial P}{\partial \rho} \right|_S \nabla^2 \rho = 0 \qquad (11.12)$$

to represent the hydrodynamic effect of a small disturbance on a fluid in equilibrium. In common with the interpretation of other wave equations in physical systems, we conclude that the perturbation propagates unchanged with a propagation speed given by

$$w = \sqrt{\frac{1}{M_U M_r} \left. \frac{\partial P}{\partial \rho} \right|_S} \qquad (11.13)$$

Equation 11.13 works remarkably well for a wide variety of fluid acoustic situations despite the approximations made in its derivation. This common expression for sound speed is often not used directly, because derivatives at constant entropy are not often tabulated. Using a standard definition of the adiabatic compressibility, $\kappa_S = (1/\rho)\partial \rho / \partial P|_S$, we can write

$$w^2 = \frac{1}{M_U M_r \rho \kappa_S} \qquad (11.14)$$

while standard thermodynamic relations allow us to write

$$w^2 = \frac{C_P}{M_U M_r C_V} \left. \frac{\partial P}{\partial \rho} \right|_T \qquad (11.15)$$

and the relations with the Helmholtz free energy then give us the expression in Eq. 11.1. Hence we conclude that a small amplitude disturbance in an equilibrium ideal fluid will propagate as an isentropic wave with wave speed given by Eq. 11.1; this is a sound wave.

11.2.2. The Speed of Sound

In statistical mechanics, an ideal gas is often considered to be a system of noninteracting point particles; the monatomic noble gases at low densities approximate this ideal condition. In this case, the equation of state can be written as $P = \rho RT$, the molar isochoric heat capacity is generated from the three translational degrees of freedom as $3R/2$, and the isobaric heat capacity is $5R/2$. For this system, Eq. 11.15 immediately gives us a value of 322.59 m/s for the speed of sound in argon ($M_r = 39.948$) at 300 K. The current standard reference value at atmospheric pressure is 322.67 m/s, from which we can conclude that this simple theory and model works remarkably well. Returning to Eq. 11.1, note that the Helmholtz free energy for such a monatomic ideal gas can be written as

$$A^{ig}(\rho, T) = RT[A^0 + \ln \rho - 3/2 \ln T] \qquad (11.16)$$

where A^0 is an integration constant with which we need not be concerned [3]. With this equation for A, the derivatives required for Eq. 11.1 are easily calculable with the same result

$$w^2 = \frac{5RT}{3M_U M_r} \qquad (11.17)$$

as previously seen for an ideal monatomic fluid.

More generally, an ideal gas can have additional degrees of freedom and internal structure. This is reflected in different values or temperature dependences in the ideal gas heat capacities of structured molecules, or, equivalently, in additional temperature dependence in the expression for the ideal gas Helmholtz free energy. Thus with estimates for the translational, rotational, vibrational, and electronic energies based on spectroscopic information, and the ideal gas heat capacities, based on such models as the rigid-rotor harmonic oscillator, can be calculated from statistical mechanics. From this information, values for the speed of sound of low-density systems can be calculated. Alternatively, accurate measurements of the speed of sound at low density can be used to provide estimates of the ideal gas properties of a molecular system. However, if the frequency associated with the sound is not commensurate with the relaxation time of the modes contributing to the ideal gas calculations, corrections must be considered when connecting measured sound speeds to ideal gas behavior [4].

Interactions between the particles in a fluid are extremely important when considering the speed of sound, or any other thermodynamic properties, outside the limited region in which the ideal gas approximation is valid. In this case, Eq. 11.16 is not an appropriate representation of the fluid's Helmholtz free energy, and Eq. 11.17 cannot be used to determine the speed of sound. Equations 11.1 and 11.13 through 11.15 are appropriate for compressed gases at elevated pressures, for saturated or compressed liquids, and for supercritical fluids. Explicit examples for the use of these equations in determining the speed of sound in a variety of fluids and state variables are given Section 11.4.

There are two general classes of approximations that we have made in order to derive the wave equation and the equivalent speed of sound expressions of Eqs. 11.1 and 11.13 through 11.15. These require that (1) the amplitude of the wave be small so that the equations can be linearized and 2 reversibility must be maintained so that isentropic behavior can be assumed. We shall briefly explore each of these requirements in this Section.

The small amplitude condition is easily fulfilled in most common examples of sound propagation. The restriction can be made that the maximum relative deviation of the density, the acoustic "condensation," is very small, $|\rho - \rho_o|/\rho_o << 1$; alternatively, one can require that $|P - P_o| << \rho_o w^2$. This condition can be satisfied if the pressure perturbations are much smaller than the equilibrium value of the pressure, which is typically the case.

To further quantify the small amplitude restriction, we note that the sound intensity level can be defined as the logarithmic ratio of an average energy flux, approximated for a plane wave as $w^3 M_U M_r (\rho - \rho_o)^2/(2\rho_o)$, to a reference value (typically 10^{-12} W/m^2). In air at ambient conditions, the small amplitude approximation breaks down (i.e., $|\rho - \rho_o|/\rho = 1$) at about 194 decibels (dB); for comparison, the range of hearing at 1000 Hz is about 0 to 140 dB. For liquid water at ambient conditions, this condition would hold up to about 240 dB. Thus we can conclude that, except in the cases of extraordinarily large "shock" pressures, the linear approximations hold for sound propagation.

The more significant restriction on our derivation of the sound equation is associated with the irreversibility caused by the viscous and heat flux effects that were ignored despite their importance in Eqs. 11.3 and 11.4. These effects lead to both attenuation of sound and dispersion—the dependence of the speed of sound on the frequency. We will not go into details here, but the inability of a fluid to instantaneously return to an equilibrium state when perturbed can be related to the relaxation times that govern the fluid's behavior. One can make various approximations to the full hydrodynamic equations to obtain corrections to the speed of sound equations given previously.

The equations can be solved in the complex plane, and to lowest order, the viscous and thermal effects are found to leave the speed of sound unchanged, and contribute only to a frequency-dependent absorption [5]. More generally, one can obtain an expansion of the form

$$w^2(f) = w^2(0)[1 + aX + O(X^2)] \quad (11.18)$$

for the real part of the speed of sound as a function of the frequency, f; the expansion parameter, X, is related to the relaxation times and the frequency, and the coefficient a can be evaluated for certain models or treated as an empirical parameter. The solution given by Eq. 11.1 represents the zero frequency limit, and, depending on the particular relaxation mechanisms present, some dispersion effects may be seen at frequencies above 1 GHz. Note that the high frequency signals correspond to small wavelengths, and it is at small wavelengths that the continuum approximation, essential in the development of hydrodynamic approaches to the fluid state, inappropriate. Finally, we note that the fluctuation-driven phenomena near the critical point in any fluid system also contribute to dispersion of the speed of sound. Details concerning these aspects of the acoustic approximation can be found in the texts listed in "Additional Reading."

11.3. EQUATIONS OF STATE FOR FLUIDS

From the previous discussion, we conclude that the speed of sound in a fluid can be determined from its Helmholtz free energy surface; see Eq. 11.1) In general, an expression for A is formed from the sum of ideal gas and residual contributions: $A(\rho T) = A^{ig}(\rho, T) + A^r(\rho, T)$. The ideal gas term represents energy associated with noninteracting particles and their internal degrees of freedom—i.e., from the statistical partition function involved with translational, electronic, vibrational, and rotational motions. This quantity can be computed from models of the molecular architecture (i.e., rigid-rotator, harmonic oscillator model for nonlinear polyatomic molecules) and spectroscopic data. Following the format of Eq. 11.16, the ideal gas portion of the Helmholtz energy can be written in general as

$$A^{ig}(\rho, T) = RT[A^0 + \ln \rho + f(T)] \quad (11.19)$$

where f gives all of the temperature dependence. It is often sufficient to consider only the A^{ig} contribution for gas phase systems at very low density; however, the residual contribution becomes important as the pressure is increased. The residual contribution to A describes the effects of interparticle interactions.

An equation of state, typically an expression for the pressure as a function of temperature and density (or volume), can also be decomposed into the ideal gas and residual contributions. The ideal gas term in the equation of state is known theoretically as $P^{ig} = \rho RT$, and it is the ideal gas heat capacity (or related thermodynamic quantity) that is used to obtain A^{ig}. Accurate expressions for the residual contributions to A or P are generally based on empirical thermodynamic data. The same decomposition of the Helmholtz free energy into ideal and residual parts holds also for mixtures; in this case, the ideal gas contribution is simply related to a mole fraction average of the pure components' ideal gas values, and cross-interaction effects make the residual values more difficult to determine. In Section 11.4, we see how these two parts of A contribute to the speed of sound in fluids.

In Figures 11.1 and 11.2 we sketch the basic phase topology of simple fluid systems to understand the scope of the Helmholtz energy formulation for fluid systems. All pure fluids have phase diagrams similar to Figure 11.1; however, quantum fluids may look somewhat different at low temperatures, and some fluids may decompose at temperatures within the area depicted. The solid-fluid phase boundary intersects the vapor-liquid equilibrium (VLE) phase boundary at the triple point, a single value of temperature and pressure at which the solid, liquid, and vapor phases coexist. The

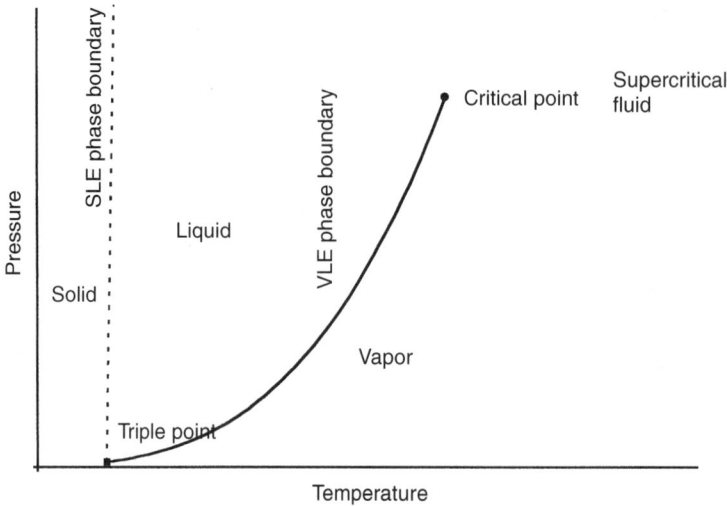

Fig. 11.1. Pressure vs temperature phase diagram of a simple pure fluid. SLE is solid-liquid equilibrium and VLE is vapor-liquid equilibrium.

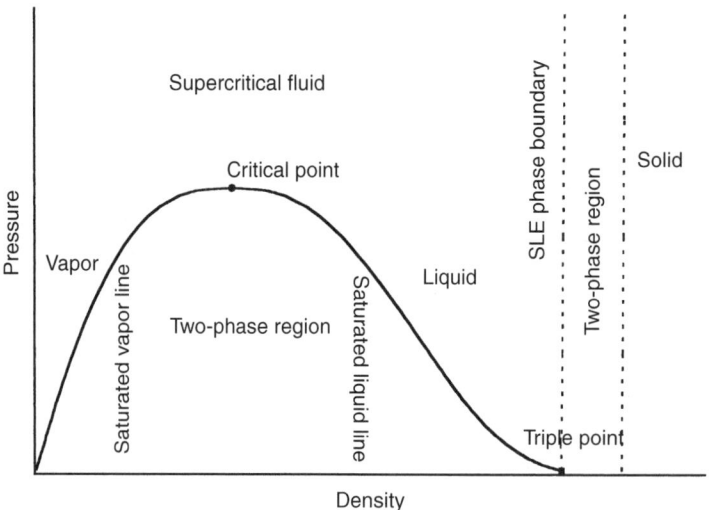

Fig. 11.2. Pressure vs density phase diagram for a simple pure fluid.

melting line extends, in principle, to infinite pressures (although there are, of course, other phenomena at extreme conditions), but the VLE phase boundary terminates at a critical point. For temperatures above the critical-point temperature, the fluid is said to be in the supercritical phase; for lower temperatures, the fluid is a vapor at pressures below the VLE phase boundary and a liquid at higher pressures (to the solid-fluid boundary). Along the VLE phase boundary, the fluid is said to be in the saturated liquid or saturated vapor state; inside the two-phase region of Figure 11.2, the equilibrium state consists of two phases, and the speed of sound is not uniquely defined.

The phase topology for mixtures can become significantly more complex. In general, for a system at fixed composition, the VLE phase boundary is no longer a simple line. The liquid will vaporize at a bubble point, and the bubbles may have a composition different from that of the coexisting liquid. The vapor condenses at a dew point, and the coexisting liquid may also have a different composition. The locus of dew points and bubble points for a fixed composition may be joined by a critical point. In addition, there may be liquid-liquid immiscibility in portions of the phase diagram. The speed of sound is well defined for each one-phase point, and the same equations given previously can be used for mixtures.

11.3.1. Simple Equations of State

There are several ways to represent the residual portion of the thermodynamic surfaces of fluids. The virial series is a density expansion for low densities based on statistical-mechanics cluster approaches [3]. The pressure can be written as

$$P = \rho RT[1 + B(T)\rho + C(T)\rho^2 + \ldots = \rho RT[1 + \sum_{i=1} B_i(T)\rho^i] \quad (11.20)$$

In this notation, B is the second virial coefficient, but B_i is the $(i+1)$ virial coefficient. Although the second and, perhaps, third virial coefficients can be determined from experimental data, virial coefficients of higher order are difficult to measure, and the series itself does not converge at higher densities. For gases at low and moderate density (i.e., below the saturation boundary or at pressures less than about 1 MPa), this type of representation can be quite accurate.

Equation 11.20 yields

$$A^r(\rho, T) = RT \sum_{i=1} \frac{1}{i} B_i(T) \rho^i] \quad (11.21)$$

and, using Eq. 11.1, the speed of sound for the virial equation of state is given by

$$w^2 = \frac{1}{M_U M_r} \left[RT + RT \sum_{i=1} (i+1) B_i \rho^i \right. $$
$$\left. - \frac{R + 2RT \sum_{i=1} \left(\frac{B_i}{T} + B_i' \right) \rho^i + RT^2 \sum_{i,j=1} \left(\frac{B_i}{T} + B_i' \right) \left(\frac{B_j}{T} + B_j' \right) \rho^{i+j}}{2f' + Tf'' + T\sum_{i=1} \frac{\rho^i}{i} \left(\frac{2B_i'}{T} + B_i'' \right)} \right] \quad (11.22)$$

where primes and double primes indicate the first and second temperature derivatives, respectively, of the virial coefficients, B_i and the temperature term in A^{ig}, f from Eq. 11.19. It is straightforward but tedious to expand the denominator in Eq. 11.22 to write a density series of the form

$$w^2 = \frac{RT\gamma^o}{M_U M_r}[1 + \sum_{i=1} K_i(T)\rho^i] \quad (11.23)$$

where γ^o is the ratio of the ideal gas heat capacities, $1 - (2Tf + T^2 f'')^{-1}$, in the current notation. The first two terms are given by [6].

$$K_1(T) = 2B_1 + 2(\gamma^o - 1)TB_1' + \frac{(1-\gamma^o)^2}{\gamma^o} T^2 B_1'' \quad (11.24)$$

and

$$K_2(T) = \frac{\gamma^o - 1}{\gamma^o}[B_1 + (2\gamma^o - 1)TB_1' - (1-\gamma^o)T^2 B_1'']^2 + \frac{2\gamma^o + 1}{\gamma^o} B_2$$
$$+ \frac{\gamma^{o2} - 1}{\gamma^o} TB_2' + \frac{(\gamma^o - 1)^2}{2\gamma^o} T^2 B_2'' \quad (11.25)$$

Thus, if the virial coefficients are known, they can be used to calculate the speed of sound in a gas.

In these expressions, we see the intimate connection between the ideal gas value of the speed of sound and the full expression for this property. In Eq. 11.23, the ideal gas contribution is identified with the zero-density value, $w^{ig} = (RT\gamma^o/M_U M_r)^{1/2}$, but this contribution from internal degrees of freedom, γ^o, also affects each term in the expression for the speed of sound as seen in Eqs. 11.24 and 11.25. It is generally true that the ideal gas speed of sound is defined by the low-density limit, and that the internal structure also is coupled to the propagation of sound at high densities.

The family of cubic equations of state represent another simple form to describe the thermodynamic surface. In general these equations can be written as [7]

$$P = \frac{\rho RT}{1 - b\rho} - \frac{a\rho^2}{1 + ub\rho + vb^2\rho^2} \tag{11.26}$$

where u and w serve to distinguish between the type of cubic equation (e.g., van der Waals, Peng–Robinson, etc.), b is typically a fluid dependent constant, and a is also fluid dependent, but may be constant or a function of temperature, depending on the model being considered. The residual Helmholtz free energy can then be written as

$$A^r(\rho, T) = -RT \ln(1 - b\rho) - \frac{a^2}{\rho\sqrt{u^2 - 4v}} \ln \frac{2 + b\rho(u + \sqrt{u^2 - 4v})}{2 + b\rho(u - \sqrt{u^2 - 4v})} \tag{11.27}$$

and the speed of sound becomes

$$w^2 = \frac{1}{M_U M_r} \left[\frac{RT}{(1 - b\rho)^2} - \frac{a\rho(2 + bu\rho)}{[vb^2\rho^2 + bu\rho + 1]^2} \right.$$
$$\left. - \frac{\left[\frac{R}{1 - b\rho} - \frac{a'\rho}{vb^2\rho^2 + bu\rho + 1} \right]^2}{2Rf' + RTf'' - \frac{a''}{b\sqrt{u^2 - 4v}} \ln \frac{2 + b\rho(u + \sqrt{u^2 - 4v})}{2 + b\rho(u + \sqrt{u^2 - 4v})}} \right] \tag{11.28}$$

The primes indicate temperature derivatives. The expression simplifies considerably for the original van der Waals equation (where $u = v = 0$ and a is constant), but the general expression of Eq. 11.28 can be used for any of the common cubic equations of state. (Note that in the case of $u = w = 0$, the logarithmic expression can be expanded, and the full equation is still well defined.)

Cubic equations are quite often used to describe important fluids in engineering applications. These equations can provide reasonably good descriptions of a thermodynamic surface, (except near the critical point), and liquid-vapor phase boundary information can be calculated directly from this approach. The sound speeds calculated from expressions such as that given previously may be adequate approximations. For more accurate descriptions of the speed of sound (and other properties), the reference quality equations, as discussed in Section 11.3.2., are required.

In addition to these simple equations that can be used to approximate the properties of real fluid systems, we mention that there are also model systems for which properties can be calculated. For instance, the hard sphere fluid, consisting of (structureless) spherical particles of diameter σ and relative mass M_r, interacts with a potential function which is infinite when the particles collide and zero otherwise. This model forms the basis for many computer simulations and theoretical results. The equation of state for this fluid has been shown empirically to be well described by the Carnahan–Starling expression [8]

$$P = \rho RT \frac{1 + \eta + \eta^2 - \eta^3}{(1 - \eta)^3} \tag{11.29}$$

where $\eta = \pi\sigma^3 \rho N_A/6$ and N_A is the Avogadro number. This leads to a Helmholtz energy expression of the form

$$A = RT \left\{ A^o + \ln \rho - \frac{3}{2} \ln T + \frac{4\eta - 3\eta^2}{(1-\eta)^2} \right\} \tag{11.30}$$

and Eq. 11.1 immediately gives

$$w^2 = \frac{RT}{M_u M_r} \left\{ \frac{5 + 10\eta - 3\eta^2 - 24\eta^3 + 37\eta^3 - 22\eta^5 + 5\eta^6}{(1-\eta)^6} \right\} \tag{11.31}$$

Unfortunately, we are not aware of any computer simulations which have generated the speed of sound of the hard sphere system, but the model can be used for real fluids with an effective hard sphere diameter regressed from available data. Of course, alternative ideal gas behavior can also be incorporated into such a hard sphere model for real fluids.

11.3.2. Other Equation of State Descriptions

The most accurate equations of state are those that seek to represent all available thermodynamic data to within their experimental uncertainty. These are generally expressed by the pressure as a function of temperature and density together with a correlation for the ideal gas properties, or directly as the Helmholtz free energy as a function of temperature and density. Such equations are available for industrially important fluids and are suitable for purposes such as instrument calibration and custody transfer applications. These equations are developed by regression of large numbers of data over a broad range of state variables and may have 30 to 40 adjustable coefficients to describe the vapor, liquid, and supercritical phases, including the phase boundary information. Of course there are also thermodynamic surfaces of intermediate complexity compared with the simple cubic or virial coefficient representations and the reference quality equations being considered here.

Examples of the accurate reference quality equations of state include the Benedict–Webb–Rubin family of equations, which have the form of

$$P = \rho RT + \sum_{i=1}^{N_1} G_i \rho^{n_i} T^{m_i} + \sum_{i=N_1}^{N} G_i \rho^{n_i} T^{m_i} \exp(\gamma \rho^2) \tag{11.32}$$

Typical equations of this form [9] have 32 coefficients G_i, with the exponents of density ranging from 2 to 13 and the exponents of temperature ranging from -4 to $\frac{1}{2}$; several important fluids have been described by this functional form with fixed exponents, n_i and m_i by regressing only the coefficients G_i based on the available experimental data on density, heat capacity, phase boundary, speed of sound, etc. This type of equation can also be converted into an expression for the Helmholtz free energy, and the speed of sound can be calculated from Eq. 11.1 as before.

Recently, several equations of the form

$$A(\rho, T) = A^{ig}(\rho, T) + \sum_{i=1}^{N_1} G_i \delta^{n_i} \tau^{m_i} + \sum_{i=N_1}^{N_2} G_i \delta^{n_i} \tau^{m_i} \exp(\delta^{l_i}) + \sum_{i=N_2}^{N} G_i \delta^{n_i} \tau^{m_i} \exp[g(\delta, \tau)] \tag{11.33}$$

have been developed for important fluids, where $\delta = \rho/\rho_c$, $\tau = T_c/T$, the subscript "c" indicates a value at the critical point, and g is a function that improves the description of the thermodynamic surface in the critical region. These equations are also determined empirically from the best available experimental data; in this case, the numbers of terms in each sum, the exponents n_i, m_i, l_i, and the function g may be optimized from the data. The resultant "structurally optimized" thermodynamic surface generally provides the best general representation over the full range of fluid states, and the speed of sound determined using Eqs. 11.1 and 11.33 provide an accurate estimate of the property. The sounds speed tabulated Section 11.6 were generated from models such as those of Eqs. 11.32 and 11.33

When reference quality thermodynamic surfaces are not available for a fluid of interest, predictive models for the surface can be used to determine the speed of sound. The residual Helmholtz energy surfaces of similar fluids generally obey a corresponding states principle [3, 7]; thus, with two-scale factors, representing changes in energy and distance scales, one can generate an approximate surface for a fluid based on the thermodynamic surface of a more well-studied fluid. This approach allows reasonable estimates of the speed of sound with a minimal amount of information. The cubic equations of state of Eq. 11.26 can be implemented when values for the critical temperature and pressure are known; the more complex cubic equations require additional information, and other extensions of the corresponding states principle can be implemented that make use of available experimental information. The molecular structure must be considered explicitly to approximate the ideal gas effects, which are quite significant; group additivity and related methods can be useful for this problem when spectroscopic data are not available. Direct predictions of the heat capacities and compressibility can also be made; these enter into the speed of sound calculation using Eq. 11.15. Alternatively, corresponding states or related models can be developed for the speed of sound itself, and these can be considered independently of the full Helmholtz energy surface.

Fluid mixtures generally obey the same thermodynamic and hydrodynamic principles as pure fluids, and thus equations of state can be constructed that look similar to those discussed previously, and the speed of sound can be computed using the same equations, with derivatives being taken at fixed composition. Among the special considerations, we note that the continuity equation, Eq. 11.2, must be revised to reflect conservation of each species present, and the irreversibilities associated with demixing upon passage of a sound perturbation must be considered at high frequencies. In addition, as mentioned previously, the phase topology of mixtures generally differs from that of pure fluids, and care must be taken with any calculation of the speed of sound that the appropriate phase has been identified. Thermodynamic surfaces for mixture can be calculated from a virial equation of state if the pure fluid coefficients are known and the relevant "cross" virial coefficients, representing interactions between the distinct species in the fluid mixture, can be estimated. Cubic equations of state can also be extended to mixtures using well-defined mixing rules and, if available, a set of interaction parameters representing the cross-species effects. Reference quality equations can also be extended to mixtures with appropriate mixing rules; available experimental data over a range of compositions can be used to regress additional correction terms and hence improve the mixture calculations.

Near the critical point of a pure fluid or a mixture, none of the equations of state discussed earlier are adequate. At this singularity in the thermodynamic surface, the behavior is governed by fluctuations that have very long length scales rather than by the comparatively short-ranged correlations observed in other regions of the phase diagram [10]. The phenomenon of "critical slowing" affects the hydrodynamics of a near-critical system, so the derivation of speed of sound as a thermodynamic property must be reexamined [6]. The thermodynamic speed of sound is strictly zero at the critical point, which can be seen by Eq. 11.15 with the additional observation that the strong critical divergence of the isobaric heat capacity is compensated by a similar decrease in the isothermal derivative; thus, the weak divergence of the isochoric heat capacity in the denominator of the expression ensures that the speed of sound approaches zero at the critical point. This value of the speed of sound can be obtained from a classical Helmholtz free energy surface, although a correct description of the approach at the critical point requires a nonanalytical, scaling theory expression for the free energy that is outside the scope of this chapter. The dispersion of sound (i.e., the dependence of the speed of sound on frequency) is very pronounced in the critical region, and it is the low-frequency limit that is governed by these thermodynamic considerations.

In summary, the thermodynamic speed of sound of a fluid system at any temperature, density (or pressure), and composition can be obtained by using Eq. 11.1 if the Helmholtz free energy surface for the fluid is known. The quality of the speed of sound calculated in this manner of course depends on the quality of the thermodynamic surface that is used. The reference quality equations can reliably be used as they should reflect the uncertainty of the available speed of sound data.

11.4. SPEED OF SOUND IN FLUIDS

Argon is one of the simplest fluid systems because of its location on the periodic table, and it is also an important commodity chemical. The system has been well studied, and the reference quality Helmholtz energy equation of Tegeler *et al.* [11] provides an excellent basis for examining the behavior of speed of sound in argon. The equation of [11] is of the form described by Eq. 11.33; in addition, a modified Benedict–Webb–Rubin equation as in Eq. 11.32 is available in [9].

In Figures 11.3 to 11.5, we show the behavior of the speed of sound over a large range of the state variables in various projections. In Figure 11.3, the saturation boundary is shown in bold; at a fixed temperature, the two lines represent the sounds speed in the coexisting vapor and liquid. Note that the sounds speeds approach zero at the critical point. Note from Figures 11.3 and 11.4 that the speed of sound isotherms cross in these pressure projections. The isochores (lines of constant density) are also seen to cross in Figure 11.5; for supercritical fluids, the speed of sound tends to increase with density and with temperature along the isochores. In Figure 11.6, we compare the calculated sound speeds with the data of Estrada-Alexanders and Trusler [12]. The 173 data were obtain between 110 K and 450 K for pressures ranging from 6.83 kPa to 19.26 MPa using a spherical acoustic resonator with experimental uncertainties from

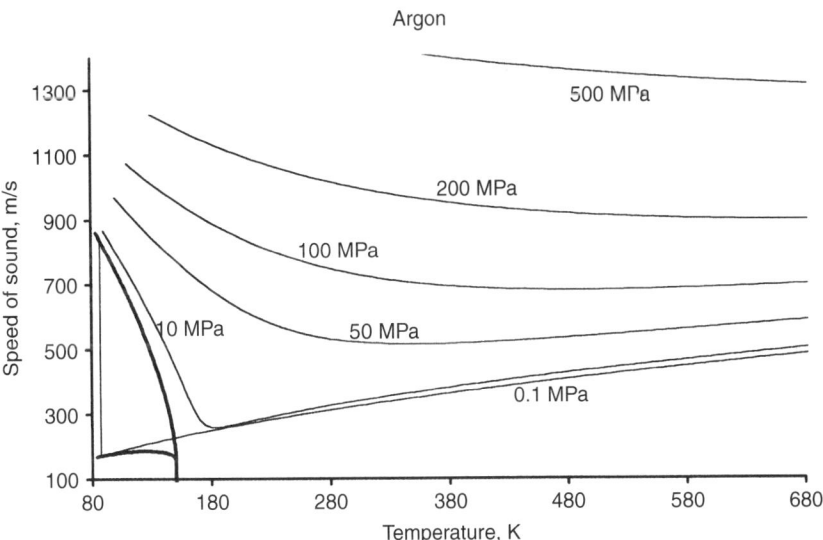

Fig. 11.3. The speed of sound in argon vs temperature along isobars. The heavy lines give values on the saturated liquid and saturated vapor curves.

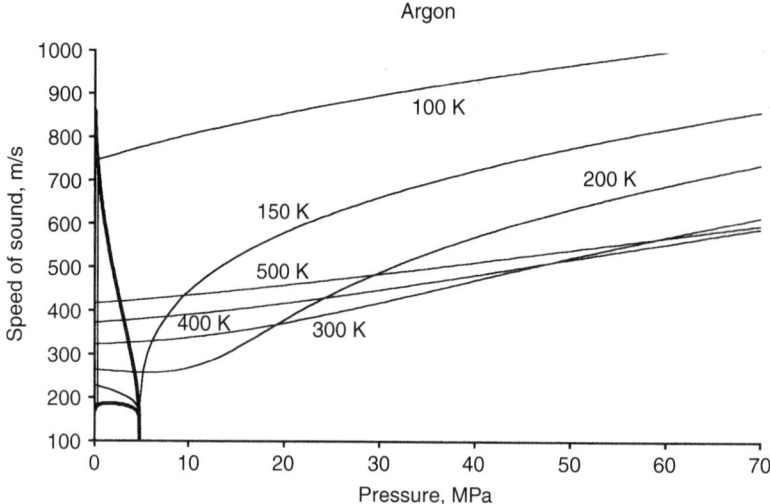

Fig. 11.4. The speed of sound in argon vs pressure along isotherms. The heavy lines give values on the saturated liquid and saturated vapor curves.

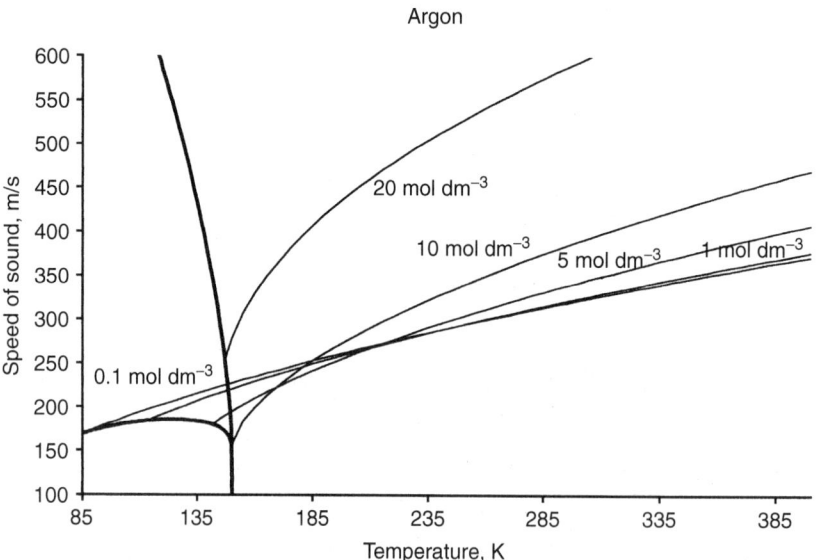

Fig. 11.5. The speed of sound in argon vs temperature along isochores. The heavy lines give values on the saturated liquid and saturated vapor curves.

0.01 to 0.007%. These data were used in the regression of the thermodynamic surface, and the deviations between the data and this reference quality surface are consistent with the experimental uncertainty assessment.

In general, as discussed previously the reference quality equations of state provide calculations of the speed of sound that are consistent quality equations of state provide calculations of the speed of sound that are consistent with the experimental data. In Figure 11.7, we compare the 300 K isotherm for argon as calculated from the equation of state of [11] with the results of various models. A density of 30 mol dm^{-3} for this temperature corresponds to a pressure of about 190 MPa. The dashed line gives the ideal gas value for argon at 300 K, as calculated from Eq. 11.17. This quantity is independent of density, and agrees quite well with the reference equation of state at zero density. In fact, all of the models agree with Eq. 11.17 in this limit. We have

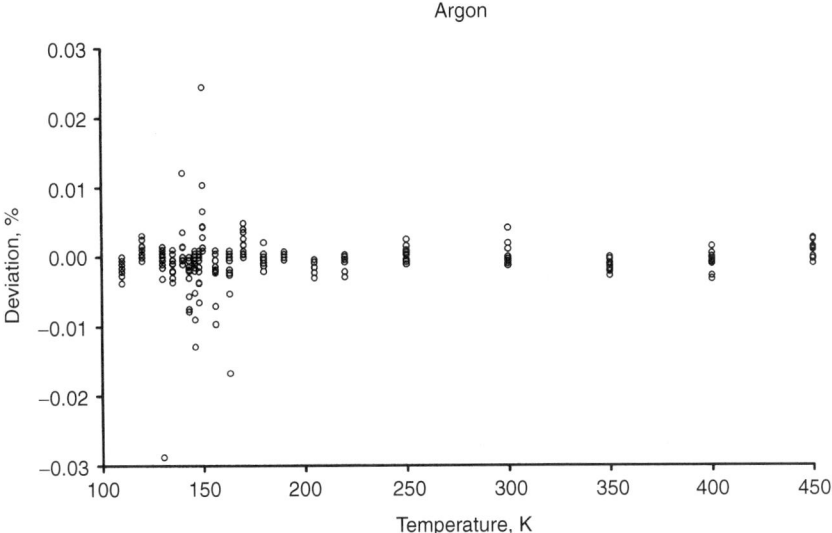

Fig. 11.6. Sample deviations between experimental and calculated values for the speed of sound in argon. Two points (with deviations of −0.26% and 0.54%) are not shown.

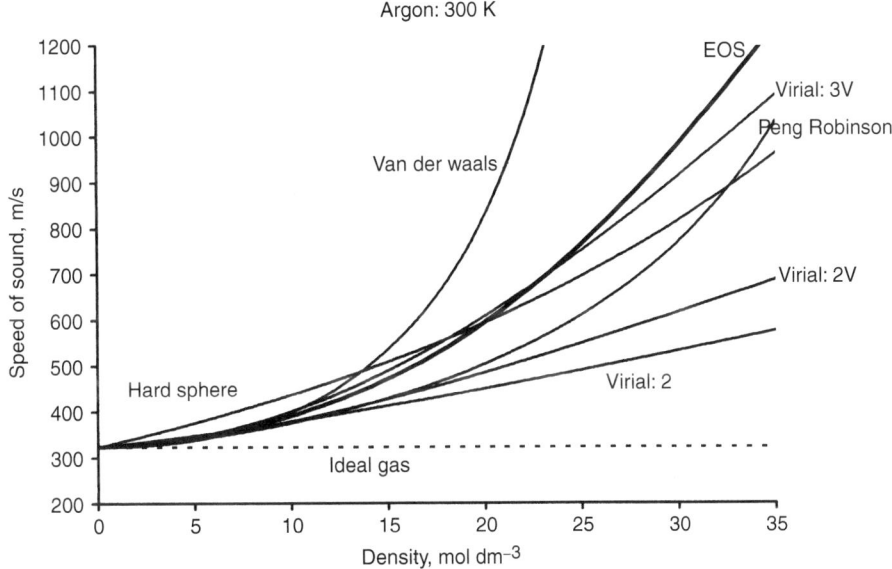

Fig. 11.7. Speed of sound of argon vs density at 300 K. The bold line labeled EOS represents values calculated from the standard reference equation of state. The other lines are values calculated from various models discussed in the text.

included three lines to represent the virial expressions of Eqs. 11.20 to 11.25. The line labeled "Virial: 2" used the first two acoustic virial coefficients [Eqs. 11.24 and 11.25], where the pressure virial coefficients and their derivatives were calculated numerically from a computer implementation of the reference thermodynamic surface of Tegeler et al. [11]. The line "Virial: 2V" used the first two acoustic virial coefficients tabulated by Estrada-Alexanders and Trusler [12] that had been determined from an interatomic potential energy function. The differences in these two lines comes about because of the difficulty in numerically determining the second temperature derivatives of the pressure virial coefficients and the sensitivity of the acoustic virial coefficients of these quantities. The model ("Virial: 3V"), using three virial coefficients as tabulated by

Estrada-Alexanders and Trusler [12], is seen to provide the best representation of the reference line even to the high pressures which are shown.

The van der Waals and Peng–Robinson equations of state are two forms of the family of cubic equations of state. The parameters of both of these equations are determined from the critical parameters of argon and, in the case of the Peng–Robinson equation, the acentric factor for argon. For the van der Waals equations, both u and v of Eq. 11.28 are zero; for the Peng–Robinson equation of state $u = 2$ and $v = -1$. These equations may have problems reproducing experimental densities at high values of the pressure, and the speeds of sound are seen to deviate significantly from the reference line. Finally, Figure 11.7 shows the results for the speed of sound of the hard sphere fluid as given in Eq. 11.31. Only a value of the effective hard sphere diameter, σ is needed in order to evaluated the parameter η. We have determined $\sigma = 2.88 \times 10^{-10}$ m by requiring that the hard sphere result be consistent with the sound speed of argon at 20 mol dm^{-3}. The hard sphere expression seems to give somewhat less curvature than the reference line for this isotherm.

Nitrogen provides our second example, and in Figures 11.8 to 11.10 we indicate the behavior of the speed of sound based on the Helmholtz energy formulation of Span *et al.* [13]. The general behaviors of the curves shown are seen to be quit similar to those of argon; for similar state points, nitrogen generally has a larger value of the speed of sound. Figure 11.11 compares the calculated speed of sound with high-quality experimental measurements on argon of Costa Gomes and Trusler [4]. The data were obtained in a spherical resonate operating at frequencies between 5 kHz and 26 kHz, and the experimental uncertainty were estimated [4] as less than 0.01%. A difference between the measured results at these frequencies and the thermodynamic speed of sound, corresponding to zero frequency, was noted by Costa Gomes and Trusler. Figure 11.11 also shows the data as adjusted to zero frequency in [4]. These zero-frequency data were considered in the regression of Span *et al.* [13] and, as can be seen, the reference equation of state agrees with the data to within their uncertainty.

The properties of water are well described by the standard formulation adopted by the International Association for the Properties of Water and Steam (IAPWS) [14].

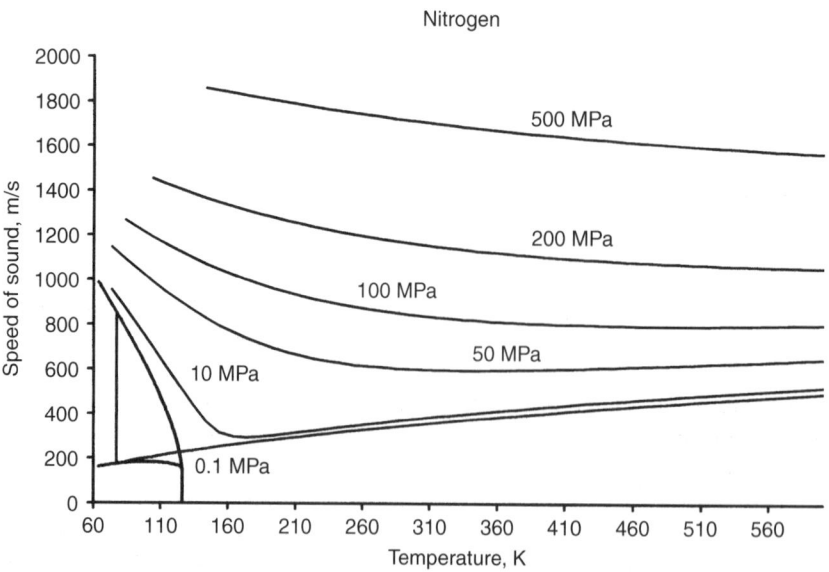

Fig. 11.8. The speed of sound in nitrogen vs temperature along isobars. The heavy lines give values on the saturated liquid and saturated vapor curves.

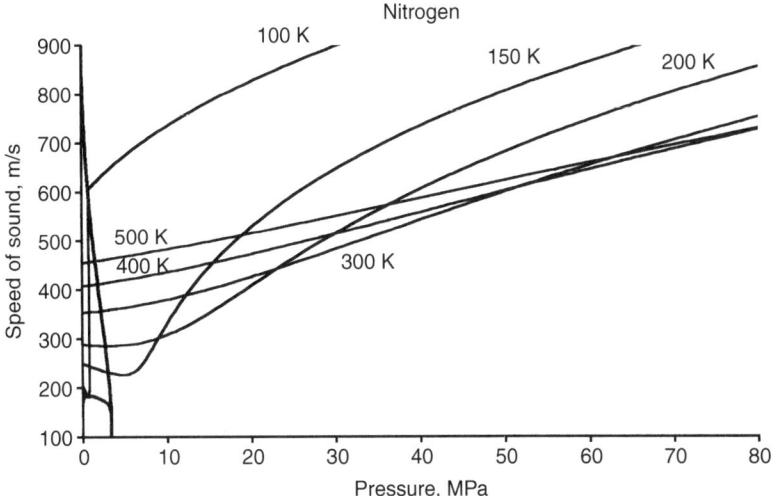

Fig. 11.9. The speed of sound in nitrogen vs pressure along isotherms. The heavy lines give values on the saturated liquid and saturated vapor curves.

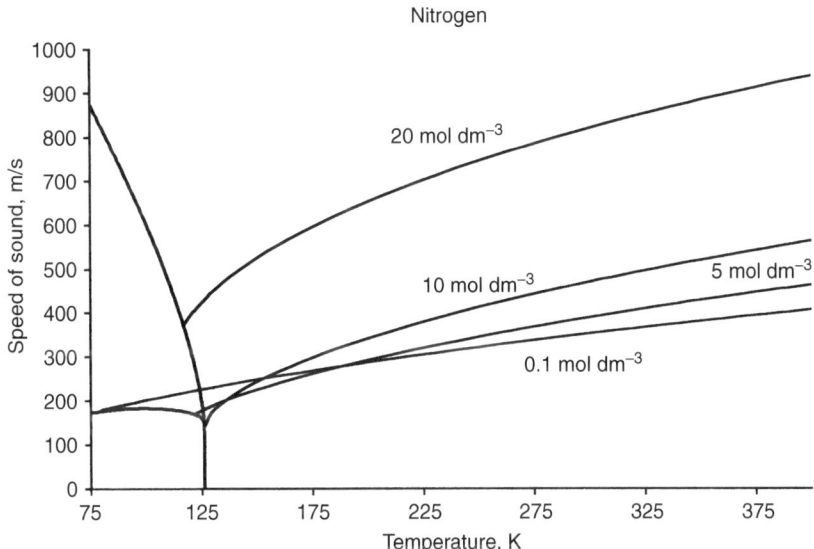

Fig. 11.10. The speed of sound in nitrogen vs temperature along isochores. The heavy lines give values on the saturated liquid and saturated vapor curves.

Figures 11.12 to 11.14 show the behavior of the speed of sound of fluid water. Because of the relatively high critical temperature of water (647.096 K) relative to that of argon and nitrogen, the region of phase equilibria is expanded in these figures. The vertical line segments in Figures 11.12 and 11.13 connected the isobars and isotherms, respectively, across the two-phase region. This is the region of liquid-vapor phase separation, and the thermodynamic speed of sound is not defined for the two-phase systems. Figure 11.12 includes a line representing the critical isobar ($P_c = 22.064$ MPa); this isobar intersects the phase boundary at the critical temperature, whereas all subcritical isobars meet the saturation boundary at liquid and vapor points, and supercritical isobars do not intersect the phase boundary. As mentioned earlier the thermodynamic speed of sound should vanish at the critical point, however, the classical equation for water gives a small but non-zero value. For some equations of state, this value may depend

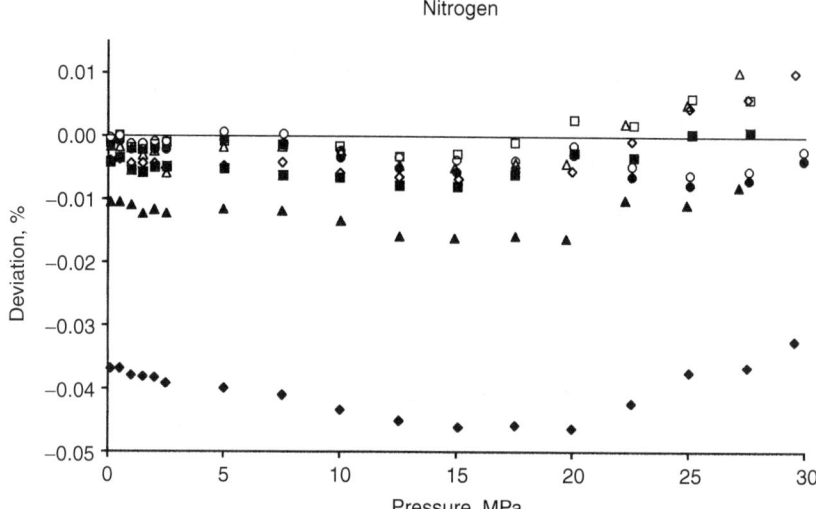

Fig. 11.11. Sample deviations between experimental and calculated values for the speed of sound in nitrogen. The closed symbols represent data obtained at frequencies between 5 kHz and 26 kHz; the open symbols are data corrected to zero frequency. Circles: 250 K; squares: 275 K; triangles: 300 K; diamonds: 350 K.

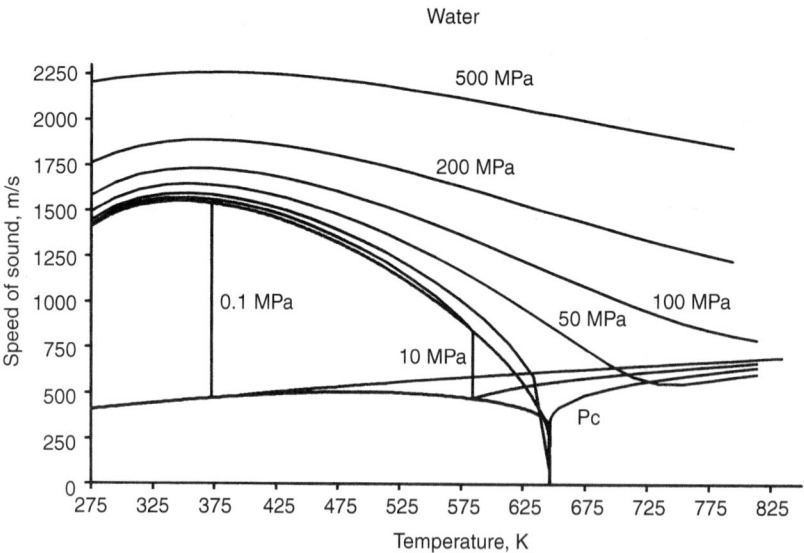

Fig. 11.12. The speed of sound in water vs temperature along isobars. The heavy lines give values on the saturated liquid and saturated vapor curves.

on the computer implementation and the particular calculation performed: a value of about 35 m/s can be obtained on the phase boundary at the critical temperature for the IAPWS equation.

Several features of the behavior of the speed of sound in water can be observed in Figure 11.13. Note that there is a small region where the speed of sound in the saturated vapor (the lower branch of the phase boundary in most of the figures) is greater than that in the coexisting liquid. This occurs for temperatures between about 643 K and the critical temperature. Near the pure-fluid critical point, the vapor and liquid states become quite similar; although the vapor has a smaller density, the speed of sound in the vapor is larger than that of the liquid in this region. The subcritical

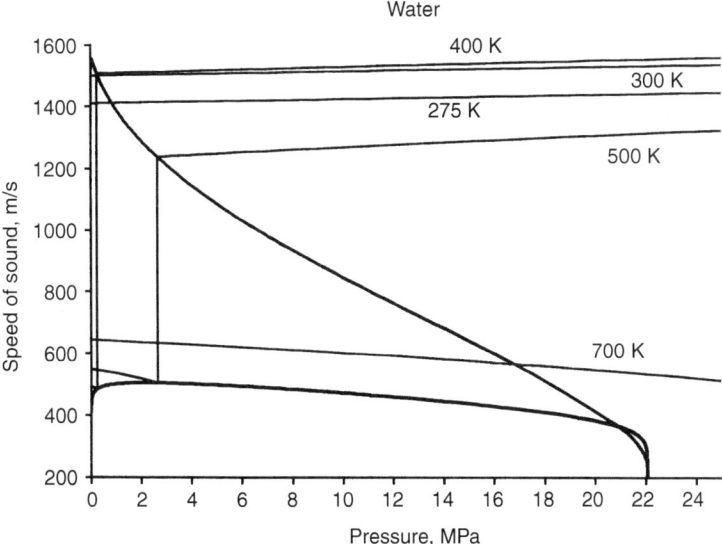

Fig. 11.13. The speed of sound in water vs pressure along isotherms. The heavy lines give values on the saturated liquid and saturated vapor curves.

isotherms in Figure 11.13 illustrate another feature: note that for fixed pressure (at about 18 MPa, to be specific), the speed of sound at 300 K and 400 K is significantly higher than that at 275 K, and the 500-K isotherm is inverted compared to the trend of the others. This behavior is related to the maxima seen in the isobars in Figure 11.12. Figure 11.14 shows the clear maximum of the speed of sound in the saturated liquid and the broader maximum in that saturated vapor phase.

The behavior of the speed of sound in air is shown in Figures 11.15 to 11.17. These values were calculated from the thermodynamic surface of Lemmon *et al.* [15] which provides a reference quality equation of state for dry air considered at the fixed composition of 0.7812 nitrogen, 0.2096 oxygen, and 0.0092 argon by mole fraction. We see that the general behavior of the mixture isobars (Fig. 11.15), isotherms

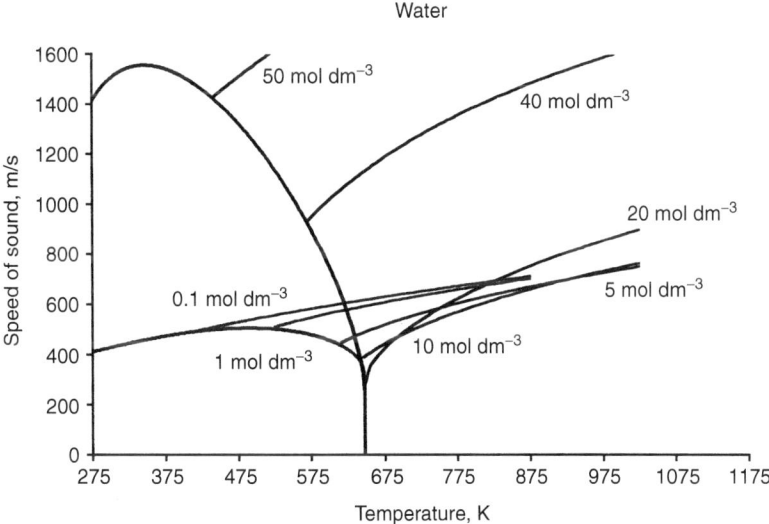

Fig. 11.14. The speed of sound in water vs temperature along isochores. The heavy lines give values on the saturated liquid and saturated vapor curves.

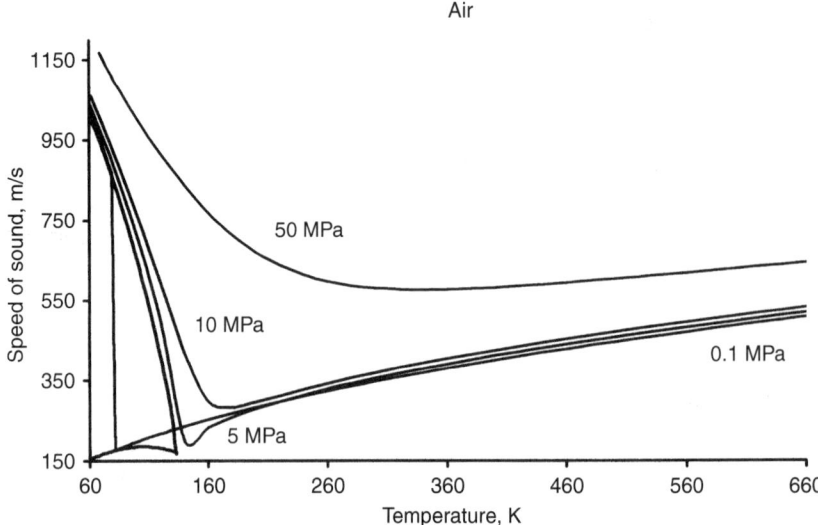

Fig. 11.15. The speed of sound in air vs temperature along isobars. The heavy lines give values on the saturated liquid and saturated vapor curves (bubble-point curve and dew-point curve).

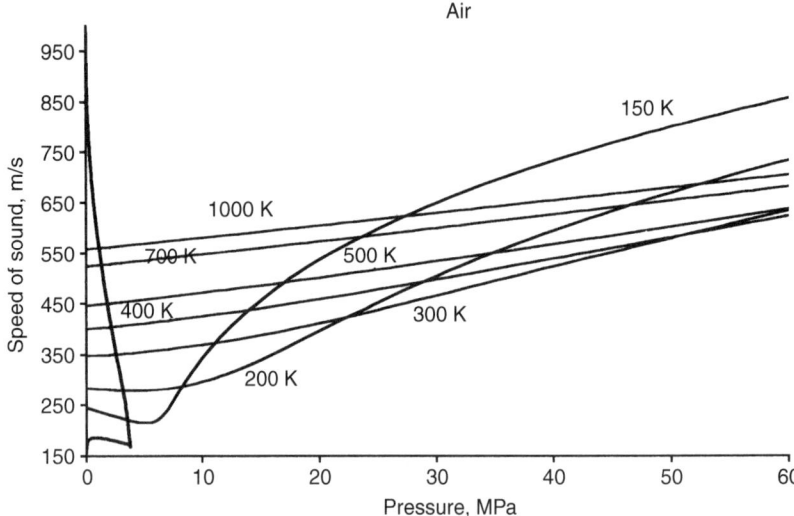

Fig. 11.16. The speed of sound in air vs pressure along isotherms. The heavy lines give values on the saturated liquid and saturated vapor curves (bubble-point curve and dew-point curve).

(Fig. 11.16), and isochores (Fig. 11.17) is quite similar to the trends seen for the pure fluids. Note that the subcritical isobar at 0.1 MPa also shows a segment in the two-phase region. This line is not vertical, however, and does not connect states in equilibrium. The lower saturation boundary is the line of dew points. At 0.1 MPa, the dew-point temperature for the composition specified is at about 81.6 K. The bubble-point temperature is at about 78.8 K. The bubbles in equilibrium with the bulk liquid and the condensate in equilibrium with the bulk vapor have different compositions than the bulk fluid. Again, the thermodynamic speed of sound is not defined in the two-phase region.

Finally, in Figure 11.18 we show some deviations between experimental sound speeds for a natural gas fluid and calculations from a model for the mixture Helmholtz free energy. The data were obtained with a cylindrical cavity for a methane-rich

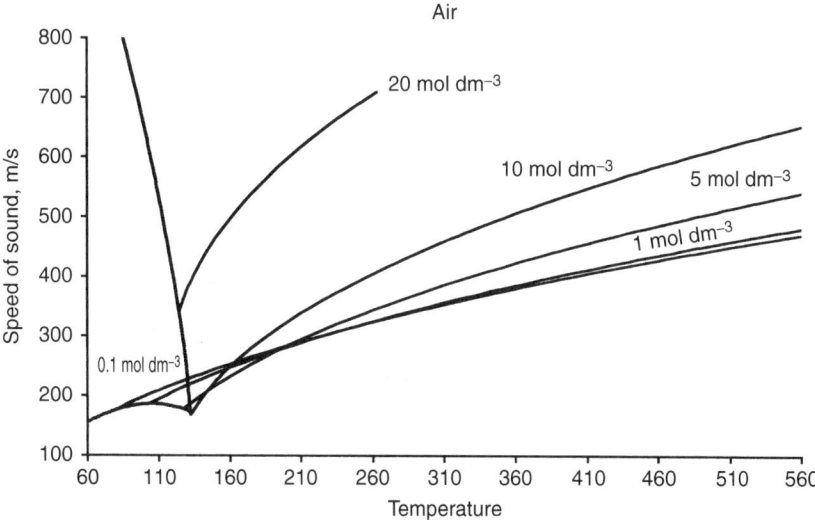

Fig. 11.17. The speed of sound in air vs temperature along isochores. The heavy lines give values on the saturated liquid and saturated vapor curves (bubble-point curve and dew-point curve).

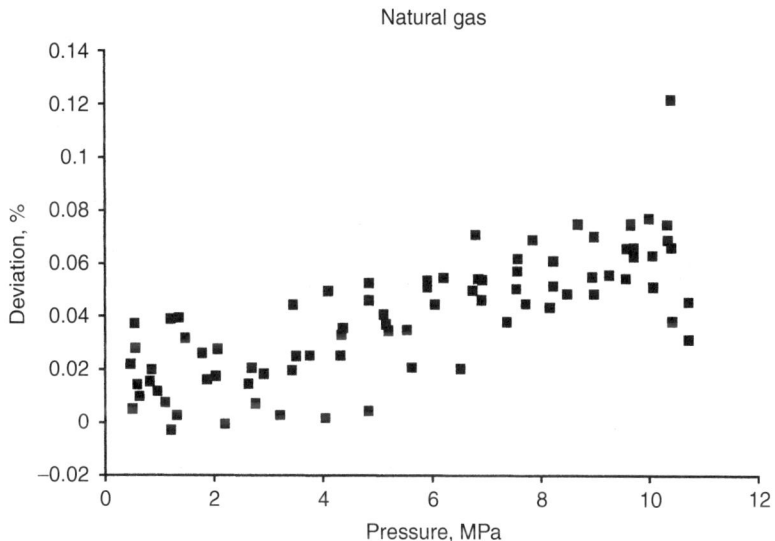

Fig. 11.18. Sample deviations between experimental and calculated values for the speed of sound in a natural gas.

ten-component mixture representing Gulf Coast natural gas [16]. The experimental uncertainty is about 0.05%. The generalized mixture model representing the base line [17] uses the reference quality equation of state for each pure component and a few mixture parameters based primarily on density and phase equilibrium data. The model is seen to agree with the data to nearly within the experimental uncertainty.

11.5. CONCLUSIONS

The thermodynamic properties of a fluid system are completely determined by the Helmholtz free energy surface of the system. We have seen that the speed of sound in such a system can be considered a thermodynamic property, and that it can be calculated from Eq. 11.1 The available sources for the speed of sound in fluid systems can be considered to be of three types: (1) experimental measurements and compilations of experimental measurements can provide the most accurate values for the speed of sound in well-defined systems; (2) reference quality Helmholtz free energy surfaces have been developed for important fluids and can describe the speed of sound essentially to within the uncertainty of available data; (3) models or predictive schemes can be used for systems for which measurements are not available, including more complex molecular systems and mixtures.

Among the sources for experimental data, the tables [18] of the Thermodynamics Research Center provide an extensive compilation for both hydrocarbon and non-hydrocarbon systems. These tables are available in both printed and electronic form. Researchers at the National Institute of Standards and Technology have developed several reference thermodynamic surfaces based on experimental measurements at NIST and elsewhere, and have compiled reference quality surfaces from several other sources. Such information is available in the form of PC-based computer programs such as the NIST Thermophysical Properties of Pure Fluids Database [19]. NIST also provides predictive computer packages that can be used for pure fluids with limited measurements and for mixtures. The catalog of computer packages from the Standard Reference Data office of NIST is available on-line at ***http://www.nist.gov/srd***. Some information on the thermophysical properties of fluids is available in the NIST chemistry webbook at **http://webbook.nist.gov,** although information about the speed of sound in fluids is not included as of the version dated February, 2000. The material in Tables 11.1 through 11.4 (see Appendix) was calculated from these computer programs.

11.6. TABLES OF THE SPEED OF SOUND IN IMPORTANT FLUIDS

Table 11.1. Argon — saturation boundary.

		Liquid		Vapor	
T (K)	P (MPa)	ρ (mol/dm^3)	w (m/s)	ρ (mol/dm^3)	w (m/s)
84	0.07	35.44	861.10	0.104	168.28
85	0.08	35.28	854.24	0.115	169.08
86	0.09	35.13	847.35	0.127	169.87
87	0.10	34.98	840.43	0.140	170.64
88	0.11	34.82	833.47	0.155	171.39
89	0.12	34.67	826.48	0.170	172.12
90	0.13	34.51	819.45	0.186	172.83
91	0.15	34.35	812.39	0.204	173.52
92	0.16	34.19	805.28	0.222	174.20
93	0.18	34.04	798.14	0.242	174.85
94	0.19	33.87	790.95	0.264	175.49
95	0.21	33.71	783.72	0.286	176.11
96	0.23	33.55	776.45	0.310	176.71
97	0.25	33.39	769.14	0.336	177.29
98	0.28	33.22	761.78	0.363	177.85
99	0.30	33.05	754.37	0.392	178.39

Table 11.1. (continued)

T (K)	P (MPa)	Liquid ρ (mol/dm³)	Liquid w (m/s)	Vapor ρ (mol/dm³)	Vapor w (m/s)
100	0.32	32.89	746.91	0.422	178.91
101	0.35	32.72	739.40	0.454	179.41
102	0.38	32.54	731.83	0.488	179.89
103	0.41	32.37	724.21	0.524	180.36
104	0.44	32.20	716.54	0.562	180.80
105	0.47	32.02	708.81	0.601	181.22
106	0.51	31.84	701.02	0.643	181.62
107	0.54	31.66	693.16	0.687	182.01
108	0.58	31.48	685.24	0.733	182.37
109	0.62	31.30	677.26	0.782	182.71
110	0.67	31.11	669.20	0.833	183.03
111	0.71	30.92	661.08	0.887	183.33
112	0.76	30.73	652.87	0.943	183.61
113	0.81	30.54	644.60	1.002	183.87
114	0.86	30.34	636.24	1.065	184.11
115	0.91	30.14	627.79	1.130	184.33
116	0.97	29.94	619.26	1.198	184.52
117	1.02	29.74	610.64	1.269	184.70
118	1.08	29.53	601.93	1.344	184.85
119	1.15	29.32	593.11	1.423	184.98
120	1.21	29.11	584.19	1.506	185.09
121	1.28	28.89	575.16	1.592	185.17
122	1.35	28.67	566.02	1.683	185.23
123	1.43	28.45	556.75	1.778	185.27
124	1.50	28.22	547.36	1.878	185.29
125	1.58	27.98	537.83	1.982	185.28
126	1.66	27.75	528.16	2.092	185.24
127	1.75	27.50	518.34	2.208	185.18
128	1.84	27.25	508.36	2.330	185.10
129	1.93	27.00	498.21	2.458	184.98
130	2.03	26.74	487.88	2.592	184.85
131	2.12	26.47	477.36	2.735	184.68
132	2.23	26.20	466.63	2.885	184.49
133	2.33	25.91	455.69	3.043	184.27
134	2.44	25.62	444.51	3.211	184.02
135	2.55	25.32	433.10	3.389	183.74
136	2.67	25.01	421.42	3.579	183.41
137	2.79	24.68	409.47	3.781	183.03
138	2.91	24.35	397.21	3.996	182.60
139	3.04	23.99	384.62	4.228	182.09
140	3.17	23.62	371.63	4.477	181.50
141	3.30	23.23	358.17	4.747	180.81
142	3.44	22.82	344.14	5.041	180.00
143	3.59	22.38	329.41	5.363	179.05
144	3.74	21.90	313.80	5.719	177.93
145	3.89	21.38	297.06	6.119	176.57
146	4.05	20.81	278.88	6.574	174.89
147	4.21	20.16	258.79	7.106	172.74
148	4.38	19.40	236.08	7.750	169.82
149	4.55	18.45	209.33	8.587	165.39
150	4.73	17.03	174.74	9.875	157.01

Table 11.2. Argon—isobar at 0.101 325 MPa.

T (K)	ρ (mol/dm^3)	w (m/s)
90	0.140	173.81
100	0.125	184.14
110	0.113	193.74
120	0.103	202.78
130	0.095	211.38
140	0.088	219.59
150	0.082	227.49
160	0.077	235.09
170	0.072	242.44
180	0.068	249.56
190	0.064	256.48
200	0.061	263.21
210	0.058	269.76
220	0.056	276.15
230	0.053	282.39
240	0.051	288.50
250	0.049	294.48
260	0.047	300.33
270	0.045	306.07
280	0.044	311.70
290	0.042	317.24
300	0.041	322.67
310	0.039	328.02
320	0.038	333.27
330	0.037	338.45
340	0.036	343.55
350	0.035	348.57
360	0.034	353.52
370	0.033	358.40
380	0.032	363.21
390	0.031	367.96
400	0.030	372.65
410	0.030	377.28
420	0.029	381.86
430	0.028	386.38
440	0.028	390.85
450	0.027	395.27
460	0.026	399.63
470	0.026	403.95
480	0.025	408.23
490	0.025	412.46
500	0.024	416.65
510	0.024	420.79
520	0.023	424.90
530	0.023	428.96
540	0.023	432.99
550	0.022	436.98
560	0.022	440.93

Table 11.2. (continued)

T (K)	ρ (mol/dm³)	w (m/s)
570	0.021	444.85
580	0.021	448.73
590	0.021	452.58
600	0.020	456.40
610	0.020	460.19
620	0.020	463.94
630	0.019	467.67
640	0.019	471.36
650	0.019	475.03
660	0.018	478.67
670	0.018	482.28
680	0.018	485.87
690	0.018	489.42
700	0.017	492.96

Table 11.3. Argon—isobar at 1 MPa.

T (K)	ρ (mol/dm³)	w (m/s)
90	34.579	824.17
100	32.955	751.59
110	31.157	672.31
116.5981	29.821	614.12
116.5981	1.240	184.63
120	1.182	189.35
130	1.046	201.51
140	0.945	212.09
150	0.865	221.67
160	0.799	230.54
170	0.744	238.88
180	0.697	246.79
190	0.656	254.34
200	0.619	261.59
210	0.587	268.58
220	0.558	275.34
230	0.532	281.89
240	0.509	288.26
250	0.488	294.47
260	0.468	300.52
270	0.450	306.44
280	0.433	312.22
290	0.418	317.88
300	0.403	323.44
310	0.390	328.88
320	0.377	334.23
330	0.366	339.49
340	0.355	344.66
350	0.344	349.74
360	0.335	354.75

continues

Table 11.3. (continued)

T (K)	ρ (mol/dm³)	w (m/s)
370	0.325	359.68
380	0.317	364.54
390	0.309	369.33
400	0.301	374.06
410	0.293	378.72
420	0.286	383.33
430	0.280	387.87
440	0.273	392.36
450	0.267	396.80
460	0.261	401.19
470	0.256	405.52
480	0.250	409.81
490	0.245	414.06
500	0.240	418.25
510	0.235	422.41
520	0.231	426.52
530	0.226	430.59
540	0.222	434.63
550	0.218	438.62
560	0.214	442.58
570	0.211	446.50
580	0.207	450.39
590	0.203	454.24
600	0.200	458.06
610	0.197	461.85
620	0.194	465.60
630	0.190	469.33
640	0.187	473.02
650	0.185	476.69
660	0.182	480.33
670	0.179	483.94
680	0.176	487.52
690	0.174	491.08
700	0.171	494.61

Table 11.4. Argon — isobar at 5 MPa.

T (K)	ρ (mol/dm³)	w (m/s)
90	34.879	844.78
100	33.343	777.34
110	31.682	705.93
120	29.828	628.46
130	27.655	541.00
140	24.831	433.54
150	19.159	248.19
160	6.090	206.39
170	4.947	223.58
180	4.307	236.49
190	3.867	247.45

Table 11.4. (continued)

T (K)	ρ (mol/dm^3)	w (m/s)
200	3.534	257.20
210	3.269	266.11
220	3.051	274.40
230	2.865	282.20
240	2.706	289.60
250	2.566	296.66
260	2.443	303.44
270	2.332	309.98
280	2.232	316.29
290	2.142	322.42
300	2.060	328.37
310	1.984	334.17
320	1.914	339.82
330	1.849	345.34
340	1.789	350.74
350	1.734	356.03
360	1.681	361.22
370	1.632	366.31
380	1.586	371.31
390	1.543	376.23
400	1.502	381.06
410	1.464	385.82
420	1.427	390.51
430	1.392	395.13
440	1.359	399.69
450	1.328	404.18
460	1.298	408.62
470	1.270	412.99
480	1.242	417.32
490	1.216	421.59
500	1.191	425.81
510	1.167	429.99
520	1.145	434.12
530	1.123	438.20
540	1.101	442.25
550	1.081	446.25
560	1.061	450.21
570	1.043	454.13
580	1.024	458.02
590	1.007	461.87
600	0.990	465.68
610	0.973	469.46
620	0.958	473.21
630	0.942	476.92
640	0.928	480.60
650	0.913	484.26
660	0.899	487.88
670	0.886	491.47
680	0.873	495.04
690	0.860	498.58
700	0.848	502.09

Table 11.5. Argon—isobar at 10 MPa.

T (K)	ρ (mol/dm^3)	w (m/s)
90	35.23	868.40
100	33.78	805.91
110	32.25	741.46
120	30.59	674.25
130	28.76	603.33
140	26.67	527.43
150	24.15	445.10
160	20.87	357.76
170	16.44	284.47
180	12.31	259.27
190	9.89	260.68
200	8.45	267.70
210	7.49	275.72
220	6.78	283.74
230	6.23	291.51
240	5.79	298.98
250	5.42	306.16
260	5.11	313.06
270	4.83	319.72
280	4.60	326.15
290	4.38	332.37
300	4.20	338.42
310	4.02	344.29
320	3.87	350.02
330	3.73	355.60
340	3.60	361.05
350	3.48	366.38
360	3.37	371.60
370	3.26	376.72
380	3.17	381.74
390	3.07	386.67
400	2.99	391.51
410	2.91	396.27
420	2.83	400.96
430	2.76	405.58
440	2.70	410.12
450	2.63	414.60
460	2.57	419.02
470	2.51	423.38
480	2.46	427.69
490	2.41	431.94
500	2.36	436.14
510	2.31	440.29
520	2.26	444.39
530	2.22	448.45
540	2.18	452.46
550	2.13	456.43
560	2.10	460.36
570	2.06	464.25
580	2.02	468.10
590	1.99	471.91

Table 11.5. (continued)

T (K)	ρ (mol/dm^3)	w (m/s)
600	1.95	475.69
610	1.92	479.44
620	1.89	483.15
630	1.86	486.83
640	1.83	490.47
650	1.80	494.09
660	1.77	497.67
670	1.75	501.23
680	1.72	504.76
690	1.70	508.26
700	1.67	511.73

Table 11.6. Argon—isobar at 50 MPa.

T (K)	ρ (mol/dm^3)	w (m/s)
100	36.26	969.74
110	35.19	927.77
120	34.12	887.55
130	33.05	849.02
140	31.97	812.28
150	30.90	777.44
160	29.83	744.66
170	28.76	714.08
180	27.70	685.82
190	26.65	659.98
200	25.63	636.63
210	24.62	615.80
220	23.65	597.47
230	22.71	581.56
240	21.81	567.93
250	20.96	556.40
260	20.14	546.77
270	19.38	538.85
280	18.66	532.42
290	17.98	527.30
300	17.34	523.31
310	16.75	520.31
320	16.19	518.16
330	15.67	516.73
340	15.18	515.91
350	14.72	515.62
360	14.28	515.78
370	13.88	516.32
380	13.50	517.19
390	13.14	518.35
400	12.80	519.74
410	12.48	521.33
420	12.17	523.11
430	11.88	525.03
440	11.61	527.09
450	11.35	529.26

continues

Table 11.6. (continued)

T (K)	ρ (mol/dm³)	w (m/s)
460	11.10	531.52
470	10.87	533.87
480	10.64	536.29
490	10.43	538.77
500	10.22	541.31
510	10.02	543.89
520	9.83	546.50
530	9.65	549.15
540	9.48	551.83
550	9.31	554.54
560	9.15	557.26
570	9.00	560.00
580	8.85	562.75
590	8.71	565.51
600	8.57	568.28
610	8.43	571.05
620	8.30	573.83
630	8.18	576.62
640	8.06	579.40
650	7.94	582.18
660	7.83	584.97
670	7.72	587.75
680	7.61	590.52
690	7.51	593.30
700	7.41	596.07

Table 11.7. Argon — isobar at 100 MPa.

T (K)	ρ (mol/dm³)	w (m/s)
110	37.41	1074.29
120	36.57	1043.43
130	35.74	1014.38
140	34.93	987.00
150	34.13	961.21
160	33.35	936.96
170	32.58	914.20
180	31.83	892.89
190	31.09	873.00
200	30.37	854.48
210	29.66	837.28
220	28.98	821.35
230	28.31	806.65
240	27.66	793.12
250	27.03	780.71
260	26.42	769.35
270	25.82	759.00
280	25.25	749.60
290	24.69	741.08
300	24.15	733.39
310	23.63	726.46
320	23.13	720.25
330	22.65	714.70

Table 11.7. (continued)

T (K)	ρ (mol/dm^3)	w (m/s)
340	22.18	709.76
350	21.73	705.38
360	21.30	701.52
370	20.88	698.13
380	20.47	695.17
390	20.09	692.61
400	19.71	690.42
410	19.35	688.56
420	19.00	687.02
430	18.66	685.75
440	18.34	684.74
450	18.02	683.98
460	17.72	683.43
470	17.43	683.08
480	17.15	682.91
490	16.87	682.92
500	16.61	683.08
510	16.35	683.38
520	16.10	683.81
530	15.86	684.37
540	15.63	685.03
550	15.40	685.80
560	15.19	686.67
570	14.97	687.62
580	14.77	688.65
590	14.57	689.76
600	14.37	690.94
610	14.18	692.17
620	14.00	693.47
630	13.82	694.83
640	13.65	696.23
650	13.48	697.68
660	13.31	699.17
670	13.15	700.71
680	13.00	702.28
690	12.84	703.88
700	12.70	705.52

Table 11.8. Nitrogen — saturation boundary.

		Liquid		Vapor	
T (K)	P (MPa)	ρ (mol/dm^3)	w (m/s)	ρ (mol/dm^3)	w (m/s)
64	0.015	30.83	986.57	0.028	162.08
65	0.017	30.69	976.36	0.033	163.20
66	0.021	30.54	966.18	0.038	164.30
67	0.024	30.39	956.04	0.044	165.37
68	0.028	30.24	945.93	0.051	166.41
69	0.033	30.09	935.83	0.059	167.43
70	0.039	29.93	925.74	0.068	168.42
71	0.045	29.78	915.66	0.077	169.39
72	0.051	29.62	905.58	0.088	170.32
73	0.059	29.47	895.49	0.100	171.23

continues

Table 11.8. Nitrogen — saturation boundary.

T (K)	P (MPa)	Liquid		Vapor	
		ρ (mol/dm^3)	w (m/s)	ρ (mol/dm^3)	w (m/s)
74	0.067	29.31	885.39	0.112	172.10
75	0.076	29.15	875.28	0.126	172.95
76	0.086	28.99	865.15	0.142	173.77
77	0.097	28.83	855.00	0.158	174.55
78	0.109	28.67	844.82	0.176	175.31
79	0.122	28.51	834.61	0.196	176.03
80	0.137	28.34	824.36	0.217	176.72
81	0.153	28.17	814.07	0.240	177.38
82	0.169	28.01	803.74	0.265	178.00
83	0.188	27.84	793.36	0.292	178.60
84	0.208	27.66	782.93	0.320	179.16
85	0.229	27.49	772.44	0.351	179.68
86	0.252	27.32	761.89	0.383	180.17
87	0.276	27.14	751.28	0.418	180.63
88	0.303	26.96	740.60	0.456	181.05
89	0.331	26.78	729.84	0.496	181.43
90	0.360	26.60	719.01	0.538	181.78
91	0.392	26.41	708.09	0.584	182.10
92	0.426	26.22	697.09	0.632	182.38
93	0.462	26.03	685.99	0.683	182.62
94	0.500	25.84	674.80	0.737	182.82
95	0.541	25.64	663.50	0.795	182.99
96	0.583	25.44	652.09	0.856	183.12
97	0.628	25.24	640.57	0.921	183.21
98	0.676	25.03	628.92	0.990	183.26
99	0.726	24.82	617.14	1.063	183.28
100	0.778	24.61	605.23	1.141	183.25
101	0.834	24.39	593.17	1.223	183.18
102	0.892	24.17	580.96	1.310	183.08
103	0.953	23.94	568.58	1.403	182.93
104	1.016	23.71	556.03	1.501	182.74
105	1.083	23.47	543.30	1.605	182.51
106	1.153	23.23	530.37	1.716	182.24
107	1.226	22.98	517.24	1.833	181.93
108	1.303	22.72	503.88	1.958	181.58
109	1.383	22.46	490.29	2.092	181.19
110	1.466	22.18	476.44	2.234	180.76
111	1.553	21.90	462.32	2.386	180.28
112	1.643	21.61	447.92	2.549	179.75
113	1.737	21.31	433.19	2.724	179.15
114	1.835	20.99	418.12	2.912	178.49
115	1.937	20.66	402.67	3.116	177.75
116	2.043	20.31	386.80	3.337	176.93
117	2.153	19.94	370.43	3.579	176.01
118	2.268	19.55	353.49	3.844	175.00
119	2.387	19.13	335.85	4.137	173.87
120	2.511	18.68	317.33	4.465	172.61
121	2.639	18.19	297.68	4.838	171.17
122	2.773	17.63	276.54	5.270	169.49
123	2.912	17.00	253.32	5.785	167.43
124	3.056	16.23	227.00	6.430	164.67
125	3.207	15.21	195.48	7.324	160.26
126	3.365	13.28	150.97	9.111	148.38

Table 11.9. Nitrogen—isobar at 0.101 325 MPa.

T (K)	ρ (mol/dm³)	w (m/s)
70	29.94	926.19
77.355	28.77	851.39
77.355	0.1646	174.82
90	0.1392	190.42
110	0.1125	212.07
130	0.0946	231.36
150	0.0817	248.99
170	0.0720	265.37
190	0.0643	280.74
210	0.0581	295.28
230	0.0530	309.11
250	0.0488	322.33
270	0.0452	335.02
290	0.0420	347.22
310	0.0393	359.00
330	0.0369	370.38
350	0.0348	381.40
370	0.0329	392.08
390	0.0312	402.44
410	0.0297	412.51
430	0.0283	422.31
450	0.0271	431.84
470	0.0259	441.12
490	0.0249	450.16
510	0.0239	458.99
530	0.0230	467.60
550	0.0221	476.02
570	0.0214	484.25
590	0.0206	492.31
610	0.0200	500.20
630	0.0193	507.93
650	0.0187	515.51
670	0.0182	522.95
690	0.0177	530.27
710	0.0172	537.45
730	0.0167	544.53
750	0.0162	551.49
770	0.0158	558.34
790	0.0154	565.10
810	0.0150	571.76
830	0.0147	578.33
850	0.0143	584.82
870	0.0140	591.22
890	0.0137	597.55
910	0.0134	603.80
930	0.0131	609.99
950	0.0128	616.10
970	0.0126	622.15
990	0.0123	628.14
1010	0.0121	634.07
1030	0.0118	639.94
1050	0.0116	645.75
1070	0.0114	651.51

continues

Table 11.9. (continued)

T (K)	ρ (mol/dm³)	w (m/s)
1090	0.0112	657.22
1110	0.0110	662.87
1130	0.0108	668.48
1150	0.0106	674.04
1170	0.0104	679.56
1190	0.0102	685.03
1210	0.0101	690.45
1230	0.0099	695.84
1250	0.0097	701.18
1270	0.0096	706.48
1290	0.0094	711.75
1310	0.0093	716.97
1330	0.0092	722.16
1350	0.0090	727.31
1370	0.0089	732.43
1390	0.0088	737.51
1410	0.0086	742.56
1430	0.0085	747.57
1450	0.0084	752.56
1470	0.0083	757.51
1490	0.0082	762.43
1510	0.0081	767.31
1530	0.0080	772.17
1550	0.0079	777.00
1570	0.0078	781.80
1590	0.0077	786.57
1610	0.0076	791.31
1630	0.0075	796.03
1650	0.0074	800.72
1670	0.0073	805.38
1690	0.0072	810.02
1710	0.0071	814.63
1730	0.0070	819.21
1750	0.0070	823.78
1770	0.0069	828.31
1790	0.0068	832.82
1810	0.0067	837.31
1830	0.0067	841.78
1850	0.0066	846.22
1870	0.0065	850.64
1890	0.0064	855.04
1910	0.0064	859.42
1930	0.0063	863.77
1950	0.0062	868.10
1970	0.0062	872.42
1990	0.0061	876.71

Table 11.10. Nitrogen—isobar at 1 MPa.

T (K)	ρ (mol/dm³)	w (m/s)
70	30.00	932.51
90	26.69	727.06
103.747	23.77	559.22
103.747	1.48	182.79

Table 11.10. (continued)

T (K)	ρ (mol/dm^3)	w (m/s)
110	1.32	194.09
130	1.02	221.53
150	0.8523	243.20
170	0.7359	261.98
190	0.6498	278.91
210	0.5829	294.53
230	0.5291	309.13
250	0.4848	322.92
270	0.4475	336.03
290	0.4158	348.56
310	0.3883	360.58
330	0.3643	372.16
350	0.3432	383.33
370	0.3244	394.13
390	0.3076	404.58
410	0.2925	414.73
430	0.2788	424.57
450	0.2663	434.15
470	0.2549	443.46
490	0.2445	452.53
510	0.2349	461.37
530	0.2260	469.99
550	0.2178	478.42
570	0.2101	486.65
590	0.2030	494.70
610	0.1963	502.58
630	0.1901	510.30
650	0.1843	517.88
670	0.1788	525.31
690	0.1736	532.61
710	0.1687	539.79
730	0.1641	546.84
750	0.1597	553.79
770	0.1556	560.63
790	0.1516	567.37
810	0.1479	574.01
830	0.1443	580.57
850	0.1410	587.04
870	0.1377	593.43
890	0.1346	599.74
910	0.1317	605.98
930	0.1289	612.14
950	0.1262	618.24
970	0.1236	624.27
990	0.1211	630.25
1010	0.1187	636.16
1030	0.1164	642.01
1050	0.1142	647.81
1070	0.1120	653.55
1090	0.1100	659.24
1110	0.1080	664.89
1130	0.1061	670.48
1150	0.1043	676.02

continues

Table 11.10. (continued)

T (K)	ρ (mol/dm³)	w (m/s)
1170	0.1025	681.52
1190	0.1008	686.98
1210	0.0991	692.39
1230	0.0975	697.76
1250	0.0959	703.09
1270	0.0944	708.38
1290	0.0930	713.63
1310	0.0916	718.84
1330	0.0902	724.02
1350	0.0888	729.16
1370	0.0876	734.26
1390	0.0863	739.33
1410	0.0851	744.37
1430	0.0839	749.37
1450	0.0827	754.34
1470	0.0816	759.28
1490	0.0805	764.18
1510	0.0795	769.06
1530	0.0784	773.91
1550	0.0774	778.72
1570	0.0764	783.51
1590	0.0755	788.27
1610	0.0745	793.00
1630	0.0736	797.71
1650	0.0727	802.39
1670	0.0719	807.04
1690	0.0710	811.67
1710	0.0702	816.27
1730	0.0694	820.84
1750	0.0686	825.39
1770	0.0678	829.92
1790	0.0670	834.42
1810	0.0663	838.90
1830	0.0656	843.36
1850	0.0649	847.79
1870	0.0642	852.20
1890	0.0635	856.59
1910	0.0628	860.96
1930	0.0622	865.31
1950	0.0616	869.63
1970	0.0609	873.94
1990	0.0603	878.22
2010	0.0597	882.48

Table 11.11. Nitrogen — isobar at 5 MPa.

T (K)	ρ (mol/dm³)	w (m/s)
70	30.283	959.16
90	27.202	772.11
110	23.385	565.77
130	16.433	288.81
150	6.029	226.09

Table 11.11. (continued)

T (K)	ρ (mol/dm³)	w (m/s)
170	4.387	254.67
190	3.605	277.24
210	3.108	296.46
230	2.753	313.52
250	2.483	329.07
270	2.266	343.47
290	2.089	356.96
310	1.940	369.71
330	1.812	381.84
350	1.702	393.42
370	1.605	404.54
390	1.519	415.24
410	1.442	425.55
430	1.373	435.52
450	1.311	445.18
470	1.254	454.54
490	1.202	463.63
510	1.154	472.48
530	1.111	481.09
550	1.070	489.48
570	1.032	497.67
590	0.997	505.67
610	0.965	513.49
630	0.934	521.15
650	0.906	528.65
670	0.879	536.01
690	0.853	543.24
710	0.830	550.33
730	0.807	557.31
750	0.786	564.17
770	0.765	570.93
790	0.746	577.59
810	0.728	584.15
830	0.711	590.62
850	0.694	597.01
870	0.678	603.32
890	0.663	609.55
910	0.649	615.70
930	0.635	621.79
950	0.622	627.81
970	0.609	633.77
990	0.597	639.66
1010	0.585	645.49
1030	0.574	651.27
1050	0.563	657.00
1070	0.553	662.67
1090	0.543	668.29
1110	0.533	673.86
1130	0.524	679.38
1150	0.515	684.86
1170	0.506	690.29
1190	0.498	695.68
1210	0.490	701.02

continues

Table 11.11. (continued)

T (K)	ρ (mol/dm^3)	w (m/s)
1230	0.482	706.33
1250	0.474	711.60
1270	0.467	716.82
1290	0.460	722.01
1310	0.453	727.16
1330	0.446	732.28
1350	0.439	737.36
1370	0.433	742.40
1390	0.427	747.42
1410	0.421	752.40
1430	0.415	757.34
1450	0.409	762.26
1470	0.404	767.14
1490	0.399	772.00
1510	0.393	776.82
1530	0.388	781.62
1550	0.383	786.38
1570	0.378	791.12
1590	0.374	795.83
1610	0.369	800.52
1630	0.365	805.17
1650	0.360	809.81
1670	0.356	814.41
1690	0.352	818.99
1710	0.348	823.55
1730	0.344	828.08
1750	0.340	832.59
1770	0.336	837.07
1790	0.332	841.53
1810	0.329	845.97
1830	0.325	850.38
1850	0.322	854.78
1870	0.318	859.15
1890	0.315	863.50
1910	0.312	867.82
1930	0.309	872.13
1950	0.305	876.42
1970	0.302	880.68
1990	0.299	884.93

Table 11.12. Nitrogen — isobar at 10 MPa.

T (K)	ρ (mol/dm^3)	w (m/s)
70	30.60	989.62
90	27.75	819.44
110	24.49	647.61
130	20.41	472.34
150	14.96	331.25
170	10.28	292.70
190	7.87	300.47
210	6.53	315.42
230	5.65	330.91

Table 11.12. (continued)

T (K)	ρ (mol/dm^3)	w (m/s)
250	5.02	345.77
270	4.54	359.83
290	4.15	373.13
310	3.84	385.75
330	3.57	397.78
350	3.35	409.27
370	3.15	420.29
390	2.98	430.88
410	2.82	441.09
430	2.69	450.94
450	2.56	460.47
470	2.45	469.71
490	2.35	478.67
510	2.26	487.38
530	2.17	495.85
550	2.09	504.10
570	2.02	512.15
590	1.95	520.01
610	1.89	527.69
630	1.83	535.21
650	1.77	542.57
670	1.72	549.80
690	1.67	556.88
710	1.62	563.84
730	1.58	570.69
750	1.54	577.42
770	1.50	584.05
790	1.46	590.58
810	1.43	597.01
830	1.39	603.36
850	1.36	609.63
870	1.33	615.82
890	1.30	621.93
910	1.27	627.97
930	1.25	633.95
950	1.22	639.86
970	1.20	645.71
990	1.17	651.49
1010	1.15	657.23
1030	1.13	662.90
1050	1.11	668.53
1070	1.09	674.10
1090	1.07	679.62
1110	1.05	685.10
1130	1.03	690.53
1150	1.01	695.92
1170	1.00	701.26
1190	0.981	706.56
1210	0.965	711.82
1230	0.950	717.05
1250	0.935	722.23
1270	0.920	727.37
1290	0.906	732.48

continues

Table 11.12. (continued)

T (K)	ρ (mol/dm^3)	w (m/s)
1310	0.893	737.56
1330	0.880	742.59
1350	0.867	747.60
1370	0.855	752.57
1390	0.843	757.51
1410	0.831	762.42
1430	0.820	767.29
1450	0.809	772.14
1470	0.798	776.96
1490	0.787	781.74
1510	0.777	786.50
1530	0.767	791.23
1550	0.758	795.93
1570	0.748	800.61
1590	0.739	805.26
1610	0.730	809.88
1630	0.721	814.48
1650	0.713	819.05
1670	0.704	823.60
1690	0.696	828.12
1710	0.688	832.62
1730	0.680	837.10
1750	0.673	841.55
1770	0.665	845.98
1790	0.658	850.38
1810	0.651	854.77
1830	0.644	859.13
1850	0.637	863.47
1870	0.630	867.79
1890	0.624	872.09
1910	0.617	876.37
1930	0.611	880.63
1950	0.605	884.87
1970	0.599	889.09
1990	0.593	893.29

Table 11.13. Nitrogen — isobar at 50 MPa.

T (K)	ρ (mol/dm^3)	w (m/s)
90	30.52	1061.02
110	28.54	963.85
130	26.61	878.83
150	24.76	806.60
170	22.99	747.47
190	21.35	701.00
210	19.83	665.92
230	18.45	640.44
250	17.22	622.64
270	16.12	610.73
290	15.14	603.25
310	14.28	599.08
330	13.50	597.38

Table 11.13. (continued)

T (K)	ρ (mol/dm^3)	w (m/s)
350	12.81	597.50
370	12.19	598.98
390	11.64	601.45
410	11.13	604.67
430	10.67	608.44
450	10.25	612.61
470	9.87	617.08
490	9.51	621.78
510	9.18	626.63
530	8.88	631.59
550	8.60	636.63
570	8.33	641.73
590	8.08	646.85
610	7.85	651.99
630	7.64	657.14
650	7.43	662.28
670	7.24	667.41
690	7.05	672.53
710	6.88	677.63
730	6.72	682.70
750	6.56	687.76
770	6.41	692.79
790	6.27	697.80
810	6.14	702.78
830	6.01	707.74
850	5.88	712.67
870	5.77	717.58
890	5.65	722.46
910	5.54	727.31
930	5.44	732.14
950	5.34	736.95
970	5.24	741.73
990	5.15	746.49
1010	5.06	751.22
1030	4.97	755.93
1050	4.89	760.62
1070	4.81	765.28
1090	4.73	769.92
1110	4.66	774.54
1130	4.59	779.14
1150	4.51	783.71
1170	4.45	788.26
1190	4.38	792.79
1210	4.32	797.30
1230	4.25	801.79
1250	4.19	806.26
1270	4.13	810.70
1290	4.08	815.13
1310	4.02	819.54
1330	3.97	823.92
1350	3.92	828.29
1370	3.87	832.64
1390	3.82	836.96

continues

Table 11.13. (continued)

T (K)	ρ (mol/dm^3)	w (m/s)
1410	3.77	841.27
1430	3.72	845.56
1450	3.67	849.83
1470	3.63	854.09
1490	3.59	858.32
1510	3.54	862.54
1530	3.50	866.73
1550	3.46	870.91
1570	3.42	875.08
1590	3.38	879.22
1610	3.34	883.35
1630	3.31	887.46
1650	3.27	891.56
1670	3.24	895.63
1690	3.20	899.70
1710	3.17	903.74
1730	3.13	907.77
1750	3.10	911.78
1770	3.07	915.78
1790	3.04	919.76
1810	3.01	923.73
1830	2.98	927.68
1850	2.95	931.61
1870	2.92	935.53
1890	2.89	939.44
1910	2.86	943.33
1930	2.84	947.20
1950	2.81	951.06
1970	2.78	954.91
1990	2.76	958.74

Table 11.14. Nitrogen — at 100 MPa.

T (K)	ρ (mol/dm^3)	w (m/s)
90	32.57	1244.55
110	31.03	1173.11
130	29.57	1110.14
150	28.19	1055.27
170	26.90	1008.09
190	25.69	967.97
210	24.57	934.20
230	23.52	906.03
250	22.54	882.74
270	21.63	863.64
290	20.78	848.09
310	19.99	835.56
330	19.26	825.55
350	18.58	817.66
370	17.94	811.52
390	17.35	806.84
410	16.79	803.39
430	16.27	800.95
450	15.79	799.36

Table 11.14. (continued)

T (K)	ρ (mol/dm^3)	w (m/s)
470	15.33	798.49
490	14.90	798.21
510	14.49	798.44
530	14.11	799.10
550	13.75	800.12
570	13.41	801.46
590	13.09	803.07
610	12.78	804.91
630	12.49	806.95
650	12.21	809.16
670	11.95	811.53
690	11.69	814.04
710	11.45	816.67
730	11.22	819.40
750	11.00	822.23
770	10.79	825.14
790	10.58	828.12
810	10.39	831.18
830	10.20	834.29
850	10.02	837.46
870	9.85	840.68
890	9.68	843.94
910	9.52	847.25
930	9.36	850.58
950	9.21	853.95
970	9.07	857.35
990	8.93	860.78
1010	8.79	864.23
1030	8.66	867.69
1050	8.53	871.18
1070	8.41	874.68
1090	8.29	878.20
1110	8.17	881.73
1130	8.06	885.27
1150	7.95	888.82
1170	7.84	892.38
1190	7.74	895.95
1210	7.64	899.52
1230	7.54	903.10
1250	7.44	906.68
1270	7.35	910.26
1290	7.26	913.84
1310	7.17	917.43
1330	7.08	921.01
1350	7.00	924.60
1370	6.92	928.18
1390	6.84	931.77
1410	6.76	935.35
1430	6.68	938.93
1450	6.61	942.50
1470	6.54	946.07
1490	6.46	949.64
1510	6.39	953.21

continues

Table 11.14. (continued)

T (K)	ρ (mol/dm³)	w (m/s)
1530	6.33	956.76
1550	6.26	960.32
1570	6.19	963.87
1590	6.13	967.41
1610	6.07	970.95
1630	6.01	974.48
1650	5.95	978.01
1670	5.89	981.53
1690	5.83	985.04
1710	5.77	988.55
1730	5.72	992.05
1750	5.66	995.55
1770	5.61	999.03
1790	5.56	1002.51
1810	5.51	1005.98
1830	5.46	1009.45
1850	5.41	1012.91
1870	5.36	1016.36
1890	5.31	1019.80
1910	5.27	1023.23
1930	5.22	1026.66
1950	5.18	1030.08
1970	5.13	1033.49
1990	5.09	1036.90

Table 11.15. Water — saturation boundary.

T (K)	P (MPa)	Liquid		Vapor	
		ρ (mol/dm³)	w (m/s)	ρ (mol/dm³)	w (m/s)
275	0.0007	55.50	1411.35	0.0003	410.33
277	0.0008	55.50	1420.78	0.0003	411.77
279	0.0009	55.50	1429.78	0.0004	413.21
281	0.0011	55.50	1438.35	0.0005	414.64
283	0.0012	55.49	1446.51	0.0005	416.06
285	0.0014	55.48	1454.29	0.0006	417.48
287	0.0016	55.47	1461.68	0.0007	418.89
289	0.0018	55.45	1468.71	0.0008	420.29
291	0.0020	55.43	1475.39	0.0008	421.69
293	0.0023	55.41	1481.72	0.0010	423.08
295	0.0026	55.38	1487.72	0.0011	424.46
297	0.0030	55.36	1493.41	0.0012	425.84
299	0.0033	55.33	1498.78	0.0013	427.21
301	0.0038	55.30	1503.86	0.0015	428.57
303	0.0042	55.27	1508.64	0.0017	429.93
305	0.0047	55.23	1513.14	0.0019	431.28
307	0.0053	55.20	1517.37	0.0021	432.62
309	0.0059	55.16	1521.33	0.0023	433.96
311	0.0066	55.12	1525.04	0.0026	435.29
313	0.0073	55.08	1528.49	0.0028	436.61
315	0.0081	55.03	1531.70	0.0031	437.93
317	0.0090	54.99	1534.68	0.0034	439.24
319	0.0100	54.94	1537.42	0.0038	440.54

Table 11.15. (continued)

T (K)	P (MPa)	Liquid		Vapor	
		ρ (mol/dm^3)	w (m/s)	ρ (mol/dm^3)	w (m/s)
321	0.0111	54.90	1539.95	0.0042	441.83
323	0.0123	54.85	1542.25	0.0046	443.11
325	0.0135	54.80	1544.35	0.0050	444.39
327	0.0149	54.74	1546.24	0.0055	445.66
329	0.0164	54.69	1547.93	0.0060	446.92
331	0.0180	54.63	1549.43	0.0066	448.17
333	0.0198	54.58	1550.73	0.0072	449.41
335	0.0217	54.52	1551.85	0.0078	450.64
337	0.0238	54.46	1552.80	0.0085	451.87
339	0.0260	54.40	1553.56	0.0093	453.08
341	0.0284	54.34	1554.16	0.010	454.28
343	0.031	54.28	1554.59	0.011	455.48
345	0.034	54.21	1554.85	0.012	456.67
347	0.037	54.15	1554.96	0.013	457.84
349	0.040	54.08	1554.91	0.014	459.01
351	0.043	54.01	1554.71	0.015	460.16
353	0.047	53.95	1554.36	0.016	461.30
355	0.051	53.88	1553.86	0.017	462.44
357	0.055	53.81	1553.22	0.019	463.56
359	0.060	53.73	1552.44	0.020	464.67
361	0.065	53.66	1551.52	0.022	465.77
363	0.070	53.59	1550.47	0.023	466.86
365	0.075	53.51	1549.29	0.025	467.93
367	0.081	53.44	1547.98	0.027	469.00
369	0.087	53.36	1546.54	0.029	470.05
371	0.094	53.28	1544.97	0.031	471.09
373	0.101	53.20	1543.29	0.033	472.12
375	0.108	53.12	1541.48	0.035	473.13
377	0.116	53.04	1539.56	0.038	474.14
379	0.125	52.96	1537.52	0.040	475.12
381	0.133	52.88	1535.36	0.043	476.10
383	0.143	52.79	1533.09	0.046	477.06
385	0.153	52.71	1530.71	0.049	478.01
387	0.163	52.62	1528.22	0.052	478.95
389	0.174	52.53	1525.62	0.055	479.87
391	0.186	52.45	1522.92	0.058	480.77
393	0.198	52.36	1520.11	0.062	481.67
395	0.211	52.27	1517.20	0.066	482.54
397	0.224	52.18	1514.18	0.070	483.41
399	0.238	52.08	1511.06	0.074	484.25
401	0.253	51.99	1507.85	0.078	485.09
403	0.269	51.90	1504.53	0.083	485.91
405	0.286	51.80	1501.12	0.088	486.71
407	0.303	51.71	1497.61	0.092	487.50
409	0.321	51.61	1494.00	0.098	488.27
411	0.340	51.51	1490.30	0.103	489.02
413	0.360	51.42	1486.51	0.109	489.76
415	0.381	51.32	1482.62	0.115	490.49
417	0.403	51.22	1478.64	0.121	491.19
419	0.426	51.11	1474.57	0.127	491.88
421	0.449	51.01	1470.41	0.134	492.56
423	0.474	50.91	1466.17	0.141	493.22

continues

Table 11.15. (continued)

T (K)	P (MPa)	Liquid ρ (mol/dm^3)	Liquid w (m/s)	Vapor ρ (mol/dm^3)	Vapor w (m/s)
425	0.500	50.81	1461.83	0.148	493.86
427	0.527	50.70	1457.40	0.156	494.48
429	0.556	50.59	1452.89	0.164	495.09
431	0.585	50.49	1448.29	0.172	495.68
433	0.616	50.38	1443.60	0.180	496.25
435	0.648	50.27	1438.83	0.189	496.81
437	0.681	50.16	1433.97	0.198	497.35
439	0.716	50.05	1429.03	0.208	497.87
441	0.752	49.94	1424.01	0.218	498.37
443	0.789	49.82	1418.90	0.228	498.86
445	0.828	49.71	1413.71	0.239	499.32
447	0.869	49.60	1408.44	0.250	499.77
449	0.911	49.48	1403.08	0.261	500.20
451	0.954	49.36	1397.64	0.273	500.62
453	0.999	49.24	1392.12	0.285	501.01
455	1.05	49.13	1386.52	0.298	501.39
457	1.10	49.01	1380.84	0.311	501.74
459	1.15	48.88	1375.08	0.325	502.08
461	1.20	48.76	1369.24	0.339	502.40
463	1.25	48.64	1363.32	0.354	502.70
465	1.31	48.51	1357.32	0.369	502.98
467	1.37	48.39	1351.24	0.385	503.24
469	1.42	48.26	1345.08	0.401	503.49
471	1.49	48.13	1338.84	0.418	503.71
473	1.55	48.01	1332.53	0.435	503.91
475	1.62	47.88	1326.13	0.453	504.09
477	1.68	47.74	1319.66	0.471	504.25
479	1.75	47.61	1313.10	0.491	504.39
481	1.83	47.48	1306.47	0.510	504.51
483	1.90	47.34	1299.76	0.531	504.61
485	1.98	47.21	1292.98	0.552	504.68
487	2.06	47.07	1286.11	0.574	504.74
489	2.14	46.93	1279.16	0.596	504.77
491	2.23	46.79	1272.14	0.619	504.78
493	2.31	46.65	1265.04	0.643	504.77
495	2.40	46.51	1257.85	0.668	504.74
497	2.50	46.36	1250.59	0.693	504.68
499	2.59	46.22	1243.25	0.719	504.60
501	2.69	46.07	1235.83	0.746	504.49
503	2.79	45.92	1228.33	0.774	504.36
505	2.89	45.77	1220.76	0.803	504.21
507	3.00	45.62	1213.10	0.833	504.03
509	3.11	45.47	1205.36	0.863	503.82
511	3.22	45.32	1197.53	0.895	503.59
513	3.34	45.16	1189.63	0.927	503.34
515	3.46	45.00	1181.65	0.961	503.06
517	3.58	44.84	1173.58	0.995	502.75
519	3.71	44.68	1165.43	1.031	502.41
521	3.83	44.52	1157.19	1.068	502.05
523	3.97	44.36	1148.88	1.105	501.65
525	4.10	44.19	1140.47	1.144	501.23
527	4.24	44.02	1131.98	1.185	500.78
529	4.38	43.86	1123.41	1.226	500.30

Table 11.15. (continued)

T (K)	P (MPa)	Liquid		Vapor	
		ρ (mol/dm^3)	w (m/s)	ρ (mol/dm^3)	w (m/s)
531	4.53	43.68	1114.74	1.269	499.79
533	4.68	43.51	1105.99	1.313	499.25
535	4.84	43.34	1097.15	1.358	498.68
537	4.99	43.16	1088.21	1.405	498.07
539	5.16	42.98	1079.19	1.453	497.43
541	5.32	42.80	1070.07	1.503	496.76
543	5.49	42.61	1060.86	1.554	496.06
545	5.66	42.43	1051.55	1.607	495.32
547	5.84	42.24	1042.15	1.662	494.54
549	6.02	42.05	1032.65	1.718	493.73
551	6.21	41.86	1023.04	1.776	492.88
553	6.40	41.66	1013.34	1.836	491.99
555	6.60	41.46	1003.53	1.898	491.07
557	6.80	41.26	993.61	1.962	490.10
559	7.00	41.06	983.59	2.028	489.10
561	7.21	40.85	973.46	2.096	488.05
563	7.43	40.64	963.21	2.167	486.96
565	7.64	40.43	952.85	2.240	485.83
567	7.87	40.21	942.38	2.315	484.65
569	8.10	40.00	931.78	2.393	483.42
571	8.33	39.77	921.06	2.473	482.15
573	8.57	39.55	910.22	2.556	480.83
575	8.81	39.32	899.25	2.643	479.46
577	9.06	39.08	888.14	2.732	478.04
579	9.32	38.85	876.90	2.824	476.56
581	9.58	38.60	865.53	2.920	475.03
583	9.85	38.36	854.00	3.020	473.44
585	10.12	38.11	842.34	3.123	471.80
587	10.39	37.85	830.52	3.230	470.09
589	10.68	37.59	818.54	3.342	468.32
591	10.97	37.32	806.40	3.458	466.48
593	11.26	37.05	794.09	3.579	464.58
595	11.56	36.77	781.60	3.705	462.60
597	11.87	36.49	768.93	3.836	460.55
599	12.19	36.20	756.07	3.973	458.42
601	12.51	35.90	743.01	4.116	456.21
603	12.83	35.59	729.72	4.265	453.91
605	13.17	35.28	716.21	4.422	451.52
607	13.51	34.95	702.44	4.587	449.04
609	13.86	34.62	688.40	4.760	446.45
611	14.21	34.28	674.06	4.942	443.75
613	14.57	33.92	659.38	5.134	440.93
615	14.94	33.56	644.33	5.337	437.99
617	15.32	33.18	628.86	5.552	434.90
619	15.71	32.78	612.91	5.781	431.67
621	16.10	32.37	596.41	6.025	428.26
623	16.50	31.93	579.28	6.286	424.67
625	16.91	31.48	561.45	6.566	420.87
627	17.33	31.00	542.82	6.869	416.82
629	17.75	30.48	523.31	7.198	412.49
631	18.19	29.93	502.83	7.558	407.83
633	18.63	29.33	481.38	7.956	402.77

continues

Table 11.15. (continued)

T (K)	P (MPa)	Liquid		Vapor	
		ρ (mol/dm^3)	w (m/s)	ρ (mol/dm^3)	w (m/s)
635	19.09	28.68	459.03	8.401	397.19
637	19.55	27.97	435.99	8.907	390.93
639	20.02	27.17	412.53	9.496	383.74
641	20.51	26.26	388.55	10.210	375.08
643	21.01	25.15	362.10	11.120	363.84
645	21.52	23.59	325.47	12.460	346.60
647	22.04	19.84	251.20	15.900	285.31

Table 11.16. Water — isobar at 0.101 325 MPa.

T (K)	ρ (mol/dm^3)	w (m/s)
275	55.51	1411.51
295	55.39	1487.89
315	55.04	1531.87
335	54.52	1552.00
355	53.88	1553.96
373.10	53.20	1543.18
373.10	0.0332	472.18
375	0.0330	473.52
395	0.0312	487.03
415	0.0296	499.70
435	0.0282	511.80
455	0.0270	523.44
475	0.0258	534.70
495	0.0247	545.62
515	0.0238	556.23
535	0.0229	566.58
555	0.0220	576.68
575	0.0213	586.55
595	0.0205	596.21
615	0.0199	605.68
635	0.0192	614.95
655	0.0186	624.05
675	0.0181	632.98
695	0.0176	641.75
715	0.0171	650.37
735	0.0166	658.84
755	0.0162	667.18
775	0.0157	675.39
795	0.0153	683.47
815	0.0150	691.43
835	0.0146	699.27
855	0.0143	707.00
875	0.0139	714.63
895	0.0136	722.16
915	0.0133	729.59
935	0.0130	736.92
955	0.0128	744.17
975	0.0125	751.33
995	0.0123	758.41
1015	0.0120	765.41

Table 11.16. (continued)

T (K)	ρ (mol/dm³)	w (m/s)
1035	0.0118	772.33
1055	0.0116	779.17
1075	0.0113	785.95
1095	0.0111	792.66
1115	0.0109	799.30
1135	0.0107	805.88
1155	0.0106	812.40
1175	0.0104	818.86
1195	0.0102	825.26
1215	0.0100	831.61
1235	0.0099	837.90
1255	0.0097	844.14

Table 11.17. Water — isobar at 1 MPa.

T (K)	ρ (mol/dm³)	w (m/s)
275	55.53	1412.95
295	55.41	1489.38
315	55.06	1533.44
335	54.54	1553.68
355	53.90	1555.76
375	53.15	1543.41
395	52.29	1519.06
415	51.34	1484.24
435	50.28	1439.85
453	49.24	1392.05
453	0.286	501.02
455	0.284	502.85
475	0.268	518.73
495	0.255	532.48
515	0.243	545.19
535	0.232	557.16
555	0.223	568.56
575	0.214	579.48
595	0.207	590.01
615	0.199	600.2
635	0.193	610.09
655	0.186	619.72
675	0.181	629.11
695	0.175	638.29
715	0.170	647.26
735	0.165	656.05
755	0.161	664.67
775	0.156	673.12
795	0.152	681.42
815	0.149	689.58
835	0.145	697.61
855	0.141	705.51
875	0.138	713.29
895	0.135	720.95
915	0.132	728.5
935	0.129	735.95

continues

Table 11.17. (continued)

T (K)	ρ (mol/dm³)	w (m/s)
955	0.126	743.3
975	0.124	750.56
995	0.121	757.72
1015	0.119	764.8
1035	0.117	771.8
1055	0.114	778.72
1075	0.112	785.56
1095	0.110	792.33
1115	0.108	799.03
1135	0.106	805.66
1155	0.104	812.22
1175	0.103	818.73
1195	0.101	825.17
1215	0.099	831.56
1235	0.097	837.89
1255	0.096	844.16

Table 11.18. Water—isobar at 5 MPa.

T (K)	ρ (mol/dm³)	w (m/s)
275	55.64	1419.38
295	55.51	1496.03
315	55.15	1540.44
335	54.64	1561.11
355	54.00	1563.72
375	53.25	1552.01
395	52.40	1528.44
415	51.46	1494.56
435	50.41	1451.34
455	49.27	1399.29
475	48.02	1338.56
495	46.64	1268.87
515	45.09	1189.36
535	43.35	1098.15
537.1	43.15	1087.81
537.1	1.41	498.04
555	1.30	520.81
575	1.22	540.41
595	1.15	557.18
615	1.09	572.16
635	1.04	585.84
655	1.00	598.56
675	0.956	610.51
695	0.921	621.84
715	0.889	632.65
735	0.859	643.03
755	0.832	653.03
775	0.807	662.71
795	0.784	672.09
815	0.762	681.21
835	0.741	690.09
855	0.722	698.76

Table 11.18. (continued)

T (K)	ρ (mol/dm³)	w (m/s)
875	0.704	707.23
895	0.686	715.52
915	0.670	723.65
935	0.655	731.62
955	0.640	739.45
975	0.626	747.14
995	0.612	754.7
1015	0.600	762.15
1035	0.587	769.48
1055	0.576	776.72
1075	0.565	783.85
1095	0.554	790.88
1115	0.543	797.83
1135	0.533	804.69
1155	0.524	811.47
1175	0.515	818.18
1195	0.506	824.81
1215	0.497	831.37
1235	0.489	837.86
1255	0.481	844.29

Table 11.19. Water — isobar at 10 MPa.

T (K)	ρ (mol/dm³)	w (m/s)
275	55.78	1427.48
295	55.63	1504.34
315	55.27	1549.15
335	54.76	1570.33
355	54.12	1573.56
375	53.38	1562.62
395	52.54	1539.98
415	51.60	1507.23
435	50.58	1465.38
455	49.45	1415.01
475	48.23	1356.38
495	46.88	1289.39
515	45.38	1213.47
535	43.70	1127.35
555	41.78	1028.37
575	39.47	910.76
584.1	38.21	847.33
584.1	3.08	472.51
595	2.85	494.96
615	2.56	524.55
635	2.36	547.42
655	2.21	566.64
675	2.09	583.48
695	1.99	598.63
715	1.90	612.52
735	1.82	625.42
755	1.75	637.54

continues

Table 11.19. (continued)

T (K)	ρ (mol/dm^3)	w (m/s)
775	1.69	649.02
795	1.63	659.95
815	1.58	670.42
835	1.53	680.49
855	1.48	690.21
875	1.44	699.62
895	1.40	708.74
915	1.37	717.62
935	1.33	726.27
955	1.30	734.71
975	1.27	742.96
995	1.24	751.03
1015	1.21	758.95
1035	1.19	766.71
1055	1.16	774.33
1075	1.14	781.83
1095	1.12	789.2
1115	1.10	796.46
1135	1.07	803.6
1155	1.05	810.65
1175	1.04	817.6
1195	1.02	824.46
1215	1.00	831.24
1235	0.98	837.93
1255	0.97	844.54

Table 11.20. Water—isobar at 50 MPa.

T (K)	ρ (mol/dm^3)	w (m/s)
275	56.82	1494.62
295	56.59	1570.91
315	56.19	1617.39
335	55.67	1641.44
355	55.05	1648.59
375	54.34	1642.61
395	53.56	1626.02
415	52.70	1600.54
435	51.77	1567.36
455	50.76	1527.34
475	49.69	1481.12
495	48.53	1429.22
515	47.29	1372.03
535	45.96	1309.84
555	44.52	1242.87
575	42.95	1171.23
595	41.23	1094.94
615	39.32	1013.97
635	37.17	928.41
655	34.70	838.86
675	31.78	746.93
695	28.24	659.20
715	24.09	590.51

Table 11.20. (continued)

T (K)	ρ (mol/dm^3)	w (m/s)
735	19.85	556.44
755	16.46	556.75
775	14.09	571.69
795	12.44	590.63
815	11.24	609.70
835	10.34	627.78
855	9.62	644.65
875	9.04	660.37
895	8.55	675.07
915	8.13	688.87
935	7.77	701.91
955	7.45	714.26
975	7.17	726.03
995	6.91	737.29
1015	6.68	748.08
1035	6.47	758.46
1055	6.28	768.47
1075	6.10	778.15
1095	5.94	787.53
1115	5.78	796.64
1135	5.64	805.49
1155	5.51	814.11
1175	5.38	822.53
1195	5.26	830.74
1215	5.15	838.77
1235	5.04	846.64
1255	4.94	854.34

Table 11.21. Water — isobar at 100 MPa.

T (K)	ρ (mol/dm^3)	w (m/s)
275	58.00	1583.41
295	57.67	1654.36
315	57.24	1699.71
335	56.71	1724.94
355	56.10	1734.86
375	55.43	1732.90
395	54.69	1721.44
415	53.90	1702.11
435	53.05	1676.13
455	52.14	1644.40
475	51.18	1607.67
495	50.17	1566.57
515	49.11	1521.67
535	47.99	1473.50
555	46.81	1422.55
575	45.58	1369.34
595	44.27	1314.37
615	42.90	1258.17
635	41.45	1201.31
655	39.92	1144.39
675	38.31	1088.12

continues

Table 11.21. (continued)

T (K)	ρ (mol/dm^3)	w (m/s)
695	36.62	1033.35
715	34.84	981.04
735	32.99	932.30
755	31.08	888.40
775	29.15	850.76
795	27.24	820.40
815	25.40	797.37
835	23.67	781.16
855	22.08	771.04
875	20.65	766.04
895	19.37	765.04
915	18.24	767.02
935	17.24	771.13
955	16.37	776.73
975	15.59	783.33
995	14.90	790.61
1015	14.28	798.30
1035	13.72	806.26
1055	13.22	814.35
1075	12.76	822.49
1095	12.35	830.64
1115	11.96	838.74
1135	11.61	846.78
1155	11.28	854.73
1175	10.98	862.60
1195	10.70	870.36
1215	10.44	878.01
1235	10.19	885.56
1255	9.96	893.00

Table 11.22. Air — saturation boundary.

T (K)	P (Bubble) (MPa)	Liquid ρ (mol/dm^3)	w (m/s)	P (Dew) (MPa)	Vapor ρ (mol/dm^3)	w (m/s)
60	0.006	33.03	1028.29	0.0026	0.005	155.14
61	0.007	32.89	1020.26	0.0033	0.006	156.38
62	0.008	32.74	1012.16	0.0041	0.008	157.61
63	0.010	32.60	1004.00	0.0051	0.010	158.81
64	0.012	32.46	995.77	0.0063	0.012	159.99
65	0.014	32.31	987.48	0.0078	0.014	161.16
66	0.017	32.17	979.13	0.0094	0.017	162.30
67	0.020	32.02	970.72	0.011	0.021	163.42
68	0.024	31.87	962.24	0.014	0.024	164.53
69	0.027	31.72	953.70	0.016	0.029	165.61
70	0.032	31.58	945.10	0.019	0.034	166.66
71	0.037	31.43	936.43	0.023	0.039	167.70
72	0.042	31.28	927.70	0.027	0.046	168.70
73	0.049	31.13	918.90	0.031	0.053	169.69
74	0.056	30.97	910.04	0.037	0.061	170.65
75	0.063	30.82	901.11	0.042	0.069	171.58
76	0.072	30.67	892.11	0.049	0.079	172.49
77	0.081	30.51	883.05	0.056	0.090	173.37
78	0.091	30.36	873.91	0.064	0.101	174.23

Table 11.22. (continued)

T (K)	P (Bubble) (MPa)	Liquid ρ (mol/dm^3)	w (m/s)	P (Dew) MPa	Vapor ρ (mol/dm^3)	w (m/s)
79	0.10	30.20	864.71	0.073	0.114	175.05
80	0.11	30.04	855.44	0.082	0.128	175.85
81	0.13	29.88	846.09	0.093	0.143	176.62
82	0.14	29.72	836.67	0.10	0.160	177.36
83	0.16	29.56	827.18	0.12	0.178	178.07
84	0.17	29.40	817.61	0.13	0.198	178.75
85	0.19	29.23	807.96	0.15	0.219	179.40
86	0.21	29.07	798.24	0.16	0.241	180.02
87	0.23	28.90	788.44	0.18	0.266	180.61
88	0.26	28.73	778.56	0.20	0.292	181.17
89	0.28	28.56	768.59	0.22	0.320	181.69
90	0.30	28.38	758.55	0.24	0.351	182.19
91	0.33	28.21	748.42	0.27	0.383	182.65
92	0.36	28.03	738.20	0.29	0.417	183.08
93	0.39	27.85	727.90	0.32	0.454	183.48
94	0.42	27.67	717.51	0.35	0.493	183.84
95	0.46	27.49	707.03	0.38	0.535	184.17
96	0.50	27.30	696.46	0.41	0.580	184.46
97	0.53	27.12	685.80	0.45	0.627	184.72
98	0.57	26.92	675.05	0.49	0.677	184.95
99	0.62	26.73	664.20	0.53	0.730	185.14
100	0.66	26.53	653.26	0.57	0.786	185.30
101	0.71	26.33	642.22	0.61	0.846	185.42
102	0.76	26.13	631.08	0.66	0.909	185.51
103	0.81	25.92	619.84	0.71	0.976	185.55
104	0.87	25.71	608.50	0.76	1.047	185.57
105	0.93	25.50	597.06	0.81	1.122	185.54
106	0.99	25.28	585.51	0.87	1.201	185.48
107	1.05	25.06	573.85	0.93	1.285	185.38
108	1.12	24.83	562.09	0.99	1.374	185.24
109	1.18	24.60	550.21	1.06	1.468	185.07
110	1.26	24.36	538.21	1.13	1.568	184.85
111	1.33	24.12	526.10	1.20	1.674	184.60
112	1.41	23.87	513.86	1.28	1.786	184.30
113	1.49	23.61	501.48	1.35	1.905	183.97
114	1.58	23.35	488.97	1.44	2.032	183.59
115	1.66	23.08	476.31	1.52	2.166	183.17
116	1.75	22.80	463.48	1.61	2.310	182.71
117	1.85	22.51	450.49	1.70	2.463	182.21
118	1.95	22.22	437.29	1.80	2.626	181.66
119	2.05	21.91	423.89	1.90	2.801	181.08
120	2.16	21.59	410.23	2.01	2.989	180.45
121	2.27	21.25	396.30	2.12	3.191	179.78
122	2.38	20.90	382.04	2.23	3.410	179.06
123	2.50	20.53	367.40	2.35	3.648	178.31
124	2.62	20.14	352.31	2.47	3.908	177.52
125	2.74	19.73	336.67	2.59	4.193	176.68
126	2.87	19.28	320.36	2.73	4.510	175.81
127	3.01	18.79	303.21	2.86	4.865	174.91
128	3.14	18.24	285.00	3.01	5.270	173.96
129	3.28	17.62	265.37	3.15	5.740	172.98
130	3.43	16.86	243.75	3.31	6.307	171.93
131	3.58	15.87	219.07	3.47	7.034	170.79
132	3.72	14.20	189.12	3.65	8.127	169.40

Table 11.23. Air — isobar at 0.101 325 MPa.

T (K)	ρ (mol/dm^3)	w (m/s)
100	0.124	198.21
200	0.061	283.45
300	0.041	347.36
400	0.030	400.50
500	0.024	446.40
600	0.020	487.07
700	0.017	523.89
800	0.015	557.85
900	0.014	589.60
1000	0.012	619.60
1100	0.011	648.15
1200	0.010	675.47
1300	0.0094	701.72
1400	0.0087	727.03
1500	0.0081	751.50
1600	0.0076	775.19
1700	0.0072	798.19
1800	0.0068	820.54
1900	0.0064	842.30
2000	0.0061	863.51
2100	0.0058	884.21
2200	0.0055	904.43
2300	0.0053	924.20
2400	0.0051	943.56
2500	0.0049	962.52
2600	0.0047	981.10
2700	0.0045	999.34
2800	0.0044	1017.25
2900	0.0042	1034.84
3000	0.0041	1052.13
3100	0.0039	1069.14
3200	0.0038	1085.87
3300	0.0037	1102.35
3400	0.0036	1118.58
3500	0.0035	1134.57
3600	0.0034	1150.34
3700	0.0033	1165.89
3800	0.0032	1181.23
3900	0.0031	1196.37
4000	0.0030	1211.31
4100	0.0030	1226.07
4200	0.0029	1240.65
4300	0.0028	1255.05
4400	0.0028	1269.29
4500	0.0027	1283.36
4600	0.0026	1297.27
4700	0.0026	1311.03
4800	0.0025	1324.65
4900	0.0025	1338.12
5000	0.0024	1351.45

Table 11.24. Air—isobar at 1 MPa.

T (K)	ρ (mol/dm^3)	w (m/s)
100	26.59	658.25
106.22	25.23	582.97
108.10	1.38	185.23
200	0.616	281.74
300	0.402	348.45
400	0.300	402.33
500	0.240	448.46
600	0.200	489.16
700	0.171	525.96
800	0.150	559.86
900	0.133	591.54
1000	0.120	621.47
1100	0.109	649.95
1200	0.100	677.21
1300	0.092	703.40
1400	0.086	728.66
1500	0.080	753.07
1600	0.075	776.72
1700	0.071	799.67
1800	0.067	821.98
1900	0.063	843.70
2000	0.060	864.87
2100	0.057	885.53
2200	0.055	905.72
2300	0.052	925.46
2400	0.050	944.79
2500	0.048	963.72
2600	0.046	982.28
2700	0.044	1000.49
2800	0.043	1018.37
2900	0.041	1035.94
3000	0.040	1053.21
3100	0.039	1070.20
3200	0.038	1086.91
3300	0.036	1103.37
3400	0.035	1119.59
3500	0.034	1135.56
3600	0.033	1151.31
3700	0.032	1166.84
3800	0.032	1182.17
3900	0.031	1197.29
4000	0.030	1212.22
4100	0.029	1226.97
4200	0.029	1241.53
4300	0.028	1255.92
4400	0.027	1270.15
4500	0.027	1284.21
4600	0.026	1298.11
4700	0.026	1311.86
4800	0.025	1325.46
4900	0.025	1338.92
5000	0.024	1352.24a

Table 11.25. Air — isobar at 5 MPa.

T (K)	ρ (mol/dm^3)	w (m/s)
100	27.22	710.56
200	3.38	279.42
300	2.02	355.63
400	1.49	411.69
500	1.18	458.30
600	0.983	498.93
700	0.843	535.45
800	0.738	569.01
900	0.657	600.34
1000	0.592	629.93
1100	0.539	658.09
1200	0.494	685.05
1300	0.457	710.97
1400	0.424	735.96
1500	0.396	760.13
1600	0.372	783.56
1700	0.350	806.30
1800	0.331	828.42
1900	0.314	849.96
2000	0.298	870.96
2100	0.284	891.46
2200	0.271	911.50
2300	0.259	931.10
2400	0.249	950.29
2500	0.239	969.10
2600	0.230	987.54
2700	0.221	1005.64
2800	0.213	1023.41
2900	0.206	1040.88
3000	0.199	1058.05
3100	0.193	1074.95
3200	0.187	1091.57
3300	0.181	1107.95
3400	0.176	1124.08
3500	0.171	1139.98
3600	0.166	1155.65
3700	0.162	1171.12
3800	0.158	1186.37
3900	0.154	1201.43
4000	0.150	1216.30
4100	0.146	1230.98
4200	0.143	1245.49
4300	0.139	1259.82
4400	0.136	1273.99
4500	0.133	1288.00
4600	0.130	1301.85
4700	0.127	1315.55
4800	0.125	1329.10
4900	0.122	1342.52
5000	0.120	1355.79

Table 11.26. Air — isobar at 10 MPa.

T (K)	ρ (mol/dm³)	w (m/s)
100	27.86	763.47
200	7.39	296.30
300	4.04	369.50
400	2.92	425.59
500	2.32	471.81
600	1.93	511.90
700	1.65	547.83
800	1.45	580.83
900	1.29	611.64
1000	1.17	640.74
1100	1.06	668.46
1200	0.975	695.01
1300	0.902	720.56
1400	0.839	745.21
1500	0.784	769.07
1600	0.736	792.20
1700	0.693	814.68
1800	0.655	836.55
1900	0.621	857.85
2000	0.591	878.64
2100	0.563	898.94
2200	0.538	918.78
2300	0.515	938.21
2400	0.494	957.23
2500	0.474	975.87
2600	0.456	994.17
2700	0.440	1012.12
2800	0.424	1029.76
2900	0.410	1047.10
3000	0.396	1064.15
3100	0.384	1080.92
3200	0.372	1097.44
3300	0.361	1113.71
3400	0.350	1129.74
3500	0.340	1145.54
3600	0.331	1161.12
3700	0.322	1176.49
3800	0.306	1206.63
3900	0.306	1206.63
4000	0.298	1221.42
4100	0.291	1236.03
4200	0.284	1250.46
4300	0.277	1264.72
4400	0.271	1278.82
4500	0.265	1292.76
4600	0.260	1306.55
4700	0.254	1320.19
4800	0.249	1333.68
4900	0.244	1347.04
5000	0.239	1360.25

Table 11.27. Air—isobar at 50 MPa.

T (K)	ρ (mol/dm^3)	w (m/s)
100	30.94	1012.41
200	21.58	670.59
300	15.23	581.66
400	11.68	582.86
500	9.54	604.06
600	8.10	629.50
700	7.06	655.45
800	6.28	680.98
900	5.66	705.86
1000	5.15	730.05
1100	4.74	753.58
1200	4.38	776.48
1300	4.08	798.80
1400	3.82	820.55
1500	3.59	841.78
1600	3.38	862.52
1700	3.20	882.80
1800	3.04	902.63
1900	2.89	922.05
2000	2.76	941.07
2100	2.64	959.73
2200	2.53	978.04
2300	2.43	996.01
2400	2.33	1013.67
2500	2.24	1031.02
2600	2.16	1048.10
2700	2.09	1064.90
2800	2.02	1081.44
2900	1.95	1097.73
3000	1.89	1113.79
3100	1.83	1129.62
3200	1.78	1145.23
3300	1.73	1160.63
3400	1.68	1175.83
3500	1.64	1190.84
3600	1.59	1205.66
3700	1.55	1220.30
3800	1.51	1234.77
3900	1.48	1249.06
4000	1.44	1263.20
4100	1.41	1277.18
4200	1.37	1291.01
4300	1.34	1304.69
4400	1.32	1318.22
4500	1.29	1331.62
4600	1.26	1344.88
4700	1.23	1358.01
4800	1.21	1371.01
4900	1.19	1383.88
5000	1.16	1396.64

Table 11.28. Air — isobar at 100 MPa.

T (K)	ρ (mol/dm^3)	w (m/s)
100	33.16	1192.37
200	26.18	932.01
300	21.14	818.47
400	17.60	779.80
500	15.09	772.41
600	13.23	778.24
700	11.80	790.14
800	10.67	805.13
900	9.75	821.78
1000	8.98	839.35
1100	8.33	857.39
1200	7.77	875.65
1300	7.29	893.95
1400	6.86	912.18
1500	6.48	930.28
1600	6.15	948.19
1700	5.85	965.90
1800	5.57	983.39
1900	5.32	1000.64
2000	5.10	1017.65
2100	4.89	1034.43
2200	4.70	1050.98
2300	4.52	1067.31
2400	4.36	1083.41
2500	4.21	1099.29
2600	4.06	1114.97
2700	3.93	1130.44
2800	3.81	1145.72
2900	3.69	1160.81
3000	3.58	1175.72
3100	3.48	1190.45
3200	3.38	1205.01
3300	3.29	1219.40
3400	3.20	1233.63
3500	3.12	1247.71
3600	3.04	1261.63
3700	2.97	1275.41
3800	2.90	1289.04
3900	2.83	1302.54
4000	2.76	1315.90
4100	2.70	1329.13
4200	2.64	1342.24
4300	2.59	1355.21
4400	2.53	1368.07
4500	2.48	1380.81
4600	2.43	1393.43
4700	2.38	1405.95
4800	2.34	1418.35
4900	2.29	1430.64
5000	2.25	1442.83

ADDITIONAL READING

Thompson, P.A. *Compressible-Fluid Dynamics* New York: McGraw Hill. (1972).

Trusler, J.P.M. (1991). *Physical Acoustics and Metrology of Fluids* Bristol: Adam Hilger.

Van Dael, W. (1975). In *Experimental Thermodynamics of Non-Reacting Systems* (International Union of Pure and Applied Chemistry, Commission on Thermodynamics and Thermochemistry; Le Neindre, B. and Vodar, B. eds. London: Butterworths. chapter 11, *Thermodynamic Properties and the Velocity of Sound*, pp. 527–575.

References

1. Slattery, J.C. *Momentum, Energy, and Mass Transfer in Continua*, New York: Krieger Publishing. (1981).
2. Thompson, P.A. *Compressible-Fluid Dynamics* New York: McGraw Hill. (1972).
3. McQuarrie, D.A. (1976). *Statistical Mechanics* New York: Harper & Row.
4. Costa Gomes, M.F., and Trusler, J.P.M. (1998). The speed of sound in nitrogen at temperatures between $T = 250$ K and $T = 350$ K and at pressures up to 30 Mpa, *J. Chem. Thermodynamics* **30**: 527.
5. Trusler J.P.M. (1991). *Physical Acoustics and Metrology of Fluids* Bristol: Adam Hilger.
6. Van Dael, W. In *Experimental Thermodynamics of Non-Reacting Systems* (International Union of Pure and Applied Chemistry, Commission on Thermodynamics and Thermochemistry Le Neindre, B. and Vodar, B, eds. London: Butterworths, chapter 11: *Thermodynamic properties and the velocity of sound*, pp. 527–575.
7. Reid, R.C., Prausnitz, J.M., and Poling, B.E. (1987). *The Properties of Gases and Liquids* New York: McGraw Hill.
8. Carnahan, N.F., and Starling, K.E. (1969). Equation of state for nonattracting rigid spheres, *J. Chem. Phys.* **51**: 635.
9. Younglove, B.A. (1982). Thermophysical properties of fluids. I. argon, ethylene, parahydrogen, nitrogen, nitrogen trifluoride, and oxygen, *J. Phys. Chem. Ref. Data* **11**, pp. 1–1 to 1–353. (368 pages).
10. Sengers, J.V., and Levelt Sengers, J.M.H. (1978). In *Progress in Liquid Physics* (Croxton, C.A., ed., Chichester, U.K: Wiley. Chapter 4: *Critical phenomena in classical fluids*, pp. 103–174.
11. Tegeler, Ch., Span, R., and Wagner, W. (1999). A new equation of state for argon covering the fluid region for temperatures from the melting line to 700 K at pressures up to 1000 MPa, *J. Phys. Chem. Ref. Data*, **28**: 779.
12. Estrada-Alexanders, A.F., and Trusler, J.P.M. (1995). The speed of sound in gaseous argon at temperatures between 110 K and 450 K and at pressures up to 19 MPa. *J. Chem. Thermodyn.* **27**: 1075.
13. Span, R., Lemmon, E.W., Jacobsen, R.T., and Wagner, W. (1998). A reference quality equation of state for nitrogen, *Int. J. Thermophys.* **19**: 1121.
14. IAWPS. (1995). Release on the Formulation for the Thermodynamic Properties of Ordinary Water Substance for General and Scientific Use; copies of this and other IAPWS releases can be obtained from the Executive Secretary of IAPWS, Dr. R.B. Dooley, Electric Power Research Institute, 3412 Hillview Ave., Palo Alto, CA 94304.
15. Lemmon, E.W., Jacobsen, R.T, Penoncello, S.G., and Friend, D.G. (2000). Thermodynamic properties of air and mixtures of nitrogen, argon, and oxygen from 60 to 2000 K at pressures to 2000 MPa. *J. Phys. Chem. Ref. Data* **29**, in press.
16. Younglove, B.A., Frederick, N.V., and McCarty, R.D. (1993). *Speed to Sound Data and Related Models for Mixtures of Natural Gas Components*, NIST Monograph 178.
17. Lemmon, E.W., and Jacobsen, R.T. (1999). *A Generalized Model for the Themodynamic Properties of Mixtures,* Int. J. Thermophys. **20**: 825.
18. Thermodynamics Research Center; Texas Engineering Experiment Station, College Station, TX (1961–2000): Natl. Inst. Stand. Tech., Boulder, CO (2000–).
19. NIST Thermophysical Properties of Pure Fluids Database (NIST12, Version 5.0). (2000). Natl. Inst. Stand. Tech., Gaithersburg, MD.

CHAPTER 12

ACOUSTIC MEASUREMENTS IN GASES: APPLICATIONS TO THERMODYNAMIC PROPERTIES, TRANSPORT PROPERTIES, AND THE TEMPERATURE SCALE

M. R. Moldover, K. A. Gillis, J. J. Hurly, J. B. Mehl, and J. Wilhelm[1]
Process Measurements Division, National Institute of Standards and Technology, Gaithersburg, Maryland, USA

Contents

Abstract		329
12.1.	Introduction and Scope	330
12.2.	Acoustic Measurements and Thermodynamic Properties of Dilute Gases	331
12.3.	Measuring the Speed of Sound	335
12.4.	Resonance Measurements of Transport Properties	340
12.5.	Acoustic Thermometry	345
12.6.	Acoustic Determination of the Universal Gas Constant R	349
12.7.	Concluding Remarks	351
12.8.	Preface to Tables	351
Acknowledgments		369
References		369

ABSTRACT

Cylindrical acoustic resonators developed at the National Institute of Standards and Technology (NIST) are routinely used to measure the speed of sound in gases with uncertainties of 0.01% or less. The pressure dependence of the data is fitted with model intermolecular potentials to obtain virial coefficients as well as gas densities ρ and heat capacities C_p with uncertainties of 0.1%. The model intermolecular potentials are also used to estimate the viscosity η and thermal conductivity λ of gases with uncertainties of less than 10%. These techniques have been applied to numerous gases, and the results are tabulated. The gases include candidate replacement refrigerants, helium-xenon mixtures used in thermoacoustic machinery, and very reactive gases used in semiconductor processing. The viscosity can be measured directly with uncertainties of less than 1% using the Greenspan acoustic viscometer, a novel acoustic resonator

---

[1] Guest Scientist from Fachbereich Chemie, Universität Rostock, D-18051 Rostock, Germany.

Handbook of Elastic Properties of Solids, Liquids, and Gases, edited by Levy, Bass, and Stern
Volume IV: Elastic Properties of Fluids: Liquids and Gases

ISBN 0-12-445764-9 / $35.00

developed at NIST. A second novel resonator is used to measure the Prandtl number ($Pr \equiv \eta C_p/\lambda$) with uncertainties on the order of 2%. The thermal conductivity is determined by combining the Prandtl number with the acoustically determined density, viscosity, and heat capacity. Spherical acoustic resonators are used to measure the speed of sound with the highest possible accuracy. An argon-filled spherical resonator was used to redetermine the universal gas constant R with a fractional standard uncertainty of 1.7×10^{-6}. The same resonator was used to measure imperfections of the internationally accepted temperature scale (ITS-90) in the range 217 to 303 K. This work is being extended to 800 K. In effect, very accurate speed-of-sound measurements will be used to calibrate thermometers.

12.1. INTRODUCTION AND SCOPE

We describe a research program at the National Institute of Standards and Technology (NIST) that uses gas-filled acoustic resonators to obtain thermodynamic and transport property data for solving both engineering and standards problems. The NIST program has determined the thermodynamic properties of more than 20 gases used in engineering; tables of these properties are in the database associated with this chapter. Acoustic measurements to determine the transport properties of these gases are proceeding. In the area of standards, this program has used an argon-filled spherical acoustic resonator to determine the internationally recognized [1] value of the universal gas constant R [2]. Subsequently, the same resonator was used to measure the imperfections of the internationally accepted temperature scale (ITS-90) in the range 216 to 303 K [3]. The study of ITS-90 is now being extended to temperatures as high as 800 K. In addition to data, this research program has led to a thorough understanding of spherical acoustic resonators [4]; cylindrical acoustic resonators [5]; Greenspan acoustic viscometers [6, 7]; and practical, reliable methods of interpolating and extrapolating speed-of-sound data [8, 9]. (We adapted several ideas for interpolation and extrapolation from Trusler and his collaborators [10–12].)

This NIST program began in 1978 when one of the present authors (M.R.M.) was asked to find an alternative method of determining the universal gas constant R. The request was stimulated by the presence of mutually inconsistent values of R in the literature [13]. In 1976, a group at Britain's National Physical Laboratory (NPL) reported a "new" value for R that had a standard uncertainty of 20 ppm (1 ppm = 1 part in 10^6) [14] and was 159 ppm larger than the value of R that had been accepted in 1972 with an uncertainty of 31 ppm [15]. The new NPL value of R was based on speed-of-sound data obtained with a cylindrical acoustic resonator and extrapolated to zero pressure using an empirical, quadratic function. Guided by the theory of the virial equation of state, Rowlinson and Tildesley [16] used a linear extrapolation of the NPL data to obtain a value of R in agreement with the 1972 value. Three years later, the NPL group reported a revised value of R and stated that its 1976 value of R was erroneous because an unsuspected nonlinearity in an acoustic transducer had led them to use a quadratic function for extrapolating [13]. Thus, NPL unintentionally demonstrated the value of the theoretical considerations that we present in Section 12.2. By the time of NPL's revision, this NIST program had begun. Ultimately, NIST confirmed the revised value of R with a much smaller standard uncertainty, 1.7 ppm.

The present review begins by describing the connections between the thermodynamic properties of dilute gases and the speed of sound. These connections lead us to recommend the use of model intermolecular potentials for analyzing speed-of-sound data (and for extrapolation to zero pressure). The data tables are based on such model potentials. Then, we describe four gas-filled resonators: (1) a cylinder used to measure the speed of sound in industrially significant gases, (2) a dumbbell-shaped Helmholtz resonator used to measure the viscosity, (3) a cylinder with sheet-metal inserts to determine the Prandtl number, and (4) the large spherical shell that was used to redetermine R and to measure the imperfections in ITS-90.

12.2. ACOUSTIC MEASUREMENTS AND THERMODYNAMIC PROPERTIES OF DILUTE GASES

Rigorous statistical mechanical arguments lead to the convenient virial expansion of the pressure p of gases as a function of density ρ and the thermodynamic temperature T

$$p = \rho RT(1 + B\rho + C\rho^2 + D\rho^3 + \cdots) \tag{12.1}$$

Here, the density virial coefficients $B(T)$, $C(T)$, $D(T)$, etc., are functions of the temperature only. All of the thermodynamic properties of dilute gases can be computed by straightforward differentiations and integrations from the density virial coefficients and the zero-pressure limit of the constant-pressure heat capacity $C_p^0(T)$. Often, $B(T)$, $C(T)$, $D(T)$, etc., are parametrized with arbitrary functions (e.g., polynomials in $1/T$) and the parameters are determined from some limited set of thermodynamic data. If this procedure is followed using speed-of-sound data, unacceptable results may be obtained, especially when the functions are extrapolated beyond the range of the data. Much better results can be obtained by using virial coefficients generated from model intermolecular potentials to represent the speed of sound $u(p,T)$.

In the limit of zero frequency, the speed of sound is a thermodynamic quantity:

$$u^2 = \left(\frac{\partial p}{\partial \rho}\right)_S = \frac{1}{\rho \beta_S} = \frac{\gamma}{\rho \beta_T} \tag{12.2}$$

Here, S is the entropy, β_S is the adiabatic compressibility, β_T is the isothermal compressibility, and $\gamma \equiv C_p/C_V$ is the ratio of the molar heat capacity at constant pressure to the molar heat capacity at constant volume. In this work, we consider audio frequencies and gases at pressures of 50 kPa or higher. Under these circumstances, the zero-frequency limit is an excellent approximation, except for those few gases composed of symmetric molecules in which the conversion of kinetic energy into vibrational energy requires unusually many collisions (e.g., very dry carbon dioxide).

Usually, one measures the speed of sound $u(p,T)$ as a function of the temperature and the pressure (instead of the density). The results may be expressed as the power series

$$u^2 = \frac{\gamma^0 RT}{M}\left(1 + \frac{\beta_a p}{RT} + \frac{\gamma_a p^2}{RT} + \frac{\delta_a p^3}{RT} + \frac{\varepsilon_a p^4}{RT} + \cdots\right) \tag{12.3}$$

Here, $\gamma^0(T) = C_p^0(T)/C_V^0(T)$ is the zero-pressure limit of the heat capacity ratio and M is the molecular mass. The temperature-dependent "acoustic virial coefficients" $\beta_a(T)$, $\gamma_a(T)$, $\delta_a(T)$, and $\varepsilon_a(T)$ can be calculated explicitly from the density virial coefficients $B(T)$, $C(T)$, etc., using rigorous thermodynamic relationships [8]. The relationships for $\beta_a(T)$ and $\gamma_a(T)$ are

$$\beta_a = 2B + 2(\gamma^0 - 1)B' + \frac{(\gamma^0 - 1)^2}{\gamma^0}B'' \tag{12.4}$$

and

$$\gamma_a = \frac{(\gamma^0 - 1)}{RT\gamma^0}[B + (2\gamma^0 - 1)B' + (\gamma^0 + 1)B'']^2 \\ + \frac{1}{RT\gamma^0}\left[(2\gamma^0 + 1)C + [(\gamma^0)^2 - 1]C' + \frac{(\gamma^0 - 1)^2}{2}C''\right] \tag{12.5}$$

In Eqs. 12.4 and 12.5, the primed functions B' and B'' denote $T(dB/dT)$ and $T^2(d^2B/dT^2)$, respectively. As illustrated by Eqs. 12.4 and 12.5, $\beta_a(T)$ and $\gamma_a(T)$ are related to the density virial coefficients by second-order differential equations. Similar, more complicated equations for $\delta_a(T)$ and $\varepsilon_a(T)$ appear in [8]. These equations can be integrated numerically provided that the acoustic virial coefficients and $C_p^0(T)$ [or equivalently $\gamma^0(T)$] are known from speed-of-sound measurements and provided that the initial values for the integration are determined by additional thermodynamic data. For many gases, the initial values are not available. When the initial values are at low temperatures and the integrations proceed toward higher temperatures, the integrations are stable in the sense that the effects of errors in the initial values decrease as the temperature increases. When the initial values are at high temperatures and the integration proceeds to lower temperatures, the uncertainties grow [11].

We recommend another way of analyzing $u^2(p,T)$. We begin with an algebraic expression for the ideal-gas heat capacity $C_p^0(T)$ that contains several parameters and following Trusler [10] a model intermolecular pair potential $\varphi(r)$ that also contains several parameters. (Here, r is the radial coordinate that measures the distance between pairs of molecules.) To obtain the temperature-dependent density virial coefficients and their derivatives, the model potential is weighted by the Boltzmann factor, integrated with respect to r, and differentiated with respect to T. Then, using Eqs. 12.4 and 12.5, etc., we compute $u^2(p,T)$ for those states where the data were taken. These computations are repeated to find the best-fit values of the parameters in $C_p^0(T)$ and $\varphi(r)$. This procedure does not require additional thermodynamic data and it solves the differential equations relating $B(T)$ and $C(T)$ to $\beta_a(T)$ and $\gamma_a(T)$ exactly. The procedure is not perfect because the true functional form of $\varphi(r)$ is not known.

For analyzing $u^2(p,T)$ data, we have used two model potentials for $B(T)$: (1) hard-core square-well potentials and (2) the hard-core Lennard-Jones potential. When we used a square-well potential for $B(T)$, we used a separate square-well function for $C(T)$. When we used a Lennard-Jones potential for $B(T)$, we included an Axilrod-Teller triple dipole term in $C(T)$. We note that Trusler [10] has used yet another potential introduced by Maitland and Smith.

The hard-core square-well potentials can be integrated to obtain very convenient, closed-form expressions for $B(T)$ and $C(T)$; we reproduce these expressions here:

$$B(T) = b_0[1 - (\lambda^3 - 1)\Delta] \tag{12.6}$$

$$C(T) = \tfrac{1}{8}b_0^2(5 - c_1\Delta - c_2\Delta^2 - c_3\Delta^3)$$

$$c_1 = \lambda^6 - 18\lambda^4 + 32\lambda^3 - 15 \tag{12.7}$$

$$c_2 = 2\lambda^6 - 36\lambda^4 + 32\lambda^3 + 18\lambda^2 - 16$$

$$c_3 = 6\lambda^6 - 18\lambda^4 + 18\lambda^2 - 6$$

where $\Delta \equiv \exp(\varepsilon/k_BT) - 1$, k_B is Boltzmann's constant, $b_0 \equiv (2/3)\pi N_A \sigma^3$ is the molar volume excluded by the hard cores, and N_A is Avogadro's constant. The adjustable parameters are ε, λ, and σ, where ε is the well depth, σ is the radius of the hard core, and λ is the ratio of the width of the well to σ. We allowed each virial coefficient to have its own values for ε, λ, and σ; thus, we do not attribute physical significance to the values that resulted from fitting the data. When the data extended to high enough pressures, we also used square-well functions for $D(T)$ and an empirical form for $E(T)$ [8].

The form of the hard-core Lennard-Jones 6-12 potential is:

$$\varphi(r_{ij}) = 4\varepsilon \left\{ \left(\frac{\sigma - 2a}{r_{ij} - 2a}\right)^{12} - \left(\frac{\sigma - 2a}{r_{ij} - 2a}\right)^6 \right\} \tag{12.8}$$

where r_{ij} is the intermolecular separation between molecules i and j, ε is the well depth, σ is the value of r_{ij} where $\varphi(r)$ crosses zero, and a is the radius of the hard core. This potential has three adjustable parameters: ε, σ, and a. The same values for these parameters are used to compute both $B(T)$ and $C(T)$.

As an example of what can be done with the hard-core Lennard-Jones potential, we consider the data for WF_6. We calculated $B(T)$ and $C(T)$ and their temperature derivatives classically using references [17,18]. The calculation of $C(T)$ required the inclusion of three-body contributions. This added a fourth adjustable parameter, ν_{123}. Following Trusler [10], we used the Axilrod-Teller triple-dipole term [19],

$$\varphi(r_{123}) = \frac{\nu_{123}(1 + \cos\theta_1 \cos\theta_2 \cos\theta_3)}{(r_{12}^3 r_{13}^3 r_{23}^3)} \tag{12.9}$$

where ν_{123} is the dispersion coefficient and θ_i is defined as the angle subtended at molecule i by molecules j and k. This is the first term in the three-body corrections to the dispersion energy for monatomic species.

Figure 12.1 displays speed-of-sound data for WF_6 and their deviations from a seven-parameter fitting function. Three of the parameters were used to represent the ideal-gas heat capacity:

$$C_p^0/R = 5.207 + 0.04251(T/K) - 3.931 \times 10^{-5}(T/K)^2 \quad 290\,K \leq T \leq 420\,K \tag{12.10}$$

The remaining four parameters were ε, σ, a, and ν_{123}. These seven parameters define the density virial equation up to $C(T)$; however, the derivatives of $B(T)$ and $C(T)$ do make noticeable contributions to the acoustic virials $\delta_a(T)$, and $\varepsilon_a(T)$, and these terms

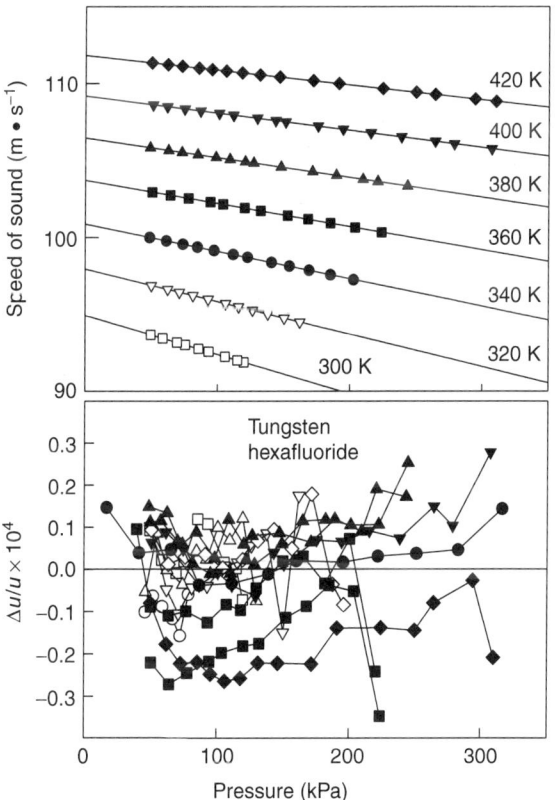

Fig. 12.1. Top, Speed-of-sound data for WF_6. Bottom, Deviations of all of the data in the top panel from a fit to a model hard-core Lennard-Jones potential.

were included when fitting the speed-of-sound data for WF_6. Figure 12.1 shows that the data are represented almost to within their noise. We also analyzed these data with the Maitland-Smith potential and obtained slightly larger deviations.

Because of experimental problems, the data for WF_6 did not extend to pressures above 300 kPa. For several other gases, the $u^2(p,T)$ data extend to higher densities. For such gases, we used hard-core square-well expressions for $D(T)$ and empirical expressions for $E(T)$. In these cases, we first completed a preliminary analysis using only the low-pressure data to find the best values of the parameters in $C_p^0(T)$. Then, we held the parameters in $C_p^0(T)$ fixed and found the best values of the remaining parameters.

To a small extent, the results for $B(T)$, $C(T)$, etc., depend on the functional forms of the model potentials. For several gases, (e.g., light alkanes), the published $u^2(p,T)$ data are accurate and span a very wide temperature range. In such cases, we found that the more realistic potentials (e.g., hard-core Lennard Jones, Maitland-Smith, etc.) yield better results than the hard core square-well potentials, especially when extrapolated to very high temperatures. Nevertheless, even the hard-core square-well potentials yield equations of state $p(\rho,T)$ with uncertainties of less than 0.1% over a wide range of conditions [8].

The tables of data attached to this chapter were obtained from model intermolecular potentials. For each gas, we have tabulated values of $C_p^0(T)$, $B(T)$, and $C(T)$, and for those gases where our data extend to high enough densities, we have tabulated $D(T)$. We have also tabulated the acoustic virial coefficients $\beta_a(T)$, $\gamma_a(T)$, and $\delta_a(T)$.

Figure 12.2 is another demonstration of the utility of fitting model potentials to $u^2(p,T)$ data. The solid curves represent the viscosity of three gases (HBr, SF_6, and BCl_3) at low pressures. These curves were calculated from model hard-core Lennard-Jones potentials that were obtained from $u^2(p,T)$ data [20]. For SF_6, the viscosities from the model potential are, on average, 4% smaller than published viscosity measurements (filled symbols). Also, every published measurement falls within 10% of the calculated curve. Equally good agreement has been obtained for CF_4, C_2F_6 [9] and C_3H_8 [10]. From these experiences, we conclude that one can use $u^2(p,T)$ data to predict the viscosity of difficult-to-handle gases such as HBr and BCl_3 over a wide range of temperatures with uncertainties on the order of 10% or less.

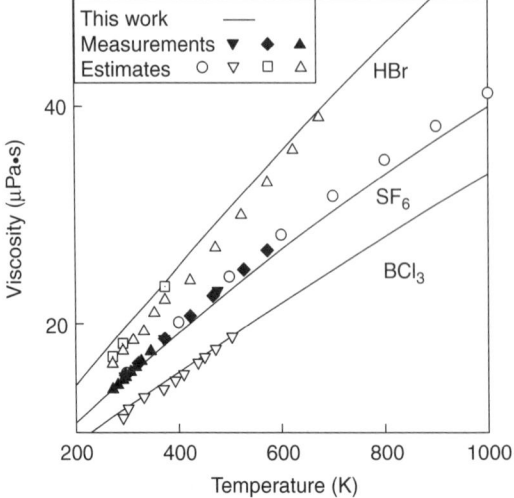

Fig. 12.2. Viscosity of three gases at zero pressure. The curves are estimates from speed-of-sound data. The filled symbols are direct measurements. For SF_6, the direct measurements differ by less than 10% from the curve. The open symbols are other estimates. For references and details, see [20].

In concluding this section, we emphasize that model potentials provide physically motivated functions for extrapolating $u^2(p,T)$ data to zero pressure. Such extrapolations to zero pressure are exactly what is required to determine $C_p^0(T)$, R, and the thermodynamic temperature from acoustic data. Indeed, as mentioned in the Section 12.1, Rowlinson and Tildesley [16] detected the extrapolation error in the earlier NPL determination of R [14] by using a semiempirical model potential to extrapolate the $u^2(p,T)$ data from NPL to zero pressure.

12.3. MEASURING THE SPEED OF SOUND

12.3.1. Under What Conditions Are Resonance Methods Most Useful?

Gas-filled acoustic resonators are well suited for measuring $u^2(p,T)$ with fractional uncertainties of less than 10^{-4} throughout the approximate density range $50 \text{ mol} \cdot \text{m}^{-3} < \rho < 10^4 \text{ mol} \cdot \text{m}^{-3}$. Under these conditions, $u^2(p,T)$ can be determined from measurements of resonance frequencies and the application of very small corrections that depend on other properties of the gas and the properties of the resonator. When the density of the gas is reduced below the lower end of this range (which corresponds to a density near that of ambient air), the corrections from viscous and thermal dissipation at the boundary between the gas and the walls of the resonator grow as $1/\rho^{1/2}$ and the signal-to-noise ratio decreases in proportion to ρ^2. (For certain gases at low pressures, the attenuation of sound throughout the volume of the gas is important and the signal-to-noise ratio decreases as ρ^3.) These effects set the lower density limit. When the density of the gas is raised above $10^4 \text{ mol} \cdot \text{m}^{-3}$ (corresponding approximately to the critical density for most gases), the uncertainties increase because the coupling between oscillations of the gas within the resonator to oscillations of the shell enclosing the gas increases. A measure of the coupling is the frequency perturbation:

$$\Delta f_{\text{shell}}/f = \mathcal{F}(\text{geometry, frequency}) \times (\rho u^2)/(\rho_{\text{shell}} u_{\text{shell}}^2) \qquad (12.11)$$

where ρ_{shell} and u_{shell} are the density and the longitudinal speed of sound in the shell. In Eq. 12.11, the function $\mathcal{F}$ depends on the shape of the resonator (spherical, cylindrical, etc.) and on the resonance mode under consideration. The magnitude of $\mathcal{F}$ becomes large when the frequency of the gas's oscillations is close to one of the resonance frequencies of the empty shell, which themselves depend on u_{shell}. For radially symmetric oscillations of a gas within a spherical shell of radius a and thickness t, the function of frequency and geometry in Eq. 12.11 is approximately

$$\mathcal{F}(\text{geometry, frequency}) = \frac{a/(2t)}{1 - (f/f_{\text{breathing}})^2} \qquad (12.12)$$

where $f_{\text{breathing}} \approx 0.22 \times u_{\text{shell}}/a$ is the frequency of the breathing oscillation of the empty shell. For resonators of other shapes, this function will have a larger zero-frequency limit. At high densities, where $\Delta f_{\text{shell}}/f$ is always large, the speed of sound can be measured more accurately by using high-frequency time-of-flight techniques than by using resonance techniques. Even at low densities, reliable measurements cannot be obtained from any mode of a gas oscillation that has a frequency that happens to be close to a resonance frequency of the empty shell.

12.3.2. Resonators for Speed-of-Sound Measurements

The resonators designed to obtain accurate values of $u^2(p,T)$ should have (1) walls that are thick and stiff, (2) transducers that are small and stiff, (3) a small surface-to-volume ratio, (4) smooth interior surfaces, and (5) a carefully designed port for admitting and removing gas. The first two qualities are desirable because they reduce hard-to-calculate corrections to the resonance frequencies from the compliance of the walls ($\Delta f_{\text{shell}}/f$) and the transducers. The third and fourth qualities ensure that the corrections from viscous and thermal dissipation can be calculated and are small. Ideally, a valve is built into the wall of the resonator to seal any port used to admit gas. When this is done, the port introduces negligible perturbations to the resonance frequencies. A very simple and frequently used alternative is to use a long, narrow capillary tube to admit gas. However, a capillary slows the pumping of gas out of the resonator so much that flushing must be used whenever samples are changed. A more sophisticated alternative to a valve is to choose the length of the port and its exterior termination to simultaneously achieve a high acoustic impedance at the resonance frequencies and a low impedance to the quasi-steady flows used to fill and empty the resonator [21].

In "routine" measurements of $u^2(p,T)$ in industrially significant gases at NIST, we have used cylindrical resonators with a circular cross section. (Typical dimensions are 14 cm long and 3 cm in radius.) When results of the highest possible accuracy are needed, we have used spherical resonators with internal volumes of 1/4, 1, or 3 liters. For spherical resonators, one can model Δf_{shell} most accurately and one can attain the lowest possible viscous and thermal attenuation at the gas-shell boundary by using the radially symmetric modes. Then, one has the smallest possible frequency perturbations from the boundary and the highest possible signal-to-noise ratio.

In principle, the measurement of the resonance frequency of a single acoustic mode is sufficient to determine the speed of sound. In practice, the redundancy that comes with measuring the frequencies f and the halfwidths g of several resonance modes provides checks of the calculated corrections. These checks are practical only if the modes under study are nondegenerate, i.e., the resonance frequencies are well separated from each other. For a spherical resonator, only the radially symmetric modes are nondegenerate; thus, they are the only ones used for practical purposes. For a cylindrical resonator, we have typically used the four lowest-frequency longitudinal modes together with the lowest-frequency radial and azimuthal modes.

For studies of gases where intramolecular relaxation is slow and for studies of the strong attenuation of sound that occurs extremely close to all liquid-vapor critical points, compact resonators with low-resonance frequencies are desirable. For such studies, Garland and Williams [22] used a toroidal resonator. Recently, we measured $u^2(p,T)$ with uncertainties on the order of 0.02% using a double Helmholtz resonator designed as an acoustic Greenspan viscometer (Section 12.4). At the frequency of the Helmholtz resonance, the wavelength of sound was 1.1 m, even though the resonator was only 6 cm long. We plan to use this resonator for measurements near the critical point of xenon.

12.3.3. Calibration Using a Gas of Known Properties

The resonance method of measuring $u^2(p,T)$ requires a resonator, frequency standard, pressure gauge, and thermometer that are stable. In practice, exacting temperature and dimensional measurements are avoided by "calibrating" the resonator with a gas, such as argon, for which the speed of sound is accurately known. The calibration is conducted by measuring the resonance frequencies of several acoustic modes when the resonator is filled with argon at the temperatures and pressures of interest. These frequencies are corrected for the thermal and viscous boundary layers and are then used to determine the relevant dimensions of the resonator. The resonator is then filled with the test gas, and the resonance frequencies of exactly the same modes are measured. After making the boundary layer corrections, these frequencies are used to determine the speed of sound in the test gas, $u_{\text{test}}(p,T)$. To a high degree, the dimensions of the resonator drop out of the determination. In effect, one determines $u_{\text{test}}(p,T)/u_{\text{argon}}(p,T)$ from the frequency ratios $f_{\text{test}}/f_{\text{argon}}$ for several modes.

A numerical example illustrating the benefits of calibration is instructive. Consider an acoustic determination of $C_p^0(T)$ for propane in the temperature range 220–460 K. In this range, γ^0 decreases from 1.18 to 1.09 and $(u_{\text{propane}})^2$ increases by 105%; however, $(u_{\text{argon}}/u_{\text{propane}})^2 = (f_{\text{argon}}/f_{\text{propane}})^2$ increases by only 16%. To determine $C_p^0(T)$ directly from u_{propane}^2 (without calibration), the data are analyzed using the first equality in Eq. 12.13.

$$\frac{C_p^0(T)}{R} = \lim_{p \to 0} \left(\frac{(M \cdot u^2)_{\text{propane}}}{(M \cdot u^2)_{\text{propane}} - RT} \right) = \frac{\gamma^0}{\gamma^0 - 1}$$
$$= \lim_{p \to 0} \left(1 - \frac{3}{5} \frac{M_{\text{argon}}}{M_{\text{propane}}} \left(\frac{f_{\text{argon}}}{f_{\text{propane}}} \right)^2 \right)^{-1} \quad (12.13)$$

To attain a fractional uncertainty of 0.1% of $C_p^0(T)$, the combined fractional uncertainties of $M \cdot u^2$ and T must be less than 9×10^{-5}. Thus, the uncertainty of T must be less than 0.04 K at 460 K. In contrast, if one calibrates with argon using the zero-pressure result $(M \cdot u^2(0,T))_{\text{argon}} = (5/3)RT$ and the last equality in Eq. 12.13, then $C_p^0(T)$ can be determined to within 0.1% if the uncertainty of T is as large as 0.56 K. Of course the temperature of the calibration must be reproduced within 0.04 K when making the test measurements; however, this is a much easier task than knowing the temperature to 0.04 K. Indeed, temperature differences on the order of 0.5 K across parts of the resonator can be tolerated, provided that the calibration is conducted in the same thermal environment as the test measurements.

12.3.4. Effects of Impurities

In our experience, the presence of unknown impurities is often the most important contribution to the uncertainties of $u^2(p,T)$ and $C_p^0(T)$ deduced from measurements of acoustic resonance frequencies. The importance of impurities is implied by the ratio of masses in the last equality in Eq. 12.13. To make an explicit estimate of the importance of impurities, we use a thermodynamic result that can be derived from the virial equation of state for a mixture [23]. If the virial expansion is truncated after the term $B\rho$, one finds

$$u_{\text{mix}}^2 = (C_{p,\text{mix}}/C_{V,\text{mix}})RT/M_{\text{mix}} \quad (12.14)$$

where the subscript "mix" denotes a mole-fraction average for each property of the mixture. Consider the limit of an ideal gas (where $C_p^0 - C_V^0 = R$), with properties denoted by the subscript "1." An undetected impurity of component "2" with a small mole fraction x will change u_1 to u_{mix}. We consider the value of $C_{p,\text{mix}}^0$ computed from $(u_{\text{mix}})^2$ via the first equality in Eq. 12.13. The fractional change in $C_{p,\text{mix}}^0$ in the limit of x approaching zero is

$$\frac{1}{C_{p1}^0} \frac{\partial C_{p1}^0}{\partial x} = (C_{p2}^0/C_{p1}^0 - 1) - \frac{(1 - M_2/M_1)}{(\gamma_1^0 - 1)} + \cdots . \quad (12.15)$$

Usually, the first term on the right hand side of Eq. 12.15 is less than the second term. For propane, where $(\gamma_1^0 - 1) \approx 0.1$, the mole fraction of ethane must be less than 3×10^{-4} if its effect on C_p^0 is to be less than 0.1%. Impurities with very large values of M_2/M_1 are particularly troublesome.

12.3.5. Transducers

At gas densities greater than approximately 50 mol·m^{-3}, it is possible to make accurate measurements of $u^2(p,T)$ using acoustic pressures as small as 10^{-6} times the pressure of the test gas. Large, efficient acoustic transducers are not needed to generate and detect such small acoustic pressures when a gas in a modest volume is resonating with a Q on the order of 10^3 to 10^4. Recognizing this, we have placed small transducers within the resonator or we have used external transducers that are coupled very weakly to the resonator. These arrangements ensure that the transducers make only very small perturbations to the resonance frequencies.

We have always used two independent transducers, one functioning as a sound generator and a second functioning as a sound detector. When measuring $u^2(p,T)$ in inert gases, we used commercially manufactured transducers that were embedded within the walls of resonators in such a way as to be flush with an interior surface. To work with reactive test gases, we built thin, stainless steel or monel diaphragms into one or two walls of the resonator to separate the test gas within the resonator from an inert gas at the same pressure outside the resonator (see Fig. 12.3). In some cases, we placed transducers immediately behind the diaphragms. In other cases, acoustic wave guides filled with inert gas conducted the sound between two remote transducers and the diaphragms. This wave guide arrangement is particularly versatile because the transducers can be at ambient temperature while the resonator is at either a very high or a very low temperature [24]. Near the diaphragms, a short length of each wave guide is filled with a screen formed of fine stainless steel wires. The screen attenuates sound, thereby preventing the occurrence of high-Q, longitudinal resonances within the wave guides. Such unintentional resonances would have interfered with measurements of the resonances of the test gas within the resonator.

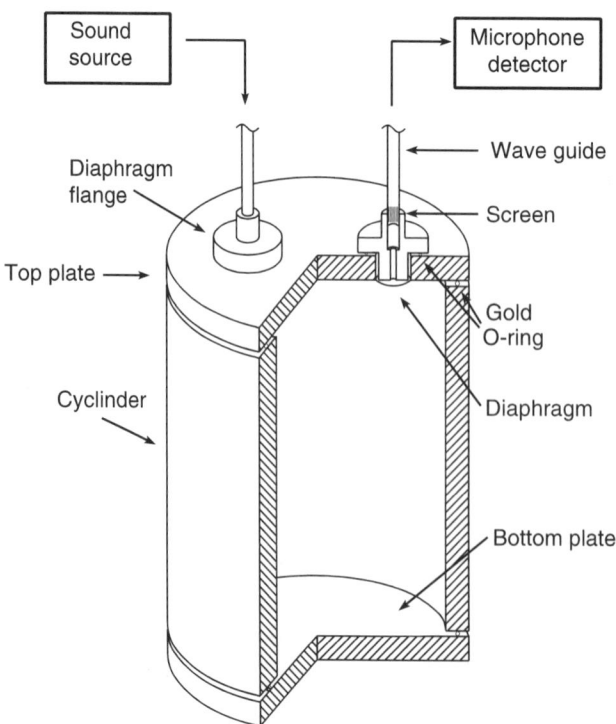

Fig. 12.3. Cylindrical acoustic resonator designed for speed-of-sound measurement.

12.3.6. Measurements of Resonance Frequencies and Halfwidths

At each state point (p,T), the frequency of the sound generator is stepped through each resonance while recording in-phase (x) and quadrature signals (y) from the detector. Typically, measurements are made at 11 frequencies separated by steps of $g_m/5$ spanning the range $f_m \pm g_m$, where g_m is the halfwidth of the resonance and the subscript "m" denotes the indices that identify the mode under study. Then, the theoretically expected, complex function of the frequency is fitted to the values $x + iy$ to obtain both f_m and g_m and also their uncertainties. The fitting function is

$$x + iy = \frac{if \mathbf{A_m}}{(f^2 - \mathbf{F_m^2})} + \mathbf{B} + \mathbf{C}(f - f_1) \tag{12.16}$$

Here, $\mathbf{A_m}$, $\mathbf{B}$, and $\mathbf{C}$ are complex constants, $\mathbf{F_m} = f_m + ig_m$ is the complex resonance frequency, and f_1 is an arbitrary constant. In Eq. 12.16, the parameters $\mathbf{B}$ and $\mathbf{C}$ account for possible crosstalk and for the effects of the "tails" of modes other than the one under study. The quality factor of a typical resonance was in the range $10^3 < Q \equiv f_m/2g_m < 10^4$ and the standard deviation of f_m was less than $10^{-5} f_m$. Under ideal conditions, the statistical uncertainty of f_m is on the order of $10^{-7} f_m$.

12.3.7. Determination of the Speed of Sound

The speed of sound u in the test gas is determined from the measured resonance frequencies f_m by $u = 2\pi f_m/k_m$. Expressions for the wave number k_m for resonances in a cylindrical cavity with a radius a and length l appear in [5]. Expressions for a spherical cavity appear in [4]. If the resonator is calibrated (Section 12.3.3) using a gas with a known speed of sound, accurate values of the dimensions of the resonator and the wave number are not needed.

The measured resonance frequencies are corrected for the thermal and viscous losses at the boundaries as well as for the small effects of the duct used to move the sample into and out of the resonance cavity [9, 10].

12.3.8. Corrections for Boundary Losses and Fill Duct

Dissipative heat exchange between the test gas and the walls of the resonator occurs within a gas layer adjacent to the walls. The thickness of this thermal boundary layer is characterized by the thermal penetration length δ_t. Similarly, momentum exchange is characterized by the viscous penetration length δ_v. These lengths are defined in terms of the acoustic frequency and the gas's properties by

$$\delta_t^2 \equiv \frac{\lambda}{\rho C_p \pi f}; \quad \delta_v^2 \equiv \frac{\eta}{\rho \pi f} \tag{12.17}$$

The thermal and viscous boundary layers perturb the resonance frequencies and halfwidths by $\Delta \mathbf{F_t} = -\Delta f_t + ig_t$ and $\Delta \mathbf{F_v} = \Delta f_v + ig_v$, respectively. For a typical resonator, such as the one shown in Fig. 12.3, these perturbations have magnitudes on the order of the volume of the gas in the boundary layers divided by the volume of the gas inside the resonator, i.e., $\Delta f_t/f_m \sim g_t/f_m \sim (\gamma - 1) \times \delta_t \times A_{\text{resonator}}/V_{\text{resonator}} \sim (\gamma 1)10^{-4}$ and $f_v/f_m \sim g_v/f_m \sim \delta_v \times A_{\text{resonator}}/V_{\text{resonator}} \sim 10^4$. For the monatomic gases at low pressures, the factor $(\gamma - 1)$ is 2/3; it is much smaller for polyatomic gases (e.g., 0.09 for propane at 460 K). Therefore, the temperature changes associated with the sound wave and the corresponding frequency perturbations $\Delta \mathbf{F_t}$ are much smaller in polyatomic gases than in monatomic gases.

Expressions for $\Delta \mathbf{F_t}$ and $\Delta \mathbf{F_v}$ for cylindrical resonators appear in [5]; the corresponding expressions for spherical resonators appear in [4]. For other geometries, the corrections may be calculated using boundary perturbation theory. As in [4], boundary perturbation theory may also be used to account for any fill duct that may be present.

12.4. RESONANCE MEASUREMENTS OF TRANSPORT PROPERTIES

Cylindrical resonators, such as the one sketched in Fig. 12.3, are not well suited to determine the viscosity η and the thermal conductivity λ. In the following paragraph, we discuss the reasons for this to motivate the design of the Greenspan acoustic viscometer and the Prandtl number resonator. The Prandtl number is defined by $Pr \equiv \eta C_p/\lambda$ and is widely used to correlate the heat transfer from a solid to a moving fluid. We plan to measure the Prandtl number in many gases and to deduce the thermal conductivity from it, together with the acoustically determined values of C_p and η.

As mentioned in Section 12.3.8, typical perturbations from viscous and thermal losses are on the order of (volume of the boundary layer)/(volume of the resonator). For the cylindrical resonators designed to measure $u^2(p,T)$, this ratio is 10^{-3} or even smaller. The halfwidths g_m of the resonances are measured with relative uncertainties on the order of 10^{-2} or less. In principle, one could use these measured values of g_m for several modes to determine the lengths δ_t and δ_v and from these lengths one could determine η and λ. However, several factors would lead to uncertainties in η and λ of several percent or more. One factor is the difficult-to-calculate contributions to g_m. Such contributions come from motion of the shell at high pressures, sound attenuation throughout the volume of the gas at low pressures, and dissipation within the transducer assemblies. A second factor is the difficulty in determining the effective internal area of a resonator within 1%. Surface roughness on the scale of δ_t and δ_v may cause the effective area to increase with pressure (as δ_t and δ_v decrease with pressure). A pressure-dependent contribution to the effective area may result from crevices that were formed where the parts of a resonator were joined during its assembly. Finally, accurate measurements of small values of g_m require a high signal-to-noise ratio and very fine temperature control during the measurement. These considerations led us to consider using a Greenspan acoustic viscometer to measure η and another specially designed resonator to measure the Prandtl number.

12.4.1. Greenspan Acoustic Viscometer

The Greenspan acoustic viscometer is a dumbbell-shaped double Helmholtz resonator that has a low-frequency, low-Q resonance in which the gas oscillates through a duct connecting two chambers (see Fig. 12.4). The low frequency leads to relatively large values of δ_v that reduce the requirements for a fine surface finish. The low Q reduces the relative importance of the difficult-to-estimate contributions to g_m and also reduces the need to maintain very high temperature stability. In the lowest-order approximation, the Helmholz resonance frequency f_m is

$$f_m^2 \approx \frac{u^2 r_d^2}{2\pi L_d V} \qquad (12.18)$$

where V is the volume of each of the chambers and L_d and r_d are the length and radius of the duct that connects the chambers. In many important cases, the halfwidth of the Helmholtz resonance can be expressed as the sum of four terms:

$$\frac{2g_m}{f_m} = \frac{1}{Q} \approx \frac{\delta_v}{r_d} + 2\varepsilon_{\text{orif}}\frac{\delta_v}{L_d} + (\gamma-1)\frac{\delta_t A}{\pi V} + (\gamma-1)\frac{C_{\text{relax}}}{C_p}\frac{2\pi f_m \tau_{\text{relax}}}{1+(2\pi f_m \tau_{\text{relax}})^2} \qquad (12.19)$$

The first term on the right-hand side (r.h.s.) of Eq. 12.19 is the desired term that is used to determine the viscosity from the radius of the duct r_d and ρ (that appears in δ_v). The

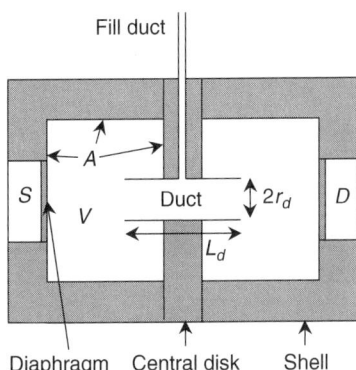

Fig. 12.4. Schematic cross section of a Greenspan acoustic viscometer [7]. This viscometer had an outer diameter of 6.2 cm and a length of 7.2 cm. The source and detector transducers (denoted "S" and "D," respectively) were PZT stacks. The volume of the chamber on the left is indicated by "V" and its inner surface area is indicated by "A."

second term on the r.h.s. of Eq. 12.19 accounts for flows that converge into and diverge out of each end of the duct. This term contains the "orifice resistance parameter" $\varepsilon_{\text{orif}}$ that Mehl calculated numerically [25]. For the resonator described in [7] and sketched in Fig. 12.4, $\varepsilon_{\text{orif}} \approx 0.97$. The third term on the r.h.s. of Eq. 12.19 accounts for the heat exchange between the gas and the walls of the resonator; it is proportional to the ratio of the volume of the thermal boundary layer in one chamber $\delta_t A$ to the volume V of that chamber. The last term on the r.h.s. of Eq. 12.19 applies to certain gases (e.g., CH_4, CO_2) that have symmetries such that many intermolecular collisions are required for their internal degrees of freedom to adjust to the temperature change associated with the acoustic oscillation. In such gases, the acoustic dissipation throughout the volume of the gas is characterized by the product $C_{\text{relax}} \cdot \tau_{\text{relax}}(\rho)$ where C_{relax} is the heat capacity associated with the slowly relaxing degrees of freedom and $\tau_{\text{relax}}(\rho)$ is the relaxation time, which is proportional to ρ^{-1}.

For the viscometer described by Wilhelm et al. [7], the transducers were piezoelectric "stacks" that were separated from the test gas by 0.1-mm-thick stainless steel diaphragms. In tests with five gases (He, Ar, Xe, N_2, CH_4), the $Q \equiv f_m/2g_m$ was in the range $20 < Q < 200$ for pressures in the range 0.2 MPa $< p < 3.2$ MPa. Less than 10% of Q^{-1} came from heat transfer between the gas and the walls of the chambers; thus, λ had to be known to 10% to determine η to 2%. The relaxation term for methane contributed less than 0.3% to Q^{-1}.

Figure 12.5 compares the viscosities determined with the Greenspan acoustic viscometer with data from the literature obtained by other methods. To obtain the results in the upper panel of Fig. 12.5, the Greenspan viscometer was used as an absolute instrument, i.e., the viscosity was calculated from measurements of its dimensions and its frequency response. The root mean square (r.m.s.) deviation of the Greenspan viscometer data from the base line is 0.47%.

The bottom panel of Fig. 12.5 shows the smaller deviations (0.12% r.m.s.) that remained after the data from the literature were used to calibrate the Greenspan viscometer. This illustrates how the instrument would perform if it were used as a relative instrument. For the calibration, we assumed that the literature values of η were correct and we replaced the values of $\varepsilon_{\text{orif}}$ and A with effective values. On reducing the calculated value of $\varepsilon_{\text{orif}}$ by 3%, all of the deviations on Fig. 12.5 were shifted upward by 0.5%. This small change in $\varepsilon_{\text{orif}}$ would result if the right angles (that were assumed in the calculations) at the ends of the duct were rounded with a radius of only 2.5 μm [25]. Such rounding is consistent with the way in which the duct was manufactured.

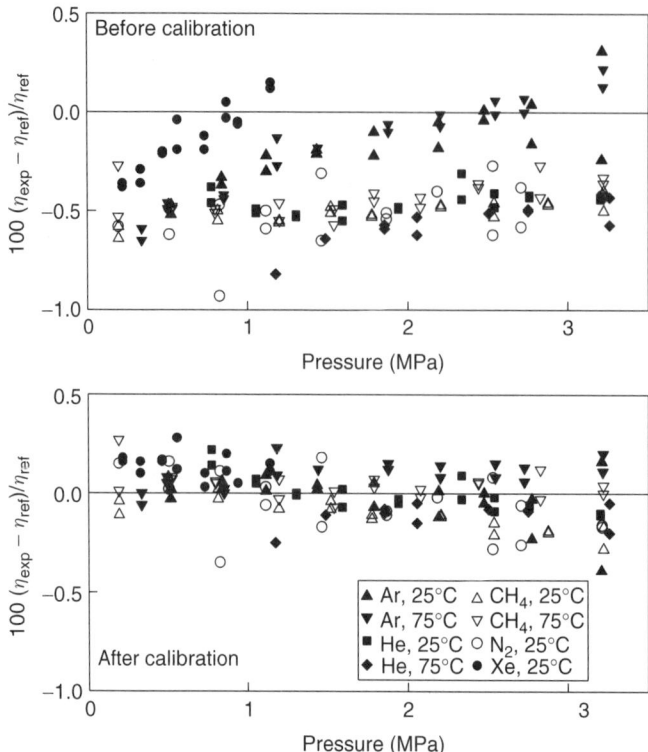

Fig. 12.5. Comparison of viscosities determined with a Greenspan acoustic viscometer with literature values. Top, Results obtained using the viscometer as an absolute instrument. Bottom, Results obtained after calibration.

The data for the noble gases in the top panel of Fig. 12.5 have a positive slope before calibration. This slope was eliminated by replacing A with effective values that were 1–5% larger than the value calculated from dimensional measurements and by assuming that A varied inversely as δ_t. The cause for this δ_t-dependent increase in A was not determined; it may be related to rough surfaces where the diaphragms and the duct were soldered into place. Alternatively, there may have been a narrow crevice where metal O-rings were used to seal the chambers to the central disk (Fig. 12.4).

The frequency of the Helmholtz mode of the Greenspan viscometer can be used to determine the speed of sound $u(p,T)$. This is a by-product of the viscosity measurement and was not expected to be particularly accurate. It was a pleasant surprise to find that the values of $u(p,T)$ for He, Ar, Xe, N_2, and CH_4 were consistently between 0.16 and 0.20% larger than values from the literature. If the volumes of the chambers were replaced with effective volumes that were 0.36% smaller, all of the speed of sound data would fall within ±0.02% of published values (see Fig. 12.6). The difference between the effective volume and the geometric volume is on the order of the volume displaced by the solder used to hold the diaphragms and the duct in place.

Apparently, a calibrated Greenspan viscometer can be used for simultaneously measuring the viscosity and the speed of sound in a gas. However, if one were to use only the Helmholtz mode of such a resonator, the redundancy that is usually available in resonance speed-of-sound measurements would be missing and the possibility of undetected systematic errors would be greatly increased. Alternatively, one could design a Greenspan viscometer with three chambers connected by two ducts. Such a resonator would have two low-frequency modes that could provide some redundancy.

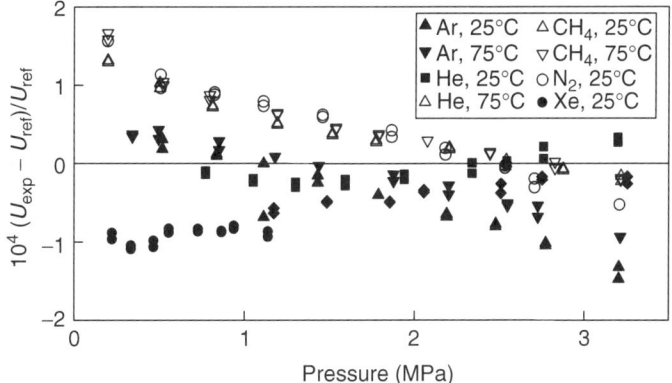

Fig. 12.6. Comparison of speed-of-sound results from a Greenspan viscometer with results from the literature. See [7] for details.

12.4.2. Resonators for Measuring the Prandtl Number

Figure 12.7 is a sketch of a prototype resonator that was designed to measure the Prandtl number. This resonator was a cylinder with an internal length, $L = 13.2$ cm, and an internal radius of $R = 4.8$ cm. A honeycomb-shaped array of hexagonal ducts was installed into the center of the resonator. Ignoring the honeycomb and the viscous and thermal boundary layers, the longitudinal modes of the cylindrical resonator have frequencies f_n

$$f_n = un/(2L); \quad n = 1, 2, 3, \ldots \quad (12.20)$$

The surface area of the honeycomb was approximately five times larger than the surface area of the cylinder even though the length of the honeycomb was much less than that of the cylinder. Thus, most of the damping of the longitudinal modes occurred within the honeycomb.

The odd longitudinal modes ($n = 1, 3, 5, \ldots$) had velocity antinodes in the plane of symmetry bisecting the honeycomb and the axis of the cylinder. In each hexagonal duct, viscosity damped the gas motion within the boundary layer (of thickness δ_v)

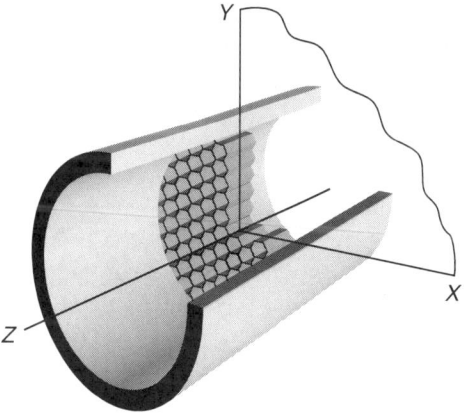

Fig. 12.7. Schematic section of the resonator used to measure the Prandtl number. The hexagonal ducts are oriented parallel to the axis of the cylindrical resonator and located midway between the ends of the cylinder. The odd longitudinal modes have a velocity antinode in the plane bisecting the resonator and the ducts. The even longitudinal modes have a temperature antinode in the same plane.

surrounding the duct's surfaces. The quality factor of these odd modes was approximately $Q_{odd} \approx$ (volume of resonator)/[$\delta_v \times$ (area of honeycomb)]. In the same plane bisecting the honeycomb, the even longitudinal modes ($n = 2, 4, 6, \ldots$) had a velocity node and a corresponding antinode in the pressure and the temperature. Heat exchange between the honeycomb and the gas within the thermal boundary layer (of thickness δ_t) strongly damped the even longitudinal modes. The quality factor of the even modes was approximately $Q_{even} \approx$ (volume of resonator)/[$(\gamma - 1)\delta_t \times$ (area of honeycomb)]. In this crude approximation, the ratio Q_{even}/Q_{odd} is independent of the volume of the resonator and of the area of the honeycomb, and the Prandtl number can be calculated from measurements of the frequencies and halfwidths of an even and odd mode:

$$Pr = \frac{\eta C_p}{\lambda} = \frac{f_{odd}}{f_{even}} \left(\frac{\delta_v}{\delta_t}\right)^2 = \frac{f_{odd}}{f_{even}} \left(\frac{(\gamma - 1)Q_{even}}{Q_{odd}}\right)^2 = \frac{f_{even}}{f_{odd}} \left(\frac{(\gamma - 1)g_{even}}{g_{odd}}\right)^2 \quad (12.21)$$

The prototype Prandtl number resonator was assembled by installing a stainless steel honeycomb into the center of a cylindrical resonator. The honeycomb was similar to the "stack" built into thermoacoustic refrigerators and heat engines. Each hexagonal duct was 2.6 cm long and had a diameter (from flat side to flat side) of 1 mm. The stainless steel sheets comprising the honeycomb were 0.05-mm thick, and the average hydraulic radius r_h of the hexagonal ducts was 0.430 mm. [$r_h \equiv 2\times$ (cross-sectional area of duct)/(perimeter of duct)] Piezoelectric (PZT) "stack" transducers were placed behind diaphragms in each end of the resonator, much like the arrangement for the Greenspan viscometer sketched in Fig. 12.4.

The frequency response of the resonator was measured as a function of pressure on an isotherm for argon, xenon, and sulfur hexafluoride. The acoustic amplitudes used for these measurements were always small enough that nonlinear contributions to the Q were negligible.

The measured frequency response was compared with predictions that account for several effects not included in Eq. 12.21, namely, the viscous and thermal losses at the cylindrical boundaries of the resonator, the "end effects" at both ends of each hexagonal duct, the reduction of the resonator's cross section by the honeycomb, and the duct used to admit and remove gas from the resonator [26].

Figure 12.8 displays preliminary results obtained with the Prandtl number resonator. For several modes and for the three gases, the measured halfwidths are compared with halfwidths calculated using data from the literature. For xenon the data span the pressure range 0.2 MPa $< p <$ 1.2 MPa, corresponding to the range $50 < Q < 300$; for argon, the ranges are 0.1 MPa $< p <$ 1 MPa and $25 < Q < 160$; for SF_6, the ranges are 0.15 MPa $< p <$ 0.4 MPa and $250 < Q < 1000$.

The results in Fig. 12.8 for the odd modes are plotted as a function of δ_v/r_h and the results for the even modes are plotted as a function of δ_t/r_h. Nearly all the results in the bottom panel of Fig. 12.8 fall within $\pm 2\%$ of a smooth curve, indicating that the effective area of the honeycomb is a smooth function of the length scale (δ_v or δ_t) used to determine the area. Probably, some of this dependence results from the sensitivity of the end corrections to the presence of sharp corners or burrs. Another contribution to scale dependence of the effective area may result from the partially obstructed and oddly shaped ducts that were located at the boundary between the honeycomb and the cylindrical shell of the resonator.

One can fit the data in the bottom panel of Fig. 12.8 by an empirical equation that represents the effective area as a function of δ_v or δ_t. If this fit is used as a calibration, the Prandtl number of another gas can be determined with a standard uncertainty on the order of 2%. This is a very encouraging result from a prototype instrument. We conjecture that the importance of the calibration function would be reduced if the ends of the honeycomb were rounded, perhaps by electropolishing.

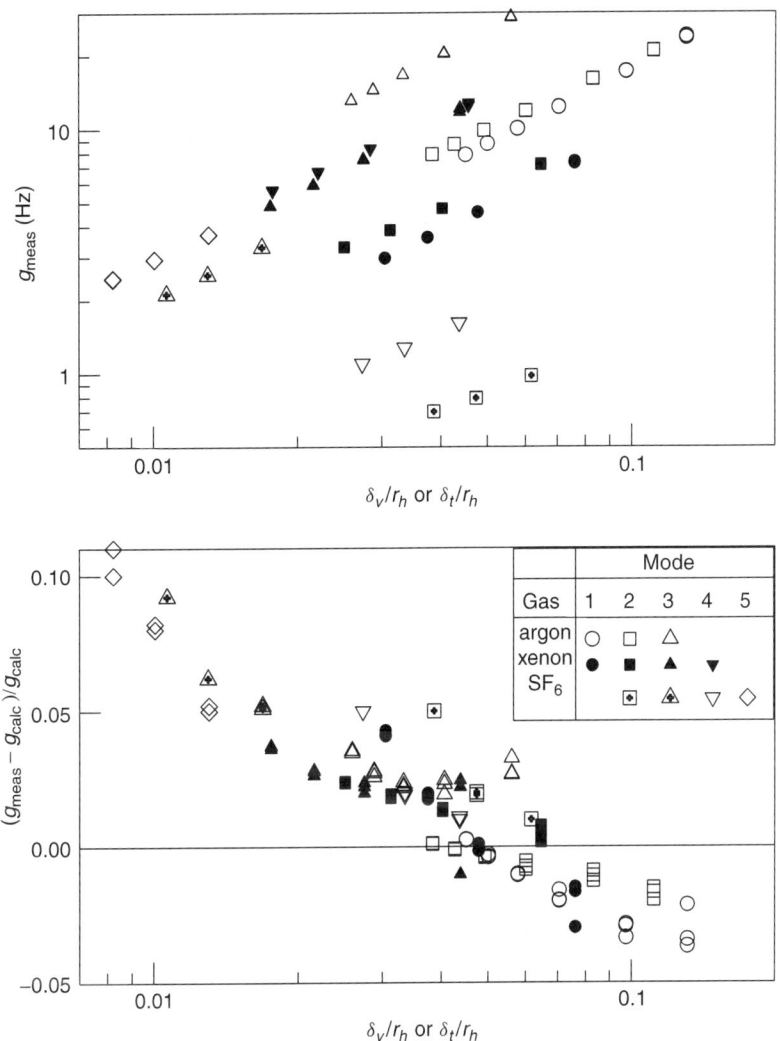

Fig. 12.8. Top, Halfwidths g of the longitudinal resonances of the Prandtl number resonator. Because SF_6 has a small value of $(\gamma - 1)$, the even, thermally damped modes for this gas have conspicuously smaller values of g. Bottom, Fractional differences between measured and calculated halfwidths. If a smooth curve were drawn through the differences, most of that data would fall within $\pm 2\%$ of it.

12.5. ACOUSTIC THERMOMETRY

12.5.1. Principles of Acoustic Thermometry

Primary acoustic thermometry relies on the connection between the speed of sound in a gas and the thermodynamic temperature of that gas. Hydrodynamics and the kinetic theory of dilute gases relate the thermodynamic temperature T, the average kinetic energy E in one degree of freedom, and the speed of sound u. In the simplest approximation,

$$3E = \frac{1}{2}mv_{\text{rms}}^2 = \frac{3}{2}k_B T, \quad u^2 = \frac{\gamma}{3}v_{\text{rms}}^2 \qquad (12.22)$$

Here, v_{rms} is the root mean square speed of a gas molecule, m is its mass, and k_B is the Boltzmann constant. For monatomic gases, $\gamma \to 5/3$ as $p \to 0$. The International System of Units assigns the exact value 273.16 K to the temperature of the triple point of water T_w. From this assignment and from Eq. 12.22, the Kelvin thermodynamic temperature T of a gas can be determined from the zero-pressure limit of the ratio of speed of sound measurements at T and T_w using the equation

$$\frac{T}{273.16\ K} = \lim_{p\to 0}\left(\frac{u^2(p,T)}{u^2(p,T_w)}\right) \quad (12.23)$$

In the NIST program, the speed-of-sound ratios were determined using the spherical, stainless steel shell that had been used to acoustically redetermine the universal gas constant R. This shell had a 2-cm-thick wall bounding a spherical cavity with a radius of 9 cm. A schematic drawing of the shell appears in Fig. 12.9. We measured the frequencies $f_\mu(T)$ of the microwave resonances of the cavity while it was evacuated to deduce the thermal expansion of the cavity. (Here, the subscripts α and μ distinguish

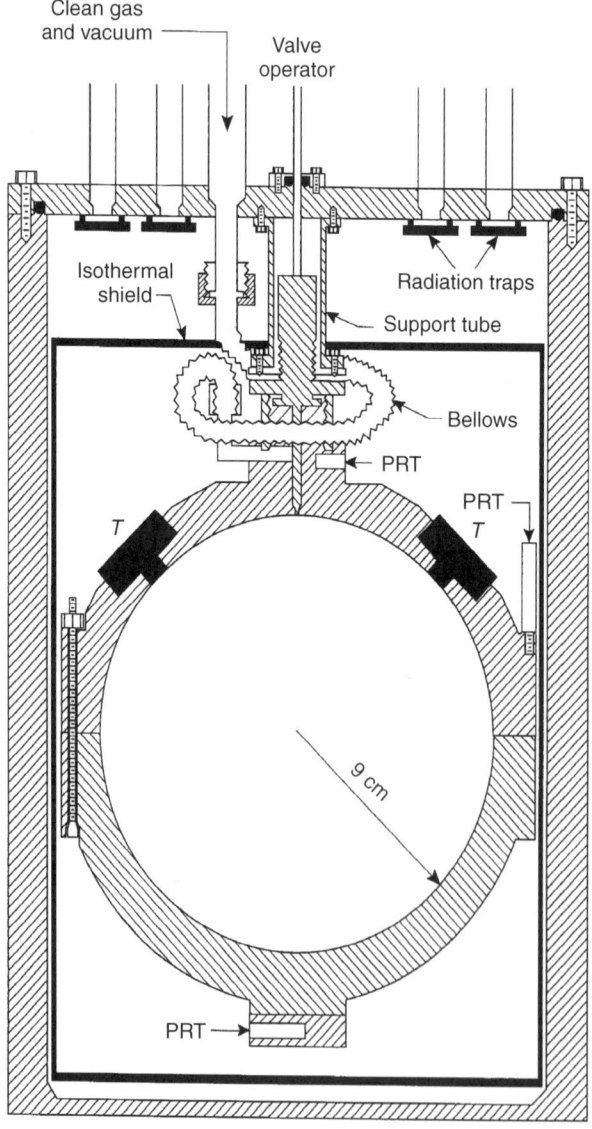

Fig. 12.9. Schematic cross section of the resonator used to measure the universal gas constant R and the difference between the international temperature scale ITS-90 and the Kelvin thermodynamic temperature. The resonator is shown within a pressure vessel. The pressure vessel was immersed in a stirred, thermostated bath that is not shown. The transducers are denoted "T" and the platinum resistance thermometers are denoted "PRT."

acoustic frequencies from microwave frequencies.) We also measured, as a function of the temperature and the pressure, the resonance frequencies $f_\alpha(p,T)$ of the radially symmetric acoustic modes of the cavity while it was filled with argon. To deduce $u^2(p,T)$, the acoustic frequencies were combined with the volume of the resonator at T_w from [2] and the thermal expansion deduced from $f_\mu(T)$. To obtain zero-pressure ratios of $u^2(p,T)/u^2(p,T_w)$, the values of $u^2(p,T)$ on each isotherm were fitted by the virial expansion Eq. 12.3 with an additional term proportional to p^{-1} that accounted for a mean free-path effect at low pressures.

The Kelvin thermodynamic temperature was determined approximately from the measured quantities by

$$\frac{T}{273.16\text{ K}} = \left(\frac{<f_\mu(T_w) + \Delta f_\mu(T_w)>}{<f_\mu(T) + \Delta f_\mu(T)>}\right)^2 \times \lim_{p \to 0} \left(\frac{f_\alpha(p,T) + \Delta f_\alpha(p,T)}{f_\alpha(p,T_w) + \Delta f_\alpha(p,T_w)}\right)^2$$

(12.24)

Because there are no radially symmetric microwave modes, it was necessary to use appropriately weighted averages of the microwave frequencies; this is indicated by the brackets "< ... >" in Eq. 12.24 [9]. (Partially resolved triplets of microwave modes were used.) In Eq. 12.24, the terms $\Delta f_\alpha(p,T)$ and $\Delta f_\mu(T)$ represent small corrections that must be applied to the acoustic and microwave frequencies, respectively. For the acoustic frequencies, these corrections depend on the thermal conductivity of the argon and the mechanical response of the shell to the acoustic pressure, as discussed in Section 12.3.1 in connection with other acoustic resonators. For the microwave frequencies, the electrical resistivity of the stainless steel shell was needed to account for the penetration of the microwave fields into a thin layer of metal at the inner boundary of the resonator. The microwave penetration length δ_μ is analogous to the thermal and viscous lengths δ_t and δ_v insofar as it leads to a Q for the microwave modes on the order of (radius of resonator)/δ_μ and a fractional frequency shift on the order of Q^{-1}.

The derivation of Eq. 12.24 requires that the same acoustic and microwave modes be used for the frequency measurements at the temperatures T and T_w and that the gas has the same value of γ at both temperatures. Maintaining the purity of the argon was a major concern and is discussed in [3]. As implied by the absence of the resonator's dimensions and eigenvalues in Eq. 12.24, and as emphasized by Mehl and Moldover [27], the spherical cavity plays a limited role in measuring u/c, the ratio of the speed of sound in a monatomic gas to the speed of microwaves (light). One may view the cavity as a temporary artifact that must remain dimensionally stable just long enough to measure $f_\alpha(p)$ and $<f_\mu>$ at the temperature T and that must not change its shape too much when the frequency measurements are repeated at T_w.

12.5.2 Why Spherical Resonators?

For thermometry, we require speed-of-sound measurements of the highest possible precision. To meet this requirement, the radially symmetric acoustic modes within a spherical cavity have several advantages compared with the acoustic modes of other cavities, including those of more easily manufactured cylindrical cavities.

For radially symmetric acoustic oscillations in a spherical cavity, the gas's motion is perpendicular to the shell everywhere; thus, it is not subject to viscous damping where the gas contacts the shell. The small surface-to-volume ratio of the spherical shell leads to the smallest possible thermal dissipation in the thermal boundary layer. The absence of viscous losses and the small thermal losses lead to higher Qs and correspondingly higher signal-to-noise ratios. The latter permits measurements at lower pressures, thereby reducing the uncertainties in extrapolating $u^2(p,T)$ to zero pressure.

The high Qs of the radially symmetric modes permit the use of small transducers embedded in the shell. The transducers were selected because they had a very small effective volume that produced only weak perturbations to the measured frequencies.

For the radially symmetric modes, the frequency perturbation from the thermal boundary layer is as small as possible for a resonator of a given volume. Thus, the dependence of the measured frequencies (and the temperatures deduced from them) on the thermal conductivity is as small as possible. For the NIST resonator, a 0.3% uncertainty in the thermal conductivity of argon propagates into a 0.2 mK uncertainty in the temperatures determined in the range 216 to 303° K. This uncertainty is a fraction of the overall standard uncertainty of the results: 0.6 mK.

As mentioned in Section 12.3.1, the coupling of the motion of the gas to the motion of the shell leads to pressure-dependent frequency perturbations. In comparison with other geometries, these perturbations are smallest for spherical shells. Also, they have been modeled quantitatively.

12.5.3. Measurements of the Thermodynamic Temperature

In the NIST acoustic thermometry (and in the acoustic redetermination of R discussed in Section 12.6), five nondegenerate acoustic modes spanning the frequency range 2.5–9.5 kHz were used. Also, three microwave triplets spanning the frequency range 1.5–5.0 GHz were used. The redundant acoustic and microwave measurements helped determine some components of the uncertainty in measuring $(T - T_{90})$, the difference between the Kelvin thermodynamic temperature T and the temperature measured on ITS-90. The redundancy also tested the theories for the corrections $\Delta f_\alpha(T)$ and $\Delta f_\mu(T)$. We now consider, in turn, the measurements of the four variables that appear in Eq. 12.24: f_μ, f_α, T, and p.

The microwave resonances had quality factors of more than 5000. The values of $< f_\mu(T) >$ could be determined with uncertainties of a few parts in 10^7 using off-the-shelf instruments and a relatively simple analysis program. The uncertainties of $< f_\mu(T) + \Delta f_\mu(T) >$ made a negligible contribution to the uncertainty of $(T - T_{90})$.

For argon, the acoustic measurements were made on near to six isotherms: 217.0950, 234.3156, 253.1500, 273.1600, 293.1300, and 302.9166 K. These measurements spanned the density range 10–200 mol·m^{-3}. At these densities, the correction terms $\Delta f_\alpha(p,T)$ were small, and in some cases they tended to cancel out the ratios in Eq. 12.24. At low pressures, the contribution to $\Delta f_\alpha(p,T)$ from the thermal boundary layer in the argon where it contacts the shell grows as $p^{-1/2}$ and the signal-to-noise ratio for measuring $f_\alpha(T)$ declines as p^2. At high pressures, the term in $\Delta f_\alpha(p,T)$ from the mechanical response of the shell to the acoustic pressure grows as p and the terms resulting from the nonideality of the argon grow as p, p^2, etc. Thus, the argon data were taken over the limited range of intermediate densities where they were most useful for determining $(T - T_{90})$.

Three capsule-style standard platinum resistance thermometers were calibrated at the triple points of argon, mercury, water, and gallium and then installed in the resonator. The uncertainty of the thermometry resulted from the uncertainty of each calibration measurement and from drifts in the system (thermometers + resistance bridge + standard resistor) during the weeks between calibrations. ITS-90 is inherently nonunique because different platinum resistance thermometers meeting ITS-90 specifications will indicate different temperatures when they are compared at temperatures not used in their calibration [28]. In this work, the nonuniqueness of ITS-90 contributed an uncertainty of approximately 0.2 mK to the temperature measurements on the isotherms at 217, 253, and 293 K. An additional uncertainty resulted from the small temperature difference ($\leq$0.5 mK) between the thermometers embedded in the top and in the bottom of the spherical shell. We estimate that our imperfect knowledge of the volume average

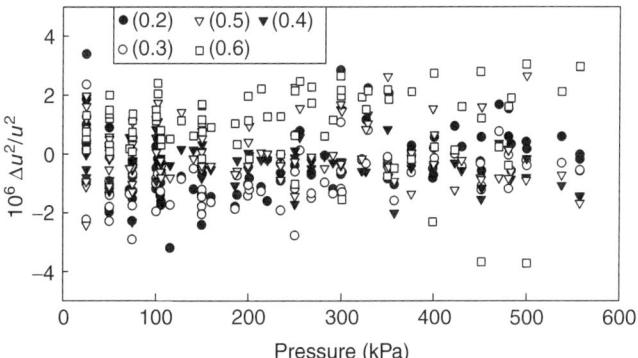

Fig. 12.10. Deviations of measurements of the speed of sound in argon from an augmented virial equation of state ($\Delta u^2 \equiv u^2_{\text{measured}} - u^2_{\text{fitted}}$). The various symbols show the results for different radially symmetric acoustic modes, which are denoted (0,2), (0,3), etc.

of the temperature within the argon was less than 0.1 mK relative to the thermometer calibrations.

We used a calibrated, fused-quartz, bourdon tube gage to measure the pressure of the argon. On the isotherm at 303 K, the worst case, u^2 changes only 0.3% over the range of pressure spanned by the data. The uncertainty of the pressure measurement made a negligible contribution to the uncertainty of $(T - T_{90})$.

The analysis of the acoustic data is a complicated subject that is described in detail in [3]. Here, we emphasize the quality of the results, as illustrated in Fig. 12.10. Note that all the fractional deviations of u^2 are within $\pm 4 \times 10^{-6}$. These results are approximately a factor of 10 more precise than those obtained with the cylindrical resonator (Fig. 12.1, bottom panel) and approximately a factor of 100 more precise that those obtained with the dumbbell-shaped Greenspan viscometer (Fig. 12.6).

With a few well-understood exceptions, $u^2(p,T)$ results obtained from resonance frequency measurements spanning a factor of 20 in pressure, a factor of 4 in frequency, and the temperature range $216\ K \leq T \leq 303$ K, could all be represented with a relative standard deviation of only 1.1×10^{-6}. This representation had 11 parameters of which six were the determined thermodynamic temperatures. The other 5 parameters were used to represent the departures of the acoustic virial coefficients $\beta_a(T)$ and $\gamma_a(T)$ in Eq. 12.3 from a model argon-argon potential.

The values of $(T - T_{90})$ ranged from (-3.6 ± 0.6) mK at 217 K to (4.6 ± 0.6) mK at 303 K. The resonator that was used at NIST to produce the data in Fig. 12.10 has been remanufactured and is now being used to determine values of $(T - T_{90})$ up to 800 K. This study may lead to replacing ITS-90 with a new temperature scale based, in part, on speed-of-sound data. If that happens, the calibration of practical thermometers will be based on acoustic measurements and the nonuniqueness inherent in ITS-90 may be reduced.

12.6. ACOUSTIC DETERMINATION OF THE UNIVERSAL GAS CONSTANT R

The most accurate measurement of the speed of sound ever made in a gas had a relative standard uncertainty of only 0.6×10^{-6}. This measurement (made at NIST) led to the value of the universal gas constant $R = (8.314\,472 \pm 0.000\,015)$ J·mol^{-1}·K^{-1}, which has a relative standard uncertainty of 1.7×10^{-6} and was internationally accepted in 1998 [1]. The physics underlying this acoustic redetermination of R is the same physics underlying the acoustic determination of the thermodynamic temperature; however, the redetermination of R had two additional complexities. First, only ratios of u^2 are needed to determine the temperature; however, an absolute measurement of u^2 is required to determine R. Any absolute measurement of u^2 requires difficult dimensional measurements. Second, any monatomic gas may be used to determine the temperature; however, a gas of known molecular mass M must be used to determine R. The mass m of a single atom (of a known isotope) is extremely well known on a scale of atomic mass units; however, the uncertainty of the average value of m for any pure gas is much larger because of the uncertainties of the relative isotopic abundances and the uncertainty of the Avogadro constant N_A. (The Avogadro constant relates the Boltzmann constant to the gas constant $R = N_A k_B$ and it relates the molecular mass to the molar mass $M = N_A m$.)

Equations 12.22 imply that a measurement of the speed of sound in a dilute, monatomic gas at the temperature T_w (which, by definition, is exactly 273.16 K) will determine R through the relations

$$R = N_A k_B = \frac{mN_A}{T_w} \lim_{p \to 0}(u^2/\gamma) = \frac{3M}{5T_w} \lim_{p \to 0}(f_m V^{1/3}/z_m)^2 \qquad (12.25)$$

To obtain the last equality of Eq. 12.25, we replaced γ with its zero-pressure limit 5/3 for the monatomic gases, and we replaced u with $(f_m V^{1/3}/z_m)$, where z_m is the exactly known eigenvalue associated with the radially symmetric mode of a gas in a perfect spherical shell of volume V, and f_m is the unperturbed resonance frequency of that mode.

The eigenvalues z_m have a very weak sensitivity to smooth departures from a perfect spherical geometry, provided that the departures do not change the volume. For acoustic measurements, this symmetry is discussed in [4] and [29]; for microwave measurements, a detailed discussion appears in [27]. This symmetry allowed NIST to use careful weighing in place of very difficult and less accurate measurements of the internal dimensions of the assembled spherical shell.

The NIST redetermination of R used a spherical shell assembled from two hemispherical parts that had been manufactured in NIST's machine shop. The shell had departures from sphericity on the order of a few parts in 10^4. The volume of the shell was determined by weighing the mercury required to fill it and by using the known density of that lot of mercury. The weighing was done with a relative standard uncertainty of 0.3×10^{-6} at ambient temperature; however, the uncertainty of $V^{2/3}$ at T_W was larger (0.7×10^{-6}) because of the uncertainty of the integrated thermal expansion of the mercury between ambient temperature and T_W.

Commercially purchased argon gas was used for the measurements; however, the relative abundances of the isotopes Ar^{36}, Ar^{38}, and Ar^{40} were not known accurately enough to calculate the mean molar mass M. Instead, a small sample of isotopically enriched Ar^{40} was purchased and chemically purified and then used as a standard for M. The spherical acoustic resonator was used to compare the speed of sound in this standard to the speed of sound in the commercial argon used for the main measurements. The relative standard uncertainty of the comparisons was 0.2×10^{-6} of u and it contributed a relative standard uncertainty of 0.4×10^{-6} to the value of R.

As in the case of acoustic thermometry, a significant contribution to the uncertainty of R (0.7×10^{-6}) resulted from the multiparameter fits used to extrapolate the speed-of-sound data to zero pressure. A smaller uncertainty resulted from the uncertainty of the thermal conductivity of argon. The thermal conductivity was needed to correct the resonance frequencies for the thermal boundary layer.

Today, the constants R and k_B are best known from acoustic measurements. They appear in relationships connecting energy to temperature. For example, the power per unit area radiated by a black body is $\pi^2(k_B T)^4/(60\hbar^3 c^2)$, where c is the speed of light and $2\pi\hbar$ is Planck's constant. Thus, one can use the acoustically established value of k_B and a measurement of the power radiated to determine the temperature of the radiator.

12.7. CONCLUDING REMARKS

We have described gas-filled acoustic resonators with four different shapes (1) cylindrical, (2) dumbbell, (3) spherical, and (4) cylindrical with an internal honeycomb of parallel ducts. These resonators compose a tool kit for accurately measuring the thermophysical properties of pure gases and gas mixtures. To use these tools effectively, we used model intermolecular potentials to obtain thermodynamic properties from the resonance frequency data. In the future, we will also use model intermolecular potentials to obtain transport properties from the halfwidth's of the resonances. The model intermolecular potentials obtained by simultaneously fitting virial coefficients and transport properties will permit reliable calculations of the thermophysical properties of gases beyond the temperature range of the acoustic data.

We have used our resonance frequency measurements to determine the ideal-gas heat capacity and the parameters in model intermolecular potentials for more than 20 gases. From these model potentials, we have computed and tabulated the density virial coefficients and the acoustic virial coefficients. The tables include the ideal-gas heat capacity; thus, they can be used to calculate any thermodynamic property of the gases. We plan to add the viscosity and the thermal conductivity to this database, and we plan to extend the database to include most of the gases for which industry requires accurate values of thermophysical properties.

12.8. PREFACE TO TABLES

The following tables contain the thermodynamic properties of 24 gases. For each gas, we provide ideal-gas heat capacities, density virial coefficients, and acoustic virial coefficients. The units for each property are shown in the heading of each column. The tabulated values should be multiplied by the power of 10 given in parentheses immediately following the value. Thus, $-6.674(-4)$ is to be understood as -6.674×10^{-4}.

All of the virial coefficients were computed from model intermolecular potential functions containing parameters that were fitted to speed-of-sound data. The temperature span of each table corresponds to the temperature span of the underlying acoustic data. Each table includes the maximum pressure of the acoustic data and a reference to a publication containing the acoustic data. The tables are ordered by the area of the industrial application.

For modeling vapor-compression refrigerators, we present data for 3 ethers, 11 partially halogenated hydrocarbons, and methane. For these gases, the chemical formulas and the refrigeration idustry's abbreviation (in parentheses) are CF_3-O-CF_2H (E125), CHF_2-O-CHF_2(E134), CF_3CH_2-O-CHF_2(E245), CCl_2HCF_3(R123), $CHFCl$-CF_3(R124), C_2HF_5(R125), CF_3-CH_2F(R134a), $CFCl_2$-CH_3(R141b), CF_2-CH_3(R143a), CHF_2-CH_3(R152a) CF_3-CHF-CHF_2(R236ea), CF_3-CH_2-CF_3(R236fa), CHF_2-CF_2-CH_2F(R245ca), CHF_2-CF_2-CF_2-CH_2F(R338mccq), and CH_4(R50).

For calibrating mass flow controllers used in the semiconductor processing industry and for process modeling, we present data for three surrogate gases — CF_4, C_2F_6, and SF_6 — and data for three process gases — WF_6, HBr, and BCl_3.

The data for three noble gases — Xe, He, and Ar — may be useful for modeling thermoacoustic machinery and carrier gases used in semiconductor processing.

The uncertainty of the tabulated values of C_p^0 is on the order of 0.1%. It is difficult to provide a statement of the uncertainty of densities calculated from the acoustically derived virial coefficients that is both concise and accurate. Instructive examples are provided in [8]. In one example, the acoustic data for pentafluoroethane (R125) in the ranges. $0.71 < T/T_c < 1.18$ and $0.05 < p/\text{MPa} < 1$ was analyzed. ($T_c = 339$ K is the critical temperature.) The resulting densities were all within 0.07% of independently measured values of the density at pressures up to the lesser of 1 MPa or 80% of the vapour pressure. Larger density differences occurred very close to the vapour pressure curve. See [8] for examples in different temperature and pressure ranges.

The tabulated values of the acoustic virial coefficients can be used to calculate the speed of sound from which they were derived with uncertainties on the order of 0.01%.

Name: pentafluorodimethyl ether (CAS # 3822-68-2)
Formula: CF_3–O–CF_2H
M.W.: 0.13602 kg/mol
Max. Pressure: 1000 kPa
Reference: Hurly, J.J., Schmidt, J.W., Gillis, K.A. (1997). Virial equation of state and ideal-gas heat capacities of pentafluoro-dimethyl ether. *Int. J. Thermophys.* **18**: 137–159.

T (K)	$C_p^0(T)/R$	$B(T)$ (m$^3\cdot$mol^{-1})	$C(T)$ (m$^6\cdot$mol^{-2})	$D(T)$ (m$^9\cdot$mol^{-3})	$\beta_a(T)$ (m$^3\cdot$mol^{-1})	$\gamma_a(T)$ (m$^3\cdot$mol$^{-1}\cdot$Pa^{-1})	$\delta_a(T)$ (m$^3\cdot$mol$^{-1}\cdot$Pa^{-2})
260	11.645	−6.674(−4)	2.090(−9)	−1.34(−10)	−1.066(−3)	−2.59(−10)	−2.3(−16)
265	11.780	−6.358(−4)	1.621(−8)	−1.11(−10)	−1.019(−3)	−2.21(−10)	−1.8(−16)
270	11.915	−6.063(−4)	2.639(−8)	−9.25(−11)	−9.748(−4)	−1.90(−10)	−1.4(−16)
275	12.050	−5.788(−4)	3.357(−8)	−7.68(−11)	−9.331(−4)	−1.64(−10)	−1.1(−16)
280	12.183	−5.531(−4)	3.848(−8)	−6.37(−11)	−8.940(−4)	−1.42(−10)	−9.1(−17)
285	12.316	−5.290(−4)	4.168(−8)	−5.27(−11)	−8.571(−4)	−1.23(−10)	−7.3(−17)
290	12.448	−5.063(−4)	4.358(−8)	−4.35(−11)	−8.223(−4)	−1.07(−10)	−5.9(−17)
295	12.579	−4.850(−4)	4.451(−8)	−3.59(−11)	−7.894(−4)	−9.40(−11)	−4.7(−17)
300	12.709	−4.650(−4)	4.472(−8)	−2.94(−11)	−7.583(−4)	−8.23(−11)	−3.8(−17)
305	12.838	−4.461(−4)	4.440(−8)	−2.40(−11)	−7.289(−4)	−7.23(−11)	−3.0(−17)
310	12.967	−4.283(−4)	4.370(−8)	−1.94(−11)	−7.010(−4)	−6.36(−11)	−2.4(−17)
315	13.094	−4.114(−4)	4.273(−8)	−1.56(−11)	−6.745(−4)	−5.60(−11)	−1.9(−17)
320	13.221	−3.954(−4)	4.158(−8)	−1.24(−11)	−6.493(−4)	−4.94(−11)	−1.5(−17)
325	13.346	−3.803(−4)	4.031(−8)	−9.67(−12)	−6.253(−4)	−4.35(−11)	−1.2(−17)
330	13.471	−3.659(−4)	3.898(−8)	−7.38(−12)	−6.025(−4)	−3.84(−11)	−9.5(−18)
335	13.594	−3.523(−4)	3.762(−8)	−5.47(−12)	−5.807(−4)	−3.38(−11)	−7.3(−18)
340	13.716	−3.393(−4)	3.626(−8)	−3.86(−12)	−5.600(−4)	−2.97(−11)	−5.6(−18)
345	13.838	−3.270(−4)	3.492(−8)	−2.51(−4)	−5.401(−4)	−2.61(−11)	−4.1(−18)
350	13.958	−3.152(−4)	3.363(−8)	−1.38(−12)	−5.212(−4)	−2.29(−11)	−3.0(−18)
355	14.077	−3.040(−4)	3.238(−8)	−4.43(−13)	−5.030(−4)	−2.00(−11)	−2.0(−18)
360	14.194	−2.933(−4)	3.118(−8)	3.37(−13)	−4.856(−4)	−1.73(−11)	−1.2(−18)
365	14.311	−2.831(−4)	3.005(−8)	9.82(−13)	−4.690(−4)	−1.50(−11)	−5.8(−19)
370	14.426	−2.733(−4)	2.899(−8)	1.51(−12)	−4.530(−4)	−1.28(−11)	−6.6(−20)
375	14.541	−2.639(−4)	2.799(−8)	1.94(−12)	−4.376(−4)	−1.09(−11)	3.5(−19)
380	14.653	−2.550(−4)	2.705(−8)	2.29(−12)	−4.229(−4)	−9.11(−12)	6.9(−19)
385	14.765	−2.464(−4)	2.618(−8)	2.57(−12)	−4.087(−4)	−7.49(−12)	9.5(−19)
390	14.875	−2.381(−4)	2.537(−8)	2.79(−12)	−3.951(−4)	−6.01(−12)	1.2(−18)
395	14.984	−2.302(−4)	2.463(−8)	2.96(−12)	−3.819(−4)	−4.65(−12)	1.3(−18)
400	15.091	−2.226(−4)	2.394(−8)	3.09(−12)	−3.693(−4)	−3.41(−12)	1.5(−18)

Name:	tetrafluorodimethyl ether (CAS # 1691-17-4)	
Formula:	$CHF_2-O-CHF_2$	
M.W.	0.118031 kg/mol	
Max. Pressure:	90 kPa	
Reference:	Gillis, K.A. (1994). Thermodynamic properties of two halogenated ethers from speed-of-sound measurements: Difluoromethoxy-difluoromethane and 2-difluoromethoxy-1,1,1-trifluoroethane. *Int. J. Thermophys.* **15**: 821–847.	

T (K)	$C_p^0(T)/R$	$B(T)$ ($m^3 \cdot mol^{-1}$)	$C(T)$ ($m^6 \cdot mol^{-2}$)	$D(T)$ ($m^9 \cdot mol^{-3}$)	$\beta_a(T)$ ($m^3 \cdot mol^{-1}$)	$\gamma_a(T)$ ($m^3 \cdot mol^{-1} \cdot Pa^{-1}$)	$\delta_a(T)$ ($m^3 \cdot mol^{-1} \cdot Pa^{-2}$)
255	10.482	−1.229(−3)	—	—	−1.826(−3)	−1.05(−9)	−1.3(−15)
260	10.672	−1.156(−3)	—	—	−1.731(−3)	−8.57(−10)	−9.6(−16)
265	10.852	−1.089(−3)	—	—	−1.644(−3)	−7.08(−10)	−7.4(−16)
270	11.022	−1.028(−3)	—	—	−1.563(−3)	−5.89(−10)	−5.7(−16)
275	11.184	−9.726(−4)	—	—	−1.488(−3)	−4.93(−10)	−4.4(−16)
280	11.338	−9.213(−4)	—	—	−1.418(−3)	−4.15(−10)	−3.4(−16)
285	11.483	−8.741(−4)	—	—	−1.353(−3)	−3.52(−10)	−2.7(−16)
290	11.622	−8.305(−4)	—	—	−1.293(−3)	−3.00(−10)	−2.2(−16)
295	11.753	−7.902(−4)	—	—	−1.236(−3)	−3.66(−10)	−2.4(−16)
300	11.879	−7.528(−4)	—	—	−1.183(−3)	−3.29(−10)	−2.0(−16)
305	11.998	−7.182(−4)	—	—	−1.134(−3)	−2.96(−10)	−1.7(−16)
310	12.113	−6.859(−4)	—	—	−1.087(−3)	−2.67(−10)	−1.5(−16)
315	12.222	−6.559(−4)	—	—	−1.044(−3)	−2.42(−10)	−1.2(−16)
320	12.328	−6.278(−4)	—	—	−1.003(−3)	−2.19(−10)	−1.1(−16)
325	12.429	−6.016(−4)	—	—	−9.641(−4)	−1.99(−10)	−9.1(−17)
330	12.527	−5.770(−4)	—	—	−9.277(−4)	−1.81(−10)	−7.8(−17)
335	12.622	−5.540(−4)	—	—	−8.934(−4)	−1.65(−10)	−6.7(−17)
340	12.716	−5.323(−4)	—	—	−8.610(−4)	−1.51(−10)	−5.8(−17)
345	12.807	−5.120(−4)	—	—	−8.304(−4)	−1.38(−10)	−5.0(−17)
350	12.897	−4.928(−4)	—	—	−8.015(−4)	−1.27(−10)	−4.4(−17)
355	12.986	−4.747(−4)	—	—	−7.741(−4)	−1.16(−10)	−3.8(−17)
360	13.075	−4.577(−4)	—	—	−7.481(−4)	−1.07(−10)	−3.3(−17)
365	13.164	−4.415(−4)	—	—	−7.235(−4)	−9.86(−11)	−2.9(−17)
370	13.254	−4.263(−4)	—	—	−7.001(−4)	−9.10(−11)	−2.6(−17)
374	13.327	−4.146(−4)	—	—	−6.822(−4)	−8.54(−11)	−2.3(−17)

Name:	2-(difluoromethoxy)-1,1,1-trifluoroethane (CAS # 1885-48-9)	
Formula:	$CF_3CH_2-O-CHF_2$	
M.W.	0.150048 kg/mol	
Max. Pressure:	50 kPa	
Reference:	Gillis, K.A. (1994). Thermodynamic properties of two halogenated ethers from speed-of-sound measurements: Difluoromethoxy-difluoromethane and 2-difluoromethoxy-1,1,1-trifluoroethane. *Int. J. Thermophys.* **15**: 821–847.	

T (K)	$C_p^0(T)/R$	$B(T)$ ($m^3 \cdot mol^{-1}$)	$C(T)$ ($m^6 \cdot mol^{-2}$)	$D(T)$ ($m^9 \cdot mol^{-3}$)	$\beta_a(T)$ ($m^3 \cdot mol^{-1}$)	$\gamma_a(T)$ ($m^3 \cdot mol^{-1} \cdot Pa^{-1}$)	$\delta_a(T)$ ($m^3 \cdot mol^{-1} \cdot Pa^{-2}$)
278	15.395	−1.499(−3)	—	—	−2.430(−3)	−1.42(−9)	−1.9(−15)
280	15.474	−1.464(−3)	—	—	−2.379(−3)	−1.35(−9)	−1.8(−15)
285	15.670	−1.383(−3)	—	—	−2.259(−3)	−1.20(−9)	−1.4(−15)
290	15.864	−1.309(−3)	—	—	−2.149(−3)	−1.06(−9)	−1.2(−15)
295	16.054	−1.241(−3)	—	—	−2.046(−3)	−9.46(−10)	−9.9(−16)
300	16.242	−1.178(−3)	—	—	−1.952(−3)	−8.45(−10)	−8.2(−16)
305	16.427	−1.120(−3)	—	—	−1.864(−3)	−7.57(−10)	−6.9(−16)
310	16.610	−1.066(−3)	—	—	−1.782(−3)	−6.80(−10)	−5.8(−16)
315	16.789	−1.017(−3)	—	—	−1.706(−3)	−6.13(−10)	−4.9(−16)
320	16.966	−9.706(−4)	—	—	−1.635(−3)	−5.53(−10)	−4.1(−16)
325	17.141	−9.278(−4)	—	—	−1.568(−3)	−5.01(−10)	−3.5(−16)
330	17.312	−8.880(−4)	—	—	−1.506(−3)	−4.55(−10)	−3.0(−16)
335	17.481	−8.509(−4)	—	—	−1.448(−3)	−4.14(−10)	−2.6(−16)
340	17.647	−8.162(−4)	—	—	−1.393(−3)	−3.77(−10)	−2.2(−16)
345	17.810	−7.837(−4)	—	—	−1.342(−3)	−3.44(−10)	−1.9(−16)
350	17.971	−7.533(−4)	—	—	−1.294(−3)	−3.15(−10)	−1.7(−16)
355	18.129	−7.247(−4)	—	—	−1.248(−3)	−2.89(−10)	−1.4(−16)
360	18.284	−6.979(−4)	—	—	−1.205(−3)	−2.66(−10)	−1.3(−16)
365	18.436	−6.726(−4)	—	—	−1.164(−3)	−2.44(−10)	−1.1(−16)
370	18.586	−6.489(−4)	—	—	−1.126(−3)	−2.25(−10)	−9.6(−17)
375	18.733	−6.264(−4)	—	—	−1.090(−3)	−2.08(−10)	−8.4(−17)
380	18.877	−6.052(−4)	—	—	−1.055(−3)	−1.92(−10)	−7.4(−17)
384	18.991	−5.891(−4)	—	—	−1.029(−3)	−1.81(−10)	−6.7(−17)

Name: 2,2-dichloro-1,1,1-trifluoroethane (CAS # 306-83-2)
Formula: CCl_2-HCF_3
M.W. 0.15293 kg/mol
Max. Pressure: 80 kPa
Reference: Goodwin, A.R.H., and Moldover, M.R. (1991). Thermophysical properties of gaseous refrigerants from speed-of-sound measurements, III: Results for 1,1-dichloro-2,2,2-trifluoroethane ($CHCl_2-CF_3$) and 1,2,-dichloro-1,2,2-trifluoroethane ($CHClF-CClF_2$). *J. Chem. Phys.* **95**: 5236–5242.

T (K)	$C_p^0(T)/R$	$B(T)$ ($m^3 \cdot mol^{-1}$)	$C(T)$ ($m^6 \cdot mol^{-2}$)	$D(T)$ ($m^9 \cdot mol^{-3}$)	$\beta_a(T)$ ($m^3 \cdot mol^{-1}$)	$\gamma_a(T)$ ($m^3 \cdot mol^{-1} \cdot Pa^{-1}$)	$\delta_a(T)$ ($m^3 \cdot mol^{-1} \cdot Pa^{-2}$)
260	11.410	−1.518(−3)	−1.099(−6)	—	−2.373(−3)	−2.23(−9)	−3.3(−15)
265	11.541	−1.440(−3)	−8.814(−7)	—	−2.261(−3)	−1.90(−9)	−2.6(−15)
270	11.670	−1.367(−3)	−7.057(−7)	—	−2.156(−3)	−1.62(−9)	−2.1(−15)
275	11.796	−1.300(−3)	−5.634(−7)	—	−2.059(−3)	−1.40(−9)	−1.7(−15)
280	11.920	−1.238(−3)	−4.479(−7)	—	−1.969(−3)	−1.20(−9)	−1.3(−15)
285	12.042	−1.180(−3)	−3.539(−7)	—	−1.884(−3)	−1.04(−9)	−1.1(−15)
290	12.161	−1.126(−3)	−2.772(−7)	—	−1.805(−3)	−9.05(−10)	−8.7(−16)
295	12.278	−1.076(−3)	−2.145(−7)	—	−1.730(−3)	−7.89(−10)	−7.0(−16)
300	12.393	−1.030(−3)	−1.632(−7)	—	−1.661(−3)	−6.90(−10)	−5.8(−16)
305	12.506	−9.859(−4)	−1.213(−7)	—	−1.595(−3)	−6.05(−10)	−4.7(−16)
310	12.617	−9.450(−4)	−8.690(−8)	—	−1.533(−3)	−5.32(−10)	−3.9(−16)
315	12.725	−9.066(−4)	−5.876(−8)	—	−1.475(−3)	−4.69(−10)	−3.2(−16)
320	12.831	−8.705(−4)	−3.574(−8)	—	−1.420(−3)	−4.14(−10)	−2.7(−16)
325	12.935	−8.366(−4)	−1.694(−8)	—	−1.368(−3)	−3.67(−10)	−2.2(−16)
330	13.037	−8.046(−4)	−1.614(−9)	—	−1.318(−3)	−3.25(−10)	−1.8(−16)
335	13.316	−7.744(−4)	−1.084(−8)	—	−1.272(−3)	−2.89(−10)	−1.5(−16)

Name: 1-chloro-1,2,2,2-tetrafluoroethane (CAS # 2837-89-0)
Formula: $CHFCl-CF_3$
M.W. 0.136477 kg/mol
Max. Pressure: 900 kPa
Reference: Gillis, K.A. (1997). Thermodynamic properties of seven gaseous halogenated hydrocarbons from acoustic measurements: $CHClFCF_3$, CHF_2CF_3, CF_3CH_3, CHF_2CH_3, $CF_3CHFCHF_2$, $CF_3CH_2CF_3$ and $CHF_2CF_2CH_2F$. *Int. J. Thermophys.* **18**: 73–135.

T (K)	$C_p^0(T)/R$	$B(T)$ ($m^3 \cdot mol^{-1}$)	$C(T)$ ($m^6 \cdot mol^{-2}$)	$D(T)$ ($m^9 \cdot mol^{-3}$)	$\beta_a(T)$ ($m^3 \cdot mol^{-1}$)	$\gamma_a(T)$ ($m^3 \cdot mol^{-1} \cdot Pa^{-1}$)	$\delta_a(T)$ ($m^3 \cdot mol^{-1} \cdot Pa^{-2}$)
250	10.664	−9.906(−4)	−8.375(−9)	—	−1.537(−3)	−6.61(−10)	−6.4(−16)
255	10.791	−9.395(−4)	−9.785(−10)	—	−1.465(−3)	−5.79(−10)	−5.2(−16)
260	10.916	−8.924(−4)	5.278(−9)	—	−1.398(−3)	−5.08(−10)	−4.2(−16)
265	11.040	−8.487(−4)	1.057(−8)	—	−1.335(−3)	−4.47(−10)	−3.4(−16)
270	11.163	−8.083(−4)	1.504(−8)	—	−1.276(−3)	−3.94(−10)	−2.8(−16)
275	11.285	−7.708(−4)	1.880(−8)	—	−1.222(−3)	−3.48(−10)	−2.3(−16)
280	11.406	−7.358(−4)	2.198(−8)	—	−1.171(−3)	−3.08(−10)	−1.9(−16)
285	11.526	−7.032(−4)	2.464(−8)	—	−1.123(−3)	−2.73(−10)	−1.6(−16)
290	11.645	−6.727(−4)	2.686(−8)	—	−1.077(−3)	−2.42(−10)	−1.3(−16)
295	11.762	−6.442(−4)	2.871(−8)	—	−1.035(−3)	−2.15(−10)	−1.1(−16)
300	11.879	−6.175(−4)	3.023(−8)	—	−9.949(−4)	−1.91(−10)	−9.0(−17)
305	11.995	−5.924(−4)	3.148(−8)	—	−9.571(−4)	−1.70(−10)	−7.5(−17)
310	12.109	−5.688(−4)	3.250(−8)	—	−9.214(−4)	−1.51(−10)	−6.2(−17)
315	12.223	−5.466(−4)	3.331(−8)	—	−8.876(−4)	−1.35(−10)	−5.2(−17)
320	12.335	−5.256(−4)	3.394(−8)	—	−8.556(−4)	−1.20(−10)	−4.3(−17)
325	12.446	−5.058(−4)	3.442(−8)	—	−8.252(−4)	−1.07(−10)	−3.6(−17)
330	12.557	−4.871(−4)	3.477(−8)	—	−7.963(−4)	−9.52(−11)	−3.0(−17)

(continued)

T (K)	$C_p^0(T)/R$	$B(T)$ (m$^3 \cdot$ mol^{-1})	$C(T)$ (m$^6 \cdot$ mol^{-2})	$D(T)$ (m$^9 \cdot$ mol^{-3})	$\beta_a(T)$ (m$^3 \cdot$ mol^{-1})	$\gamma_a(T)$ (m$^3 \cdot$ mol$^{-1} \cdot$ Pa^{-1})	$\delta_a(T)$ (m$^3 \cdot$ mol$^{-1} \cdot$ Pa^{-2})
335	12.666	−4.694(−4)	3.501(−8)	—	−7.689(−4)	−8.48(−11)	−2.5(−17)
340	12.774	−4.526(−4)	3.515(−8)	—	−7.428(−4)	−7.54(−11)	−2.1(−17)
345	12.881	−4.367(−4)	3.522(−8)	—	−7.180(−4)	−6.71(−11)	−1.7(−17)
350	12.987	−4.216(−4)	3.521(−8)	—	−6.943(−4)	−5.96(−11)	−1.4(−17)
355	13.092	−4.072(−4)	3.514(−8)	—	−6.717(−4)	−5.29(−11)	−1.2(−17)
360	13.196	−3.935(−4)	3.501(−8)	—	−6.500(−4)	−4.69(−11)	−9.9(−18)
365	13.299	−3.804(−4)	3.485(−8)	—	−6.294(−4)	−4.14(−11)	−8.1(−18)
370	13.400	−3.680(−4)	3.465(−8)	—	−6.096(−4)	−3.65(−11)	−6.6(−18)
375	13.501	−3.561(−4)	3.441(−8)	—	−5.907(−4)	−3.21(−11)	−5.3(−18)
380	13.601	−3.448(−4)	3.415(−8)	—	−5.725(−4)	−3.20(−11)	−5.0(−18)
385	13.699	−3.339(−4)	3.387(−8)	—	−5.551(−4)	−2.81(−11)	−4.0(−18)
390	13.797	−3.236(−4)	3.357(−8)	—	−5.384(−4)	−2.46(−11)	−3.2(−18)
395	13.893	−3.136(−4)	3.326(−8)	—	−5.223(−4)	−2.14(−11)	−2.5(−18)
400	13.989	−3.041(−4)	3.294(−8)	—	−5.069(−4)	−1.86(−11)	−1.9(−18)

Name: pentafluoroethane (CAS # 354-33-6)
Formula: C$_2$HF$_5$
M.W.: 0.12002 kg/mol
Max. Pressure: 1000 kPa
Reference: Gillis, K.A., and Moldover, M.R. (1996). Practical determination of gas densities from the speed of sound using square-well potentials. *Int. J. Thermophys.* **17**: 1305–1324.

T (K)	$C_p^0(T)/R$	$B(T)$ (m$^3 \cdot$ mol^{-1})	$C(T)$ (m$^6 \cdot$ mol^{-2})	$D(T)$ (m$^9 \cdot$ mol^{-3})	$\beta_a(T)$ (m$^3 \cdot$ mol^{-1})	$\gamma_a(T)$ (m$^3 \cdot$ mol$^{-1} \cdot$ Pa^{-1})	$\delta_a(T)$ (m$^3 \cdot$ mol$^{-1} \cdot$ Pa^{-2})
240	9.967	−6.330(−4)	−8.029(−8)	−1.35(−10)	−9.776(−4)	−2.88(−10)	−2.7(−16)
245	10.090	−6.011(−4)	−4.822(−8)	−1.12(−10)	−9.319(−4)	−2.40(−10)	−2.1(−16)
250	10.212	−5.715(−4)	−2.450(−8)	−9.31(−11)	−8.891(−4)	−2.01(−10)	−1.6(−16)
255	10.333	−5.440(−4)	−7.043(−9)	−7.74(−11)	−8.492(−4)	−1.69(−10)	−1.3(−16)
260	10.453	−5.184(−4)	5.710(−9)	−6.44(−11)	−8.117(−4)	−1.44(−10)	−1.0(−16)
265	10.573	−4.945(−4)	1.492(−8)	−5.37(−11)	−7.766(−4)	−1.22(−10)	−7.9(−17)
270	10.691	−4.722(−4)	2.145(−8)	−4.47(−11)	−7.435(−4)	−1.05(−10)	−6.3(−17)
275	10.810	−4.513(−4)	2.597(−8)	−3.73(−11)	−7.124(−4)	−9.00(−11)	−5.0(−17)
280	10.927	−4.317(−4)	2.896(−8)	−3.11(−11)	−6.830(−4)	−7.77(−11)	−4.0(−17)
285	11.044	−4.132(−4)	3.082(−8)	−2.59(−11)	−6.553(−4)	−6.73(−11)	−3.2(−17)
290	11.160	−3.959(−4)	3.183(−8)	−2.15(−11)	−6.291(−4)	−5.84(−11)	−2.5(−17)
295	11.275	−3.795(−4)	3.221(−8)	−1.79(−11)	−6.042(−4)	−5.09(−11)	−2.0(−17)
300	11.390	−3.641(−4)	3.212(−8)	−1.48(−11)	−5.807(−4)	−4.44(−11)	−1.6(−17)
305	11.504	−3.495(−4)	3.171(−8)	−1.22(−11)	−5.583(−4)	−3.88(−11)	−1.3(−17)
310	11.617	−3.357(−4)	3.105(−8)	−1.00(−11)	−5.370(−4)	−3.39(−11)	−1.0(−17)
315	11.730	−3.226(−4)	3.024(−8)	−8.19(−12)	−5.168(−4)	−2.97(−11)	−8.2(−18)
320	11.841	−3.102(−4)	2.932(−8)	−6.63(−12)	−4.975(−4)	−2.59(−11)	−6.5(−18)
325	11.952	−2.984(−4)	2.834(−8)	−5.32(−12)	−4.792(−4)	−2.27(−11)	−5.1(−18)
330	12.063	−2.872(−4)	2.733(−8)	−4.20(−12)	−4.616(−4)	−1.98(−11)	−3.9(−18)
335	12.173	−2.765(−4)	2.632(−8)	−3.27(−12)	−4.449(−4)	−1.72(−11)	−3.0(−18)
340	12.281	−2.664(−4)	2.532(−8)	−2.48(−12)	−4.288(−4)	−1.49(−11)	−2.2(−18)
345	12.390	−2.567(−4)	2.435(−8)	−1.81(−12)	−4.135(−4)	−1.29(−11)	−1.6(−18)
350	12.497	−2.474(−4)	2.341(−8)	−1.25(−12)	−3.988(−4)	−1.10(−11)	−1.1(−18)
355	12.604	−2.386(−4)	2.252(−8)	−7.74(−13)	−3.847(−4)	−9.36(−12)	−6.5(−19)
360	12.710	−2.301(−4)	2.168(−8)	−3.78(−13)	−3.712(−4)	−7.87(−12)	−3.1(−19)
365	12.816	−2.220(−4)	2.088(−8)	−4.77(−14)	−3.582(−4)	−6.51(−12)	−2.9(−20)
370	12.920	−2.143(−4)	2.013(−8)	2.28(−13)	−3.458(−4)	−5.28(−12)	2.0(−19)
375	13.024	−2.069(−4)	1.944(−8)	4.56(−13)	−3.338(−4)	−4.16(−12)	3.8(−19)
380	13.128	−1.997(−4)	1.879(−8)	6.44(−13)	−3.223(−4)	−3.14(−12)	5.2(−19)
385	13.230	−1.929(−4)	1.820(−8)	7.97(−13)	−3.112(−4)	−2.20(−12)	6.4(−19)
390	13.332	−1.863(−4)	1.765(−8)	9.22(−13)	−3.005(−4)	−1.34(−12)	7.3(−19)
395	13.433	−1.800(−4)	1.714(−8)	1.02(−12)	−2.901(−4)	−5.53(−13)	8.0(−19)
400	13.534	−1.739(−4)	1.668(−8)	1.10(−12)	−2.802(−4)	1.74(−13)	8.5(−19)

Name: 1,1,1,2-tetrafluoroethane (CAS # 811-97-2)
Formula: CF_3-CH_2F
M.W.: 0.10203 kg/mol
Max. Pressure: 600 kPa
Reference: Gillis, K.A., and Moldover, M.R. (1996). Practical determination of gas densities from the speed of sound using square-well potentials. *Int. J. Thermophys.* **17**: 1305–1324.

T (K)	$C_p^0(T)/R$	$B(T)$ ($m^3 \cdot mol^{-1}$)	$C(T)$ ($m^6 \cdot mol^{-2}$)	$D(T)$ ($m^9 \cdot mol^{-3}$)	$\beta_a(T)$ ($m^3 \cdot mol^{-1}$)	$\gamma_a(T)$ ($m^3 \cdot mol^{-1} \cdot Pa^{-1}$)	$\delta_a(T)$ ($m^3 \cdot mol^{-1} \cdot Pa^{-2}$)
233	8.728	−1.013(−3)	4.006(−8)	—	−1.429(−3)	−6.39(−10)	−7.0(−16)
235	8.776	−9.863(−4)	3.908(−8)	—	−1.396(−3)	−6.01(−10)	−6.3(−16)
240	8.894	−9.238(−4)	3.681(−8)	—	−1.318(−3)	−5.17(−10)	−5.0(−16)
245	9.012	−8.673(−4)	3.482(−8)	—	−1.246(−3)	−4.46(−10)	−3.9(−16)
250	9.129	−8.158(−4)	3.310(−8)	—	−1.181(−3)	−3.87(−10)	−3.1(−16)
255	9.245	−7.690(−4)	3.163(−8)	—	−1.121(−3)	−3.36(−10)	−2.5(−16)
260	9.360	−7.261(−4)	3.040(−8)	—	−1.065(−3)	−2.93(−10)	−2.0(−16)
265	9.475	−6.868(−4)	2.937(−8)	—	−1.014(−3)	−2.56(−10)	−1.6(−16)
270	9.588	−6.507(−4)	2.853(−8)	—	−9.664(−4)	−2.24(−10)	−1.3(−16)
275	9.701	−6.174(−4)	2.786(−8)	—	−9.223(−4)	−1.96(−10)	−1.1(−16)
280	9.813	−5.867(−4)	2.734(−8)	—	−8.812(−4)	−1.72(−10)	−8.5(−17)
285	9.924	−5.583(−4)	2.695(−8)	—	−8.428(−4)	−1.51(−10)	−6.9(−17)
290	10.034	−5.319(−4)	2.669(−8)	—	−8.070(−4)	−1.33(−10)	−5.6(−17)
295	10.143	−5.074(−4)	2.653(−8)	—	−7.735(−4)	−1.16(−10)	−4.6(−17)
300	10.252	−4.846(−4)	2.646(−8)	—	−7.421(−4)	−1.02(−10)	−3.7(−17)
305	10.360	−4.634(−4)	2.648(−8)	—	−7.126(−4)	−8.95(−11)	−3.0(−17)
310	10.467	−4.435(−4)	2.657(−8)	—	−6.849(−4)	−7.83(−11)	−2.4(−17)
315	10.573	−4.249(−4)	2.672(−8)	—	−6.588(−4)	−6.83(−11)	−2.0(−17)
320	10.678	−4.075(−4)	2.693(−8)	—	−6.342(−4)	−5.93(−11)	−1.6(−17)
325	10.782	−3.912(−4)	2.719(−8)	—	−6.109(−4)	−5.13(−11)	−1.2(−17)
330	10.885	−3.758(−4)	2.749(−8)	—	−5.889(−4)	−4.41(−11)	−9.7(−18)
335	10.988	−3.613(−4)	2.782(−8)	—	−5.681(−4)	−3.77(−11)	−7.5(−18)
340	11.090	−3.477(−4)	2.819(−8)	—	−5.484(−4)	−3.19(−11)	−5.6(−18)

Name: 1,1-dichloro-1-fluoroethane (CAS # 1717-00-6)
Formula: $CFCl_2-CH_3$
M.W.: 0.11695 kg/mol
Max. Pressure: 70 kPa
Reference: Goodwin, A.R.H., and Moldover, M.R. (1991). Thermophysical properties of gaseous refrigerants from speed-of-sound measurements, II: Results for 1,1-dichloro-1-fluoroethane (CCl_2FCH_3). *J. Chem. Phys.* **95**: 5230–5235.

T (K)	$C_p^0(T)/R$	$B(T)$ ($m^3 \cdot mol^{-1}$)	$C(T)$ ($m^6 \cdot mol^{-2}$)	$D(T)$ ($m^9 \cdot mol^{-3}$)	$\beta_a(T)$ ($m^3 \cdot mol^{-1}$)	$\gamma_a(T)$ ($m^3 \cdot mol^{-1} \cdot Pa^{-1}$)	$\delta_a(T)$ ($m^3 \cdot mol^{-1} \cdot Pa^{-2}$)
260	9.792	−1.388(−3)	—	—	−2.118(−3)	−1.26(−9)	−1.7(−15)
265	9.897	−1.321(−3)	—	—	−2.023(−3)	−1.12(−9)	−1.4(−15)
270	10.003	−1.259(−3)	—	—	−1.934(−3)	−1.01(−9)	−1.2(−15)
275	10.108	−1.201(−3)	—	—	−1.851(−3)	−9.01(−10)	−9.7(−16)
280	10.213	−1.147(−3)	—	—	−1.773(−3)	−8.09(−10)	−8.1(−16)
285	10.319	−1.096(−3)	—	—	−1.700(−3)	−7.29(−10)	−6.9(−16)
290	10.424	−1.049(−3)	—	—	−1.631(−3)	−6.57(−10)	−5.8(−16)
295	10.530	−1.004(−3)	—	—	−1.566(−3)	−5.94(−10)	−5.0(−16)
300	10.635	−9.623(−4)	—	—	−1.504(−3)	−5.37(−10)	−4.2(−16)
305	10.740	−9.228(−4)	—	—	−1.446(−3)	−4.87(−10)	−3.6(−16)
310	10.846	−8.856(−4)	—	—	−1.391(−3)	−4.42(−10)	−3.1(−16)
315	10.951	−8.505(−4)	—	—	−1.338(−3)	−4.02(−10)	−2.7(−16)

Name: 1,1,1-trifluoroethane (CAS # 420-46-2)
Formula: CF_3-CH_3
M.W. 0.0840412 kg/mol
Max. Pressure: 1000 kPa
Reference: Gillis, K.A. (1997). Thermodynamic properties of seven gaseous halogenated hydrocarbons from acoustic measurements: $CHClFCF_3$, CHF_2CF_3, CF_3CH_3, CHF_2CH_3, $CF_3CHFCHF_2$, $CF_3CH_2CF_3$ and $CHF_2CF_2CH_2F$. *Int. J. Thermophys.* **18**: 73–135.

T (K)	$C_p^0(T)/R$	$B(T)$ ($m^3 \cdot mol^{-1}$)	$C(T)$ ($m^6 \cdot mol^{-2}$)	$D(T)$ ($m^9 \cdot mol^{-3}$)	$\beta_a(T)$ ($m^3 \cdot mol^{-1}$)	$\gamma_a(T)$ ($m^3 \cdot mol^{-1} \cdot Pa^{-1}$)	$\delta_a(T)$ ($m^3 \cdot mol^{-1} \cdot Pa^{-2}$)
250	8.324	−6.466(−4)	8.478(−9)	−4.48(−11)	−9.408(−4)	−2.09(−10)	−1.5(−16)
255	8.444	−6.140(−4)	2.271(−8)	−3.73(−11)	−8.985(−4)	−1.76(−10)	−1.2(−16)
260	8.561	−5.838(−4)	3.301(−8)	−3.11(−11)	−8.590(−4)	−1.49(−10)	−9.2(−17)
265	8.676	−5.557(−4)	4.031(−8)	−2.61(−11)	−8.220(−4)	−1.26(−10)	−7.2(−17)
270	8.790	−5.297(−4)	4.533(−8)	−2.20(−11)	−7.873(−4)	−1.07(−10)	−5.7(−17)
275	8.901	−5.054(−4)	4.860(−8)	−1.86(−11)	−7.547(−4)	−9.15(−11)	−4.5(−17)
280	9.011	−4.828(−4)	5.055(−8)	−1.58(−11)	−7.241(−4)	−7.82(−11)	−3.5(−17)
285	9.119	−4.616(−4)	5.152(−8)	−1.35(−11)	−6.952(−4)	−6.69(−11)	−2.8(−17)
290	9.226	−4.418(−4)	5.173(−8)	−1.15(−11)	−6.679(−4)	−5.73(−11)	−2.2(−17)
295	9.331	−4.233(−4)	5.140(−8)	−9.88(−12)	−6.422(−4)	−4.91(−11)	−1.7(−17)
300	9.434	−4.058(−4)	5.066(−8)	−8.50(−12)	−6.178(−4)	−4.21(−11)	−1.4(−17)
305	9.537	−3.894(−4)	4.964(−8)	−7.34(−12)	−5.947(−4)	−3.60(−11)	−1.1(−17)
310	9.638	−3.740(−4)	4.841(−8)	−6.35(−12)	−5.728(−4)	−3.08(−11)	−8.5(−18)
315	9.738	−3.594(−4)	4.706(−8)	−5.51(−12)	−5.521(−4)	−2.63(−11)	−6.7(−18)
320	9.837	−3.456(−4)	4.563(−8)	−4.80(−12)	−5.323(−4)	−2.23(−11)	−5.2(−18)
325	9.935	−3.326(−4)	4.417(−8)	−4.18(−12)	−5.135(−4)	−1.88(−11)	−3.9(−18)
330	10.032	−3.202(−4)	4.270(−8)	−3.65(−12)	−4.956(−4)	−1.58(−11)	−2.9(−18)
335	10.128	−3.085(−4)	4.124(−8)	−3.20(−12)	−4.786(−4)	−1.31(−11)	−2.2(−18)
340	10.223	−2.974(−4)	3.982(−8)	−2.81(−12)	−4.623(−4)	−1.07(−11)	−1.5(−18)
345	10.318	−2.869(−4)	3.844(−8)	−2.47(−12)	−4.468(−4)	−8.64(−12)	−9.7(−19)
350	10.412	−2.768(−4)	3.712(−8)	−2.18(−12)	−4.320(−4)	−6.76(−12)	−5.4(−19)
355	10.505	−2.673(−4)	3.586(−8)	−1.92(−12)	−4.178(−4)	−5.09(−12)	−2.0(−19)
360	10.599	−2.582(−4)	3.466(−8)	−1.70(−12)	−4.042(−4)	−3.58(−12)	−8.8(−20)
365	10.691	−2.495(−4)	3.352(−8)	−1.50(−12)	−3.912(−4)	−2.23(−12)	−3.2(−19)
370	10.784	−2.412(−4)	3.246(−8)	−1.33(−12)	−3.787(−4)	−1.01(−12)	−5.0(−19)
375	10.876	−2.332(−4)	3.145(−8)	−1.18(−12)	−3.667(−4)	9.85(−14)	−6.5(−19)
380	10.968	−2.256(−4)	3.052(−8)	1.31(−15)	−3.552(−4)	−9.11(−13)	9.6(−18)
385	11.061	−2.184(−4)	2.964(−8)	1.31(−15)	−3.442(−4)	1.28(−13)	1.0(−18)
390	11.153	−2.114(−4)	2.883(−8)	1.31(−15)	−3.336(−4)	1.08(−12)	1.1(−18)
395	11.245	−2.048(−4)	2.808(−8)	1.31(−15)	−3.233(−4)	1.94(−12)	1.1(−18)
400	11.338	−1.983(−4)	2.738(−8)	1.31(−15)	−3.135(−4)	2.74(−12)	1.1(−18)

Name: 1,1-difluoroethane (CAS # 75-37-6)
Formula: CHF_2-CH_3
M.W. 0.0660508 kg/mol
Max. Pressure: 1000 kPa
Reference: Gillis, K.A. (1997). Thermodynamic properties of seven gaseous halogenated hydrocarbons from acoustic measurements: $CHClFCF_3$, CHF_2CF_3, CF_3CH_3, CHF_2CH_3, $CF_3CHFCHF_2$, $CF_3CH_2CF_3$ and $CHF_2CF_2CH_2F$. *Int. J. Thermophys.* **18**: 73–135.

T (K)	$C_p^0(T)/R$	$B(T)$ ($m^3 \cdot mol^{-1}$)	$C(T)$ ($m^6 \cdot mol^{-2}$)	$D(T)$ ($m^9 \cdot mol^{-3}$)	$\beta_a(T)$ ($m^3 \cdot mol^{-1}$)	$\gamma_a(T)$ ($m^3 \cdot mol^{-1} \cdot Pa^{-1}$)	$\delta_a(T)$ ($m^3 \cdot mol^{-1} \cdot Pa^{-2}$)
243	7.119	−9.318(−4)	−1.850(−7)	−2.99(−10)	−1.243(−3)	−5.46(−10)	−7.3(−16)
245	7.154	−9.092(−4)	−1.573(−7)	−2.76(−10)	−1.216(−3)	−5.05(−10)	−6.5(−16)
250	7.242	−8.561(−4)	−1.004(−7)	−2.27(−10)	−1.154(−3)	−4.19(−10)	−4.9(−16)
255	7.330	−8.076(−4)	−5.754(−8)	−1.88(−10)	−1.096(−3)	−3.49(−10)	−3.8(−16)
260	7.420	−7.631(−4)	−2.544(−8)	−1.56(−10)	−1.043(−3)	−2.92(−10)	−2.9(−16)
265	7.510	−7.223(−4)	−1.513(−9)	−1.31(−10)	−9.938(−4)	−2.46(−10)	−2.3(−16)
270	7.601	−6.848(−4)	1.618(−8)	−1.10(−10)	−9.483(−4)	−2.08(−10)	−1.8(−16)

continues

(continued)

T (K)	$C_p^0(T)/R$	$B(T)$ ($m^3 \cdot mol^{-1}$)	$C(T)$ ($m^6 \cdot mol^{-2}$)	$D(T)$ ($m^9 \cdot mol^{-3}$)	$\beta_a(T)$ ($m^3 \cdot mol^{-1}$)	$\gamma_a(T)$ ($m^3 \cdot mol^{-1} \cdot Pa^{-1}$)	$\delta_a(T)$ ($m^3 \cdot mol^{-1} \cdot Pa^{-2}$)
275	7.692	−6.501(−4)	2.911(−8)	−9.24(−11)	−9.059(−4)	−1.77(−10)	−1.4(−16)
280	7.784	−6.181(−4)	3.480(−8)	−7.82(−11)	−8.665(−4)	−1.51(−10)	−1.1(−16)
285	7.876	−5.884(−4)	4.489(−8)	−6.65(−11)	−8.296(−4)	−1.29(−10)	−9.0(−17)
290	7.968	−5.608(−4)	4.926(−8)	−5.67(−11)	−7.952(−4)	−1.11(−10)	−7.2(−17)
295	8.061	−5.352(−4)	5.200(−8)	−4.85(−11)	−7.629(−4)	−9.57(−11)	−5.8(−17)
300	8.154	−5.113(−4)	5.352(−8)	−4.17(−11)	−7.327(−4)	−8.27(−11)	−4.7(−17)
305	8.247	−4.889(−4)	5.411(−8)	−3.59(−11)	−7.042(−4)	−7.17(−11)	−3.8(−17)
310	8.341	−4.680(−4)	5.401(−8)	−3.10(−11)	−6.773(−4)	−6.22(−11)	−3.1(−17)
315	8.434	−4.485(−4)	5.340(−8)	−2.69(−11)	−6.520(−4)	−5.41(−11)	−2.5(−17)
320	8.528	−4.301(−4)	5.243(−8)	−2.33(−11)	−6.281(−4)	−4.71(−11)	−2.1(−17)
325	8.621	−4.128(−4)	5.120(−8)	−2.03(−11)	−6.055(−4)	−4.10(−11)	−1.7(−17)
330	8.715	−3.996(−4)	4.980(−8)	−1.77(−11)	−5.841(−4)	−3.57(−11)	−1.4(−17)
335	8.808	−3.813(−4)	4.829(−8)	−1.55(−11)	−5.638(−4)	−3.11(−11)	−1.1(−17)
340	8.901	−3.668(−4)	4.672(−8)	−1.36(−11)	−5.445(−4)	−2.70(−11)	−9.3(−18)
345	8.993	−3.532(−4)	4.513(−8)	−1.20(−11)	−5.262(−4)	−2.34(−11)	−7.6(−18)
350	9.086	−3.403(−4)	4.355(−8)	−1.05(−11)	−5.087(−4)	−2.03(−11)	−6.2(−18)
355	9.178	−3.281(−4)	4.200(−8)	−9.31(−12)	−4.921(−4)	−1.74(−11)	−5.0(−18)
360	9.269	−3.165(−4)	4.050(−8)	−8.23(−12)	−4.762(−4)	−1.49(−11)	−4.0(−18)
365	9.360	−3.055(−4)	3.905(−8)	−7.28(−12)	−4.611(−4)	−1.27(−11)	−3.2(−18)
370	9.450	−2.950(−4)	3.767(−8)	−6.46(−12)	−4.466(−4)	−1.07(−11)	−2.5(−18)
375	9.540	−2.851(−4)	3.635(−8)	−5.74(−12)	−4.327(−4)	−8.83(−12)	−2.0(−18)
380	9.629	−2.756(−4)	3.511(−8)	3.53(−15)	−4.195(−4)	−9.96(−12)	−2.6(−19)
385	9.717	−2.666(−4)	3.394(−8)	3.53(−15)	−4.068(−4)	−8.27(−12)	−2.9(−20)
390	9.804	−2.580(−4)	3.285(−8)	3.53(−15)	−3.946(−4)	−6.73(−12)	1.6(−19)
395	9.891	−2.498(−4)	3.183(−8)	3.53(−15)	−3.829(−4)	−5.32(−12)	3.2(−19)
400	9.976	−2.419(−4)	3.088(−8)	3.53(−15)	−3.717(−4)	−4.04(−12)	4.5(−19)

Name:	1,1,1,2,3,3-hexafluoropropane (CAS # 431-63-0)
Formula:	$CF_3-CHF-CHF_2$
M.W.	0.152039 kg/mol
Max. Pressure:	600 kPa
Reference:	Gillis, K.A. (1997). Thermodynamic properties of seven gaseous halogenated hydrocoarbons from acoustic measurements: $CHClFCF_3$, CHF_2CF_3, CF_3CH_3, CHF_2CH_3, CH_3CHFCH_2, $CF_3CH_2CF_3$ and $CHF_2CF_2CH_2F$. *Int. J. Thermophys.* **18**: 73–135.

T (K)	$C_p^0(T)/R$	$B(T)$ ($m^3 \cdot mol^{-1}$)	$C(T)$ ($m^6 \cdot mol^{-2}$)	$D(T)$ ($m^9 \cdot mol^{-3}$)	$\beta_a(T)$ ($m^3 \cdot mol^{-1}$)	$\gamma_a(T)$ ($m^3 \cdot mol^{-1} \cdot Pa^{-1}$)	$\delta_a(T)$ ($m^3 \cdot mol^{-1} \cdot Pa^{-2}$)
267.15	14.593	−1.032(−3)	−1.576(−6)	−5.60(−9)	−1.741(−3)	−1.53(−9)	−4.2(−15)
270	14.678	−1.006(−3)	−1.286(−6)	−4.75(−9)	−1.699(−3)	−1.31(−9)	−3.5(−15)
275	14.826	−9.622(−4)	−8.900(−7)	−3.58(−9)	−1.628(−3)	−1.01(−9)	−2.5(−15)
280	14.973	−9.214(−4)	−6.032(−7)	−2.73(−9)	−1.561(−3)	−7.89(−10)	−1.9(−15)
285	15.119	−8.830(−4)	−3.957(−7)	−2.10(−9)	−1.498(−3)	−6.23(−10)	−1.4(−15)
290	15.264	−8.467(−4)	−2.458(−7)	−1.63(−9)	−1.438(−3)	−4.97(−10)	−1.0(−15)
295	15.408	−8.125(−4)	−1.380(−7)	−1.27(−9)	−1.382(−3)	−4.02(−10)	−7.8(−16)
300	15.552	−7.801(−4)	−6.109(−8)	−1.00(−9)	−1.328(−3)	−3.30(−10)	−6.0(−16)
305	15.694	−7.495(−4)	−6.785(−9)	−7.95(−10)	−1.277(−3)	−2.73(−10)	−4.6(−16)
310	15.835	−7.204(−4)	−3.095(−8)	−6.35(−10)	−1.229(−3)	−2.30(−10)	−3.6(−16)
315	15.976	−6.929(−4)	−5.654(−8)	−5.11(−10)	−1.182(−3)	−1.95(−10)	−2.8(−16)
320	16.116	−6.667(−4)	−7.327(−8)	−4.13(−10)	−1.139(−3)	−1.68(−10)	−2.2(−16)
325	16.254	−6.418(−4)	−8.356(−8)	−3.36(−10)	−1.097(−3)	−1.45(−10)	−1.8(−16)
330	16.392	−6.180(−4)	−8.920(−8)	−2.75(−10)	−1.057(−3)	−1.27(−10)	−1.4(−16)
335	16.529	−5.954(−4)	−9.153(−8)	−2.26(−10)	−1.018(−3)	−1.12(−10)	−1.2(−16)
340	16.665	−5.738(−4)	−9.152(−8)	−1.87(−10)	−9.819(−4)	−9.97(−11)	−9.6(−17)
345	16.800	−5.532(−4)	−8.992(−8)	−1.55(−10)	−9.469(−4)	−8.89(−11)	−7.9(−17)
350	16.934	−5.335(−4)	−8.726(−8)	−1.30(−10)	−9.134(−4)	−7.97(−11)	−6.5(−17)
335	17.067	−5.147(−4)	−8.394(−8)	−1.09(−10)	−8.812(−4)	−7.16(−11)	−5.4(−17)
360	17.199	−4.967(−4)	−8.024(−8)	−9.15(−11)	−8.504(−4)	−6.44(−11)	−4.5(−17)
365	17.330	−4.794(−4)	−7.638(−8)	−7.74(−11)	−8.208(−4)	−5.80(−11)	−3.7(−17)
370	17.460	−4.628(−4)	−7.251(−8)	−6.57(−11)	−7.923(−4)	−5.22(−11)	−3.1(−17)
375	17.590	−4.469(−4)	−6.874(−8)	−5.59(−11)	−7.649(−4)	−4.70(−11)	−2.6(−17)
377	17.641	−4.407(−4)	−6.727(−8)	−1.47(−11)	−7.543(−4)	−5.25(−11)	−9.6(−18)

Name: 1,1,1,3,3,3-hexafluoropropane (CAS # 690-39-1)
Formula: $CF_3-CH_2-CF_3$
M.W. 0.152039 kg/mol
Max. Pressure: 1000 kPa
Reference: Gillis, K.A. (1997). Thermodynamic properties of seven gaseous halogenated hydrocarbons from acoustic measurements: $CHClFCF_3$, CHF_2CF_3, CF_3CH_3, CHF_2CH_3, $CF_3CHFCHF_2$, $CF_3CH_2CF_3$ and $CHF_2CF_2CH_2F$. *Int. J. Thermophys.* **18**: 73–135.

T (K)	$C_p^0(T)/R$	$B(T)$ ($m^3 \cdot mol^{-1}$)	$C(T)$ ($m^6 \cdot mol^{-2}$)	$D(T)$ ($m^9 \cdot mol^{-3}$)	$\beta_a(T)$ ($m^3 \cdot mol^{-1}$)	$\gamma_a(T)$ ($m^3 \cdot mol^{-1} \cdot Pa^{-1}$)	$\delta_a(T)$ ($m^3 \cdot mol^{-1} \cdot Pa^{-2}$)
275	14.292	−9.136(−4)	−7.494(−7)	−3.63(−10)	−1.536(−3)	−9.69(−10)	−1.0(−15)
280	14.442	−8.748(−4)	−5.534(−7)	−2.97(−10)	−1.473(−3)	−7.88(−10)	−7.7(−16)
285	14.592	−8.384(−4)	−4.033(−7)	−2.45(−10)	−1.414(−3)	−6.45(−10)	−5.9(−16)
290	14.743	−8.040(−4)	−2.884(−7)	−2.02(−10)	−1.358(−3)	−5.31(−10)	−4.6(−16)
295	14.893	−7.716(−4)	−2.004(−7)	−1.68(−10)	−1.305(−3)	−4.41(−10)	−3.5(−16)
300	15.044	−7.409(−4)	−1.330(−7)	−1.41(−10)	−1.255(−3)	−3.68(−10)	−2.8(−16)
305	15.194	−7.119(−4)	−8.153(−8)	−1.18(−10)	−1.207(−3)	−3.09(−10)	−2.2(−16)
310	15.344	−6.844(−4)	−4.237(−8)	−9.95(−11)	−1.162(−3)	−2.61(−10)	−1.7(−16)
315	15.495	−6.582(−4)	−1.274(−8)	−8.42(−11)	−1.118(−3)	−2.22(−10)	−1.4(−16)
320	15.645	−6.334(−4)	9.514(−9)	−7.16(−11)	−1.077(−3)	−1.89(−10)	−1.1(−16)
325	15.796	−6.098(−4)	2.604(−8)	−6.11(−11)	−1.038(−3)	−1.62(−10)	−8.9(−17)
330	15.946	−5.874(−4)	3.812(−8)	−5.23(−11)	−1.000(−3)	−1.40(−10)	−7.2(−17)
335	16.096	−5.659(−4)	4.676(−8)	−4.50(−11)	−9.645(−4)	−1.21(−10)	−5.9(−17)
340	16.247	−5.455(−4)	5.274(−8)	−3.88(−11)	−9.302(−4)	−1.06(−10)	−4.8(−17)
345	16.397	−5.260(−4)	5.667(−8)	−3.35(−11)	−8.974(−4)	−9.21(−11)	−4.0(−17)
350	16.548	−5.074(−4)	5.902(−8)	−2.91(−11)	−8.660(−4)	−8.06(−11)	−3.3(−17)
355	16.698	−4.895(−4)	6.017(−8)	−2.53(−11)	−8.358(−4)	−7.08(−11)	−2.7(−17)
360	16.848	−4.725(−4)	6.041(−8)	−2.21(−11)	−8.069(−4)	−6.22(−11)	−2.2(−17)
365	16.999	−4.561(−4)	5.997(−8)	−1.93(−11)	−7.791(−4)	−5.48(−11)	−1.9(−17)
370	17.149	−4.404(−4)	5.903(−8)	−1.70(−11)	−7.525(−4)	−4.83(−11)	−1.6(−17)
375	17.300	−4.253(−4)	5.773(−8)	−1.49(−11)	−7.268(−4)	−4.26(−11)	−1.3(−17)
380	17.450	−4.109(−4)	5.618(−8)	2.68(−16)	−7.021(−4)	−4.41(−11)	−7.5(−18)
385	17.600	−3.970(−4)	5.448(−8)	2.66(−16)	−6.783(−4)	−3.93(−11)	−6.1(−18)
390	17.751	−3.836(−4)	5.268(−8)	2.68(−16)	−6.554(−4)	−3.50(−11)	−5.0(−18)
395	17.901	−3.707(−4)	5.083(−8)	2.68(−16)	−6.333(−4)	−3.11(−11)	−4.0(−18)
400	18.052	−3.584(−4)	4.898(−8)	2.68(−16)	−6.120(−4)	−2.76(−11)	−3.2(−18)

Name: 1,1,2,2,3-pentafluoropropane (CAS # 679-86-7)
Formula: $CHF_2-CF_2-CH_2F$
M.W. 0.134049 kg/mol
Max. Pressure: 1000 kPa
Reference: Gillis, K.A. (1997). Thermodynamic properties of seven gaseous halogenated hydrocoarbons from acoustic measurements: $CHClFCF_3$, CHF_2CF_3, CF_3CH_3, CHF_2CH_3, $CF_3CHFCHF_2$, $CF_3CH_2CF_3$ and $CHF_2CF_2CH_2F$. *Int. J. Thermophys.* **18**: 73–135.

T (K)	$C_p^0(T)/R$	$B(T)$ ($m^3 \cdot mol^{-1}$)	$C(T)$ ($m^6 \cdot mol^{-2}$)	$D(T)$ ($m^9 \cdot mol^{-3}$)	$\beta_a(T)$ ($m^3 \cdot mol^{-1}$)	$\gamma_a(T)$ ($m^3 \cdot mol^{-1} \cdot Pa^{-1}$)	$\delta_a(T)$ ($m^3 \cdot mol^{-1} \cdot Pa^{-2}$)
310	14.696	−8.264(−4)	−2.319(−7)	−5.00(−10)	−1.394(−3)	−4.68(−10)	−5.2(−16)
315	14.885	−7.950(−4)	−1.520(−7)	−4.34(−10)	−1.344(−3)	−3.93(−10)	−4.2(−16)
320	15.071	−7.653(−4)	−9.126(−8)	−3.78(−10)	−1.296(−3)	−3.32(−10)	−3.4(−16)
325	15.252	−7.370(−4)	−4.531(−8)	−3.30(−10)	−1.250(−3)	−2.82(−10)	−2.8(−16)
330	15.430	−7.102(−4)	−1.078(−8)	−2.88(−10)	−1.206(−3)	−2.42(−10)	−2.3(−16)
335	15.604	−6.846(−4)	1.494(−8)	−2.52(−10)	−1.165(−3)	−2.08(−10)	−1.9(−16)
340	15.774	−6.603(−4)	3.385(−8)	−2.21(−10)	−1.125(−3)	−1.80(−10)	−1.6(−16)
345	15.940	−6.371(−4)	4.749(−8)	−1.95(−10)	−1.086(−3)	−1.57(−10)	−1.3(−16)
350	16.102	−6.150(−4)	5.708(−8)	−1.71(−10)	−1.050(−3)	−1.38(−10)	−1.1(−16)
355	16.260	−5.939(−4)	6.355(−8)	−1.51(−10)	−1.015(−3)	−1.21(−10)	−9.4(−17)
360	16.414	−5.737(−4)	6.763(−8)	−1.33(−10)	−9.808(−4)	−1.07(−10)	−8.0(−17)

continues

(continued)

T (K)	$C_p^0(T)/R$	$B(T)$ (m³·mol⁻¹)	$C(T)$ (m⁶·mol⁻²)	$D(T)$ (m⁹·mol⁻³)	$\beta_a(T)$ (m³·mol⁻¹)	$\gamma_a(T)$ (m³·mol⁻¹·Pa⁻¹)	$\delta_a(T)$ (m³·mol⁻¹·Pa⁻²)
365	16.564	−5.544(−4)	6.989(−8)	−1.18(−10)	−9.484(−4)	−9.53(−11)	−6.8(−17)
370	16.711	−5.359(−4)	7.078(−8)	−1.05(−10)	−9.173(−4)	−8.50(−11)	−5.8(−17)
375	16.853	−5.182(−4)	7.065(−8)	−9.27(−11)	−8.874(−4)	−7.59(−11)	−4.9(−17)
380	16.991	−5.012(−4)	6.976(−8)	1.64(−13)	−8.586(−4)	−7.65(−11)	−1.8(−17)
385	17.126	−4.848(−4)	6.833(−8)	1.64(−13)	−8.309(−4)	−6.91(−11)	−1.5(−17)
390	17.256	−4.692(−4)	6.652(−8)	1.64(−13)	−8.042(−4)	−6.24(−11)	−1.3(−17)
395	17.383	−4.541(−4)	6.446(−8)	1.64(−13)	−7.785(−4)	−5.65(−11)	−1.1(−17)
400	17.506	−4.396(−4)	6.225(−8)	1.64(−13)	−7.537(−4)	−5.12(−11)	−9.2(−18)

Name:	1,1,1,2,2,3,3,4-octafluorobutane (CAS # 662-35-1)
Formula:	$CF_3-CF_2-CF_2-CH_2F$
M.W.	0.202047 kg/mol
Max. Pressure:	400 kPa
Reference:	Defibaugh, D.R., Carrillo, N.E., Hurly, J.J., Moldover, M.R., Schmidt, J.W., Weber, L.A. (1997). Thermodynamic properties of HFC-338mccq, $CF_3CF_2CF_2CH_2F$, 1,1,1,2,2,3,3,4-octafluorobutane. *J. Chem. Eng. Data* **42**: 488–496.

T (K)	$C_p^0(T)/R$	$B(T)$ (m³·mol⁻¹)	$C(T)$ (m⁶·mol⁻²)	$D(T)$ (m⁹·mol⁻³)	$\beta_a(T)$ (m³·mol⁻¹)	$\gamma_a(T)$ (m³·mol⁻¹·Pa⁻¹)	$\delta_a(T)$ (m³·mol⁻¹·Pa⁻²)
300	19.748	−1.162(−3)	−6.284(−7)	—	−2.037(−3)	−1.30(−9)	−1.3(−15)
305	19.899	−1.114(−3)	−4.808(−7)	—	−1.955(−3)	−1.11(−9)	−1.0(−15)
310	20.053	−1.069(−3)	−3.632(−7)	—	−1.878(−3)	−9.44(−10)	−8.0(−16)
315	20.208	−1.026(−3)	−2.695(−7)	—	−1.805(−3)	−8.09(−10)	−6.4(−16)
320	20.366	−9.861(−4)	−1.948(−7)	—	−1.736(−3)	−6.96(−10)	−5.1(−16)
325	20.528	−9.480(−4)	−1.353(−7)	—	−1.671(−3)	−6.01(−10)	−4.2(−16)
330	20.692	−9.120(−4)	−8.793(−8)	—	−1.609(−3)	−5.21(−10)	−3.4(−16)
335	20.860	−8.778(−4)	−5.031(−8)	—	−1.550(−3)	−4.54(−10)	−2.8(−16)
340	21.033	−8.453(−4)	−2.053(−8)	—	−1.494(−3)	−3.96(−10)	−2.3(−16)
345	21.209	−8.145(−4)	2.934(−9)	—	−1.441(−3)	−3.48(−10)	−1.9(−16)
350	21.390	−7.852(−4)	2.130(−8)	—	−1.390(−3)	−3.06(−10)	−1.5(−16)
355	21.576	−7.573(−4)	3.556(−8)	—	−1.342(−3)	−2.69(−10)	−1.3(−16)
360	21.768	−7.308(−4)	4.651(−8)	—	−1.296(−3)	−2.38(−10)	−1.1(−16)
365	21.965	−7.054(−4)	5.478(−8)	—	−1.252(−3)	−2.11(−10)	−8.8(−17)
370	22.168	−6.812(−4)	6.090(−8)	—	−1.210(−3)	−1.88(−10)	−7.4(−17)
375	22.378	−6.581(−4)	6.529(−8)	—	−1.169(−3)	−1.67(−10)	−6.2(−17)
380	22.594	−6.359(−4)	6.828(−8)	—	−1.131(−3)	−1.49(−10)	−5.2(−17)
385	22.818	−6.148(−4)	7.016(−8)	—	−1.094(−3)	−1.33(−10)	−4.4(−17)
390	23.049	−5.945(−4)	7.116(−8)	—	−1.059(−3)	−1.19(−10)	−3.7(−17)
395	23.287	−5.750(−4)	7.146(−8)	—	−1.025(−3)	−1.07(−10)	−3.1(−17)
400	23.534	−5.564(−4)	7.120(−8)	—	−9.920(−4)	−9.58(−11)	−2.6(−17)

Name: methane (CAS # 74-82-8)
Formula: CH_4
M.W.: 0.016043 kg/mol
Max. Pressure: 3000 kPa
Reference: Gillis, K.A and Moldover M.R. (1996). Practical determination of gas densities from the speed of sound using square-well potentials. *Int. J. Thermophys.* **17**: 1305–1324.

(continued)

T (K)	$C_p^0(T)/R$	$B(T)$ ($m^3 \cdot mol^{-1}$)	$C(T)$ ($m^6 \cdot mol^{-2}$)	$D(T)$ ($m^9 \cdot mol^{-3}$)	$\beta_a(T)$ ($m^3 \cdot mol^{-1}$)	$\gamma_a(T)$ ($m^3 \cdot mol^{-1} \cdot Pa^{-1}$)	$\delta_a(T)$ ($m^3 \cdot mol^{-1} \cdot Pa^{-2}$)
125	4.003	−2.565(−4)	1.923(−9)	2.64(−13)	−3.100(−4)	−6.74(−11)	−3.2(−17)
130	4.004	−2.379(−4)	2.357(−9)	3.44(−13)	−2.869(−4)	−5.43(−11)	−2.2(−17)
135	4.004	−2.214(−4)	2.667(−9)	3.77(−13)	−2.662(−4)	−4.39(−11)	−1.6(−17)
140	4.005	−2.065(−4)	2.885(−9)	3.82(−13)	−2.476(−4)	−3.56(−11)	−1.1(−17)
145	4.005	−1.932(−4)	3.034(−9)	3.70(−13)	−2.307(−4)	−2.90(−11)	−8.2(−18)
150	4.006	−1.811(−4)	3.132(−9)	3.50(−13)	−2.155(−4)	−2.37(−11)	−5.9(−18)
155	4.006	−1.702(−4)	3.190(−9)	3.25(−13)	−2.015(−4)	−1.94(−11)	−4.3(−18)
160	4.007	−1.602(−4)	3.219(−9)	2.98(−13)	−1.887(−4)	−1.58(−11)	−3.1(−18)
165	4.009	−1.511(−4)	3.227(−9)	2.72(−13)	−1.770(−4)	−1.29(−11)	−2.2(−18)
170	4.010	−1.426(−4)	3.217(−9)	2.47(−13)	−1.662(−4)	−1.05(−11)	−1.6(−18)
175	4.012	−1.349(−4)	3.196(−9)	2.24(−13)	−1.562(−4)	−8.50(−12)	−1.1(−18)
180	4.015	−1.277(−4)	3.165(−9)	2.03(−13)	−1.470(−4)	−6.84(−12)	−7.8(−19)
185	4.018	−1.211(−4)	3.128(−9)	1.84(−13)	−1.384(−4)	−5.44(−12)	−5.2(−19)
190	4.021	−1.149(−4)	3.087(−9)	1.67(−13)	−1.304(−4)	−4.28(−12)	−3.2(−19)
195	4.026	−1.092(−4)	3.042(−9)	1.52(−13)	−1.229(−4)	−3.30(−12)	−1.8(−19)
200	4.031	−1.038(−4)	2.996(−9)	1.38(−13)	−1.160(−4)	−2.47(−12)	−7.4(−20)
205	4.036	−9.882(−5)	2.948(−9)	1.27(−13)	−1.094(−4)	−1.77(−12)	3.7(−21)
210	4.042	−9.412(−5)	2.900(−9)	1.17(−13)	−1.033(−4)	−1.18(−12)	6.1(−20)
215	4.049	−8.970(−5)	2.852(−9)	1.08(−13)	−9.754(−5)	−6.80(−13)	1.0(−19)
220	4.057	−8.554(−5)	2.805(−9)	1.00(−13)	−9.212(−5)	−2.56(−13)	1.3(−19)
225	4.066	−8.162(−5)	2.759(−9)	9.33(−14)	−8.701(−5)	1.04(−13)	1.5(−19)
230	4.075	−7.792(−5)	2.713(−9)	8.76(−14)	−8.219(−5)	4.09(−13)	1.6(−19)
235	4.085	−7.442(−5)	2.669(−9)	8.28(−14)	−7.763(−5)	6.68(−13)	1.7(−19)
240	4.097	−7.110(−5)	2.626(−9)	7.86(−14)	−7.332(−5)	8.87(−13)	1.7(−19)
245	4.109	−6.796(−5)	2.585(−9)	7.52(−14)	−6.923(−5)	1.07(−12)	1.8(−19)
250	4.122	−6.498(−5)	2.544(−9)	7.23(−14)	−6.535(−5)	1.23(−12)	1.7(−19)
255	4.136	−6.214(−5)	2.506(−9)	6.99(−14)	−6.167(−5)	1.36(−12)	1.7(−19)
260	4.151	−5.944(−5)	2.468(−9)	6.79(−14)	−5.817(−5)	1.47(−12)	1.7(−19)
265	4.166	−5.687(−5)	2.432(−9)	6.63(−14)	−5.484(−5)	1.56(−12)	1.6(−19)
270	4.183	−5.441(−5)	2.398(−9)	6.51(−14)	−5.166(−5)	1.64(−12)	1.6(−19)
275	4.201	−5.207(−5)	2.365(−9)	6.41(−14)	−4.864(−5)	1.70(−12)	1.5(−19)
280	4.219	−4.983(−5)	2.333(−9)	6.35(−14)	−4.575(−5)	1.75(−12)	1.5(−19)
285	4.239	−4.769(−5)	2.302(−9)	6.30(−14)	−4.299(−5)	1.79(−12)	1.4(−19)
290	4.259	−4.563(−5)	2.273(−9)	6.27(−14)	−4.034(−5)	1.82(−12)	1.4(−19)
295	4.280	−4.366(−5)	2.245(−9)	6.27(−14)	−3.782(−5)	1.84(−12)	1.3(−19)
300	4.303	−4.178(−5)	2.218(−9)	6.27(−14)	−3.539(−5)	1.86(−12)	1.2(−19)
305	4.326	−3.996(−5)	2.192(−9)	6.29(−14)	−3.307(−5)	1.87(−12)	1.2(−19)
310	4.349	−3.822(−5)	2.167(−9)	6.32(−14)	−3.084(−5)	1.88(−12)	1.1(−19)
315	4.374	−3.655(−5)	2.143(−9)	6.36(−14)	−2.870(−5)	1.88(−12)	1.1(−19)
320	4.399	−3.493(−5)	2.121(−9)	6.41(−14)	−2.664(−5)	1.88(−12)	1.0(−19)
325	4.426	−3.338(−5)	2.099(−9)	6.46(−14)	−2.465(−5)	1.87(−12)	9.9(−20)
330	4.453	−3.189(−5)	2.078(−9)	6.52(−14)	−2.274(−5)	1.87(−12)	9.4(−20)
335	4.480	−3.044(−5)	2.058(−9)	6.59(−14)	−2.090(−5)	1.86(−12)	9.0(−20)
340	4.508	−2.905(−5)	2.038(−9)	6.66(−14)	−1.913(−5)	1.84(−12)	8.6(−20)
345	4.537	−2.771(−5)	2.020(−9)	6.73(−14)	−1.741(−5)	1.83(−12)	8.2(−20)
350	4.566	−2.641(−5)	2.002(−9)	6.18(−14)	−1.575(−5)	1.81(−12)	7.9(−20)
355	4.596	−2.515(−5)	1.985(−9)	6.88(−14)	−1.415(−5)	1.80(−12)	7.5(−20)
360	4.627	−2.394(−5)	1.969(−9)	6.96(−14)	−1.260(−5)	1.78(−12)	7.2(−20)
365	4.657	−2.276(−5)	1.953(−9)	7.04(−14)	−1.109(−5)	1.76(−12)	6.9(−20)
370	4.688	−2.163(−5)	1.938(−9)	7.13(−14)	−9.634(−6)	1.74(−12)	6.6(−20)
375	4.719	−2.053(−5)	1.924(−9)	7.21(−14)	−8.219(−6)	1.72(−12)	6.3(−20)

Name: tetrafluoromethane (CAS # 75-73-0)
Formula: CF_4
M.W. 0.08805 kg/mol
Max. Pressure: 1500 kPa
Reference: Hurly, J.J. (1999). Thermophysical properties of gaseous CF_4 and H_2F_6 from speed-of-sound measurements. *Int. J. Thermophys.* **20**: 455–484.

(continued)

T (K)	$C_p^0(T)/R$	$B(T)$ (m$^3 \cdot$mol^{-1})	$C(T)$ (m$^6 \cdot$mol^{-2})	$D(T)$ (m$^9 \cdot$mol^{-3})	$\beta_a(T)$ (m$^3 \cdot$mol^{-1})	$\gamma_a(T)$ (m$^3 \cdot$mol$^{-1} \cdot$Pa^{-1})	$\delta_a(T)$ (m$^3 \cdot$mol$^{-1} \cdot$Pa^{-2})
175	5.257	−3.005(−4)	5.144(−9)	—	−3.897(−4)	−6.15(−11)	−2.5(−17)
180	5.344	−2.830(−4)	6.076(−9)	—	−3.688(−4)	−6.11(−11)	−1.9(−17)
185	5.432	−2.670(−4)	6.766(−9)	—	−3.494(−4)	−4.25(−11)	−1.4(−17)
190	5.520	−2.523(−4)	7.269(−9)	—	−3.313(−4)	−3.54(−11)	−1.1(−17)
195	5.608	−2.387(−4)	7.624(−9)	—	−3.145(−4)	−2.94(−11)	−7.9(−18)
200	5.697	−2.260(−4)	7.865(−9)	—	−2.988(−4)	−2.43(−11)	−5.9(−18)
205	5.785	−2.143(−4)	8.016(−9)	—	−2.841(−4)	−2.01(−11)	−4.4(−18)
210	5.874	−2.034(−4)	8.096(−9)	—	−2.703(−4)	−1.65(−11)	−3.2(−18)
215	5.962	−1.932(−4)	8.122(−9)	—	−2.574(−4)	−1.35(−11)	−2.4(−18)
220	6.050	−1.837(−4)	8.105(−9)	—	−2.451(−4)	−1.09(−11)	−1.7(−18)
225	6.138	−1.748(−4)	8.055(−9)	—	−2.335(−4)	−8.65(−12)	−1.2(−18)
230	6.225	−1.664(−4)	7.980(−9)	—	−2.226(−4)	−6.76(−12)	−7.7(−19)
235	6.312	−1.586(−4)	7.886(−9)	—	−2.122(−4)	−5.14(−12)	−4.6(−19)
240	6.398	−1.512(−4)	7.778(−9)	—	−2.024(−4)	−3.75(−12)	−2.3(−19)
245	6.483	−1.442(−4)	7.659(−9)	—	−1.930(−4)	−2.57(−12)	−6.1(−20)
250	6.568	−1.376(−4)	7.533(−9)	—	−1.841(−4)	−1.55(−12)	6.8(−20)
255	6.652	−1.313(−4)	7.403(−9)	—	−1.756(−4)	−6.72(−13)	1.6(−19)
260	6.735	−1.254(−4)	7.270(−9)	—	−1.675(−4)	7.83(−14)	2.3(−19)
265	6.817	−1.198(−4)	7.136(−9)	—	−1.598(−4)	7.22(−13)	2.8(−19)
270	6.899	−1.145(−4)	7.002(−9)	—	−1.524(−4)	1.27(−12)	3.1(−19)
275	6.980	−1.094(−4)	6.869(−9)	—	−1.453(−4)	1.75(−12)	3.3(−19)
280	7.060	−1.046(−4)	6.738(−9)	—	−1.386(−4)	2.15(−12)	3.4(−19)
285	7.139	−9.995(−5)	6.609(−9)	—	−1.321(−4)	2.49(−12)	3.4(−19)
290	7.217	−9.556(−5)	6.484(−9)	—	−1.258(−4)	2.79(−12)	3.4(−19)
295	7.294	−9.136(−5)	6.361(−9)	—	−1.198(−4)	3.03(−12)	3.4(−19)
300	7.370	−8.735(−5)	6.242(−9)	—	−1.140(−4)	3.24(−12)	3.3(−19)
305	7.446	−8.352(−5)	6.126(−9)	—	−1.085(−4)	3.42(−12)	3.2(−19)
310	7.520	−7.984(−5)	6.014(−9)	—	−1.032(−4)	3.56(−12)	3.0(−19)
315	7.594	−7.631(−5)	5.905(−9)	—	−9.801(−5)	3.68(−12)	2.9(−19)
320	7.667	−7.293(−5)	5.800(−9)	—	−9.305(−5)	3.77(−12)	2.7(−19)
325	7.738	−6.969(−5)	5.699(−9)	—	−8.826(−5)	3.85(−12)	2.6(−19)
330	7.809	−6.657(−5)	5.602(−9)	—	−8.364(−5)	3.91(−12)	2.4(−19)
335	7.879	−6.357(−5)	5.508(−9)	—	−7.917(−5)	3.95(−12)	2.3(−19)
340	7.948	−6.068(−5)	5.418(−9)	—	−7.486(−5)	3.98(−12)	2.2(−19)
345	8.016	−5.790(−5)	5.331(−9)	—	−7.068(−5)	4.00(−12)	2.0(−19)
350	8.083	−5.522(−5)	5.248(−9)	—	−6.665(−5)	4.01(−12)	1.9(−19)
355	8.150	−5.264(−5)	5.168(−9)	—	−6.274(−5)	4.01(−12)	1.8(−19)
360	8.215	−5.014(−5)	5.090(−9)	—	−5.895(−5)	4.00(−12)	1.6(−19)
365	8.279	−4.774(−5)	5.016(−9)	—	−5.528(−5)	3.99(−12)	1.5(−19)
370	8.343	−4.541(−5)	4.945(−9)	—	−5.172(−5)	3.96(−12)	1.4(−19)
375	8.406	−4.317(−5)	4.877(−9)	—	−4.827(−5)	3.94(−12)	1.3(−19)
380	8.467	−4.099(−5)	4.812(−9)	—	−4.492(−5)	3.91(−12)	1.2(−19)
385	8.528	−3.889(−5)	4.749(−9)	—	−4.167(−5)	3.88(−12)	1.1(−19)
390	8.588	−3.686(−5)	4.688(−9)	—	−3.851(−5)	3.84(−12)	1.0(−19)
395	8.647	−3.488(−5)	4.630(−9)	—	−3.545(−5)	3.80(−12)	9.6(−20)
400	8.706	−3.297(−5)	4.575(−9)	—	−3.246(−5)	3.76(−12)	8.8(−20)
405	8.763	−3.112(−5)	4.521(−9)	—	−2.956(−5)	3.72(−12)	8.1(−20)
410	8.820	−2.933(−5)	4.470(−9)	—	−2.674(−5)	3.67(−12)	7.4(−20)
415	8.875	−2.758(−5)	4.421(−9)	—	−2.400(−5)	3.63(−12)	6.8(−20)
420	8.930	−2.589(−5)	4.374(−9)	—	−2.132(−5)	3.58(−12)	6.2(−20)
425	8.984	−2.425(−5)	4.329(−9)	—	−1.872(−5)	3.54(−12)	5.6(−20)
430	9.038	−2.265(−5)	4.285(−9)	—	−1.619(−5)	3.49(−12)	5.1(−20)
435	9.090	−2.110(−5)	4.244(−9)	—	−1.372(−5)	3.44(−12)	4.6(−20)
440	9.142	−1.959(−5)	4.204(−9)	—	−1.131(−5)	3.39(−12)	4.1(−20)
445	9.193	−1.812(−5)	4.165(−9)	—	−8.960(−6)	3.34(−12)	3.7(−20)
450	9.243	−1.670(−5)	4.128(−9)	—	−6.670(−6)	3.29(−12)	3.3(−20)
455	9.293	−1.531(−5)	4.093(−9)	—	−4.435(−6)	3.24(−12)	2.9(−20)
460	9.341	−1.395(−5)	4.059(−9)	—	−2.255(−6)	3.20(−12)	2.6(−20)
465	9.389	−1.263(−5)	4.027(−9)	—	−1.259(−7)	3.15(−12)	2.2(−20)
470	9.437	−1.135(−5)	3.995(−9)	—	−1.953(−6)	3.10(−12)	1.9(−20)
475	9.483	−1.010(−5)	3.965(−9)	—	−3.983(−6)	3.05(−12)	1.6(−20)

Name: hexafluoroethane (CAS # 76-16-4)
Formula: C_2F_6
M.W.: 0.138012 kg/mol
Max. Pressure: 1500 kPa
Reference: Hurly, J.J. (1999). Thermophysical properties of gaseous CF_4 and H_2F_6 from speed-of-sound measurements. *Int. J. Thermophys.* **20**: 455–484.

T (K)	$C_p^o(T)/R$	$B(T)$ ($m^3 \cdot mol^{-1}$)	$C(T)$ ($m^6 \cdot mol^{-2}$)	$D(T)$ ($m^9 \cdot mol^{-3}$)	$\beta_a(T)$ ($m^3 \cdot mol^{-1}$)	$\gamma_a(T)$ ($m^3 \cdot mol^{-1} \cdot Pa^{-1}$)	$\delta_a(T)$ ($m^3 \cdot mol^{-1} \cdot Pa^{-2}$)
210	10.244	−5.772(−4)	−3.057(−8)	−2.47(−12)	−9.171(−4)	−2.84(−10)	−1.9(−16)
215	10.401	−5.472(−4)	−1.734(−8)	−2.30(−12)	−8.725(−4)	−2.40(−10)	−1.5(−16)
220	10.557	−5.194(−4)	−7.094(−9)	−2.14(−12)	−8.309(−4)	−2.03(−10)	−1.1(−16)
225	10.712	−4.936(−4)	8.205(−10)	−1.99(−12)	−7.921(−4)	−1.73(−10)	−9.0(−17)
230	10.865	−4.696(−4)	6.914(−9)	−1.84(−12)	−7.557(−4)	−1.47(−10)	−7.0(−17)
235	11.017	−4.473(−4)	1.158(−8)	−1.70(−12)	−7.216(−4)	−1.26(−10)	−5.5(−17)
240	11.168	−4.264(−4)	1.512(−8)	−1.57(−12)	−6.896(−4)	−1.08(−10)	−4.4(−17)
245	11.316	−4.069(−4)	1.777(−8)	−1.44(−12)	−6.594(−4)	−9.23(−11)	−3.4(−17)
250	11.464	−3.886(−4)	1.973(−8)	−1.31(−12)	−6.310(−4)	−7.91(−11)	−2.7(−17)
255	11.609	−3.714(−4)	2.113(−8)	−1.19(−12)	−6.042(−4)	−6.78(−11)	−2.2(−17)
260	11.753	−3.552(−4)	2.209(−8)	−1.08(−12)	−5.789(−4)	−5.81(−11)	−1.7(−17)
265	11.896	−3.400(−4)	2.272(−2)	−9.70(−13)	−5.548(−4)	−4.98(−11)	−1.3(−17)
270	12.037	−3.257(−4)	2.307(−8)	−8.64(−13)	−5.321(−4)	−4.26(−11)	−1.0(−17)
275	12.176	−3.121(−4)	2.321(−8)	−7.62(−13)	−5.105(−4)	−3.63(−11)	−8.2(−18)
280	12.313	−2.992(−4)	2.319(−8)	−6.63(−13)	−4.899(−4)	−3.09(−11)	−6.3(−18)
285	12.449	−2.871(−4)	2.303(−8)	−5.68(−13)	−4.704(−4)	−2.62(−11)	−4.8(−18)
290	12.583	−2.755(−4)	2.278(−8)	−4.76(−13)	−4.518(−4)	−2.21(−11)	−3.6(−18)
295	12.715	−2.646(−4)	2.246(−8)	−3.88(−13)	−4.340(−4)	−1.86(−11)	−2.7(−18)
300	12.846	−2.541(−4)	2.208(−8)	−3.02(−13)	−4.171(−4)	−1.54(−11)	−1.9(−18)
305	12.975	−2.442(−4)	2.166(−8)	−2.19(−13)	−4.009(−4)	−1.27(−11)	−1.3(−18)
310	13.102	−2.348(−4)	2.121(−8)	−1.39(−13)	−3.854(−4)	−1.03(−11)	−7.5(−19)
315	13.227	−2.257(−4)	2.075(−8)	−6.12(−14)	−3.706(−4)	−8.19(−12)	−3.4(−19)
320	13.351	−2.171(−4)	2.028(−8)	1.40(−14)	−3.564(−4)	−6.34(−12)	−1.7(−20)
325	13.473	−2.089(−4)	1.980(−8)	8.70(−14)	−3.428(−4)	−4.70(−12)	2.4(−19)
330	13.594	−2.010(−4)	1.933(−8)	1.58(−13)	−3.298(−4)	−3.26(−12)	4.5(−19)
335	13.712	−1.935(−4)	1.886(−8)	2.26(−13)	−3.172(−4)	−1.99(−12)	6.1(−19)
340	13.829	−1.862(−4)	1.840(−8)	2.93(−13)	−3.052(−4)	−8.69(−13)	7.4(−19)
345	13.945	−1.793(−4)	1.795(−8)	3.58(−13)	−2.936(−4)	1.21(−13)	8.4(−19)
350	14.059	−1.727(−4)	1.751(−8)	4.20(−13)	−2.824(−4)	9.95(−13)	9.1(−19)
355	14.171	−1.663(−4)	1.709(−8)	4.81(−13)	−2.717(−4)	1.77(−12)	9.7(−19)
360	14.281	−1.601(−4)	1.668(−8)	5.41(−13)	−2.613(−4)	2.45(−12)	1.0(−18)
365	14.390	−1.542(−4)	1.629(−8)	5.99(−13)	−2.513(−4)	3.05(−12)	1.0(−18)
370	14.497	−1.485(−4)	1.591(−8)	6.55(−13)	−2.417(−4)	3.58(−12)	1.1(−18)
375	14.603	−1.430(−4)	1.555(−8)	7.09(−13)	−2.324(−4)	4.05(−12)	1.1(−18)
380	14.706	−1.378(−4)	1.520(−8)	7.63(−13)	−2.234(−4)	4.46(−12)	1.1(−18)
385	14.809	−1.327(−4)	1.487(−8)	8.14(−13)	−2.147(−4)	4.83(−12)	1.1(−18)
390	14.910	−1.278(−4)	1.455(−8)	8.65(−13)	−2.063(−4)	5.15(−12)	1.1(−18)
395	15.009	−1.230(−4)	1.425(−8)	9.14(−13)	−1.982(−4)	5.43(−12)	1.0(−18)
400	15.106	−1.184(−4)	1.396(−8)	9.62(−13)	−1.903(−4)	5.67(−12)	1.0(−18)
405	15.202	−1.140(−4)	1.369(−8)	1.01(−12)	−1.827(−4)	5.88(−12)	1.0(−18)
410	15.297	−1.097(−4)	1.343(−8)	1.05(−12)	−1.753(−4)	6.07(−12)	9.9(−19)
415	15.390	−1.056(−4)	1.318(−8)	1.10(−12)	−1.682(−4)	6.23(−12)	9.7(−19)
420	15.481	−1.016(−4)	1.294(−8)	1.14(−12)	−1.613(−4)	6.36(−12)	9.5(−19)
425	15.571	−9.767(−5)	1.272(−8)	1.19(−12)	−1.545(−4)	6.48(−12)	9.3(−19)
430	15.660	−9.390(−5)	1.251(−8)	1.23(−12)	−1.480(−4)	6.58(−12)	9.1(−19)
435	15.747	−9.025(−5)	1.231(−8)	1.27(−12)	−1.416(−4)	6.66(−12)	8.9(−19)
440	15.832	−8.670(−5)	1.212(−8)	1.31(−12)	−1.355(−4)	6.73(−12)	8.7(−19)
445	15.916	−8.327(−5)	1.194(−8)	1.35(−12)	−1.295(−4)	6.78(−12)	8.5(−19)
450	15.999	−7.993(−5)	1.177(−8)	1.38(−12)	−1.236(−4)	6.83(−12)	8.3(−19)
455	16.080	−7.669(−5)	1.161(−8)	1.42(−12)	−1.180(−4)	6.86(−12)	8.1(−19)
460	16.160	−7.354(−5)	1.146(−8)	1.46(−12)	−1.125(−4)	6.88(−12)	7.9(−19)
465	16.238	−7.048(−5)	1.132(−8)	1.49(−12)	−1.071(−4)	6.90(−12)	7.7(−19)
470	16.315	−6.750(−5)	1.119(−8)	1.53(−12)	−1.019(−4)	6.91(−12)	7.5(−19)
475	16.391	−6.461(−5)	1.106(−8)	1.56(−12)	−9.678(−5)	6.91(−12)	7.3(−19)

Name: sulfur hexafluoride (CAS # 2551-62-4)
Formula: SF_6
M.W.: 0.146054 kg/mol
Max. Pressure: 1500 kPa
Reference: Hurly, J.J., Defibaugh, D.R., and Moldover, M.R. (2000). Thermodynamic properties of sulfur Hexafluoride. *Int. J. Thermophys.* **21**: 739–765.

T (K)	$C_p(T)/R$	$B(T)$ ($m^3 \cdot mol^{-1}$)	$C(T)$ ($m^6 \cdot mol^{-2}$)	$D(T)$ ($m^9 \cdot mol^{-3}$)	$\beta_a(T)$ ($m^3 \cdot mol^{-1}$)	$\gamma_a(T)$ ($m^3 \cdot mol^{-1} \cdot Pa^{-1}$)	$\delta_a(T)$ ($m^3 \cdot mol^{-1} \cdot Pa^{-2}$)
230	9.367	−4.868(−4)	−2.852(−8)	−7.24(−12)	−7.636(−4)	−1.88(−10)	−1.1(−16)
235	9.552	−4.645(−4)	−1.950(−8)	−7.65(−12)	−7.320(−4)	−1.63(−10)	−8.6(−17)
240	9.734	−4.437(−4)	−1.219(−8)	−7.96(−12)	−7.022(−4)	−1.41(−10)	−7.0(−17)
245	9.913	−4.242(−4)	−6.272(−9)	−8.18(−12)	−6.740(−4)	−1.22(−10)	−5.8(−17)
250	10.088	−4.059(−4)	−1.478(−9)	−8.32(−12)	−6.473(−4)	−1.06(−10)	−4.8(−17)
255	10.261	−3.888(−4)	2.399(−9)	−8.40(−12)	−6.221(−4)	−9.25(−11)	−4.0(−17)
260	10.430	−3.726(−4)	5.528(−9)	−8.44(−12)	−5.981(−4)	−8.06(−11)	−3.4(−17)
265	10.596	−3.574(−4)	8.042(−9)	−8.42(−12)	−5.753(−4)	−7.02(−11)	−2.8(−17)
270	10.759	−3.430(−4)	1.005(−8)	−8.38(−12)	−5.536(−4)	−6.12(−11)	−2.4(−17)
275	10.919	−3.295(−4)	1.165(−8)	−8.30(−12)	−5.330(−4)	−5.34(−11)	−2.0(−17)
280	11.077	−3.166(−4)	1.291(−8)	−8.20(−12)	−5.133(−4)	−4.65(−11)	−1.7(−17)
285	11.231	−3.045(−4)	1.388(−8)	−8.07(−12)	−4.945(−4)	−4.05(−11)	−1.5(−17)
290	11.382	−2.929(−4)	1.463(−8)	−7.93(−12)	−4.766(−4)	−3.53(−11)	−1.3(−17)
295	11.531	−2.820(−4)	1.518(−8)	−7.78(−12)	−4.595(−4)	−3.07(−11)	−1.1(−17)
300	11.676	−2.715(−4)	1.558(−8)	−7.62(−12)	−4.431(−4)	−2.66(−11)	−9.7(−18)
305	11.819	−2.616(−4)	1.585(−8)	−7.44(−12)	−4.275(−4)	−2.30(−11)	−8.5(−18)
310	11.959	−2.521(−4)	1.601(−8)	−7.27(−12)	−4.124(−4)	−1.98(−11)	−7.4(−18)
315	12.096	−2.431(−4)	1.608(−8)	−7.09(−12)	−3.981(−4)	−1.70(−11)	−6.5(−18)
320	12.230	−2.345(−4)	1.608(−8)	−6.90(−12)	−3.842(−4)	−1.46(−11)	−5.8(−18)
325	12.362	−2.262(−4)	1.603(−8)	−6.72(−12)	−3.710(−4)	−1.24(−11)	−5.1(−18)
330	12.491	−2.184(−4)	1.592(−8)	−6.53(−12)	−3.582(−4)	−1.04(−11)	−4.5(−18)
335	12.617	−2.108(−4)	1.578(−8)	−6.35(−12)	−3.460(−4)	−8.67(−12)	−4.1(−18)
340	12.741	−2.036(−4)	1.561(−8)	−6.16(−12)	−3.342(−4)	−7.13(−12)	−3.6(−18)
345	12.862	−1.966(−4)	1.541(−8)	−5.98(−12)	−3.228(−4)	−5.76(−12)	−3.3(−18)
350	12.980	−1.900(−4)	1.520(−8)	−5.80(−12)	−3.119(−4)	−4.54(−12)	−2.9(−18)
355	13.096	−1.836(−4)	1.497(−8)	−5.63(−12)	−3.013(−4)	−3.45(−12)	−2.6(−18)
360	13.210	−1.774(−4)	1.474(−8)	−5.46(−12)	−2.911(−4)	−2.48(−12)	−2.4(−18)
365	13.320	−1.715(−4)	1.449(−8)	−5.29(−12)	−2.813(−4)	−1.62(−12)	−2.2(−18)
370	13.429	−1.658(−4)	1.424(−8)	−5.13(−12)	−2.718(−4)	−8.46(−13)	−2.0(−18)
375	13.535	−1.603(−4)	1.399(−8)	−4.97(−12)	−2.627(−4)	−1.58(−13)	−1.8(−18)
380	13.638	−1.550(−4)	1.374(−8)	−4.81(−12)	−2.538(−4)	4.55(−13)	−1.6(−18)
385	13.740	−1.499(−4)	1.349(−8)	−4.66(−12)	−2.452(−4)	1.00(−12)	−1.5(−18)
390	13.839	−1.450(−4)	1.325(−8)	−4.52(−12)	−2.369(−4)	1.49(−12)	−1.4(−18)
395	13.935	−1.402(−4)	1.300(−8)	−4.38(−12)	−2.289(−4)	1.92(−12)	−1.3(−18)
400	14.029	−1.356(−4)	1.276(−8)	−4.24(−12)	−2.211(−4)	2.31(−12)	−1.2(−18)
405	14.121	−1.312(−4)	1.253(−8)	−4.11(−12)	−2.136(−4)	2.65(−12)	−1.1(−18)
410	14.211	−1.269(−4)	1.229(−8)	−3.98(−12)	−2.063(−4)	2.96(−12)	−1.0(−18)
415	14.299	−1.227(−4)	1.207(−8)	−3.86(−12)	−1.992(−4)	3.23(−12)	−9.3(−19)
420	14.384	−1.187(−4)	1.185(−8)	−3.74(−12)	−1.923(−4)	3.47(−12)	−8.7(−19)
425	14.467	−1.148(−4)	1.164(−8)	−3.63(−12)	−1.856(−4)	3.68(−12)	−8.1(−19)
430	14.548	−1.110(−4)	1.143(−8)	−3.52(−12)	−1.791(−4)	3.86(−12)	−7.6(−19)
435	14.627	−1.073(−4)	1.123(−8)	−3.42(−12)	−1.728(−4)	4.03(−12)	−7.1(−19)
440	14.704	−1.038(−4)	1.104(−8)	−3.32(−12)	−1.667(−4)	4.17(−12)	−6.7(−19)
445	14.779	−1.003(−4)	1.085(−8)	−3.22(−12)	−1.607(−4)	4.29(−12)	−6.3(−19)
450	14.852	−9.700(−5)	1.067(−8)	−3.13(−12)	−1.549(−4)	4.40(−12)	−5.9(−19)
455	14.923	−9.374(−5)	1.049(−8)	−3.04(−12)	−1.493(−4)	4.49(−12)	−5.6(−19)
460	14.992	−9.059(−5)	1.032(−8)	−2.96(−12)	−1.438(−4)	4.57(−12)	−5.3(−19)

Name: tungsten hexaflurode (CAS 7783-82-6)
Formula: WF_6
M.W.: 0.29784 kg/mol
Max. Pressure: 300 kPa
Reference: Hurly, J.J. (2000). Thermophysical properties of gaseous tunsten hexafluoride from speed-of-sound measurements. *Int. J. Thermophys.* **21**: 185–206.

(continued)

T (K)	$C_P^0(T)/R$	$B(T)$ (m^3·mol^{-1})	$C(T)$ (m^6·mol^{-2})	$D(T)$ (m^9·mol^{-3})	$\beta_a(T)$ (m·mol^{-1})	$\gamma_a(T)$ (m^3·mol^{-1}·Pa^{-1})	$\delta_a(T)$ (m^3·mol^{-1}·Pa^{-2})
290	14.229	−7.955(−4)	−1.113(−7)	—	−1.341(−3)	−4.71(−10)	−3.2(−16)
295	14.326	−7.637(−4)	−8.388(−8)	—	−1.289(−3)	−4.13(−10)	−2.6(−16)
300	14.422	−7.338(−4)	−6.120(−8)	—	−1.241(−3)	−3.63(−10)	−2.1(−16)
305	14.516	−7.057(−4)	−4.245(−8)	—	−1.196(−3)	−3.20(−10)	−1.8(−16)
310	14.607	−6.791(−4)	−2.695(−8)	—	−1.152(−3)	−2.82(−10)	−1.5(−16)
315	14.697	−6.541(−4)	−1.415(−8)	—	−1.112(−3)	−2.50(−10)	−1.2(−16)
320	14.785	−6.305(−4)	−3.597(−9)	—	−1.073(−3)	−2.21(−10)	−1.0(−16)
325	14.871	−6.081(−4)	5.087(−9)	—	−1.036(−3)	−1.97(−10)	−8.4(−17)
330	14.954	−5.869(−4)	1.221(−8)	—	−1.001(−3)	−1.75(−10)	−7.1(−17)
335	15.036	−5.668(−4)	1.803(−8)	—	−9.681(−4)	−1.56(−10)	−5.9(−17)
340	15.116	−5.477(−4)	2.277(−8)	—	−9.365(−4)	−1.39(−10)	−5.0(−17)
345	15.194	−5.295(−4)	2.658(−8)	—	−9.064(−4)	−1.24(−10)	−4.2(−17)
350	15.270	−5.123(−4)	2.964(−8)	—	−8.777(−4)	−1.11(−10)	−3.5(−17)
355	15.344	−4.958(−4)	3.205(−8)	—	−8.503(−4)	−9.98(−11)	−3.0(−17)
360	15.416	−4.801(−4)	3.393(−8)	—	−8.240(−4)	−8.96(−11)	−2.5(−17)
365	15.486	−4.652(−4)	3.536(−8)	—	−7.989(−4)	−8.04(−11)	−2.1(−17)
370	15.554	−4.509(−4)	3.641(−8)	—	−7.749(−4)	−7.23(−11)	−1.8(−17)
375	15.620	−4.372(−4)	3.715(−8)	—	−7.519(−4)	−6.50(−11)	−1.5(−17)
380	15.684	−4.242(−4)	3.762(−8)	—	−7.298(−4)	−5.84(−11)	−1.3(−17)
385	15.747	−4.116(−4)	3.788(−8)	—	−7.087(−4)	−5.26(−11)	−1.1(−17)
390	15.807	−3.996(−4)	3.795(−8)	—	−6.883(−4)	−4.73(−11)	−9.2(−18)
395	15.865	−3.881(−4)	3.788(−8)	—	−6.688(−4)	−4.26(−11)	−7.8(−18)
400	15.921	−3.771(−4)	3.768(−8)	—	−6.500(−4)	−3.83(−11)	−6.6(−18)
405	15.976	−3.665(−4)	3.739(−8)	—	−6.319(−4)	−3.45(−11)	−5.5(−18)
410	16.028	−3.563(−4)	3.701(−8)	—	−6.145(−4)	−3.10(−11)	−4.6(−18)
415	16.078	−3.465(−4)	3.656(−8)	—	−5.977(−4)	−2.78(−11)	−3.9(−18)
420	16.127	−3.371(−4)	3.607(−8)	—	−5.815(−4)	−2.50(−11)	−3.2(−18)

Name: hydrogen bromide (CAS # 10035-10-6)
Formula: HBr
M.W.: 0.080912 kg/mol
Max. Pressure: 1500 kPa
Reference: Hurly, J.J. (2000). Thermophysical properties of gaseous HBr and BCl_3 from speed-of-sound measurements. *Int. J. Thermophys.*

T (K)	$C_P^0(T)/R$	$B(T)$ (m^3·mol^{-1})	$C(T)$ (m^6·mol^{-2})	$D(T)$ (m^9·mol^{-3})	$\beta_a(T)$ (m·mol^{-1})	$\gamma_a(T)$ (m^3·mol^{-1}·Pa^{-1})	$\delta_a(T)$ (m^3·mol^{-1}·Pa^{-2})
230	3.503	−3.277(−4)	−1.608(−8)	—	−3.668(−4)	−8.14(−11)	−2.5(−17)
235	3.503	−3.118(−4)	−1.079(−8)	—	−3.482(−4)	−6.88(−11)	−2.0(−17)
240	3.503	−2.971(−4)	−6.686(−9)	—	−3.310(−4)	−5.85(−11)	−1.6(−17)
245	3.503	−2.835(−4)	−3.494(−9)	—	−3.150(−4)	−5.00(−11)	−1.3(−17)
250	3.503	−2.708(−4)	−1.020(−9)	—	−3.001(−4)	−4.29(−11)	−1.0(−17)
255	3.503	−2.590(−4)	8.906(−10)	—	−2.863(−4)	−3.70(−11)	−8.5(−18)
260	3.503	−2.480(−4)	2.358(−9)	—	−2.734(−4)	−3.21(−11)	−6.9(−18)
265	3.503	−2.377(−4)	3.476(−9)	—	−2.613(−4)	−2.79(−11)	−5.6(−18)
270	3.503	−2.281(−4)	4.318(−9)	—	−2.500(−4)	−2.43(−11)	−4.6(−18)
275	3.503	−2.190(−4)	4.941(−9)	—	−2.394(−4)	−2.13(−11)	−3.8(−18)
280	3.503	−2.105(−4)	5.392(−9)	—	−2.294(−4)	−1.87(−11)	−3.1(−18)
285	3.503	−2.025(−4)	5.707(−9)	—	−2.200(−4)	−1.64(−11)	−2.6(−18)
290	3.503	−1.949(−4)	5.914(−9)	—	−2.111(−4)	−1.45(−11)	−2.1(−18)
295	3.503	−1.878(−4)	6.036(−9)	—	−2.027(−4)	−1.28(−11)	−1.8(−18)
300	3.503	−1.810(−4)	6.092(−9)	—	−1.948(−4)	−1.13(−11)	−1.5(−18)
305	3.504	−1.746(−4)	6.096(−9)	—	−1.873(−4)	−9.96(−12)	−1.2(−18)

continues

(continued)

T K	$C_p^0(T)/R$	$B(T)$ (m³·mol⁻²)	$C(T)$ (m⁶·mol⁻²)	$D(T)$ (m⁹·mol⁻³)	$\beta_a(T)$ (m³·mol⁻¹)	$\gamma_a(T)$ (m³·mol⁻¹·Pa⁻¹)	$\delta_a(T)$ (m³·mol⁻¹·Pa⁻²)
310	3.504	−1.686(−4)	6.060(−9)	—	−1.801(−4)	−8.82(−12)	−1.0(−18)
315	3.504	−1.628(−4)	5.994(−9)	—	−1.733(−4)	−7.81(−12)	−8.6(−19)
320	3.504	−1.573(−4)	5.905(−9)	—	−1.669(−4)	−6.92(−12)	−7.2(−19)
325	3.504	−1.521(−4)	5.798(−9)	—	−1.608(−4)	−6.12(−12)	−6.0(−19)
330	3.505	−1.472(−4)	5.680(−9)	—	−1.549(−4)	−5.42(−12)	−5.0(−19)
335	3.505	−1.425(−4)	5.553(−9)	—	−1.493(−4)	−4.79(−12)	−4.2(−19)
340	3.505	−1.380(−4)	5.420(−9)	—	−1.440(−4)	−4.22(−12)	−3.4(−19)
345	3.506	−1.337(−4)	5.285(−9)	—	−1.389(−4)	−3.72(−12)	−2.8(−19)
350	3.506	−1.296(−4)	5.149(−9)	—	−1.340(−4)	−3.26(−12)	−2.3(−19)
355	3.506	−1.257(−4)	5.014(−9)	—	−1.294(−4)	−2.85(−12)	−1.9(−19)
360	3.507	−1.219(−4)	4.880(−9)	—	−1.249(−4)	−2.48(−12)	−1.5(−19)
365	3.507	−1.183(−4)	4.749(−9)	—	−1.206(−4)	−2.15(−12)	−1.2(−19)
370	3.508	−1.148(−4)	4.622(−9)	—	−1.165(−4)	−1.85(−12)	−9.6(−20)
375	3.509	−1.115(−4)	4.498(−9)	—	−1.125(−4)	−1.57(−12)	−7.4(−20)
380	3.509	−1.083(−4)	4.379(−9)	—	−1.087(−4)	−1.32(−12)	−5.5(−20)
385	3.510	−1.053(−4)	4.264(−9)	—	−1.050(−4)	−1.09(−12)	−3.8(−20)
390	3.511	−1.023(−4)	4.154(−9)	—	−1.015(−4)	−8.84(−13)	−2.4(−20)
395	3.511	−9.949(−5)	4.049(−9)	—	−9.812(−5)	−6.94(−13)	−1.3(−20)
400	3.512	−9.676(−5)	3.948(−9)	—	−9.485(−5)	−5.20(−13)	−2.5(−21)
405	3.513	−9.413(−5)	3.852(−9)	—	−9.169(−5)	−3.61(−13)	5.9(−21)
410	3.514	−9.159(−5)	3.761(−9)	—	−8.864(−5)	−2.15(−13)	1.3(−20)
415	3.515	−8.914(−5)	3.675(−9)	—	−8.570(−5)	−8.13(−14)	1.9(−20)
420	3.516	−8.678(−5)	3.593(−9)	—	−8.286(−5)	4.18(−14)	2.4(−20)
425	3.517	−8.450(−5)	3.515(−9)	—	−8.011(−5)	1.55(−13)	2.9(−20)
430	3.518	−8.229(−5)	3.442(−9)	—	−7.745(−5)	2.59(−13)	3.2(−20)
435	3.519	−8.016(−5)	3.373(−9)	—	−7.488(−5)	3.55(−13)	3.5(−20)
440	3.520	−7.810(−5)	3.308(−9)	—	−7.239(−5)	4.44(−13)	3.7(−20)

Name: boron trichloride (CAS # 10294-34-5)
Formula: BCl_3
M.W.: 0.11717 kg/mol
Max. Pressure: 1500 kPa
Reference: Hurly, J.J. (2000). Thermophysical properties of gaseous HBr and BCl_3 from speed-of-sound measurements, *Int. J. Thermophys.*

T (K)	$C_p^0(T)/R$	$B(T)$ (m³·mol⁻¹)	$C(T)$ (m⁶·mol⁻²)	$D(T)$ (m⁹·mol⁻³)	$\beta_a(T)$ (m³·mol⁻¹)	$\gamma_a(T)$ (m³·mol⁻¹·Pa⁻¹)	$\delta_a(T)$ (m³·mol⁻¹·Pa⁻²)
290	7.447	−7.649(−4)	−2.992(−7)	—	−1.216(−3)	−5.17(−10)	−3.6(−16)
295	7.490	−7.376(−4)	−2.492(−7)	—	−1.176(−3)	−4.55(−10)	−2.9(−16)
300	7.533	−7.118(−4)	−2.068(−7)	—	−1.138(−3)	−4.02(−10)	−2.4(−16)
305	7.574	−6.873(−4)	−1.707(−7)	—	−1.103(−3)	−3.55(−10)	−2.0(−16)
310	7.616	−6.641(−4)	−1.400(−7)	—	−1.068(−3)	−3.15(−10)	−1.7(−16)
315	7.656	−6.420(−4)	−1.137(−7)	—	−1.036(−3)	−2.79(−10)	−1.4(−16)
320	7.696	−6.211(−4)	−9.126(−8)	—	−1.004(−3)	−2.48(−10)	−1.2(−16)
325	7.735	−6.011(−4)	−7.207(−8)	—	−9.747(−4)	−2.21(−10)	−1.0(−16)
330	7.774	−5.821(−4)	−5.563(−8)	—	−9.462(−4)	−1.97(−10)	−8.4(−17)
335	7.812	−5.640(−4)	−4.156(−8)	—	−9.189(−4)	−1.76(−10)	−7.1(−17)
340	7.849	−5.467(−4)	−2.950(−8)	—	−8.927(−4)	−1.57(−10)	−5.9(−17)
345	7.886	−5.302(−4)	−1.917(−8)	—	−8.676(−4)	−1.40(−10)	−5.0(−17)
350	7.922	−5.144(−4)	−1.032(−8)	—	−8.435(−4)	−1.25(−10)	−4.2(−17)
355	7.957	−4.993(−4)	−2.741(−9)	—	−8.203(−4)	−1.12(−10)	−3.6(−17)
360	7.992	−4.848(−4)	3.736(−9)	—	−7.981(−4)	−1.00(−10)	−3.0(−17)
365	8.026	−4.709(−4)	9.266(−9)	—	−7.766(−4)	−8.99(−11)	−2.5(−17)
370	8.059	−4.576(−4)	1.398(−8)	—	−7.560(−4)	−8.06(−11)	−2.1(−17)
375	8.092	−4.448(−4)	1.798(−8)	—	−7.362(−4)	−7.22(−11)	−1.8(−17)
380	8.124	−4.325(−4)	2.138(−8)	—	−7.170(−4)	−6.46(−11)	−1.5(−17)
385	8.156	−4.207(−4)	2.424(−8)	—	−6.985(−4)	−5.79(−11)	−1.3(−17)
390	8.187	−4.093(−4)	2.665(−8)	—	−6.807(−4)	−5.18(−11)	−1.1(−17)
395	8.217	−3.983(−4)	2.866(−8)	—	−6.635(−4)	−4.63(−11)	−9.0(−18)
400	8.247	−3.878(−4)	3.032(−8)	—	−6.469(−4)	−4.13(−11)	−7.5(−18)
405	8.276	−3.776(−4)	3.169(−8)	—	−6.308(−4)	−3.69(−11)	−6.2(−18)
410	8.304	−3.678(−4)	3.279(−8)	—	−6.152(−4)	−3.28(−11)	−5.1(−18)
415	8.332	−3.583(−4)	3.367(−8)	—	−6.001(−4)	−2.92(−11)	−4.2(−18)
420	8.359	−3.492(−4)	3.436(−8)	—	−5.856(−4)	−2.59(−11)	−3.4(−18)

Name: xenon (CAS # 7440-63-3)
Formula: Xe
M.W.: 0.1313 kg/mol
Max. Pressure: 1500 kPa
Reference: Hurly, J.J., Schmidt, J.W., Boyes, S.J., and Moldover, M.R. (1997). Virial equation of state of helium, xenon, and helium-xenon mixtures from speed-of-sound an burnett P rho T measurements. *Int. J. Thermophys.* **18**: 579–634.

T (K)	$C_p^0(T)/R$	$B(T)$ (m$^3\cdot$mol^{-1})	$C(T)$ (m$^6\cdot$mol^{-2})	$D(T)$ (m$^9\cdot$mol^{-3})	$\beta_a(T)$ (m$^3\cdot$mol^{-1})	$\gamma_a(T)$ (m$^3\cdot$mol$^{-1}\cdot$Pa^{-1})	$\delta_a(T)$ (m$^3\cdot$mol$^{-1}\cdot$Pa^{-2})
210	2.5	−2.501(−4)	6.598(−11)	—	−2.381(−4)	−4.11(−11)	−9.6(−18)
215	2.5	−2.395(−4)	1.425(−9)	—	−2.257(−4)	−3.52(−11)	−7.8(−18)
220	2.5	−2.295(−4)	2.503(−9)	—	−2.142(−4)	−3.01(−11)	−6.3(−18)
225	2.5	−2.202(−4)	3.355(−9)	—	−2.034(−4)	−2.59(−11)	−5.2(−18)
230	2.5	−2.115(−4)	4.026(−9)	—	−1.933(−4)	−2.33(−11)	−4.2(−18)
235	2.5	−2.033(−4)	4.548(−9)	—	−1.838(−4)	−1.92(−11)	−3.4(−18)
240	2.5	−1.955(−4)	4.951(−9)	—	−1.749(−4)	−1.65(−11)	−2.8(−18)
245	2.5	−1.882(−4)	5.257(−9)	—	−1.664(−4)	−1.43(−11)	−2.3(−18)
250	2.5	−1.813(−4)	5.483(−9)	—	−1.585(−4)	−1.23(−11)	−1.9(−18)
255	2.5	−1.747(−4)	5.646(−9)	—	−1.510(−4)	−1.06(−11)	−1.5(−18)
260	2.5	−1.685(−4)	5.755(−9)	—	−1.439(−4)	−9.11(−12)	−1.2(−18)
265	2.5	−1.626(−4)	5.823(−9)	—	−1.371(−4)	−7.82(−12)	−1.0(−18)
270	2.5	−1.570(−4)	5.856(−9)	—	−1.307(−4)	−6.69(−12)	−8.1(−19)
275	2.5	−1.517(−4)	5.861(−9)	—	−1.246(−4)	−5.70(−12)	−6.6(−19)
280	2.5	−1.466(−4)	5.844(−9)	—	−1.188(−4)	−4.83(−12)	−5.2(−19)
285	2.5	−1.417(−4)	5.809(−9)	—	−1.133(−4)	−4.07(−12)	−4.2(−19)
290	2.5	−1.371(−4)	5.760(−9)	—	−1.080(−4)	−3.40(−12)	−3.3(−19)
295	2.5	−1.327(−4)	5.699(−9)	—	−1.030(−4)	−2.80(−12)	−2.5(−19)
300	2.5	−1.284(−4)	5.630(−9)	—	−9.819(−5)	−2.28(−12)	−1.9(−19)
305	2.5	−1.244(−4)	5.554(−9)	—	−9.359(−5)	−1.81(−12)	−1.4(−19)
310	2.5	−1.205(−4)	5.473(−9)	—	−8.917(−5)	−1.40(−12)	−9.8(−20)
315	2.5	−1.168(−4)	5.388(−9)	—	−8.494(−5)	−1.03(−12)	−6.3(−20)
320	2.5	−1.132(−4)	5.301(−9)	—	−8.088(−5)	−7.06(−13)	−3.5(−20)
325	2.5	−1.098(−4)	5.213(−9)	—	−7.699(−5)	−4.16(−13)	−1.1(−20)
330	2.5	−1.064(−4)	5.123(−9)	—	−7.324(−5)	−1.58(−13)	7.6(−21)
335	2.5	−1.033(−4)	5.034(−9)	—	−6.963(−5)	7.14(−14)	2.3(−20)
340	2.5	−1.002(−4)	4.946(−9)	—	−6.617(−5)	2.76(−13)	3.6(−20)
345	2.5	−9.726(−5)	4.858(−9)	—	−6.282(−5)	4.59(−13)	4.6(−20)
350	2.5	−9.442(−5)	4.771(−9)	—	−5.960(−5)	6.21(−13)	5.4(−20)
355	2.5	−9.168(−5)	4.686(−9)	—	−5.649(−5)	7.66(−13)	6.0(−20)
360	2.5	−8.903(−5)	4.603(−9)	—	−5.350(−5)	8.96(−13)	6.4(−20)
365	2.5	−8.648(−5)	4.522(−9)	—	−5.060(−5)	1.01(−12)	6.8(−20)
370	2.5	−8.401(−5)	4.442(−9)	—	−4.780(−5)	1.11(−12)	7.0(−20)
375	2.5	−8.162(−5)	4.365(−9)	—	−4.509(−5)	1.20(−12)	7.2(−20)
380	2.5	−7.931(−5)	4.290(−9)	—	−4.247(−5)	1.28(−12)	7.3(−20)
385	2.5	−7.707(−5)	4.217(−9)	—	−3.994(−5)	1.36(−12)	7.3(−20)
390	2.5	−7.491(−5)	4.146(−9)	—	−3.749(−5)	1.42(−12)	7.3(−20)
395	2.5	−7.281(−5)	4.078(−9)	—	−3.511(−5)	1.48(−12)	7.3(−20)
400	2.5	−7.078(−5)	4.012(−9)	—	−3.280(−5)	1.53(−12)	7.2(−20)

Name: helium (CAS # 7440-59-7)
Formula: He
M.W.: 0.0040026 kg/mol
Max. Pressure: 1500 kPa
Reference: Hurly, J.J., Schmidt, J.W., Boyes, S.J., and Moldover, M.R. (1997). Virial equation of state of helium, xenon, and helium-xenon mixtures from speed-of-sound and Burnett P rho T measurements. *Int. J. Thermophys.* **18**: 579–634.

(continued)

T (K)	$C_p^0(T)/R$	$B(T)$ (m$^3 \cdot$mol^{-1})	$C(T)$ (m$^6 \cdot$mol^{-2})	$D(T)$ (m$^9 \cdot$mol^{-3})	$\beta_a(T)$ (m$^3 \cdot$mol^{-1})	$\gamma_a(T)$ (m$^3 \cdot$mol$^{-1} \cdot$Pa^{-1})	$\gamma_a(T)$ (m$^3 \cdot$mol$^{-1} \cdot$Pa^{-2})
210	2.5	1.216(−5)	1.197(−10)	—	2.324(−5)	3.95(−14)	−9.7(−22)
215	2.5	1.214(−5)	1.188(−10)	—	2.316(−5)	3.75(−14)	−9.1(−22)
220	2.5	1.213(−5)	1.179(−10)	—	2.308(−5)	3.57(−14)	−8.6(−22)
225	2.5	1.212(−5)	1.171(−10)	—	2.229(−5)	3.39(−14)	−8.1(−22)
230	2.5	1.210(−5)	1.162(−10)	—	2.291(−5)	3.22(−14)	−7.6(−22)
235	2.5	1.208(−5)	1.153(−10)	—	2.282(−5)	3.07(−14)	−7.2(−22)
240	2.5	1.207(−5)	1.144(−10)	—	2.274(−5)	2.92(−14)	−6.7(−22)
245	2.5	1.205(−5)	1.136(−10)	—	2.265(−5)	2.78(−14)	−6.4(−22)
250	2.5	1.203(−5)	1.127(−10)	—	2.257(−5)	2.65(−14)	−6.0(−22)
255	2.5	1.202(−5)	1.119(−10)	—	2.249(−5)	2.53(−14)	−5.7(−22)
260	2.5	1.200(−5)	1.111(−10)	—	2.240(−5)	2.42(−14)	−5.4(−22)
265	2.5	1.198(−5)	1.103(−10)	—	2.232(−5)	2.31(−14)	−5.1(−22)
270	2.5	1.196(−5)	1.095(−10)	—	2.224(−5)	2.21(−14)	−4.9(−22)
275	2.5	1.194(−5)	1.087(−10)	—	2.216(−5)	2.12(−14)	−4.6(−22)
280	2.5	1.192(−5)	1.080(−10)	—	2.209(−5)	2.03(−14)	−4.4(−22)
285	2.5	1.190(−5)	1.072(−10)	—	2.201(−5)	1.95(−14)	−4.2(−22)
290	2.5	1.188(−5)	1.065(−10)	—	2.194(−5)	1.87(−14)	−4.0(−22)
295	2.5	1.186(−5)	1.058(−10)	—	2.186(−5)	1.80(−14)	−3.8(−22)
300	2.5	1.184(−5)	1.050(−10)	—	2.180(−5)	1.73(−14)	−3.6(−22)
305	2.5	1.182(−5)	1.043(−10)	—	2.173(−5)	1.66(−14)	−3.4(−22)
310	2.5	1.180(−5)	1.037(−10)	—	2.166(−5)	1.60(−14)	−3.3(−22)
315	2.5	1.178(−5)	1.030(−10)	—	2.160(−5)	1.54(−14)	−3.1(−22)
320	2.5	1.176(−5)	1.023(−10)	—	2.154(−5)	1.49(−14)	−3.0(−22)
325	2.5	1.174(−5)	1.017(−10)	—	2.149(−5)	1.44(−14)	−2.9(−22)
330	2.5	1.172(−5)	1.011(−10)	—	2.143(−5)	1.39(−14)	−2.7(−22)
335	2.5	1.169(−5)	1.004(−10)	—	2.138(−5)	1.35(−14)	−2.6(−22)
340	2.5	1.167(−5)	9.982(−11)	—	2.134(−5)	1.31(−14)	−2.5(−22)
345	2.5	1.165(−5)	9.922(−11)	—	2.130(−5)	1.27(−14)	−2.4(−22)
350	2.5	1.163(−5)	9.863(−11)	—	2.126(−5)	1.23(−14)	−2.3(−22)
355	2.5	1.161(−5)	9.806(−11)	—	2.123(−5)	1.19(−14)	−2.2(−22)
360	2.5	1.159(−5)	9.749(−11)	—	2.120(−5)	1.16(−14)	−2.1(−22)
365	2.5	1.157(−5)	9.694(−11)	—	2.117(−5)	1.13(−14)	−2.0(−22)
370	2.5	1.155(−5)	9.639(−11)	—	2.115(−5)	1.10(−14)	−1.9(−22)
375	2.5	1.152(−5)	9.586(−11)	—	2.114(−5)	1.07(−14)	−1.9(−22)
380	2.5	1.150(−5)	9.533(−11)	—	2.113(−5)	1.05(−14)	−1.8(−22)
385	2.5	1.148(−5)	9.482(−11)	—	2.112(−5)	1.02(−14)	−1.7(−22)
390	2.5	1.146(−5)	9.431(−11)	—	2.112(−5)	1.00(−14)	−1.6(−22)
395	2.5	1.144(−5)	9.382(−11)	—	2.113(−5)	9.81(−14)	−1.6(−22)
400	2.5	1.142(−5)	9.333(−11)	—	2.114(−5)	9.62(−14)	−1.5(−22)

Name: argon (CAS # 7440-37-1)
Formula: Ar
M.W.: 0.039948 kg/mol
Max. Pressure: 1500 kPa
Reference: Aziz, R.A. (1993). A highly accurate interatomic potential for argon. *J. Chem. Phys.* **99**: 4518–4525.

T (K)	$C_P^0(T)/R$	$B(T)$ (m$^3 \cdot$ mol^{-1})	$C(T)$ (m$^6 \cdot$ mol^{-2})	$D(T)$ (m$^9 \cdot$ mol^{-3})	$\beta_a(T)$ (m$^3 \cdot$ mol^{-1})	$\gamma_a(T)$ (m$^3 \cdot$ mol$^{-1} \cdot$ Pa^{-1})	$\delta_a(T)$ (m$^3 \cdot$ mol$^{-1} \cdot$ Pa^{-2})
100	2.5	−1.824(−4)	−8.830(−10)	—	−1.804(−4)	−4.58(−11)	−1.6(−17)
110	2.5	−1.531(−4)	1.080(−9)	—	−1.449(−4)	−2.50(−11)	−7.2(−18)
120	2.5	−1.305(−4)	1.853(−9)	—	−1.182(−4)	−1.41(−11)	−3.3(−18)
130	2.5	−1.126(−4)	2.114(−9)	—	−9.735(−5)	−8.01(−12)	−1.5(−18)
140	2.5	−9.809(−5)	2.150(−9)	—	−8.061(−5)	−4.42(−12)	−6.5(−19)
150	2.5	−8.606(−5)	2.087(−9)	—	−6.688(−5)	−2.24(−12)	−2.5(−19)
160	2.5	−7.596(−5)	1.985(−9)	—	−5.542(−5)	−8.81(−13)	−6.7(−20)
170	2.5	−6.736(−5)	1.871(−9)	—	−4.573(−5)	−2.25(−14)	2.0(−20)
180	2.5	−5.995(−5)	1.760(−9)	—	−3.742(−5)	5.23(−13)	5.6(−20)
190	2.5	−5.350(−5)	1.656(−9)	—	−3.022(−5)	8.67(−13)	6.9(−20)
200	2.5	−4.785(−5)	1.561(−9)	—	−2.393(−5)	1.08(−12)	6.9(−20)
210	2.5	−4.785(−5)	1.477(−9)	—	−1.839(−5)	1.21(−12)	6.4(−20)
220	2.5	−3.840(−5)	1.402(−9)	—	−1.348(−5)	1.27(−12)	5.7(−20)
230	2.5	−3.442(−5)	1.335(−9)	—	−9.091(−6)	1.30(−12)	5.0(−20)
240	2.5	−3.084(−5)	1.276(−9)	—	−5.155(−6)	1.30(−12)	4.3(−20)
250	2.5	−2.759(−5)	1.225(−9)	—	−1.605(−6)	1.29(−12)	3.6(−20)
260	2.5	−2.465(−5)	1.179(−9)	—	−1.610(−6)	1.23(−12)	3.0(−20)
270	2.5	−2.196(−5)	1.138(−9)	—	−4.535(−6)	1.23(−12)	2.5(−20)
280	2.5	−1.950(−5)	1.102(−9)	—	7.205(−6)	1.19(−12)	2.1(−20)
290	2.5	−1.724(−5)	1.070(−9)	—	9.650(−6)	1.14(−12)	1.7(−20)
300	2.5	−1.516(−5)	1.041(−9)	—	1.190(−5)	1.10(−12)	1.4(−20)
310	2.5	−1.323(−5)	1.015(−9)	—	1.397(−5)	1.06(−12)	1.1(−20)
320	2.5	−1.145(−5)	9.921(−10)	—	1.588(−5)	1.02(−12)	8.9(−21)
330	2.5	−9.792(−6)	9.714(−10)	—	1.765(−5)	9.74(−13)	7.0(−21)
340	2.5	−8.249(−6)	9.527(−10)	—	1.929(−5)	9.34(−13)	5.4(−21)
350	2.5	−6.809(−6)	9.357(−10)	—	2.082(−5)	8.94(−13)	4.1(−21)
360	2.5	−5.462(−6)	9.204(−10)	—	2.225(−5)	8.57(−13)	3.0(−21)
370	2.5	−4.021(−6)	9.064(−10)	—	2.358(−5)	8.21(−13)	2.1(−21)
380	2.5	−3.017(−6)	8.936(−10)	—	2.482(−5)	7.87(−13)	1.3(−21)
390	2.5	−1.904(−6)	8.820(−10)	—	2.599(−5)	7.54(−13)	6.3(−22)
400	2.5	−8.558(−7)	8.713(−10)	—	2.708(−5)	7.23(−13)	8.1(−23)
410	2.5	1.326(−7)	8.164(−10)	—	2.811(−5)	6.93(−13)	−3.7(−22)
420	2.5	1.066(−6)	8.524(−10)	—	2.908(−5)	6.65(−13)	−7.5(−22)
430	2.5	1.948(−6)	8.440(−10)	—	2.999(−5)	6.39(−13)	−1.1(−21)
440	2.5	2.784(−6)	8.362(−10)	—	3.085(−5)	6.13(−13)	−1.3(−21)
450	2.5	3.576(−6)	8.290(−10)	—	3.166(−5)	5.89(−13)	−1.5(−21)
460	2.5	4.328(−6)	8.222(−10)	—	3.242(−5)	5.66(−13)	−1.7(−21)
470	2.5	5.042(−6)	8.159(−10)	—	3.315(−5)	5.44(−13)	−1.8(−21)
480	2.5	5.722(−6)	8.100(−10)	—	3.383(−5)	5.24(−13)	−1.9(−21)
490	2.5	6.368(−6)	8.045(−10)	—	3.449(−5)	4.17(−13)	−3.0(−21)
500	2.5	6.984(−6)	7.993(−10)	—	3.510(−5)	4.00(−13)	−3.0(−21)

ACKNOWLEDGMENTS

This work was supported in part by the Office of Naval Research. We thank the Deutscher Akademischer Austauschdienst (Germany), which provided a stipend to one of the authors (J.W.), thereby facilitating his participation in this work.

References

1. Mohr, P.J., and Taylor, B.N. (1999). CODATA Recommended Values of the Fundamental Physical Constants: 1998. *J. Phys. Chem. Ref. Data* **28**: 1713–1852.
2. Moldover, M.R., Trusler, J.P.M., Edwards, T.J., Mehl, J.B., and Davis, R.S. (1988). Measurement of the universal gas constant R using a spherical acoustic resonator. *J. Res. Nat. Bureau Stand. (US)* **93**: 85–144.

3. Moldover, M.R., Boyes, S.J., Meyer, C.W., and Goodwin, A.R.H. (1999). Thermodynamic temperatures of the triple points of mercury and gallium and in the interval 217 K to 303 K. *J. Res. Natl. Inst. Stand. Tech.* **104**: 11–46.
4. Moldover, M.R., Mehl, J.B., and Greenspan, M. (1986). Gas-filled spherical resonators: Theory and experiment. *J. Acoust. Soc. Am.* **79**: 253–272.
5. Gillis, K.A. (1994). Thermodynamic properties of two gaseous halogenated ethers from speed-of-sound measurement: Difluoromethoxy-difluoromethane and 2-difluoromethoxy-1,1,1-trifluoroethane. *Int. J. Thermophysics* **15**: 821–847.
6. Gillis, K.A., Mehl, J.B., and Moldover, M.R. (1996). Greenspan acoustic viscometer for gases. *Rev. Sci. Instrum.* **67**: 1850–1857.
7. Wilhelm, J., Gillis, K.A., Mehl, J.B., and Moldover, M.R. (submitted). An improved Greenspan acoustic viscometer. *Int. J. Thermophys.*
8. Gillis, K.A., and Moldover, M.R. (1996). Practical determination of gas densities from the speed of sound using square-well potentials. *Int. J. Thermophys.* **17**: 1305–1324.
9. Hurly, J.J. (1999). Thermophysical properties of gaseous CF_4 and C_2F_6 from speed-of-sound measurements. *Int. J. Thermophys.* **20**: 455.
10. Trusler, J.P.M. (1997). Equation of state for gaseous propane determined from the speed of sound. *Int. J. Thermophys.* **18**: 635–654.
11. Estrada-Alexanders, A.F., Trusler, J.P.M., and Zarari, M.P. (1995). Determination of thermodynamic properties from the speed of sound. *Int. J. Thermophys.* **16**: 663–673.
12. Trusler, J.P.M., Wakeham, W.A., and Zarari, M. (1997). Model intermolecular potentials and virial coefficients determined from the speed of sound. *Mol. Phys.* **90**: 695–703.
13. Colclough, A.R., Quinn, T.J., and Chandler, T.R.D. (1979). An acoustic redetermination of the gas constant. *Proc. R. Soc. Lond. A* **368**: 125–139.
14. Quinn, T.J., Colclough, A.R., and Chandler, T.R.D. (1976). A new determination of the gas constant by an acoustical method. *Phil. Trans. R. Soc. Lond. A* **283**: 367–420.
15. Batuecas, T. (1972). Volume normal moléculaire, V_o, des gaz á l'ètat idéal, in *Atomic Masses and Fundamental Constants*, vol. 4, Sanders, J.H., and Wapstra, A.H., eds., pp. 534–542, New York: Plenum.
16. Rowlinson, J.S., and Tildesley, D.J. (1977). The determination of the gas constant from the speed of sound. *Proc. R. Soc. Lond. A* **358**: 281–286.
17. Mason, E.A., and Spurling, T.H. (1969). *The Virial Equation of State*. Oxford: Pergamon Press.
18. Dulla, R.J., Rowlinson, J.S., and Smith, W.R. (1971). Effective pair potentials in fluids in the presence of three-body forces. II. *Mol. Phys.* **21**: 299–315.
19. Axilrod, B.M., and Teller, E.J. (1943). Interaction of the van der Waals type between three atoms. *J. Chem. Phys.* **11**: 299–300.
20. Hurly, J.J. (in press). Thermophysical properties of gaseous HBr and BCl_3 from speed-of-sound measurement. *Int. J. Thermophys.*
21. Goodwin, A.R.H. (1988). *Thermophysical Properties from the Speed of Sound*, pp. 113–117, PhD Thesis, University College London.
22. Garland, C.W., and Williams, R.D. (1974). Low-frequency sound velocity near the critical point of xenon. *Phys. Rev. A* **10**: 1328–1332.
23. Van Dael, W., (1975). Thermodynamic properties and the velocity of sound, in *Experimental Thermodynamics*, vol. II, Le Neindre, B., and Vodar, B., eds., pp. 527–577, London: Butterworths.
24. Gillis, K.A., Moldover, M.R., and Goodwin, A.R.H. (1991). Accurate acoustic measurements in gases under difficult conditions. *Rev. Sci. Instrum.* **62**: 2213–2217.
25. Mehl, J.B. (1999). Greenspan acoustic viscometer: Numerical calculations of fields and duct-end effects. *J. Acoust. Soc. Am.* **106**: 73–82.
26. Gillis, K.A., Mehl, J.B., and Moldover, M.R. (in preparation). Acoustic resonator for Prandtl number measurements.
27. Mehl, J.B., and Moldover, M.R. (1986). Measurement of the ratio of the speed of sound to the speed of light. *Phys. Rev. A* **34**: 3341–3344.
28. Strouse, G.F. (1992). NIST assessment of ITS-90 non-uniqueness for 25.5 Ω SPRTs at gallium, indium, and cadmium fixed points, in *Temperature: Its Measurement and Control in Science and Industry*, vol 6, Schooley, J.F., ed., pp. 175–178, New York: American Institute of Physics.
29. Mehl, J.B. (1986). Acoustic resonance frequencies of deformed spherical resonators: I. *J. Acoust. Soc. Am.* **71**: 1109–1113; Mehl, J.B. (1982). Acoustic resonance frequencies of deformed spherical resonators: II. *J. Acoust. Soc. Am.* **79**: 278–285.

CHAPTER 13

USE OF CYLINDRICAL ACOUSTIC RESONANCE TO MEASURE THE SPEED OF SOUND IN GASES

F. W. Giacobbe
Chicago Research Center/American Air Liquide, Inc., Countryside, Illinois, USA

Contents

Abstract		371
13.1.	Introduction	372
13.2.	Apparatus	373
13.3.	Gases	375
13.4.	Method of Operation	376
13.5.	Experimental Results and Discussion	377
13.6.	Conclusion	384
Acknowledgments		385
References		385

ABSTRACT

This chapter describes the design and use of a cylindrical acoustic resonator that has been employed to accurately measure absolute sound speeds in pure gases and gas mixtures. This device was used to make experimental sound speed measurements, at ambient temperatures, in two gaseous mixtures (air and mixtures of helium and argon) and in pure argon, carbon dioxide, helium, hydrogen, methane, neon, nitrogen, nitrous oxide, and oxygen. The apparatus and methods employed to produce and correct this laboratory data for the influences of temperature, viscosity, and thermal conductivity within the cylindrical resonator are also discussed.

Mean acoustic velocities could be resolved (using the apparatus and techniques described in this chapter) with uncertainties of approximately ±0.15% from three separate sets of experimental test trials per gas tested. Repeated determinations of the acoustic velocities in dry, carbon dioxide-free air indicated that uncertainties of less than ±0.1% could be achieved in average acoustic velocities obtained from 10 or more independent sets of test trials.

It is believed that the description, operation, and necessary corrections associated with the use of the cylindrical acoustic resonator described herein will be useful to others who may wish to construct similar devices for their own use in the measurement of sound speeds in other gases or gas mixtures in the future.

Handbook of Elastic Properties of Solids, Liquids, and Gases, edited by Levy, Bass, and Stern
Volume IV: Elastic Properties of Fluids: Liquids and Gases
Copyright © 2001 by Academic Press

ISBN 0-12-445764-9 / $35.00   All rights of reproduction in any form reserved.

13.1. INTRODUCTION

Many different experimental methods have been employed in the past to make acoustic velocity measurements in various gases and gas mixtures. Generally, the most accurate of these methods involves the use of cylindrical [4] or spherical [10] cavity resonance techniques.

Typically, cylindrical cavity resonance involves the establishment of a standing sound wave between two transducers (one acting as an emitter and the other as a receiver) at each end of a cylindrical tube. In another mode of operation involving cylindrical cavity resonance, a single transducer faces a flat but movable end plate or piston at the other end of the cylinder. When two transducers are employed, one of them is moved toward, or away from, the other while the sonic driving frequency of the emitting transducer is held constant. When the distance separating the transducers is equal to an integer number of half wavelengths, the output signal generated in the receiving transducer reaches a maximum. Therefore, the precise distance between two successive peaks in the receiving transducer's output signal is equal to one half of one sonic wavelength in the gas phase separating the two transducers. So, the speed of sound in the gas can be estimated by simply multiplying the fixed resonance frequency by the experimentally determined sonic wavelength.

When a single transducer (opposing a flat but movable end plate or piston) is employed, the operation is similar except that one usually must relate changes in the electrical behavior of the single transducer (due to the establishment of resonance within the cylindrical tube) to the movement of a flat object at the other end of the cylinder. These changes also vary periodically with each half wavelength change in the acoustic path length. And, as above, the speed of sound in the gas medium can be estimated by multiplying the fixed frequency of the emitting transducer by the experimentally determined sonic wavelength. One of the problems with both of these approaches is the need to move something (e.g., transducer, or a flat end plate, or a piston) at one end of a cylindrical tube [21].

According to some authorities, spherical resonators may be capable of more exact measurements of sonic velocities in gases than cylindrical resonance devices [10]. But, these devices are more difficult to fabricate and operate. And, typically, several rather complicated corrections (involving the effects of the thermal boundary layer at the resonator wall, the elastic compliance of the resonator, sound absorption by the gas, perturbations of the transducers, and departures from perfectly spherical geometry) must be made in determining very accurate sound speeds [9]. Often, these corrections are avoided entirely by using spherical resonators to determine relative acoustic velocities instead of absolute values [9, 52].

In any case, the purpose of this chapter is to describe the operation and practical use of a specialized fixed path length, cylindrical acoustic resonator that was designed, fabricated, and tested in the hope of obviating some of the problems indicated previously with respect to other types of cylindrical (as well as spherical) acoustic resonators. The main goal in building this device was to enable the precise, and relatively rapid, measurement of absolute acoustic velocities in unusual or atypical gas mixtures. But, prior to that activity, the operation and performance of this device was evaluated by using it to measure sound speeds in several pure gases and two gas mixtures. It is thought that a description of this work in experimentally determining sound speeds in these gases, using this apparatus, would be of interest to others having a need to make relatively precise sound speed measurements in other gases or gas mixtures in the future. It is further acknowledged that some, but not all, of the information described herein has been discussed in more detail in another publication [26]. However, the present source of elastic property information was deemed to be an exceptional location for more widely disseminating the many potentially useful aspects and results of this experimental work. In addition, earlier publications of our own and others have been referred to as frequently as possible to minimize any unnecessary duplication of information.

13.2. APPARATUS

A schematic of the acoustic resonator designed and employed during our own studies may be seen in Figure 13.1. Except for the transducers, transducer support structures, and a few other minor items, this device was fabricated from 304 and 316 stainless steel. All components that could be permanently connected were welded together (using the tungsten/inert gas [TIG] method) as indicated in Figure 13.1. The small-diameter tubing, used as a heat exchanger, was also fabricated from stainless steel. It had an outside diameter of 3.175 mm and was welded in place as shown. The overall length (uncoiled) of this section of tubing was about 6.0 m. Acetone was employed to clean all gas "wetted" metal parts of the system prior to the initial assembly. In addition, leak testing of the entire system, using nitrogen gas at approximately 10 atm, was also performed to ensure the integrity of all welded seams after the initial assembly.

Two identical acoustic transducers (Allied Electronics Type: S100RL-M) were securely mounted within machined brass support structures at each end of the resonator. These acoustic transducers were selected mainly because of their relatively flat front diaphragms. This feature permitted an accurate and direct measurement of the distance between the transducer diaphragms (i.e., 35.41 cm). These transducers also had the advantages of off-the-shelf availability, low cost, ruggedness, very good sensitivity (with amplification) to a wide range of acoustic driving frequencies, and a diaphragm diameter almost equal to the inner diameter of the cylindrical resonator (i.e., 2.804 cm).

The brass transducer supports were held firmly in place by threaded stainless steel end caps (MDC Vacuum Products Corporation: DS-112). These components are also illustrated in Figure 13.1. Fiducial marks on the end caps and central tube section were used to ensure that O-ring compression was constant. Therefore, the distance between the transducer diaphragms was also constant, even if the system was opened periodically for inspection and then closed again.

The annular cavity, between the central gas resonance tube and the outer shell of the resonator, was filled with water and sealed prior to all experimental measurements. This relatively large volume of water acted as a thermal heat sink that either preheated or cooled the entering gas stream (depending on whether the gas was slightly cooler or hotter than the water), via the coiled stainless steel heat exchanger. The water reservoir also maintained the central gas resonance chamber at a relatively constant temperature (very near, or at ambient) during each experimental trial run. No attempt was made to control this temperature. However, it was monitored continuously during each experimental trial run using a precalibrated platinum resistance thermometer (Omega: THX-700-GP-12) inserted in the water reservoir, as shown in Figure 13.1. Temperatures could be accurately read to the nearest 0.1 C using this thermometer and, generally, did not vary by more than ±0.1 C during any single trial run. Relative errors in the gas temperature due to this small variation in temperature were about ±1 part in 3000 or about ±0.033%. These average trial temperature readings were used to correct all of the individual sound speed measurements, determined from the original data, to the common temperatures of 0.0, 20.0, and/or 25.0 C.

The acoustic resonator was initially purged at relatively high gas flow rates. Then, the gas under study was passed, at a very low flow rate (ca. 3.0 liters/hour), through the coiled stainless steel tubing and then through a very small annular space (ca. 0.13 mm) between the left-hand side brass transducer support and the inner wall of

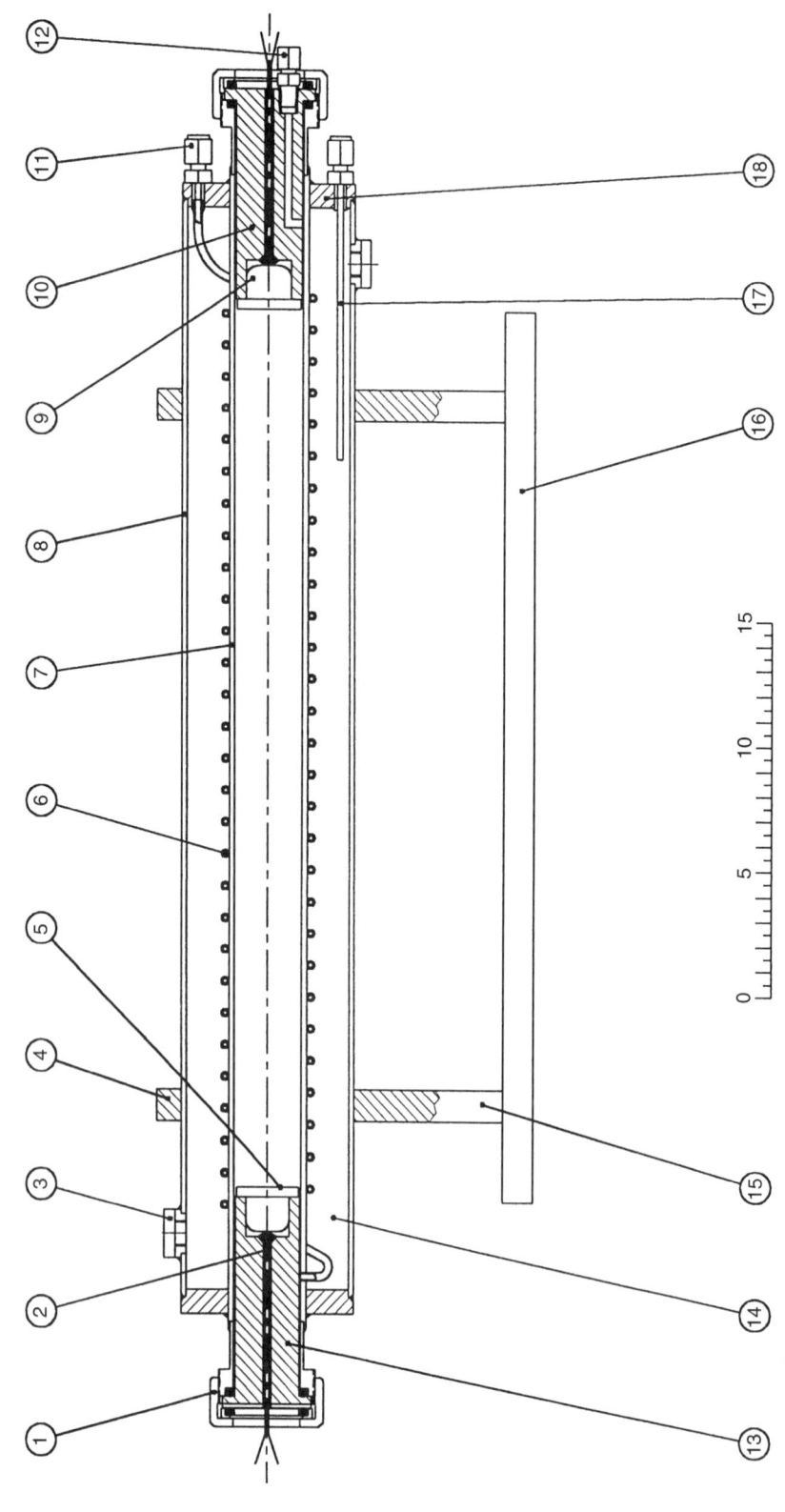

Fig. 13.1. Schematic of cylindrical acoustic resonance device and approximate scale (cm). 1 = threaded SS end cap, 2 = epoxy seal, 3 = water reservoir plug, 4 = top of vertical Al support, 5 = transducer diaphragm, 6 = coiled SS heat exchanger, 7 = central resonance cavity, 8 = outer SS shell, 9 = acoustic transducer, 10 = brass end support, 11 = gas inlet line, 12 = gas outlet, 13 = similar to 10, 14 = water reservoir, 15 = bottom of vertical Al support, 16 = horizontal Al support, 17 = Pt resistance thermometer, and 18 = SS end cap.

the central resonance chamber. The gas then entered the main section of the central resonance chamber (between the two acoustic transducers) at a pressure virtually equivalent to the prevailing atmospheric pressure, then it passed into the annular space between the right-hand side brass transducer support and the central resonance tube, and then it escaped to the atmosphere through a cavity drilled into that same brass transducer support. Flammable gases were additionally directed through a short section of stainless steel tubing, attached to the outlet gas brass transducer support, and burned as they exited into the atmosphere. This flow pattern was deliberately employed to ensure that each new gas tested would quickly and thoroughly remove (by displacement) all traces of any other gas residue remaining within the central section of the acoustic resonator after any preceding experimental test trials. The Doppler shifting of the sound speed due to this flowing gas stream was negligible, although others have made provisions to eliminate this potential source of error in other types of acoustic systems [44]. Just before the start of each test trial, barometer readings (subsequently corrected for temperature and latitude) were also taken to allow for their later use in corrections to the initially determined sound speeds.

13.3. GASES

Acoustic velocities were measured in two mixtures (air and mixtures of helium and argon) and in nine pure gases (argon, carbon dioxide, helium, hydrogen, methane, neon, nitrogen, nitrous oxide, and oxygen).

The air was obtained from a nonlubricated air compressor and storage tank maintained at 4.5 atm. Before its delivery to the resonator, the compressed air was purified by passing it (at 4.5 atm) through a highly activated bed of molecular sieve, which removed all but insignificant traces of carbon dioxide and water vapor [25]. A direct measurement of the carbon dioxide and water vapor concentrations in the purified compressed air stream indicated a residual concentration of carbon dioxide of less than 5.0 ppm (detectability limit of the analyzer) and a residual concentration of water vapor of less than 1.0 ppm. Therefore, this gas mixture has been referred to as dry, carbon dioxide-free air. The reader may wish to refer to other published experimental studies that have already addressed the influence of minor impurities, such as carbon dioxide and water vapor, on acoustic velocities in air [31].

Mixtures of helium and argon were prepared by accurately blending the pure gaseous components together within a high-pressure stainless steel sampling cylinder maintained at fixed temperatures very near ambient conditions (using a constant temperature water bath). During the blending process, gas pressures (to the nearest ± 0.34 atm [or ± 5 psi]) and temperatures (to the nearest ± 0.1 C) were accurately measured and recorded to permit precise calculations regarding the final mixture concentrations. Nine separate blended samples of the helium/argon gas mixtures, at total system pressures very near 69.0 atm (or 1000 psig), were prepared in this way. Helium/argon mole fractions within these mixtures ranged between 0.1/0.9 and 0.90/0.1, in helium mole fraction step size increases as near as possible to 0.1 per gas mixture tested.

Nitrogen gas was produced in a heat exchanger attached directly to a bulk source of highly purified cryogenic liquid nitrogen. High-purity argon, helium, and oxygen were obtained from large high-pressure cylinders (No. 49) of these gases. Carbon dioxide (liquid) and hydrogen were obtained from standard-size high-purity cylinders (No. 44). Methane, neon, and nitrous oxide (liquid) were obtained from much smaller high-purity cylinders (No. 16). All of these gases were at least 99.99% pure. However, the nitrogen and oxygen gases were equal to, or better than, 99.9998% pure.

13.4. METHOD OF OPERATION

During operation, one of the transducers (hereafter referred to as the active transducer) was excited by a sinusoidal wave generator (BK Precision: No. 3011B, 2 MHz Function Generator). Actual excitation frequencies were measured using a separate digital frequency meter (BK Precision: No. 3022, 2 MHz Sweep Function Generator and Digital Frequency Meter). This arrangement permitted the direct measurement of digital frequencies as high as 20 kHz with a resolution of ± 1 Hz. Of course, the implied reading error here is $\pm$ 1 part in 20,000, but the same resolution was obtained at lower frequencies, indicating a much higher relative error. Nevertheless, since the method of employing the experimental frequency data to determine acoustic sound speeds relied on a curve fitting routine over an extended range of frequencies, average errors due to uncertainties in the actual frequency readings were probably negligible.

The excitation frequency employed to drive the active transducer was slowly increased from zero until a condition of resonance was established between both of the transducers. This condition of resonance could be detected by measuring the output current (AC) induced in the field coil of the receiving transducer. The output current in this transducer reached a maximum each time the excitation frequency in the active transducer passed through a system resonance frequency. And, each standing wave pattern between the two transducers increased by exactly one-half of one wavelength (i.e., the wavelength of each standing wave decreased with an increase in the resonance frequency but each standing wave contained an additional half wavelength of the new wavelength generated) with each successive increase in the resonance frequency. Furthermore, since the first resonance condition produced a standing sound wave having a wavelength exactly twice the distance between the transducer diaphragms, and the second resonance condition produced a standing sound wave having a wavelength equal to this distance, and so on, a complete data set relating resonance frequency and sonic wavelength could be generated quite easily. And also, as a first approximation, it is possible to regard acoustic velocities as independent of frequency (especially for unconfined sonic disturbances propagated at frequencies below 20 kHz) [9, 48]. Therefore, a plot of resonance frequencies vs reciprocal sonic wavelengths should produce a straight line. The slope of this straight line should be nearly equal to the acoustic velocity in the residing test gas, and its intercept value should be zero (provided that there are no systematic errors associated with the wavelength or resonance frequency measurements). In fact, this type of graphical technique was actually employed to make first approximation estimates of acoustic velocities in the various gases tested in this way. A significant advantage in this approach is that the distance between the transducer diaphragms must be measured accurately only once. All wavelengths, under resonance conditions, are then calculated by multiplying this dimension by the ratio of two integers (i.e., 2/n, where n = 1, 2, 3, ...). In addition, if a suitably large distance (say, 30–50 cm) separates the transducers, slight imperfections in the flatness of their diaphragms can be reduced to insignificance. Of course, other types of transducers that really do have flat diaphragms can be used to completely eliminate this potential problem [2]. One of these types of transducers, containing perfectly flat diaphragms, is commercially available within Radio Shack retail outlet stores. It is referred to as a "Piezoelectric Speaker Element." However, we have not yet tested any of these transducers within our own sound speed measuring equipment so it is not possible to comment here on their suitability in this kind of application.

To improve on the resolution of our equipment, the induced output current from the signal-receiving transducer was amplified (Keithley: 427 Current Amplifier) and the resulting AC signal was fed to an oscilloscope (BK Precision: Model 2120) to facilitate fine tuning of the frequency input to the driving transducer. However, tuning the system to find resonance frequencies was not an exact process because the system itself was so sensitive. In fact, the errors related to the differences between the measured resonance frequencies and the true resonance frequencies (whatever they really were) were greater than the errors associated with the ability of the digital frequency meter to accurately read frequencies. But, the errors associated with this decision-making process were minimized because of the graphical technique used to analyze the frequency vs wavelength data. In other words, the random nature of the errors associated with determining the experimental resonance frequencies tended to produce both slightly high and low values. So, if enough data points were taken, and if the same experiments could be repeated often enough, errors in the initial acoustic velocities, determined from line slopes, could be reduced significantly.

13.5. EXPERIMENTAL RESULTS AND DISCUSSION

Some of the derived results of this experimental work have been listed in Table 13.1. The origin of much of this data is explained in more detail in the following paragraphs.

The "Average Initial Velocity" data listed in Table 13.1 were produced directly from three line slopes that were obtained from each of the plotted resonance frequency vs reciprocal wavelength data sets collected for each gas tested. Least squares curve fitted lines obtained from the original experimental data may be seen in Figure 13.2. To improve clarity and readability, only one line for each gas is plotted without showing any of the original data points. Additional information indicating the position of the plotted data points may be seen in our earlier publication [26]. If that article is referred to it will be seen that practically all of the experimental data points seem to lie almost exactly on the curve fitted lines. And, all repeated sets of data points lie almost exactly on top of each other for each of the tested gases. There are two reasons for this behavior. One reason for this is that there was very good (but not exact) repeatability between corresponding experimental sets of data, and the other is that the scales chosen

Table 13.1. Acoustic velocities derived directly from experimental data[a] [26].

Specific gas	Average initial velocity (m/sec)[b]	Initial corrected velocity at 25.0 C (m/sec)[c]	Final corrected velocity at 25.0 C (m/sec)[d]	Final corrected velocity at 0.0 C (m/sec)[e]
Air	345.7	345.9	346.5	331.6
Argon	321.0	320.8	321.5	307.7
Carbon dioxide	267.5	268.4	268.7	257.2
Helium	1008.4	1009.8	1013.9	970.5
Hydrogen	1309.9	1317.6	1321.5	1264.9
Methane	447.0	449.0	449.5	430.2
Neon	451.1	451.5	452.6	433.2
Nitrogen	349.1	350.9	351.3	336.3
Nitrous oxide	264.8	266.4	266.6	255.2
Oxygen	327.1	329.0	329.5	315.4

[a]Sample means only; more extensive data are given in an earlier reference [26].

[b]Average initial velocities from slopes of three sets of plotted resonance frequencies vs 1.0/wavelength for each gas. See text and other references for methods of correcting the experimental data.

[c]Obtained directly from average initial velocity/line slope data.

[d]Obtained directly from initial corrected velocity at 25.0 C.

[e]Obtained directly from final corrected velocity at 25.0 C.

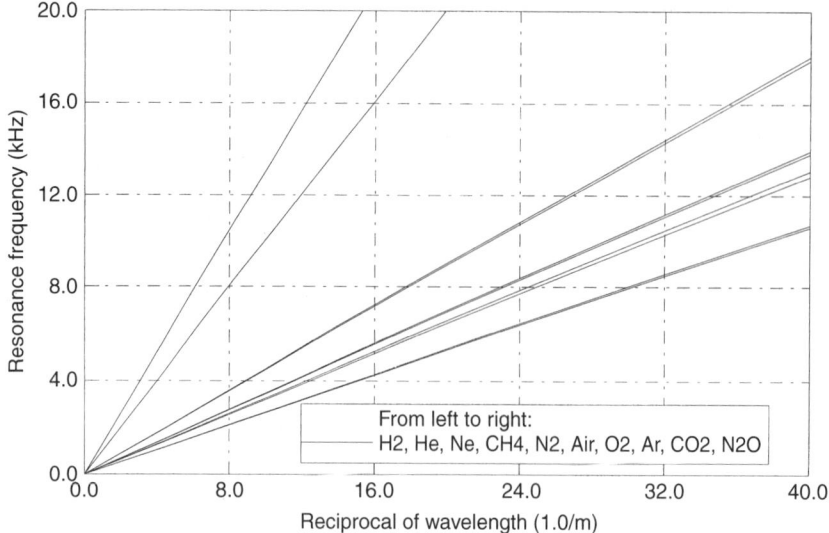

Fig. 13.2. Acoustic resonance behavior in common gases.

to allow for plotting all of the original data on a single graph was too coarse to allow for distinguishability among corresponding data points within each data set.

In addition to random errors in the resonance frequencies, there were some systematic errors in the measured resonance frequencies. These errors were not in the frequency readings themselves, due to a faulty frequency meter, but systematic errors in determining the exact position of resonance. Evidence of this behavior became apparent during the least squares curve fitting procedure employed to find the slopes (i.e., the acoustic velocities) of the best straight lines for all of the original experimental data plotted (e.g., see Figs. 2 and 4 in [26]) for all of the gases tested during these experimental studies.

During our curve fitting procedure, intercept values were not forced to zero. And, instead of random positive and negative variations in intercept values near the origin of the graphs, all averaged intercepts (generally, three per gas tested) were negative. Furthermore, the averaged negative intercept values seemed to be directly proportional to the acoustic velocity in the particular gas tested (the reader may wish to note here that light gases, such as helium and hydrogen, tend to promote high sound speeds and they also have high thermal conductivities). These effects have been more fully documented and described in the earlier publication noted previously.

In any case, ignoring the slightly nonzero intercept values and finding the line slopes by fitting the data collected during each separate trial run to $Y = mX + b$, resulted in corrected acoustic velocities that were in good agreement with many of the previously reported experimental values. This fact indicates that each of the straight lines originally plotted were uniformly shifted in the negative direction. Therefore, if the line slopes had been forced to zero, there would have been poorer agreement between the acoustic velocities determined during this work and those determined in earlier experimental studies unrelated to this work.

As noted previously, three independent acoustic velocities (i.e., one for each set of the three trial runs per gas) were initially determined for each gas tested at whatever the true gas temperature was during the data collection period. These are the "Average Initial Velocity" values listed in Table 13.1. To convert all velocity values to a common temperature of 25.0 C, the average initial acoustic velocities were multiplied by $(298.2/T)^{1/2}$, where T was the average absolute temperature (in K) of the gas during its actual trial run period. The acoustic velocities produced by this procedure have been called "Initial Corrected Velocity at 25.0 C" in Table 13.1. Often, sound speeds in gases are tabulated at 0.0 C instead of 25.0 C, but the procedure above was used because the acoustic velocities adjusted to the temperature of 25.0 C had to be further corrected due to the effects of viscosity and thermal conductivity within the resonance tube [5]. And, the parameters required to make these corrections are usually tabulated at (or near) 25.0 C. So, fewer (or less drastic) changes in these data were required prior to making corrections in the calculated velocities.

Several methods of correcting acoustic velocities (initially measured within cylindrical tubes) for gas viscosity and thermal conductivity have been described in detail in other publications (see references indicated previously). Generally, these methods require the use of the heat capacity ratio of the tested gas, the heat capacity of the gas at constant pressure, the ratio of the gas thermal conductivity and viscosity, the gas viscosity, the gas density, the resonance frequency employed to excite the gas, and the radius of the cylindrical tube used in making the experimental measurements. All of these parameters could be readily measured (e.g., the resonance tube inside diameter [2.804 cm] was directly measured), calculated, or found in suitable references at, or near, 25.0 C [14]. Values listed near, but not at, 25.0 C, were converted to the 25.0 C equivalents. However, the resonance frequencies employed throughout each trial run varied from very low values to relatively high values and a graphical technique (described in detail previously) was employed to find acoustic velocities throughout the entire range of frequencies actually employed. Therefore, the viscosity and thermal conductivity corrections were made using the average resonance frequency employed during each trial run. Employing these data, and one of the correction techniques noted earlier [43], all of the original acoustic velocities, initially converted to 25.0 C, were redetermined and listed in Table 13.1 as "Final Corrected Velocity at 25 C."

The barometric pressures existing during the time of each experimental trial run (corrected for temperature and latitude) were also employed in correcting the original experimental acoustic velocities. As noted previously, due to the very low gas flow rates employed through our cylindrical resonance device during each experimental trial, the barometric pressures that existed within our laboratory were, for all practical purposes, equivalent to the test gas pressures that existed within the cylindrical resonator. A knowledge of these gas pressures is important because the speed of sound in a real gas is a function of pressure [17].

In addition to the procedure noted previously, the corrected velocities at 25.0 C were converted to equivalent values at 0.0 C to compare them with previously determined acoustic velocities typically reported at this temperature [7]. These corrected experimental acoustic velocities, at 0.0 C, have also been listed in Table 13.1 of this chapter and have been called "Final Corrected Velocity at 0.0 C." Additional details regarding these results may be referred to within our earlier article. However, to check the potential accuracy of one possible theoretical method of calculating the speed of sound in gases, some additional theoretical sound speed calculations were made. These theoretical calculations were made using a simplified method of computing the unconfined velocity of sound in any pure gas. The classical equation [82] that applies here is:

$$c = (\sigma RT/M)^{1/2} \tag{13.1}$$

where c is the velocity of sound in the gas, σ is the ratio of the gas's molar heat capacities at constant pressure and constant temperature (i.e., Cp/Cv), R is the universal gas constant, T is the absolute temperature, and M is the gas's molecular weight. The initial parameters, and resulting calculated sound speeds, have all been listed in Table 13.2 for all of the gases originally listed within Table 13.1. The data within the "Reported Temperature" column of this table contains the corresponding temperatures that were listed along with the Cp/Cv values, within the original literature sources of

Table 13.2. Comparison of corrected experimental acoustic velocity values and sound speeds calculated using the ideal gas law, all at 0.0 C and 1.0 atm.

Specific gas[a]	Molar ratio (Cp/Cv)	Reported temperature (deg C)	Molecular mass (g/mol)	Final corrected experimental velocities at 0.0 C (m/sec)	Ideal gas calculated velocities at 0.0 C (m/sec)	Difference (%)
Air[b]	1.403	0.0	28.96	331.6	331.7	−0.03
Argon	1.670	24.2	39.95	307.7	308.1	−0.13
Carbon dioxide	1.292	29.9	44.01	257.2	258.2	−0.39
Helium	1.667	23.1	4.003	970.5	972.5	−0.20
Hydrogen	1.405	24.4	2.016	1264.9	1258.1	0.54
Methane	1.302	25.1	16.04	430.2	429.3	0.20
Neon[b]	1.640	25.0	20.18	433.2	429.6	0.83
Nitrogen	1.403	23.0	28.01	336.3	337.3	−0.29
Nitrous oxide	1.277	25.3	44.01	255.2	256.7	−0.59
Oxygen[b]	1.401	15.0	32.00	315.4	315.3	0.03

[a]Cp/Cv ratio (corrected to 1.0 atm) and temperature data from Zemansky [82], p. 130, unless gas name followed by (b).

[b]Cp/Cv ratio and temperature data from AIP Handbook, pp. 3–71.

this data [30, 82]. Using these temperatures and then correcting the sound speeds to 0.0 C produces the same calculated result that would be found by assuming that the specific heat ratio, at the listed temperature, had the same value at 0.0 C. Since all of the temperatures (corresponding to the listed specific heat ratios) were not too far from 0.0 C, this assumption was probably not very inaccurate.

For comparative purposes, the corrected experimental acoustic velocities at 0.0 C have also been relisted in Table 13.2. In addition, percentage differences between the strictly calculated sound speeds and the experimentally determined sound speeds are also indicated in Table 13.2. It is interesting to see how well the theoretical values compare with the experimentally determined sound speeds. The effect of each gas's compressibility factor (i.e., Z) on the theoretical sound speeds was also checked to see if it would make much of a difference in the calculated values. However, it was found that, at least at 0.0 C and near 1.0 atm, the differences created within the experimentally calculated sound speeds, with and without using compressibility factors, was practically negligible. Probably the main reason for this result is due to the fact that all of the compressibility factors were very near 1.0 and these factors only modified the original results by a factor of Z raised to the 1/2 power. So, all of the correction factors became even closer to 1.0 after taking square roots.

To obtain a better idea of how a more extensive series of separate sets of measurements would affect uncertainties in acoustic velocities determined using the apparatus and techniques described herein, an additional series of seven experimental test trials in dry, carbon dioxide-free air (for a total of 10 complete test trials) were performed. Each of these test trials produced 20 pairs of resonance frequency vs reciprocal wavelength values. Therefore, a total of 200 separate pairs of data points were generated after completion of these 10 test trials. All of the initial acoustic velocities (determined from the slopes of the resonance frequency vs reciprocal wavelength plots) and the corrected acoustic velocities related to these 10 test trials in air have been listed in Table IV our earlier article on this subject. The mean acoustic velocity of the corrected values, converted to their 0.0 C equivalents, was 331.5 m/sec (at nearly 1.0 atm), and the sample standard deviation was 0.4 m/sec. It is interesting to note that, according to Pierce [58], the best current value for the speed of sound in dry air at 0.0 C is also

331.5 m/sec. Other references list values ranging between 331.44 and 331.68 m/sec for the speed of sound in dry, carbon dioxide–free air [29]. In particular, Hardy *et al.* [36], reported an average sound speed (from the results of five earlier studies) of 331.45 m/sec in dry, carbon dioxide–free air. And, Wong [78] also lists similar results from many earlier experimental studies. In any case, the numerical uncertainty in the mean acoustic velocity determined during our experimental studies may be taken as ±0.3 m/sec (95% confidence interval [69]). This result indicates that the apparatus and techniques employed during this work were capable of producing mean experimental acoustic velocities, in pure gases or gas mixtures, with uncertainties on the order of ±0.09% (for 10 or more separate sets of test trials involving the same gas or gas mixture). In some cases, this error is much smaller than the differences that one can find between acoustic velocities reported for the same gas in different sources of the existing literature, e.g., see [6].

It is not possible to apply the same uncertainty estimate, obtained for 10 test trials, to the means of the acoustic velocities determined from only three independent test trials. The uncertainties associated with these means must certainly be higher. A detailed comparison of the mean experimental acoustic velocities were listed in Table III of our earlier article, and the results of even earlier experimental studies indicated that minimum uncertainties in our results were approximately ±0.15%. This error estimate was used to obtain all of the numerical uncertainty values listed in Table III in our earlier article.

An additional experimental acoustic resonance study was also made involving several different gaseous mixtures consisting of helium and argon. These gas mixtures were blended within a high-pressure sampling cylinder, as indicated earlier in this chapter. The actual blended concentrations of each gas were calculated after the blending process, using the ideal gas law, by properly taking into account the partial pressure and temperature of each gas in the mixture. More sophisticated calculations involving compressibility factors for each gas in the mixture were deemed to be unnecessary due to the uncertainty in the exact pressure of each gas in the mixture. The total system pressure of each blended gas mixture was held as near as possible to 69.0 atm (1000 psig), without exceeding this pressure. This constraint was due to the maximum pressure that the blending gauge, selected for this application, was capable of indicating. In total, nine different blended helium/argon gas mixtures were prepared in this way.

At least one day after each gas mixture was prepared, that mixture was allowed to flow through our acoustic resonance device, and three separate experimental sets of measurements of resonance frequencies, as a function of reciprocal wavelengths, were made as in our earlier experiments involving pure gases and air. After each of the three sets of new experimental measurements were made using one of the gas mixtures, the remaining portion of that mixture was discarded. The same cylinder was then evacuated and partially filled with helium (or argon, depending on the particular gas mixture to be prepared next) and reevacuated. This process was repeated at least three times to ensure complete removal of the previous gas mixture before preparation of the next one.

In addition to preparing and testing the acoustic resonance behavior of the gas mixtures described previously, samples of pure helium and pure argon were also placed within the same sampling cylinder used to store the mixtures. Then, using this sampling cylinder as a source of pure helium, and then pure argon, the acoustic resonance behavior of these pure gases was measured using the same procedures used to test the gas mixtures. This process was employed to ensure that all of the experimental results (involving the pure helium, pure argon, and helium/argon mixtures), to be compared later, would have been collected under nearly identical conditions.

The actual gas mixtures that were obtained as a result of our blending process are listed in Table 13.3. In addition, many of the pertinent experimental measurements

Table 13.3. Experimental and derived acoustic velocity data for mixtures of helium and argon.

Mole fraction helium	Initial gas temperature (deg C)	Initial gas pressure (mm Hg)	Average initial velocity (m/sec)^a	Initial corrected velocity at 20.0 C (m/sec)^b	Final corrected velocity at 20.0 C (m/sec)^b	Final corrected velocity at 0.0 C (m/sec)^b
0.0000	23.1	742.9	319.6	317.9	318.6	307.5
	23.2	742.9	319.5	317.8	318.4	307.4
	23.2	742.7	319.8	318.1	318.7	307.6
			Sample mean	317.9	318.6	307.5
0.1002	22.8	749.1	334.4	332.8	333.5	321.9
	22.9	749.1	335.0	333.3	334.0	322.4
	23.0	749.1	335.0	333.3	334.0	322.4
			Sample mean	333.1	333.8	322.2
0.2032	21.9	740.9	351.5	350.4	351.1	338.9
	22.0	740.9	352.1	350.9	351.6	339.4
	22.2	741.3	352.6	351.2	352.0	339.8
			Sample mean	350.8	351.6	339.4
0.3015	22.8	752.8	373.3	371.6	372.4	359.5
	22.7	752.8	373.8	372.1	372.9	359.9
	22.5	752.8	373.0	371.4	372.2	359.3
			Sample mean	371.7	372.5	359.6
0.4016	22.4	739.6	399.8	398.2	399.2	385.3
	22.6	739.6	399.8	398.1	399.0	385.2
	22.5	739.6	399.4	397.7	398.7	384.8
			Sample mean	398.0	399.0	385.1
0.5012	22.1	743.8	431.7	430.2	431.3	416.3
	22.3	743.8	432.2	430.5	431.7	416.7
	22.4	743.8	432.0	430.2	431.3	416.4
			Sample mean	430.3	431.4	416.5
0.6053	21.3	746.0	475.0	473.9	475.3	458.8
	21.4	746.0	474.8	473.7	475.1	458.6
	21.5	746.0	475.4	474.2	475.6	459.1
			Sample mean	473.9	475.3	458.8
0.6980	23.0	743.7	526.6	523.9	525.5	507.3
	22.9	743.7	528.2	525.6	527.2	508.9
	23.0	743.7	528.1	525.4	527.0	508.7
			Sample mean	525.0	526.6	508.3
0.7956	22.3	738.1	605.0	602.6	604.8	583.8
	22.4	738.1	605.0	602.6	604.7	583.7
	22.6	738.3	606.4	603.7	605.9	584.9
			Sample mean	603.0	605.1	584.1
0.8993	22.7	749.6	741.8	738.4	741.6	715.8
	22.8	749.6	741.7	738.2	741.3	715.6
	22.8	749.6	742.8	739.3	742.4	716.7
			Sample mean	738.6	741.8	716.0
1.0000	26.1	744.3	1010.2	999.8	1003.8	968.9
	26.7	743.7	1012.2	1000.8	1004.8	969.9
	26.9	744.6	1015.2	1003.4	1007.4	972.4
			Sample mean	1001.3	1005.3	970.4

^{a}From slope of resonance frequency vs 1.0/wavelength.

^{b}See text and earlier reference for methods of converting and correcting data [26].

and results that were obtained from our study involving these mixtures (as well as our repeated measurements involving the pure helium and argon gas samples) are also listed in Table 13.3. Our methods of data reduction and analysis of the original experimental data collected during this part of our study have already been discussed, in part, in this chapter and in our earlier publication [26].

To further exemplify relationships between the tabulated experimental results and the original experimental data, the resonance frequencies, as a function of reciprocal wavelength, for the helium/argon gas mixtures have been plotted in Figure 13.3. Again, for the sake of clarity, some of the original lines and actual data points have been deliberately omitted from this graph. The actual plotted lines are averages from the three independent sets of measurements per gas mixture tested. If each of the three individual lines (corresponding to the three separate test trials per gas mixture tested) had been plotted, it would have been difficult to distinguish between them. In addition, most of the actual experimental data points, not plotted, appeared to lie exactly on the plotted lines. Corresponding data points collected from the repeated test trials also appeared to lie exactly on top of each other.

An additional comparison of experimental and theoretical sound speeds in helium/argon gas mixtures has been made in Figure 13.4. In that graph, experimental sound speeds, as a function of gas composition in mole fraction units, have been plotted together with theoretically calculated sound speeds, all at 20.0 C. In the case of these gas mixtures, the reference temperature of 20.0 C was used because experimental data, involving the viscosity of helium/argon gas mixtures at 20.0 C [27, 41], were employed to correct the initially determined sound speed values.

In each case involving the helium/argon gas mixtures, theoretical sound speeds were obtained by assuming that the ratio of molar heat capacities (at constant pressure and constant temperature, i.e., Cp/Cv) of the mixed gases was exactly 5/3 and that the average molecular weight of the mixture was equal to the sum of the products of the mole fraction of each binary gaseous component multiplied by the molecular weight of that component. Equation 13.1 for calculating the speed of sound in pure gases was then employed (at $T = 293.2$ K) to calculate the speed of sound in the gas mixture. The same procedure was followed for pure helium and pure argon, except that the molecular weights of the pure gases were employed in these calculations. In referring

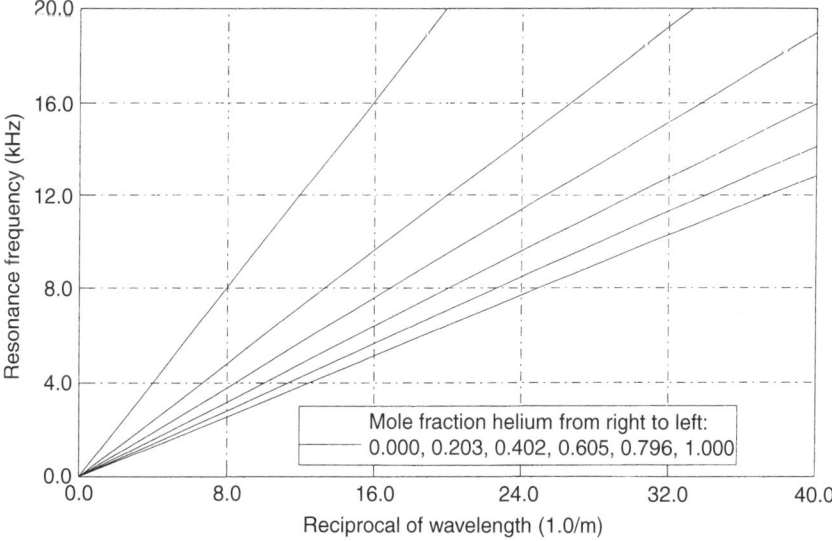

Fig. 13.3. Acoustic resonance in helium/argon mixtures (averages near 20.0 C).

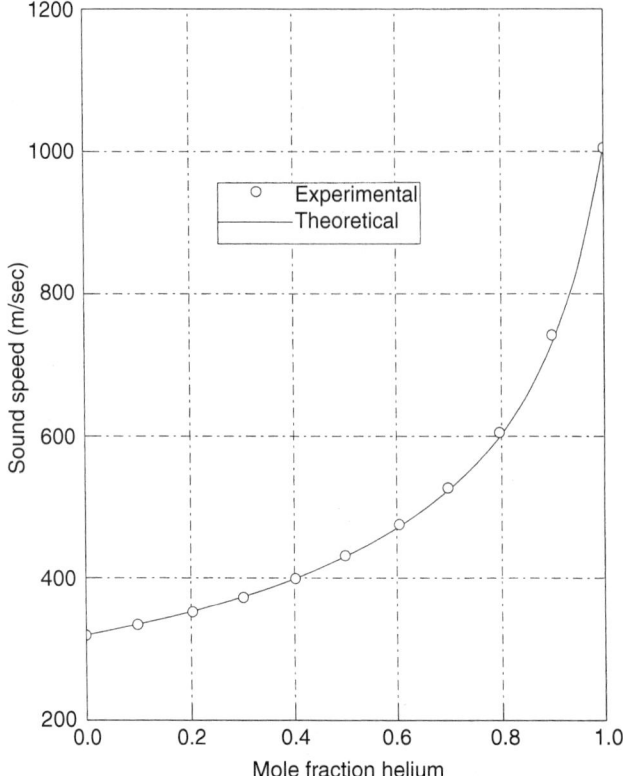

Fig. 13.4. Experimental vs theoretical sound speeds in helium/argon gas mixtures at 20.0 C.

to Figure 13.4, the agreement between the experimental and theoretical results seems very good, but not perfect. Experimental uncertainties in the true gas compositions, due mainly to uncertainties in the blended gas pressures (or other types of more significant operator errors), may be responsible for the discrepancies that are most noticeable at helium mole fractions ranging from 0.7 to 0.9. It is believed that these discrepancies are not a result of the peculiar viscosity behavior of helium/argon gas mixtures, as a function of gas composition [27, 41], because the viscosity corrections, made in the experimentally measured sound speeds, took that behavior into account.

In any case, these results for the helium/argon gas mixtures are quite interesting because they prove that, at least in the case of one binary mixture of noble gases, theoretically calculated sound speeds compare quite favorably with good experimental sound speed measurements. This kind of agreement may hold with other binary mixtures of noble gases. However, binary gaseous mixtures containing no (or, at most, one) noble gases are probably not likely to exhibit good agreement between theoretical and experimental sound speed determinations unless some rather sophisticated corrections are made in the calculated theoretical values.

13.6. CONCLUSION

The main significance of this chapter is that it describes an interesting experimental/graphical method of rapidly determining relatively accurate absolute acoustic velocities in pure gases and gas mixtures. This work also illustrates that the experimental apparatus required to perform this kind of work is relatively inexpensive to fabricate and not very difficult to operate. Using the apparatus and techniques described herein: mean, unconfined, absolute acoustic velocities could be determined, with uncertainties of approximately ±0.15% from only three separate experimental trials per gas tested. It has also been estimated that uncertainties of approximately ±0.09% (95% confidence interval) could be achieved in corrected experimental acoustic velocities obtained from 10 experimental trials per gas tested. Furthermore, it has been demonstrated that absolute acoustic sound speeds in binary mixtures of noble gases could be determined with good repeatability and in relatively good agreement with theoretically calculated values. Additional laboratory studies in the future (using the apparatus and methods of data reduction/analysis described here), involving other gas mixtures, may produce other worthwhile experimental results. These results may also lead to the development of additional useful information regarding the thermodynamic properties of selected uncommon gas mixtures. And, this goal may be achieved without the need of employing any hardware more sophisticated than that described in this chapter, especially if the expense of building and operating other types of acoustic resonance devices cannot be justified by significant improvements in precision and accuracy.

ACKNOWLEDGMENTS

The assistance of M. Pizzo, in welding and in collecting all of the raw experimental data associated with this experimental study, is gratefully acknowledged. The work of P. Vanacek, in machining the SS and brass hardware needed to fabricate the acoustic resonator employed during this study, is also gratefully acknowledged.

References

1. Ashkenazi, S. and Polturak, E. (1988). *Am. J. Phys.* **56**: 836–839.
2. Ashkenazi-Younglove (1979). See Refs [1, 8, 24, 61, 68, 80].
3. Beranek, L.L. (1988). *Acoustic measurements*, pp. 46, 313–328, Woodbury, NY: Acoustical Society of America.
4. Beranek-Younglove (1979). See Refs [3, 12, 15, 16, 18, 22, 23, 34, 35, 42, 45, 47, 57, 59, 60, 62, 66–68, 71, 80].
5. Beranek-Younglove (1980). See Refs [3, 11, 37, 38, 42, 43, 49, 56, 63–65, 76, 81].
6. Beranek-Sen (1990). See Refs [3, 13, 30, 46, 48, 64].
7. Beranek Vargaftik (1975). See Refs [3, 13, 30, 48, 64, 74].
8. Bretz, M. and Shapiro, M.L. (1989). *Am. J. Phys.* **57**: 129–133.
9. Colgate, S.O., Sona, C.F., Reed, K.R., and Sivaraman, A. (1990). *J. Chem. Eng. Data* **35**: 1–5.
10. Colgate-Moldover (1979). See Refs [9, 19, 20, 51–55].
11. Crandall, I.B. *Theory of vibrating systems and sound*, (1926). pp. 229–241, New York: Van Nostrand.
12. Crouthamel, C.E. and Diehl, H. (1948). *Anal. Chem.* **20**: 515–520.
13. Dean, J.A. ed., (1992). *Lange's handbook of chemistry*, 14th ed., pp. 5, 142–144, New York: McGraw-Hill.
14. Dean-Wark (1988). See Refs [13, 30, 46, 48, 50, 74, 75].
15. El-Hakeem, A.S., Gaggioli, R.A., and Romer, I.C. (1966). *J. Acoust. Soc. Am.* **40**: 1485–1488.
16. El-Hakeem, A.S. (1965). *J. Chem. Phys.* **42**: 3132–3133.
17. El-Hakeem-VanIterbeek (1958). See Refs [16, 18, 32, 39, 40, 67, 72, 73].
18. Ewing, M.B., McGlashan, M.L., and Trusler, J.P.M. (1985). *J. Chem. Thermo.*, **17**: 549–559.
19. Ewing, M.B., Owasu, A.A., and Trusler, J.P.M. (1989). *Physica A* **156**: 899–908.
20. Ewing, M.B. and Trusler, J.P.M. (1989). *J. Chem. Phys.* **90**: 1106–1115.
21. Ewing-VanItterbeek. (1957). See Refs [18, 22, 23, 34, 35, 60, 70, 72].
22. Gammon, B.E. (1976). *J. Chem. Phys.* **64**: 2556–2568.
23. Gammon, B.E. and Douslin, D.R. (1976). *J. Chem. Phys.* **64**: 203–218.
24. Garrett, S.L., Swift, G.W., and Packard, R.E. (1981). *Physical* **107B**: 601–602.
25. Giacobbe, F.W. (1991). *Gas Sep. Purif.* **5**: 16–20.

26. Giacobbe, F.W. (1993). *J. Acoust. Soc. Am.* **94**: 1200–1210.
27. Giacobbe, F.W. (1994). *J. Acoust. Soc. Am.* **96**: 3568–3580.
28. Grabau, M. (1933). *J. Acoust. Soc. Am.* **5**: 1–9.
29. Grabau-Smith (1963). See Refs [28, 30, 33, 36, 68].
30. Gray, D.E. eds. (1972). *American Institute of Physics Handbook*, 3rd ed., pp. 2:34, 3:60, 71–74, 4:147, New York: McGraw-Hill.
31. Gray-Wong (1987). See Refs [30, 77, 79].
32. Greenspan, M. (1956). *J. Acoust. Soc. Am.* **28**: 644–648.
33. Greenspan, M. (1987). *J. Acoust. Soc. Am.* **82**: 370–372.
34. Greenspan, M. and Thompson, M.C. (1953). *J. Acoust. Soc. Am.* **25**: 92–96.
35. Grimsrud, D.T. and Wrentz, J.H. (1967). *Phys. Rev.* **157**: 181–190.
36. Hardy, H.C., Telfair, D., and Pielemeier, W.H. (1942). *J. Acoust. Soc. Am.* **13**: 226–233.
37. Helmholtz, H.L.F. (1954). *Sensations of Tone*, p. 90, New York: Dover.
38. Henry, P.S.H. (1931). *Proc. Phys. Soc. (London)* **43**: 340–362.
39. Herzfeld, K.F. and Litovitz, T.A. (1959). *Absorption and Dispersion of Ultrasonic Waves*, pp. 185–216, New York: Academic Press.
40. Holmes, R., Smith, F.A., and Tempest, W. (1963). *Proc. Phys. Soc. (London)* **81**: 311–319.
41. Iwasaki, H. and Kestin, J. (1963). *Physica* **29**: 1345–1372.
42. Kaye, G.W.C. and Sherratt, G.G. (1933). *Proc. R. Soc. (London) Ser. A* **141**: 123–143.
43. Kinsler, L.E., Frey, A.R., Coppens, A.B., and Sanders, J.V. (1982). *Fundamentals of Acoustics*, 3rd ed., pp. 209–210, New York: Wiley.
44. Kniazuk, M. and Prediger, F.R. (1955). *Inst. Auto* **28**: 1916–1917.
45. Kortbeek, P.J., Muringer, M.J.P., Trappeniers, N.J., and Biswas, S.N. (1985). *Rev. Sci. Instrum.* **56**: 1269–1273.
46. *L'Air Liquide Gas Encyclopedia*, (1976). p. 1098, New York: Elsevier, North-Holland Pub. Co.
47. Lewis, J.L., Nierode, D.E., Gaggioli, R.A., and Romer, I.C. (1969). *J. Acoust. Soc. Am.* **45**: 1037–1308.
48. Lide, D.R. (1994). ed., *CRC Handbook of Chemistry and Physics*, 75th ed., pp. 14: 34–35, Boca Raton, FL: CRC Press.
49. Mason, W.P. (1928). *Phys. Rev.* **31**: 283–295.
50. *Matheson Gas Data Book*, 6th ed., (1980). p. 344, Lyndhurst NJ: Matheson Corp.
51. Mehl, J.B. (1982). *J. Acoust. Soc. Am.* **71**: 1109–1113.
52. Mehl, J.B. and Moldover, M.R. (1981). *J. Chem. Phys.* **74**: 4062–4077.
53. Moldover, M.R., Trusler, J.P.M., Edwards, T.J., Mehl, J.B., Davis, R.S. (1988). *J. Res. NBS* **93**: 85–144.
54. Moldover, M.R., Trusler, J.P.M., Edwards, T.J., Mehl, J.B., and Davis, R.S. (1988). *Phys. Rev. Lett.* **60**: 249–252.
55. Moldover, M.R., Waxman, M., and Greespan, M. (1979). *High Temp. High Pres.* **11**: 75–86.
56. Norton, G.A. (1935). *J. Acoust. Soc. Am.* **7**: 16–26.
57. Oberst, H. (1937). *Akus. Z.* **2**: 76–92.
58. Pierce, A.D. (1989). p. 28, *Acoustics*, Woodbury, NY: Acoustical Society of America.
59. Pierce, G.W. (1925). *Proc. Am. Acad. Arts Sci.* **60**: 271–302.
60. Plumb, H. and Cataland, G. (1966). *Metrologia* **2**: 127–139.
61. Polturak, E., Garrett, S.L., and Lipson, S.G., (1986). *Rev. Sci. Instrum.* **57**: 2837–2841.
62. Quinn, T.J., Chandler, T.R.D., and Colclough, A.R. (1974). *Nature* **250**: 218.
63. Raleigh, J.W.S. (1945). *Theory of Sound*, 2nd ed., vol. II, p. 319, New York: Dover.
64. Sen, S.N. (1990). *Acoustics, Waves, and Oscillations*, pp. 131–133, New York: Wiley.
65. Shields, F.D. and Lagemann, R.T. (1957). *J. Acoust. Soc. Am.* **29**: 470–475.
66. Shoemaker, D.P., Garland, C.W., and Nibler, J.W. (1989). *Experiments in Physical Chemistry*, pp. 113–118, New York: McGraw-Hill.
67. Sivaraman, A. and Gammon, B.E. (1986). *Thermophysical Properties of Natural Gas Components Determined by Speed of Sound Measurements*, Rep. No. Niper-142 (NTIS Acc. No. DE86009255/XAB) pp. 1–28.
68. Smith, D.H. and Harlow, R.G. (1963). *Br. J. Appl. Phys.* **14**: 102–106.
69. Taylor, J.K. (1991). *Statistical Techniques for Data Analysis*, pp. 66–68, Chelsea, MI: Lewis Pub. Co.
70. Tempest, W. (1957). PhD Thesis, University of Liverpool.
71. Van Dael, W. and Van Itterbeek, A. (1965). The velocity of sound in dense fluids, in *Physics of High Pressures and the Condensed Phase*, Van Itterbeek, A., ed., pp. 297–357, New York: Wiley.
72. Van Itterbeek, A. (1957). *J. Acoust. Soc. Am.* **29**: 584–587.
73. Van Itterbeek, A. and DeLaet, W. (1958). *Physica* **24**: 59–67.
74. Vargaftik, N.B. (1975). *Handbook of Physical Properties of Liquids and Gases*, 2nd ed., p. 28, Washington, DC: Hemisphere Pub. Corp.
75. Wark, K. (1988). *Thermodynamics*, 5th ed., pp. 837–839, New York: McGraw-Hill.
76. Weston, D.E. (1953). *Proc. Phys. Soc. (London) Ser.* **B66**: 695–709.

77. Wong, G.S.K. and Embleton, T.F.W. (1984). *J. Acoust. Soc. Am.* **76**: 555–559.
78. Wong, G.S.K. (1986). *J. Acoust. Soc. Am.* **79**: 1359–1366.
79. Wong, G.S.K. (1987). *J. Acoust. Soc. Am.* **82**: 373–374.
80. Younglove, B.A. and McCarty, R.D. (1979). *Thermodynamic Properties of Nitrogen Gas Derived from Measurements of Sound Speed*, Rep. No. NASA-RP-1051 (NTIS Acc. No. N80-14257/3) pp. 1–53.
81. Younglove, B.A. and McCarty, R.D. (1980). *J. Chem. Thermo.* **12**: 1121–1128.
82. Zemansky, M.W. (1957). *Heat and Thermodynamics*, 4th ed., pp. 130–134, New York: McGraw-Hill.

CHAPTER 14

SPEED OF SOUND IN PLANETARY ATMOSPHERES

R. D. Lorenz and W. B. Hubbard
Lunar and Planetary Laboratory, University of Arizona, Tucson, Arizona, USA

Contents

14.1. Introduction 389
14.2. Acoustic Measurements on Spacecraft 389
14.3. Instrumental Considerations 391
14.4. Speed of Sound—A Diagnostic of Composition and Transport 392
14.5. Acoustic-Gravity Waves in High Planetary Atmospheres 394
References 397

14.1. INTRODUCTION

The speed of sound in the atmospheres of the other planets can be considerably higher or lower than on Earth. Sound speed is a common proxy for temperature in terrestrial meteorology and has been long proposed for measurement in planetary atmospheres as a diagnostic for composition, temperature, and hydrogen transport. Sound speed is an important environmental parameter, e.g., in determining the flight Mach number of aerospace vehicles. Planetary atmospheres also support, in addition to sound waves, oscillations of much lower frequency, which may play a significant role in the large-scale energetics of tenuous upper layers. Large-scale gravity-acoustic waves have been observed in the upper atmospheres of most planets as well as Saturn's moon Titan.

14.2. ACOUSTIC MEASUREMENTS ON SPACECRAFT

In many terrestrial applications, such as meteorological, oceanographic, or industrial measurements, the composition of the medium (and hence its elastic properties—for ideal gases, its relative molecular mass and specific heat) is either known or assumed, so that the sound speed relates directly to temperature. For planetary applications, although temperature can be constrained by the measurement of sound speed, it is generally more useful to determine temperature separately and thence constrain the medium properties.

Acoustic measurements for spacecraft were proposed as early as 1966 (Hanel and Strange [1]), indeed before *in situ* planetary measurements of any kind had been made.

Their concept was to measure the sound speed in a temperature-controlled spiral duct, thus constraining the relative molecular mass and ratio of specific heats. The application in mind was to resolve the relative abundances of CO_2, N_2, and Ar in the then-unknown Martian atmosphere.

The instrument would also infer the density of the gas from the acoustic impedance (the product of sound speed and density) determined from the signal strength. The duct contained one transmitter and two receivers—the difference in strength for the two identical receivers gave the absorption and hence allowed the determination of the impedance losses in the acoustic circuit. Knowing the density from this measurement, and the pressure from separate pressure measurements, the relative molecular mass could be determined, and thence the ratio of specific heats. The two measurements (speed of sound and acoustic impedance) allow the solution of the two unknowns, making the assumption of an atmosphere with only three major components.

More recently, the first outer planet acoustic measurement has been sent on its way. The Surface Science Package (SSP) on the ESA Huygens probe to Saturn's moon Titan includes (Zarnecki et al. [2]) an acoustic sounder. Titan, unique among planetary satellites, has a thick nitrogen atmosphere. This sounder, acting as a SODAR (Sonic Detection and Ranging), will measure topographic roughness at the landing site prior to impact, in the same way as a radar altimeter. Additionally, it will constrain the depth of any liquid hydrocarbon deposit the probe might land in by sensing an echo from the bottom and possibly detect acoustic backscatter from raindrops during the descent. The package also includes a speed-of-sound instrument to constrain the composition of surface liquid and facilitate the measurement of lake depth. This sensor will also operate in the atmosphere, sampling every 10m in altitude or so—a better altitude resolution than the temperature sensors of the Huygens Atmospheric Structure Instrument (HASI; Fulchignoni et al. [3]). The combination of the HASI temperature measurement and the SSP speed of sound may provide useful information on the altitude variation of the methane mixing ratio (via its effect on the relative molecular mass), which varies from around 8% near the surface to about 2% at the tropopause, e.g., Yelle et al. [4].

The same Atmospheric Structure instrument carries a passive microphone to search for emissions due to possible thunder (Grard [5]), although it has been argued elsewhere that thunder is not likely to be present. This microphone may record noise due to operation of pumps or valves on the probe, as well as possible aeroacoustic noise.

Passive acoustic sensors have been flown on Soviet Venus landers, although with the intent of detecting acoustic emission due to thunder. Although events were detected (Ksanfomality et al. [6]), it is unclear whether these were indeed due to thunder or merely aeroacoustic noise caused by the turbulent airflow over the probe (both sources peak at frequencies of a few hertz). Data from these sensors was used, on the surface, to constrain the surface windspeeds (Ksanfomality et al. [7]).

More recently yet, and with rather more fanfare, a microphone was sent to the Martian surface on the unsuccessful Mars polar lander. This purely passive instrument (Delory et al. [8]) was intended principally for public outreach—simply to record the sounds of another world. The instrument itself was highly miniaturized, being provided minimal mass, power, and data volume allocations on the lander, and used a highly integrated digital signal processor derived from mobile phone technology to perform Fourier analysis and data compression on the recorded signals.

Although instruments on planetary landers to date have used hot-wire anemometry techniques (Viking, Mars Polar Lander) or windsocks (Pathfinder), at least one future mission plans to use an acoustic anemometer: the British Beagle 2 Mars lander will use an ultrasonic resonant device to determine wind speed and direction (Towner et al. [9]).

This compact device, based on a terrestrial anemometer, will measure windspeed over the range 0–40 m/s and direction to 3 degrees.

A possible future application of speed of sound measurements on Mars (Lorenz [10]) is on the polar caps. The Martian surface pressure varies by about 30% over the course of a year, as CO_2 condenses out, forming seasonal ice or frost caps. It is not well understood to what extent condensation is direct (as frost onto the surface) or indirect (as CO_2 snow at higher altitudes that then settles onto the surface). A speed-of-sound measurement at the surface would be a compact diagnostic, since direct condensation as frost would enrich the near-surface atmosphere in trace gases like argon and nitrogen, which would change the sound speed significantly.

In the future, some sort of passive sound recording instrument is likely to be included on most planetary landers, in view of the public interest in such measurements, and the minimal resource requirements (largely an effect of the rapid developments in microelectronics associated with cellular telephones). The Mars microphone mentioned previously had a mass of less than 50 g, fit in a box 2.5-cm square by 0.5-cm thick, and required only 150 mW of power. It could record telephone-quality sound for 10 sec in a data file only 20 kilobytes long.

The present NASA Planetary Exploration Roadmap includes plans for a Jupiter multiprobe mission and a Neptune orbiter (which might include an atmosphere probe). Acoustic instrumentation might be carried on these probes, although, as discussed later, to usefully constrain atmospheric properties other than temperature such as condensation and convection. Neptune and Uranus are more likely prospects for this kind of measurement.

14.3. INSTRUMENTAL CONSIDERATIONS

The threshold altitude at which it is possible to perform speed-of-sound measurements depends on the matching of the transducer to the medium. At high altitudes (i.e., low pressures, thus low density and low acoustic impedance), the transducers are not well matched to the medium, and the achievable signal to perform the measurement is too small.

As an example, when the pressure in a test chamber was reduced to Mars ambient (7.5 mbar), the signal from the microphone used in the Mars Microphone experiment dropped by some 25–28 dB, largely as a result of the poorer matching of the microphone to the atmospheric acoustic impedance.

For speed-of-sound measurements, the absorption due to the atmosphere itself is not significant, although active or passive acoustic remote sensors may need to take into account the spectral filtering by the atmosphere. Here it may be noted that in general the absorptivity or attenuation coefficient of the atmosphere increases strongly with frequency.

As an example, Williams [11] has noted that most sounds (those in the human audible range) will be attenuated in the Martian atmosphere far more than in the terrestrial atmosphere. Whereas attenuation coefficients of 9×10^{-4} and 0.572 dB/m apply on Earth at frequencies of 20 Hz and 20 kHz, respectively, the corresponding values for Mars are 0.419 and 8.98 dB/m.

Three other considerations apply to operating frequency selection. One is the measurement to be made — clearly a higher frequency permits a better time resolution. Second, the transducers to be used for the measurement will have a response function that will have a strong sensitivity variation with frequency. Most transducers are operated in a condition close to resonance, although the resonant frequency may depend on the medium (and hence on the pressure of the atmosphere, which may vary as a probe descends through the atmosphere). Finally, ambient noise is a consideration — a free-falling probe is likely to excite aeroacoustic signals, which will have a peak spectral power at a (typically low) frequency that relates to the descent speed divided by the characteristic dimension of the vehicle. At very high frequencies, thermally produced noise in the atmosphere becomes important.

The acoustic instrument on the Huygens probe (Zarnecki *et al.* [2]) uses a pulse transit time measurement across an approximately 15-cm gap: the measurement is made in both directions between the two transducers to remove any wind effects. Since accurate timing is important, and the signal losses are quite small due to the short transit distance, the speed-of-sound measurement operates at the fairly high frequency of 1 MHz. In contrast, the sounder measurement (where the weak echo from a surface 100 m or more away is received) operates at a lower frequency — 15 kHz — with better transducer performance, lower thermal noise, and lower attenuation, giving better overall signal-to-noise.

14.4. SPEED OF SOUND — A DIAGNOSTIC OF COMPOSITION AND TRANSPORT

The sound speed relates to the compressibility of the medium, $c^2 = (\partial P/\partial \rho)s$. In an ideal gas, this may be expressed as (e.g., Rogers and Mayhew [12]; Hanel and Strange [1])

$$c = \sqrt{\gamma T R_0 / M} \tag{14.1}$$

where γ is the ratio of specific heats, T is the temperature in Kelvin, and R_o is the universal gas constant (8314 J/kg/mol). M is the mean relative molecular mass (RMM) of the gas mixture, the sum of the RMMs of the component gases M_i, weighted by their mole fraction X_i.

In general, variations in temperature are the principal determinant of variations in the speed of sound in planetary atmospheres. This may be seen in Figure 14.1, comparing the speeds of sound in the Venusian and terrestrial atmospheres.

Just like water vapor on Earth, condensable constituents in planetary atmospheres (principally water on Jupiter, ammonia on Saturn, and methane on Titan, Uranus, and Neptune) have their highest concentrations at low, warm altitudes. Higher in the troposphere, their concentration becomes limited by their low vapor pressures at the low local temperature. Above this "cold-trapping" altitude — usually close to the tropopause — the concentration is fairly constant, at the level determined by the cold trap. Since the condensable species affects the relative molecular mass, and the specific heat of the gas mixture, these changes in concentration can influence the speed of sound. Lorenz [13] discusses this influence, which is small on Jupiter and Saturn owing to the small concentration of the condensable species even below the cold trap but significant on Uranus and Neptune. Figure 14.2 shows the expected speed-of-sound profiles for the four giant planets.

On Titan, the speed of sound may be used to constrain the methane mixing ratio. The temperature will be measured directly by thermometers on the HASI instrument, while the bulk composition of the atmosphere will be measured at intervals by a gas chromatograph/mass spectrometer, which will establish the mixing ratio of argon (the only gas likely to be present at the level of a few percent, but which has not yet been measured). Knowing the argon mixing ratio and the temperature, the speed-of-sound measurement will constrain the methane mixing ratio, which is known to be around 1.7% near the tropopause at 40 km but increases in an unknown profile below that altitude. In addition, since the speed-of-sound measurements are made more frequently, and have a negligible time constant, than are the direct temperature measurements, the sound speed data may be used to detect vertical temperature inhomogeneities such as gravity-acoustic waves (see Section 14.5, and Friedson [14]) or convective cells.

SPEED OF SOUND IN PLANETARY ATMOSPHERES

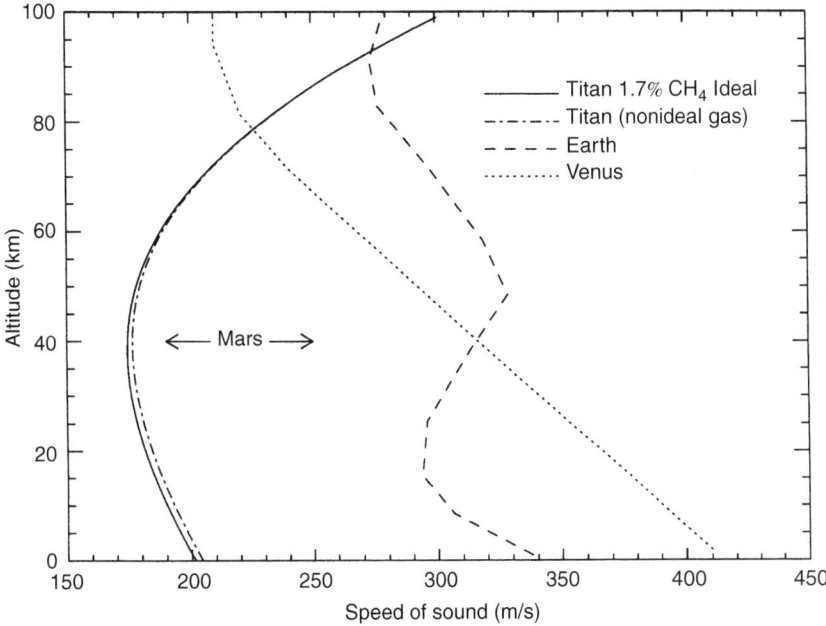

Fig. 14.1. Speed of sound on the terrestrial planets and Titan. The profiles mirror the temperature structure of the planets. The Martian atmospheric temperature and hence sound speed is highly variable due to absorption of sunlight by dust storms so is shown as a range. The profile for Titan is shown with and without a correction for nonideality at low pressures: a variable methane concentration might increase the speed near the surface by a similar amount.

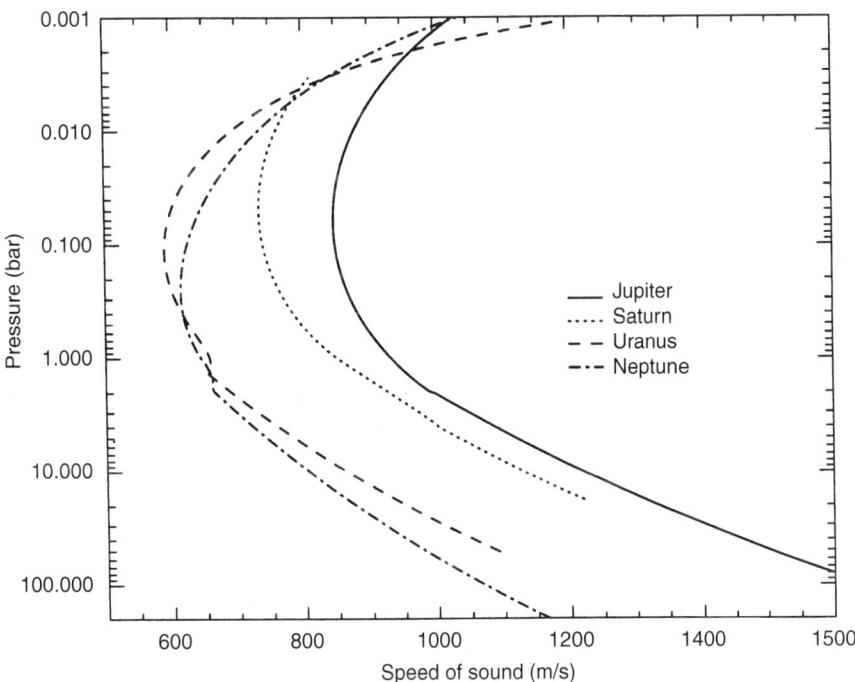

Fig. 14.2. Idealized speed-of-sound profile for the four giant planets using low-order polynomials to describe the temperature structure and assuming equilibrium para hydrogen fraction, and variable condensable abundance. The kinks in the curves are due to the sharp change in mixing ratio of condensable species (parts-per-thousand water on Jupiter and Saturn, and a few percent of ammonia and methane on Uranus and Neptune, respectively).

While the ideal gas law is sufficiently accurate in the hydrogen-rich atmospheres of the outer planets (except in their extreme deep interiors, e.g., Duffy *et al.* [15], but at pressures and temperatures far too high for *in situ* measurement), Titan's atmosphere is so cold that for accuracies better than 5%, the ideal gas equation no longer holds (see Fig. 14.1)—the atmosphere is close enough to condensation that corrections for intermolecular forces must be applied (or equivalently the effective ratio of specific heats changes; indeed, the ratio of specific heats is usually determined from laboratory speed-of-sound measurements, e.g., Younglove and McCarty [16]) Although a virial equation of state or other formulation may be used, Lindal *et al.* [17] suggested that a correction factor F_C to the temperature should be applied, of the form

$$F_C = 1 + A \frac{p}{T^B} \quad (14.2)$$

where p is the local pressure in Pa, T is the temperature in K, and $A = 0.0563$, $B = 2.75$.

An additional factor, beyond the molecular composition, is the spin state of hydrogen molecules (see, e.g., Massie and Hunten [18], and Appendix VIII of Chamberlain and Hunten [19]). The two spin states, ortho and para, have significantly different specific heats at constant pressure (and hence ratios of specific heats) at temperatures between 50 and 350 K: The two states have an equilibrium ratio that depends on temperature: in the high temperature limit, the fraction of hydrogen f_p in the para form is 25%; at temperatures below 100 K (and thus relevant to the tropopause regions in the outer planet atmospheres), the para fraction rises above 50%. Because hydrogen is the dominant constituent, this otherwise obscure physical effect assumes considerable importance and can modify the lapse rate (i.e., the rate of change of temperature with altitude) measurably (Massie and Hunten [19]), and substantially influence convective processes (Conrath and Gierasch [20]; Smith and Gierasch [21]).

In a uniform, steady atmosphere, the para hydrogen would ultimately assume its equilibrium value everywhere—25% in the deep warm interior, closer to 50% near the cold tropopause, and lower again in the warm stratosphere. However, it takes a significant time (hours, if no sites to catalyze the conversion are available) for equilibrium to be attained, so if there are transport processes such as convection that have timescales faster than this, significant disequilibrium concentrations of para hydrogen can occur. As an example, upwelling from the interior would bring 25% para hydrogen from the deep to the tropopause, where 50% might be expected.

If the molecular composition and temperature of the atmosphere can be independently determined, the speed of sound can be used in giant planet atmospheres (particularly that of Neptune) to constrain the para hydrogen fraction, and thus the relative timescales of convective and equilibration processes. The para hydrogen fraction can be measured by other means (e.g., infrared spectroscopy; Conrath and Gierasch [20]) but speed of sound is one of the most convenient *in situ* measurements.

14.5. ACOUSTIC-GRAVITY WAVES IN HIGH PLANETARY ATMOSPHERES

Figure 14.3(a) shows atmospheric pressure-temperature profiles of the four giant planets and Titan; Figure 14.3(b) shows the corresponding altitudes. The broad temperature minimum occurring in all four planets at about 100 mbar mirrors the sound-speed minimum shown in Figure 14.2. The general increase of temperatures with altitude (i.e., with decreasing pressure) above this minimum means that all of these atmospheres are stably stratified and can support gravity waves at pressures below about

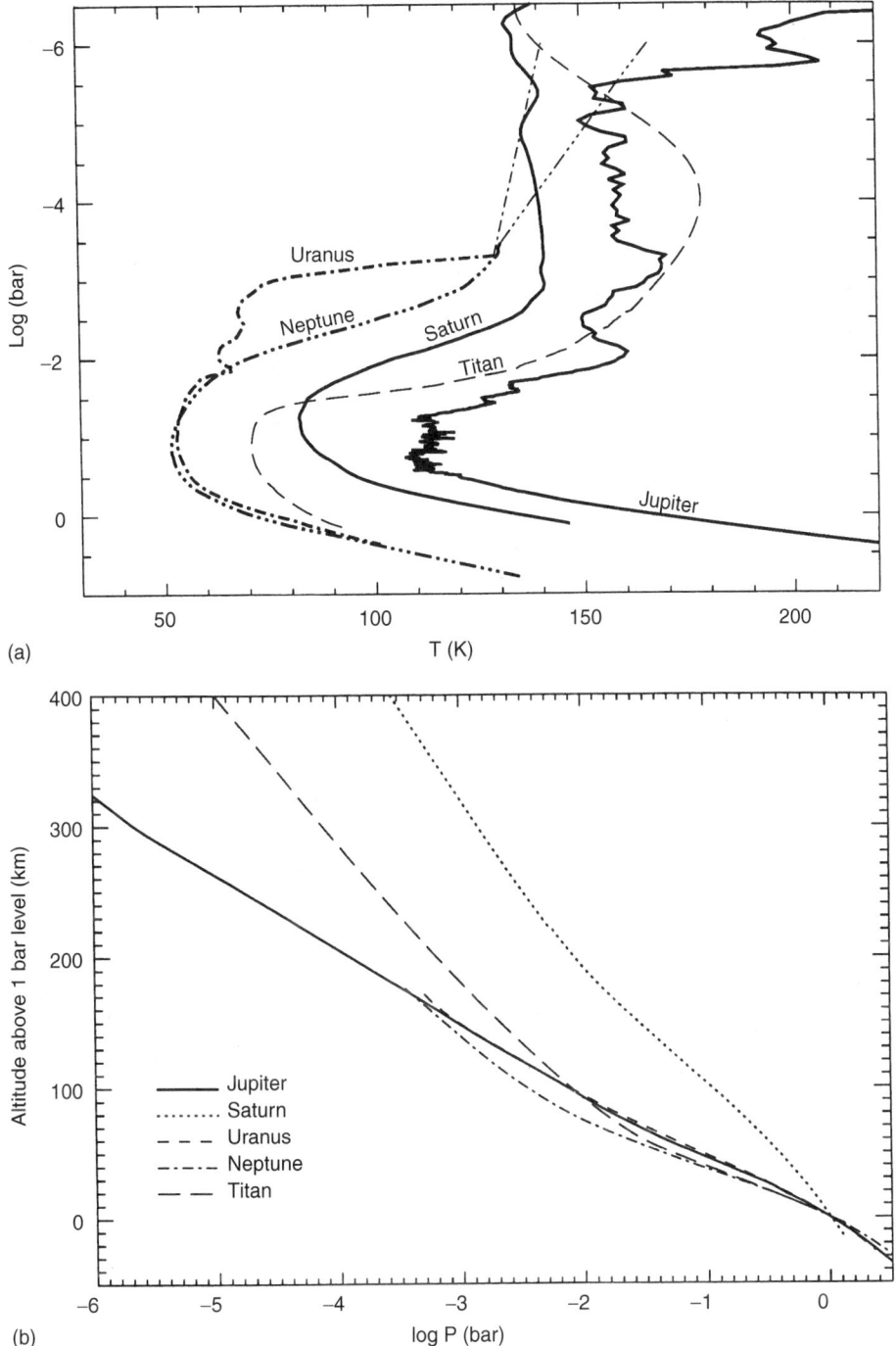

Fig. 14.3. (a) Real pressure-temperature profiles of the four giant planet atmospheres: the higher-order structure in these data would be apparent in sound-speed measurements. The Jupiter profile shows *in situ* data obtained by the Galileo entry probe experiment (Huber [25]); oscillations in temperature at pressures below 100 mbar are probably acoustic-gravity waves. The Saturn profile at pressures above 1 mbar is obtained from radio-occultation data (Lindal *et al.* [26]), while the profile at pressures below 1 mbar is derived from stellar occultation observations (Hubbard *et al.* [27]). Profiles for Uranus and Neptune are obtained from radio-occultation data (Lindal *et al.* [28]; Lindal [29]), for Titan from [Yelle *et al.* [4]]. (b) Pressure-altitude profiles of the giant planets and Titan.

100 mbar. Because such waves involve changes in pressure and density as well as periodic displacements against the force of gravity, they are frequently called acoustic-gravity waves. The spectrum of these waves includes sufficiently low frequencies and long wavelengths that they can be directly detected by remote observations such as occultation of spacecraft radio links and occultations of stars.

If the high planetary atmosphere is taken to be isothermal at temperature T and with constant scale height H and gravitational acceleration g, the dispersion relation for acoustic-gravity waves takes the form (Houghton [22])

$$k_v^2 = k_h^2[(\omega_B^2/\omega^2) - 1] + (\omega^2 - \omega_A^2)/c^2 \tag{14.3}$$

where k_v is the vertical wave number, k_h is the horizontal wave number, ω is the temporal frequency, ω_B is the Brunt-Vaisala or buoyancy frequency, and $\omega_A = c/2H$ is the acoustic cutoff frequency.

Figure 14.4 shows values of the buoyancy period ($= 2\pi/\omega_B$) vs pressure in the high atmospheres of the giant planets. As pressures increase above a few hundred millibars, ω_B^2 approaches zero and then goes negative, implying that gravity waves cannot exist at the higher pressures in the deeper atmosphere.

Our best indication that such oscillations actually exist in high planetary atmospheres comes from observations of scintillations of natural (stars) and artificial (spacecraft radio links) sources as they are occulted by the atmospheres. The occultation geometry is such that the slant path length through the atmosphere is $\sim(rH)^{0.5}$, where r is the planetary radius; the path length comprises several thousand kilometers for a giant planet. As a result, the fluctuations in the occulted signal primarily preserve the effects of density fluctuations with very small values of k_h. According to the dispersion relation, if $k_h \ll 1/H$ and $k_v \sim 1/H$, we have $\omega^2 \ll \omega_B^2$, and the waves are effectively stationary in time for an hour or longer. Figure 14.5 shows an example of strong

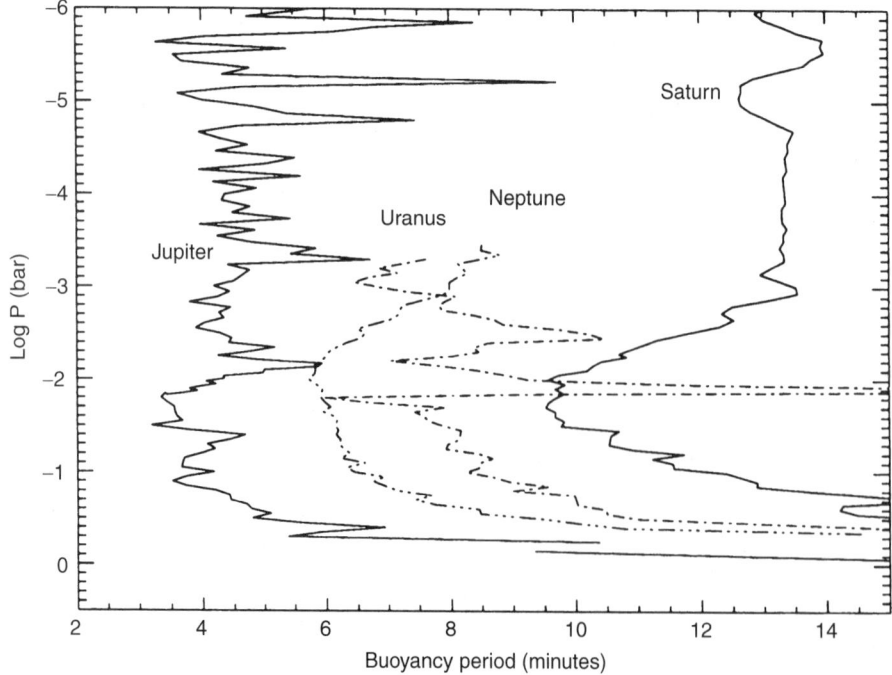

Fig. 14.4. Buoyancy periods in the high atmospheres of the giant planets. Jupiter's periods are shorter than those of the other planets because of higher gravity.

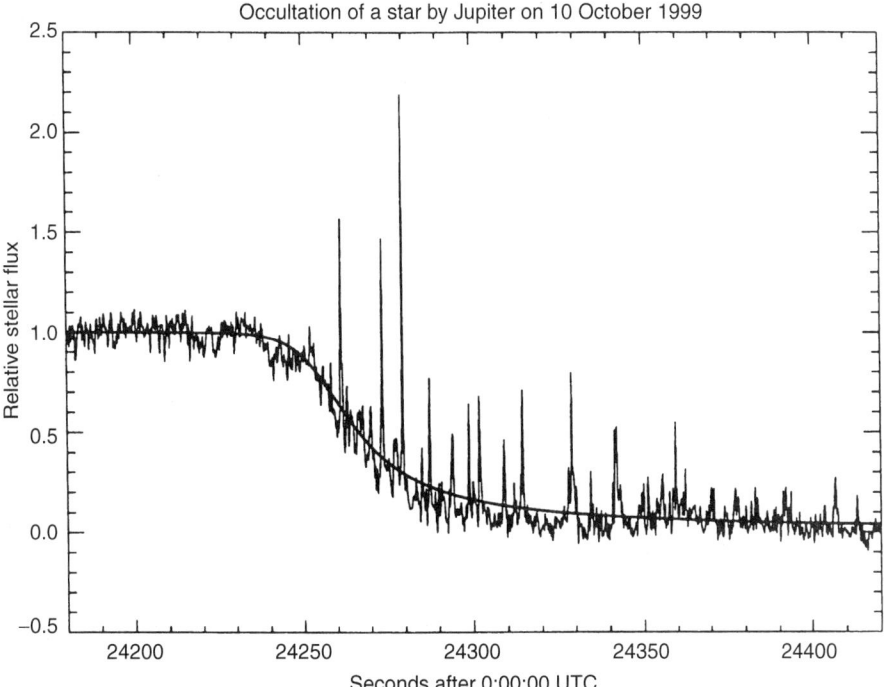

Fig. 14.5. A stellar occultation showing strong scintillations as the stellar image sinks into Jupiter's high atmosphere. The smooth curve shows the predicted stellar signal for an atmosphere with a constant scale height H and no fluctuations due to superimposed acoustic-gravity waves.

scintillations probably caused by acoustic-gravity waves in Jupiter's high north-polar atmosphere at pressures of a few microbars.

The power spectrum of acoustic-gravity waves in high planetary atmospheres is anisotropic, with power-law indices of about -3, corresponding to pancakelike structures with aspect ratios of tens to several hundred (Sicardy *et al.* [23]; Friedson [14]; Narayan and Hubbard [24]).

References

1. Hanel, R.A. and Strange, M.G. (1966). Acoustic experiment to determine the composition of an unknown planetary atmosphere. *J. Acoust. Soc. Am.* **40**: 896–905.
2. Zarnecki, J.C., Banaszkiewicz, M., Bannister, M., Boynton, W.V., Challenor, P., Clark, B., Daniell, P.M., Delderfield, J., English, M.A., Fulchignoni, M., Garry, J.R.C., Geake, J.E., Green, S.F., Hathi, B., Jaroslawski, S., Leese, M.R., Lorenz, R.D., McDonnell, J.A.M., Merryweather-Clarke, N., Mill, C.S., Miller, R.J., Newton, G., Parker, D.J., Svedhem, L.H., Turner, R.F., and Wright, M.J. (1997). The surface science package, in *Huygens: Science, Payload and Mission, ESA SP-1177*, Wilson, A., ed., pp. 177–195, Noordwijk, The Netherlands: European Space Agency.
3. Fulchignoni, M., Angrilli, F., Bianchini, G., Bar-Nun, A., Barucci, M., Borucki, W., Coradini, M., Coustenis, A., Ferri, F., Grard, R.J., Hamelin, M., Harri, A.M., Leppelmeier, G.W., Lopez-Moreno, J., McDonnell, J.A.M., McKay, C., Neubauer, F.M., Pedersen, A., Picardi, G., Pirronello, V., Pirjola, R., Rodrigo, R., Schwingenschuh, C., Seiff, A., Svedhem, H., Thrane, E., Vanzani, V., Visconti, G., and Zarnecki, J. (1997). The Huygens Atmospheric Structure Instrument (HASI), *Huygens: Science, Payload and Mission*, ESA SP-1177, Wilson, A., ed., pp. 163–176. Noordwijk, The Netherlands: European Space Agency.
4. Yelle, R.V., Strobell, D.F., Lellouch, E., and Gautier, D. (1997). Engineering models for Titan's atmosphere, in *Huygens: Science, Payload and Mission*, ESA SP-1177, Wilson A., ed., pp. 253–256, Noordwijk, The Netherlands: European Space Agency.
5. Grard, R.J. (1997). Atmospheric electricity and lightning activity models for titan, in *Huygens: Science, Payload and Mission*, ESA SP-1177, Wilson, A., ed., pp. 257–265, Noordwijk, The Netherlands: European Space Agency.

6. Ksanfomality, L.V., Scarf, F.L., and Taylor, W.L.. (1983a). *The Electrical Activity of the Atmosphere of Venus*, in Venus. Hunten, D.M., *et al*., eds. University of Arizona.
7. Ksanfomality, L.V., Goroshkova, N.V., and Khondryev, V.K. (1983b). Wind velocity near the surface of venus from acoustic measurements, *Cosmic Res.* **21**: 161–167 (translated from *Kosmicheskii Issledovania* **21**: 218–224).
8. Delory, G.T., Luhmann, J.G., Curtis, D.W., Primbsch, J.H., Mozer, F.S., Sircar, I., Friedman, I.D., Linkin, S., and Lipatov, S. (2000). An audio microphone for the '98 Mars Polar Lander. *J. Geophys. Res.*
9. Towner, M.C., Zarnecki, J.C., Leese, M.R., Patel, M.R., Ringrose, T.J., Hathi, B., Pullan, D., and Sims, M.R. (2000). The Beagle 2 Environmental Sensors: Instrument Measurements and Capabilities, *Lunar and Planetary Science Conference XXXI*, Houston, March 2000 (CD ROM Proceedings, paper 1028).
10. Lorenz, R.D. (2000). Compact Polar Instrumentation for Studying In-Situ the Martian CO_2 cycle, *2nd International Conference on Mars Polar Science and Exploration*, Reykjavik, August 2000.
11. Williams, J-P. (2000). The acoustic environment of the martian surface. *J. Geophys. Res.*
12. Rogers, G.F.C. and Mayhew, Y.C. (1967). *Engineering Thermodynamics: Work and Heat Transfer*. London: Longman.
13. Lorem, R.D. (1999). Speed of sound in outer planet atmospheres *Planetary & Space Science*, **47**, 67–77.
14. Friedson, A.J. (1994). Gravity waves in titan's atmosphere. *Icarus* **109**: 40–57.
15. Duffy, T.S., Vos, W.L., Zha, C-S. Hemley, R.J., and Mao, H. (1994). Sound velocities in dense hydrogen and the interior of jupiter. *Science* **263**: 1590–1593.
16. Younglove, B. and McCarty, R.D. (1979). *Thermodynamic Properties of Nitrogen Gas Derived from Measurements of Sound Speed*. NASA Reference Publication 1051 (NBSIR 79-1611).
17. Lindal, G.F., Wood, G.E., Hotz, H.B., Sweetnam, D.N., Eshleman, V.R., and Tyler, G.L. (1983). The atmosphere of Uranus: An analysis of the Voyager 1 radio occultation measurements. *Icarus* **53**: 348–356.
18. Massie, S.T. and Hunten, D.M. (1982). Conversion of para and ortho hydrogen in the jovian planets. *Icarus* **49**: 213–226.
19. Chamberlain, J.W. and Hunten, D.M. (1987). *Theory of Planetary Atmospheres*, 2nd ed., San Diego: Academic Press.
20. Conrath, B.J. and Gierasch, P.J. (1984). Global variation of the para hydrogen fraction in jupiter's atmopshere and implications for dynamics on the outer planets, *Icarus* **57**: 184–204.
21. Smith, M.D. and Gierasch, P.J. (1995). Convection in the outer planet atmospheres including ortho-para hydrogen conversion. *Icarus* **116**: 159–179.
22. Houghton, J.T. (1977). *The Physics of Atmospheres*. Cambridge: Cambridge University Press.
23. Sicardy, B., Ferri, F., Roques, F., Lecacheux, J., Pau, S., Brosch, N., Nevo, Y., Hubbard, W.B., Reitsema, H.J., Blanco, C., Carreira, E., Beisker, W., Bittner, C., Bode, H.-J., Bruns, M., Appleby, G. Forrest, R.W., Nicolson, I.K.M., Hollis, A.J., Miles, R. (1999). The structure of titan's atmosphere from the 28 Sgr occultation. *Icarus* **142**: 357–390.
24. Narayan, R. and Hubbard, W.B. (1988). Theory of anisotropic refractive scintillation: application to stellar occultation by neptune. *Astrophys. J.* **325**: 503–518.
25. Huber, L.F. (2000). Galileo Probe Data Set Archive. Available at: *http://atmos.nmsu.edu/PDS/review/gp_0001/* (Planetary Atmospheres PDS web site).
26. Lindal, G.F., Sweetnam, D.N., and Eshleman, V.R. (1985). The atmosphere of Saturn: An analysis of the Voyager radio occultation measurements. *Astron. J.* **90**: 1136–1146.
27. Hubbard, W.B., Porco, C.C., Hunten, D.M., Rieke, G.H., Rieke, M.J., McCarthy, D.W., Haemmerle, V., Haller, J., McLeod, B., Lebofsky, L.A., Marcialis, R., Holberg, J.B., Landau, R., Carrasco, L., Elias, J., Buie, M.W., Dunham, E.W., Persson, S.E., Boroson, T., West, S., French, R.G., Harrington, J., Elliot, J.L., Forrest, W.J., Pipher, J.L., Stover, R.J., Brahic, A., and Grenier, I. (1997). Structure of Saturn's mesosphere from the 28 Sgr occultations. *Icarus* **130**: 404–425.
28. Lindal, G.F., Lyons, J.R., Sweetnam, D.N., Eshleman, V.R., Hinson, D.P., Tyler, G.L. (1987). The atmosphere of Uranus: Results of radio occultation measurements with Voyager 2. *J. Geophys. Res.* **92**: 14987–15001.
29. Lindal, G.F. (1992). The atmosphere of Neptune: An analysis of radio occultation data acquired with voyager 2. *Astron. J.* **103**: 967–982.

CHAPTER 15

SOUND SPEED IN NORMAL STARS

Sarbani Basu
Department of Astronomy, Yale University, New Haven, Connecticut, USA

Contents

Abstract	399
15.1. Introduction	399
15.2. Determining the Sound-Speed Profiles in Stars	400
15.3. Sound Speed in the Sun	407
15.4. Implications of Helioseismology Results	419
Acknowledgment	420
References	420

ABSTRACT

We discuss how the sound-speed profiles in stars are determined with the help of solar models. We give a brief overview of the observational constraints that the models must satisfy before discussing how the models are constructed and what the sound-speed results look like. We also discuss the special case of the Sun in detail. Helioseismology, or the study of solar oscillations, has provided an almost direct and very precise measurement of the sound speed in the solar interior. We give a brief introduction to helioseismology to discuss how the solar sound-speed profile is determined before going on to discuss what the solar sound-speed profile is.

15.1. INTRODUCTION

Stars can be described as spheres of gas. The temperature near and at the center of the sphere is high enough to initiate nuclear fusion, which releases a tremendous amount of energy. Of interest to the astrophysical community are the sound-speed profiles in the stars, which are related to the compressibility of the material forming the star.

Stars are too far away to be able to measure the sound speed and other properties with currently available techniques — the closest star (other than the Sun), Proxima Centauri, is 4.3 light years (4.1×10^{16} m) away. Except for the Sun, our knowledge of stellar sound-speed profiles comes through modeling stellar structure using basic physical principles and inputs and ensuring that the models satisfy the few observational constraints that we have. In recent years we have been able to measure the sound-speed profile in the Sun throughout the interior using seismic techniques.

In this chapter we discuss how one can estimate the sound speed inside stars. Section 15.2 deals with stars in general. Section 15.3 deals with the special case of the Sun. The implications of the solar results are discussed in Section 15.4.

15.2. DETERMINING THE SOUND-SPEED PROFILES IN STARS

15.2.1. Observational Constraints

Our considerable knowledge of stars is a result of observing stellar luminosities, colors, and spectra. In the case of binary stars, masses can be determined through observations of orbital parameters of the binary system.

The color of a star is an indication of its surface temperature (since it indicates that one part of the spectrum emits more than another), and assuming stars are black-bodies can be used to obtain an effective temperature of the surface of the star. The luminosity of the star can be measured if the distance to the star is known. The luminosity and temperature can be used to determine the radius of the star. This process is fairly accurate. The luminosity can also be used to estimate the mass of the star using the empirical mass–luminosity relationship determined from binary stars. Figure 15.1 shows the empirical mass–luminosity relation for stars on the "main-sequence," i.e., stars that fuse ("burn") hydrogen in their cores; the data were obtained from Popper [1]. In the figure and in all further discussions, the mass, radius, and luminosity of a star are

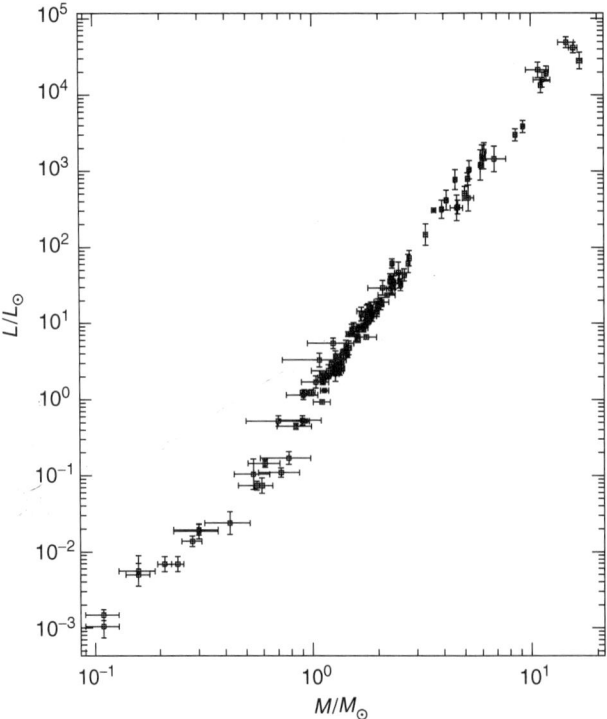

Fig. 15.1. The observed mass–luminosity relation of main-sequence stars (i.e., stars that are young enough to be burning hydrogen in the core). The data were obtained from Popper [1]. The data are for stars in binary systems for which mass can be determined from orbital parameters.

expressed in terms of the solar mass ($M_\odot$), luminosity ($L_\odot$), and radius ($R_\odot$). The mass of the Sun is 1.989×10^{33} g (from [2]), the luminosity is 3.846×10^{33} erg s^{-1} [3], and the radius is approximately 6.96×10^{10} cm [4] (see Antia [5], Schou *et al.* [6], and Brown and Christensen-Dalsgaard [7] for recent discussion about the exact value of the solar radius).

Stellar spectra — lines and continua — tell us about the state of the star's atmosphere. The nature of a star's atmosphere is controlled by the temperature and density of the gases and the acceleration due to gravity (g), which determines the stratification of the photosphere. The energy escaping from the interior of the star, which depends on the mass and age of the star, and to a smaller extent on the composition, also plays a role.

A plot of stellar luminosity against temperature or color of stars is called the Hertzsprung-Russell (HR) diagram. The most remarkable feature of the HR diagram is that stars are not distributed over the diagram at random, exhibiting all combinations of temperature and luminosity. Instead, the stars form a well-defined pattern. The position of a star on the HR diagram is an indication of the evolutionary state of the star. Figure 15.2 shows the HR diagram for main-sequence stars. As stars get older, they move along this diagonal till they exhaust hydrogen in the core. Once hydrogen is exhausted and helium combustion begins in the core, the stars move to the right of the main sequence and the star is said to have moved off the main sequence. At the very least, the model of a star of a given mass should fall in the appropriate region of the HR diagram and satisfy the mass–luminosity constraint.

In recent years, another type of observation has been gaining in importance, and that is the normal modes of oscillations of stars. In the case of the Sun, where observations are easy to make, the study of solar oscillation ("helioseismology") has enabled very precise determinations of solar structure, and is discussed in Section 15.3. The field is in its infancy for other stars where the main problem is observational difficulty

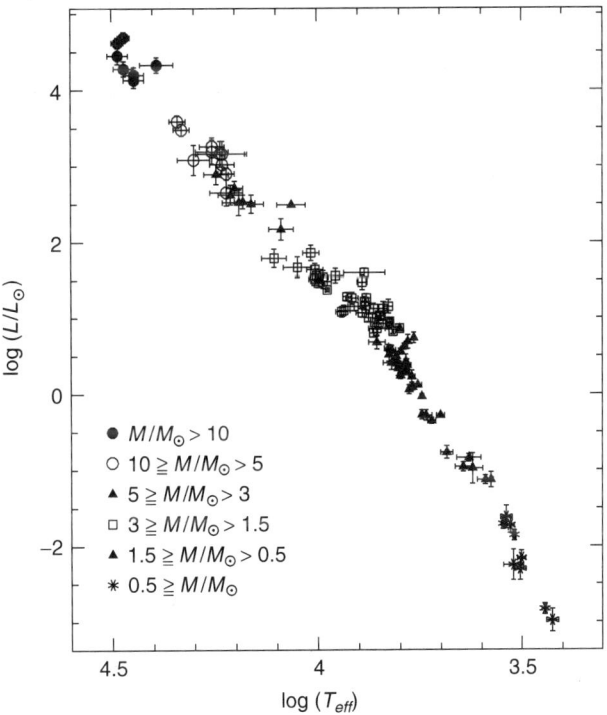

Fig. 15.2. The Hertzsprung-Russell diagram for the main-sequence stars shown in Figure 15.1. Note that as per convention, temperature decreases along the abscissa. The different symbols denote different mass ranges of the stars.

because of turbulence in the earth's atmosphere. Pulsations of specific kinds of stars like cepheid variables, etc., have been observed for some time now, but these have only one or two modes. Although these data do constrain some properties of these stars, they are not enough to determine the sound speed directly. White dwarfs are known to oscillate in many modes, and a number of these stars have been studied.

15.2.2. Equations Describing Stellar Structure

The equations governing stellar structure have been discussed in detail by Cox and Giuli [8], Schwarzschild [9], Chandrasekhar [10], and Clayton [11]. More recent developments are discussed by Hansen and Kawaler [12] and Kippenhahn and Weigert [13]. We give a quick overview.

The first assumption involved in making stellar models is that stars are spherical. This implies that rotation and magnetic fields do not unduly change stellar structure. Spherical symmetry implies that the structure of a star can be expressed as a function of just one variable, r, the radial distance from the center. Thus, if m is the mass enclosed within a sphere of a given radius r, then

$$\frac{dm}{dr} = 4\pi r^2 \rho \tag{15.1}$$

where ρ is the density. Since stars expand or contract over their lifetime, it is generally easier to use the Lagrangian form of the equations, i.e., with mass m as the independent variable. Thus, Eq. 15.1 can be rewritten as

$$\frac{dr}{dm} = \frac{1}{4\pi r^2 \rho} \tag{15.2}$$

The second assumption is that of hydrostatic equilibrium, i.e., that pressure balances gravity. This holds true during all the long-lived phases of a star's life. The equation of hydrostatic equilibrium is

$$\frac{dp}{dr} = -\frac{Gm}{r^2} \rho \tag{15.3}$$

where p is the pressure and G is the gravitational constant. In the Lagrangian form,

$$\frac{dp}{dm} = -\frac{Gm}{4\pi r^4} \tag{15.4}$$

Since stars are not just passive spheres of gas but produce energy through nuclear reactions at the core, an energy equation needs to be considered. In a stationary state energy l flows through a shell of radius r per unit time as a result of nuclear reactions in the interior. If ϵ is the energy released per unit mass per second by nuclear reactions, then

$$\frac{dl}{dm} = \epsilon \tag{15.5}$$

The last, and perhaps most difficult to derive, is the equation of energy transport, which determines the temperature at any point. In general terms, and with the help of Eq. 15.4, the equation can be written quite trivially as

$$\frac{dT}{dm} = -\frac{GmT}{4\pi r^4 p} \nabla \tag{15.6}$$

where ∇ is the dimensionless "temperature gradient" $d \ln T / d \ln p$. The difficulty lies in determining what ∇ is.

If energy transport is through radiation, then one can show that in the limit of diffusive radiative transfer (i.e., when the mean free path of the photons is smaller than the characteristic length scales in the star)

$$\nabla = \nabla_{\text{rad}} = \frac{3}{16\pi a \tilde{c} G} \frac{\kappa l\, p}{m T^4} \qquad (15.7)$$

where, a is the usual radiation-density constant, $\tilde{c}$ the speed of light, and κ the opacity. The opacity is an external input to stellar models. The calculation of opacity is generally extremely complex and modern calculations are available as tables rather than in any analytic form (see, e.g., tables from the OPAL project—Iglesias and Rogers [14], the low temperature opacities from Alexander and Ferguson [15] and Kurucz [16]). Generally, opacity is expressed as a function of temperature, density, and composition. Equation 15.7 does not hold near the stellar surface where the diffusion approximation breaks down and the condition of local thermal equilibrium does not apply. In those regions one needs to solve the full equations of radiative transfer. Instead of doing the calculations themselves, modelers often use the relationship between temperature and optical depth, τ, (the "T-τ" relationship) obtained by detailed calculations done by others (e.g., the Krishna-Swamy T-τ relationship [17]).

If energy transport is by convection, i.e., through the exchange of macroscopic mass elements ("convective elements," "eddies") one has to use the convective temperature gradient. Deep inside the star this is usually the adiabatic temperature gradient $\nabla_{\text{ad}} \equiv (d \ln T / d \ln p)_s$, ($s$ being the specific entropy), which is determined by the equation of state. In the outer layers one needs to use some treatment of convection, e.g., the mixing-length formulation, or the formulation of Canuto and Mazzitelli [18] to calculate the convective flux. It should be noted that there is no "theory" of convection as such, and one has to use the approximate formalisms mentioned. The mixing-length theory was first proposed by Prandtl [19]. His model of convection was analogous to heat transfer by particles; the transporting particles are the macroscopic eddies, and their mean free path is the "mixing length." This was applied to stars by Biermann [20], Vitense [21], and Böhm-Vitense [22]. Different mixing-length formalisms have different assumptions about what the mixing length is. The mixing length is generally parameterized by a dimensionless parameter α, the mixing-length parameter. The lack of a proper theory has led people to try numerical simulations to study convection (e.g., Stein and Nordlund [23]). Some of these are producing general and interesting results about mixing lengths in the Sun and other stars (see e.g., Ludwig et al. [24]; Trampedach et al. [25]).

Whether a portion of a star transports energy by radiation or convection depends on the relative magnitudes of the radiative temperature gradient ∇_{rad} and the adiabatic temperature gradient ∇_{ad}. If ∇_{rad} required to transport energy by radiation is less than ∇_{ad}, then energy is transferred by radiation, otherwise a part of the energy is transported by convection. This condition is often called the "Schwarzschild criterion" for the onset of convection. Figure 15.3 shows the different temperature gradients in two stellar models.

Equations 15.2, 15.4, 15.5, and 15.6 (along with the equations to actually determine the temperature gradient) form the basic equations of stellar structure. However, they cannot be solved without external inputs. As seen from these equations, there are four equations for five unknowns, r, ρ, P, T and l. We thus need some prescription to write at least one of them in terms of the others. This prescription is given by the equation of state, where the density ρ can be written as a function of P, T and the composition, thus $\rho = \rho(P, T, X_i)$, X_i being the mass fraction of species i and denotes the composition of the material inside the star. Like the opacity, the equation of state of stellar material is determined through relatively complex calculations (see Eggleton et al. [26], Däppen et al. [27], and Rogers et al. [28]). The equation of state has to

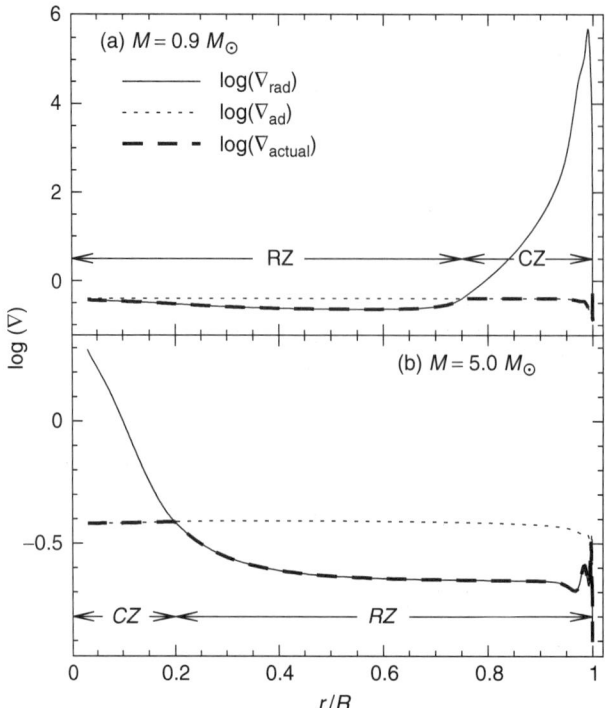

Fig. 15.3. The dimensionless temperature gradient $\nabla \equiv d\ln T / d\ln P$ plotted as a function of fractional radius for two stellar models, one with mass of $0.9\,M_\odot$ and one with mass of $5.0 M_\odot$. The models were constructed with the zero-age main-sequence code of Hansen and Kawaler [12]. In the figure, $\nabla_{\rm ad}$ is the adiabatic temperature gradient, $\nabla_{\rm rad}$ is the radiative temperature gradient. The actual temperature gradient in the models (the one used in Eq. 15.6) is shown by the thick dashed lines. The regions with convective energy transport (the convection zone, or CZ) and radiative energy transport (the radiative zone or RZ) are indicated. In addition to the CZ in the core, the $5M_\odot$ model has a very thin convection zone very close to the surface that is not marked in the figure.

account for processes like ionization and recombination at the correct temperature ranges. Because densities in stellar cores can be very high (150 g cm^{-3} in the center of the Sun and higher for stars of lower mass than the Sun), the equation of state has to deal with possible electron degeneracy, and since central temperatures can be very high in high mass stars, it has to deal with radiation pressure, etc. However, to first approximation, one can make order of magnitude estimates by assuming that stellar matter is a fully ionized, ideal gas.

If the composition profile of a star is known, one can solve for the structure of a star and determine its sound-speed profile. For a star that has just started evolving, this is not much of a problem; since nuclear reactions have barely begun, the composition is homogeneous and it is enough to specify the mix of hydrogen and helium. This stage is called the "zero age main sequence" or ZAMS. (In most usual stellar calculations, all elements heavier than helium are lumped together and referred to as "metals." The hydrogen mass fraction is denoted as X, helium mass fraction as Y, and fraction of metals as Z and follow the normalization $X + Y + Z = 1$.) The structure of a ZAMS star is completely determined by its mass and composition.

There is no *a priori* way to determine the composition profile within a star that has evolved to some extent. Nuclear reactions occurring deep inside the star continually change the composition within. The way the problem is dealt with is to "evolve" the star from a state when it had a homogeneous composition and follow the changes of composition as well. This obviously means that Eqs. 15.2, 15.4, 15.5, and 15.6 need to be modified to take the time variations into account. Equations 15.2 and 15.6

do not change. Equation 15.4 needs to be changed to take into account the possible expansion or contraction of the star. Equation 15.5 has to be changed to account for energy released or absorbed due to contraction or expansion. These additional terms are important only for very specific stages of stellar evolution. In addition, there are equations for the time evolution of the composition. Thus, the equations for stellar structure and *evolution* are

$$\frac{\partial r}{\partial m} = \frac{1}{4\pi r^2 \rho} \tag{15.8}$$

$$\frac{\partial p}{\partial m} = -\frac{Gm}{4\pi r^4} - \frac{1}{4\pi r^2}\frac{\partial^2 r}{\partial t^2} \tag{15.9}$$

$$\frac{\partial l}{\partial m} = \epsilon - c_p \frac{\partial T}{\partial t} + \frac{\delta}{\rho}\frac{\partial p}{\partial t} \tag{15.10}$$

$$\frac{\partial T}{\partial m} = -\frac{GmT}{4\pi r^4 p}\nabla \tag{15.11}$$

and

$$\frac{\partial X_i}{\partial t} = \frac{m_i}{\rho}\left(\sum_j r_{ji} - \sum_k r_{ik}\right), \quad i = 1, \ldots, I \tag{15.12}$$

In Eq. 15.10, c_p is the specific heat at constant pressure and $\delta \equiv -(\partial \ln \rho / \partial \ln T)_p$ are quantities defined by the equation of state. Equation 15.12 is a set of I equations (the normalization $\sum_i X_i = 1$ can replace one of the equations) for change in the mass-fraction X_i of nucleus $i = 1, \ldots, I$. The quantity $r_{ji} = r_{ji}(P, T, X_i)$ is the rate at which species j is destroyed to produce species i. Similarly, r_{ik} is the rate at which i is destroyed producing species k, m_i is the atomic mass of species i. In case there is a convection-zone present, additional equations are required to regulate mixing of elements (see Kippenhahn and Weigert [13], Chapter 8). Also, additional equations are required if one considers diffusion and gravitational settling of helium and heavy elements toward the core (see, e.g., Thoul *et al.* [29] and Proffitt and Michaud [30]).

To get physical solutions to the previous equations we need to impose boundary conditions on the variables. At the center, mass $m = 0$, and two conditions follow trivially, i.e., $r_c = 0$ and $l_c = 0$, the subscript "c" denoting center. Unfortunately, boundary conditions on T_c and p_c are not straightforward, hence the remaining required boundary conditions are applied at the surface.

The boundary conditions at the surface are, if anything, more complex than the central boundary conditions. One requirement is that the mass of a star remains constant over its lifetime. This is not strictly correct. High-mass stars are known to lose a lot of mass through stellar winds. Mass loss in the Sun, however, is quite small — only about $2 \times 10^{-14} M_\odot$ yr^{-1} and does not vary by more than a factor of two over a solar cycle (see Zirker [31]). Thus, mass loss can be ignored in stars of low mass. Though very important for high-mass stars, the process of mass loss and how it affects the structure and evolution of stars is not properly understood. Assuming that the total mass remains constant, an extreme boundary condition would be to assume that as $m \to M$, $p \to 0$, and $T \to 0$, M being the mass of the star. Although these conditions reflect the fact that the pressure and temperature at the surface are many orders of magnitude less than the interior pressure and temperature, they are by no means correct because pressure and temperature have a rather extended transition to interstellar values. In practice, more complicated relations are applied, taking account of the fact that one can define a radius R of the star from where the bulk of radiation is emitted. Details of the boundary conditions and their effect on the structure of the star are discussed by Kippenhahn and Weigert [13], Hansen and Kawaler [12], etc. Models of the Sun must satisfy two extra constraints: the radius should be 1 $R_\odot$ and the luminosity 1 $L_\odot$ at the current age of the Sun, which is about 4.6 Gyr [32]. This implies that two free parameters are needed to satisfy the equations, and these are generally taken to be the initial helium mass-fraction Y and the mixing-length parameter α.

15.2.3. Sound Speed in Stars

Once we know how to build stellar models, finding the sound-speed profile is simple. Figure 15.4 shows the sound-speed profiles in a number of different stellar models plotted as a function of fractional radius. All the models are homogeneous, zero-age main-sequence models constructed with the ZAMS code of Hansen and Kawaler [12] and have a composition of $X = 0.74$ and $Y = 0.24$. Note that the models of the lower-mass stars ($M \leq 1.1\ M_\odot$) have higher sound speeds in the outer layers than the higher-mass stars because the low-mass stars have convective envelopes and convection zones have higher temperature gradients than radiative zones. Higher temperature gradients result in higher temperature, and hence higher sound speeds in the envelope of the low-mass stars.

The structure of stars change as they continue to burn hydrogen in their core. This causes the sound-speed profile to change. Figure 15.5 shows the sound-speed profile in two stars that have envelope convective zones at three different epochs in their evolution. These models have been constructed by Christensen-Dalsgaard [33]. Note that the sound speed in the core decreases sharply with time. This is a result of accumulation of helium, which increases the local mean molecular weight, thereby

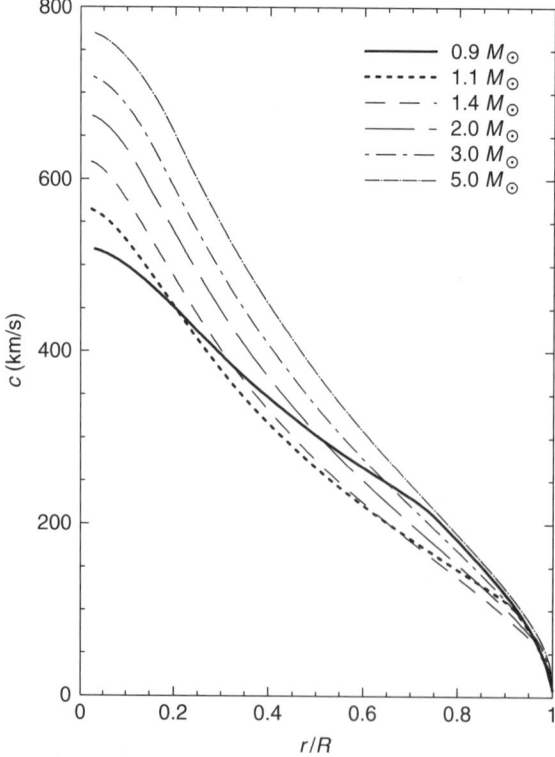

Fig. 15.4. The sound-speed profiles of several zero-age main-sequence stellar models plotted as a function of fractional radius. The thick lines indicate models that have an extended convection zone in the outer layers only. The others are models with convection zones in the core.

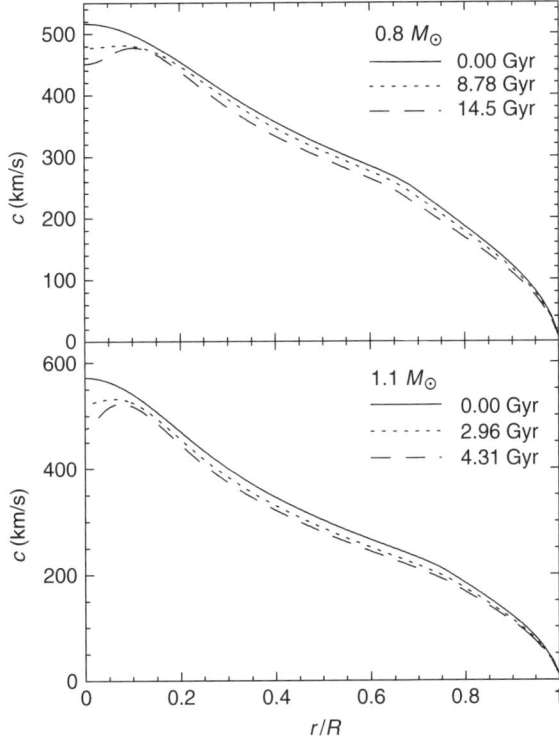

Fig. 15.5. The time variation of the sound-speed profile in two stellar models that have convection zones in the envelope of the star. These models were constructed by Christensen-Dalsgaard [33]. Time $t = 0$ corresponds to the zero age main sequence.

decreasing sound speed (for an ideal gas $c^2 \propto T/\mu$, c being the sound speed, T the temperature, and μ the mean molecular weight).

In Figure 15.6 we show the evolution of the sound-speed profile of two stellar models that have convection zones in their cores. The change in the sound speed in the core with age is quite dramatic. The presence of the convection zone causes the helium produced in the center to be mixed with surrounding material, increasing the mean molecular weight in a larger region than in models shown in Figure 15.5. The mixing in the convective core results in a discontinuous abundance profile, which results in a discontinuous sound-speed profile. The core also shrinks with age, thus introducing a sharp composition gradient even in that part of the radiative region where nuclear reactions do not take place.

15.3. SOUND SPEED IN THE SUN

The Sun is the only star for which the sound-speed profile has been determined observationally. Modern solar models tell us that solar sound speed varies from a few kilometers per second in the outer layers to slightly greater than 500 km/s in the energy-generating core (see Christensen-Dalsgaard *et al.* [34] for a discussion of modern solar models). The models also predict that the energy is transferred by radiation in the inner 70% of the Sun and by convection in the outer 30%. The change between the convection zone and the radiative zone causes a discontinuity in the second derivative of the temperature and hence of the sound speed. The sound speed increases inward till about $0.07 R_\odot$. Below this, the sound speed decreases despite increasing temperatures. This decrease is caused by the accumulation of helium as a result of hydrogen fusion.

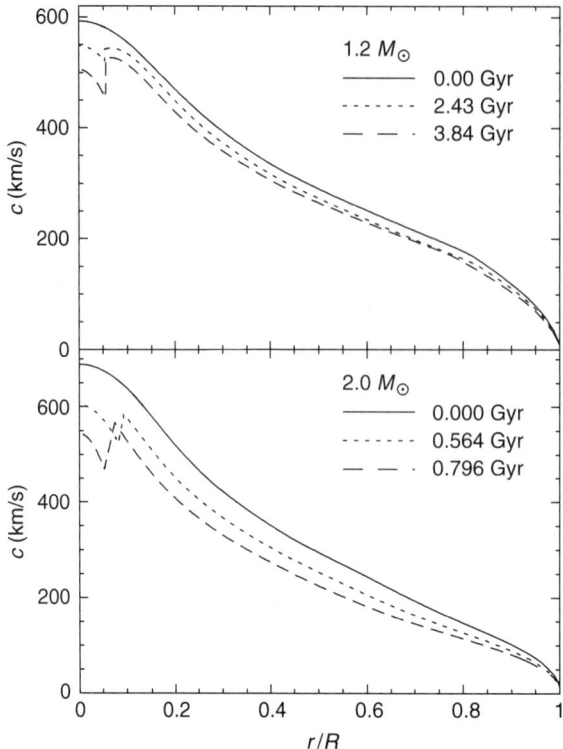

Fig. 15.6. The time variation of the sound-speed profile in two stellar models that have convection zones in the core of the star. These models were constructed by Christensen-Dalsgaard [33].

Increase in the local helium abundance increases the mean molecular weight, causing the sound speed to decrease.

Direct determination of the speed of sound inside the Sun has been made possible by the relatively young field of helioseismology. Helioseismology is the study of solar interior using the frequencies of solar oscillations. The Sun is oscillating simultaneously in many millions of global normal modes of oscillation. These resonant acoustic eigenmodes of the Sun are excited in the subsurface turbulent layers of the Sun. The frequencies of the oscillations can be measured to extremely high precision. A brief history of the subject and a straightforward description of observing techniques can be found in the article by Hill *et al.* [35]. The frequencies of these oscillations depend on the properties of the Sun, viz., structure, rotation, magnetic fields, etc. Hence, if the frequencies are known, these properties can be determined.

There are a number of long-term observational projects to observe the Sun to determine solar oscillation frequencies. Among these are the Global Oscillation network Group (GONG), which is a network of six solar telescopes situated at six different places on the earth to enable almost continuous observation of the Sun [36]. In addition there are smaller ground-based networks, such as the Birmingham Solar Oscillation Network (BiSON [37]), the IRIS network [38], and the Low-ℓ (LOWL [39]) instrument. Among space-based programs are the three helioseismology instruments (GOLF, VIRGO, and MDI) on board the Solar and Heliospheric Observatory (SOHO), a spacecraft that observes the Sun continuously from its vantage point at the L1 Lagrange point between the Sun and the Earth to obtain continuous observations of the Sun [40].

To a good approximation the oscillations can be described as linear and adiabatic. Each mode of solar oscillation has a velocity amplitude of about 10–20 cm/s, which is very small compared with the sound speed as determined from solar models. They are also adiabatic. The period of the oscillations is of the order of 5 min, which is much smaller than the Kelvin-Helmholtz time scale for the Sun ($\simeq 10^7$ years), and hence, heat transfer during each oscillation period can be neglected in most of the solar interior.

Stellar oscillation frequencies are characterized by three numbers, (n, ℓ, m). The quantity n, referred to as the radial order of the mode, is generally the number of nodes in the radial direction. The angular distribution is generally described by spherical harmonics Y_ℓ^m, ℓ the degree of the mode is roughly speaking the number of wavelengths along the circumference, and m, the azimuthal order is the number of nodes along the equator. For a spherically symmetric system the position of the equator is arbitrary and depends on the choice of the orientation of the coordinate system. This implies that all modes with the same n and ℓ have the same frequency regardless of m. The frequency of the (n, ℓ) multiplet is a function of the structure of the star and hence can be used to probe the structure of the star. The presence of rotation, magnetic fields, or other large-scale flows breaks the spherical symmetry of the system and lifts the frequency degeneracy between modes with the same radial order and degree but different m.

The Sun is a slow rotator. The solar equator rotates with a period of about 25 days at the surface, while the rotation period at a latitude of 60 degrees is about 32 days [41, 42]. Thus, the ratio of the centrifugal force to the gravitational force at the equator is only about 2 parts in 10^5 and, hence, to the first order, rotation does not affect the mean frequency of an (n, ℓ) multiplet. The deviation from sphericity in the Sun has been measured to be very small; the relative difference between equatorial and polar radii is about a part in 10^5 [43]. Thus, the mean frequencies of the different (n, ℓ) multiplets can be used to probe the structure of the Sun.

15.3.1. Equations Describing Solar Oscillations

Since the Sun is a gaseous object, we begin with the basic equations of hydrodynamics, i.e., the equations of continuity and motion:

$$\frac{\partial \rho}{\partial t} + \nabla \cdot (\rho \vec{v}) = 0 \tag{15.13}$$

$$\rho \frac{d\vec{v}}{dt} = -\nabla p + \rho \vec{g} \tag{15.14}$$

where,

$$\vec{g} = \nabla \Phi, \quad \text{and,} \quad \nabla^2 \Phi = -4\pi G \rho \tag{15.15}$$

Here, ρ is the density, t is time, $\vec{v}$ is the velocity, p is the pressure, $\vec{g}$ is acceleration due to gravity, and Φ is the gravitational potential. Also needed is the energy equation, which can be written in the form

$$\frac{dq}{dt} = \frac{1}{\rho(\Gamma_3 - 1)} \left(\frac{dp}{dt} - \frac{\Gamma_1 p}{\rho} \frac{d\rho}{dt} \right) \tag{15.16}$$

where

$$\Gamma_1 = \left(\frac{\partial \ln p}{\partial \ln \rho} \right)_{ad} \quad \text{and,} \quad \Gamma_3 - 1 = \left(\frac{\partial \ln T}{\partial \ln \rho} \right)_{ad} \tag{15.17}$$

We assume that the equilibrium structure is static and the oscillations are small perturbations to the equilibrium structure. Thus, all time derivatives of equilibrium quantities vanish. Hence, the continuity equation is satisfied trivially, while the equation of motion reduces to the equation of hydrostatic support, i.e.,

$$\nabla p = \rho \vec{g} = \rho \nabla \Phi \tag{15.18}$$

Note that this is the full three-dimensional form of Eq. 15.3 discussed in Section 15.2.2.

As has been said before, solar oscillations have very small amplitudes, thus we can consider linear perturbations only. Thus, for example, pressure at any time t can be written as

$$p(\vec{r}, t) = p(\vec{r}) + p'(\vec{r}, t) \tag{15.19}$$

where $p(\vec{r})$ is the time-independent equilibrium pressure and $p'(\vec{r}, t)$ is a small Eulerian perturbation. The Lagrangian perturbation on the other hand is

$$\delta p(r, t) = p(\vec{r} + \vec{\delta r}(\vec{r}, t)) - p(\vec{r}) = p'(\vec{r}, t) + \vec{\delta r}(\vec{r}, t) \cdot \nabla p \tag{15.20}$$

where, $\vec{\delta r}$ is the displacement from the equilibrium position. The perturbations to the other quantities can be written in exactly the same way. By definition, the equilibrium state has no velocities — the velocity $\vec{v}$ in the context of the Lagrangian perturbation is simply the time derivative of the displacement. Since we are looking for oscillatory solutions, we search for solutions with a time dependence of the form $\exp(-i\omega t)$, ω being the frequency of the oscillation. Such a choice is possible because the equations do not explicitly depend on t.

Substituting the (Eulerian) perturbed quantities in the continuity equation, and keeping only quantities to the first order in perturbation (and integrating with respect to time), we get

$$\rho' + \nabla \cdot (\rho \vec{\delta r}) = 0 \tag{15.21}$$

The equation of motion gives us

$$\rho \frac{\partial^2 \vec{\delta r}}{\partial t^2} = -\nabla p' + \rho \vec{g}' + \rho' \vec{g} \tag{15.22}$$

where, $\vec{g}' = \nabla \Phi'$, and Φ' is the solution of

$$\nabla^2 \Phi' = -4\pi G \rho' \tag{15.23}$$

For the heat equation on the other hand, it is easier to consider the Lagrangian perturbation because under the assumptions that have been made, the time derivative of the various quantities is simply the time derivative of the Lagrangian perturbation in those quantities. Thus from Eq. 15.16 one obtains

$$\frac{\partial \delta q}{\partial t} = \frac{1}{\rho(\Gamma_3 - 1)} \left(\frac{\partial \delta p}{\partial t} - \frac{\Gamma_1 p}{\rho} \frac{\partial \delta \rho}{\partial t} \right) \tag{15.24}$$

or, in the adiabatic limit where energy loss is negligible,

$$p' + \vec{\delta r} \cdot \nabla p = \frac{\Gamma_1 p}{\rho} (\rho' + \vec{\delta r} \cdot \nabla \rho) \tag{15.25}$$

Since the Sun is a sphere, it is desirable to cast the equations in spherical polar coordinates. Thus, the displacement $\vec{\delta r}$ can be decomposed as

$$\vec{\delta r} = \xi_r \hat{a}_r + \vec{\xi}_t \tag{15.26}$$

where, $\hat{a}_r$ is the unit vector in the radial direction, ξ_r is the radial component of the displacement vector, and ξ_t is the transverse component. An added advantage of using spherical polar coordinates is the fact that tangential gradients of the equilibrium quantities do not exist. Thus, e.g., the heat equation (Eq. 15.25) becomes

$$\rho' = \frac{\rho}{\Gamma_1 p} p' + \rho \xi_r \left(\frac{1}{\Gamma_1 p} \frac{dp}{dr} - \frac{1}{\rho} \frac{d\rho}{dr} \right) \tag{15.27}$$

The tangential component of the equation of motion (Eq. 15.22) is

$$\rho \frac{\partial^2 \vec{\xi}_t}{\partial t^2} = -\nabla_t p' + \rho \nabla_t \Phi' \qquad (15.28)$$

or (taking the tangential divergence of both sides)

$$\rho \frac{\partial^2}{\partial t^2}(\nabla_t \cdot \vec{\xi}_t) = -\nabla_t^2 p' + \rho \nabla_t^2 \Phi' \qquad (15.29)$$

The continuity equation (Eq. 15.21) after decomposition can be used to eliminate the term $\nabla_t \cdot \vec{\xi}_t$ from Eq. 15.29 to obtain

$$-\frac{\partial^2}{\partial t^2}\left[\rho' + \frac{1}{r^2}\frac{\partial}{\partial r}(\rho r^2 \xi_r)\right] = -\nabla_t^2 p' + \rho \nabla_t^2 \Phi' \qquad (15.30)$$

The radial component of the equation of motion gives

$$\rho \frac{\partial^2 \xi_r}{\partial t^2} = -\frac{\partial p'}{\partial r} - \rho' g + \rho \frac{\partial \Phi'}{\partial r} \qquad (15.31)$$

where gravity acts in the negative r direction. Finally, the Poisson's equation becomes

$$\frac{1}{r^2}\frac{\partial}{\partial r}\left(r^2 \frac{\partial \Phi'}{\partial r}\right) + \nabla_t^2 \Phi' = -4\pi G \rho' \qquad (15.32)$$

Analysis of these equations shows that the angular dependence of the perturbed variables can be written in terms of spherical harmonics (see, e.g., Christensen-Dalsgaard [44]). Thus,

$$\xi_r(r, \theta, \phi, t) \equiv \xi_r(r) Y_\ell^m(\theta, \phi) \exp(-i\omega t) \qquad (15.33)$$

$$p'(r, \theta, \phi, t) \equiv p'(r) Y_\ell^m(\theta, \phi) \exp(-i\omega t) \qquad (15.34)$$

and so on.

Once the description of the variables in terms of spherical harmonics is substituted in Eqs. 15.30, 15.31, and 15.32, Eq. 15.27 can be used to eliminate the quantity ρ' to obtain

$$\frac{d\xi_r}{dr} = -\left(\frac{2}{r} + \frac{1}{\Gamma_1 p}\frac{dp}{dr}\right)\xi_r + \frac{1}{\rho c^2}\left(\frac{S_\ell^2}{\omega^2} - 1\right)p' - \frac{\ell(\ell+1)}{\omega^2 r^2}\Phi' \qquad (15.35)$$

from Eq. 15.30, where $c^2 = \Gamma_1 p/\rho$ is the squared sound speed and S_ℓ^2 is the Lamb frequency defined by

$$S_\ell^2 = \frac{\ell(\ell+1)c^2}{r^2} = k_t^2 c^2 \qquad (15.36)$$

Equation 15.31 and the equation of hydrostatic equilibrium give

$$\frac{dp'}{dr} = \rho(\omega^2 - N^2)\xi_r + \frac{1}{\Gamma_1 p}\frac{dp}{dr}p' + \rho \frac{d\Phi'}{dr} \qquad (15.37)$$

where, N is the *buoyancy frequency* defined as

$$N^2 = g\left(\frac{1}{\Gamma_1 p}\frac{dp}{dr} - \frac{1}{\rho}\frac{d\rho}{dr}\right) \qquad (15.38)$$

This is the frequency with which a small element of fluid will oscillate when it is disturbed from its equilibrium position. When $N^2 < 0$, the fluid is unstable to convection, and in such regions part of the energy will be transported by convection. And

Eq. 15.32 becomes

$$\frac{1}{r^2}\frac{d}{dr}\left(r^2\frac{d\Phi'}{dr}\right) = -4\pi G\left(\frac{p'}{c^2} + \frac{\rho\xi_r}{g}N^2\right) + \frac{\ell(\ell+1)}{r^2}\Phi' \quad (15.39)$$

Equations 15.30, 15.37, and 15.39 need to be solved only to find the radial dependence and the frequency. The transverse component of displacement vector can be written in terms of $p'(r)$ and $\Phi'(r)$ and one can show that

$$\vec{\xi}_t(r,\theta,\phi,t) = \xi_t(r)\left(\frac{\partial Y_\ell^m}{\partial \theta}\hat{a}_\theta + \frac{1}{\sin\theta}\frac{\partial Y_\ell^m}{\partial \phi}\hat{a}_\phi\right)\exp(-i\omega t) \quad (15.40)$$

where $\hat{a}_\theta$ and $\hat{a}_\phi$ are the unit vectors in the θ and ϕ directions, respectively, and

$$\xi_t(r) = \frac{1}{r\omega^2}\left(\frac{1}{\rho}p'(r) - \Phi'(r)\right) \quad (15.41)$$

15.3.2. Properties of Solar Oscillations

Once the equations describing solar oscillations are derived one can go ahead and study the properties of such oscillations. Detailed derivations and analyses of the properties of solar (and stellar) oscillations can be found in Unno et al. [45], Cox [46], Christensen-Dalsgaard and Berthomieu [47], and Gough [48]. We shall try to find some of the broad properties, particularly those that enable us to use solar oscillations to study the solar interior.

To study the main properties of solar oscillations, we use the so-called Cowling approximation, where the perturbation to the gravitation field Φ' is ignored. From a spherical-harmonic expansion of Φ' one can see that it is small in two situations, first when ℓ is large, i.e., when the perturbations are restricted to the outer layers where ρ (and hence ρ') is small. The second is when radial order n is large. In this case, ρ' changes sign rapidly and the positive and negative contributions from ρ' cancel each other, leaving a very small Φ'.

When Φ' is neglected, Eqs. 15.35–15.39 reduce to a system of two first-order equations,

$$\frac{d\xi_r}{dr} = -\left(\frac{2}{r} - \frac{1}{\Gamma_1}H_p^{-1}\right)\xi_r + \frac{1}{\rho c^2}\left(\frac{S_\ell^2}{\omega^2} - 1\right)p' \quad (15.42)$$

$$\frac{dp'}{dr} = \rho(\omega^2 - N^2)\xi_r - \frac{1}{\Gamma_1}H_p^{-1}p' \quad (15.43)$$

where $H_p = -dr/d\ln p$ is the pressure scale height. For high-order oscillations, the eigenfunctions vary much more rapidly than the coefficients in the equations. Thus, the left-hand side of Eq. 15.42 is much larger than the first term of the right-hand side. Hence, in the crudest approximation, we can neglect the first term in Eq. 15.42 and the second term in Eq. 15.43 and combine the two resulting equations to obtain a wave equation of the form

$$\frac{d^2\xi}{dr^2} = \frac{\omega^2}{c^2}\left(1 - \frac{N^2}{\omega^2}\right)\left(\frac{S_\ell^2}{\omega^2} - 1\right)\xi_r \quad (15.44)$$

This is the simplest possible approximation to equations of nonradial oscillations but is enough to illustrate some of the key properties. One can see immediately that the equation does not always have an oscillatory solution. The solution is oscillatory when (1) $\omega^2 < S_\ell^2$, and $\omega^2 < N^2$, or (2) $\omega^2 > S_\ell^2$, and $\omega^2 > N^2$. The solution is exponential otherwise.

In Fig. 15.7 we show N^2 and S_ℓ^2 plotted as a function of depth for a standard solar model. The figure shows that modes for which the first condition is true are trapped mainly in the core (since N^2 is negative in the convection zone and the sun has an envelope convection zone). These are the *g-modes* since the restoring force is gravity through buoyancy. No g-modes have been observed so far in the case of the Sun. Modes that satisfy the second condition are oscillatory in the outer regions, though low-degree modes can penetrate right to the center. These are the *p-modes* since the restoring force is predominantly pressure. Note that *p*-modes of different degrees penetrate to different depths within the Sun. High-degree modes penetrate to shallower depths than low-degree modes. For a given degree, modes of higher frequency penetrate to deeper depths. Thus, modes of different degrees sample different layers of the Sun. Hence, if we have frequencies of modes with a wide range in degrees, we could probe the structure of the Sun as a function of radius.

15.3.3 Inverting Oscillation Frequencies

We now concentrate on putting the equations in a form that will allow us to derive the structure of the Sun from the frequencies. We start from the perturbed form of the equation of motion, i.e., Eq. 15.22. In the previous discussion we mentioned that the displacement vector can be written as $\vec{\delta r}(\vec{r})\exp(-i\omega t)$. Substituting this in the equation, we get

$$-\omega^2 \rho \vec{\delta r} = -\nabla p' + \rho \vec{g}' + \vec{g}\rho' \qquad (15.45)$$

Substituting for ρ' from the continuity equation (Eq. 15.21) and for p' from Eq. 15.27, we get

$$-\omega^2 \rho \vec{\delta r} = \nabla(c^2 \rho \nabla \cdot \vec{\delta r} + \nabla p \cdot \vec{\delta r}) - \vec{g}\nabla \cdot (\rho \vec{\delta r}) - G\rho \nabla \left(\int_V \frac{\nabla \cdot (\rho \vec{\delta r})dV}{|\vec{r}-\vec{r}'|} \right) \qquad (15.46)$$

In Eq. 15.46, ω is the observed quantity, and we would like to find c^2 and ρ (and hence p assuming hydrostatic equilibrium). However, $\vec{\delta r}$ is not observed and therefore

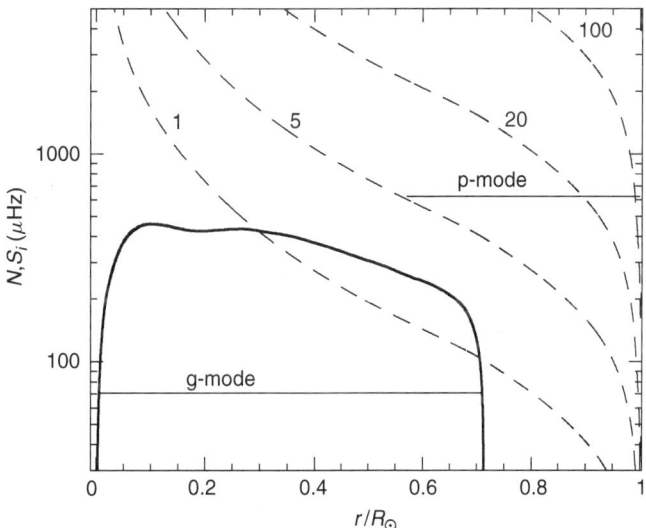

Fig. 15.7. The buoyancy frequency N (Eq. 15.38) shown by the continuous line, and the characteristic acoustic frequency S_ℓ (Eq. 15.36) shown by the dashed lines labeled with the value of ℓ, for a standard solar model plotted as a function of radius. This figure indicates the depth to which different modes of different degrees penetrate. The horizontal lines indicate the trapping region of a g-mode and of a degree 5 p-mode.

the equations cannot be inverted directly. The way out of this is to recognize that Eq. 15.46 defines an eigenvalue problem of the form

$$\mathcal{L}\vec{\delta r} = -\omega^2 \vec{\delta r} \qquad (15.47)$$

$\mathcal{L}$ being a differential operator. Under specific boundary conditions, namely, $\rho(R_\odot) = p(R_\odot) = 0$, the eigenvalue problem defined by Eq. 15.46 is Hermitian [49]. Thus, the variational principle can be used to linearize Eq. 15.46 around a known solar model to obtain

$$\frac{\delta\omega^2}{\omega^2} = -\frac{\int_V \rho \vec{\delta r}^* \cdot \delta\mathcal{L}\vec{\delta r}\, dV}{\int_V \rho \vec{\delta r}^* \cdot \vec{\delta r}\, dV} \qquad (15.48)$$

where $\delta\omega$ is the difference in frequency between the known solar model (or the "reference model") and the Sun, $\delta\mathcal{L}$ contains information about the difference between the reference model and the Sun, and $\vec{\delta r}$ is the displacement eigenfunction for the known solar model and thus can be calculated. One such equation can be written for each mode of oscillation, and the set of equations can be used to calculate the difference in structure between the solar model and the Sun, and thus determine the structure of the Sun.

The denominator of the right-hand side of Eq. 15.48 is often called the mode inertia, I, since it can be shown that the time-averaged kinetic energy of a mode is proportional to $\omega^2 I$. Equation 15.48 implies that for a given difference in structure, the resulting differences in frequencies of modes with a high inertia are less than those of modes with lower inertia. For modes of a given frequency, lower-degree (i.e., deeply penetrating) modes have higher mode inertias than higher-degree (i.e., shallow) modes.

There is an additional complication that comes in the way of inverting Eq. 15.48, and that is we really do not know how to model the layers just below the solar surface properly. Equation 15.48 implies that we can invert the solar frequencies provided we know how to model the Sun. However, the approximations that are used to calculate the convective flux (e.g., the mixing-length formalisms of various kinds) are not good enough to model the solar surface properly. Also the adiabatic approximation certainly breaks down in the outer layers. This implies that the right-hand side of Eq. 15.48 does not account for the frequency difference $\delta\omega/\omega$. Fortunately, we know the rough form of the frequency differences due to errors in surface layers of the model. For modes that are not of very high degree ($\ell \simeq 200$ and lower), the structure of the wavefront near the surface is almost independent of the degree, the wave-vector being almost completely radial. This implies that any errors in frequency due to errors in the surface structure have to be a function of frequency alone once the mode inertia has been taken into account. It can also be shown (e.g., Gough [50]) that surface perturbations cause the error in frequency to be a slowly varying function of frequency, which can be modeled as a sum of low-degree polynomials. This effect is shown in Fig. 15.8, which shows the scaled frequency difference between two models, which differ only near the surface due to differences in their convection formalisms. The scaling factor $Q_{n\ell}$ is given by

$$Q_{nl} = \frac{I_{n\ell}}{\bar{I}_0(\omega_{n,\ell})} \qquad (15.49)$$

here, $I_{n\ell}$ is the inertia of a mode of degree ℓ and order n, and $\bar{I}_0(\omega_{n,\ell})$ is the inertia of a mode of degree 0 interpolated to have the same frequency as the mode of degree ℓ and order n. This ratio takes into account the fact that high-degree modes that do not penetrate deep inside the Sun are perturbed more easily than low-degree modes of the same frequency.

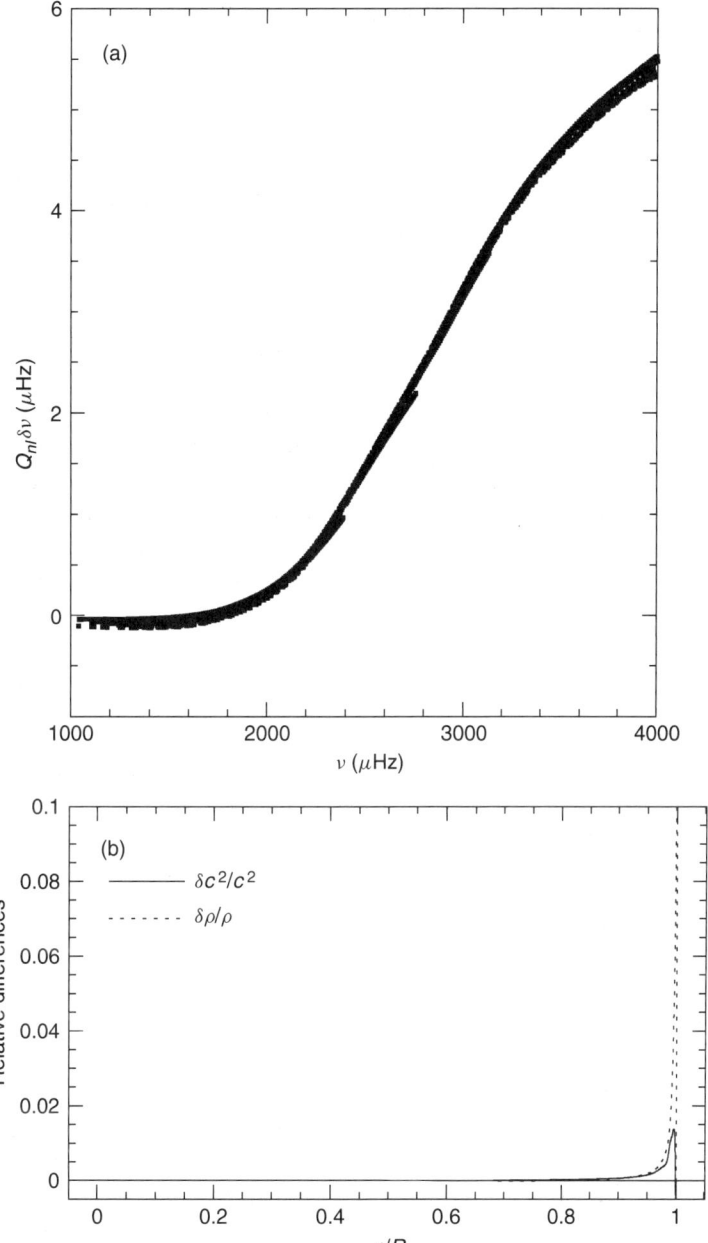

Fig. 15.8. (a) The scaled frequency differences between two solar models, one constructed with the standard mixing-length formalism for convection, the other with the Canuto and Mazzitelli [18] formalism. Frequency differences for p-modes of degree 0 to 200 are shown in the figure. (b) The relative difference in squared sound speed and density between the two solar models. Note that the differences between the two solar models are concentrated in the near-surface layers.

The modified Eq. 15.48 can be rewritten to isolate the difference between the model and the Sun to obtain

$$\frac{\delta \omega_i}{\omega_i} = \int K^i_{c^2,\rho}(r) \frac{\delta c^2}{c^2}(r) dr + \int K^i_{\rho,c^2}(r) \frac{\delta \rho}{\rho}(r) dr + \frac{F(\omega_i)}{I_i} \qquad (15.50)$$

where the index i denotes any mode. The terms $\delta c^2/c^2$ and $\delta \rho/\rho$ are the relative squared sound speed and relative density differences between the reference model and the Sun, the kernels $K_{c^2,\rho}$ and K_{ρ,c^2} are known functions of the reference model, and $F(\omega)$ is the function of frequency that accounts for near-surface errors. Kernels for a few modes are shown in Fig. 15.9. Note that kernels for lower-degree modes penetrate deeper in radius than kernels of higher-degree modes. Details of the derivations of the kernels can be found in Antia and Basu [51].

15.3.4. Inversion Techniques

There are two complementary methods of using Eq. 15.50 to determine $\delta c^2/c^2$ and $\delta\rho/\rho$: the regularized least squares (RLS) [51, 52] and the optimally localized averages (OLA) [53–55]. In the former, the three unknown functions are parameterized on a set of basis functions like splines, and the spline-coefficients are found by fitting the given data under the constraint that the solution is smooth. The latter involves finding a linear combination of the kernels such that the combination is localized in space. The linear combination is called the averaging kernel. The solution obtained is then an average of the true solution weighted by the averaging kernel. Some details of inversion theory can be found in the article by Gough and Thompson [56]. We give a brief discussion of the OLA method as used in helioseismic inversions since this is the method used to obtain the solar sound-speed results discussed in Section 15.3.5.

Usually a variant of the OLA [57] method called "Subtractive" Optimally Localized Averages (SOLA) [58] is used. The aim here is to construct explicitly well-localized resolution kernels (generally called the *averaging* kernel) $\mathcal{K}$ such that

$$<f> = \int \mathcal{K} f(r) dr \qquad (15.51)$$

represents the average of the quantity f over a sufficiently narrow range in r. Thus, if

$$\mathcal{K} = \sum_i c_i K^i_{c^2,\rho}, \quad \text{then,} \quad \left\langle \frac{\delta c^2}{c^2} \right\rangle = \sum_i c_i \frac{\delta\omega_i}{\omega_i} \qquad (15.52)$$

From Eq. 15.50 we see that this is possible only if $\int \mathcal{K} dr = 1$, and if $\mathcal{C} = \sum_i c_i K^i_{\rho,c^2}$ and $\mathcal{F} = \sum_i c_i F(\omega_i)$ are small. Let $\mathcal{T}(r, r_0)$ be a sufficiently localized function positioned at r_0 that we would like the averaging kernels to resemble, then the coefficients

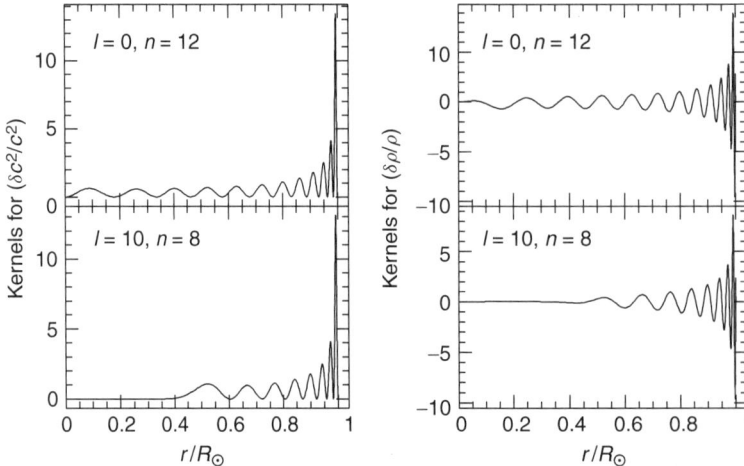

Fig. 15.9. Kernels for relative sound-speed difference and relative density difference for two modes. Note that the kernels for the lower-degree mode penetrate deeper than the kernel for the higher-degree mode.

c_i for the position r_0 can be determined by minimizing

$$\int \left(\sum_i c_i K^i_{c^2,\rho}(r, r_0) - \mathcal{T}(r, r_0) \right)^2 dr + \beta \int \left(\sum_i c_i K^i_{\rho,c^2}(r, r_0) \right)^2 dr + \mu \sum_{i,j} c_i c_j E_{ij} \quad (15.53)$$

subject to constraints given in Eqs. 15.55 and 15.56. In the previous equation, E_{ij} is the error-covariance matrix of the observations, β is a tradeoff parameter designed to minimize $\mathcal{C}$, and μ is a tradeoff parameter that controls the propagated error in the solution. Normally, $\mathcal{T}$ is chosen to be a Gaussian centered at r_0 and is referred to as the target kernel. The minimization is done under the constraint that the resulting averaging kernel is unimodular, i.e.,

$$\int \sum_i c_i K^i_{c^2,\rho} dr = 1 \quad (15.54)$$

To suppress the surface term, we add additional constraints:

$$\sum_i c_i \frac{\psi_j(\omega_i)}{I_i} = 0, \quad j = 1, \ldots, m \quad (15.55)$$

where $\psi_j(\omega_i)$ are basis functions, like for instance Legendre polynomials, such that $F(\omega) = \sum_{j=1}^{m} a_j \psi_j(\omega)$.

15.3.5. Results With Solar Data

We use the solar oscillation frequencies as determined from data obtained by the Michelson Doppler Interferometer (MDI) on board SOHO during the first year of its mission [59]. Figure 15.10 shows a plot of the modes as a ℓ-ν diagram. Note the amazing precision of the data.

The "reference solar model" used in the inversions is Model S of Christensen-Dalsgaard et al. [34]. This model is used because many helioseismological results in literature are based on this reference model. Studies have shown that changing the reference model has negligible effect on the solar sound speeds obtained from the inversion results as long as the solar model is reasonable [60].

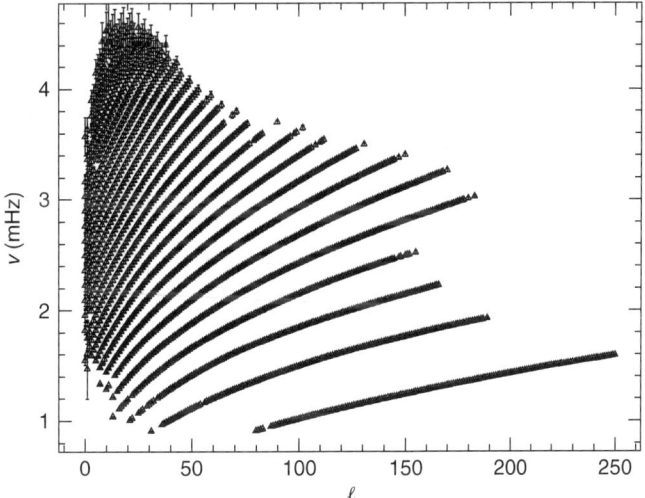

Fig. 15.10. Oscillation frequencies of the Sun plotted as a function of the degree of the modes. 1000σ error-bars are shown. The ridges correspond to different values of the order n of the modes, starting from $n = 0$ at the bottom.

Model S was constructed with the Livermore (OPAL) equation of state [28]. For temperatures higher than 10^4 K, an early version of the OPAL opacities was used [61]; opacities from the tables of Kurucz [16] were used at lower temperatures. The model incorporates the diffusion of helium and heavy elements below the convection zone. The ratio between heavy elements and hydrogen abundance at surface is $Z/X = 0.0245$ [62]. The model has an age of 4.6 Gyr without pre-main-sequence. The model has been described in detail by Christensen-Dalsgaard *et al.* [34].

The inversion results are shown in Figure 15.11, which is a figure of the relative difference in the sound speed between the Sun and the reference solar model plotted as a function of radius. The results are remarkable in that the sound speed of the solar model is very close to that of the Sun. The maximum difference in the interior is 0.2%. The bump in the sound-speed differences around $0.7R_\odot$ is believed to be the result of accumulation of helium below the convection zone of the model. This points to the existence of mechanisms for changing the composition profile below the convection zone base in the Sun, the most plausible being mixing induced by rotation.

Because the sound speed of the reference solar model is known, these results can be used to determine the solar sound speed as a function of radius, and the solar sound speed thus calculated is tabulated in Table 15.1. Figure 15.12 shows the results plotted as a function of radius. The inversion results do not go deep enough into the core to be able to see the decrease in sound speed toward the center.

Despite finite resolution we can see the changeover from the convection zone to the radiative zone at around $0.7R_\odot$. The inversion results can in fact be used to determine the position of the change (see, e.g., Christensen-Dalsgaard *et al.* [63], Basu and Antia [64], and Basu [65]) accurately.

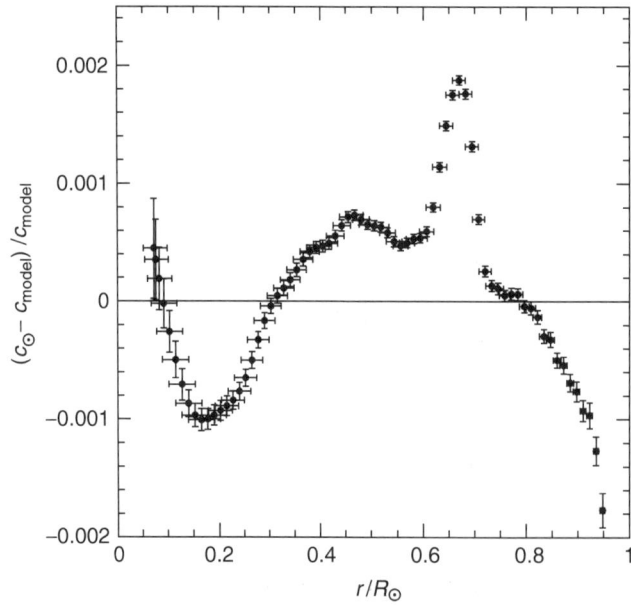

Fig. 15.11. The relative sound-speed difference between the Sun and Model S as obtained by inverting the relative frequency differences between the Sun and the model. The vertical error-bars represent 1σ errors because of errors in the data. The horizontal error-bars are a measure of the resolution of the inversion. Because the sound speed of the model is known, these results shown can be used to determine the solar sound-speed profile.

Table 15.1. Solar sound speed as a function of radius.

Radius ($R_\odot$)	Sound speed (km s^{-1})	Error (km s^{-1})
0.0725	511.270	0.217
0.0762	511.131	0.175
0.0828	510.630	0.135
0.1029	506.943	0.089
0.1276	498.114	0.067
0.1528	485.079	0.048
0.1773	469.888	0.042
0.2020	453.262	0.038
0.2273	435.963	0.035
0.2400	427.407	0.032
0.2651	411.036	0.029
0.2903	395.546	0.026
0.3030	388.150	0.025
0.3410	367.385	0.022
0.3789	348.688	0.019
0.4043	337.128	0.017
0.4295	326.307	0.016
0.4549	316.108	0.015
0.4802	306.408	0.014
0.5055	297.159	0.013
0.5308	288.299	0.012
0.5688	275.547	0.011
0.5814	271.406	0.011
0.6067	263.157	0.011
0.6320	254.922	0.010
0.6573	246.422	0.010
0.6826	237.197	0.010
0.6952	232.088	0.010
0.7078	226.389	0.010
0.7204	219.634	0.010
0.7331	212.749	0.010
0.7457	205.876	0.010
0.7584	198.981	0.010
0.7710	192.071	0.010
0.7837	185.135	0.010
0.7963	178.128	0.009
0.8217	163.925	0.009
0.8470	149.303	0.010
0.8723	134.016	0.010
0.8976	117.733	0.010
0.9102	109.070	0.010
0.9355	90.077	0.011
0.9481	79.332	0.012
0.9735	52.590	0.016

15.4 IMPLICATIONS OF HELIOSEISMOLOGY RESULTS

The results of helioseismology have shown that the solar models are very like the real Sun. This is true for solar-density profile and adiabatic index as well (e.g., Basu et al. [60]). Helioseismic results also indicate that the inputs to stellar models, such as the equation of state and opacity, are good representations of the actual equation of

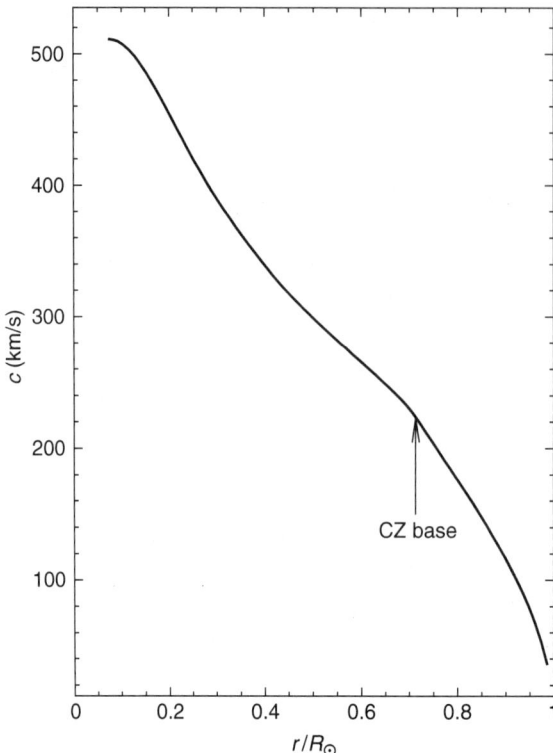

Fig. 15.12. Sound-speed profile of the Sun as determined from the results shown in Figure 15.11. The continuous line is the sound-speed profile; the 1σ error limits are too small to be seen because of the scale of the plot.

state and opacity of stellar matter. Thus, we are successful in building solar models. Because the equations governing solar evolution are the same as those governing the evolution of other stars, helioseismic results have largely validated our model of stellar structure and evolution in general. This gives us confidence that our estimates of the sound speed in other stars based on models are very close to the truth.

ACKNOWLEDGMENT

The author would like to thank J. Christensen-Dalsgaard for permission to use unpublished stellar models.

References

1. Popper, D.M. (1980). *Ann. Rev. Astron. Astroph.* **18**: 115–164.
2. Cohen, E.R. and Taylor, B.N. (1987). *Rev. Mod. Phys.* **59**: 1121–1148.
3. Willson, R.C. and Hudson, H.S. (1988). *Nature* **332**: 810–812.
4. Allen, C.W. (1973). *Astrophys. Quantities* London: Athlone Press.
5. Antia, H.M. (1998). *Astron. Astrophys.* **330**: 336–340.
6. Schou, J., Kosovichev, A.G., Goode, P.R., and Dziembowski, W.A. (1997). *Astrophys. J.*, **489**, L197–L200.
7. Brown, T.M. and Christensen-Dalsgaard, J. (1998). *Astrophys. J.* **500**: L195–L198.
8. Cox, J.P. and Giuli, R.T. (1968). *Principles of Stellar Structure*. New York: Gordon Breach.
9. Schwarzschild, M. (1958). *Structure and Evolution of the Stars*. Princeton: Princeton University Press.
10. Chandrasekhar, S. (1939). *An Introduction to the Study of Stellar Structure*. Chicago: University of Chicago Press (reissued in 1957 by Dover Publ.).
11. Clayton, D.D. (1968). *Principles of Stellar Evolution and Nucleosynthesis*. New York: McGraw-Hill.

12. Hansen, C.J. and Kawaler, S.D. (1994). *Stellar Interiors. Physical Principles, Structure, and Evolution.* New York: Springer.
13. Kippenhahn, R. and Weigert, A. (1990). *Stellar Structure and Evolution.* Berlin: Springer-Verlag.
14. Iglesias, C.A. and Rogers, F.J. (1996). *Astrophys. J.* **464**: 943–953.
15. Alexander, D.R. and Ferguson, J.W. (1994). *Astrophys. J.* **437**: 879–891.
16. Kurucz, R.L. (1991). In *NATO ASI Series, Stellar Atmospheres: Beyond Classical Models*, Crivellari, L., Hubeny, I., and Hummer, D.G., eds., pp. 441–448, Doedrecht: Kluwer.
17. Krishna-Swamy, K.S. (1966). *Astrophys. J.* **145**: 174–194.
18. Canuto, V.M. and Mazzitelli, I. (1991). *Astrophys. J.* **370**: 295–311.
19. Prandtl, L. (1925). *Z. Agnew. Math. Mech.* **5**: 136
20. Biermann, L. (1951). *Z. Astrophys.* **28**: 304.
21. Vitense, E. (1953). *Zs. Astrophys.* **32**: 108.
22. Böhm-Vitense, E. (1958). *Zs. Astrophys.* **46**: 108.
23. Stein, R.F. and Nordlund, Å. (1998). *Astrophys. J.* **499**, 914–933.
24. Ludwig, H.-G., Freytag, B., and Steffen, M. (1999). *Astron. Astrophys.* **346**: 111–124.
25. Trampedach, R., Stein, R.F., Christensen-Dalsgaard, J., and Nordlund, Å. (1999). In *Theory and Tests of Convection in Stellar Structure,* Giménez, A., Guinan, E.F., and Montesinos, B., eds., Astronomical Society of the Pacific Conference Series, San Francisco, **173**: 233–236.
26. Eggleton, P.P., Faulkner, J., and Flannery, B.P. (1973). *Astron. Astrophys.* **23**: 325–330.
27. Däppen, W., Mihalas, D., Hummer, D.G., and Mihalas, B.W. (1988). *Astrophys. J.* **332**: 261–270.
28. Rogers, F.J., Swenson, F.J., and Iglesias, C.A. (1996). *Astrophys. J.* **456**: 902–908.
29. Thoul, A.A., Bahcall, J.N., and Loeb, A. (1994). *Astrophys. J.* **421**: 828–842.
30. Proffitt, C.R., and Michaud, G. (1991). *Astrophys. J.* **380**: 238–250.
31. Zirker, J.B. (1981). In *Solar Phenomena in Stars and Stellar Systems*, Bonnet, R.M. and Dupree, A.K., eds., pp. 301–318, Boston: Reidel.
32. Wasserberg, G.J., in Bahcall, J.N. and Pinsonneault, M.H. (1995). *Rev. Mod. Phys.* **67**: 781–808.
33. Christensen-Dalsgaard, J. (1993). In *Proc. IAU Colloq. 137: Inside the Stars*, Baglin, A. and Weiss, W.W., eds., Astronomical Society of the Pacific Conference Series, San Francisco, **40**: 483–496.
34. Christensen-Dalsgaard, J., *et al.* (1996). *Science* **272**: 1286–1292.
35. Hill, F., Deubner, F.-L., and Isaak, G. (1991). In *Solar Interior and Atmosphere*, Cox, A.N., Livingston, W.C., and Matthews, M., eds., pp. 329–400. Space Science Series, Tucson University of Arizona Press.
36. Harvey, J.W., *et al.* (1996). *Science* **272**: 1284–1286.
37. Elsworth, Y., *et al.* (1994). *Astrophys. J.* **434**: 801–806.
38. Fossat, E. (1991). *Solar Phys.* **133**: 1–12.
39. Tomczyk, S., Streander, K., Card, G., Elmore, D., Hull, H., and Caccani, A. (1995). *Sol. Phys.* **159**: 1–21.
40. Domingo, V., Fleck, B., and Poland, A.I. (1995). *Solar Phys.* **162**: 1–37.
41. Antia, H.M., Basu, S., and Chitre, S.M. (1998) *Mon. Not. Roy. Astron. Soc.* **298**: 543–556.
42. Schou, J., Antia, H.M., Basu, S., *et al.* (1998). *Astrophys. J.* **505**: 390–390.
43. Kuhn, J.R., Bush, R.I., Scheick, X., and Scherrer, P.H., (1998). *Nature* **392**: 155–157.
44. Christensen-Dalsgaard, J. (1997). *Lecture Notes on Stellar Oscillations, 4th ed.* Aarhus: Aarhus University.
45. Unno, W., Osaki, Y., Shibahashi, H. (1989). *Non-radial Oscillations in Stars*, 2nd ed. Tokyo: Tokyo University Press.
46. Cox, J.P. (1980). *Theory of Stellar Pulsation*, Princeton: Princeton University Press.
47. Christensen-Dalsgaard, J. and Berthomieu, G. (1991). In *Solar Interior and Atmosphere*, Cox, A.N., Livingston, W.C., and Matthews, M., eds., Space Science Series, pp. 401–478, Tucson: University of Arizona Press.
48. Gough, D.O. (1993). In *Astrophysical Fluid Dynamics, Les Houches Session XLVII*, Zahn, J.-P. and Zinn-Justin, J., eds., pp. 399–560, Amsterdam: Elsevier.
49. Chandrasekhar, S. (1964). *Astrophys. J.* **139**: 664–674.
50. Gough, D.O. (1990). *Lecture Notes in Physics*, vol **367**: pp. 283–318, Osaki, Y., and Shibahashi, H., eds. Berlin: Springer.
51. Antia, H.M. and Basu, S. (1994). *Astron. Astrophys. Suppl.* **107**: 421.
52. Dziembowski, W.A., Pamyatnykh, A.A., and Sienkiewicz, R. (1990). *Mon. Not. Roy. Astron. Soc.* **244**: 542–550.
53. Kosovichev, A.G., *et al.* (1992). *Mon. Not. Roy. Astron. Soc.* **259**: 536–558.
54. Christensen-Dalsgaard, J., and Thompson, M.J. (1995). In *Proc. GONG'94: Helio- and Astero-seismology from Earth and Space*, Ulrich, R.K., Rhodes Jr., E.J., and Däppen, W., eds., Astronomical Society of the Pacific Conference Series, **76**: 144–147.
55. Basu, S., Chaplin, W.J., Christensen-Dalsgaard, J., Elsworth, Y., Isaak, G.R., New, R., Schou, J., Thompson, M.J., and Tomczyk, S. (1997). *Mon. Not. Roy. Astron. Soc.* **291**: 243–251.

56. Gough, D.O., and Thompson, M.J. (1991). In *Solar Interior and Atmosphere*, Cox, A.N., Livingston, W.C., and Matthews, M. eds., pp. 519–561, Space Science Series, Tucson: University of Arizona Press.
57. Backus, G. and Gilbert, F. (1967). *Geophys. J. R. Astr. Soc.* **13**: 247–276.
58. Pijpers, F.P. and Thompson, M.J. (1992). *Astron. Astrophys.* **262**: L33–L36.
59. Schou, J., *et al.* (1998). In *Proc. SOHO6/GONG98 Workshop, Structure and Dynamics of the Interior of the Sun and Sun-like Stars*, Korzennik, S.G. and Wilson, A., eds., ESA SP-418, vol 2, pp. 845–849, Noordwijk: European Space Agency.
60. Basu, S., Bahcall, J.N., and Pinsonneault, M. (2000). *Astrophys J.* **529**: 1084–1100
61. Rogers, F.J. and Iglesias, C.A. (1992). *Astrophys. J.* **401**: 361–366.
62. Grevesse, N., and Noels, A. (1993). In *Origin and Evolution of the Elements*, Prantzos, N., Vangioni-Flam, E., and Cassé, M., eds., pp. 15–25, Cambridge: Cambridge University Press.
63. Christensen-Dalsgaard, J., Gough, D.O., Thompson, M.J. (1991). *Astrophys. J.* **378**: 413–437.
64. Basu, S., Antia, H.M. (1997). *Mon. Not. Roy. Astron. Soc.* **287**: 189–198.
65. Basu, S. (1998). *Mon. Not. Roy. Astron. Soc.* **298**: 719–728.

CHAPTER 16

THE PROPERTIES OF CONDENSED MATTER IN WHITE DWARFS AND NEUTRON STARS

Shmuel Balberg and Stuart L. Shapiro[1]
Department of Physics, University of Illinois at Urbana-Champaign, Urbana, Illinois, USA

Contents

Abstract		423
16.1.	Introduction	423
16.2.	The Cold, High-Density Equation of State	425
16.3.	White Dwarf and Neutron Star Structure	433
16.4.	Observations of Condensed Matter in White Dwarfs	437
16.5.	Observations of Condensed Matter in Neutron Stars	439
16.6.	Thermal Properties of Matter in White Dwarfs and Neutron Stars	441
16.7.	Concluding Remarks	444
References		445

ABSTRACT

White dwarfs and neutron stars are stellar objects with masses comparable to that of our sun. However, at the end-point stages of stellar evolution, these objects do not sustain any thermonuclear burning and therefore can no longer support the gravitational load of their own mass by generating thermal pressure. Rather, matter in their interiors is compressed to much higher densities than commonly found in normal stars, and pressure is created by degenerate fermion kinetic energy and particle interactions. As a result, white dwarfs and neutron stars offer unique cosmic laboratories for studying matter at very high densities. In this chapter we discuss the basic properties of condensed matter at extreme densities and summarize the extent to which these properties can be examined by observations of compact objects.

16.1. INTRODUCTION

Astronomical phenomena provide many examples where matter exists in extreme conditions not found in terrestrial environments. One example is the high density of

---

[1] Also from the Department of Astronomy and the National Center for Supercomputing Applications, University of Illinois at Urbana–Champaign, Urbana, IL 61801.

Handbook of Elastic Properties of Solids, Liquids, and Gases, edited by Levy, Bass, and Stern
Volume IV: Elastic Properties of Fluids: Liquids and Gases
Copyright © 2001 by Academic Press
All rights of reproduction in any form reserved.

ISBN 0-12-445764-9 / $35.00

degenerate matter in "compact objects" — the relics of stars that have ceased burning thermonuclear fuel, and thereby no longer generate thermal pressure to support themselves against gravitational collapse. By contracting appreciably from their original sizes, the interiors of compact objects reach sufficiently high densities to produce nonthermal pressure via degenerate fermion pressure and particle interactions. Compact objects provide cosmic laboratories for studying the properties of matter at high densities.

Firm observational evidence and well-founded theoretical understanding both exist for two classes of compact objects that support themselves against collapse by cold, degenerate fermion pressure: *white dwarfs*, whose interiors resemble a very dense solid, with an ion lattice surrounded by degenerate electrons, and *neutron stars*, whose cores resemble a giant atomic nucleus — a mixture of interacting nucleons and electrons, and possibly other elementary particles and condensates. White dwarfs are supported by the pressure of degenerate electrons, whereas neutron stars are supported by pressure due to a combination of nucleon degeneracy and nuclear interactions. These unique states of matter are achieved by significant compression of stellar material. Table 16.1 compares the principal physical quantities of a typical white dwarf and neutron star with those of the sun.[2]

Condensed matter in compact objects spans an enormous range of densities, which we loosely refer to as "high densities." These extend from about 7 gm cm^{-3} (e.g., the density of terrestrial $^{56}_{26}$Fe), at the surface of a cold neutron star or white dwarf, to as much as $\rho \approx 10^{15}$ gm cm^{-3}, several times the density in atomic nuclei, in the cores of neutron stars. Matter at the various densities found in compact objects exhibits a variety of novel properties. Electromagnetic, strong, and weak interactions all play an important role in determining the character of compact objects. Because these objects are bound by gravity, they are a meeting point of all four of the fundamental forces of nature. Correspondingly, the astrophysics of white dwarfs and neutron stars incorporates a wide variety of physics, including nuclear, particle, solid state, and gravitation physics, to name a few areas.

In this chapter we briefly survey the theory of condensed matter at high densities in compact objects and illustrate how the basic theory is tested through astronomical observations. Because we must cover 15 orders of magnitude in density, our presentation is at most introductory in nature, and we encourage the interested reader to pursue the cited references. A detailed introduction to the physics of high-density matter and compact objects can be found in the textbook *Black Holes, White Dwarfs and Neutron Stars*, by Shapiro and Teukolsky [1].

We begin by considering the fundamental nature of cold ($T = 0$) high-density matter in Section 16.2, and distinguish between different regimes of high density.

Table 16.1. Parameters for the sun and a typical white dwarf and neutron star.

Object	Mass ($M_\odot^a$)	Radius (km)	Mean density (gm cm^{-3})	Mean pressure (dyne cm^{-2})	GM/Rc^{2b}
Sun	1	$\sim 7 \times 10^5$	~ 1	$\sim 10^{12}$	$\sim 10^{-6}$
White dwarf	≤ 1.4	$\sim 5 \times 10^3$	$\sim 10^7$	$\sim 10^{24}$	$\sim 10^{-4}$
Neutron star	1–3	~ 10	$\sim 10^{15}$	$\sim 10^{35}$	$\sim 10^{-1}$

[a]$M_\odot \equiv 1.989 \times 10^{33}$ gm = one solar mass.

[b]This ratio measures the importance of relativistic gravitation, i.e., general relativity.

[2] Throughout this chapter we use units that are the standard in astrophysical research: cgs for microscopic properties of matter and solar units (denoted by $\odot$) for macroscopic properties of astronomical objects.

In Section 16.3 we connect these microscopic properties with the fundamental macroscopic parameters of a compact object through the hydrostatic equilibrium dependence of mass and radius on central density. We summarize the fundamental predictions regarding the structure of white dwarfs and neutron stars. In Sections 16.4 and 16.5 we examine how observations of white dwarfs and neutron stars can be used to probe the properties high-density matter. We briefly discuss the perturbative effects of a finite temperature in Section 16.6.

16.2. THE COLD, HIGH-DENSITY EQUATION OF STATE

The pressure and energies in white dwarfs and neutron stars are nonthermal; thermal effects due to a finite temperature can be treated as a perturbation. We may therefore treat high-density matter as having a zero temperature to very good approximation. The equation of state (EoS) of the matter then reduces to a single-parameter function, $P(\rho_0)$ and $\rho(\rho_0)$, or $P(\rho)$, where P is the pressure, ρ_0 is the rest–mass density and ρ is the total mass–energy density, which accounts for internal (possibly relativistic) particle energies as well as rest–mass energy.

There exist two main regimes of high density, distinguished as follows. As long as all nucleons are confined to nuclei, their contribution to the total pressure is negligible compared with that of the degenerate electrons. At some threshold density, $\rho_{n-\text{drip}}$, it becomes favorable for the nuclei to disintegrate, i.e., neutrons "drip" out of the nuclei and form a "nucleon gas." The standard EoS of Baym *et al.* [2] suggests that $\rho_{n-\text{drip}} \approx 4 \times 10^{11}$ gm cm^{-3}. We may distinguish between the EoS for $\rho \leq \rho_{n-\text{drip}}$, which characterizes matter in white dwarfs and in the outermost layers of neutron stars, and $\rho > \rho_{n-\text{drip}}$, which describes matter in the interior of neutron stars.

16.2.1. The Equation of State below Neutron Drip Density: $\rho \lesssim 4 \times 10^{11}$ gm cm^{-3}

In matter below the neutron drip density, the ions provide a Coulomb lattice of pointlike charges, which is (to good approximation) independent of the properties of the surrounding electrons. The EoS of such matter is governed mainly by the electron gas. To lowest approximation, we may treat the electrons as an ideal fermion gas, incorporating some Coulomb corrections at relatively low density ($\rho \leq 10^4$ gm cm^{-3}) and corrections due to inverse β—decay just below the neutron drip density (10^9 gm cm$^{-3} \leq \rho \leq \rho_{n-\text{drip}}$). The EoS of condensed matter below neutron drip density is well understood. The standard equations for cold, degenerate matter in white dwarfs (helium-, carbon-, oxygen-, and, possibly, iron-dominated models) have been derived by Chandrasekar [3] and Salpeter [4], whereas models for equilibrated matter,[3] are based on the works of Dirac [5] and Feynman *et al.* [6] for $\rho \leq 10^4$ gm cm^{-3} and, e.g., of Harrison and Wheeler [7] and Baym *et al.* [2] for higher densities.

16.2.1.1. The Ideal Fermion Gas

For almost the entire range of high densities, the electrostatic energy associated with the structure of matter is much smaller than the Fermi energies. Consequently, Coulomb forces are generally negligible in a first-order treatment of the high density EoS. The electron component of high-density matter can therefore be described by a cold, single-species gas of noninteracting fermions. At zero temperature the fermions fill all the

---

[3]The equilibrium isotope of matter is the nucleus of highest binding energy per nucleon. At low densities this isotope is normal $^{56}_{26}$Fe, but as density increases so do the atomic mass and neutron to proton ratio [1].

states with momentum $p \leq p_F$ and none of the states with $p > p_F$, where p_F is the Fermi momentum. The corresponding Fermi energy of the particle species is

$$E_F \equiv \left((p_F c)^2 + (mc^2)^2\right)^{1/2} \qquad (16.1)$$

where $c = 3 \times 10^{10}$ cm sec^{-1} is the speed of light in vacuum and m is the fermion rest mass.

For electrons, their number density, n_e, is directly related to their Fermi momentum, $p_{F,e}$ by integrating over all occupied phase space ($h = 6.63 \times 10^{-27}$ ergs sec is Plank's constant):

$$n_e = \int_0^{p_{F,e}} n_e(p) dp \equiv 2\frac{1}{h^3} \int_0^{p_{F,e}} 4\pi p^2 dp = \frac{8\pi p_{F,e}^3}{3h^3} \qquad (16.2)$$

The factor of 2 in the second equation arises from the electron spin degeneracy.

The pressure the electrons supply is calculated through the mean momentum flux of the electron gas,

$$P = \frac{1}{3}\int_0^{p_{F,e}} v_e(p) n_e(p) p\, dp = \frac{2}{h^3}\int_0^{p_{F,e}} \frac{p^2 c^2}{(p^2 c^2 + (mc^2)^2)^{1/2}} 4\pi p^2 dp = \frac{m_e c^2}{\lambda_e^3}\phi(x) \qquad (16.3)$$

where $x_e \equiv p_{F,e}/m_e c$ is the electron "relativity parameter," $\lambda_e \equiv h/(2\pi m_e c)$ is the electron Compton wavelength, $v_e = p_e c^2/E_e$ is the electron velocity and

$$\phi(x) = \frac{1}{8\pi^2}\{x(1+x^2)^{1/2}(2x^2/3 - 1) + \ln[x + (1+x^2)^{1/2}]\} \qquad (16.4)$$

The mass–energy density of the free electrons is also uniquely related to the Fermi momentum as

$$\varepsilon_e = \int_0^{p_{F,e}} E_e(p) n_e(p) dp = \frac{2}{h^3}\int_0^{p_{F,e}} (p^2 c^2 + (mc^2)^2)^{1/2} 4\pi p^2 dp = \frac{mc^2}{\lambda_e^3}\chi(x_e) \qquad (16.5)$$

where

$$\chi(x) = \frac{1}{8\pi^2}\{x(1+x^2)^{1/2}(1 + 2x^2) - \ln[x + (1+x^2)^{1/2}]\} \qquad (16.6)$$

However, even when the degenerate electrons contribute most of the pressure, the mass–energy density is dominated by the rest mass of the ions, which are very nonrelativistic at these densities. Thus, the density of the matter may be simply taken as

$$\rho = \rho_0 = \frac{n_e m_B}{Y_e} \qquad (16.7)$$

where Y_e is the mean number of electrons per nucleon and m_B is the mean nucleon mass. In the case of white dwarfs we may set $Y_e = Z/A = 0.5$ ($Z =$ atomic number, $A =$ atomic weight), which is appropriate for fully ionized helium, carbon, or oxygen, the most abundant constituents in a white dwarf, and $m_B = 1.66 \times 10^{-24}$ gm. Combining Eqs. 16.2, 16.3, and 16.7 provides the basic EoS of electron pressure–dominated high-density condensed matter in this regime.

There exist opposite limits to the cold fermion gas equation of state: the low-density, nonrelativistic ($x \ll 1$) and the high-density, extremely relativistic ($x \gg 1$) limits. From Eqs. 16.2 and 16.7 we find that in cold matter with $Y_e = 0.5$, the electron relativity parameter satisfies $x_e = 1$ at $\rho \approx 10^6$ gm cm^{-3}, which therefore marks the transition density between a nonrelativistic (NR) and extremely relativistic (ER) electron gas. It is convenient to write the EoS in the two limits in a *polytropic* form,

$$P(\rho) = K\rho^\Gamma \qquad (16.8)$$

where

$$\text{NR } x_e \ll 1, \rho \ll 10^6 \text{ gm cm}^{-3} : \Gamma = \tfrac{5}{3}, K = 1.0036 \times 10^{13} Y_e^{5/3} \quad (16.9)$$

$$\text{ER } x_e \gg 1, \rho \gg 10^6 \text{ gm cm}^{-3} : \Gamma = \tfrac{4}{3}, K = 1.2435 \times 10^{15} Y_e^{4/3} \quad (16.10)$$

The constant K is in cgs units, yielding a pressure in dyne cm^{-2} for a density in gm cm^{-3}. Note that in this approximation the composition of the matter enters only through Y_e. Correspondingly, helium, carbon, and oxygen, which all have $Y_e = 0.5$, have identical ideal equations of state, slightly stiffer than that of iron ($Y_e = 26/56 \approx 0.43$). Fully equilibrated matter (often referred to as "catalyzed") has a Y_e that decreases with density and is therefore softer than matter composed of a single element.

The free, degenerate electron pressure EoS outlined here is a good approximation for the equation of state below neutron-drip density. It was employed by Chandrasekar in his pioneering analysis of equilibrium white dwarfs [3], for which he received the Nobel prize in 1983. More exact treatments were introduced in later years, which included the two main required corrections — electrostatic effects at low densities and neutronization (or inverse β — decay) at higher densities.

16.2.1.2. Electrostatic Corrections to the Cold Equation of State: $\rho \lesssim 10^4$ gm cm^{-3}

There exists a net electrostatic correction to the ideal equation of state due to the fact that the local distribution of charge is very nonuniform. The fact that positive charge is concentrated in pointlike ions causes the average electron–ion separation to be smaller than the average distance between electrons. The net electrostatic potential felt by the electrons is thus an attractive one, which effectively reduces the pressure for a given density.

Electrostatic corrections to the cold equation state are mostly important at relatively low densities. Electrostatic energies are inversely proportional to the average separation between particles, $\langle r \rangle$ which is naturally proportional to $n_e^{-1/3}$. The relative importance of electrostatic energy, E_C, between a degenerate, nonrelativistic electron and an ion of charge Z can be estimated through

$$1\frac{E_C}{E'_F} \equiv \frac{Ze^2/<r>}{p_{F,e}^2/2m_e} \propto n_e^{-1/3} \quad (16.11)$$

where $E'_{F,e} = p_{F,e}^2 \propto n_e^{2/3}$ is the Fermi kinetic energy of the nonrelativistic electrons. Unlike the case of hot matter (where the mean electron kinetic energy is $\sim k_B T$), the relative importance of electrostatic corrections *decreases* with density as $n_e^{-1/3}$.

A rough estimate of the quantitative electrostatic correction can be performed by using the Wigner-Seitz approximation, which describes the lattice as neutral sphere with a central pointlike ion and an ambient uniform electron gas.[4] For a carbon or oxygen lattice, the correction to the pressure is typically of the order of a few percent. At lower densities the electron distribution deviates from uniformity, and more sophisticated approaches, namely, the *Thomas-Fermi* and *Thomas-Fermi-Dirac* models, must be invoked [4, 6].

16.2.1.3. Neutronization Corrections to the Cold Equation of State: 10^9 gm cm$^{-3} \rho \lesssim 4 \times 10^{11}$ gm cm^{-3}

Stable high-density matter must be in chemical equilibrium to all types of reactions, including the weak interactions that drive β-decay and electron capture ("inverse

---

[4] We note that such an approximation is actually better suited for white dwarfs than for laboratory solids, where the electron distribution is much more nonuniform.

β — decay"):

$$n \to p + e + \bar{\nu}_e, \qquad p + e \to n + \nu_e \qquad (16.12)$$

where n and p denote a neutron and a proton, respectively, e denotes an electron, and $\nu_e(\bar{\nu}_e)$ denote an electron neutrino (anti-neutrino). If the matter's composition is out of β — equilibrium, it will adjust through β — decays or electron capture. Both types of reactions change the electron per nucleon fraction, Y_e, and thus affect the EoS (Eqs. 16.9 and 16.10).

In cold white dwarfs and neutron stars the weakly interacting neutrinos freely escape the system: a zero neutrino abundance implies a zero neutrino chemical potential. The condition of chemical equilibrium is then stated as:

$$\mu_n = \mu_p + \mu_e \qquad (16.13)$$

where μ_x denotes the chemical potential of species x. The condition of chemical equilibrium is the fundamental origin of the stable existence of neutrons in nuclei and in uniform $n-p-e$ matter. For free, single particles the chemical potential is identical to the rest mass. The masses in MeV of the three elementary particles of Eq. 16.13 are $m_n = 939.6$, $m_p = 938.3$, and $m_e = 0.511$ (1 MeV is equivalent to $\sim 1.78 \times 10^{-27}$ gm). It is energetically allowed to have a *free* neutron decay in the reaction $n \to p + e + \bar{\nu}_e$ ($m_n - m_p - m_e \approx 0.8$ MeV). Indeed, the lifetime of a free neutron is only about 1000 sec before it undergoes a β-decay. By contrast, in a cold, degenerate, noninteracting $n-p-e$ gas neutron decay can be "blocked": because the protons and electrons must obey the Pauli principle, the decays will be suppressed if the energy available to the newly formed electron and proton is insufficient to place them above their respective Fermi levels. At such high densities, the equilibrium state of the gas includes a finite fraction of neutrons.

In bulk matter, the electron chemical potential, which is equal to the electron Fermi energy, rises with electron number density. Hence, maintaining chemical equilibrium (Eq. 16.13) may require some protons to capture electrons and covert to neutrons. In matter below the neutron drip density these conversions occur in nuclei, so their neutron fraction increases—they "neutronize." The result is a net decrease in the electron abundance (Y_e) at high densities, which lowers the pressure and "softens" the EoS.

The fact that "neutronization" becomes energetically favorable at densities below the neutron drip must be taken into account when formulating an exact EoS for this range. In matter composed of nuclei and free electrons, β — decay is limited to relatively high densities ($\rho \gtrsim 10^9$ gm cm^{-3}), and the exact threshold for the onset of neutronization depends on the nature of the nucleus, since the nuclear interactions must be taken into account in addition to Fermi kinetic energies. For example, the reaction $^{12}_{6}C + e \to ^{12}_{5}B$ requires the nucleus to absorb an energy of ~ 14 MeV, which an electron can supply if the density of a pure carbon lattice reaches $\sim 3 \times 10^{10}$ gm cm^{-3}. The EoS of matter close to the neutron drip density thus depends on its chemical composition even for elements of equal Y_e, such as helium and carbon [4].

16.2.2. The Equation of State above Neutron Drip Density: $\rho \gtrsim 4 \times 10^{11}$ gm cm^{-3}

As neutronization proceeds, the nuclei become increasingly neutron rich. The so-called "tensor-force" component of the nuclear interactions causes like nucleons to repel one another, so that the binding energy of a neutron-rich nucleus is smaller than in one where $Z/A = 0.5$. Fully equilibrated matter reaches the last stable isotope ^{118}Kr($Y_e \approx 0.31$) just before the density of $\rho \equiv \rho_{n-\text{drip}} \approx 4.3 \times 10^{11}$ gm cm^{-3}. At higher densities it becomes favorable for some neutrons to "drip" out of their parent nuclei. The nucleon component of the matter can no longer be confined to pointlike objects and the nuclei begin to dissolve. As density is increased, a larger fraction of the nucleons exists as "free" particles, outside the nuclei. This is a gradual transition, until the matter approaches the density of atomic nuclei, where all nuclei have essentially dissolved, and the distribution of nucleons becomes uniform. The EoS above the neutron drip density must be formulated by consistently including the effects of nuclear physics, which become the governing component of the properties of matter as the density of atomic nuclei, $\rho_{nuc} \approx 2.8 \times 10^{14}$ gm/cm^3 is approached.

16.2.3. Subnuclear Densities: 4×10^{11} gm cm^{-3} $\lesssim \rho \lesssim 2.8 \times 10^{14}$ gm cm^{-3}

An analysis of cold matter in the range $\rho_{n-\text{drip}} \leq \rho \leq \rho_{nuc}$ is complicated by requiring chemical equilibrium between nucleons inside the nuclei and those that have dripped outside. One must account for the effects of the surrounding gas of free nucleons on the nuclei, as well as other effects such as nuclei surface and Coulomb energies. In addition, the nuclei are expected to be very neutron rich, deviating from the $Z/A = 0.4$–0.5 found in terrestrial nuclei. Nonetheless, the properties of matter in this range of densities can still be derived by a natural extrapolation from ordinary nuclei, and, indeed, the EoS for such matter is believed to be well understood. The principal studies of matter (e.g., [8]) are based on the nuclear liquid-drop model and are suitable for most applications regarding neutron star structure. For problems where a more accurate description of the neutron star inner crust is required, attention must be given to lattice effects, as the nuclei and free nucleons arrange in a distinct spatial structure (where the nuclei settle into bubbles, slabs, or rods, depending on density [9]). It is expected that at $\rho \approx \frac{1}{2}\rho_{nuc}$ all nuclei will have dissolved so that the matter is completely uniform.

16.2.4. Supernuclear Densities: $\rho \gtrsim 2.8 \times 10^{14}$ gm cm^{-3}

The upper end of the high-density regime is referred to as "supernuclear," where $\rho \geq \rho_{nuc}$. As matter is compressed to such densities the EoS becomes gradually dominated by the degenerate nucleons and the nucleon interactions.

It is instructive to begin by considering an ideal uniform mixture of neutrons, protons, and electrons. For any given nucleon number density, n_N, solving the equilibrium composition n_n, n_p, and n_e (the neutron, proton, and electron number densities, respectively) requires three equations. The first is baryon number conservation,

$$n_N = n_n + n_p \tag{16.14}$$

and the other two arise from imposing chemical equilibrium (Eq. 16.13) and charge neutrality.

The condition for charge neutrality is

$$n_p = n_e \tag{16.15}$$

For a noninteracting mixture of nucleons and electrons, all chemical potentials are simply the Fermi energies (Eq. 16.1). Electrons are extremely relativistic at nuclear densities so their electron chemical potential is

$$\mu_e \approx pF_{,e}c = \hbar c (3\pi^2 Y_e n_N)^{1/3} \approx 100 \left(\frac{Y_e}{0.03}\right)^{1/3} \left(\frac{n_N}{n_{nuc}}\right)^{1/3} \text{MeV} \tag{16.16}$$

where $n_{nuc} = 0.16$ fm^{-3} is the number density of nuclei at the saturation density (and 1 fm $= 10^{-13}$ cm). Even if the electron fraction per baryon is only a few percent, the electron chemical potential exceeds the mass difference between neutrons and protons by two orders of magnitude. The only way to maintain a finite electron fraction (which is required to balance the protons for charge neutrality) *and* satisfy chemical equilibrium is by having a significantly larger neutron than proton fraction (note that for nonrelativistic fermions $E_F \propto n^{2/3}$ and for extremely relativistic fermions $E_F \propto n^{1/3}$). Equilibrium matter at nuclear densities must be very neutron dominated, and objects composed of such matter are thus "neutron stars." For a noninteracting gas, the ratio of neutrons to protons in equilibrium must be about 8:1, which is also representative of more realistic models of supernuclear densities.

The noninteracting gas approximation is not reliable for deriving the EoS at supernuclear densities. Unlike electrostatic perturbations, nucleon–nucleon interactions are not negligible, and the interaction energies are comparable to the Fermi energies of the degenerate nucleons (electrons do not feel the strong interaction and may still be treated as noninteracting). Modeling of the nucleon–nucleon interaction is one of the longest-standing problems in nuclear physics, still only partially solved. Profound difficulties exist due to the absence of a comprehensive theory of the interactions and the difficulty of obtaining experimental data for $\rho > \rho_{nuc}$. Further complications arise due to the fact that the Fermi and interaction energies at $\rho \sim 2-3\rho_{nuc}$ reach a sizable fraction of the rest mass, and relativistic effects must be taken into account as well. It is not yet possible to apply quantum chromodynamics (QCD), the fundamental theory of strong interactions, to the many-body nuclear domain at $\rho \approx \rho_{nuc}$. Instead, the most useful approaches are still based on phenomenological potential formalisms, and many-body Schrödinger-like systems of equations [1].

The EoS of supernuclear matter remains to date a field of active research. Current approaches include variational methods based on deduced two and three nucleon interactions ([10], see [11] for a recent study), and relativistic mean field approximations ([12], see [13] for a review). The models are based on fitting parameters to reproduce the empirically determined properties of finite nuclei with $n = n_{nuc}$. A rough approximation of the properties of the nucleon component of supernuclear matter can be obtained through the effective, nonrelativistic model of [14],

$$\rho(n, x) = E_{\text{gas}}(n, x) + an^2 + b(n_n - n_p)^2 + cn^{\delta+1} \qquad (16.17)$$

The first term E_{gas} is the mass–energy density of a noninteracting gas of nucleons at density n and composition n_n, n_p. The coefficient a is negative, representing the long-range attractive component of the inter-nucleon force, while c is positive, representing the short-range repulsive component. The power δ is larger than unity, so that short-range repulsion dominates at high density. The *symmetry* term includes the positive coefficient b, which describes the "tensor-force" that repels like nucleons, and its contribution is therefore minimized for symmetric nuclear matter ($n_n = n_p$). The values of a, b, c, and δ are derived by requiring Eq. 16.17 to reproduce the assumed properties of symmetric matter at $n_N = n_{nuc}$. See [14, 15] for typical (model-dependent) values of these parameters.

Finally, we note that it is quite possible that other particles, besides neutrons, protons, and electrons, coexist in stable equilibrium at supernuclear matter. The most obvious example is the muon, which is a lepton similar to the electron but having a rest mass of $m_\mu \approx 105$ MeV. The equilibrium condition for the muons will simply be $\mu_\mu = \mu_e$, so from Eq. 16.16 it is evident that at densities $n \gtrsim n_{nuc}$ it is energetically favorable to convert some electrons into muons through the weak interactions. At densities of $n \gtrsim 2n_{nuc}$, it is possible that other exotic particles will appear, such as hyperons (baryons that are heavier than nucleons), Bose-Einstein condensates of mesons (i.e., pions or kaons), or even conversion of the nucleons into an uniform mixture of quarks (see [13, 16, 17] for recent reviews of the possible presence and roles of such particles in neutron stars).

16.2.5. Basic Properties of High-Density Matter

In Table 16.2 we present an EoS of cold, fully catalyzed matter ranging from normal iron at zero pressure to supernuclear densities. The EoS in the subnuclear regime is compiled from [2, 6, 8] and are "standard" for studying catalyzed high-density matter. For realistic white dwarf models, which are presumably composed mostly of helium or carbon and oxygen, and never reached sufficiently high temperatures to catalyze their nuclei, a slightly different EoS must be used, based on a single species; see [4]. We also tabulate one "state-of-the-art" EoS for the supernuclear range [11].

The EoS $P(\rho)$ is plotted in Figure 16.1(a). It is especially instructive to examine corresponding values of the adiabatic index, $\Gamma \equiv d\ln P/d\ln\rho$, plotted in Figure 16.1(b), along with the sound speed ($c_s = \sqrt{dP/d\rho}$). Note that around $\rho \approx 10^6$ gm cm^{-3}, the EoS transforms, as expected, from a nonrelativistic $\Gamma = 5/3$ polytrope to a relativistic $\Gamma = 4/3$ one as the Fermi energy of the electrons, which dominate the pressure, gradually becomes relativistic. There is a sharp drop in the adiabatic index around the neutron drip density, since at first the dripped neutrons contribute to mass density but not to pressure. Only as matter approaches nuclear densities does the nucleon pressure become important, pushing the adiabatic index back up to values in the range 2–3.

Along with the EoS, we also examine some thermodynamic quantities of cold, high-density matter. In Table 16.2 we also list the adiabatic index, sound speed, bulk modulus Y, and incompressibility K, defined as

$$Y = n\frac{dP}{dn} \quad \text{and} \quad K = 9\frac{dP}{dn} \equiv 9n\frac{d^2\rho}{dn^2} \tag{16.18}$$

Note that the bulk modulus (which has units of pressure) is the reciprocal of the quantity usually defined as the compressibility ($\chi \equiv Y^{-1}$), whereas the "incompressibility" (units of energy) is more commonly used in nuclear physics applications (and is indeed measurable in nuclei). In Table 16.3 we list typical values of Γ, c_s, Y, K,

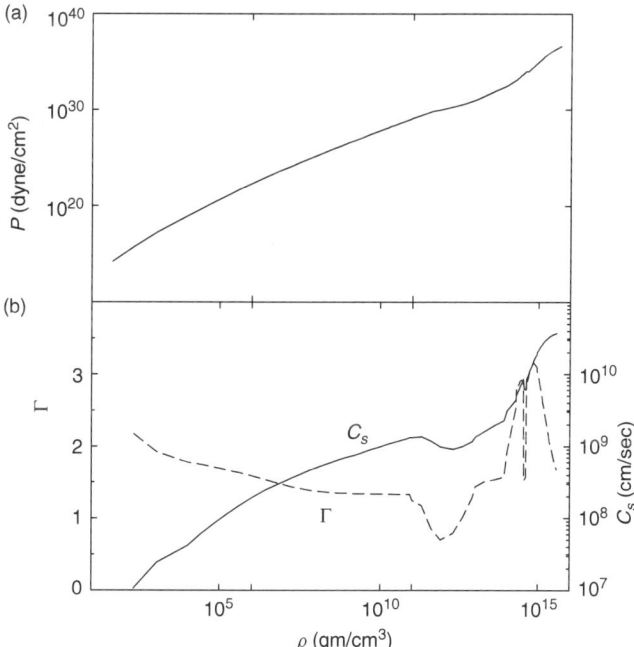

Fig. 16.1. The equation of state of cold, catalyzed high-density matter, based on [2, 6, 8, 11]. (a) The pressure density relation. (b) Adiabatic index and sound speed.

Table 16.2. The equation of state[a] of cold, catalyzed high-density matter[b].

ρ (gm cm^{-3})	n_B (cm^{-3})	P (dyne cm^{-2})	Γ	Y (dyne cm^{-2})	K (MeV)	c_s (cm sec^{-1})
1.00E+01	6.02E+24	5.13E+11	7.03E+00	3.20E+12	7.68E−06	5.67E+05
2.00E+01	1.21E+25	2.59E+13	3.09E+00	7.32E+13	3.41E−05	1.91E+06
5.00E+01	3.01E+25	2.21E+14	2.50E+00	5.51E+14	1.03E−04	3.32E+06
1.00E+02	6.02E+25	1.13E+15	2.23E+00	2.88E+15	2.69E−04	5.37E+06
2.00E+02	1.20E+26	5.14E+15	2.14E+00	1.10E+16	5.15E−04	7.43E+06
5.00E+02	3.01E+26	3.54E+16	2.05E+00	7.27E+16	1.36E−03	1.21E+07
1.00E+03	6.02E+26	1.44E+17	1.90E+00	2.95E+17	2.75E−03	1.72E+07
2.00E+03	1.20E+27	4.91E+17	1.74E+00	8.56E+17	3.99E−03	2.07E+07
5.00E+03	3.01E+27	2.53E+18	1.82E+00	4.61E+18	8.61E−03	3.04E+07
1.00E+04	6.02E+27	9.01E+18	1.80E+00	1.75E+19	1.63E−02	4.18E+07
2.00E+04	1.20E+28	3.07E+19	1.78E+00	5.45E+19	2.54E−02	5.22E+07
5.00E+04	3.01E+28	1.51E+20	1.73E+00	2.61E+20	4.87E−02	7.22E+07
1.00E+05	6.02E+28	4.90E+20	1.69E+00	8.91E+20	8.32E−02	9.44E+07
2.00E+05	1.20E+29	1.57E+21	1.65E+00	2.59E+21	1.21E−01	1.14E+08
5.00E+05	3.01E+29	7.07E+21	1.60E+00	1.13E+22	2.11E−01	1.50E+08
1.00E+06	6.02E+29	2.16E+22	1.59E+00	3.63E+22	3.39E−01	1.90E+08
2.00E+06	1.20E+30	6.38E+22	1.57E+00	9.97E+22	4.65E−01	2.23E+08
5.00E+06	3.01E+30	2.63E+23	1.48E+00	3.90E+23	7.29E−01	2.79E+08
1.00E+07	6.02E+30	6.81E+23	1.44E+00	1.03E+24	9.64E−01	3.21E+08
2.00E+07	1.20E+31	1.88E+24	1.44E+00	2.68E+24	1.25E+00	3.66E+08
5.00E+07	3.01E+31	6.84E+24	1.40E+00	9.55E+24	1.78E+00	4.37E+08
1.00E+08	6.02E+31	1.78E+25	1.38E+00	2.57E+25	2.40E+00	5.07E+08
2.00E+08	1.20E+32	4.67E+25	1.36E+00	6.29E+25	2.94E+00	5.61E+08
5.00E+08	3.01E+32	1.53E+26	1.39E+00	2.11E+26	3.95E+00	6.50E+08
1.00E+09	6.02E+32	3.91E+26	1.17E+00	4.57E+26	4.27E+00	6.76E+08
2.00E+09	1.20E+33	8.83E+26	1.36E+00	1.19E+27	5.57E+00	7.72E+08
5.00E+09	3.01E+33	3.03E+27	1.36E+00	4.12E+27	7.71E+00	9.07E+08
1.00E+10	6.01E+33	7.24E+27	1.35E+00	1.01E+28	9.47E+00	1.01E+09
2.00E+10	1.20E+34	1.83E+28	1.24E+00	2.24E+28	1.05E+01	1.06E+09
5.00E+10	3.00E+34	5.63E+28	1.13E+00	6.37E+28	1.19E+01	1.13E+09
1.00E+11	5.99E+34	1.40E+29	1.26E+00	1.80E+29	1.69E+01	1.34E+09
2.00E+11	1.20E+35	3.13E+29	1.17E+00	3.69E+29	1.73E+01	1.36E+09
5.00E+11	2.99E+35	8.28E+29	5.60E−01	4.65E+29	8.75E+00	9.63E+08
1.00E+12	5.97E+35	1.26E+30	8.48E−01	1.05E+30	9.86E+00	1.02E+09
2.00E+12	1.19E+36	2.12E+30	7.53E−01	1.61E+30	7.60E+00	8.97E+08
5.00E+12	2.98E+36	4.79E+30	1.11E+00	5.34E+30	1.01E+01	1.03E+09
1.00E+13	5.95E+36	1.16E+31	1.44E+00	1.76E+31	1.66E+01	1.33E+09
2.00E+13	1.19E+37	3.26E+31	1.52E+00	4.94E+31	2.34E+01	1.57E+09
5.00E+13	2.97E+37	1.32E+32	1.53E+00	2.03E+32	3.84E+01	2.01E+09
1.00E+14	5.92E+37	3.93E+32	1.87E+00	8.36E+32	7.94E+01	2.88E+09
2.00E+14	1.18E+38	1.93E+33	2.75E+00	5.54E+33	2.64E+02	5.24E+09
5.00E+14	2.89E+38	1.77E+34	3.00E+00	5.54E+34	1.08E+03	1.03E+10
1.00E+15	5.50E+38	1.52E+35	2.87E+00	6.52E+35	6.68E+03	2.30E+10
2.00E+15	9.36E+38	9.20E+35	2.23E+00	3.01E+36	1.81E+04	[c]
5.00E+15	1.57E+39	4.87E+36	1.50E+00	1.44E+37	5.14E+04	[c]

[a]The equation of state up to $\rho \sim \rho_{nuc} = 2.8 \times 10^{14}$ is based on Table 5 of [8] (© 1971, The American Astronomical Society). The supernuclear equation of state is adapted from [11].

[b]Shortened notation: $2.00E+10$ is 2.00×10^{10}.

[c]The APR [11] equation of state is not relativistic, and at extremely high densities it has an unphysical superluminal (greater than light speed) speed of sound.

Table 16.3. Typical values of thermodynamic quantities in the sun, white dwarfs, and neutron stars.

Object	Γ	c_s (cm sec^{-1})	Y (dyne cm^{-2})	K (MeV)	C_v ergs deg^{-1} gm^{-1}
Sun	5/3	$\sim 10^7$	$\sim 10^{12}$	$\sim 10^{-4}$	$\sim \frac{3}{2}(1/\mu + 1/m_p)\, k_B$
White dwarf (C/O)	4/3	$\sim 5 \times 10^8$	$\sim 10^{24}$	~ 10	$\frac{3}{2}\, k_B/\mu$
Neutron star	2–3	$\sim 10^{10}$	$\sim 10^{35}$	~ 200	$\sim 10^{-3}\, k_B/m_p$[a]

[a] For an ideal nonrelativistic neutron gas and a central temperature of $\sim 10^9\,°K \rightarrow k_B T/E_F \approx 10^{-3}$.

and the specific heat capacity, c_v, for condensed matter found in white dwarfs and neutron stars. For the purpose of comparison, we also give values for the sun, where the pressure roughly follows an ideal Maxwell-Boltzmann law $P \sim n k_B T$. Note how the extreme conditions of high densities lead to properties that are very different than those found for terrestrial materials. For example, the speed of sound in a neutron star is expected to be several tens of percent of the speed of light!

16.3. WHITE DWARF AND NEUTRON STAR STRUCTURE

Compact objects are self-gravitating equilibria with a mass comparable to that of the sun, $1\,M_\odot \equiv 1.989 \times 10^{33}$ g. Both white dwarfs and neutron stars are *centrally condensed* objects — most of their mass is located in a high-density core, which is limited to a fraction of the volume; furthermore, the radius of the objects *decreases* with increasing mass. These traits are characteristic of a configuration supported by degenerate-fermion pressure.

16.3.1. Construction of a Self-gravitating, Equilibrium Star

The equilibrium structure of a self-gravitating object is derived from the equations of hydrostatic equilibrium. The simplest case is that of a spherical, nonrotating, static configuration, where for a given EoS all macroscopic properties are parameterized by a single parameter, e.g., the central density. In the case of compact objects, the gravitational fields are strong enough that calculations must be performed in the context of general relativistic (rather than Newtonian) gravity. The fundamental equation of hydrostatic equilibrium in its general relativistic form has been derived by Tolman [18] and Oppenheimer and Volkoff [19] and is known as the "TOV" equation:

$$\frac{dP(r)}{dr} = -\frac{Gm(r)\rho(r)}{r^2}\left(1 + \frac{P(r)}{c^2\rho(r)}\right)\left(1 + \frac{4\pi r^3 P(r)}{c^2 m(r)}\right)\left(1 - \frac{2Gm(r)}{c^2 r}\right)^{-1} \quad (16.19)$$

This equation simply states that at any radial distance r, the gravitational pull by the mass interior to r, $m(r)$, is balanced by the gradient of the pressure $P(r)$; $G = 6.67 \times 10^{-8}$ cm^3 gm^{-1} sec^{-2} is the gravitational constant. Note that the first term on the right-hand side is the only term in the nonrelativistic case, where $P/(\rho c^2) \ll 1$ and $2Gm/(rc^2) \ll 1$. The second and third factors arise from the pressure being a form of energy density ("regeneration of pressure" effect), and the last term includes the correction due to the curvature of space in the strong gravitational field of the star.

16.3.2. Stable Configurations of High-Density Matter

Given an EoS, Eq. 16.19 can be integrated simultaneously with the mass equation

$$\frac{dm(r)}{dr} = 4\pi r^2 \rho(r) \quad (16.20)$$

to determine the entire profile of the object. The actual integration is typically solved by numerical means, assuming a central density $\rho(r = 0) = \rho_c$, and integrating outward from the center until reaching the surface, where $P(r) = 0$, which identifies $r = R$ as the radius of the star and $M(R)$ as its mass. We emphasize that the TOV equations provide the *gravitational* mass (or total mass–energy) of the star, which includes the effect of the gravitational binding energy, $E_{GB} \sim GM^2/R$, as well as the internal and rest–mass energy of the stellar constituents. Note that in the case of neutron stars, $(GM^2/R)/(Mc^2) \sim 10 - 20\%$; indeed, the gravitational mass of a neutron star is measurably lower than the total rest mass of its constituents.

Varying ρ_c produces a *sequence* of models for the given EoS, yielding $M(\rho_c)$ and $R(\rho_c)$ along the sequence. The resulting sequences for stars composed of pure carbon (based on [18]) and of cold, catalyzed condensed matter (with the EoS tabulated in Table 16.2) are both shown in Figure 16.2 (note that carbon-dominated matter beyond the neutron drip density is unphysical and therefore omitted). The most distinct feature about the equilibrium sequences is the existence of local maxima in the $M(\rho_c)$ curves. A hydrostatic equilibrium configuration is dynamically *unstable* to catastrophic gravitational collapse if $dM/d\rho_c < 0$, since a radial perturbation would cause it to collapse on itself [1]. Therefore, stable carbon white dwarfs exist only if the central density is $\rho_c \leq 6 \times 10^9$ gm cm^{-3}, while cold, catalyzed matter has two distinct regimes of stable configurations: one with central densities below $\sim 10^{10}$ gm cm^{-3} (equivalent to white dwarfs) and one where the central density lies roughly in the range $10^{13} \leq \rho_c \leq 10^{15}$ gm cm^{-3} (neutron stars). Configurations where $\rho_c \gtrsim \rho_{n-\text{drip}}$ are unstable and cannot be found in nature. This situation is a direct consequence of the nature of the EoS as shown in Figure 16.1. Quantitatively, it can be shown that a hydrostatic configuration is stable only if its mass-averaged adiabatic index, $\overline{\Gamma}$, satisfies [1]

$$\overline{\Gamma} - \frac{4}{3} \gtrsim \frac{2GM}{Rc^2} \tag{16.21}$$

It is evident, therefore, that the effect of neutronization and neutron drip, which cause $\overline{\Gamma}$ to drop below 4/3, is to separate astrophysical objects with high-density matter into two distinct classes of stable configurations, which are shown schematically in Figure 16.3.

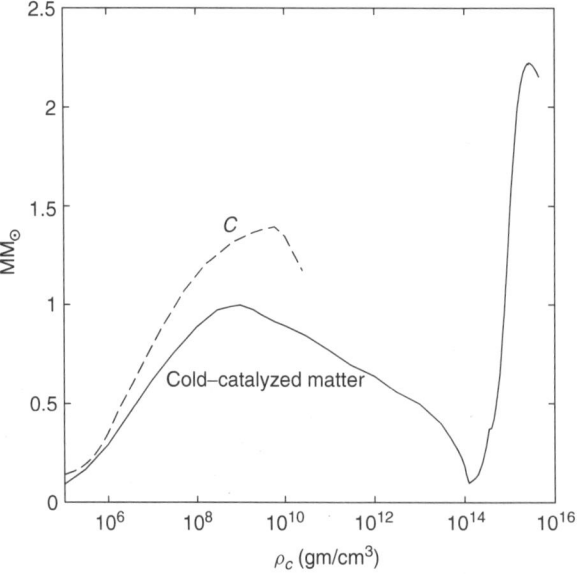

Fig. 16.2. Mass vs central density of a cold, carbon star [18] (dashed line) and of a star composed of cold, catalyzed matter (solid line).

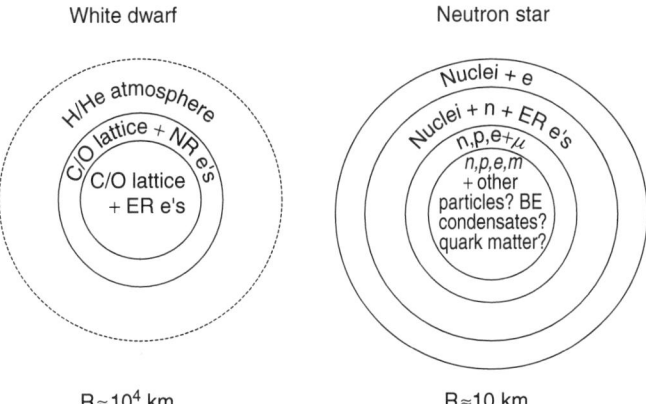

Fig. 16.3. Schematic illustration of the structure of a typical white dwarf (left) and a neutron star (right).

16.3.2.1. *White Dwarfs*

White dwarfs are composed of matter below the neutronization density. The astrophysical scenario that creates white dwarfs, i.e., quasi-static contraction of a progenitor star at the end point of thermonuclear burning, does not allow matter to reach high enough temperatures to achieve equilibrium composition through further burning. Accordingly, the composition of white dwarfs is mostly dominated by its nuclear ashes. In the standard scenario, white dwarfs are composed mostly of carbon and oxygen, but there is observational evidence that helium white dwarfs (and perhaps even iron-core white dwarfs, see Section 16.4.2.) exist as well. The pressure throughout the star is due to degenerate electrons, which are nonrelativistic in the outer layers but relativistic in the interior of the more massive stars with $M \sim M_\odot$.

By approximating the EoS as a Newtonian polytrope (Eq. 16.8), Chandrasekar [3] derived the basic features of white dwarfs:

$$\rho_c \lesssim 10^6 \text{ gm cm}^{-3} \; \Gamma = \frac{5}{3}: \quad R \simeq 1.12 \times 10^9 \times \left(\frac{\rho_c}{10^6 \text{ gm cm}^{-3}}\right)^{-1/6} \left(\frac{Y_e}{0.5}\right)^{5/6} \text{ cm}$$

$$M \simeq 0.496 \times \left(\frac{\rho_c}{10^6 \text{ gm cm}^{-3}}\right)^{1/2} \left(\frac{Y_e}{0.5}\right)^{5/2} M_\odot \quad (16.22)$$

$$\rho_c \gtrsim 10^6 \text{ gm cm}^{-3} \; \Gamma = \frac{4}{3}: \quad R \simeq 3.35 \times 10^9 \times \left(\frac{\rho_c}{10^6 \text{gm cm}^{-3}}\right)^{-1/3} \left(\frac{Y_e}{0.5}\right)^{2/3} \text{ cm}$$

$$M \simeq 1.46 \times \left(\frac{Y_e}{0.5}\right)^2 M_\odot \quad (16.23)$$

White dwarfs are expected to have radii of the order of 10^4 km—roughly the size of the Earth, or about 1% of that of the sun (which has a radius $R_\odot \approx 7 \times 10^5$ km).

As long as the EoS is approximated as a pure polytrope, so that electrostatic and neutronization corrections are ignored, helium, carbon, and oxygen white dwarfs ($Y_e = 0.5$) are all identical, and an iron-dominated white dwarf is only slightly different ($Y_e \approx 0.43$). Cold, catalyzed matter, which is used for the mass sequence in Figure 16.2, has a significantly softer EoS (due to Y_e decreasing with increasing density), and its sequence lies, therefore, lower than those of stars composed of a single species. The most significant result in Eq. 16.23 is that there is no dependence of M on the central density. The mass $M_{Ch} \simeq 1.46 \, M_\odot$, known as the Chandrasekar mass, is the asymptotic value a white dwarf can reach if it achieves sufficiently high density, so that its entire structure is governed by relativistic fermions (in practice this mass cannot be reached, since the outermost layers have nonrelativistic electrons). Nonrotating white dwarfs cannot have a mass exceeding M_{Ch}, and the precise limit on their mass is lower by several percent due to neutronization at high densities (Section 16.2.1.3.3.) *and* general relativistic effects. Despite the negligible effect on the hydrostatic equilibrium *profile* of white dwarfs, general relativity is required for a full analysis of white dwarf *stability* because an exact $\Gamma = \frac{4}{3}$ star is unstable to gravitational collapse (Eq. 16.21).

16.3.2.2. Neutron Stars

Neutron stars are composed mostly of matter at nuclear densities, including a core with supernuclear densities, and is topped by a thin crust at subnuclear densities. The crust, composed of cold, catalyzed matter, may also be divided into an inner part with $\rho_{n-\mathrm{drip}} \leq \rho \leq \rho_{nuc}$ and an outer part where $\rho \leq \rho_{n-\mathrm{drip}}$.

Immediately after the neutron was discovered (1932), Landau modeled a neutron star as a gas of noninteracting, degenerate neutrons [21]. He found that neutron stars would have a maximum mass of about $1 M_\odot$ and a radius of several kilometers. Although this was a very crude approximation, it does suggest the correct orders of magnitude regarding the structure of these objects: a neutron star has a mass comparable to that of the sun compressed to the size of a medium city! If supported by degenerate fermion pressure, a neutron star cannot have a mass that exceeds the Chandrasekar limit by much, and its radius is inversely proportional to the mass of the pressure-providing fermion. A neutron star should indeed have radius about $m_n/m_e \sim 10^3$ times smaller that of a white dwarf.

A realistic treatment of neutron star structure requires general relativity [20], whose effects are appreciable in this case. Furthermore, quantitative estimates must be based on a realistic EoS at all densities and especially a realistic model for the supernuclear regime, where particle interactions are most important. In particular, these interactions are repulsive at short distances and oppose compression, thereby stiffening the EoS in comparison to a free particle gas. Indeed, plausible equations of state predict that $M_{\max}(NS) \sim 2\, M_\odot$, ($\sim 2.2\, M_\odot$ for [11]), whereas the star's radius will lie in the range $R \sim 10\text{--}15$ km. As an example, we show in Figure 16.4 the M vs R relation

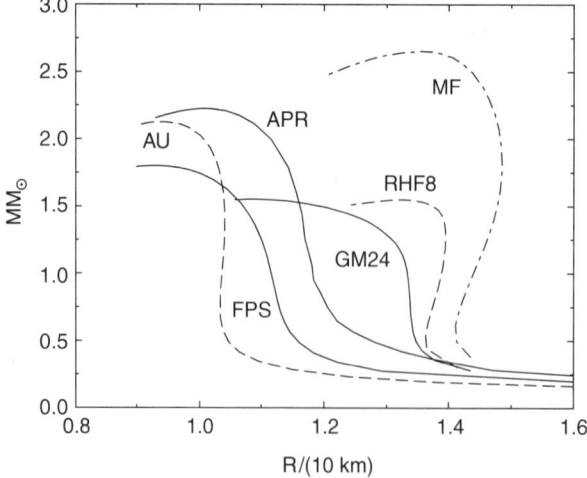

Fig. 16.4. The mass vs radius relation for a nonrotating neutron star for several supernuclear equations of state: Huber *et al.* (RHF8, [74]), Pandharipande and Smith (MF, [75]), Glendenning (GM24, [13], p. 244), Friedman and Pandharipande (FPS, [76]), Wiringa *et al.* (AU, [10]), and Akmal *et al.* (APR, [11]).

found for a few representative equations of state of supernuclear densities. The *exact* value of $M_{\max}(NS)$ depends on the assumed EoS and therefore provides an integral measure of the properties of matter at supernuclear densities [17, 22]. We also note that the existence of a maximum mass for neutron stars has important astrophysical implications because it suggests that a larger mass cannot be sustained by cold pressure and must inevitably collapse to a black hole.

16.4. OBSERVATIONS OF CONDENSED MATTER IN WHITE DWARFS

More than 2000 white dwarfs have been discovered to date, based on the spectroscopic properties of observed stars [23]. Although the thermal energy in white dwarfs is small compared with the Fermi energies of the electrons in the interiors, it is still sufficient to generate an observable surface luminosity for several billion years. The luminosity of a star depends on its radius, R, and its effective surface temperature, T_{eff}, according to

$$L = 4\pi\sigma R^2 T_{\text{eff}}^4 \tag{16.24}$$

where $\sigma = 5.67 \times 10^{-5}$ erg cm^{-2} sec^{-1} deg^{-4} is the Stephan-Boltzmann constant. The luminosities of white dwarfs are typically 10^{-3} to 10^{-2} of that of the sun, $L_\odot \approx 4 \times 10^{33}$ ergs sec^{-1}, but their radii are also smaller than those of ordinary stars by a factor of ~ 100. As a result, white dwarfs have apparent surface temperature of several times 10^4 °K, unusually high in comparison with most observable stellar objects, so they do appear very "white." The existence of white dwarfs was established spectroscopically (by determining the surface temperature of sources) as early as 1910.

White dwarfs are believed to be remnants of stars with initial masses in the range 0.1–8 $M_\odot$, which are not massive enough to complete the thermonuclear burning process all the way to iron. They are mostly composed of carbon and oxygen, but in some cases thermonuclear burning ceased before these elements were produced, and such stars are dominated be helium.

16.4.1. Radii

Observational determination of a white dwarf radius is straightforward if its flux, F, is measured and its distance from the Earth, D, is known:

$$F(D) = \frac{L}{4\pi D^2} \rightarrow R^2 = \frac{FD^2}{\sigma T_{\text{eff}}^4} \tag{16.25}$$

The stellar radius is derived only after the effective surface temperature is obtained spectroscopically [24]. Estimated radii of white dwarfs reside in the range of 0.007–0.013 $R_\odot$, which is consistent with an object supported by degenerate electron pressure (Section 16.3.2.) and thereby confirms the basic nature of white dwarfs.

16.4.2. Mass-Radius Relations

The most significant test of the nature of matter in a white dwarf is obtained by comparing observed mass-radius relations with the theoretical predictions. Once the radius of a white dwarf is determined, details of the surface emission, namely effects of gravitational acceleration on line emission [25] (which scales as M/R^2), or gravitational redshift [26] (which scales as M/R), are then used to estimate the surface gravity. An independent (and usually more accurate) estimate of the mass can be obtained directly for white dwarfs in binary star systems, through Kepler's third law [28]. Through these

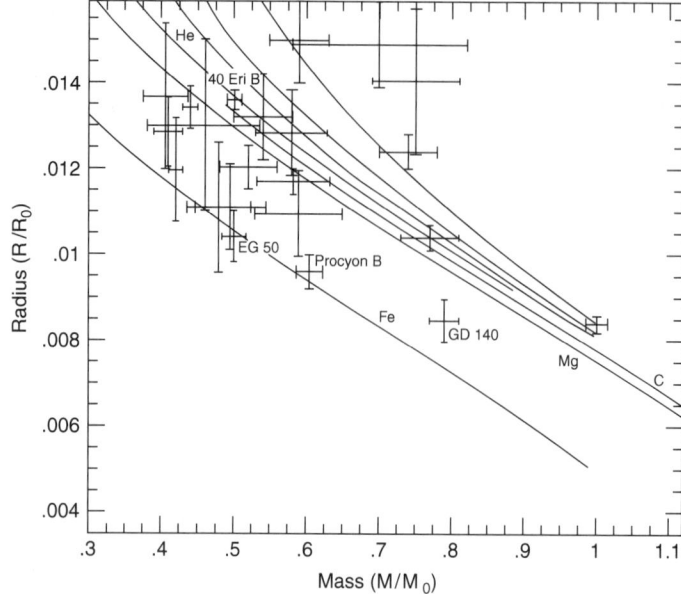

Fig. 16.5. The mass–radius relation for white dwarfs. Solid lines labeled He, C, Mg, and Fe denote the zero-temperature mass-radius relation for a star composed of each element of [18]. Models of white dwarfs with a hydrogen atmosphere [32] are for an effective atmosphere temperature of 30,000 °K (dotted line) and 15,000 °K and 8,000 °K (dashed lines). Estimates for observed white dwarfs are shown with a 1σ error (from Provencal *et al.* [31], © 1998, The American Astronomical Society).

methods, masses of a few white dwarfs have been estimated with reasonable accuracy, and most seem to cluster around $\sim 0.6\,M_\odot$ [27]. Some larger-mass white dwarfs are also known, including the most famous Sirius B [29], with $M \approx 1\,M_\odot$. Note that these masses are smaller than M_{Ch}: it is believed that stellar evolution, and especially periods of mass loss, limit the masses of most white dwarfs to $M \leq 1\,M_\odot$, even though the EoS could sustain somewhat higher masses.

Some examples of observationally determined radii and mass for white dwarfs are presented in Table 16.4. Note that larger-mass stars have smaller radii, as is expected

Table 16.4. Mass–radius relations of selected white dwarfs with 1σ errors.

Object	Mass $(M_\odot)$	Radius $(R_\odot)$
Masses from Binary Motion		
Sirius B	1.000 ± 0.016	0.0084 ± 0.0002
Procyon B	0.604 ± 0.018	0.0096 ± 0.0004
40 Eri B	0.501 ± 0.011	0.0136 ± 0.0002
Masses from Surface Acceleration		
EG 50	0.50 ± 0.06	0.0104 ± 0.0006
GD 140	0.79 ± 0.09	0.0085 ± 0.0005
Masses from Surface Gravitational Redshift		
CD-38 10980	0.74 ± 0.04	0.01245 ± 0.0004
W485A	0.59 ± 0.04	0.0150 ± 0.001

Adapted from Provencal *et al.* [31] © 1998, The American Astronomical Society.

(Eqs. 16.22 and 16.23). It is interesting to note that for some time, there was a nonnegligible discrepancy between the observations and theory, where observed radii seemed to be ~10–20% smaller than estimated by theory for carbon white dwarfs [25]. Only improved estimates of distances to several white dwarfs with the *Hipparcos* satellite and better modeling of white dwarf atmospheres [30] have allowed this discrepancy to be mostly resolved [31]. A comparison of current observed mass-radii determinations with theoretical curves is presented in Figure 16.5, and in general, the fit is indeed very good: these new results seem to confirm that the composition of most white dwarfs is indeed dominated by medium-weight elements (carbon and oxygen). However, they also imply that a small minority of white dwarfs do have relatively small radii and may therefore contain iron cores, which presents an intriguing puzzle from the point of view of stellar evolution.

16.5. OBSERVATIONS OF CONDENSED MATTER IN NEUTRON STARS

Neutron stars are often identified observationally as very accurately pulsating sources, where the pulsation is attributed to the star's rotation. The period of pulsation limits the size of the source to no more than a few tens of kilometers, thus indicating the presence of a very compact object. Since the first discovery in 1967, close to 800 pulsars have been identified, more than 700 of those in radio waves [33]. About 30 x-ray pulsars have also been observed, and a smaller number of nonpulsating x-ray sources of various types are also most likely to contain a neutron star [34].

Theory suggests that neutron stars are formed when a massive star is disrupted in a "supernova" explosion. As the massive star evolves, matter in its core undergoes thermonuclear burning all the way to iron, which is the most tightly bound nucleus and therefore cannot burn further. When the iron core is massive enough it collapses under its own weight, until the collapse is halted when nuclear densities are reached. Most of the gravitational binding energy is released in the form of neutrinos, but some is transformed into an outgoing shock wave that expels the star's envelope and leaves behind the newly formed neutron star. Indeed, several dozen young neutron stars have been found in sites of past supernovae, confirming this scenario.

16.5.1. Neutron Star Masses

The masses of more than 20 neutron stars have been determined observationally through their gravitational pull on a binary companion star. Within the errors of measurement, the masses of these stars appear to cluster in a narrow range of $1.35 \pm 0.1\,M_\odot$ [34]. This is especially true of the masses of neutron stars in the four known binary-pulsar systems (where the companion object is also a neutron star), listed in Table 16.5., where mass determinations are most accurate. Another example is the neutron star - white dwarf binary (once believed to be a double binary neutron star system) B2303 + 46. The properties of the orbit allow to limit the mass of the neutron star to $1.2 < M_{NS} < 1.4\,M_\odot$ [35]. In particular, the Hulse-Taylor binary pulsar 1913 + 16 (the coordinates of its location in the sky) has been very accurately determined to have $M_{1913+16} = 1.4411 \pm 0007\,M_\odot$. Clearly, a high density EoS must satisfy $M_{max} \geq 1.44\,M_\odot$ to be consistent with this observation.

This mass limit does not impose a serious constraint on realistic models for the supernuclear EoS and is satisfied by all but the softest equations. There has been much debate whether the narrow range of observed neutron star masses is evidence for the maximum mass being rather low ($M_{max}(NS) \approx 1.5\,M_\odot$) [36], or whether it is due to astrophysical effects, which could restrict the range of masses of observable stars. Recently, there is growing evidence that heavier neutron stars exist: one in the Vela

Table 16.5. Mass determinations with errors of neutron stars in known binary pulsar systems.

Star	Median mass ($M_\odot$)	68% Central limits	95% Central limits
J1518 + 4904 pulsar	1.56	+0.13/−0.44	
J1518 + 4904 companion	1.05	+0.45/−0.11	
B1534 + 12 pulsar	1.339	±0.003	±0.006
B1534 + 12 companion	1.339	±0.003	±0.006
B1913 + 16 pulsar	1.4411	±0.00035	±0.0007
B1913 + 16 companion	1.3874	±0.00035	±0.0007
B2127 + 11C pulsar	1.349	±0.040	±0.080
B2127 + 11C companion	1.363	±0.040	±0.080

Adapted from Thorsett and Chakrabarty [34] (© 1999, The American Astronomical Society).

X-1 binary is now estimated to have a mass of $\sim 1.9\,M_\odot$ [37], and the oscillations observed in some galactic x-ray sources seem to indicate that they include neutron stars with masses larger than $2\,M_\odot$ [38]. If these estimates are confirmed with improved observations in the future, they will provide a much more stringent test that would be consistent only with the stiffer models for the supernuclear EoS.

16.5.2. Radii

The mass and radius relation of a nonrotating neutron star is uniquely defined for any given EoS through the solution of the TOV equations (Figure 16.4). However, to date, observational methods for estimating neutron star radii lack the accuracy required to critically differentiate between realistic equations of state. Hopefully, data from a new generation of x-ray satellites will provide substantial constraints on the compactness (M/R) of neutron stars (either through their thermal emission [39] or from emission by material accreting on them [40]).

16.5.3. Rotation Periods

The observed pulsation period of a neutron star is attributed to rotation. Most measured periods are in the range 0.25–1 sec, but a subclass of *millisecond* pulsars is also known to exist. The fastest spinning neutron star observed to date has a period of $P = 1.56$ msec (so that it spins faster than an egg-beater!).

Rotation provides an additional centrifugal barrier that assists the internal pressure in supporting the gravitational load. The resulting configuration then depends on both central density and rotation period and must be solved self-consistently including the effects of general relativity. The maximum mass a given EoS can support increases with respect to the static (TOV) value as the star is allowed to rotate faster [41]. There must exist a lower limit on the rotation *period* (upper limit on the rotation *rate*) because too rapid rotation would cause the star to shed mass at the equator. The general trend is that a softer EoS predicts a smaller lower limit on the rotation period because a more compact star is more tightly bound gravitationally and so is better suited to resist the centrifugal force. At present, the limit of ~ 1.56 msec is not very restrictive and is consistent with basically all realistic models of high-density matter. Discovery of a submillisecond ($P \gtrsim 0.5$ msec) pulsar would allow distinguishing more directly between competing supernuclear EoSs [1, 42, 43].

16.5.4. Pulsar Glitches

Pulsar periods are observed to increase gradually with time, implying that rotational energy of the star is being lost (primarily by electromagnetic dipole radiation). However, several pulsars have been observed to undergo sudden *decreases* in the rotation period glitches, which are believed to originate from the transfer of angular momentum between different parts of the star. The current leading model [44, 45] suggests that the source of excess angular momentum is the inner curst, and thus the magnitude of glitch phenomena can be used to set a lower limit on the fraction of the total moment of inertia the inner crust must carry. Recent results [46] suggest the crust of the Vela pulsar must carry at least 1.4% of the total moment of inertia of the star, which is sufficient to rule out EoSs that are very soft in the range $1 \leq \rho/\rho_{nuc} \leq 2$. Possible progress in theoretical modeling of glitch phenomena could provide further limits on the properties of high-density matter near the nuclear saturation density.

16.6. THERMAL PROPERTIES OF MATTER IN WHITE DWARFS AND NEUTRON STARS

Fermions in both white dwarfs and neutron stars are expected to have degenerate energies that are much larger than their thermal energies. However, compact objects do have finite temperatures that are a relic of their progenitors. Because degenerate matter has very long scattering lengths for individual particles, the cores of both white dwarfs and neutron stars are expected to be excellent thermal conductors, and practically isothermal.

Finite temperature effects on the properties of condensed matter in compact objects has become a very active theoretical and observational field in recent years, especially in the case of white dwarfs. A finite temperature leads to cooling through thermal emission processes that are potentially observable. Examining the cooling history (i.e., temperature vs age) of compact objects thus provides empirical evidence regarding the thermal properties of high-density matter. Here we can only briefly mention some of the topics of current interest.

16.6.1. Thermal Effects in White Dwarf Structure

The regions most affected by a finite temperature are the low-density outer layers of a white dwarf, which may include a thin atmosphere composed of helium and possibly also hydrogen. For lower densities the electrons are not degenerate and a surface temperature of 10^4 °K will severely alter the EoS, especially in the regime where electrostatic corrections are important. Modern EoSs for a hydrogen/helium atmosphere [47] have been incorporated in several studies of white dwarf structure (see [48] for helium stars, [32] for carbon-oxygen, and [49] for a recent compilation of various compositions). The thermal pressure in the outer layers is generally found to be very effective in inflating the stellar radius. As seen from Figure 16.5, a surface temperature of $T_{eff} = 10^4$ °K increases the radius by a significant factor [32, 49] with respect to the cold [18] models.

There has been a recent revival of studies of white dwarf thermal evolution and its dependence on the properties of matter at the relevant high densities. We briefly review here the two main aspects of these studies, namely, cooling and pulsations.

16.6.1.1. White Dwarf Cooling

The basic theory of white dwarf cooling was established by Mestel in 1952 [50]. The heat is originally stored in the nondegenerate ion lattice, and the dominant cooling mechanism is photon diffusion to the surface from the isothermal core through the nondegenerate outer layers. The simplest results are based on Kramer's approximation for the photon opacities in the nondegenerate regime:

$$\kappa = \kappa_0 \rho T^{-3.5} \qquad (16.26)$$

where κ is the mean opacity (in $cm^2 gm^{-1}$) and $\kappa_0 = 4.34 \times 10^{24} Z(1 + X_H)$ $cm^2 gm^{-1}$ is a composition-dependent constant, X_H and Z being the mass-fraction of hydrogen and heavy (not hydrogen or helium) elements, respectively. The luminosity through the outer layers, $L(r)$, depends on the temperature gradient according to the radiative diffusion equation

$$L(r) = -4\pi r^2 \frac{c}{3\kappa\rho} \frac{d}{dr}(aT^4) \qquad (16.27)$$

(where $a = 7.7565 \times 10^{-15}$ erg deg^{-4} is the radiation constant). Combining the above thermal temperature gradient with the pressure gradient required for hydrostatic stability (Eq. 16.19, see [1]), one finds that the surface luminosity satisfies

$$L = 5.7 \times 10^6 \mu Y_e^2 \frac{1}{Z(1+X)} \left(\frac{M}{M_\odot}\right) T_c^{3.5}, \qquad (16.28)$$

where μ is the mean molecular weight in the nondegenerate atmosphere and T_c is the core temperature in the star. For typical white dwarfs ($M \sim 1 M_\odot$) with no hydrogen in the atmosphere, $X = 0$, $Z \approx 0.1$, $\mu = 1.4$, $Y_e = 0.5$, an observed luminosity in the range 10^{-5}–$10^{-2} L_\odot$ corresponds to a central temperature of $T_c \sim 10^6$–10^7 °K.

The key elements of Eq. 16.28 are that the surface luminosity (which is observable) is clearly related to the mass of the star, the composition and the internal temperature. If the mass is also determined separately (as discussed previously), luminosity can then provide a direct test of the star's thermal and composition profiles. Furthermore, the age of the white dwarf can also be estimated based on its cooling time scale, which depends on the luminosity function (Eq.16.28), the stellar mass, and the specific heat capacity of white dwarf material. The latter (in ergs $deg^{-1} gm^{-1}$) if roughly that of the ion lattice,

$$C_v(\text{ions}) = 2\frac{3}{2}\frac{k_B}{\mu} \qquad (16.29)$$

where the factor of 2 comes from the three degrees of collective vibration and rotation, in addition to the three single ion degrees of motion. Due to their degeneracy, the specific heat capacity of the electrons is suppressed by a factor of $\sim k_B T/E_F$ [51], which for a density of 10^8 gm cm^{-3} and an internal temperature of $\sim 10^7$ °K is 10^{-3}.

Although models that include more detailed microphysics do reproduce the general results of the Mestel model (see, e.g. [1, 29, 52, 53]), accurate work that is compatible with the quality of current observations must include several important perturbations. Most notably, accurate low-temperature opacities, semidegenerate electron thermal conductivity, and pressure-induced ionization must be accounted for in modeling the white dwarf atmosphere. Crystallization of the ion gas may also have a significant effect on the thermal history of the star, both as a transient source of (latent) heat and by effectively reducing the heat capacity. In particular, below the *Debye temperature*, $\Theta_D \simeq 4 \times 10^3 \rho^{1/2}$ °K, the specific heat capacity becomes temperature dependent [1],

$$C_v = \frac{12\pi^4}{5} k_B \left(\frac{T}{\Theta_D}\right)^3 \qquad (16.30)$$

The specific effects of the finite temperature on the thermodynamic properties of matter in white dwarf atmosphere are most notable through studies of pulsations (see Section 16.6.1.2).

It is noteworthy that for several years there seemed to be a paucity of low-luminosity (and therefore, old) observed white dwarfs, which posed a nontrivial puzzle in terms of models for the stellar evolution. Only rather recently it was realized [55] that the properties of the finite temperature atmospheres would make old white dwarfs dimmer than previously anticipated (especially if their atmosphere included only a helium layer and no hydrogen).

16.6.1.2. Pulsation of White Dwarfs

The basic theory of stellar pulsation is reviewed in detail by Sarbani Basu in this volume. We note that studies of pulsations in general allow one to estimate the sound speed profile of the star, its temperature gradient, and even the mean molecular weight in its core. It has been well established [56] that stellar pulsation provides an effective probe for the properties of matter in a star. Commonly referred to as "asteroseismology," this field is rapidly emerging due to significant advances in observational instrumentation and theoretical modeling.

White dwarfs, like many other stars, may undergo stable pulsations which leave a detectable imprint on the time dependence of the star's luminosity [57, 58]. Since pulsations tend to damp over time (or the star evolves out of the instability strip, if it is overstable to the pulsations), they are most easily observable in younger, and therefore hotter, white dwarfs. Modeling observed pulsations of young white dwarfs is particularly important for probing the EoS of semidegenerate matter [59, 60]. With the aid of theoretical models, studies of white dwarfs' pulsations have been used for independently restricting the mass-radius relationships and dimensions of the hydrogen envelope [58, 61], and in general seem to hold much potential for future studies of the details of white dwarf physics. For example, some pulsation modes are theoretically predicted to be more sensitive to the extent of crystallization of the atmosphere material. Recent analysis of the periods and relative magnitudes of various pulsation modes in the white dwarf BPM 37093 [60] helps determine the extent of crystallization in the atmosphere (about 50% in mass), which compares favorably with theory. In another recent example [62], it is possible to identify, for the first time, the contraction rate of the pre–white dwarf star PG 1159–035 by measuring the secular changes over time of the periods of various modes.

16.6.2. Thermal Effects in Neutron Star Structure

Although finite temperature effects are practically negligible when considering the overall structure of evolved neutron stars, they are important in assessing their thermal history and as probes of matter in the interior of these objects. Some thermal effects on structure do exist in newly born neutron stars (often called "proto-neutron stars").

16.6.2.1. Proto-Neutron Stars

Neutron stars are born in supernovae with initial internal temperatures of several times 10^{11} °K and initial entropies of $\sim 2k_B$ per nucleon (k_B is the Boltzmann constant). These values are large enough to impose nontrivial effects on the composition and structure of the neutron star compared with the $T = 0$ case [16]. At such high temperatures the matter is *not* transparent to the thermal neutrinos [1, 62], and the proto-neutron star must cool through neutrino diffusion to the surface, which occurs over a time scale of several seconds (much longer than a dynamical time of milliseconds) [63]. This fundamental prediction was dramatically confirmed by the observed neutrino pulse from the supernova 1987A (the closest supernova observed from the Earth in 400 years), which lasted for ~ 12 seconds. (See [64] for a review). Further studies of the properties of high-temperature, supernuclear-density matter are required to model this brief early cooling epoch and its possible observational signatures [65].

16.6.2.2. Cooling of Neutron Stars

After the initial rapid cooling stage, which lasts several days, a neutron star settles on a slower cooling curve [1, 66]. Once neutrinos escape the system freely, they must be continuously produced to fuel the emission process. Subsequent neutrinos are mostly produced by the so-called URCA (β and inverse β decays as in Eq. 16.12) and other processes. The isothermal core at this stage is expected to have a temperature of several times $10^8 \, °K$, whereas at the surface the temperature is down to several times $10^6 \, °K$, which provides for continuous surface thermal luminosity in soft x-rays. This soft x-ray luminosity is much more difficult to detect than the optical ultraviolet luminosity of white dwarfs. Current observations are generally limited to young (age $\lesssim 10^6$ years) neutron stars, where the main cooling mechanism is still the neutrino emission from the core. An important consequence of this situation is that estimates of thermal emission from neutron stars serve as a probe of the properties of matter in the core [67].

More than 20 compact, isolated soft x-ray sources observed with x-ray imaging telescopes have been identified as neutron stars [68]. In most cases the source is too faint to allow for a spectroscopic determination of the surface temperature or the emission is dominated by other phenomena (most likely magnetospheric emission). However, in a handful of cases, surface thermal emission has been strong enough to be detected and the surface temperature has been deduced to within a factor of two [68]. The measurements are still not accurate enough to determine whether the observed layer is the actual surface or a possible hydrogen atmosphere, which would have different emission properties, but they are sufficient to place some constraints on the rates of neutrino emitting processes in the core [69].

Perhaps the most notable conclusion to date of observed neutron star cooling rates is that they support the theoretical prediction that the nucleons in the core couple to a *superfluid* state. Theoretical models suggest that the strong interactions will pair the neutrons in the core in a $^3P_1$ superfluid and the protons to a $^1S_0$ superconductor with critical temperatures of the order of $10^9 \, °K$ [70]. Neutrino emitting processes, such as β-decays, must break a coupled Cooper pair before its constituents can participate in the decay. The main effect of nucleon superfluidity is therefore damping the efficiency of neutrino emission from the core (there is also a secondary effect due to the modulation of the heat capacity, which first increases discontinuously as the star cools to the critical temperature, and then decreases exponentially at lower temperatures). Observed neutron star surface temperatures are apparently too high to be consistent with the cooling rates predicted for normal core, but they agree with the suppressed cooling rates when the nucleons couple to a superfluid state [71, 72]. Further progress in analyses of the matter in the interiors of neutron stars is expected through measurements by NASA's recently launched Chandra x-ray satellite.

16.7. CONCLUDING REMARKS

The theoretical and observational study of compact objects remains one of the most exciting fields in modern astronomy. In essence, this research is also an exploration of the properties of condensed matter at extreme densities. Predictions regarding the properties of white dwarfs and neutron stars serve to test our understanding of matter at these high densities, and theories of high-density matter serve as a basis for interpreting observational results regarding these objects. Most exciting, these objects bring together all four of the fundamental forces of nature and probe regimes not accessible in the terrestrial laboratory. They provide the most numerous and accessible sample of objects where relativistic gravitation — general relativity — plays a role in determining their physical properties.

In this chapter we described the tight interconnection between the microscopic (local) properties of condensed matter at high densities and the macroscopic (global) properties of white dwarfs and neutron stars. Although the fundamental principles of cold, high-density matter are believed to be well understood, and are generally consistent with observations, key questions still remain, and new observations may give rise to new puzzles. The current boom in capabilities of Earth-bound telescopes and satellite instrumentation promises that many more puzzles — and, hopefully, answers — are in store regarding the nature of cosmic matter at high densities.

References

1. Shapiro, S.L., and Teukolsky, S.A. (1983). *Black Holes, White Dwarfs and Neutron Stars*. Wiley, New York: NY.
2. Baym, G., Pethick, C.J., and Sutherland, P. (1971). The ground state of matter at high densities: Equation of state and stellar models. *Astrophys. J.* **170**: 299.
3. Chandrasekar, S. (1931). The density of white dwarf stars. *Phil. Mag.* **11**: 592; The maximum mass of ideal white dwarfs. *Astrophys. J.* **74**: 81.
4. Salpeter, E.E. (1961). Energy and pressure of zero temperature plasma. *Astrophys. J.* **134**: 669.
5. Dirac, P.A.M. (1926). On the theory of quantum mechanics. *Proc. Roy. Soc. London Ser. A* **112**: 661.
6. Feynman, R.P., Metropolis, N., and Teller, E. (1949). Equations of state of elements based on the generalized Fermi-Thomas Theory. *Phys. Rev.* **75**: 1561.
7. Harrison, B.K., and Wheeler, J.A. (1958). See in Harrison, B.K., Thorne, K.S., and Wheeler, J.A. (1965). Gravitation theory and gravitational collapse. Chicago: University of Chicago Press.
8. Baym, G., Bethe, H.A., and Pethick, C.J. (1971). Neutron star matter. *Nucl. Phys. A* **175**: 225.
9. Ravenhall, D.G., and Pethick, C.J. (1995). Matter at large neutron excess and the physics of neutron-star crusts. *Ann. Rev Nuc. Pat. Sci.* **45**: 429.
10. Wiringa, R.B., Fiks, V., and Fabrocini, A. (1988). Equation of state for dense nuclear matter. *Phys. Rev.* **C38**: 1010.
11. Akmal, A., Pandharipande, V.R., and Ravenhall, D.G. (1998). The equation of state for nucleon matter and neutron star structure. *Phys. Rev.* **C58**: 1804.
12. Serot, B.D., and Walecka, J.D. (1979). Properties of finite nuclei in a relativistic quantum field theory. *Phys. Lett. B* **87**: 172.
13. Glendenning, N.K. (1996). *Compact Stars*, Springer, New York: NY.
14. Lattimer, J.M., and Swesty, D.F. (1991). An effective equation of state for hot dense matter. *Nucl. Phys. A* **535**: 331.
15. Balberg, S., and Gal, A. (1997). An effective equation of state for dense matter with strangeness. *Nucl. Phys. A* **625**: 435.
16. Prakash, M. *et al.* (1997). Composition and structure of proto-neutron stars. *Phys. Rep.* **280**: 1.
17. Balberg, S., Lichtenstadt, I., and Cook, G.B. (1999). Roles of hyperons in neutron stars. *Astrophys. J. Supp.* **121**: 515.
18. Tolman, R.C. (1939). Static solutions of einstein's filed equations for spheres of fluids. *Phys. Rev.* **55**: 364.
19. Oppenheimer, J.R., and Volkoff, G.M. (1939). On massive neutron cores. *Phys. Rev.* **55**: 374.
20. Hamada, T., and Salpeter, E.E. (1961). Models for zero-temperature stars, *Astrophys. J.* **134**: 683.
21. Landau, L.D. (1932). On the theory of stars. *Phys. Z. Sowjwtunion* **1**: 285.
22. Pandharipande, V.R., Pines, D., and Smith, R.A. (1976). Neutron star structure: theory, observation and speculation. *Astrophys. J.* **208**: 550.
23. McCook, G.P., and Sion, E.M. (1999). A catalog of spectroscopically identified white dwarfs. *Astrophys. J. Supp.* **121**: 1.
24. Shipman, H.L. (1979). Masses and radii of white dwarfs III. Results for 110 hydrogen-rich and 28 Helium-rich stars. *Astrophys. J.* **228**: 240.
25. Schmidt, H. (1996). The empirical white dwarf mass-radius relation and its possible improvement by HIPPARCOS. *Astron. Astrophys.* **311**: 852.
26. Shapiro, S.L., and Teukolsky, S.A. (1976). On the maximum gravitational redshift of white dwarfs. *Astrophys. J.* **203**: 697.
27. Weidermann, V. (1990) Masses and evolutionary status of white dwarfs and their progenitors. *Ann. Rev Astron. Astrophys.* **28**: 103.
28. Greenstein J.L. (1986). White dwarfs in wide binaries. I - Physical properties. II - Double degenerates and composites. *Astron. J.* **92**: 859.

29. Gatewood, G.D., and Gatewood, C.V. (1978). A study of sirius. *Astrophys. J.* **225**: 191.
30. Wood, M.A. (1995). Theoretical white dwarf luminosity functions: DA models, in *Proc. 9th European Workshop on White Dwarfs* pp. 41, Koster, D., and Werner, K. eds., Berlin: Springer.
31. Provencal, J.L., Shpiman, H.L., Høg, E., and Thejll, P. (1998). Testing the white dwarf mass-radius relation with *HIPPARCOS*. *Astrophys. J.* **494**: 759.
32. Taylor, J.H., Manchester, R.N., and Lyne, A.G. (1993). Catalog of 558 pulsars. *Astrophys. J. Supp.* **88**: 529. See the Princeton Pulsar-Group website for an updated catalog, http://www.pulsar.princeton.edu.
33. *X-Ray Binaries*, Lewin, W.H.G., van Paraijs, J., and van den Heuvel, E.P.J. Eds. (1995). Cambridge: Cambridge University Press.
34. Thorsett, S.E., and Chakrabarty, D. (1999). Neutron star mass measurements. I. Radio pulsars. *Astrophys. J.* **512**: 288.
35. van Kerkwijk, M.H., and Kulkarni, S.R. (1999). A massive white dwarf companion to the eccentric binary pulsar system PSR B2303 + 46. *Astrophys. J. Lett.* **516**: L25.
36. Bethe, H.E., and Brown, G.E. (1995). Observational constraints on the maximum neutron star mass. *Astrophys. J. Lett.* **445**: L29.
37. van Kerkwijk, M.H. (2000). Neutron star mass determinations. To appear in *Proc. ESO Workshop on Black Holes in Binaries and Galactic Nuclei*, Kaper, L., van den Heuvel, E.P.J., and Woudt, P.A. eds., Springer-Verlag, Berlin available on the Los-Alamos pre-print server, astro-ph/0001077.
38. van der Klis, M. (1998). KlioHertz quasi-periodic oscillations in low mass x-ray binaries, in *The Many Faces of Neutron Stars*, pp. 337. NATO ASI series Buccheri, R., van Paradijs, J., and Alpar, M.A. eds., Dordrecht: Kluwer.
39. Ögelman, H. (1995). X-ray observations of cooling neutron stars, in *The Lives of the Neutron Stars* pp. 101. Alpar, M.A., Kizilogu, Ü., and van Paradijs, J. eds., Kluwer: Dordechet.
40. Miller, M.C., and Lamb, F.K. (1998). Bounds on the compactness of neutron stars from brightness oscillations during x-ray bursts. *Astrophys. J. Lett.* **499**: L37.
41. Cook, G.B., Shapiro, S.L., and Teukolsky, S.A. (1994). Rapidly rotating neutron stars in general relativity: Realistic equations of state *Astrophys. J.* **424**: 823.
42. Shapiro, S.L., Teukolsky, S.A., and Wasserman, I. (1983). Implications of the millisecond pulsar for neutron star models. *Astrophys. J.* **272**: 702.
43. Lattimer, J.M., Prakash, M., Masak, D., and Yahil, A. (1990). Rapidly rotating pulsars and the equation of state. *Astrophys. J.* **355**: 241.
44. Pines, D., and Alpar, M.A. (1985). Superfluidity in neutron stars. *Nature* **316**: 27.
45. Alpar, M.A., Chau, H.F., Cheng, K.S., and Pines, D. (1993). Postglitch relaxation of the VELA pulsar after its first eight large glitches - A reevaluation with the vortex creep model. *Astrophys. J.* **409**: 345.
46. Link, B., Epstein, R.I., and Lattimer, J.M. (1999). Pulsar constraints on neutron star structure and equation of state. *Phys. Rev. Lett.* **83**: 3362.
47. Saumon, D., Chabrier, G., and van Horn, H.M. (1995). An equation of state for low-mass stars and giant planets. *Astrophys. J. Supp.* **99**: 713.
48. Hansen, B.M.S., and Sterl Phinney, E. (1998). Stellar forensics. I - Cooling curves. *Mon. Not. Roy. Astron. Soc.* **294**: 557.
49. Panei, J.L., Althaus, L.G., and Benvenuto, O.G. (2000). Mass-radius relations for white dwarfs of different internal compositions. *Astron. Astrophys.* **353**: 970.
50. Mestel, L. (1952). On the theory of white dwarfs stars I. The energy sources of white dwarfs. *Mon. Not. Roy. Astron. Soc.* **112**: 583.
51. Pathria, R.K. (1972). *Statistical Mechanics*. Oxford: Pergamon Press.
52. Iben, I., Jr., and Tutukov, A.V. (1984). Cooling of low-mass carbon-oxygen dwarfs from the planetary nucleus stage through the crystallization stage. *Astrophys. J.* **282**: 615.
53. D'Antona, F., and Mazzitelli, I. (1990). Cooling of white Dwarfs. *Ann. Rev Astron. Astrophys.* **28**: 139.
54. Hansen, B.M.S. (1999). Cooling models for old white dwarfs. *Astrophys. J.* **520**: 680.
55. Hansen, B.M.S. (1998). Old and blue white-dwarf stars as a detectable source of microlensing events. *Nature* **394**: 860.
56. Cox, J.P. (1980). *Theory of Stellar Pulsation*. Princeton: Princeton University Press, NJ.
57. Bradley, P.A., Winget, D.E., and Wood, M.A. (1993). The potential for asteroseismology of DB white dwarf stars. *Astrophys. J.* **406**: 661.
58. Bradley, P.A., and Winget, D.E. (1991). Asteroseismology of white dwarf stars. I - Adiabatic results. *Astrophys. J. Supp.* **75**: 463.
59. Bradley, P.A., Winget, D.E., and Wood, M.A. (1993). Maximum rates of period change for DA white dwarf models with carbon and oxygen cores. *Astrophys. J. Lett.* **391**: 33.
60. Montgomery, M.H., and Winget, D.E. (1999). The effect of crystallization on the pulsations of white dwarf stars. *Astrophys. J.* **526**: 976.
61. Goldreich, P., and Wu, Y. (1999). Gravity modes in ZZ Ceti stars. I. Quasi-adiabatic analysis of overstability. *Astrophys. J.* **511**: 904; Gravity modes in ZZ Ceti stars. II. eigenvalues and eigenfunctions. *Astrophys. J.* **519**: 783.

62. Costa, J.E.S., Kepler, S.O., and Winget, D.E. (1999). Direct measurement of a secular pulsation period change in the pulsating hot pre-white dwarf PG 1159–035. *Astrophys. J.* **522**: 973.
63. Reddy, S., and Prakash, M. (1997). Neutrino scattering in a newly born neutron star. *Astrophys. J.* **423**: 689.
64. Burrows, A., and Lattimer, J.M. (1986). The birth of neutron stars, *Astrophys. J.* **307**: 178.
65. Bethe, H.E. (1990). Supernovae. *Rev. Mod. Phys* **62**: 801.
66. Pons, J.A., Reddy, S., Prakash, M., Lattimer, J.M., and Miralles, J.A. (1999). Evolution of Proto-Neutron Stars. *Astrophys. J.* **513**: 780.
67. Pethick, C.J. (1992). Cooling of neutron stars. *Rev. Mod. Phys* **64**: 1133.
68. Schaab, Ch., Weber, F., Weigel, M.K., and Glendenning, N.K. (1996). Thermal evolution of compact stars. *Nucl. Phys. A* **605**: 531.
69. Schaab, Ch., Weber, F., and Weigel, M.K. (1998). Neutron superfluidity in strongly magnetic interiors of neutron stars and its effect on thermal evolution. *Astron. Astrophys.* **335**: 596.
70. Page D. (1998). Thermal evolution of isolated neutron stars, in *The Many Faces of Neutron Stars*, pp. 539, NATO ASI series Buccheri, R., van Paradijs, J., and Alpar, M.A. eds., Dordrecht: Kluwer.
71. Takatsuka, T., and Tamagaki, R. (1993). Superfluidity in neutron star matter and symmetrical nuclear matter. *Prog. Theo. Phys. Suppl.* **112**: 27.
72. Schaab, Ch., Voskresensky, D., Sedrakian, A.D., Weber, F., and Weigel, M.K. (1997). Impact of medium effects on the cooling of nonsuperfluid and superfluid neutron stars. *Astron. Astrophys.* **321**: 591.
73. Schaab, Ch., Balberg, S., and Schaffner-Bielich, J. (1998). Implications of hyperon superfluidty for neutron star cooling. *Astrophys. J. Lett.* **504**: L99.
74. Huber, H., Weber, F., Weigel, M.K., and Schaab, Ch. (1998). Neutron star properties with relativistic equations of state. *Int. J. Mod. Phys. E* **7**: 301.
75. Pandharipande, V.R., and Smith, R.A. (1975). Nuclear matter calculations with mean stellar fields. *Phys. Lett. B* **59**: 15.
76. Friedman, B., and Pandharipande, V.R. (1981). Hot and cold, nuclear and neutron matter. *Nucl. Phys. A* **361**: 502.

Index

A

Acoustic thermometry
  gases, 345–349
  planetary atmospheres, 389–398
Air
  saturation boundary, 320–321
  sound speed, 287–288, 320–327
Alkenes
  in crude oil, 208–209
  velocity measurements, 212, 214
Analytic signal, 10–11
Anderson-Brinkman-Morel (ABM) state, 121
Argon
  cylindrical acoustic resonance measurements, 375–384
  ideal gas value, 281–282
  saturation boundary, 269–270
  sound speed, 280–283, 290–297
    measurement deviations, 349
  thermodynamic properties, 368–369
Aromatic hydrocarbons, 210–211

B

BCS theory of superconductivity, 120–121
Benedict-Webb-Rubin equations, 279
Benzene, sound speed measurement, 14–15
Beudant, François S., 28
Birmingham Solar Oscillation Network (BiSON), 408
Boltzmann Equation, 235, 236
  rarified gas, 260–263
Boric acid, and sound absorption in water, 100, 103–106, 110–111
Boron trichloride, thermodynamic properties, 366
Bouyancy frequency, 396
Brillouin scattering, 178–179
Brunt-Vaisala frequency, 396
Bubbly liquid, 183–205. *See also* Liquid
  bubble dynamics, 191–197
    bubble shape, 191–192
    near-field effects, 192
    oscillation amplitude, 193–194
    pressure and bubble volume, 192–193, 194–195
  bubbles as sound sources, 203
  distortions, 190–191
  linear waves, 184–191
    effective medium approach, 187–188
    Foldy's multiple scattering theory, 188–191
    Wood's method, 184–187
  phase velocity, 197–201
  plane wave propagation, 201–203
  sound speed, 197–201
    equilibrium, 186
    frozen, 186

C

Carnahan-Starling expression, 278
Chemical warfare agents, sound speed determination in, 3–21
  experimental, 13–14
  frequency domain analysis, 7–8
  introduction to, 3–5
  liquid density, 8–10
  loss mechanisms, 9
  precursor chemicals, 17
  results, 14–16
  shear wave coupling, 6
  swept frequency data, 11–13
  theory behind, 5–13
  time domain analysis, 10–13
  wall resonance modulation, 14–15
  wave interactions, 18–20
Chemical Weapons Convention, 4
Colladon, Jean-Daniel, 28
Compact objects, 424. *See also* Stars, neutron; Stars, white dwarf
Condensed matter, 423–447
  below neutron drip density, 425–428
  cold, high-density equation of state, 425–427, 425–433
  densities of, 424
  neutron stars, 439–441
  properties of, 431–433
  stable configurations of, 433–437
  subnuclear densities, 429
  supernuclear densities, 429–430
  in white dwarfs, 437–439
Cowling approximation, 412–413
Cylindrical acoustic resonators, 329–330, 371–387
  acoustic thermometry, 345–349
  boundary losses corrections, 339–340
  calibration, 336–337

Cylindrical acoustic resonators, *continued*
   conditions most useful for, 335
   fill duct corrections, 339–340
   impurities and, 337
   sound speed measurements, 335–340
   transducers, 338
   transport properties measurements, 340–345
   universal gas constant determination, 349–351

D

De Gennes elastic energy, 169
Detection methods, underwater, 28–29
Dichloro-1-fluoroethane, 1,1-, thermodynamic properties, 356
Diffusion, gas mixture, 256–257
Difluoroethane, 1,1-, thermodynamic properties, 357–358
Difluoromethoxy-1,1,1-trifluoroethane, 2-, thermodynamic properties, 353

E

Eigenmodes, as sound sources, 203
Elastic constants
   deduced from first sound velocity, 172–176
   gases, 235–265
   liquid crystals, 159–181
Equation of state
   cold, high-density, 425–433
     above neutron drip density, 428–429
     below neutron drip density, 425–428
     electrostatic corrections to, 427
     neutronized corrections to, 427–428
     subnuclear densities, 429
     supernuclear densities, 429–430
   fluid, 274–280
     critical point and, 280
     cubic, 277–278, 283
     hard sphere, 278
     reference quality, 279, 281–283
     simple, 276–278
     thermodynamic, 271–272
     virial, 276–277, 282–283
   gas
     ideal, 237
     virial, 238, 330
Ethylene glycol, sound speed, 14–15
Euler equation, 271

F

Fermi-Dirac statistics, 119–120
Fluid
   continuity equation, 270
   equations of state, 274–280
     critical point and, 280
     cubic, 277–278, 283
     hard sphere, 278
     reference quality, 279, 281–283
     simple, 276–278
     thermodynamic, 271–272
     virial, 276–277, 282–283
   Euler equation, 271
   hard sphere, 278
   hydrodynamic equations, 270–273
   Navier-Stokes equation, 270–271
   particle interactions, 273–274
   phase topology, 275–276
   sound as thermodynamic property of, 267–328
   sound speed, 273–274, 280–289
     air, 320–327
     argon, 290–297
     nitrogen, 297–310
     water, 310–320
   vapor-liquid equilibrium phase boundary, 275–276
   wave equation, 268–274
Foldy approximation, 194–195
Fourier transform
   chemical warfare agent, 10–11
Frank-Oseen elastic energy, 165
Freedericksz transition, 171–172

G

Gas
   acoustic measurements, 329–370
   acoustic propagation constant, 243–244
   acoustic thermometry, 345–349
     principles of, 345–347
     spherical resonators in, 346–348
     thermodynamic temperature measurements, 348–349
   acoustic virial coefficients, 331–332
   collisions, 248–249
   compressibility, 380
   convective energy transport, 259
   cylindrical acoustic resonance measurements, 371–387
     apparatus, 373–375
     average initial velocity data, 377–379
     curve fitting, 378
     errors in, 378
     final corrected velocity, 379
     gases used, 375, 381–384
     initial corrected velocity, 378–379
     method of operation, 376–377
     resolution improvements, 376–377
     uncertainty in, 380–381
   diatomic, 244–245
   diffusion, 256–257
   distribution function, 258, 259
   eigenvalues, 350
   elastic constants, 235–265
     acoustics measurements, 235–236
     relaxation processes, 241–246, 254–255
   equation of motion, 239
   equations of state, virial, 238, 330
   halogen, 247–249
   hard-core Lennard-Jones potential, 332–333, 334
   hard-core square-well potential, 334
   hard-core square-well potentials, 332
   ideal, 236–238
     equation of continuity, 236
     equation of state, 237
     sound speed, 237–238
   ideal Fermion, 425–427
   inverse collisions, 259–260

INDEX

low pressure measurements, 257–263
Maxwell distribution, 260–261
measurement results, 247–256
model potential forms, 334–335
moderate pressure measurements, 245–247
polyatomic, 244–245
rotation, 242
rotational energy levels, 252–256
SO_2, 247, 249–256
sound speed, 333–340, 371–387
   calibration, 336–337
   corrections for boundary losses and fill duct, 339–340
   cylindrical acoustic resonance measurements, 371–387
   determining, 339
   halfwidths measurements, 339
   ideal gas, 237–238
   impurities, effects of, 337
   real gas corrections, 238
   resonance frequencies measurements, 339
   resonance methods, 335, 340–345
   resonators for, 335–336
   transducers, 338
   transport processes contributions, 239–241
Stokes-Navier equation, 239, 257
systems not in equilibrium, 263
tables of thermodynamic properties, 351–369
   boron trichloride, 366
   helium, 367–368
   hexafluoroethane, 363
   hydrogen bromide, 365–366
   methane, 360–361
   1-chloro-1,2,2,2-tetrafluoroethane, 354–355
   1,1-dichloro-1-fluoroethane, 356
   1,1-difluoroethane, 357–358
   1,1,1,2,3,3-hexafluoropropane, 358
   1,1,1,3,3,3-hexafluoropropane, 359
   1,1,1,2,2,3,3,4-octafluorobutane, 360
   1,1,2,2,3-pentafluoropropane, 359–360
   1,1,1,2-tetrafluoroethane, 356
   1,1,1-trifluoroethane, 357
   pentafluorodimethyl ether, 352
   pentafluoroethane, 355
   sulfur hexafluoride, 364
   tetrafluorodimethyl ether, 352–353
   tetrafluoromethane, 361–362
   tungsten hexafluoride, 364–365
   2-dichloro-1,1,1-trifluoroethane, 354
   2-difluoromethoxy-1,1,1-trifluoroethane, 353
   xenon, 367
temperature and absorption, 251–252
thermodynamic properties of dilute, 331–335
translational motion, 241–242
transport properties resonance measurements, 340–345
   Greenspan acoustic viscometer, 340–343
   Prandtl number, 343–345
ultrasonic absorption/dispersion, 243
universal gas constant, 330, 349–351
vibration, 242–244
virial coefficients tables, 351–369
virial equation of state, 238, 330
viscosity of, 329–330, 334, 341–342
v-t transitions, 248–249
Gaussian envelope, 10
Ginzburg criterion, 170, 176
Global Oscillation Network Group (GONG), 408
Glycerine, sound speed, 14–15
G-modes, 413
Greenspan acoustic viscometer, 340–343

H

Halogen, 247–249
Helioseismology, 401–402, 408–417
   solar models and, 419–420
Helium
   cylindrical acoustic resonance measurements, 375–384
   superfluid ^{3}HE, 117–146
      anisotropy, 121–122
      boundary condition, 135, 136
      coherence effects, 121
      first sound, 125–126
      fourth sound, 129–137
      normal liquid state, 117–122
      phases, 117–118
      properties of, 119
      second sound, 126–129
      spin-entropy wave, 137–144
      temperature dependence, 119–120, 128–129, 132–134, 135
      third sound, 125
      wave propagation, 122–125
   superfluid helium four (He II), 147–157
      acoustic properties, 148–150
      fifth sound, 150
      first sound, 148, 154
      fourth sound, 149, 155–156
      second sound, 149, 155
      sound speed measurement, 150–154
      third sound, 150
   thermodynamic properties, 367–368
Helmholtz free energy, 268
   fluid systems, 275–276
   ideal gas, 274–275
   monatomic ideal gas, 273
   natural gas fluid deviations, 288–289
Hexafluoroethane, thermodynamic properties, 363
Hexafluoropropane, 1,1,1,2,3,3-, thermodynamic properties, 358
Hexafluoropropane, 1,1,1,3,3,3-, thermodynamic properties, 359
Hilbert transform, 10–11, 16
Huygens Atmospheric Structure Instrument (HASI), 390, 392
Hydrodynamics, Smectic liquid crystal, 163–164
Hydrogen
   rotational energy level, 242
   spin state, 394
Hydrogen bromide, thermodynamic properties, 365–366

I

Impedance anlyzer, 13

INDEX

Interferometry
   gas velocity measurement techniques, 245–247
   Swept-Frequency Acoustic, 3–21
   water sound speed measurements, 30–31, 38–39
International Association for the Properties of Water and Steam (IAPWS), 284
Internationally accepted temperature scale (ITS-90), 330
IRIS network, 408
Isopropanol, sound speed measurement, 14–15

J

Jupiter, atmospheric measurements, 393, 394–396

L

Lagrangian perturbation, 410
Landau Fermi liquid theory, 120, 122
Landau-Ginzburg energy, 165, 169
Landau-Peierls form, Smectic deformations, 161–162, 164
Langmuir circulation cells, 201
Liquid
   bubbly, 183–205
   causes of sound absorption, 85–92
   density in chemical warfare agents, 8–10
   petroliferous, 207–231
Liquid crystals, 159–181
   anisotropic critical point, 175
   elastic constants
      critical behavior, 166, 170
      deduced from Rayleigh and Brillouin scattering, 178–179
      deduced from velocity of first sound, 172–176
      Ginzburg criterion, 170
      heat behavior, 171
      layer-compression modulus B, 176–178
      SmC order parameter fluctuations and, 169–170
   elastic properties, 159–181
   free energy, 162–163, 165–166
   layer-compression energy, 162
   Nematic phases, 160
      distortions, 161, 162
      elastic energy, 161
      Freedericksz transition, 171–172
   N-SmA transition, 165–169
      elastic constant behavior, 173–175
   sample preparation, 171
   Smectic phases, 160
      elastic energy, 161–163
      hydrodynamics, 163–164
      layer-compression modulus B deduction, 176–178
      parameter fluctuations, 166–168
      parameter modulus relaxation, 168–169
      SmA-SmC transition, 169–171, 176
      sound propagation, 163–164

M

Magnesium sulfate ($MgSO_4$), and sound absorption in water, 93, 96, 97–100, 109
Mars, atmospheric measurements, 390–391, 394–396
Mathcad program, 7, 16–18
Maxwell distribution, 260–261
Methane, thermodynamic properties, 360–361
Michelson Doppler Interferometer (MDI), 417–418
Mode inertia, 414
Multiple scattering theory, Foldy's, 188–191

N

N-Alkanes
   in crude oil, 208
   velocity measurements, 212, 213
Napthenes
   in crude oil, 209–210
   velocity measurements, 212, 215
NASA Planetary Exploration Roadmap, 391
National Institute of Standards and Technology (NIST)
   sound speed in gases measurements, 329–370
   Thermophysical Properties of Pure Fluids Database, 289–290
   universal gas constant value, 349–351
Navier-Stokes Equation, 239–240, 257, 270–271
Neptunian waters, 32–33, 42–44. *See also* Seawater; Water
   definition of, 33, 42–44
   sound speed equations, recommended, 77–78
   temperature-salinity-depth volume, 42–44
Network analyzer, 13
Newton, Isaac, 25, 28
Nitrogen
   gas, cylindrical acoustic resonance measurements, 375–384
   relaxation, 254–255
   saturation boundary, 297–298
   sound speed, 283–285, 299–310

O

Octafluorobutane, 1,1,1,2,2,3,3,4-, thermodynamic properties, 360
Optimally localized averages (OLA) method, 416–417
Optoacoustic effect, 247
Oscillating superleak transducer (OST), 138–139
Oxygen, vibrational relaxation, 254

P

Pauli exclusion principle, 119
Pentafluorodimethyl ether, 352
Pentafluoroethane, thermodynamic properties, 355
Pentafluoropropane, 1,1,2,2,3-, thermodynamic properties, 359–360
Petroleum classification, 208
Petroliferous liquid, 207–231
   acoustic properties, 207–218
   crude oil classification, 211–212
      acoustic velocity, 212–218
      density/specific gravity, 211–212
      indirect methods, 212–218
      viscosity, 211–212
   crude oil components, 208–211
      alkenes, 208–209
      aromatic hydrocarbons, 210–211
      n-Alkanes, 208
      napthenes, 209–210

crude oil products, 211
   rocks saturated with, 219–231
      heavy oil saturants, 224–230
      light oil saturants, 220–224
      pore fluid effects, 219–220
Planck-Einstein relation, 253–254
Planetary atmospheres, 389–398
   acoustic-gravity waves, 394–397
   acoustic measurements on spacecraft, 389–391
   hydrogen molecule spin state and, 394
   instrument considerations, 391–392
   pressure-temperature profiles, 394–396
   temperature and, 392–394
Plane wave interaction
   glass/quartz interface, 18–19
   glass/water interface, 19–20
   input PSI field, 20
   water/glass interface, 19
P-modes, 413
Poiseulle's Law, 212
Prandtl number, 343–345
Pulse-echo technique
   gas measurements, 245
   time-of-flight, 10–11
   transform, 11–13

Q
Quasi-particle, 120

R
Rayleigh scattering, elastic constant deduction, 178–179
Regularized least squares (RLS) method, 416–417
Relaxation effects
   gases elastic properties, 241–245, 245–246
   liquid crystal, 177–178
   Maxwellian, 177
   water sound absorption, 83–84, 85–92, 86–87, 95, 104–106, 109–110

S
Saturn, atmospheric measurements, 393, 394–396
Saybolt viscosity, 212
Schwarzschild criterion, 403
Seawater, *see also* Bubbly liquid; Water
   chemical relaxation of salts, 87–89
   low frequency sound absorption mechanisms, 103–106
   Neptunian, 32–33, 42–44
      definition of, 33, 42–44
      sound speed equations, recommended, 77–78
      temperature-salinity-depth volume, 42–44
   sound absorption, 94–106, 95–96, 108–110
      equation, 109
      techniques at sea, 100–103
   sound speed, 26–27, 46–49
      at atmospheric pressure, 46, 47, 48, 60–64, 65–66
      definition of, 26–27
      under pressure, 47–49, 64.66–72
      salinity and, 26, 27, 60–64
      standard, 26–27
Seismic survey, 207

Seismic wave velocity
   heavy oil saturated consolidated rocks, 224–228
   heavy oil saturated unconsolidated sand, 228–230
   light oil saturated consolidated rocks, 220–224
   light oil saturated unconsolidated sand, 224, 225
Solar and Heliospheric Observatory (SOHO), 408
Sound
   as thermodynamic property of fluids, 267–328
   wave propagation in superfluids, 122–125
Sound attenuation
   bubbly liquid, 198–201
   chemical warfare agents, 8–10
   fourth sound, 134–135
   superfluid, 125–126
Sound dispersion, liquid crystal, 166–169
Sound intensity, 84–85
Sound propagation
   liquid crystal, 163–164
   low pressure gas, 257–263
   multilayered medium, 6
   Smectic liquid crystal, 163–164
   systems not in equilibrium, 263
   as thermodynamic property, 268, 270
   wave equation in fluids, 268–274
Sound speed
   bubbly liquid, 186, 197–201
   chemical warfare agents, 3–22
   equilibrium, 186
   fluid, 273–274, 279–289
   frozen, 186
   gases, 329–370, 371–387
   planetary atmospheres, 389–398
   pure and Neptunian water, 23–81
   stars, normal, 399–422
   stars, white dwarf/neutron, 423–447
   Sun, 407–419
   superfluid ^{3}HE, 117–146
   superfluid helium four, 147–157
Space, low frequency sound waves, 257
Spectrophone, 247
Spin-entropy wave propagation, 137–144
   attenuation, 143–144
   nonlinear response, 141–143
   oscillating superleak transducer, 138–139
   superfluid density and, 139–141
Stars, 399–400, 423–447
   atmosphere of, 401
   boundary conditions, 405–406
   color of, 400
   condensed matter in, 423–447
   construction of self-gravitating, equilibrium, 433
   energy transport equation, 402–403
   equation of state, 403–404
      cold, high-density, 425–433
   hydrostatic equilibrium equation, 402
   luminosity of, 400–401
   mass-luminosity relation, 400–401
   neutron, 423, 424
      condensed matter in, 439–441
      cooling, 444
      formation of, 439
      mass, 439–440
      proto-, 443–444

Stars, neutron, *continued*
    pulsar glitches, 441
    radii, 440
    rotation periods, 440
    structure, 433–434, 436–437
    structure of, 436–437
    thermal properties of matter, 441, 443–444
  normal, 399–422
  observational constraints, 400–402
  oscillations, 401–402
  sound speed, 399–422
  stable configurations of, 433–437
  structure, changes in, 406–407
  structure, equations describing, 402–408
  structure and evolution equations, 404–405
  Sun, 407–421
  white dwarf, 423, 424
    basic features of, 435–436
    condensed matter in, 437–439
    cooling, 441–443
    mass-radius relations, 437–439
    pulsation, 443
    radii, 437
    structure, 433–436
    thermal properties of matter, 441–443
  zero age main sequence, 404, 406
Stokes-Kirchoff coefficient, 86, 92, 108
Stokes-Navier equation, 239–240, 257, 270–271
Sturm, Charles, 28
Subtractive optimally localized averages (SOLA), 416–417
Sulfur dioxide (SO_2), 247, 249–256
  relaxation, 250–251
  vibrational modes, 249–250
Sulfur hexafluoride, thermodynamic properties, 364
Sun
  bouyancy frequency, 411
  continuity and motion equations, 409
  continuity equation, 411
  energy equation, 409
  heat equation, 410
  hydrostatic equilibrium equation, 411
  hydrostatic support equation, 409
  Lagrangian perturbation, 410
  mass loss, 405
  models, 405–406, 414–415, 419–420
  motion equation, 411, 413
  oscillation of, 401–402, 408–417
    equations, 409–412
    frequencies, 409
    frequency inversions, 413–417, 418
    linear and adiabatic, 408–409
    properties, 412–413
  Poisson's equation, 411
  rotation, 409
  sound speed, 407–419
Superconductivity, BCS theory, 120–121
Superfluid
  anisotropic, 123–125
  boundary condition, 135, 136
  density, 139–141
  density fraction, 130–134
  first sound, 125–126
  fourth sound, 129–137
  helium four, 147–157
  in neutron stars, 444
  nonlinear response, 141–143
  phases of, 117–118
  second sound, 126–129
  spin-entropy wave, 137–144
  third sound, 125
  ^{3}HE, sound speed, 117–146
  two-fluid model, 122–123, 148–149
  wave propagation in, 122–125
  zero temperature limit, 132–134
Superleak, 129
  oscillating superleak transducer, 138–139
Surface Science Package (SSP), 390
Swept-Frequency Acoustic Interferometry (SFAI), 3–21
  electronic system, 13–14
  results, 14–16
  theory behind, 5–13
  transducer, 7

T

Tetrafluorodimethyl ether, thermodynamic properties, 352–353
Tetrafluoroethane, 1,1,1,2-, thermodynamic properties, 356
Tetrafluoroethane, 1-chloro-1,2,2,2-, thermodynamic properties, 354–355
Tetrafluoromethane, thermodynamic properties, 361–362
Thermodynamics Research Center, 289
Time domain analysis, 10–13
  signal-to-noise ratio, 11–12
Titan, atmospheric measurements, 392–394, 394–396
Toluene, sound speed measurement, 14–15
Transducer
  gas measurements, 338
  oscillating superleak, 138–139
  Peshkov, 127–128
  Swept-Frequency Acoustic Interferometry, 7, 13–14
Trifluoroethane, 1,1,1-, thermodynamic properties, 357
Trifluoroethane, 2-dichloro-1,1,1-, thermodynamic properties, 354
Tungsten hexafluoride, thermodynamic properties, 364–365
Two-fluid model, 122–123, 148–149

U

U.S. National Bureau of Standards (NBS)
  velocimeter, 36–38
  water speed measurements, 30–32, 36–38
U.S. Naval Ordnance Laboratory (NOL)
  velocimeter, 36–38
  water sound speed measurements, 31–32, 36–38

V

Vapor-liquid equilibrium phase boundary (VLE), 275–276

INDEX

Velocimeters
   pulse-type, 36–38
   sing-around, 37
   type A1, 36–38
   type A2, 39–41
   type B, 38–39
   type C, 41–42
   water sound speed measurements, 30, 32, 35–42
      method for absolute, 35–36
Viscosity
   crude oil, 212
   gas, 329–330, 334, 341–342
   Saybolt, 212

W

Water, 87–88. *See also* Seawater
   compressibility, 28
   fresh/pure, 25, 93–94
   polar, 103
   saturation boundary, 310–314
Water, sound absorption in, 83–115
   above 20 kHZ, 92–96
   boron and, 100, 103–106, 110–111
   causes of in liquids, 85–92
   chemical relaxation of salts, 87–89
   classical measurements, 98, 99
   equations, 96, 106–109
   fresh/pure, 93–94
   laboratory measurement techniques, 89–92
      direct, 90–91
      resonator, 91
      T-jump, 91–92, 100
   low frequency, 96–108
      measurement techniques, 100–103
   magnesium sulfate and, 93, 96, 97–100, 109
   mechanisms of, 83–84, 95–96, 103–106
   nonlinear effects, 98, 99
   pressure and, 94, 95–96, 110–111
   relaxation processes, 86–87, 95, 109–110
      buffer effect, 104–105
      catalysis effect, 105–106
   seawater, 94–106, 95–96, 108–110
   single path technique, 110–111
   Stokes-Kirchoff coefficient, 86, 92, 108
   temperature and, 94, 98, 99–100, 107
   transmission loss, 84
Water, sound speed in, 23–81, 284–287
   available measurements, 44–49
   Chen and Millero's research, 52, 53–54, 71–72, 73–74
   closed basins, 42–44
   Colladon and Sturm experiment, 28
   common salinity, 60–62
   deep sound channel, 29
   Del Grosso and Mader's research, 33, 34–35, 38–39, 51, 64, 70–71, 72
   equations, 29–30, 49–56
      author key, 57
      compared, 56–74
      direct, 49, 55
      formulation of, 50–54
      NRL, 67, 69, 73
      recommended, 56–76
      revised, 49, 54, 56
      simplified, 49–50, 53, 56
      summarized, 54–56
      validity, 72–74
   interferometry, 30–31
   intermediate salinity, 24, 62–64, 65–66
   Kuwahara's research, 29–31
   measurements, 14–15
      NBS, 30–31
      NOL, 31–32
   modern, 30–35
   Neptunian, 32–33, 42–44
   overview, 25–27
   physics of, 25–27
   pressure and, 25–26
   pure/fresh, 24–25, 46, 47
      at atmospheric pressure, 57–60
      recommended equation for, 77
   research
      calculated values and, 29–30
      early, 27–29
      history, 24–25
   seawater, 26–27, 46–49
      at atmospheric pressure, 46, 47, 48, 60–64, 65–66
      definition of, 26–27
      under pressure, 47–49, 64.66–72
      salinity and, 26, 27, 60–64
      standard, 26–27
   tables, 310–320
   velocimeters and, 30, 32, 35–42
   Wilson's research, 50, 52, 68–70

X

Xenon, thermodynamic properties, 367

Z

Zero age main sequence (ZAMS), 404, 406

CONVERSION FACTORS BETWEEN SI AND CGS

Quantity	Multiply SI	by	to obtain CGS
Length	meter (m)	10^2	centimeter (cm)
Mass	kilogram (kg)	10^3	gram (g)
Time	second (s)	1	second (s)
Force	newton (N)	10^5	dyne
Energy	joule (J)	10^7	erg
Power	watt (W)	10^7	erg/s
Volume density	kg/m^3	10^{-3}	g/cm^3
Pressure	pascal (Pa)	10	dyne/cm^2
Speed	m/s	10^2	cm/s
Energy density	J/m^3	10	erg/cm^3
Elastic modulus	Pa	10	dyne/cm^2
Coefficient of viscosity	Pa·s	10	dyne·s/cm^2
Volume velocity	m^3/s	10^6	cm^3/s
Acoustic intensity	W/m^2	10^3	erg/(s·cm^2)
Mechanical impedance	N·s/m	10^3	dyne·s/cm
Specific acoustic impedance	Pa·s/m	10^{-1}	dyne·s/cm^3
Acoustic impedance	Pa·s/m^3	10^{-5}	dyne·s/cm^5
Mechanical stiffness	N/m	10^3	dyne/cm

Some Useful Conversions:

1 in. = 2.5400 cm
1 ft = 0.3048 m
1 yd = 0.9144 m
1 fathom = 1.829 m
1 mi (statute) = 1.609 km
1 mi (nautical international) = 1 nm = 6076 ft = 1.852 km
1 mph = 1.6093 km/h = 0.4470 m/s
1 knot = 1 nm/h = 1.1508 mph = 1.852 km/h = 0.5144 m/s
1 atm = 14.70 psi = 29.92 in. Hg (32°C) = 33.90 ft H$_2$O = 1.033×10^4 kgf/m^2 = 1.0133 bar
1 bar = 1×10^6 dyne/cm^2 = 14.50 psi = 1×10^5 Pa
1 neper/m = 8.7 dB/m

Metric (SI) Multipliers

Prefix	Abbreviation	Value
Tera	T	10^{12}
Giga	G	10^9
Mega	M	10^6
Kilo	k	10^3
Hecto	h	10^2
Deka	da	10^1
Deci	d	10^{-1}
Centi	c	10^{-2}
Milli	m	10^{-3}
Micro	μ	10^{-6}
Nano	n	10^{-9}
Pico	p	10^{-12}
Femto	f	10^{-15}